Crustal Permeability

Crustal Permeability

Edited by

Tom Gleeson

University of Victoria
Victoria
British Columbia
Canada

Steven E. Ingebritsen

U.S. Geological Survey
Menlo Park
California
USA

WILEY

We dedicate this book to our families who support and inspire us, and to Henry Darcy whose legacy of solving both scientific and practical problems continues to guide the discipline of hydrogeology.

Conversion factors for permeability and hydraulic-conductivity units

In this book we emphasize the use of permeability (k) and SI units (m²) as the measure of ease of fluid flow under unequal pressure. However hydraulic conductivity (K) and a variety of other units are used in practice. Permeability is a rock property, whereas hydraulic conductivity reflects both rock and fluid properties (fluid viscosity and density) − see Chapter 1. The approximate conversion from k to K here assumes that the fluid is water at standard temperature and pressure. Water viscosity varies by a factor of ~26 and water density by a factor of ~3 between 0° C and the critical point of water. Other fluids such as hydrocarbons can exhibit much larger viscosity ranges. In the table below, we show the unit conversion for 1 m² as well as 10^{-15} m² which is a more realistic permeability for geological materials.

		Permeability, k		Hydraulic conductivity, K		
	cm^2	Darcy		$m\ s^{-1}$	$m\ d^{-1}$	$ft\ d^{-1}$
1 m² =	10^4	10^{12}		10^7	$9x10^{11}$	$3x10^{12}$
10^{-15} m² =	10^{-11}	0.001 (1 mD)		10^{-8}	$9x10^{-4}$	$3x10^{-3}$

Contents

List of contributors

Jennifer Arrigo
CUAHSI,
Boston, MA, USA

Hiroshi Asanuma
Fukushima Renewable Energy Institute,
National Institute of Advanced Industrial Science and Technology,
Koriyama, Japan

Amlan Banerjee
Indian Statistical Institute,
Kolkata, India

R. Sky Bristol
U.S. Geological Survey,
Denver, CO, USA

Kurt Bucher
Mineralogy and Petrology,
University of Freiburg,
Freiburg, Germany

Erick R. Burns
U.S. Geological Survey,
Portland, OR, USA

Andrew Campbell
Department of Earth & Environmental Science,
New Mexico Institute of Mining and Technology,
Socorro, NM, USA

Johnson R. Cann
School of Earth and the Environment,
University of Leeds,
Leeds, UK

Michael Cardiff
Department of Geoscience,
University of Wisconsin-Madison,
Madison, WI, USA

Calum Chamberlain
School of Geography, Environment and Earth Sciences,
Victoria University of Wellington,
Wellington, New Zealand

James A. D. Connolly
Department of Earth Sciences,
Swiss Federal Institute of Technology,
Zurich, Switzerland

Simon C. Cox
GNS Science,
Dunedin, New Zealand

Laura Crossey
Department of Earth and Planetary Sciences,
University of New Mexico,
Albuquerque, NM, USA

Hugh Daigle
Department of Petroleum and Geosystems Engineering,
University of Texas at Austin,
Austin, TX, USA

Jacob DeAngelo
U.S. Geological Survey,
Menlo Park, CA, USA

Jean Desroches
Services Pétroliers Schlumberger,
Paris, France

Paul H. Denys
School of Surveying,
University of Otago,
Dunedin, New Zealand

Russell L. Detwiler
Department of Civil and Environmental Engineering,
University of California,
Irvine, CA, USA

Damien Duff
CEMI - Centre for Excellence in Mining Innovation,
Sudbury, ON, Canada

Erik Eberhardt
Geological Engineering, EOAS,
The University of British Columbia,
Vancouver, BC, Canada

Jean E. Elkhoury
Department of Civil and Environmental Engineering,
University of California,
Irvine, CA, USA;
Schlumberger-Doll Research,
Cambridge, MA, USA

Ying Fan
Department of Earth and Planetary Sciences,
Rutgers University,
New Brunswick, NJ, USA

Michael Fienen
U.S. Geological Survey,
Middleton, WI, USA

Arianne Ford
Centre for Exploration Targeting,
The University of Western Australia,
Crawley, WA, Australia

Carl W. Gable
Los Alamos National Laboratory,
Los Alamos, NM, USA

Valentin Gischig
Swiss Competence Centre on Energy Research (SCCER-SoE),
ETH Zurich,
Zurich, Switzerland

Tom Gleeson
Department of Civil Engineering,
University of Victoria,
Victoria, BC, Canada

David Gochis
NCAR,
Boulder, CO, USA

Kazuhiko Goto
Graduate School of Science and Engineering,
Kagoshima University,
Kagoshima, Japan

Akira Hasegawa
Research Center for Prediction of Earthquakes and Volcanic Eruptions,
Graduate School of Science, Tohoku University,
Sendai, Japan

Stephen Hickman
U.S. Geological Survey,
Menlo Park, CA, USA

Naoshi Hirata
Earthquake Research Institute,
University of Tokyo, Tokyo, Japan

Albert Hofstra
U.S. Geological Survey,
Denver, CO, USA

Richard Hooper
CUAHSI,
Boston, MA, USA

Shinichiro Horikawa
Earthquake and Volcano Research Center,
Graduate School of Environmental Studies,
Nagoya University, Nagoya, Japan

TreVor Howald
Department of Earth & Environmental Science,
New Mexico Institute of Mining and Technology,
Socorro, NM, USA

Christian Huber
School of Earth and Atmospheric Sciences,
Georgia Institute of Technology,
Atlanta, GA, USA

Toshihiro Igarashi
Earthquake Research Institute,
University of Tokyo,
Tokyo, Japan

Takashi Iidaka
Earthquake Research Institute,
University of Tokyo,
Tokyo, Japan

Steven E. Ingebritsen
U.S. Geological Survey,
Menlo Park, CA, USA

Takuya Ishibashi
Fukushima Renewable Energy Institute,
National Institute of Advanced Industrial Science and Technology,
Koriyama, Japan;
Department of Energy and Mineral Engineering, EMS Energy Institute,
Pennsylvania State University,
University Park, PA, USA

Takaya Iwasaki
Earthquake Research Institute,
University of Tokyo,
Tokyo, Japan

Norman Jones
Civil and Environmental Engineering,
Brigham Young University,
Provo, UT, USA

Peter K. Kaiser
CEMI - Centre for Excellence in Mining Innovation,
Sudbury, ON, Canada

Karl Karlstrom
Department of Earth and Planetary Sciences,
University of New Mexico,
Albuquerque, NM, USA

Hiroshi Katao
Disaster Prevention Research Institute,
Kyoto University,
Kagoshima, Japan

Aitaro Kato
Earthquake Research Institute,
University of Tokyo,
Tokyo, Japan

Shari Kelley
Geologic Mapping,
New Mexico Bureau of Geology and Mineral Resources,
Socorro, NM, USA

Masahiro Kosuga
Graduate School of Science and Technology,
Hirosaki University,
Hirosaki, Japan

Atsuki Kubo
Kochi Earthquake Observatory, Faculty of Science,
Kochi University,
Kochi, Japan

Robert Lowther
Newcrest Mining Limited, Cadia Valley Operations,
South Orange, NSW, Australia

Virgil Lueth
New Mexico Bureau of Geology and Mineral Resources,
Socorro, NM, USA

Elco Luijendijk
Department of Structural Geology and Geodynamics,
Georg-August-Universität Göttingen,
Göttingen, Germany

Michael Manga
Department of Earth and Planetary Science,
University of California,
Berkeley, CA, USA

Takeshi Matsushima
Institute of Seismology and Volcanology, Faculty of Sciences,
Kyushu University,
Shimabara, Japan

Toru Matsuzawa
Research Center for Prediction of Earthquakes and Volcanic Eruptions,
Graduate School of Science, Tohoku University,
Sendai, Japan

Andrew M. McCaig
School of Earth and the Environment,
University of Leeds,
Leeds, UK

Ricardo Medina
Department of Civil and Environmental Engineering,
University of California,
Irvine, CA, USA

Catriona D. Menzies
Ocean and Earth Science, National Oceanography Centre Southampton,
University of Southampton,
Southampton, UK

Steven Micklethwaite
Centre for Exploration Targeting,
The University of Western Australia,
Crawley, WA, Australia

Stephen A. Miller
Center for Hydrogeology and Geothermics (CHYN),
University of Neuchâtel,
Neuchâtel, Switzerland

Tsutomu Miura
Disaster Prevention Research Institute,
Kyoto University,
Kagoshima, Japan

Hiroki Miyamachi
Graduate School of Science and Engineering,
Kagoshima University,
Kagoshima, Japan

Nils Moosdorf
University of Hamburg,
Hamburg, Germany;
Leibniz Center for Marine Tropical Ecology,
Bremen, Germany

Joseph P. Morris
Schlumberger-Doll Research,
Cambridge, MA, USA;
Computational Geosciences,
Lawrence Livermore National Laboratory,
Livermore, CA, USA

Lawrence Murdoch
Environmental Engineering and Earth Sciences,
Clemson University,
Clemson, SC, USA

Haruhisa Nakamichi
Disaster Prevention Research Institute,
Kyoto University,
Kagoshima, Japan

Kazushige Obara
Earthquake Research Institute,
University of Tokyo,
Tokyo, Japan

Tomomi Okada
Research Center for Prediction of Earthquakes and Volcanic Eruptions,
Graduate School of Science, Tohoku University,
Sendai, Japan

Takashi Okuda
Earthquake and Volcano Research Center, Graduate School of
Environmental Studies,
Nagoya University,
Nagoya, Japan

J. Olson
Civil and Environmental Engineering,
Utah State University,
Logan, UT, USA

Lara Owens
Ormat Technologies, Inc.,
Reno, NV, USA

Aaron Packman
Department of Civil and Environmental Engineering,
Northwestern University,
Evanston, IL, USA

Scott Peckham
INSTAAR, University of Colorado,
Boulder, CO, USA

Jeff D. Pepin
Department of Earth & Environmental Science,
New Mexico Institute of Mining and Technology,
Socorro, NM, USA

Mark Person
Department of Earth & Environmental Science,
New Mexico Institute of Mining and Technology,
Socorro, NM, USA

Shanan E. Peters
Department of Geoscience,
University of Wisconsin-Madison,
Madison, WI, USA

Fred Phillips
Department of Earth & Environmental Science,
New Mexico Institute of Mining and Technology,
Socorro, NM, USA

Yury Y. Podladchikov
Earth Sciences Department,
University of Lausanne,
Lausanne, Switzerland

Giona Preisig
Swiss Geological Survey,
Bundesamt für Landestopografie Swisstopo,
Wabern bei Bern, Switzerland

Romain Prioul
Schlumberger-Doll Research,
Cambridge, MA, USA

Mark Ranjram
Civil Engineering Department,
McGill University,
Montreal, QC, Canada

Stephen Richard
Arizona Geological Survey,
Tucson, AZ, USA

Vincent Roche
School of Earth Sciences,
University College Dublin,
Dublin, Ireland

Jonny Rutqvist
Lawrence Berkeley National Laboratory (LBNL),
Berkeley, CA, USA

Demian M. Saffer
Department of Geosciences, Center for Geomechanics, Geofluids, and
Geohazards,
Pennsylvania State University,
University Park, PA, USA

Atsushi Saiga
Tono Research Institute of Earthquake Science,
Mizunami, Japan

Shinichi Sakai
Earthquake Research Institute,
University of Tokyo,
Tokyo, Japan

Elizabeth J. Screaton
Department of Geological Sciences,
University of Florida,
Gainesville, FL, USA

A. P. S. Selvadurai
Department of Civil Engineering and Applied Mechanics,
McGill University,
Montréal, QC, Canada

Heather A. Sheldon
Australian Resources Research Centre,
Commonwealth Scientific and Industrial Research Organisation (CSIRO),
Kensington, WA, Australia

Zheming Shi
School of Water Resources and Environment,
China University of Geosciences,
Beijing, China;
Department of Earth and Planetary Science,
University of California,
Berkeley, CA, USA

Frank A. Spane
Pacific Northwest National Laboratory,
Richland, WA, USA

Ingrid Stober
Institute of Applied Geosciences,
Karlsruhe Institute of Technology (KIT),
Karlsruhe, Germany

Yanqing Su
School of Earth and Atmospheric Sciences,
Georgia Institute of Technology,
Atlanta, GA, USA

Rupert Sutherland
GNS Science,
Lower Hutt, New Zealand

Donald Sweetkind
U.S. Geological Survey,
Denver, CO, USA

Hiroaki Takahashi
Institute of Seismology and Volcanology,
Hokkaido University,
Sapporo, Japan

Tetsuya Takeda
National Research Institute for Earth Science and Disaster Prevention,
Tsukuba, Japan

David Tarboton
Civil and Environmental Engineering,
Utah State University,
Logan, UT, USA

Joshua Taron
U.S. Geological Survey,
Menlo Park, CA, USA

Damon A. H. Teagle
Ocean and Earth Science, National Oceanography Centre Southampton,
University of Southampton,
Southampton, UK

Toshiko Terakawa
Earthquake and Volcano Research Center, Graduate School of
Environmental Studies,
Nagoya University,
Nagoya, Japan

Stacy Timmons
Aquifer Mapping Program,
New Mexico Bureau of Geology and Mineral Resources,
Socorro, NM, USA

Noriyoshi Tsuchiya
Graduate School of Environmental Studies,
Tohoku University,
Sendai, Japan

Noriko Tsumura
Faculty of Science,
Chiba University,
Chiba, Japan

Norihito Umino
Research Center for Prediction of Earthquakes and Volcanic Eruptions,
Tohoku University,
Sendai, Japan

Benoît Valley
Center for Hydrogeology and Geothermics (CHYN),
University of Neuchâtel,
Neuchâtel, Switzerland

Mirko van der Baan
Department of Physics,
University of Alberta,
Edmonton, AB, Canada

Clifford I. Voss
U.S. Geological Survey,
Menlo Park, CA, USA

Chi-Yuen Wang
Department of Earth and Planetary Science,
University of California,
Berkeley, CA, USA

Guang-Cai Wang
School of Water Resources and Environment,
China University of Geosciences,
Beijing, China

Noriaki Watanabe
Graduate School of Environmental Studies,
Tohoku University,
Sendai, Japan

Philipp Weis
Institute of Geochemistry and Petrology,
ETH Zürich,
Zürich, Switzerland
GFZ German Research Centre for Geosciences,
Potsdam, Germany

Colin F. Williams
U.S. Geological Survey,
Menlo Park, CA, USA

James Witcher
Witcher and Associates,
Las Cruces, NM, USA

Walter Witt
Centre for Exploration Targeting,
The University of Western Australia,
Crawley, WA, Australia

David Wolock
U.S. Geological Survey,
Lawrence, KS, USA

Takuji Yamada
Institute of Seismology and Volcanology,
Hokkaido University,
Sapporo, Japan

Yoshiko Yamanaka
Earthquake and Volcano Research Center, Graduate School of
Environmental Studies,
Nagoya University,
Nagoya, Japan

Bruce W. D. Yardley
School of Earth and the Environment,
University of Leeds,
Leeds, UK

Keisuke Yoshida
Research Center for Prediction of Earthquakes and Volcanic Eruptions,
Tohoku University,
Sendai, Japan

Ilya Zaslavsky
San Diego Supercomputer Center,
San Diego, CA, USA

About the companion websites

This book is accompanied by two companion websites:

One website includes:
- Powerpoints of all figures from the book for downloading

www.wiley.com/go/gleeson/crustalpermeability/

The other website includes:
- A persistent data portal for sharing crustal-permeability data

http://crustalpermeability.weebly.com/

CHAPTER 1

Introduction

TOM GLEESON[1] AND STEVEN E. INGEBRITSEN[2]

[1] *Department of Civil Engineering, University of Victoria, Victoria, BC, Canada;* [2] *U.S. Geological Survey, Menlo Park, CA, USA*

Permeability is the primary control on fluid flow in the Earth's crust. Thus, characterization of permeability is a central concern of many Earth scientists; hydrogeologists and petroleum engineers recognize it as their most essential parameter. More broadly considered, permeability is the key to a surprisingly wide range of geological processes, because it also controls the advection of heat and solutes and generation of anomalous pore pressures (Fig. 1.1). The practical importance of permeability – and the potential for large, dynamic changes in permeability – is highlighted by ongoing issues associated with hydraulic fracturing for hydrocarbon production ("fracking"), enhanced geothermal systems, and geologic carbon sequestration.

The measured permeability of the shallow continental crust is so highly variable that it is often considered to defy systematic characterization. Nevertheless, some order has been revealed in globally compiled data sets, including postulated relations between permeability and depth on a whole-crust scale (i.e., to approximately 30 km depth; e.g., Manning & Ingebritsen 1999; Ingebritsen & Manning 2010) and between permeability and lithology in the uppermost crust (to approximately 100 m depth: Gleeson *et al.* 2011). The recognized limitations of these empirical relations helped to inspire this book.

Although there are many thousands of research papers on crustal permeability, this is the first book-length treatment. Here, we have attempted to bridge the historical dichotomy between the hydrogeologic perspective of permeability as a static material property that exerts control on fluid flow and the perspective of economic geologists, crustal petrologists, and geophysicists who have long recognized permeability as a dynamic parameter that changes in response to tectonism, fluid production, and geochemical reactions.

This book is based in large part on a special thematic issue of the *Geofluids* journal published in early to mid-2015 (*Geofluids* **15**:1–2). Several changes and improvements differentiate the book from the thematic issue: the authors of the 22 original *Geofluids* papers have had the opportunity to revise and update their respective chapters, and three additional chapters

have been added to fill gaps in the topical coverage (Ishibashi *et al.*, this book; Taron *et al.*, this book; Yardley, this book); the introductory material has been revised and expanded; the reference list has been consolidated and updated; an index has been added; and a complementary website (http://crustalpermeability.weebly.com/) has been built to house permeability data and other supporting information. Much of this introduction, and much of the bridging material between topical sections of the book, is derived from the introduction to the *Geofluids* thematic issue, with changes and additions where appropriate.

MOTIVATION AND BACKGROUND

This book is motivated by the controlling effect of permeability on diverse geologic processes; by practical challenges associated with emerging technologies such as hydraulic fracturing, enhanced geothermal systems, and geologic carbon sequestration; and by the historical dichotomy between the hydrogeologic concept of permeability as a static material property that exerts control on fluid flow and the perspective of other Earth scientists who have long recognized permeability as a dynamic parameter. Issues associated with hydraulic fracturing, enhanced geothermal systems, and geologic carbon sequestration have already begun to promote a constructive dialog between the static and dynamic views of permeability, and here we have made a conscious effort to include both viewpoints. We focus on the quantification of permeability, encompassing both direct measurement of permeability in the uppermost crust and inferential permeability estimates, mainly for the deeper crust.

The directly measured permeability (k) of common geologic media varies by approximately 16 orders of magnitude, from values as low as 10^{23} m^2 in intact crystalline rock, intact shales, and fault gouge, to values as high as 10^{-7} m^2 in well-sorted gravels. Permeability can be regarded as a process-limiting parameter in that it largely determines the feasibility of advective solute transport ($k \gtrsim 10^{-20}$ m^2), advective heat transport

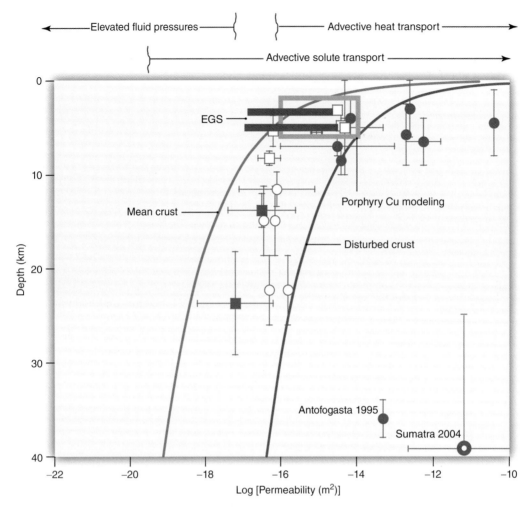

Fig. 1.1. Crustal-scale permeability (k) data. Arrows above the graph indicate approximate ranges of k over which certain geologically significant processes are likely. The "mean crust" k curve is based on k estimates from hydrothermal modeling and the progress of metamorphic reactions (Manning and Ingebritsen 1999). However, on geologically short timescales, k may reach values significantly in excess of these mean crust values (Ingebritsen and Manning 2010). The power-law fit to these high-k data – exclusive of the Sumatra datum (Waldhauser *et al.* 2012) – is labeled "disturbed crust." The evidence includes rapid migration of seismic hypocenters (solid circles), enhanced rates of metamorphic reaction in major fault or shear zones (open circles), recent studies suggesting much more rapid metamorphism than had been canonically assumed (solid squares), and anthropogenically induced seismicity (open squares); bars depict the full permissible range for a plotted locality and are not Gaussian errors. Red lines indicate k values before and after enhanced geothermal systems reservoir stimulation at Soultz (upper line) (Evans *et al.* 2005) and Basel (lower line) (Häring *et al.* 2008) and green rectangle is the k-depth range invoked in modeling the formation of porphyry-copper ores (Weis *et al.* 2012). (*See color plate section for the color representation of this figure.*)

($k \geq 10^{-16}\,\mathrm{m}^2$), and the generation of elevated fluid pressures ($k \lesssim 10^{-17}\,\mathrm{m}^2$) (Fig. 1.1) – processes which in turn are essential to ore deposition, hydrocarbon migration, metamorphism, tectonism, and many other fundamental geologic phenomena.

In the brittle upper crust, topography, magmatic heat sources, and the distribution of recharge and discharge dominate patterns of fluid flow, and externally derived (meteoric) fluids are common (e.g., Howald *et al.*, this book). In contrast, the hydrodynamics of the ductile lower crust are dominated by devolatilization reactions and internally derived fluids (e.g., Connolly & Podladchikov, this book). The brittle–ductile transition between these regimes occurs at 10–15 km depth in typical continental crust. Permeability below the brittle–ductile transition is non-negligible, at least in active orogenic belts (equivalent to mean bulk k of order 10^{-19} to $10^{-18}\,\mathrm{m}^2$) so that the underlying ductile regime can be an important fluid source to the brittle regime (e.g., Ingebritsen & Manning 2002).

The objective of this book is to synthesize the current understanding of static and dynamic permeability through representative contributions from multiple disciplines. In this introduction, we define crucial nomenclature, discuss the "static" and "dynamic" permeability perspectives, and very briefly summarize the contents of the book. Additional summary and synthesis can be found before and after the three main sections of the book, which are labeled "the physics of permeability," "static permeability," and "dynamic permeability."

NOMENCLATURE: POROSITY, PERMEABILITY, HYDRAULIC CONDUCTIVITY, AND RELATIVE PERMEABILITY

Here, we define some of the key hydrogeologic parameters that are repeatedly used in this book, namely porosity, permeability, hydraulic conductivity, and relative permeability. These are conceptually related but distinct concepts.

First, we note that all of these parameters are continuum properties that are only definable on a macroscopic scale. Perhaps most obviously, at any microscopic point in a domain, porosity ($V_{\text{void}}/V_{\text{total}} = n$) will be either 0 in the solid material or 1 in a pore space. As one averages over progressively larger volumes, the computed value of n will vary between 0 and 1 and, if the medium is sufficiently homogeneous, the volume-averaged value of n will eventually become nearly constant over a volume range, which has been termed the representative elementary volume (REV) (Bear 1972, 1979). Figure 1.2 shows, for example, a hypothetical section of volcanic ash-flow tuff; note the distinctly different porosity of the flow center relative to the flow top and bottom.

The concept of permeability – the ability of a material to transmit fluid – also applies only at an REV scale and can be regarded as reflecting detailed solid–fluid geometries that we cannot map and thus wish to render as macroscale properties. Exact analytical expressions for permeability can be obtained for simple geometries such as bundles of capillary tubes or parallel plates (constant-aperture fractures), but actual pore-fracture geometries are never known.

Porosity (n)–permeability (k) relations have been the subject of many studies (e.g., Luijendijk & Gleeson, this book), and there is often a positive correlation between these two essential quantities. However, even in the case of classical porous media, a correlation between n and k cannot be assumed for mixed-size grains, or when comparing media with greatly different grain sizes. For instance, although there is a positive correlation between n and k for clays themselves, clays are 10^4–10^{10} times less permeable than well-sorted sands (e.g., Freeze & Cherry 1979), despite having generally higher porosities. Furthermore, positive correlation between n and k cannot be assumed in more complex media. Consider again our ash-flow tuff example (Fig. 1.2): the top and bottom of an ash flow cool relatively rapidly, retaining their original high porosities (approximately 0.50), but the permeability of this "unwelded" material is relatively low, because the pores are small and not well connected. If the ash flow is sufficiently thick, pores deform and collapse in the slowly cooling interior, where the final value of porosity can be quite low (<0.05). However, the flow interior also tends to fracture during cooling, and the interconnected fractures transmit water very effectively despite the low overall porosity. The net result of the cooling history is that flow interiors typically have up to 10^4 times higher permeability than "unwelded" flow tops and bottoms, despite their much lower porosities (0.05 vs. 0.50).

Both laboratory and *in situ* (borehole) testing normally return values of hydraulic conductivity (K) rather than permeability (k), and this parameter reflects both rock and fluid properties:

$$K = \frac{k\rho_f g}{\mu_f},$$

where $\rho_f g$ is the specific weight of the fluid and μ_f is its dynamic viscosity. In order to compare rock properties among different geothermal conditions, or different fluids (e.g., hydrocarbons vs. aqueous fluids), it is necessary to convert measured values of K to values of k (e.g., Stober & Bucher, this book). Considering once again our ash-flow tuff example: if the surficial outcrop depicted in Figure 1.2 could somehow be translated from standard temperature and pressure (STP = 15°C, 1 bar) to 300°C and approximately 1000 bars (approximately 10 km depth), without any changes in its physical morphology, its permeability k would not change, but its hydraulic conductivity would be approximately 10 times larger because of the increase in the ρ_f/μ_f ratio.

Finally, the empirically based concept of relative permeability is used to extend the linear flow law for viscous fluids (i.e., Darcy's law) to multiphase systems. Relative permeability (k_r) represents the reduction in the mobility of one fluid phase due to the interfering presence of another fluid phase in the pore space and is treated as a scalar varying from 0 to 1, usually as some function of volumetric fluid saturation (e.g., $V_{\text{liquid}}/V_{\text{void}}$, where for instance $[V_{\text{vapor}} + V_{\text{liquid}}]/V_{\text{void}} = 1$). This concept is widely invoked in the context of hydrocarbon migration and production (oil–gas–liquid water) and unsaturated flow above the water table (air–liquid water), but is also applied to multiphase flow in hydrothermal systems – for instance by Weis (this book), who allows for the presence of three distinct phases in the void space (vapor + liquid + solid NaCl). Because

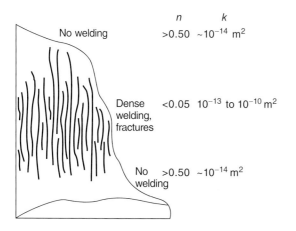

Fig. 1.2. Cross section through a hypothetical ash-flow tuff unit showing typical values of porosity (n) and permeability (k). The thickness of individual ash-flow tuff sheets ranges from a few meters to more than 300 m. Tertiary ash-flow tuffs are widespread in the western United States, particularly in the Basin and Range province. (Adapted from Winograd 1971.)

methane-saturated shales can have very low permeabilities to basinal brines, some studies have used relative permeability effects to explain anomalous pressure in mature sedimentary basins (e.g., Deming *et al.* 2002).

STATIC VERSUS DYNAMIC PERMEABILITY

Some economic geologists, geophysicists, and metamorphic petrologists have long recognized permeability as a dynamic parameter that changes in response to dewatering, fluid production, and seismicity (e.g., Sibson *et al.* 1975; Walder & Nur 1984; Yardley 1986; Hanson 1995; Connolly 1997). For the purposes of this book, we consider "dynamic permeability" to include any transient variation in permeability, regardless of timescale. However, as pointed out by Huber & Su (this book), "dynamic permeability" also has a traditional and much narrower technical definition as frequency-dependent permeability.

The view of permeability as a dynamic parameter varying with time is in stark contrast to the hydrogeologic concept of permeability as a static material property that exerts control on fluid flow. Indeed, the term "intrinsic permeability," widely used in the hydrogeologic and petroleum engineering literature, seems to imply an immutable property.

However, there is abundant evidence that permeability varies in time as well as space, and that temporal variability in permeability is particularly pronounced in environments characterized by strong chemical and thermal disequilibrium. Laboratory experiments involving hydrothermal flow in crystalline rocks under pressure, temperature, and chemistry gradients often result in order-of-magnitude permeability decreases over daily to subannual timescales due to water–rock interaction (e.g., Morrow *et al.* 1981; Moore *et al.* 1994; Yasuhara *et al.* 2006), and field observations of continuous, cyclic, and episodic hydrothermal-flow transients at various timescales also suggest transient variations in permeability (e.g., Baker *et al.* 1987; Hill *et al.* 1993; Haymon 1996; Fornari *et al.* 1998; Sohn 2007). The occurrence of active, long-lived (10^3–10^6 years) hydrothermal systems (Cathles *et al.* 1997), despite the tendency for permeability to decrease with time due to water–rock interaction, implies that other processes such as hydraulic fracturing and earthquakes regularly create new flow paths (e.g., Rojstaczer *et al.* 1995). Indeed, in the past decade, coseismic permeability enhancement and subsequent permeability decay have been directly observed (Elkhoury *et al.* 2006; Kitagawa *et al.* 2007; Xue *et al.* 2013). It is also clear that sufficiently overpressured fluids cannot be contained in the crust and will create the permeability necessary to escape (e.g., Cathles & Adams 2005; Connolly & Podladchikov 2015; Weis, 2015). These various observations have inspired suggestions that

crustal-scale permeability is a dynamically self-adjusting or even emergent property (e.g., Townend & Zoback 2000; Rojstaczer *et al.* 2008; Weis *et al.* 2012), reflecting a dynamic competition between permeability creation by processes such as fluid sourcing and tectonic fracturing and permeability destruction by processes such as compaction, diagenesis, hydrothermal alteration, and retrograde metamorphism. An important caveat is that there is likely a fundamental difference between permeability structure and evolution between prograde and retrograde metamorphism (Yardley, this book). Whereas pervasively wet rocks and near-lithostatic fluid pressures may accompany prograde metamorphism, localized hydration of dry rocks by fluid flow under near-hydrostatic fluid pressures is likely characteristic of retrograde metamorphism.

CONTENTS OF THIS BOOK

The following chapter of this book proposes a data structure to embrace and extend the existing knowledge of crustal permeability. The remainder of this book can be broadly categorized as dealing with the physics of permeability (5 chapters), static permeability (6 chapters), and dynamic permeability (13 chapters). Additional summary and synthesis sections are provided before and after these three main sections of the book.

DATA STRUCTURES TO INTEGRATE AND EXTEND EXISTING KNOWLEDGE

We live in an era of exploding information technology. Thus, the initial chapter in this book, by Fan *et al.*, outlines a vision for the "DigitalCrust": a community-governed, four-dimensional data system of the Earth's crustal structure. The DigitalCrust concept calls for a particular emphasis on crustal permeability and porosity, which have not been synthesized elsewhere and play an essential role in crustal dynamics. The Crustal Permeability data portal associated with this book at http://crustalpermeability .weebly.com/ is a complementary effort intended to unearth permeability data currently tucked away in many dusty corners of the web and in even dustier reports, books, and theses. The intent is to provide links to online, peer-reviewed permeability data that are globally accessible. In contrast to DigitalCrust, the Crustal Permeability data portal will not host data, and data do not have to be spatially located. Data requirements are simply that the data be peer reviewed (published in a peer-reviewed journal, book, or report); include permeability or other related fluid flow and transport parameters; and be hosted and publicly available on an online data repository such as figshare or institutional web pages such as the those of the United States Geological Survey (USGS).

ACKNOWLEDGMENTS

We thank the USGS Powell Center for hosting two workshops that led to the *Geofluids* thematic issue and this book; the USGS Geothermal and Volcano Hazards programs (SEI) and the Natural Sciences and Engineering Research Council (TG); and Erick Burns, Hedeff Essaid, Paul Hsieh, Christian Huber, Jennifer Lewicki, Michael Manga, Craig Manning, Mark Person, and Richard Worden for thoughtful comments that helped to improve the introductory and bridging material.

Mark Ranjram compiled the permeability–depth relations depicted in Figure 1.1 of Chapter 30. We also thank all the authors for their persistence and for their substantial contributions to the *Geofluids* thematic issue and this book; the 65 or so referees who helped to evaluate and improve those contributions; and the Editors and staff of *Geofluids* for their advice and support. Chapters coauthored by TG and SEI were handled with editorial independence.

CHAPTER 2

DigitalCrust – a 4D data system of material properties for transforming research on crustal fluid flow

YING FAN[1], STEPHEN RICHARD[2], R. SKY BRISTOL[3], SHANAN E. PETERS[4], STEVEN E. INGEBRITSEN[5], NILS MOOSDORF[6,7], AARON PACKMAN[8], TOM GLEESON[9], ILYA ZASLAVSKY[10], SCOTT PECKHAM[11], LAWRENCE MURDOCH[12], MICHAEL FIENEN[13], MICHAEL CARDIFF[4], DAVID TARBOTON[14], NORMAN JONES[15], RICHARD HOOPER[16], JENNIFER ARRIGO[16], DAVID GOCHIS[17], J. OLSON[14] AND DAVID WOLOCK[18]

[1] Department of Earth and Planetary Sciences, Rutgers University, New Brunswick, NJ, USA; [2] Arizona Geological Survey, Tucson, AZ, USA; [3] U.S. Geological Survey, Denver, CO, USA; [4] Department of Geoscience, University of Wisconsin-Madison, Madison, WI, USA; [5] U.S. Geological Survey, Menlo Park, CA, USA; [6] University of Hamburg, Hamburg, Germany; [7] Leibniz Center for Marine Tropical Ecology, Bremen, Germany; [8] Department of Civil and Environmental Engineering, Northwestern University, Evanston, IL, USA; [9] Department of Civil Engineering, University of Victoria, Victoria, BC, Canada; [10] San Diego Supercomputer Center, San Diego, CA, USA; [11] INSTAAR, University of Colorado, Boulder, CO, USA; [12] Environmental Engineering and Earth Sciences, Clemson University, Clemson, SC, USA; [13] U.S. Geological Survey, Middleton, WI, USA; [14] Civil and Environmental Engineering, Utah State University, Logan, UT, USA; [15] Civil and Environmental Engineering, Brigham Young University, Provo, UT, USA; [16] CUAHSI, Boston, MA, USA; [17] NCAR, Boulder, CO, USA; [18] U.S. Geological Survey, Lawrence, KS, USA

ABSTRACT

Fluid circulation in the Earth's crust plays an essential role in surface, near-surface, and deep-crustal processes. Flow pathways are driven by hydraulic gradients but controlled by material permeability, which varies over many orders of magnitude and changes over time. Although millions of measurements of crustal properties have been made, including geophysical imaging and borehole tests, this vast amount of data and information has not been integrated into a comprehensive knowledge system. A community data infrastructure is needed to improve data access, enable large-scale synthetic analyses, and support representations of the subsurface in Earth system models. Here, we describe the motivation, vision, challenges, and an action plan for a community-governed, four-dimensional data system of the Earth's crustal structure, composition, and material properties from the surface down to the brittle–ductile transition. Such a system must not only be sufficiently flexible to support inquiries in many different domains of Earth science, but it must also be focused on characterizing the physical crustal properties of permeability and porosity, which have not yet been synthesized at a large scale. DigitalCrust is envisioned as an interactive virtual exploration laboratory where models can be calibrated with empirical data and alternative hypotheses can be tested at a range of spatial scales. It must also support a community process for compiling and harmonizing models into regional syntheses of crustal properties. Sustained peer review from multiple disciplines will allow constant refinement in the ability of the system to inform science questions and societal challenges and to function as a dynamic library of our knowledge of Earth's crust.

Key words: data integration, deep-crustal dynamics, earth system models, groundwater, groundwater–surface water interaction, permeability

MOTIVATION

Fluid flow in the Earth's crust depends strongly on material permeability, which varies in space and through time. As data and knowledge accumulate, and as we increasingly tackle interdisciplinary questions (Bodnar *et al.* 2013), a georeferenced, time-evolving data system of crustal structure and properties is needed to address a wide range of scientific and societal questions.

Understanding Earth's critical zone

The Earth's critical zone is the region from the top of the terrestrial biosphere to the depth of active groundwater circulation

Crustal Permeability, First Edition. Edited by Tom Gleeson and Steven E. Ingebritsen.
© 2017 John Wiley & Sons, Ltd. Published 2017 by John Wiley & Sons, Ltd.
Companion Websites: www.wiley.com/go/gleeson/crustalpermeability/
http://crustalpermeability.weebly.com/

(National Research Council 2001). Critical zone science focuses on understanding the physical, chemical, and biological processes regulating critical zone evolution, determining its role in sustaining human society and terrestrial ecosystems, and predicting responses to anthropogenic, climatic, and tectonic forcing (Banwart *et al.* 2013). Fluid circulation plays a central role in critical zone processes, regulating chemical weathering, soil formation, ecosystem evolution, and biogeochemical cycling (Berner & Berner 1996; Jones & Mulholland 2000; Brantley *et al.* 2011; Boano *et al.* 2014). Carbon cycle research has focused on the Earth's atmosphere and surface, but 99.9% of all carbon is stored in the lithosphere (Kempe 1979). Thus, even small changes in fluxes from the crust can have major consequences for the ocean–atmosphere system. Chemical weathering, a primary driver of global biogeochemical cycling, depends strongly on subsurface water residence times (Berner 1978; Maher & Chamberlain 2014), which is primarily controlled by 3D hydrological flow paths and material rock properties (McGuire *et al.* 2005). Weathering depth is unknown (West 2012) yet critical to understanding global biogeochemical fluxes. Existing predictions of material fluxes are based on 2D bedrock geological maps and, therefore, neglect deeper rock strata and geothermal waters (e.g., Becker *et al.* 2008). A major advance in overcoming these and many other limitations would be a 4D knowledge system for managing and synthesizing existing and newly acquired data on the Earth's crust.

Assessing resource sustainability

Groundwater is the largest freshwater resource and primary source of drinking water for two billion people (Morris *et al.* 2003a). It also plays a central role in agriculture (Foster & Chilton 2003; Giordano 2009) and sustains the health of many ecosystems (Alley *et al.* 2002). Nevertheless, groundwater is not adequately managed to ensure sustainability (Danielopol *et al.* 2003; Foster & Chilton 2003; Brunner & Kinzelbach 2005; Konikow & Kendy 2005; Fogg & LaBolle 2006; Gleeson *et al.* 2010; Sophocleous 2010), and nearly a quarter of humanity lives in areas of groundwater stress (Gleeson *et al.* 2012a). A key factor in sustainability is groundwater residence time related to the renewal rate, which can be many millennia, well beyond the typical time horizon of human policies (Gleeson *et al.* 2012b). Residence time has been modeled assuming a consistent decrease in permeability with depth (Jiang *et al.* 2010b), but a single low permeability layer can control groundwater age (Gassiat *et al.* 2013).

Fluid hydrocarbons in the upper crust also currently play a vital role in the energy budget for society. Knowledge of subsurface structures and properties is a prerequisite for addressing many of the energy issues surrounding energy resources, including harvesting of geothermal energy (Mortensen & Axelsson 2013), carbon sequestration (Shrag 2007; Benson & Cole 2008), exploitation of unconventional oil/gas reservoirs,

and fluid-injection-induced seismicity associated with all of these activities (Hitzman *et al.* 2012).

Understanding deeper crustal dynamics

Hydrogeologists, geologists, and geophysicists have begun to actively explore the role of groundwater and other subsurface fluids in fundamental geologic processes, such as crustal heat transfer, ore deposition, hydrocarbon migration, seismicity, tectonic deformation, and diagenesis and metamorphism (e.g., Burns *et al.* 2015; Connolly & Podladchikov 2015; Howald *et al.* 2015; Micklethwaite *et al.* 2015; Miller 2015; Okada *et al.* 2015; Weis 2015). The permeability of the Earth's crust is of particular interest because it largely determines the feasibility of important physicochemical processes, such as advective solute/heat transport (Burns *et al.* 2015; Saffer 2015) and the generation of elevated fluid pressures by processes such as physical compaction, heating, mineral dehydration, and fluid injection (Connolly & Podladchikov 2015; Miller 2015; Weis 2015).

Current understanding supports a general distinction between the hydrodynamics of the brittle upper crust, where hydrostatic fluid pressures are the norm and meteoric fluids are common, and those of the ductile lower crust, where metamorphic reactions and internally derived fluids dominate hydrodynamic behavior. The brittle–ductile transition between these regimes depends on temperature, strain rate, and rheology, but occurs at 10–15 km depth in typical continental crust. In tectonically active regions, high permeability episodically exists below the brittle–ductile transition (Connolly & Podladchikov 2015), such that fluid input from the ductile regime can be important to the cycling of some elements, and perhaps even to the balancing of the global water cycle over geologic time (Ingebritsen & Manning 2002).

This book highlights the historical dichotomy between the hydrogeologic concept of permeability as a static material property that exerts control on fluid flow and the perspective of other Earth scientists who have long recognized permeability as a dynamic parameter that changes in response to tectonism, devolatilization, and geochemical reactions. The dynamic view of crustal permeability is consistent with indications that fluid pressure is close to the lithostatic load during prograde metamorphism below the brittle–ductile transition (e.g., Fyfe *et al.* 1978); sufficiently overpressurized fluids cannot be contained in the crust, leading to fracturing and other processes that create permeability. More recently, it has been suggested that the permeability of the brittle crust may also be dynamically self-adjusting, responding to tectonism and external fluid sources much as the deeper crust responds to the magnitude of internal fluid sources (cf., Cathles & Adams 2005; Rojstaczer *et al.* 2008; Weis *et al.* 2012). The temporal evolution of permeability can be abrupt or gradual: stream-flow responses to moderate-to-large earthquakes demonstrate that dynamic stresses can instantaneously change permeability by factors of up to 20 on a regional scale, whereas a 10-fold decrease in the permeability of a package of shale in a compacting basin may require 10^7 years (Ingebritsen & Gleeson 2015). Thus, in the absence of seismicity, assuming that permeability is a static parameter can be reasonable for low-temperature hydrogeologic investigations with timescales of days to decades. Data compilations of deeper crustal material properties are likely to lead to a markedly better understanding of deeper crustal dynamics.

Supporting earth system modeling

There is an urgent need for large-scale data synthesis to support the development of integrated earth system models, which account for material and energy fluxes and key abiotic–biotic interactions in the atmosphere, lithosphere, and hydrosphere. Earth system models are critical tools for predicting future global environmental change, such as that addressed by the Intergovernmental Panel on Climate Change. However, even well-understood groundwater–surface water interactions in the top tens of meters of the crust are poorly represented in current earth system models, and most do not include subsurface processes at depths >2–3 m. Efforts to extend earth system models deeper into the crust have been hindered by deficiencies in subsurface data. Global, realistic 3D gridded permeability and porosity fields for continental crust do not yet exist, but

recent efforts to map near-surface permeability and porosity (Gleeson *et al.* 2014) provide an important starting point.

DATA INTEGRATION TO TRANSFORM SCIENCE

Table 2.1 is a partial list of ongoing data integration efforts that have impacted our views of Earth systems interactions in many different ways. One example is the Macrostrat database (Peters 2006), which integrates existing stratigraphic information and aims to represent the Earth's upper crust as surface polygons that extend from the surface downward as stacks of lithostratigraphic and chronostratigraphic units. Macrostrat has integrated more than 36,000 rock units in North America, New Zealand, and the deep sea and is being augmented with the DeepDive machine reading system (Peters *et al.* 2014). Interactions between biotic and abiotic processes leave signatures in the rock record, and Macrostrat puts these signatures back into stratigraphic context, allowing them to be quantified in a space–time framework. Fossil records in the Paleobiology Database and the GPlates paleogeographic reconstructions are integrated with these data to produce a 4D model of the evolving Earth. Global-scale, deep-time syntheses of biological, geochemical, and sedimentary data have allowed new quantitative tests of long-standing hypotheses. For example, large-scale

Table 2.1 Examples of ongoing data integration efforts and the starting point of DigitalCrust

Data	Source	Format
World Topography, Bathymetry	CSDMS: http://csdms.colorado.edu/wiki/Topography_data	Gridded
FAO World Harmonized Soil Map	IIASA: http://webarchive.iiasa.ac.at/Research/LUC/External-World-soil-database/HTML/	Global gridded, polygons for countries
Global Lithologic Map	University of Hamburg: http://www.clisap.de/research/b:-climate-manifestations-and-impacts/crg-chemistry-of-natural-aqueous-solutions/global-lithological-map/	Surface polygons
World Geologic Maps	USGS WMS and ESRI map services: http://energy.usgs.gov/OilGas/AssessmentsData/WorldPetroleumAssessment/WorldGeologicMaps.aspx	Surface polygons
World Tectonic Stress Map	GFZ Potsdam: http://dc-app3-14.gfz-potsdam.de/pub/introduction/introduction_frame.html	Gridded, with points and lines
Global Sediment Thickness	UCSD: http://igppweb.ucsd.edu/~gabi/sediment.html	Gridded
Global Map of Surface Heat Flow	Map: Cardiff U: http://onlinelibrary.wiley.com/doi/10.1002/ggge.20271/abstract, Point: IHFC: http://www.heatflow.und.edu/	Gridded and points
National Geothermal Data System, SMU Geothermal Map	http://geothermaldata.org/, http://www.smu.edu/Dedman/Academics/Programs/GeothermalLab/DataMaps	Gridded maps and points, thermal profiles, thermal conductivity
Continental Stratigraphy	University of Wisconsin: http://macrostrat.org/. Data set of polygons tessellating North America, with associated time-stratigraphy description	Polygons with vertical sequence of layers
Global Aquifer Maps	BGR and UNESCO: http://www.whymap.org/whymap/EN/Home/whymap_node.html	Surface polygons
US Aeromagnetic Survey	USGS: http://mrdata.usgs.gov/magnetic/	Gridded, resolutions vary from region to region
US Gravity Anomaly	USGS: http://mrdata.usgs.gov/geophysics/gravity.html	Gridded
Groundwater Atlas, 25 US Aquifers	USGS: http://pubs.usgs.gov/ha/ha730/gwa.html	Surface polygons with thickness (isopachs)
Global Permeability and Porosity	McGill University (GLHYMPS): http://onlinelibrary.wiley.com/doi/10.1002/2014GL059856/abstract	Surface polygons

compilations of sedimentary data have played an important role in modeling biogeochemical cycling (e.g., Ronov 1978; Berner 2004), and Macrostrat has been used to calibrate sulfate burial fluxes and to better constrain the role of the sulfur cycle in regulating atmospheric oxygen (Halevy *et al.* 2012; Canfield & Kump 2013). Spatial–temporal patterns of sedimentation in Macrostrat have also been shown to quantitatively reproduce many major features in the macroevolutionary history of marine animals (Peters 2005, 2008b; Finnegan *et al.* 2011) and planktonic foraminifera (Peters *et al.* 2013). Combined with stable isotopic proxy records of biogeochemical cycling, global temperature, and rates of volcanism and crustal weathering, it appears likely that the correlations between paleobiological and macrostratigraphic data reflect common biological and stratigraphic responses to Earth system changes (e.g., Peters 2005; Hannisdal & Peters 2011), a hypothesis that emerges from, and can only be adequately tested with, integrated data deriving from the Earth's crust.

A second example is the UN-FAO Global Harmonized Soil Database. Large amounts of soil survey data from multiple nations and continents, often built using different soil taxonomies, horizon definitions and attributes, and compiled at different scales of resolution and with different formats, were harmonized through an international partnership, which defined a new set of soil attributes critical to agriculture and recommended methodologies for developing taxo-transfer rules. The result was a global data set at 30 arc-sec grids with 20 soil physical, chemical, and biological attributes. This data set (and predecessors) has been the sole basis for deriving soil hydraulic parameters necessary for calculating soil water fluxes in all global land models and servers as the primary resource for constraining global soil organic carbon stocks and fluxes (e.g., Batjes 1996; Hiederer & Kochy 2011).

THE DIGITALCRUST VISION

We envision a 4D space–time (xyz–t) data infrastructure designed to accommodate the structure and properties of the upper crust, from the surface down to the brittle–ductile transition, which occurs at 10–15 km depth in continental crust with a geothermal gradient of ~25–30°C km^{-1} but can be as shallow as 4–5 km in regions of high heat flow. In regions with adequate seismic networks, the brittle–ductile transition can be crudely mapped on the basis of the distribution of earthquakes with depth (e.g., Nazareth & Hauksson 2004; Tanaka & Ishikawa2005).

DigitalCrust must be a web-oriented, data-service-enabled, and spatially and temporally referenced workspace where the geosciences community can contribute and register data and model outputs, visualize, explore, and synthesize existing data to test hypotheses across space–time in ways that account for uncertainties. This is a daunting task and will require support from the broader Earth science community, including from initiatives such as EarthScope, national and regional geologic surveys, and funding agencies. Later, we describe some of the key elements required in DigitalCrust.

A geologic scaffolding

The foundation of DigitalCrust is a geologic scaffolding that describes the basic geologic fabric of the Earth's upper crust, from the critical zone to the brittle–ductile transition, and includes data spanning its full range of physical, chemical, and biological properties (Fig. 2.1). To accomplish this, DigitalCrust must receive contributions from all disciplinary domains involving the lithosphere, the hydrosphere, and the biosphere. Thus, despite the fact that it was originally motivated by the need to better understand and model crustal fluid flow, it must be an integrative data infrastructure that spans multiple domains of expertise in the Earth sciences. This broad vision is an attempt both to express the actual level of Earth systems integration that we believe occurs in nature and to respond to a common scientific and data infrastructure need that has been expressed in many Earth science communities. Because the most relevant intersection for many different types of geoscientists is defined by the common field location and rocks that they work on, regardless of whether or not they share any scientific expertise or disciplinary knowledge, DigitalCrust stands to promote both data discovery and interdisciplinary cross-fertilization by proactively connecting scientists on the basis of their intersection in the Earth's crust.

Hydrogeologic properties as key data content and service

Within the foundational geologic scaffolding, DigitalCrust will support multiscale integration of fluid-relevant properties. Improved description and synthesis of these properties, particularly permeability and porosity, have been a driving force behind DigitalCrust. Although millions of soil and aquifer analyses and measurements have been made, the data are dispersed and unstructured in the scientific literature, government archives, and myriad online web pages and repositories. Scales, standards, and formats also vary. We face several major challenges, including discovering this vast amount of information and organizing it within the geologic scaffolding and developing automated methods and algorithms for deriving meaningful hydrogeologic properties based on multiple data types.

Community knowledge repository and management system

As a community knowledge repository, DigitalCrust will integrate existing large-scale data sets (e.g., Table 2.1) and leverage current visualization tools to allow scientists to view what data already exist at given xyz–t coordinate and within a domain context, and what data/knowledge gaps remain to be filled. It will then allow scientists to contribute data sets to the growing knowledge base through a DigitalCrust node, with support for placing the data in an archival repository, obtaining an identifier

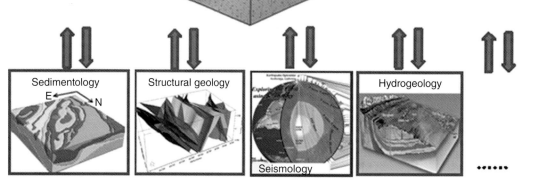

Fig. 2.1. The geologic scaffolding of DigitalCrust from the critical zone to the brittle–ductile transition (A), receiving contribution from and delivering service to a wide range of Earth science disciplines (B–E). (Image source: (A) Adapted from Winter *et al.* (1998), (B) McInerney *et al.* (2005), (C) Hinz *et al.* (2012), (D) IRIS (http://www.iris.edu/hq/), and (E) Paschke *et al.* (2011)). (*See color plate section for the color representation of this figure.*)

for data, and releasing it for community use. Contributors can view how their new entries fit into or impact the framework and receive a response from the system with recommendations on related data that they may not be aware of, as well as recognition of their data/knowledge contribution.

As a knowledge management system, DigitalCrust will index geoscience data sources from raw observation, through multiple levels of processing, interpretation, integration, and synthesis into models that are also incorporated into the repository. Linkage between observations and derived data sets through this chain should allow tracing provenance of information. The system should also include tools for social interactions such as review, discussion, correction, and updates to observations and interpretations at all levels. The resources in this system are accessed using simple web protocols and interchange formats that are documented, tested, and adopted by the DigitalCrust community. The data/information at a given geographic reference point will be delivered via an open application programming interface (API) that will support the

development of specialized third-party applications as well as the DigitalCrust online resource itself.

Central to the vision is the use of a branching and versioning system, such as "Git" and "GitHub" in software development, which supports a common repository of best available data and most proven models, while allowing any researcher to create their own development fork. Formal peer review and community consensus will integrate branches back into the master DigitalCrust branch. Borrowing from the genomics community, which allows microcitation to unambiguously reference discrete data on organisms (Patrinos *et al.* 2012), DigitalCrust will provide a capacity for citing and referencing data and data products.

Given the anticipated scope, DigitalCrust must be governed by the community it intends to serve. It differs from many common crowdsourcing models in that contributions will be attributed to specific members of the scientific community, allowing the community to regulate itself by, for example, trusting or not trusting the contributions based on individually demonstrated knowledge and expertise. A community governance model, to help sustain the integrity of the system as a whole, is being advanced as part of the NSF EarthCube initiative, which seeks to establish transformative cyber-infrastructure in support of the geosciences. Key features of organizational governance will likely involve standards for adoption and verification, organizational commitment through a membership process, impartial advisory boards, and other tested mechanisms to ensure system viability and sustainability.

Flexible information architecture

The DigitalCrust platform must provide a modular, configurable data storage and access component that is not only sufficiently flexible to interface with existing databases and technologies but also structured in such a way as to provide a useful synthetic resource. "Standard" database designs have been developed for community use, but users inevitably find that there are missing entities and properties. The emergence of no-schema, document-type databases, such as CouchDb and MongoDb (e.g., Sadalage & Fowler 2012), provides technologies for hybrid fixed-schema and open-world information exchange models. The basic idea is to define a schema for common information items that are broadly shared, such as geologic unit descriptions. Document-type databases allow unlimited addition of new properties to any entity as key–value pairs, or more complex multivalue data structures; thus, a standard schema can be readily extended by any individual or group. If the properties in new schema or schema extensions are mapped to properties in the existing information model, it then becomes possible to automate integration between data using the different schema. If new entities and properties emerge that many users find useful, they are documented and registered for consistent reuse and greater interoperability. This approach has been deployed in the National Geothermal Data System as a basis for information exchange using web services (Anderson *et al.* 2013) and in USGS ScienceBase as a method for continually expanding data capabilities with new access, analysis, and visualization parameters. DigitalCrust will extend this concept, using content models as "document templates" in a no-schema database that will provide the open-world flexibility and extensibility required by geoscientists, while also promoting standardization of commonly used entities and properties, such as lithostratigraphically defined local and regional rock units.

For geofluids applications, the DigitalCrust architecture needs to assimilate observational data and interpretations from all available sources, including geologic maps, cross sections and structural contours, hydrogeologic unit delineations, soil tests, slug tests, aquifer pump tests, and indirect property estimates obtained through model inversions. The DigitalCrust architecture should be flexible such that researchers can upload any data and create products or models at any scales, choosing from a variety of automated methods, while supporting uncertainty propagation in derived products. Close disciplinary engagements are required to assure that data are used and interpreted properly in syntheses.

AN ACTION PLAN

To ensure the success of DigitalCrust, we must reach out to the broader Earth sciences community, tapping common visions, synergistic efforts, and funding support to build the next generation of Earth science data infrastructure in a distributed, loosely coupled architecture. The NSF EarthCube program, along with the USGS John Wesley Powell Center for Earth System Analysis and Synthesis, is poised to support these activities, bringing together Earth scientists and computer scientists to tackle some of the biggest data challenges. The first step to be taken is to use available collaborative mechanisms to engage additional disciplinary experts, data owners, and use case testers as we begin bringing together architectural and data components.

The second step is to set up the basic system architecture and integrate the existing community data systems, such as those listed in Table 2.1. This will allow us to demonstrate the concept immediately and expose data availability and gaps. Some simple visualization capabilities will be developed leveraging the development in other Earth science communities. This will prepare us to develop community-sourcing capabilities that allow data uploading, indexing, and editing, as well as a discussion forum for testing multiple interpretations or models.

The third step, of immediate interest to the geofluids community, is to develop and test the capabilities of the system to generate 3D gridded data sets of crustal permeability, porosity, and other relevant properties, integrating multiple data types, scales, and levels of uncertainty. Research is needed to define models, standards, and rules of data harmonization, and a working group will be formed to help guide technical development of these standards. This will connect DigitalCrust with the science and society motivations discussed in Section "Motivation" and facilitate the longer term process of building a coherent data system in support of crustal fluid investigations.

CONCLUDING REMARKS

The current need for Earth system–level syntheses related to crustal fluid dynamics, the explosion of information on crustal structure and material properties, and rapid advances in computing and information science and technology have all converged to both enable and require the development of DigitalCrust. It is a nontrivial task, one that requires transdiscipline, transcommunity, and transagency collaboration in a sustained effort. The NSF EarthCube program presents one opportunity to construct DigitalCrust, primarily because both are aligned by their need to engage a much broader swath of the geoscience community than typically routinely collaborates. The potential utility of DigitalCrust as a community resource for hydrogeologists to better understand fluid flow in the Earth system and its role in Earth's material and energy cycles at multiple scales and to more broadly reach the geoscience community, does, however, provide ample motivation.

ACKNOWLEDGMENTS

This project is supported by the joint NSF–USGS John Wesley Powell Center for Earth System Analysis and Synthesis working group and an NSF EarthCube Geo-Domain Community Workshop grant (EAR-1251557). We thank Jeanne DiLeo at the USGS for graphic support. The authors declare no conflict of interests. Any use of trade, product, or firm names is for descriptive purposes only and does not imply endorsement by the US Government.

PART I

The physics of permeability

CHAPTER 3

The physics of permeability

TOM GLEESON[1] AND STEVEN E. INGEBRITSEN[2]

[1] *Department of Civil Engineering, University of Victoria, Victoria, BC, Canada;* [2] *U.S. Geological Survey, Menlo Park, CA, USA*

Key words: Darcy's law, dynamic permeability, inertial forces, non-Newtonian fluids, fracture permeability

Darcy's law is an expression of conservation of momentum that describes viscous fluid flow through a porous medium. It provides the scientific basis for the concept of permeability. The first two chapters of this section explore limits to the validity of Darcy's law. For porous media subjected to harmonic pressure forcing, the effective permeability is frequency dependent, an effect that has classically been represented as a dynamic correction to the effective fluid viscosity and has been termed "dynamic permeability" (a term that in this book we define more broadly to include all temporal changes in permeability). This results from the fact that Darcy's law (and therefore permeability) was originally developed for cases where inertial forces are negligible. When inertial forces are significant, the ratio of fluid flux to pressure gradient may not be constant. The seminal model of Johnson *et al.* (1987) (JKD) defined a critical frequency that represents the transition from viscous- to inertia-dominated momentum balance in homogeneous porous media. Lattice Boltzmann simulations by Huber & Su (this book) exhibit good agreement with JKD except for their most heterogeneous example, which exhibits a resonance behavior (their Fig. 4.7). They consider adding a transient term to the flux equation, which results in a hyperbolic (rather than parabolic) model for mass conservation in porous media (their Eq. 4.21) but fails to explain the entire suite of numerical results. Laboratory experiments by Medina *et al.* (this book)

also explore limits to Darcy's law; they show that, for fluids containing suspensions of solids, the linear relationship between fluid flux and pressure gradient that applies for Newtonian fluids breaks down, and strongly heterogeneous velocity fields develop within fractures. Small (3%) variations in solid volume fraction can cause twofold velocity variation.

The next three chapters address several aspects of permeability characterization in fractured rocks, long recognized as a particularly challenging hydrogeologic problem. Selvadurai (this book) explores the influence of stress state (axial normal stress) on the permeability of a single fracture through cylinders of Barre Granite and demonstrates $>10^2$-fold variation in fracture k in response to axial stress ranging from 0 to $7+$ MPa, with k hysteresis ($<10^1$-fold) between loading and unloading cycles. Ishibashi *et al.* (this book) qualitatively and quantitatively demonstrate channeling of flow by fracture roughness, and the evolution of channeling with shear displacement, relating the moment magnitude of microearthquakes to enhancement of fracture permeability. Finally, Rutqvist (this book) reviews a variety of field data on stress-induced permeability changes in fractured rock and also discusses the effects of thermally and chemically induced fracture closure; his Figure 8.14 summarizes fractured rock k to 0.6 km depth at Gideå, Sweden, and his Figure 8.16 summarizes crystalline bedrock k to $7+$ km depth from a variety of sites.

CHAPTER 4

A pore-scale investigation of the dynamic response of saturated porous media to transient stresses

CHRISTIAN HUBER AND YANQING SU

School of Earth and Atmospheric Sciences, Georgia Institute of Technology, Atlanta, GA, USA

ABSTRACT

The dynamical response of saturated porous media to transient stresses is complex because of the coupling between the solid and fluid phases. Over the last three decades, theoretical models have emerged and they predict that the transient response of porous media to pore-pressure fluctuations depends only on a single dimensionless number. This single parameter represents the ratio of the forcing frequency to a characteristic frequency of the medium. Although theoretical models for the frequency dependence of the effective permeability of the medium have successfully predicted the response of porous media at high frequency observed in laboratory and numerical experiments, they rely on assumptions that limit their applicability to homogeneous media and narrow pore-size distributions. We use pore-scale flow simulations with four different porous media topologies to study the effect of pore geometry and pore-size distribution on the dynamic response to transient pore-pressure forcing. We find a good agreement with published theoretical work for all but one medium that exhibits the broadest pore-size distribution and, therefore, the largest degree of pore-scale heterogeneity. Our results suggest the presence of a resonance peak at high frequency where the discharge, and therefore the effective permeability, is significantly amplified compared with their value around the resonant frequency. We suggest two interpretations to explain resonance. At the continuum scale, a finite speed of pore-pressure propagation during transients requires the addition of a correction term to Darcy's law. We derive a hyperbolic mass conservation equation that admits resonance under certain conditions. This model can explain the peak observed for one medium but fails to explain the absence of resonance displayed by the three other media. The second interpretation, motivated by the different pore-size distributions for the texture where resonance is observed, calls for pore-scale processes and heterogeneous pore-pressure distribution among primary and secondary flow pathways.

Key words: dynamic permeability, pore-scale modeling, saturated porous media

INTRODUCTION

The Earth's crust is porous and heterogeneous over a wide range of length scales (Ingebritsen & Manning 2010). In the upper few kilometers, the pore space is generally saturated with aqueous fluids that play a significant role in water–rock chemical reactions. Pore fluids also affect the physical properties of the heterogeneous medium. One of the most important properties for mass and heat transport is the permeability of the matrix to the flow of fluids. Besides long-term changes in permeability associated with chemical reactions and pore clogging, the response to rapid stress transients (dynamic stresses) can also significantly affect the effective permeability of saturated porous media because of poroelastic effects (Terzaghi 1925; Biot 1956a,b, 1962).

Maurice Biot developed the first theoretical model of the propagation of acoustic waves in saturated porous media in the mid-1950s. Biot found two independent solutions to the propagation of acoustic waves in porous media that he referred to as waves of the first and second kind. For both types of waves, the net drag force exerted by the solid matrix on the fluid controls the dissipation of mechanical energy and by extension the attenuation of the propagating waves. At low frequency, the wave of the second kind, sometimes referred to as Biot's slow wave, consists of pressure transport by diffusion (Darcy transport) and can lead to larger attenuation over the seismic frequency band (Pride *et al.* 2002). An important concept that emerged from the work of Biot is that the drag force that couples the fluid to the solid matrix is frequency dependent and can be represented as a dynamic correction on the effective fluid viscosity.

Following the seminal work of Biot, several groups studied the discharge of pore fluids subjected to harmonic pressure forcing and found, in agreement with Biot's analysis, the effective permeability of the medium to be frequency dependent and complex valued at high frequency ($\pi/2$ phase lag between the forcing and the discharge) (Biot 1962; Johnson et al. 1987; Smeulders et al. 1992; Berryman 2003). The frequency-dependent permeability is often referred to as the dynamic permeability of a medium. The Johnson–Koplik–Dashen (JKD) dynamic permeability model offered the first scaling relationship to model the frequency dependence of permeability (Johnson et al. 1987). It relies on a single tree parameter, the rollover frequency ω_c, which depends on fluid and matrix properties. This critical frequency represents the transition from viscous- to inertia-dominated momentum balance in a homogeneous porous medium (order of kilohertz or higher for water-saturated sandstones) (Pride et al. 2002). Later, the model was verified experimentally and numerically using simple geometries (Johnson et al. 1987; Sheng & Zhou 1988; Smeulders et al. 1992; Muller & Sahay 2011; Pazdniakou & Adler 2013). In addition, Smeulders et al. (1992) provided a more rigorous mathematical validation of the model using standard homogenization methods.

For the most part, pore-scale calculations have been limited to simple and highly symmetric geometries because of computational limitations. Numerical results showed a good agreement with the existence of a single scaling function (related to the critical frequency) that was assumed in Johnson et al. (1987). Recently, a pore-scale study using a lattice Boltzmann flow solver was able to extend the range of validation to more realistic geometries (Pazdniakou & Adler 2013). Their study actually solved for the dynamic permeability with a different problem setup. The porous medium is periodic and therefore infinite (no boundary conditions), the flow is buoyancy driven, and the forcing is homogeneous and applied through a transient harmonic perturbation of the bulk force responsible for the flow. In this study, we use a lattice Boltzmann pore-scale flow model to investigate different porous media topologies and their effect on the dynamic permeability and verify the scaling of the JKD model. Each domain has finite dimensions and the flow is pressure driven. The dynamic forcing is introduced by an imposed pore-pressure oscillation at one of the boundary. Although the difference in these two models is subtle, the choice of setup can lead to significantly different outcomes, and the effect of finite domains and pressure boundary conditions needs to be studied. We observe that the dynamic permeability response is generally in good agreement with the scaling proposed by Johnson et al. (1987) and Smeulders et al. (1992). We discuss these results in analogy to well-known properties of electric circuit involving a resistor and a capacitor in parallel (Debye relaxation).

In specific cases, however, we observe a significant departure from the JKD (or Debye relaxation) model. In particular, we observe features that suggest a resonance behavior that is not consistent with the theory of Johnson et al. (1987)

and Smeulders et al. (1992). We propose two alternative explanations for the existence of resonance. First, using a continuum-scale argument, we discuss the importance of a correction term to Darcy's law for transient flows that allows us to derive a hyperbolic version of the mass conservation equation. We show that this new mass conservation equation converges to the standard parabolic diffusion of pore pressure at low frequency, but allows the propagation of damped waves and resonance at high frequency. Although this continuum-scale model offers a satisfying framework to explain the occurrence of resonance in one medium, it fails to explain the lack of resonance in the three other media. Alternatively, we suggest that pore-scale effects, such as different pore-size distributions (PSDs), can facilitate pore-pressure excitation between heterogeneous flow pathways in response to forced pore-pressure excitations.

BACKGROUND

Maurice Biot (1962) introduced a model for the propagation of acoustic waves in saturated porous media by computing the net drag force between the fluid and an oscillating matrix under simplified geometry such as Poiseuille and duct flows. He showed that the drag associated with harmonic forcing leads to an effective fluid viscosity that can become complex and that displays a frequency dependence. It is important to note that Biot's approach was conducted at the pore scale and used an unsteady version of Stokes equation for the flow in the limit of a compressible fluid. The momentum equation represented, therefore, a balance between three terms: inertia, pressure, and viscous stresses.

Johnson et al. (1987) and Smeulders et al. (1992) showed that by matching inertial forces associated with the transient forcing with viscous forces, a critical frequency emerges

$$\omega_c = \frac{\eta}{\rho_f k_0 \alpha_\infty}, \tag{4.1}$$

where η is the dynamic viscosity of the pore fluid, ρ_f its density, k_0 the static permeability, ϕ the porosity of the medium, and lastly, α_∞ is the dynamic tortuosity at infinite frequency of the medium, which is related to the formation factor $F = \phi/\alpha_\infty$ (Johnson et al. 1987).

The ratio of the forcing frequency ω to ω_c controls the dynamic response of saturated (homogeneous) porous media to transient pore-pressure forcing. It is important to note that, in isotropic homogeneous media, the effect of the microstructure of the porous medium only emerges through three independent scalar parameters: the static ($\omega \to 0$) permeability, porosity, and formation factor of the medium. As discussed by Pride et al. (2002), once corrected for the dynamic permeability response, using mass conservation, the partial differential equation that describes the evolution of the pore pressure is a diffusion equation and is therefore parabolic and dissipative. We therefore expect that the forced pressure oscillations decay with time as expected from a diffusion equation.

METHODS

Computation of the formation factor

We constructed four porous media synthetically either using a stochastic algorithm of crystal nucleation and growth following the method described in Hersum and Marsh (2006) (textures referred to as cuboids and spheres) or creating void space with simple geometrical shapes (e.g., spheres, tubes, and wavelike tubes). Figure 4.1 shows the pore structure of the four media.

The formation factor of a heterogeneous medium is defined as the effective electric conductivity of the medium normalized to that of the pore-filling fluid assuming the solid matrix is a perfect insulator. We use the analogy between the steady-state solution of the (heat) diffusion equation and the solution to Poisson's equation for the electric potential in heterogeneous media.

We apply a 3D lattice Boltzmann model for heat conduction in porous media. In the lattice Boltzmann method, the diffusion equation is modeled following a statistical approach, and it is replaced by a discrete version of Boltzmann's equation. Boltzmann's equation describes the evolution of particle probability functions $g(\mathbf{x},\mathbf{v},t)$ that represent the probability of finding a particle at position $\mathbf{x}$, traveling with velocity $\mathbf{v}$ at time t. Particles stream through the domain and collide with each other, which leads to the following equation for the evolution of the g_is:

$$g_i(\mathbf{x} + \mathbf{e}_i dt, t + dt) = g_i(\mathbf{x}, t) - \frac{1}{\tau_b}[g_i(\mathbf{x}, t) - g_i^0(\mathbf{x}, t)], \quad (4.2)$$

where it is assumed that the collision operator reduces to a single relaxation time (Bhatnagar *et al.* 1954). The index i refers the discrete set of possible trajectories $\mathbf{e}_i$ on the lattice (nearest neighbors), τ_b is the relaxation time (related to the thermal diffusivity), and g_i^0 is the local equilibrium particle probability

distribution function. The macroscopic field of interest, here temperature $T(\mathbf{x},t)$, is obtained by summing the local distribution functions

$$T(\mathbf{x}, t) = \sum_i g_i(\mathbf{x}, t). \quad (4.3)$$

The equilibrium distributions are linearly dependent on the local temperature

$$g_i^0(\mathbf{x}, t) = t_i T(\mathbf{x}, t), \quad (4.4)$$

where t_i is lattice weights, which in our model with seven discrete velocity $\mathbf{e}_i$ are $t_0 = 1/4$, $t_{1-6} = 1/8$ (see Fig. 4.2).

This method has been shown to recover the diffusion equation in 3D and allows for a simple treatment of the internal solid–fluid boundaries (no heat flux). This is done with the bounce-back rule that specifies that distributions are reflected on solid obstacles

$$g_i(\mathbf{x}, t) = g_{\bar{i}}(\mathbf{x}, t) \quad \mathbf{x} \in \text{Solid}, \quad (4.5)$$

where the overbar denotes the velocity direction opposite to i; in other words, the distributions are all reflected backward when they encounter a solid node in the physical domain (Huber *et al.* 2008, 2010). In these calculations, the temperature at the inlet ($z = 0$) and outlet ($z = L_z$) is fixed to 1 and 0, respectively. The calculations are run until a steady state is reached. For more information about lattice Boltzmann models for the diffusion equation, the reader is referred to Wolf-Gladrow (2000).

Lattice Boltzmann model for fluid flow

Our pore-scale flow simulations are also based on the lattice Boltzmann method. We use the Palabos open-source library (www.palabos.org), to compute the 3D flow field at the pore

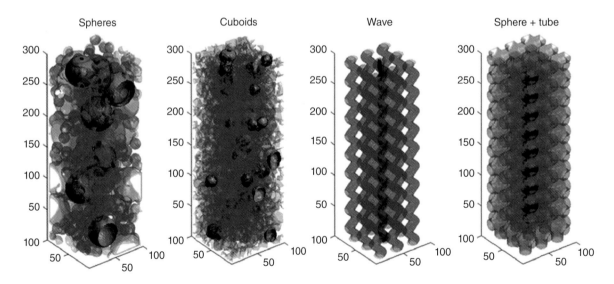

Fig. 4.1. Pore-scale representations of the four porous media. Warm colors (red) show larger pores. All textures are synthetically constructed; spheres and cuboids are constructed with a stochastic nucleation and growth algorithm following the procedure described in Hersum and Marsh (2006). (*See color plate section for the color representation of this figure.*)

Lattice velocities (7)
for the diffusion equation

Lattice velocities (19)
for the flow equation

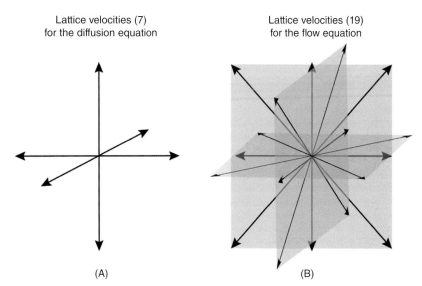

Fig. 4.2. Diagram showing the velocity discretization used for the lattice Boltzmann modeling. (A) The seven-velocity $\mathbf{e}_i$ (including a rest velocity $\mathbf{e}_0$) model for the calculation of the formation factor is shown. (B) The 19-velocity model c, used for the flow calculations at the pore scale, is illustrated. (*See color plate section for the color representation of this figure.*)

(A)

(B)

scale in each medium. Similar to the calculations of the formation factor discussed earlier, we compute the evolution of particle probability density functions $f(\mathbf{x},\mathbf{v},t)$ subjected to streaming (free flow) and collisions with other particles or boundaries where mass and momentum are conserved locally. The discretized form of the evolution equation for $f(\mathbf{x},\mathbf{v},t)$ is similar as well

$$f_i(\mathbf{x} + \mathbf{c}_i \mathrm{d}t, t + \mathrm{d}t) = f_i(\mathbf{x}, t) - \frac{1}{\tau_f}[f_i(\mathbf{x}, t) - f_i^0(\mathbf{x}, t)], \qquad (4.6)$$

where τ_f is the relaxation time to the local equilibrium distribution functions f_i^0 and i is an index that discretizes the space of available trajectories for particles. Because of the four scalar fields that are conserved locally (one for mass and three for momentum), the particle motion is now limited to 19 directions; see Figure 4.2. After assigning the first statistical moments of the particle probability distribution functions f_i to conserved macroscopic quantities such as local density and momentum

$$\rho(\mathbf{x}, t) = \sum_i f_i(\mathbf{x}, t), \quad \rho u(\mathbf{x}, t) = \sum_i \mathbf{c}_i f_i(\mathbf{x}, t), \qquad (4.7)$$

a multiscale expansion yields a compressible version of Navier–Stokes equations (Frisch *et al.* 1986, 1987; Higuera & Jimenez 1989). In the expansion, the kinematic viscosity of the fluid v is identified with

$$v = \frac{\eta}{\rho} = c_s^2 \mathrm{d}t \left(\tau_F - \frac{1}{2} \right), \qquad (4.8)$$

where c_s^2 is a constant that depends on the choice of spatial discretization (1/3 here). In each calculation, pressure boundary conditions are applied at $z = 0$ and $z = L_z$. In the first stage, steady boundary conditions are applied to obtain a steady discharge through the medium. It allows us to calculate the static permeability

$$k_0 = \frac{\eta L_z q_0}{\Delta p^0}, \qquad (4.9)$$

where Δp^0 is the steady pressure drop imposed on the medium and the discharge is computed with

$$q_0 = \frac{1}{V} \int_V \mathbf{v}(\mathbf{x}) \cdot \mathbf{n}_z \mathrm{d}x^3, \qquad (4.10)$$

where $\mathbf{v}(\mathbf{x})$ is the steady-state pore-scale velocity field and $\mathbf{n}_z$ the direction along the pressure gradient. After reaching a steady discharge, the outlet pressure is varied harmonically around its static value P_{out}^0

$$P_{\text{out}}(t) = P(L_z, t) = P_{\text{out}}^0 + \Delta p \sin(\omega t). \qquad (4.11)$$

We compute the flow field and therefore calculate the discharge with a high temporal resolution during many pressure cycles to calculate the dynamic permeability. The amplitudes of the pressure fluctuations and gradients are small enough that the compressibility effect remains limited in our lattice Boltzmann simulations.

Postprocessing of transient discharge data

In response to the dynamic pore-pressure condition, the flux q_{out} becomes

$$q_{\text{out}} = q_{\text{out}}^0 + \Delta q \sin(\omega t - \phi) \qquad (4.12)$$

when the system reaches a quasi-steady state. ϕ is the phase lag between the flux and the imposed pressure oscillations. We fit the outlet discharge with a sine function $A \sin(\omega t + B) + C$ to obtain the amplitude Δq and the phase ϕ. The dynamic permeability $k(\omega)$ is calculated as

$$k(\omega) = \frac{\eta \Delta q(\omega)}{\Delta p(\omega)/L_z}. \qquad (4.13)$$

By conducting simulations over a wide range of frequency, we can establish the spectral response of each medium to the pressure oscillations.

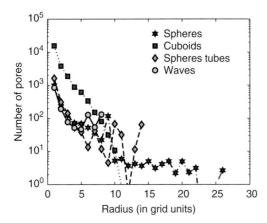

Fig. 4.3. Pore-size distribution for the four media calculated from the model of Yang *et al.* (2009). In this model, the radius of each pore is determined by the largest sphere that can be fully included into the pore.

RESULTS

Expected dynamical response

We selected four porous media structures to provide a test to the predicted self-similar nature of the dynamic response of the permeability in terms of frequency (Johnson *et al.* 1987; Smeulders *et al.* 1992). These textures range from about 30% to 60% porosity, the formation factors vary by a factor of 3, and the static permeability varies by a factor of 6. It is interesting to note that the JKD model for the dynamic permeability only depends on these three factors. We are also interested, within the limits of the number of porous media structures tested here, to investigate whether the PSD can affect the dynamical response independently from these three parameters.

The PSD for each medium is computed with the model of Yang *et al.* (2009), where in each pore, the largest sphere that can be fully included into the volume of the pore defines its effective radius (pore size). The results of these PSD calculations are shown in Figure 4.3. We note that the range of pore sizes for cuboids, sphere tubes, and waves are comparable but that spheres display a significantly greater range of pore sizes with several pores with radius >20 grid units.

The formation factors are computed according to the section Computation of the formation factor, where a 3D diffusion model is relaxed to steady state. An example of steady-state temperature distribution in the cuboids texture is shown in Figure 4.4 for reference, and the results are listed in Table 4.1.

For each porous medium, we conducted between 15 and 20 simulations with different forcing frequencies to obtain the effective permeability of the medium over a range of frequency that extends over 3 orders of magnitude.

In Figure 4.5A, we show the un-normalized results for three of the four media. We can clearly observe similar trends with a sudden decay of effective permeability as the forcing frequency

Fig. 4.4. Thermal field in the porous medium (cuboids) at steady state. Here, the solid fraction is assumed as a perfect insulator. The formation factor is computed from the effective thermal conductivity of the medium at steady state. (*See color plate section for the color representation of this figure.*)

Table 4.1 Summary of the steady flow calculations

Name	Porosity ϕ	Formation factor F	Permeability k_0 (dimensionless)	JKD critical frequency ω_c
Spheres + tubes	0.49	0.3	2.98	6.2×10^4
Waves	0.21	0.11	0.48	1.4×10^3
Spheres	0.61	0.15	0.89	1.05×10^3
Cuboids	0.61	0.39	1.16	2.04×10^3

approaches the critical frequency of each medium ω_c (see vertical lines). Once the frequency is normalized with ω_c (Fig. 4.5B), the data collapse as expected for the self-similar trend observed and documented by several authors (Johnson *et al.* 1987; Sheng & Zhou 1988; Smeulders *et al.* 1992; Müller & Sahay 2011; Pazdniakou & Adler 2013). We find an excellent agreement with the theory developed by Johnson *et al.* (1987) and Smeulders *et al.* (1992) irrespective of the PSD of the medium. The complex nature of the dynamic permeability is better portrayed by the phase lag between the harmonic pressure forcing and the computed discharge (Fig. 4.6) and again shows an excellent agreement with previous studies (Johnson *et al.* 1987; Sheng & Zhou 1988; Smeulders *et al.* 1992; Müller & Sahay 2011; Pazdniakou & Adler 2013).

Anomalous behavior

We conducted the same simulations on the last porous medium (spheres) and obtained a significantly different result.

The work remains preliminary; however, the runs were checked for reproducibility and the results are robust. The dynamic response of the permeability and phase lag with the harmonic forcing is similar to what we observe for the other media except for the existence of a peak at high frequency $\omega > \omega_c$ (see Figs 4.7 and 4.8). The phase lag displays an excursion to negative phases (or in that context, positive phase with $\phi > \pi$ to satisfy causality) that coincides with the resonant permeability peak. We observed, although not shown here, a similar behavior in another medium that displayed sharp heterogeneities at the pore scale.

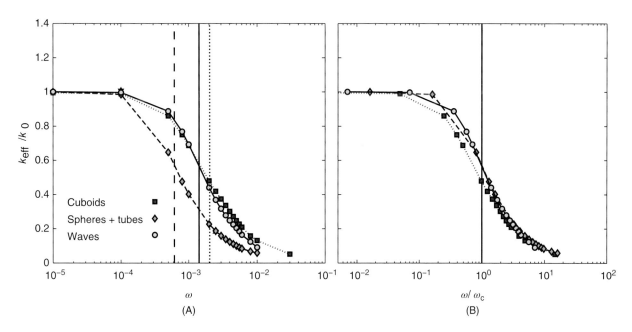

Fig. 4.5. Effective (dynamic) permeability as function of frequency for three of the four porous media used in this study. (A) How differences in porosity, static permeability k_0, and formation factor influence the response. (B) The rollover frequency ω_c provides a satisfying normalization factor to observe a self-similar behavior between the different media.

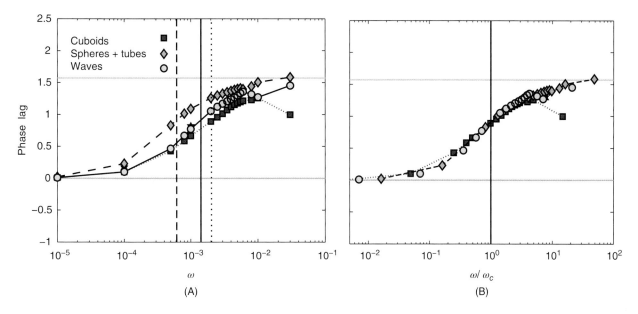

Fig. 4.6. The phase lag ϕ between the pore-pressure forcing at the outlet and the discharge in the porous medium as function of frequency (A) and normalized frequency (B). The results are consistent with previous studies (Johnson *et al.* 1987; Sheng & Zhou 1988; Smeulders *et al.* 1992; Müller & Sahay 2011; Pazdniakou & Adler 2013).

The existence of a permeability peak at high frequency is unexpected from the existing theory and suggests a resonance-like behavior over a narrow range of forcing frequencies. It is important to realize that neither Darcy's equation, even corrected for frequency-dependent permeability, nor the groundwater flow equation (diffusion) can cause a resonance-like behavior. Resonance occurs generally as a response to transient forcing of hyperbolic partial differential equations. In the Discussion section, we present an analogy between the standard JKD theory for the dynamic permeability and electric circuits and show that this theory fails to explain the peak observed in Figure 4.7. We discuss two possible causes

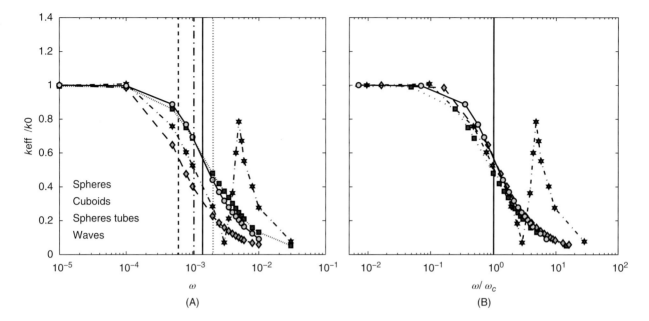

Fig. 4.7. Same as Figure 4.5 but with the last porous medium (spheres). Note that the medium referred to as spheres has the broadest pore-size distribution.

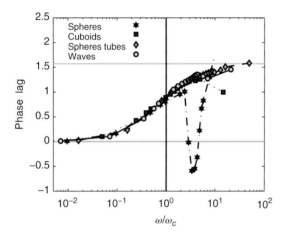

Fig. 4.8. Phase lag between the pore-pressure fluctuations and the discharge as function of normalized frequency for all media.

for the peak in dynamic permeability: the resonance is governed by the dynamics at the continuum scale (modified Darcy flux) or by microstructural properties.

DISCUSSION

Analogy to Debye relaxation

The linear theory of flow in porous media is often compared with linear electric circuit theory because of the many analogies between the two fields. First, Ohm's law is equivalent to Darcy's law with the hydraulic conductivity replacing the electrical conductivity and the head potential (or pseudopotential) as defined

by Hubbert (1940) replacing the electrical potential. It is therefore natural to yield analogies between the dynamic permeability model presented by Johnson *et al.* (1987) and complemented by Smeulders *et al.* (1992) and the spectral response of electric circuits. The two equations that define the dynamic permeability and the dynamic tortuosity in the JKD model, once written in the frequency domain, are reminiscent to Ohm's law and the dynamical response of a capacitor, respectively

$$\mathbf{v} = -\frac{k(\omega)}{\eta\phi}\nabla p, \qquad i\omega\rho_f\alpha(\omega)\mathbf{v} = -\nabla p. \qquad (4.14)$$

The second equation highlights the importance of inertia and shows that a phase lag exists at high frequency between the discharge $\phi\mathbf{v}$ and the pressure forcing. The $\pi/2$ phase between the current and the potential introduced by inertia is reminiscent of the effect of adding a capacitor to an electric circuit. If a resistor is set in parallel with a capacitor (see Fig. 4.9), the relative importance of the resistor and capacitor on the impedance of the circuit will be controlled by the imposed frequency and a Debye relaxation similar to the dynamical response of the permeability is obtained.

The transition from creep- to inertia-dominated momentum dissipation in porous media is therefore similar to the response of a parallel RC circuit. We can even push the analogy further and use it to estimate the frequency at which we expect the transition to occur. The Debye frequency is obtained by matching the imaginary part of the capacitor impedance to that of the resistor

$$\mathrm{Im}(Z_c) = Z_R,$$

$$\frac{1}{\omega C} = R, \qquad (4.15)$$

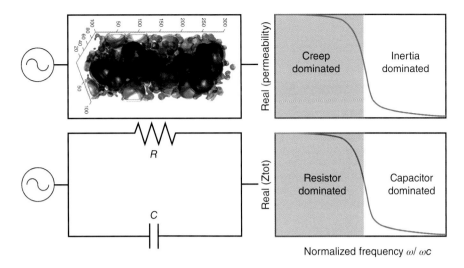

Fig. 4.9. Analogy between the dynamic permeability model of Johnson *et al.* (1987) and RC electric circuits in parallel. The spectral response of the effective permeability is similar to the spectral response of the overall impedance of the electric circuit.

and hence, $\omega_D = 1/RC$. By analogy, the resistance in the dynamic version of Darcy's law is

$$Z_R = \frac{\eta\phi}{k(\omega)} \qquad (4.16)$$

while, using Eq. 4.14, the inertia associated with the capacitor can be defined as

$$\mathrm{Im}(Z_c) = \omega\rho_f\alpha(\omega). \qquad (4.17)$$

Matching the two impedances allows us to retrieve the general form of the critical frequency of the JKD model

$$\omega_c = \frac{\eta\phi}{\rho_f\alpha_\infty k_0}. \qquad (4.18)$$

This analogy is useful because it also provides some clues as to what is problematic with the results that display resonance-like features.

Resonance and the importance of hyperbolic effects

One of the limitations of Darcy's law is that it represents a continuum approximation of a steady-state flow through a porous medium. The justification for using a time-independent momentum closure equation is that the Reynolds number that characterizes the flow is $\ll 1$. Interestingly, regardless of this approximation, one can still derive a time-dependent mass conservation equation (a simple form of the groundwater flow equation) from Darcy's law that describes the pressure or head distribution in time and space in the porous medium

$$\frac{1}{K_f}\frac{\partial p}{\partial t} = -\frac{k}{\eta}\frac{\partial^2 p}{\partial x^2}, \qquad (4.19)$$

where for simplicity, the flow is assumed perpendicular to gravity, the permeability is homogeneous, and K_f refers to the fluid's

bulk modulus. To be consistent with the theory developed by Johnson *et al.* (1987) and Smeulders *et al.* (1992), the solid matrix is viewed as incompressible for the sake of this argument. Equation 4.19 is a diffusion equation with the hydraulic diffusivity $D_h = K_f k/\eta$. The diffusion equation is parabolic, and consequently, the pore pressure can propagate at an infinite speed from the boundaries inside the domain. Consider an initial Dirac delta function of increased pore pressure centered at $\mathbf{x} = 0$ at $t = 0$. At any $t = \mathrm{d}t$, for any small $\mathrm{d}t$, the pressure profile that results from Eq. 4.19 will have finite (nonzero) pressure values except at $\mathbf{x} \to \pm\infty$. The propagation of pressure is therefore instantaneous.

In reality, we know that it will take a finite time for pressure to propagate a finite distance in a porous medium. It is therefore necessary to introduce a transient term to modify Darcy's equation (even in the frequency domain, but we will restrict the analysis to the time domain here). A similar argument has been developed for heat transfer and the development of heat waves. Cattaneo (1958) and Vernotte (1958) arrived independently at the conclusion that a proper account of heat transfer by conduction should include an additional term in Fourier's law to consider that there is a finite time τ for the propagation of heat that depends on material properties. Following the same strategy, we arrive at the following definition for the transient Darcy's equation that satisfies a finite speed of pressure propagation

$$\tau\frac{\partial q}{\partial t} + q = -\frac{k}{\eta}\frac{\partial p}{\partial z}, \qquad (4.20)$$

where q is the discharge in the z-direction. Although this equation is different from Biot's equations and the equations used by Johnson *et al.* (1987) and Smeulders *et al.* (1992), it is important to note some similarities. First, this equation is quite similar to the definition of unsteady Stokes flows and it is therefore not a novel concept. Also, in Biot's model, two separate but coupled equations were used to describe the displacement of each phase (fluid and solid) and Biot used an unsteady Stokes equation for the flow field that is not identical but quite similar to Eq. 4.20. The Cattaneo–Vernotte (CV) term in Eq. 4.20 (time derivative) introduces a phase lag for a finite value of τ, in agreement with the JKD model. The main difference here is that both Darcy's flux and the CV terms are balancing the pressure forcing in Eq. 4.19. The partitioning between the inertial term and the Darcian behavior depends on the value of the relaxation time τ and the forcing frequency ω; it is therefore not identical to Eq. 4.14.

Although the addition of a transient or unsteady term to the flux equation is not entirely novel, it is important to draw attention to its effect on the mass conservation equation. We argue that when Eq. 4.20 is introduced in the continuity equation, we retrieve a simple hyperbolic telegraph equation

$$\frac{\tau}{K_f}\frac{\partial^2 p}{\partial t^2} + \frac{1}{K_f}\frac{\partial p}{\partial t} = \frac{k}{\eta}\frac{\partial^2 p}{\partial x^2}. \qquad (4.21)$$

If the relaxation time $\tau \ll 1$, then the parabolic groundwater flow equation is retrieved. One can directly conclude that this hyperbolic equation admits the propagation of damped waves that travel with a finite velocity $c = (K_f F / \rho_f)^{1/2}$, where we identified the relaxation time τ with $\rho_f k_0 / \eta = 1/\omega$. This means that our model is consistent with that of JKD where the transition between inertia- and viscous-dominated regimes occurs around the rollover frequency ω_c. The adequate boundary conditions with respect to our study are

$$p(0, t) = p_1,$$
$$p(L_z, t) = \Delta p^0 + \Delta p \sin(\omega t), \qquad (4.22)$$

where Δp^0 is the static pressure difference across the sample and Δp is the amplitude of the harmonic pressure perturbation. Using sine transforms

$$p(z, t) = \sum_{n=1}^{\infty} S_n(t) \sin\left(\frac{n\pi z}{L_z}\right), \qquad (4.23)$$

the mass conservation equation unfolds into n second-order nonhomogeneous ODE

$$\frac{d^2 S_n}{dt^2} + \frac{1}{\tau}\frac{dS_n}{dt} + \frac{n^2\pi^2}{L_z^2}\frac{kK_f}{\eta\tau}S_n$$
$$= \frac{2n\pi}{L_z^2}\frac{kK_f}{\eta\tau}[p(0, t) + (-1)^{n+1}p(L_z, t)]. \qquad (4.24)$$

The homogeneous equations that do not include the harmonic forcing yield solutions

$$S_n^h(t) = A_n \exp\left[-\frac{\gamma}{2}(1 + \sqrt{1 - 4\omega_n^2\tau^2})t\right]$$
$$+ B_n \exp\left[-\frac{\gamma}{2}(1 - \sqrt{1 - 4\omega_n^2\tau^2})t\right], \qquad (4.25)$$

where A_n and B_n are constants that are constrained by the initial and boundary conditions, and the characteristic frequencies $\omega_n = n^2\pi^2 kK_f / \eta L_z^2$ and finally, $\gamma = 1/\tau$. These solutions are dissipative (damping), and the damping increases when $\tau \to 0$, which is consistent with the diffusive behavior of the equation in this limit.

Because the harmonic forcing has no influence over the homogeneous solution, it is more important to study the particular solution for resonance effects. Because of the harmonic forcing, it is convenient to write the particular solutions as

$$S_n^p(t) = C_n \sin(\omega t) + D_n \cos(\omega t). \qquad (4.26)$$

The coefficients C_n and D_n are obtained after substituting Eq. 4.26 into Eq. 4.24

$$C_n = \frac{2(-1)^{n+1}\tau^2[(\omega_n')^2 - \omega^2]\omega_n'}{\omega^2 + \tau^2[(\omega_n')^2 - \omega^2]^2},$$
$$D_n = \frac{2(-1)^n \tau\omega\omega_n'}{n\pi[\omega^2 + \tau^2[(\omega_n')^2 - \omega^2]^2]}, \qquad (4.27)$$

where the characteristic frequency of the medium ω_n' is defined by

$$\omega_n' = \sqrt{\frac{\omega_n}{\tau}}. \qquad (4.28)$$

The overall amplitude the waves associated with the particular solution in response to the forcing at frequency ω is

$$X_n = \sqrt{C_n^2 + D_n^2} = 2\omega_n'\tau\left[\frac{(\omega/n\pi) - \tau[(\omega_n')^2 - \omega^2]}{\omega^2 + \tau^2[(\omega_n')^2 - \omega^2]^2}\right], \qquad (4.29)$$

which implies a finite resonance when the forcing frequency ω approaches (but not equals) ω_n'. The actual solution for the resonant frequency requires finding the roots of a fifth-order polynomial (ω) that also depends on the choice of relaxation time τ and the order of the harmonic considered n. It is beyond the scope of this work to provide an analysis of this polynomial. It is, however, instructive to reflect on the amplitude of the particular solution as the forcing approaches resonance $\omega \to \omega_n'$

$$X_n(\omega_n') = \frac{2\tau}{n\pi}, \qquad (4.30)$$

which shows that the amplitude of high-frequency harmonics may become small. Only the lowermost modes may display visible resonance peaks. Moreover, in the limit where $\tau \to 0$, no resonance is observed, which is consistent with the character of the partial differential equation in that limit.

Our hyperbolic model for the mass conservation in a porous medium subjected to transient forcing therefore admits a resonant behavior if the relaxation time τ becomes important, that is, when $\omega\tau > 1$. Alternatively, resonance may occur when the forcing frequency approaches $n\pi c / L_z$, where c is the pressure wave propagation speed that depends on the compressibility of the fluid and the formation factor of the medium. There are obviously some simplifying assumptions in this model. For instance, we have assumed that τ was independent of frequency and we use the asymptotic limits for the permeability (k_0) and the dynamic tortuosity to identify what governs the relaxation time using the standard equations for dynamic permeability. We therefore assume that τ is controlled by the physical properties of the porous medium and pore fluid and that it does not depend on the applied forcing. This assumption, although not justified, is consistent with the theory developed for heat wave propagation in heterogeneous media (Ozisik & Tzou 1994; Ordonez-Miranda & Alvarado-Gil 2011; Ordonez-Miranda et al. 2012).

In light of this hyperbolic description of the mass balance in porous media, we can estimate the expected range of frequency that should display resonance for the different porous media. A satisfying model should be able to explain the peak observed with the spheres topology and be consistent with the absence of resonance observed for the three other media. We first compute ω'_n for $n = 1$ for all media. The actual resonance is not expected to take place exactly at ω'_n, but from a visual inspection of the roots of the polynomial $f(\omega)$ the actual position of the resonance is $\omega'_n < \omega^* < 2\omega'_n$. In Figure 4.10, we compare the set of forcing frequencies tested in our simulations with ω'_n. The approximate range of frequency where the continuum hyperbolic mass balance equation predicts a resonant effect is consistent with our results with the spheres medium. However, we note that our calculations should have allowed us to observe a resonant peak in each medium. This informs us that while the continuum-scale hyperbolic model may be consistent for one of our medium at high frequency, the lack of resonance in the other media indicates that it is not sufficient to explain our results.

One could argue that the absence of resonance reflects that the amplitude of the particular solution is negligible and that the resonance is therefore difficult to measure. The hyperbolic model is generally consistent with the transient Stokes equation at the base of the theory of linear poroelasticity developed by Biot; however, it does not provide a satisfying explanation for our pore-scale simulations.

Alternatively, one can argue that pore-scale processes control the existence of the resonant peak. The porous media constructed by a stochastic process (spheres and cuboids) display more heterogeneity in terms of pore-scale structures. More specifically, the spheres medium was built with a broader PSD than the other three media. In heterogeneous media, at the pore scale, one should expect the plane wave assumption for the pressure propagation at the continuum scale to fail. In media that display competing pathways with different hydraulic responses, pore-pressure gradients can become significantly perpendicular to the main direction of propagation, and mass/pressure exchanges between different pathways may significantly affect the stress propagation. The visualization of the pore-pressure distribution in the spheres medium over time during one period can yield important information about the propagation of stress transients in the porous medium. Figure 4.11 shows snapshots of the pore-pressure field. We observe pressure waves propagating from the outlet (right) into the medium. An important feature worth noting is that at a fixed distance from the dynamic boundary (outlet), the pore pressure can vary significantly spatially. A possible explanation for the resonant behavior is that a heterogeneous porous medium can be viewed as a collection of connected primary and secondary pathways for the fluid and by extension pressure wave propagation. One clearly observes from Figure 4.11 that pressure fluctuations can propagate further into the medium

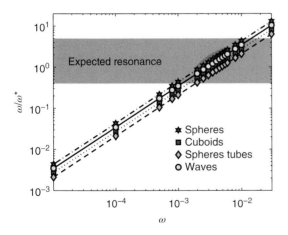

Fig. 4.10. Comparison between the sampled forcing frequency used in our simulations and the estimated range for the resonant frequency of the hyperbolic forced mass conservation equation ω^*. We observe that our sampling should have allowed us to observe resonance in all four media. The dark grey ellipse marks the region where we observe resonance for the spheres data.

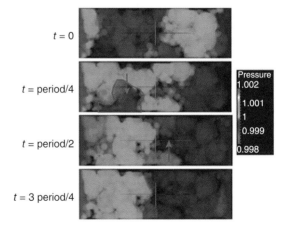

Fig. 4.11. 3D visualization of the pore-pressure field (normalized) at the forcing frequency corresponding to the maximum of the resonance peak in the spheres medium. The four images show the temporal evolution of the pressure field every quarter period. Note the regions with large lateral pore-pressure gradients (the imposed gradient is left to right) highlighted in red. It shows that flow pathways with different hydraulic connectivity have different response times to the forcing and that large pore-scale pore-pressure imbalance can emerge, violating the assumption of planar pressure wave propagating from the outlet. (See *color plate section* for the color representation of this figure.)

along the main flow pathways creating some important pressure gradients perpendicular to the main flow direction. This should lead to significant mass exchanges between connected pathways with different transient responses to stress propagation.

The resonant behavior may be associated with the existence of a finite time delay for the pore pressure to relax between primary and secondary pathways. If the response time for the pore-pressure exchange between competing but connected pathways approaches the period of the forcing, we argue that resonance may possibly occur. To test this hypothesis, we introduce a dimensionless variable ζ

$$\zeta = \frac{(1/\phi VT)\int_0^T \int_{\phi V}[(\partial p/\partial x)^2 + (\partial p/\partial y)^2]\mathrm{d}V}{(\Delta p/L_z)^2}, \quad (4.31)$$

that quantifies the amount of transverse pressure gradient cumulated over a period $T(2\pi/\omega)$. If competing pathways with different characteristic response times coexist, we argue that transverse pressure gradients will become more important and ζ value increases. We computed ζ for all media at $\omega/\omega_c = 5$, where the resonance peak maximum is found for the spheres medium. We find values of $\zeta = 0.9, 3.4, 40.1,$ and 130.1 for sphere tubes, waves, cuboids, and spheres, respectively. The medium exhibiting resonance is therefore characterized by a much greater ζ value than the other media, which suggests that transverse pore-scale pressure gradients are responsible for resonance.

Because these features in the spheres medium are sub-REV, it could explain the shortcomings of the continuum-scale interpretation to explain resonance. It is interesting to reflect on the major differences between our study and that of Pazdniakou and Adler (2013) where they do not observe resonance. Pazdniakou and Adler (2013) introduced the forcing as a homogenous oscillatory perturbation of a bulk force in the fluid and consider periodic and, therefore, infinite media. The homogeneous forcing applied in their study does not allow the buildup of significant pore-scale pressure gradients as the fluid responds uniformly to the perturbation instantaneously. There is no transport of the information on the changing pore pressure from the boundaries of the domain (there are no boundaries) and therefore no finite time response of the domain to the excitation. This could explain the different results between the two studies.

In the future, we plan on developing more highly heterogeneous structures at the pore scale and test whether resonance is promoted by pore-scale heterogeneities. In parallel, we suggest that an idealized theoretical model that decomposes the porous medium as a collection of regions according to the fluid's mobility may improve the continuum description of these heterogeneous media (dual- and multirate mass transfer models by Harvey & Gorelick 2000, Liu *et al.* 2007 and Porta *et al.* 2013). Such a framework including a finite relaxation time for pressure (leakage) between the more mobile and less mobile subsets of the medium could provide a framework to test whether pore-scale heterogeneity can influence a significant departure to the theory of JKD and even lead to resonant effects.

CONCLUSIONS

The propagation of pore-pressure transients in saturated porous media is complex at high frequency, when the short wavelength can interact with the heterogeneous structure of the medium. Over the last three decades, successful models for the dynamic response of porous media to harmonic pressure forcing have been developed and tested. They highlight that only three continuum-scale descriptors of the complex pore structure are important in homogeneous and isotropic media to characterize the spectral response of permeability. These descriptors are the porosity of the medium, its formation factor, and its static permeability. We have constructed four synthetic porous media from the pore scale with different porosity, formation factors, and permeability. Using a lattice Boltzmann method to compute the fluid flow at the pore scale, we find a very good agreement between our results and the theory for porous media with relatively narrow PSDs.

The medium that shows the broadest range of pore sizes, however, has a distinctive response to the imposed pressure transients. We observe a feature that resembles a resonance peak for the effective permeability at high frequency. Drawing from analogies with electric circuits and heat wave models, we postulate that the resonance feature is either governed by processes that operate at the continuum scale or by pore-scale processes that arise because of significant pore-scale heterogeneity. The development of a mass conservation equation at the continuum scale that corrects for the finite time required for pore pressure to propagate in the medium allows for resonance behavior to take place. It predicts resonance over the correct range of frequency for the case where our numerical results display resonant features. It, however, fails to explain the absence of resonance observed for the other media. Based on our numerical results, we argue that the resonance we observe is rather caused by pore-scale processes, whereby significant local pore-pressure gradients can form between heterogeneous flow pathways characterized by different response timescales to the imposed pressure excitations. Future studies focused on the distribution of pore pressure in heterogeneous media will potentially shed light on the dynamical processes that control the existence and the factors that govern the resonance of highly heterogeneous and saturated porous media.

ACKNOWLEDGMENTS

C.H. and Y.S. thank *Geofluids* Editor C. Manning and acknowledge the thorough reviews of two anonymous reviewers and funding from an ACS-PRF DNI grant.

CHAPTER 5

Flow of concentrated suspensions through fractures: small variations in solid concentration cause significant in-plane velocity variations

RICARDO MEDINA[1], JEAN E. ELKHOURY[1,2], JOSEPH P. MORRIS[2,3], ROMAIN PRIOUL[2], JEAN DESROCHES[4] AND RUSSELL L. DETWILER[1]

[1] *Department of Civil and Environmental Engineering, University of California, Irvine, CA, USA;* [2] *Schlumberger-Doll Research, Cambridge, MA, USA;* [3] *Computational Geosciences, Lawrence Livermore National Laboratory, Livermore, CA, USA;* [4] *Services Pétroliers Schlumberger, Paris, France*

ABSTRACT

Flow of high-concentration suspensions through fractures is important to a range of natural and induced subsurface processes where fractures provide the primary permeability (e.g., mud volcanoes, sand intrusion, and hydraulic fracturing). For these flows, the simple linear relationship between pressure gradient and flow rate, which applies for viscous-dominated flows of Newtonian fluids, breaks down. We present results from experiments in which a high concentration (50% by volume) of granular solids suspended in a non-Newtonian carrier fluid (0.75% guar gum in water) flowed through a parallel-plate fracture. Digital imaging and particle image velocimetry analysis provided detailed two-dimensional maps of velocities within the fracture. Results demonstrate the development of a strongly heterogeneous velocity field within the fracture. Surprisingly, we observed the highest velocities along the no-flow boundaries of the fracture and the lowest velocities along the centerline of the fracture. Depth-averaged simulations using a recently developed model of the rheology of concentrated suspensions of monodisperse solids in Newtonian carrier fluids reproduced experimental observations of pressure gradient versus flow rate. Results from additional simulations suggest that small (3%) variations in solid concentration within the fracture can lead to significant (factor of two) velocity variations within the fracture yet negligible changes in observed pressure gradients. Furthermore, the variations in solid concentration persist over the length of the fracture, suggesting that such heterogeneities may play a significant role in the transport of concentrated suspensions. Our results suggest that a simple fracture-averaged conductivity does not adequately represent the transport of suspended solids through fractures, which has direct implications for subsurface suspension flows where small concentration variations are likely.

Key words: fracture, granular, non-Newtonian, rheology, suspension

INTRODUCTION

Subsurface flows of fluids with high concentrations of suspended solids are important to a range of naturally occurring and applied problems. Naturally occurring phenomena such as intrusion of magmas composed of crystals suspended in silicate melts (Mader *et al.* 2013) and mobilization of suspended sediments in the shallow crust such as sand intrusion in sedimentary basins (Huuse *et al.* 2010) and mud volcanoes (Manga & Brodsky 2006) involve migration of fluidized solids through preexisting or propagating fractures. Engineering applications include environmental remediation (Murdoch *et al.* 2006), mud injection during drilling (Bittleston *et al.* 2002), and injection of slurries containing high concentrations of sands during hydraulic fracturing for both oil and gas production (Kern *et al.* 1959; Montgomery 2013). Suspended solids can alter fracture transmissivity if they become immobilized within the fracture or if their concentration is sufficiently large to change the suspension rheology.

The volumetric concentration of solids in a suspension, ϕ, strongly influences its rheology. In dilute suspensions ($\phi \lesssim 0.2$), interactions between particles are negligible, and the rheology of the suspension is similar to that of the suspending fluid, with the effective viscosity, μ, increasing with ϕ (Krieger & Dougherty 1959). Even at very low ϕ, particles may bridge or clog at pore throats leading to reduced permeability (Khilar & Fogler 1998). These permeability reductions may be reversed when the clogged particles are remobilized due to earthquake-induced shaking or pressure oscillations (Elkhoury *et al.* 2006, 2011). When ϕ approaches a critical limit, ϕ_{cr}, between the random loose and dense packing limits for the solids $(0.55 \lesssim \phi_{cr} \lesssim 0.64)$, the particles become completely jammed and abruptly change to a rigid porous medium (Haw 2004). The permeability then decreases to that of a porous medium composed of the jammed solids. Subsequent large pressure perturbations can lead to fluidization of the jammed solids and remobilization (e.g., mud volcanoes). Here, we are interested in intermediate concentrations $(0.2 \lesssim \phi \lesssim 0.55)$ where fluid and solid flow together but particle–particle interactions are non-negligible leading to frictional losses that significantly alter the rheology of the suspension from that of the suspending fluid.

Over the past decade, frictional models for the behavior of dry granular solids have been extended to represent the behavior of solids suspended in viscous fluids. Early efforts focused on flows of granular solids down inclined planes (du Pont *et al.* 2003; Cassar *et al.* 2005) and established the importance of the dimensionless "viscous number," $I_v = \mu_f \dot{\gamma}/P_s$, which relates the timescale of the movement of a single particle subjected to a force $P_s d^2$ (where P_s is the pressure acting on a solid particle of diameter, d) in a fluid with viscosity, μ_f, to the timescale of the displacement of a particle caused by the imposed shear rate, $\dot{\gamma}$. Boyer *et al.* (2011) demonstrated that relationships of the form $\tau = \mu(I_v)P_s$ can adequately quantify the shear stress and proposed corresponding constitutive relationships for the effective friction coefficient, $\mu(I_v)$, and volume fraction, $\phi(I_v)$. As shear rate (or I_v) increases, $\phi(I_v)$ decreases from a maximum of ϕ_{cr} when $I_v \to 0$ and $\mu(I_v)$ increases from a minimum when $I_v \to 0$. In pressure-driven internal flows (e.g., tubes and fractures), where the shear rate is maximum at the walls and vanishes along the centerline, the dependence of μ and ϕ on I_v gives rise to plug-flow behavior for larger values of ϕ_0 (the average solid concentration of the well-mixed suspension and the uniform concentration at the inlet). The result is a localized region of high shear rate near the walls where $\phi < \phi_0$ and a region in the center of the flow where $\phi \to \phi_{cr}$ and the fluid and solid move at the same velocity. Lecampion & Garagash (2014) extended the constitutive relationships for $\phi(I_v)$ and $\mu(I_v)$ proposed by Boyer *et al.* (2011) to develop a model for pressure-driven flows through tubes and channels.

Parallel-sided channels provide an idealized analog to fractures in geologic systems where fractures typically have rough walls that may also be permeable. However, as with early studies of Newtonian (Witherspoon *et al.* 1980) and non-Newtonian (Di Federico 1997) fluid flow in fractures, beginning with this idealized geometry provides a well-controlled step toward understanding more complicated geometries. The emphasis of previous suspension-flow studies in channels was to quantify the distribution of solids and velocity across the gap between the surfaces (or aperture). Here, we consider larger three-dimensional flow fields, where the velocity may also vary in the plane of the fracture. We are particularly interested in the influence of boundary conditions on suspension flows. In experimental and computational studies of fluid flow through fractures, uniform pressure is typically applied along two boundaries to create the pressure gradient that drives the flow. When studying suspension flows, it is also necessary to prescribe ϕ_0 at the inlet boundary. The obvious choice is to also assume uniform ϕ_0, but due to the strong dependence of the effective viscosity μ on ϕ, small variations of ϕ within the flow field can cause variations and instabilities in the velocity field. In addition, many previous studies of the rheology of concentrated suspensions focused on idealized monodisperse (Karnis *et al.* 1966; Lyon & Leal 1998a) or bimodal (Lyon & Leal 1998b) spherical solids. Here, we explore the implications of the complex rheology of a mixture of guar and silica sand representative of suspended solids encountered in the subsurface. In particular, we focus on conditions where suspended solids flow with the fluid, and we do not consider conditions under which the settling of solids within the fracture is important.

We present results from a pair of experiments in which we flowed high-solid-concentration fluid through a parallel-sided fracture with two different boundary-condition configurations. To aid in interpreting the results of the experiments, we simulated flow through the experimental system using the rheological model of Lecampion & Garagash (2014).

OVERVIEW OF EXPERIMENTS

We designed an experimental apparatus to explore the role of suspension rheology and flow geometry on fracture flow. Transparent parallel-sided fractures provide the ability to both directly measure the flow geometry under experimental conditions and visualize and quantify the velocity field within the fracture. Here, we describe the experimental apparatus, the details of the fluid–solid mixture used for the experiments, the configuration of the experimental system, and the procedure used to carry out the experiments.

Experimental apparatus

A rotating stand rigidly fixed a high-sensitivity 12-bit charge-coupled device camera (Photometrics Quantix KAF-6303e) above a monochromatic (red) light-emitting diode panel. Clamps held the fracture cell to the stand between the light source and the camera. Two ~0.3-cm-thick aluminum shims separated the two fracture surfaces (15 cm × 15 cm × 1.2 cm smooth glass plates) and served as no-flow boundaries along the fracture edges. The fracture cell secured the fracture

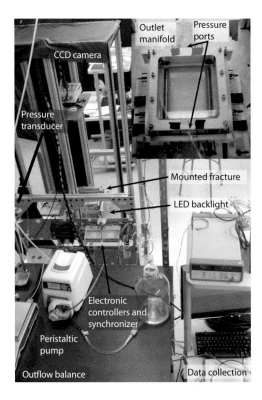

Fig. 5.1. Photograph of experimental setup and fracture cell (inset).

surfaces while allowing visualization of the entire flow field. An electronic controller synchronized 65-ms pulses of the light source with the camera exposure to provide reproducible high-resolution (76×76-μm pixels) measurements of transmitted light intensity (Fig. 5.1). Section "Image analysis" describes how we processed measured intensities to yield velocity fields.

We carried out two experiments in the same fracture with different inlet/outlet boundary conditions. Experiment A used

linear inlet and outlet manifolds (Fig. 5.2A), which included large rectangular channels that spanned both ends of the fracture. Initially, a high-capacity syringe pump pushed slurry into one end of the inlet manifold (dark gray arrow) and out a waste line at the other end of the inlet manifold (light gray arrow) filling the manifold with slurry. We then closed the waste line and opened the two outlets on either side of the outlet manifold (black arrows) to initiate flow through the fracture. Experiment B used a wedge-shaped manifold (Fig. 5.2B) that allowed us to flow directly into the fracture without prefilling the manifold. This configuration included only a single inlet and a single outlet tube. Furthermore, the wedge-shaped manifold tapers gradually from the inlet port (gray arrow) to a rectangle with the same width (W) and aperture (h) as the fracture. A differential pressure transducer connected to the ports located at the center of the inlet and outlet manifolds (marked by X in both configurations in Fig. 5.2) measured the differential fluid pressure (ΔP_f) across the fracture at high temporal resolution (0.3 Hz) during each experiment.

Fluid description and experimental configuration

For both experiments, we used a carrier fluid consisting of 0.75% by volume mixture of guar gum and water. This guar/water mixture is a shear thinning fluid that behaves as a Newtonian fluid under low shear rates and exhibits non-Newtonian behavior at higher shear rates (Fig. 5.3). We selected guar because it is a well-characterized high-viscosity fluid that can be prepared reliably and consistently as a base fluid for high-solid-concentration slurries. A laboratory-grade blender (Waring 7012G) mixed the guar/water solution. We slowly added guar with the blender operating at 6800 rpm. After adding all of the guar, we added biocide (glutaraldehyde, 0.005% by volume) and increased the blender speed to 16,900 rpm and mixed for at least 10 min to ensure complete

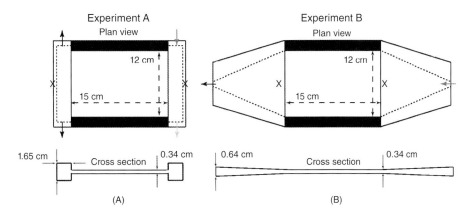

Fig. 5.2. Schematic of inlet and outlet manifold configurations for Experiments A and B. The separate schematics shown for Experiments A and B highlight the difference between the manifold geometries for the two experiments. For Experiment A, a large rectangular channel (much more conductive than the fracture) bounded each end of the fracture. For Experiment B, the manifold gradually tapered from the inlet/outlet tubing to the fracture geometry. The schematic shows the location of the inlet (dark gray arrows), outlet (black arrows), and waste (light gray arrow; Experiment A only). Symbol 'x' marks the locations of the pressure ports that were connected to the differential pressure transducer.

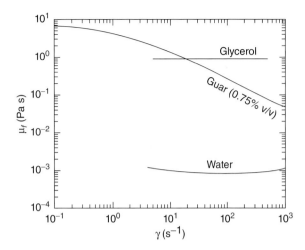

Fig. 5.3. Measured viscosities (μ_f) plotted against shear rate ($\dot{\gamma}$) for water, glycerol, and the 0.75% guar–water mixture used as the carrying fluid for the experiments.

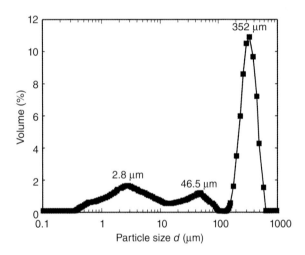

Fig. 5.4. Particle-size distribution for the solids used in the flow experiments. The solids consisted of angular silica sand grains.

hydration of the guar. Applying a vacuum for at least 12 h removed most of the trapped air bubbles from the carrier fluid.

We prepared the high-solid-concentration fluid by adding 50% by volume of silica sand to the de-aired carrier fluid. The sand had a multimodal particle-size distribution ranging from submicron to about 600 μm (Fig. 5.4). Note that the fines served to reduce the permeability of the solids, which reduced their settling rate such that negligible settling occurred within the fracture. A rotary paddle mixed the slurry as we slowly added sand. After adding all of the sand to the guar solution, we stirred the slurry under vacuum for approximately 15 min to ensure a well-mixed and bubble-free slurry. Removing bubbles was important both for the flow characteristics (entrained gas increases compressibility of the fluid) and optical quantification of the flow field (Section "Image analysis").

After mixing, we immediately transferred the slurry to a high-capacity (12.5 l) syringe pump to minimize solid particle settling prior to initiating slurry flow. The syringe pump consisted of a clear polycarbonate pipe (2.5 m long, 2.5 cm inner diameter) fitted with a plunger from a 60-ml syringe. A plastic funnel capped the bottom of the tube and provided a smooth transition from the pipe to the 3-mm-inner-diameter tubing. Water pumped through a tube using a peristaltic pump (Masterflex LS) displaced the plunger and pushed slurry through the funnel at specified flow rates. Balances recorded the mass flow rate of water into the syringe ($Q_{in}\rho_w$) and the mass flow rate of slurry from the fracture ($Q_{out}\rho_s$).

Experimental procedure

Prior to initiating flow experiments, we used light-transmission techniques to measure the fracture aperture field. This involved two steps: (i) measuring the mean fracture aperture and (ii) measuring the spatial distribution of aperture within the fracture (see Detwiler *et al.* 1999 for details). To measure the mean aperture, we oriented the fracture vertically with the inlet at the bottom and filled the inlet tubing, manifold, and about 10% of the fracture with dyed water (FD&C Blue No. 1 at 32 mg l^{-1}, Warner Jenkins) and acquired a set of images. We then measured the volume of fluid (V_d) required to nearly fill the fracture (~90%) and acquired another set of images. This provided an accurate measure of the mean fracture aperture, $\langle b \rangle = V_d/A_d$, where A_d is the area occupied by the injected fluid. We then acquired a set of images with the fracture completely filled with dyed water, flushed the fracture with ~10 pore volumes of deionized water, and acquired a set of reference images with the fracture filled with deionized water. We used these images to calculate the spatial distribution of the fracture aperture (described in Section "Aperture measurement"). After measuring the aperture field, we drained and dried the fracture and associated tubing to prepare for the flow experiment.

After drying the fracture, we filled the inlet tubing with solid-free carrier fluid while carefully preventing entrainment of air bubbles. After filling all of the inlet tubing, we slowly filled the fracture with carrier fluid and acquired a set of reference images. We then rotated the fracture to horizontal and initiated slurry injection at 6.0 ml min^{-1}. When the fracture was completely filled with slurry and the effluent mass flow rate reached steady state, we began a stepped flow experiment with image acquisition at regular intervals.

IMAGE ANALYSIS

Raw images consist of measured light intensity values, which we transformed to light absorbance, $A = \ln\left(\frac{I_r}{I_i}\right)$, where I_r is the measured intensity at a pixel in a reference image and I_i is the measured intensity at the same pixel in the measured image. Converting measured intensities to absorbance allows quantitative comparison of images between experiments by eliminating the influence of variations in camera or light-source settings. In addition, absorbance fields provide greater contrast between flowing particles and the carrier fluid.

Aperture measurement

Light absorbance can also be directly related to fracture aperture by applying the Beer–Lambert law (Christian *et al.* 2014) to measurements of the fracture filled with clear and dyed water:

$$I = I_0 e^{-\varepsilon\, Ch + \xi}, \qquad (5.1)$$

where I is the measured intensity at a location, I_0 is the incident light intensity, ε is the absorption coefficient of the solute, C is the dye concentration, h is the solute-filled gap, and ξ, is a constant that accounts for absorbance by the solvent and the glass plates (Detwiler *et al.* 1999). Though the fracture consists of two pieces of flat glass, small variations in long-wavelength aperture are common. These variations can be quantified using light-transmission techniques to measure the fracture aperture; see, for example, Detwiler *et al.* (1999). The aperture at any location is then calculated as

$$h_{i,j} = \frac{A_{i,j}}{\langle A \rangle} \langle h \rangle, \qquad (5.2)$$

where $\langle A \rangle$ is absorbance averaged over the entire field. This method of measuring fracture aperture yields measurements of $h_{i,j}$ that are accurate to within approximately $\pm 1\%$ of $\langle h \rangle$, or about $30\,\mu m$ for the fracture used in these experiments. Table 5.1 summarizes the details of the fracture aperture measurements for Experiments A and B.

Table 5.1 Parameters and geometry for both experiments

Fracture size	
$L \times W$ (cm)	15×12
Aperture (cm)	
Mean	0.34
Standard deviation	<0.01
Flow-rate range (ml min^{-1})	0.2–6.0

List of notations

Symbol	Definition
α	Tracked particle (–)
$\dot{\gamma}$	Shear rate (s^{-1})
ΔP_f	Pressure difference of fluid across fracture (psi)
Δt	Time step used in simulation (s)
Δx	Size of cell in the x-direction (cm)
Δx_α	x-Displacement of tracked particle (cm)
Δy	Size of cell in the y-direction (cm)
Δy_α	y-Displacement of tracked particle (cm)
ε	Absorption coefficient of solute (l mol^{-1} cm^{-1})
μ	Effective carrier-fluid viscosity (Pa s)
μ_f	Viscosity of carrier fluid (Pa s)
ξ	Absorbance of solvent and glass plates (–)
T	Shear stress (Pa)
Φ	Volumetric concentration of solids (–)
ϕ_0	Average solid volume concentration (–)
ϕ_{cr}	Critical volumetric concentration of solids (–)
ρ_s	Density of slurry (kg l^{-1})
ρ_w	Density of water (kg l^{-1})
Ω_{ij}	Region occupied by cell i,j (cm^2)
A	Light absorbance (–)
A_d	Area occupied by dyed water (cm^2)
C	Solute concentration (mol l^{-1})
h	Fracture aperture (cm)
I	Measured light intensity (cd)
I_0	Incident light intensity (cd)
I_v	Viscous number (–)
i,j	Coordinate positions in the x- and y-directions, respectively
L	Fracture length (cm)
P	Total pressure (Pa)
P_s	Pressure acting on a solid particle (Pa)
P_f	Pressure acting on the fluid (Pa)
Q_{in}	Influent flow rate (ml min^{-1})
Q_{out}	Effluent flow rate (ml min^{-1})
U_α	Volume of parcel of fluid occupied by tracked particle α (cm^3)
V_d	Volume occupied by dyed water (cm^3)
V_x	Velocity component in flow direction (m s^{-1})
v^x_{ij}	x-Velocity component in cell i,j (cm s^{-1})
v^y_{ij}	y-Velocity component in cell i,j (cm s^{-1})
W	Fracture width (cm)
$\langle \cdot \rangle$	Spatially averaged quantity

Particle image velocimetry

We used particle image velocimetry (PIV) analysis to calculate velocity fields from the measured absorbance fields using a modified version of the MATLAB-based software, PIVlab (Thielicke & Stamhuis 2012). A high-pass filter applied to the absorbance fields removed long-wavelength features and increased the contrast between individual sand grains and the surrounding carrier fluid. We divided the fracture image into 40×40-pixel subregions and calculated the cross-correlation between corresponding subregions in pairs of sequential images. The PIV algorithm provided a local measure of the average distance sand grains moved from one frame to the next. We note that the absorbance fields provide a high-resolution measure of the depth-integrated absorbance of the slurry-filled fracture. Thus, the resulting velocity fields indicate an average measure of the velocity of the solids at each location in the fracture. For $\phi_0 = 0.5$ (our experiments), we expected the formation of a high-concentration plug in the center of the fracture. Therefore, the measured velocity fields are probably more heavily weighted by particles traveling at the peak velocity and, thus, represent an overestimate of the average slurry velocity. We performed PIV analysis on the entire data set (1000s of images) and constructed the time series of the evolving velocity field within the cell. The supplemental material includes animations of the measured absorbance fields used for the PIV analysis.

EXPERIMENTAL RESULTS

After initializing slurry flow through the fracture, we carried out the two flow experiments by sequentially decreasing the flow rate through a sequence of steps and then increasing it through a subset of the same flow rates. At each flow rate, we attempted to allow the pressure differential (ΔP_f) and effluent mass flow rate ($Q_{out}\rho_s$) to reach steady state. Figure 5.5 shows the time series of mass flow rate of water pumped into the large injection syringe, $Q_{in}\rho_w$ (black circles), the differential pressure measured across the fracture, ΔP_f (dark gray), and the normalized effluent mass flow rate, $\frac{Q_{out}\rho_s}{Q_{in}\rho_w}$ (dashed black line). Because the density of the slurry (ρ_s) was ~1.8 kg l^{-1}, we expected the dashed black line in Figure 5.5 to be relatively constant at a value of 1.8. This was the case for Experiment A, but for Experiment B, each change in flow rate resulted in an immediate increase in $\frac{Q_{out}\rho_s}{Q_{in}\rho_w}$ during decreasing flow-rate steps (or, vice versa, decrease in $\frac{Q_{out}\rho_s}{Q_{in}\rho_w}$ during increasing flow-rate steps) followed by a gradual decrease in $\frac{Q_{out}\rho_s}{Q_{in}\rho_w}$ (or, respectively, an increase in $\frac{Q_{out}\rho_s}{Q_{in}\rho_w}$) during the following period of constant flow rate. This anomalous behavior

during Experiment B resulted from expansion/compression of a volume of about 75 ± 5 ml of air that entered the injection syringe during filling. At each flow rate, as the trapped air equilibrated with the new pressure, $\frac{Q_{out}\rho_s}{Q_{in}\rho_w}$ gradually approached the expected steady-state value of 1.8.

Despite the transient flow rates observed during Experiment B, the transients observed in ΔP_f are consistent with the changing flow rate. That is, as the flow rate gradually approached steady state after each change in flow rate, ΔP_f approached steady state at a similar rate. An exception occurred at the lowest flow rates ($t \sim 100$–150 min), where ΔP_f became relatively constant even when flow rate changed. For Experiment A, the correlation between Q_{out} and ΔP_f was not as clear, particularly for the increasing flow steps. This was likely due to changes in the distribution of solids and fluid within the inlet and outlet manifolds.

To interpret the transient behavior observed during the experiments, it is useful to plot steady-state flow rate (Q_{out}) versus the corresponding ΔP_f at each of the measured flow rates for each experiment to clarify how the slurry rheology affected the transmissivity of the constant-permeability fracture (Fig. 5.6). Despite significant differences in the behavior of the time series of the two experiments, plots of Q_{out} versus ΔP_f are surprisingly similar for both experiments. Most notably, results from both experiments suggest a yield stress ~0.05 psi (or a positive ΔP_f that must be exceeded to initiate flow) as $Q_{out} \to 0$. Also, after the flow rate had been reduced to near zero and then increased, the two experiments exhibited significantly different behaviors. This observation emphasizes the potential for hysteretic behavior in flows of high-concentration slurries.

In addition to the fracture-scale observations of Q_{out} and ΔP_f, PIV analysis of sequential pairs of images provided discrete measurements of the velocity field within the fracture. Averaging sequential velocity fields, measured during a period when the

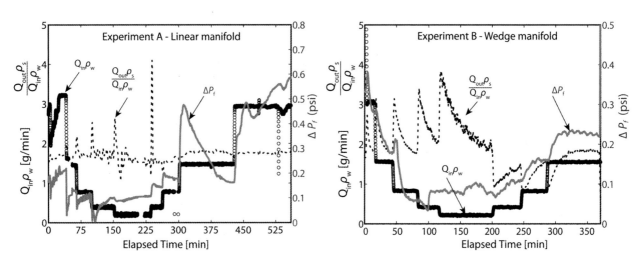

Fig. 5.5. Mass flow rate of water into the syringe pump (black circles) and differential pressure across the fracture (dark gray) plotted against time for Experiments A and B. The dashed black line is the ratio of the mass flow rate of high-ϕ effluent from the fracture over the mass flow rate of water into the syringe. At steady state, the dashed black line will be equal to the density of the high-solid-concentration fluid (~1.8 kg l^{-1}).

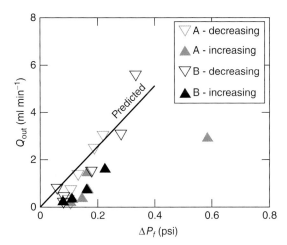

Fig. 5.6. Volumetric flow rate through the fracture (Q_{out}) plotted against differential pressure (ΔP_f) for the full range of measured flow rates for both experiments.

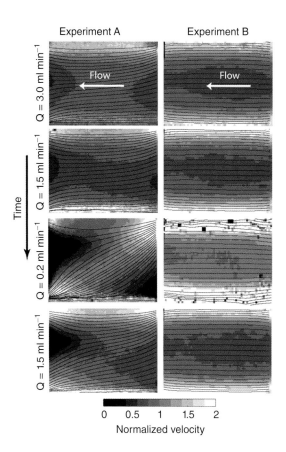

Fig. 5.7. Normalized velocity fields measured over the entire fracture for a subset of the flow rates for Experiments A and B superimposed with the corresponding streamlines. The velocity fields are normalized by the depth-averaged velocity ($V = Q_{out}/W \langle h \rangle$), during each step to facilitate the comparison of velocity fields and profiles at different flow rates. A jammed (~zero velocity) region developed in Experiment A. High-velocity regions/bands are observed in both experiments near the no-flow (top and bottom) boundaries. See the included supplemental material for animations detailing the evolution of the velocity field for both experiments.

observed flow rate was approximately constant, provided a relatively noise-free measure of velocity throughout the fracture at each flow rate. Figure 5.7 shows a representative subset of these solid velocity fields for both experiments. During Experiment A at early time, the flow field was nearly one-dimensional, with the slightly surprising result that the highest velocities occurred along the no-flow boundaries (top and bottom of each frame in Fig. 5.7), where one typically expects fluid velocity to be the lowest due to no-slip conditions on the boundaries. However, at later time, the flow field became more complex with a large region near the outflow boundary with zero velocity and a smaller region near the inlet boundary with near-zero velocity. When the flow rate returned to 1.5 ml min⁻¹ (last frame at bottom), the inlet region returned to a similar configuration to the earlier measurement at the same flow rate (second frame from top) but the outlet region remained jammed. This hysteretic response was a direct manifestation of the geometry of the boundaries and helps explain the difference in pressure response to decreasing (down triangles) and increasing (upright triangles) flow rates observed in Figure 5.6. The outlet manifold was much larger than the fracture aperture, such that pressure losses within the manifold were relatively small compared with those in the fracture. However, the flow geometry caused a stagnation point at the middle of the manifold, which led to the development of the zero velocity region when shear rate decreased. The resulting jamming of the solids was not immediately reversible when flow rate was subsequently increased.

By contrast, the inlet and outlet boundaries for Experiment B precluded a stagnation point within the flow system. The resulting flow field remained nearly one-dimensional throughout the duration of the experiment. Furthermore, because no jamming occurred, there was no evidence of hysteresis in the relationship between flow rate and differential pressure. However, as with Experiment A, the highest velocities occurred along the no-flow boundaries, which was, again, not expected.

Figure 5.8 shows a representative subset of average velocity profiles ($\overline{V}_x = \frac{1}{L} \int_0^L V_x \mathrm{d}x$, where V_x is the component of the velocity in the x or flow direction) normalized by the mean velocity ($\langle V \rangle = \frac{Q_{out}}{W \langle h \rangle}$). These normalized velocity profiles provide a more quantitative measure of the velocity distributions observed during Experiment B. The magnitude of the high-velocity channels along the edges of the fracture increases relative to $\langle V \rangle$ as the flow rate decreases. In addition, at the lowest flow rate, the velocity in the center region of the fracture also increases relative to $\langle V \rangle$. This behavior likely reflects a change in the distribution of solids across the aperture at the lowest flow rate. Because our experimental system does not measure velocity distributions across the fracture aperture, the source of this shift in measured velocities from the expected mean velocity is unclear. However, the process is readily reversible when the flow rate increases and the velocity profile at 1.5 ml min⁻¹ is almost identical to that measured during the decreasing steps.

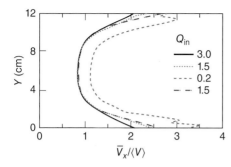

Fig. 5.8. Normalized velocity profiles corresponding to the velocity fields shown in Figure 5.7. These are average profiles (over the fracture length) measured perpendicular to the flow direction. The velocity near the no-flow boundaries is approximately twice the velocity in the middle of the fracture.

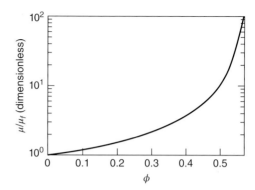

Fig. 5.9. The apparent Newtonian viscosity, μ, of fully developed slurry flow between two plates can be predicted given the Newtonian viscosity (μ_f) of the carrier fluid and the solid concentration (φ) of the slurry (Lecampion & Garagash 2014).

COMPUTATIONAL SIMULATIONS

In this section, we present results from simulations of flow through the fracture to test possible explanations for the strong velocity heterogeneities observed within the uniform-aperture fracture.

Flow of a homogeneous suspension in a Newtonian fluid

We used the rheological model developed by Lecampion & Garagash (2014) to predict the flow of concentrated suspensions. They considered the rheology of monodisperse solids suspended in a Newtonian carrier fluid. Their model reproduces experimentally observed rheologies over the entire range $0 < \phi < \phi_{cr}$ by combining an effective pressure-dependent yield stress (typical of granular media) with hydrodynamic stresses. Although they present a general model for developing pressure-driven flow, we use their simpler result for fully developed flow between parallel plates, which allows the depth integration (across the fracture aperture) of the momentum equation. In this limit, they demonstrated that the cubic law (e.g., Witherspoon *et al.* 1980)

$$Q = W \frac{h^3}{12\mu} \frac{\Delta P}{L} \qquad (5.3)$$

predicts the total flow rate, Q, between horizontal parallel plates, where μ is the apparent viscosity of the slurry, h is the aperture, W is the width, L is the length of the fracture, and ΔP is the total pressure differential measured across the length of the fracture. The total pressure, $P = P_f + P_s$, includes contributions from the pressure acting on the fluid and solid, respectively, but for the relatively free-flowing conditions considered in our experiments, we expect P_s to be small such that $P \approx P_f$. Figure 5.9 plots the unique relationship between μ/μ_f and ϕ developed by Lecampion & Garagash (2014) and used in our simulations. Note that this relationship assumes a Newtonian carrier fluid.

The measurements of carrier-fluid rheology in Figure 5.3 demonstrate that our carrier fluid was non-Newtonian, except

at the lowest shear rates. To determine the appropriate effective viscosity to use in our model, we must determine a shear rate that is representative of the experimental conditions. The characteristic shear rate, $\langle \dot{\gamma} \rangle$, of the fluid between the plates will scale with

$$\langle \dot{\gamma} \rangle \simeq \frac{\langle V \rangle}{h} \simeq \frac{Q}{Wh^2}. \qquad (5.4)$$

For our experiments, $W \simeq 10$ cm, $h \simeq 0.3$ cm, and the flow rate varied from 0.2 to 3.0 ml min^{-1}. The highest of these flow rates corresponds to $\langle \dot{\gamma} \rangle \simeq 0.05$ s^{-1}. Consequently, for our experiments, at all flow rates, $\langle \dot{\gamma} \rangle$ was below the lowest measured shear rate in Figure 5.3, and we can assume that the guar-based carrier fluid was within its Newtonian regime with $\mu_f \simeq 8$ Pa s.

A homogeneous slurry, with $\phi = 0.5$, flowing between two plates with Newtonian carrier fluid $\mu_f \simeq 8$ Pa s results in a slurry with $M \simeq 84.3$ Pa s (Fig. 5.9). Figure 5.6 compares experimental observations with the model predictions of ΔP_f for the measured range of Q_{out}. The predictions are in excellent agreement with the experimental results for decreasing flow rate (down triangles) for both Experiments A and B. However, the experimental results show evidence of a yield stress ~0.05 psi (i.e., a nonzero pressure differential when the flow rate approaches zero), which is not predicted by the model.

While we are able to obtain good agreement for the relationship between Q_{out} and ΔP_f, the experimental results indicate significant variations in the velocity field within the fracture. In the following sections, we present an approach for capturing the details of a heterogeneous flow field that explains the experimentally observed flow structure.

Flow of a heterogeneous suspension

In this section, we describe our approach for simulating the flow of variable solid-concentration fluid within a fracture, which uses a Lagrangian particle-based approach for tracking the solid concentration within the fracture. We approximate

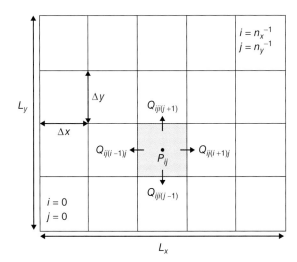

Fig. 5.10. The regular cell structure employed by the width-averaged flow solver. The domain Ω_{ij} of cell ij is highlighted in gray. The pressure, P_{ij}, is cell centered, while fluxes into neighboring elements are face centered.

the potentially three-dimensional geometry of the fracture with a two-dimensional array of locations where the local fracture aperture, $h(x,y)$ is known

$$h_{ij} = h\left[x = \left(i + \frac{1}{2}\right)\Delta x, y = \left(j + \frac{1}{2}\right)\Delta y\right], \qquad (5.5)$$

where Δx is the size of the Cartesian cells and x and y are coordinates lying in the midplane of the fracture (Fig. 5.10). Within each cell, we relate pressure gradient and slurry flow rate by assuming locally fully developed flow between the plates, which avoids the need to discretize between the fracture surfaces. This reduction from three-dimensional aperture space to a two-dimensional approximation significantly simplifies the solution of the resulting system of equations.

There are many options for tracking the interface between multiple phases (see Kothe & Rider (1995) for a review) including high-order advection (Alves *et al.* 2003), interface reconstruction (Youngs 1984), level sets (Sussman *et al.* 1994), and particle-based methods (Morris & Monaghan 1997; Monaghan 2012). We use Lagrangian particles that naturally track fluid-history-dependent variables (ϕ, time since injection, etc.) in a frame of reference tied to the fluid itself, which minimizes advection errors (Monaghan 2012). This method also makes it possible to track the evolution of variations in ϕ due to carrier-fluid loss into the matrix. Our approach resembles particle-in-cell (PIC) methods used for fluid dynamics simulations (Harlow *et al.* 1964) where Lagrangian marker particles representing parcels of fluid are placed throughout the computational domain. The particles carry information such as μ_f and ϕ for a given fluid parcel. At injector locations, source terms are introduced into the flow equations and Lagrangian marker particles are injected with the appropriate volume fractions of the components being injected at that time. Within each time step of length Δt, the Lagrangian particles contribute to

the volume fractions of the various components within the Cartesian cell in which they are located. For example, ϕ_{ij}, the solid concentration in cell i, j, can be estimated using a volume-weighted average across the particles present within the cell:

$$\phi_{ij} = \frac{\sum_{\alpha \in \Omega_{ij}} U_\alpha \phi_\alpha}{\sum_{\alpha \in \Omega_{ij}} U_\alpha}, \qquad (5.6)$$

where Ω_{ij} is the region occupied by cell i, j (highlighted in gray in Fig. 5.10):

$$i\Delta x < x < (i+1)\Delta x$$
$$i\Delta x < y < (i+1)\Delta x \qquad (5.7)$$

and U_α is the volume of the parcel of fluid tracked by particle α and ϕ_α is the solid concentration of the parcel of fluid tracked by particle α. Equation (5.6) provides solid concentrations, ϕ_{ij}, at all points on the grid for each time step. Combining these values of ϕ_{ij} with the relationship plotted in Figure 5.9 provides calculations of the effective slurry viscosity, μ_{ij} within each cell.

Using Equation 5.3, we relate the flow rate Q_{ijkl} from cell i, j into cell k, l to the geometry and differential pressure between the two cells:

$$Q_{ijkl} = \Delta x \frac{h_{ijkl}^3}{12\mu_{ijkl}} \frac{P_{ij} - P_{kl}}{\Delta x} = -\frac{h_{ijkl}^3}{12\mu_{ijkl}}(P_{kl} - P_{ij}), \qquad (5.8)$$

where h_{ijkl} and μ_{ijkl} are the aperture and effective slurry viscosity averaged between cell i, j and cell k, l. We can now assemble a set of linear equations to be solved for the unknown pressures, P_{ij}, by considering the total flux into each cell from its neighbors (see Fig. 5.10):

$$q_{i,j} = \sum_{\substack{(k,l) \in [(i-1,j),(i+1,j), \\ (i,j-1),(i,j+1)]}} Q_{ijkl}, \forall i, j, \qquad (5.9)$$

where q_{ij} is the local injection (positive) or withdrawal rate (negative) of fluid from cell i, j. In addition, depending on the flow geometry considered, there will be a number of cells with prescribed pressure according to applied pressure boundary conditions. The new flow field can then be used to update the location of all Lagrangian particles within the fracture. For example, if particle α occupies cell i, j, its change in position, Δx_α Δy_α is given by

$$\Delta x_\alpha = v_{ij}^x \Delta t$$
$$\Delta y_\alpha = v_{ij}^y \Delta t, \qquad (5.10)$$

where v^x_{ij}, v^y_{ij} are the current velocity components in cell i, j and Δt is the discrete time step used for integration. At this point, the model updates other history-dependent variables such as the solid concentration at each particle due to the rate of local carrier-fluid loss to the matrix. Although in our experiments there was no loss of carrier fluid into the matrix, this is not necessarily the case for natural systems, either because the fracture walls are not impermeable or because small fissures may take some fluid.

For simplicity, we have presented the equations for the case where buoyancy effects are neglected (horizontal fracture). This is sufficient for this study where gravity is acting perpendicular to the plane of the fracture and the timescale for settling of solids is considerably longer than the residence time of the slurry in the fracture. We have also implemented the more general case that accounts for potential slumping of denser fluids.

Simulations of heterogeneous flow in a fracture

Both experiments developed nonuniform flow fields, despite the homogeneous composition of the injected fluid (Fig. 5.7). In the case of Experiment A, clear stagnant zones developed at the lowest flow rates. However, both experiments clearly exhibited high-velocity zones along the edges of the fracture. In the case of Experiment B, these features were present and stable at all injection rates. In this section, we investigate what mechanisms might explain the development and stability of such high-velocity channels within the fracture. Using the numerical model described in the previous section, we consider the following possible sources of velocity-field heterogeneity within the fracture: (i) aperture variability; (ii) blockages in the manifold at the inlet and outlet; or (iii) heterogeneity of ϕ within the inlet manifold.

Our aperture-field measurements indicate variations within a few percent of the average at most and generally much less. Such small variations in aperture cannot explain doubling of the fluid velocity. Furthermore, the measured variations in aperture do not correlate with the high-velocity flow channels, thus ruling out the first hypothesis.

Figure 5.11 shows the results from simulations in which we introduced blockages within the upstream and downstream manifolds and assumed a constant-ϕ slurry throughout the fracture. This simulation captures the details of the velocity field resembling the experiment near the inlet and the outlet. However, the dissipative nature of flow between two plates results in the flow becoming essentially uniform (across the fracture width, W) in the middle of the fracture (Fig. 5.11).

Finally, we considered the possibility that variations in upstream solid concentration can lead to stable heterogeneity in the flow field. This hypothesis assumes that changes induced either within the upstream tubing or within the upstream manifold induce systematic changes in the upstream solid concentration pumped into the fracture. While the precise mechanism controlling this segregation has not been identified, we can investigate the implications for flow within the observed portion of the fracture. For this hypothesis to be plausible, the induced changes should be relatively small. Furthermore, because the velocity distribution for Experiment B was independent of distance from the upstream inlet, the imposed changes in the upstream solid concentration must propagate downstream through the fracture without undergoing significant change.

We performed simulations assuming a prescribed distribution of ϕ at the upstream end of the fracture corresponding to approximately 2 cm width channels on either side (Fig. 5.12). Our simulations indicate that the imposed changes in the upstream ϕ values are indeed preserved during flow, leading to sustained variations along the entire length of the fracture, including the outlet manifold (see Fig. 5.12). In addition, our analysis indicates that a reduction of as little as 3% in ϕ can lead to a factor-of-two increase in velocity within the low-ϕ channels over that in the higher ϕ central flow region (see Figs 5.13 and 5.14).

Our simulations show that only the third hypothesis is consistent with the experimental observations. In our experiments, heterogeneity in the solid-concentration field induced within the upstream tubing and/or inlet manifold likely resulted in variations in ϕ within the fracture. However, in natural and engineered systems, particularly at larger scales, heterogeneities in ϕ are likely to be prevalent. Our results suggest that even small variations in ϕ are both stable and sufficient to induce large (factor of 2) velocity variations in flow channels within

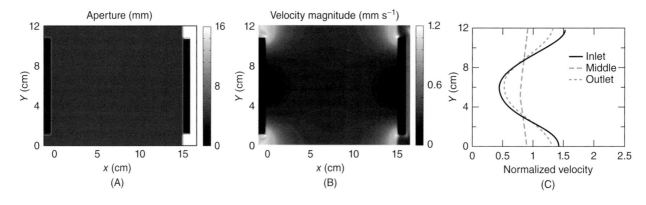

Fig. 5.11. Numerical simulation of fully developed flow of high-solid-volume slurry in an attempt to match the experimentally observed heterogeneities by introducing blockages within the upstream and downstream manifolds through changes to the aperture field (A) while assuming the fluid remains homogeneous. The velocity field (B) and profiles (C) indicate that the flow midway along the fracture is essentially homogeneous due to the dissipative nature of flow within the fracture.

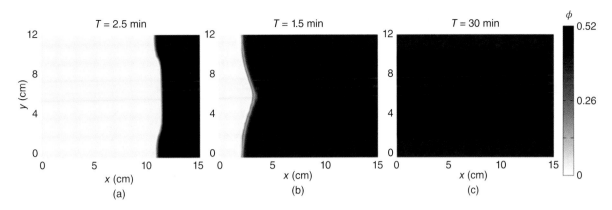

Fig. 5.12. Numerical simulation of fully developed flow of high-solid-volume slurry in the fracture with imposed heterogeneity in ϕ at the upstream boundary. The gray scale represents the evolving values of ϕ within the fracture. In this example, we impose $\phi = 0.47$ at the upstream boundary (at right) within high-speed channels of approximately 2 cm width. The central region has $\phi = 0.515$, resulting in $\phi = 0.5$ overall. Our numerical model indicates that prescribed upstream variations in ϕ will propagate from inlet to outlet in a stable manner.

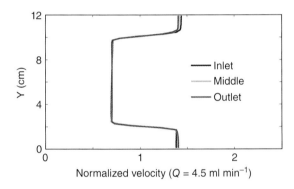

Fig. 5.13. Steady-state velocity profiles corresponding to the heterogeneous-solid-concentration model shown in Fig. 5.12. The model predicts that the slurry velocities are approximately doubled in regions of approximately 2 cm width, which is in good agreement with the experiment (Fig. 5.8)

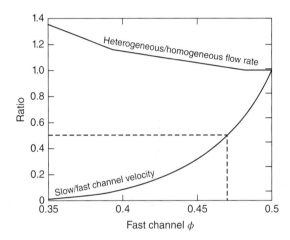

Fig. 5.14. This plot explores the influence of increasing solid-concentration contrast in the fracture on the heterogeneity in the flow field using the same geometry shown in Figure 5.12. The lower curve shows the predicted ratio of the velocity in the central portion of the fracture to the velocity in the two 2-cm-wide low-ϕ channels. As ϕ approaches 0.34 in the fast channel, the flow in the central region stagnates. A value of $\phi = 0.47$ approximates the experimental observation of velocity doubling in the fast channel zones (a ratio of 0.5 on this plot). The upper curve shows the ratio of total flow rate in the heterogeneous scenario compared with the flow of a homogeneous solid concentration of 0.5 for the same pressure drop across the fracture. We see that even as the central region stagnates, the total flow rate differs from the homogeneous solution by only tens of percent.

the fracture. Figure 5.14 explores the relationship between the velocity of the central region and the fast channels as we introduce progressively greater differences between the values of ϕ in the two regions, while maintaining the same ϕ_0. For the same pressure drop across the fracture, we also calculated the corresponding total flow rate obtained for these heterogeneous scenarios compared with the total flow rate for a homogeneous slurry with $\phi = 0.5$ (upper curve in Fig. 5.14). Even as the velocity ratio between the slow and fast channels approaches zero, the difference from the total flow rate predicted by the homogeneous theory is only tens of percent. We also investigated the effect of varying the width of fast channels assumed to have a solid concentration of 0.47 (Fig. 5.15) upon the flow and confirm that the homogeneous theory predicts a total flow rate in close agreement with the heterogeneous flow field. Even when the fast channels are enlarged to 4 cm and $\phi \rightarrow \phi_{cr}$ in the central channel (corresponding to 2/3 of the total area of the fracture) the total flow rate predicted for the homogeneous assumption is within 10% of the heterogeneous result. These

two sets of simulations indicate that the reduction in flow within the slower central channel has offset the impact of the fast channels on the average flow field. As the central region stagnates, nonlinear effects become stronger and the homogeneous solution becomes less accurate. These results suggest that bulk measurements, such as ΔP_f and Q_{out}, provide only weak constraints upon the nature of the flow within the fracture. Specifically, very high-velocity channels may develop within the fracture while the total flow rate changes only slightly.

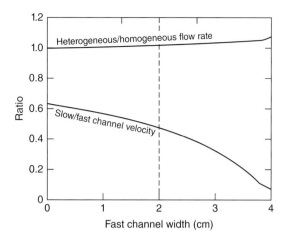

Fig. 5.15. We explore the influence of varying the width of fast channels with $\phi = 0.47$, while maintaining the same average value of ϕ across the entire fracture using the same geometry as shown in Fig. 5.12. The upper curve shows the predicted ratio of the flow rate in the slower central portion of the fracture to two channels with $\phi = 0.47$. As the width of the fast channels increases, the flow in the central region progressively slows due to increased solid concentration. Also, two channels of width 2 cm approximate the experimental observation of velocity doubling in the fast channel zones (a ratio of 0.5 on this plot). The upper curve shows the ratio of total flow rate in the heterogeneous scenario compared with the flow of a homogeneous solid concentration of 0.5 for the same pressure drop across the fracture. Even as the central region stagnates, the total flow rate is well approximated by the homogeneous solution.

However, other transport properties of the heterogeneous system, such as initial breakthrough of slurry and dispersion, will differ greatly from the homogeneous scenario.

CONCLUDING REMARKS

We have demonstrated that the experimentally observed relationship between pressure drop and total flow rate through an idealized smooth-walled fracture is predicted well by the cubic law using a recently developed rheological model to represent the effective viscosity of the concentrated slurry. However, our experimental results revealed significant variations in the velocity field within the fracture though the cubic law assumes uniform velocity in the fracture plane. Further analyses demonstrated that small variations (~3%) in the spatial distribution of ϕ were sufficient to induce large (factor of 2) velocity variations in channels within the fracture. As $\phi \to \phi_{cr}$, very small changes in ϕ can result in large differences in the effective viscosity of the slurry and induce both high-velocity channels and immobile or jammed regions. Furthermore, once

established, ϕ heterogeneity and the resulting velocity variations persisted through the length of the fracture; and, once formed, jammed regions persist, even after increasing the shear rate. Finally, while the actual flow within the fracture/fault may be highly heterogeneous, the average pressure drop and total flow rate through the system will remain close to that predicted for homogeneous flow. Consequently, while the assumption of uniform flow may match observations of pressure drop and total flow rates, it may greatly underestimate both the time of breakthrough and degree of dispersion of the slurry transported within the fracture.

In our experiments, the variations in ϕ that caused velocity-field heterogeneity were induced by upstream boundary conditions, despite efforts to maintain a uniform boundary condition of $\phi = \phi_0$. In the subsurface, heterogeneity is ubiquitous and the potential for uniform-ϕ flows to persist in natural or engineered systems seems unlikely. For example, in real fractures, the rock matrix bounding the fracture may have a non-negligible permeability and the resulting loss of fluid to the matrix can cause nonuniform changes in ϕ. In addition, aperture variability caused by fracture surface roughness leads to velocity-field variability, even in the absence of solids. In the presence of high concentrations of suspended solids, we expect plug-flow behavior similar to that observed in our experiments when the amplitude of the surface roughness is significantly smaller than the fracture aperture. In fractures with regions with apertures that are on the order of the largest particle diameters, the likelihood of jamming will increase, resulting in relationships between flow and pressure gradient that are more difficult to predict than in our experiments.

ACKNOWLEDGMENTS

The authors would like to thank Schlumberger for sponsoring this research and Luke Shannon for his assistance with developing the experimental system.

SUPPORTING INFORMATION

Additional supporting information may be found in the online version of this chapter, *Geofluids* (2015) 15, p. 24–36:

Data S1. Animation of the absorbance fields calculated for the two experiments (Exp_A.avi)

Data S2. Animation of the absorbance fields calculated for the two experiments (Exp_B.avi)

CHAPTER 6

Normal stress-induced permeability hysteresis of a fracture in a granite cylinder

A. P. S. SELVADURAI

Department of Civil Engineering and Applied Mechanics, McGill University, Montréal, QC, Canada

ABSTRACT

This chapter examines the influence of axial stress-induced closure of a fracture on its permeability. The experiments were conducted on a cylinder of Barre Granite measuring 457 mm in diameter and 510 mm in height, containing a central cylindrical cavity of diameter 75 mm. Radial flow hydraulic pulse tests were conducted in a previous research investigation (Selvadurai APS, Boulon MJ, Nguyen TS (2005) The permeability of an intact granite. *Pure and Applied Geophysics*, **162**, 373–407) to determine the permeability characteristics of the intact granite. In the continuation of the research, a fracture was introduced in the cylinder with its nominal plane normal to the axis of the cylinder. Axial compressive stress was applied normal to the plane of the fracture. An increase in the compressive normal stress acting on the fracture caused a reduction in the aperture of the fracture, which resulted in the reduction in its permeability. Steady-state radial flow tests were conducted on the fractured axially stressed sample to determine the variation of fracture permeability with axial normal stress. The analytical developments also take into account the flow through the matrix region as the normal stress increases. The results of the experimental investigations indicate that the complete stress relief of a fracture previously subjected to a normal stress of 7.5 MPa can result in a permeability increase of approximately three orders of magnitude. These findings are relevant to shallow depth geotechnical construction activities where enhanced fluid flow can be activated by stress relief. As the fracture aperture closes with high normal stress, the flow through the matrix can be appreciable and if this factor is not taken into consideration the interpretation of fracture permeability can be open to error. This factor can be of interest to the interpretation of permeability of fractures in deep crustal settings where the stresses acting normal to the fracture surface can inhibit flow in the fracture.

Key words: axially stressed fractures, flow in fractures, permeability hysteresis, stress-induced fracture closure

INTRODUCTION

Fractures are geological features that can dominate the fluid transport characteristics of geologic media. Fractures are important to groundwater hydrology, environmental earth science and, particularly, environmental geomechanics problems associated with energy resources recovery from gas-bearing shales, geologic sequestration of greenhouse gases and deep geologic disposal of hazardous material including heat-emitting radioactive wastes. The literature covering these areas is vast, and no attempt is made to provide a comprehensive literature review. Articles covering these topics and those with specific relevance to the role of fractures on the containment strategy are given by Noorishad *et al.* (1984), Boulon *et al.* (1993); Selvadurai & Nguyen (1995, 1997), Nguyen & Selvadurai (1995, 1998), Rutqvist & Tsang (2002), Chan *et al.* (2005),

Selvadurai & Yu (2005), and Selvadurai et al. (2015). The importance of fractures and other defects will invariably depend on the scale at which such defects enter the description of the flow processes in naturally occurring geological media. For example, in laboratory-scale testing of granitic rocks, the fractures can be intergranular cracks in the three major minerals, coincident grain boundary cracks and transgranular cracks. In contrast, at regional scales, the fractures are largely well defined, distinct defects with geometrical characteristics that enable the definition of features such as gouge and debris fill, which can be treated as separate material regions. An intermediate scale can refer to conditions that can be encountered at a fracture intersecting a borehole and packer system installed for the measurement of hydraulic properties. A hydraulically created fracture will, in general, involve originally nearly mated surfaces; these can experience either closure, opening or relative

Crustal Permeability, First Edition. Edited by Tom Gleeson and Steven E. Ingebritsen.
© 2017 John Wiley & Sons, Ltd. Published 2017 by John Wiley & Sons, Ltd.
Companion Websites: www.wiley.com/go/gleeson/crustalpermeability/
http://crustalpermeability.weebly.com/

movement depending on the state of stress existing in the vicinity of the borehole. The scale of the problem is such that a typical situation involving a borehole intersecting a fracture can be investigated at near-field scale in the laboratory. Investigations involving both laboratory and field determinations of the permeability of mechanically inactive or stationary fractures are numerous, and references to these studies can be found in the volumes and articles by Snow (1968b), Louis (1974), Raven & Gale (1985), Gale (1990), Makurat et al. (1990a,b), Boulon et al. (1993), Yeo et al. (1998), Pyrak-Nolte & Morris (2000), Bart et al.(2004), Hans & Boulon (2003), Sausse & Genter (2005), Selvadurai et al. (2005), Giacomini et al. (2007) and Selvadurai & Selvadurai (2010).

A striking feature in the modeling of fractures in a hydrogeological context is that, by and large, the fractures are treated as stationary features that experience no response to either mechanical or thermal phenomena or changes to the fluid pressure. This is at variance with the well-established approaches in geomechanics of rocks where response of the fracture to mechanical effects and failure is generally the norm in terms of modeling the mechanical performance of the fracture. The earliest investigations in this area are due to Patton (1966) who conducted experiments on artificial fractures, and the focus was primarily on evaluating the strength of the joint. Processes such as dilatancy during shear, which can contribute to drastic alterations in the permeability of fractures, were left uninvestigated. The investigations by Ladanyi & Archambault (1970), Jaeger (1971), Barton & Choubey (1977) and Bandis et al. (1983) proposed strength criteria, which were departures from the bilinear form of the shear strength criterion for the fracture. A great deal of attention has been focused on the mechanical characterization of the performance of the fracture, and these efforts are also described in the literature (Selvadurai & Boulon 1995; Nguyen & Selvadurai 1998; Jafari et al. 2004). Permeability alterations during shear were investigated by Witherspoon et al. (1979), Barton (1982), Elliot et al. (1985), Benjelloun (1993) and Adler (1997). The work of Nguyen & Selvadurai (1998) dealing with dilatant fractures builds on the model proposed by Plesha (1987), incorporating the influences not only of empirical relationships based on commonly used fracture topography measures such as the *joint roughness coefficient* (JRC) and *joint compressive strength* (JCS) but also a mathematically consistent elastoplasticity model that accounts for asperity degradation (Makurat et al. 1990a,b). The transformation of asperity degradation in fractures to gouge development, which ultimately has a significant influence on mechanically induced alterations to the permeability of fractures, requires alternative approaches that can involve continuum to discrete transformation of a fracture zone and continual degradation of the fragments with mechanical action (Selvadurai & Boulon 1995; Selvadurai & Sepehr 1999a,b; Selvadurai 2009; Massart & Selvadurai 2012, 2014). Such procedures, without provision for a proper algorithm that transforms a continuum region to a discrete system, will be of marginal value to modeling gouge development in fractures. Recent investigations of the influence of permeability evolution during stressing of rocks have been prompted by an interest in identifying the development of excavation damage zones around openings that are planned for the deep geologic disposal of heat-emitting radioactive wastes (Hudson et al. 2008; Nguyen & Jing 2008; Rejeb et al. 2008; Rutqvist et al. 2008; Najari & Selvadurai, 2014; Liu et al. 2013). The evolution of micromechanical damage in rocks and the consequent alteration in the deformability characteristics have been documented by Zoback & Byerlee (1975), Shiping et al. (1994) and Kiyama et al. (1996), but the extension of the concepts to permeability evolution is largely theoretical (Mahyari & Selvadurai 1998; Selvadurai 2004; Selvadurai & Shirazi 2004, 2005). Recent investigations by Shao et al. (1999), Souley et al. (2001), Zhou et al. (2006), Hu et al. (2010) and Massart & Selvadurai (2012) have developed permeability evolution concepts that match experimental data derived from testing of rock cores. Other examples of permeability evolution, particularly under isotropic compression, have been discussed by several investigators including Zhu & Wong (1997), Selvadurai & Glowacki (2008), Selvadurai et al. (2011) and Selvadurai & Jenner (2012). In these studies, the emphasis is on micromechanical defect generation under complex stress states, which contributes to either permeability enhancement or permeability reduction depending on the state of stress.

Investigations that examine the influence of stress states on the permeability characteristics of fractures are rare. Major obstacles for incorporating the influences of mechanical actions on the fluid transport characteristics of fractures can arise from the following: (i) the size of samples or scale effects that are representative of the fractures (Witherspoon et al. 1979), (ii) the precise identification of the alteration in the stress states relevant to problems in hydrogeomechanics (Haimson 1975; Stephansson 1985; Ingebritsen et al. 2006), (iii) the physical arrangements of a test large enough to apply, with a degree of control, the variety of relative motions that can result from alterations in the ambient stress state, (iv) the development of sound experimental procedures that can be used to observe permeability alterations in the fracture regions as the stress state is altered and (iv) the relationship of the evolving permeability to processes such as fracture closure and gouge generation, particularly as a function of the attainment of the failure threshold in regions of the fracture. Simultaneous consideration of all these factors is an almost impossible task in experimental geomechanics research and to date attention has focused on extremely small-scale investigations that may not completely capture the *in situ* characteristics of fractures at scales of interest to hydrogeomechanical applications.

This chapter presents the results from a series of fracture flow experiments conducted on a large cylinder of Barre Granite measuring 457 mm in diameter and 510 mm in height containing a 57 mm diameter cylindrical cavity located at the centre of the cylinder over its entire height (Fig. 6.1). To the

Fig. 6.1. Cylindrical sample of the Barre Granite used in the experimental investigations containing a V-notch for fracture creation. (Adapted from Selvadurai et al. 2005.)

author's knowledge, this is one of the few instances where a large cylinder of the scale indicated has been used in a laboratory experiment. The experimental facility was designed in such a way that transient radial flow pulse tests could first be conducted on the intact cylinder through pressurization of the central borehole. In a previous study (Selvadurai *et al.* 2005), the experimental facility was used to determine the intact permeability of the Barre Granite, which corresponds to the matrix permeability of the material. In this chapter, we consider the situation where the cylinder contains a fracture at a central plane perpendicular to its axis. The alteration in the permeability of the fracture due to an axial normal stress is determined by conducting steady-state radial flow experiments. The influence of the radial flow through the intact segments of the fractured cylinder is included in calculation of the fracture permeability.

THEORETICAL ASPECTS

Radial flow hydraulic pulse testing of the intact cylinders

The theoretical background and modeling of the radial flow hydraulic pulse tests conducted on the intact Barre Granite

was presented by Selvadurai *et al.* (2005). The conventional approach for estimating the permeability from hydraulic pulse tests relies on the piezo-conduction equation, and derivations of the equations are given in standard texts (Bear 1972; Phillips, 1991; Selvadurai 2000; Ichikawa & Selvadurai 2012). The pressure transients are assumed to occur in a fluid-saturated porous medium that can be described by the classical theory of poroelasticity proposed by Biot (1941) (see also Selvadurai & Yue 1994; Selvadurai 1996, 2007) that takes into account the coupling between the deformations of the porous skeleton and the compressible pore fluid. In conventional treatments of hydraulic pulse tests, it is implicitly assumed that the pressure transients can be described by the uncoupled piezo-conduction equation, which accounts for the compressibility of the pore fluid, the compressibility of the porous skeleton and the compressibility of the material constituting the porous fabric. A comparison between results for the one-dimensional hydraulic pulse test derived from the piezo-conduction equation, and Biot's theory of poroelasticity was performed by Wang (2000) and generally, the results for the two approaches compare favourably for low-permeability materials . More recently, Selvadurai & Najari (2013) conducted similar studies and arrived at the conclusion that the piezo-conduction equation can be used satisfactorily for estimating the permeability characteristics of low-permeability materials such as Westerly Granite, Indiana Limestone and Stanstead Granite from Quebec, Canada. Additional influences of air fraction in the pressurized cavity in a hydraulic pulse test are also discussed by Selvadurai & Najari (2015). The simplest analytical result that can be used to interpret two-dimensional axisymmetric radial flow hydraulic pulse tests assumes that the pulse test is conducted in a cylindrical cavity located in a fluid-saturated porous medium of infinite extent. Selvadurai & Carnaffan (1997) have investigated the validity of this assumption particularly when hydraulic pulse tests are conducted in annular fluid-saturated regions and provide constraints that allow the application of the theoretical results to infinite domains. The influence of air voids within the pressurized cavity on the interpretation of the conventional hydraulic pulse test was examined more recently by Selvadurai & Ichikawa (2013). Avoiding details, it can be shown that the pressure decay in a fluid-filled cylindrical cavity located in a fluid-saturated porous medium due to transient radial flow in the infinite region is given by

$$p(t) = p_0 \exp(4\alpha\beta)\mathrm{erfc}(2\sqrt{\alpha\beta}). \qquad (6.1)$$

In (6.1), $p(r,t)$ is the transient pressure potential in the porous region, t is the time variable, r is the radial coordinate, $\mathrm{erfc}(x)$ is the complementary error function, p_0 is the initial pressure, and

$$\alpha = \left(\frac{\pi a^2 S}{V_w C_w \gamma_w}\right); \quad \beta = \left(\frac{\pi T_R t}{V_w C_w \gamma_w}\right), \qquad (6.2)$$

where

$$S = l\gamma_w(nC_w + C_{\mathrm{eff}}); \quad T_R = k_i l = K_i l\gamma_w\eta. \qquad (6.3)$$

Also, in (6.2) and (6.3), S is the storage coefficient (nondimensional); T_R is the transmissivity of the cylinder of height l (L^2/T); n is the porosity (nondimensional); C_w is compressibility of the pore water (LT^2/M); C_{eff} is the compressibility porous skeleton (LT^2/M); K_i is the permeability of the intact rock (L^2); k_i is the hydraulic conductivity of the intact rock (L/T); V_w is the volume of the pressurized region (L^3); η is the dynamic viscosity (M/LT) and γ_w is the unit weight of the fluid (M/L^2T^2).

Hydraulic pulse tests conducted previously by Selvadurai *et al.* (2005) estimate that the permeability of the intact granite is between $0.40 \times 10^{-18}\,\text{m}^2$ and $1.20 \times 10^{-18}\,\text{m}^2$. Steady-state flow tests conducted on the intact cylinder gave values that ranged from $1.21 \times 10^{-18}\,\text{m}^2$ to $1.46 \times 10^{-18}\,\text{m}^2$. Considering these ranges, the permeability of the intact granite is taken to be approximately $1.20 \times 10^{-18}\,\text{m}^2$

Radial steady flow testing of fractured cylinders

In this series of experiments, the intact granite cylinder containing the central cavity was fractured by diametral compression along a pre-cut groove located on the outer surface of the cylinder (Fig. 6.1). The fracture is obtained by inducing a near tensile stress field at a diametral plane. When dealing with gouge-free plane fractures, the issue of fracture topography, fracture roughness and so on, and their influences on the fluid transport characteristics have been investigated extensively in the literature (Raven & Gale 1985; Cook 1992; Pyrak-Nolte & Morris 2000; Méheust & Schmittbuhl 2003). The objective,

here, is to use the parallel-plate analogue as an elementary model and to establish the pattern of reduction in the permeability of the fracture, as the stress normal to the fracture plane is subjected to quasi-static stress cycling. Considering the parallel-plate model and fluid flow through the fracture with an aperture 2λ it can be shown that

$$K_f = \frac{\eta Q_f \log_e(b/a)}{4\pi\lambda(p_a - p_b)} \quad [\text{Units(length)}^2], \qquad (6.4)$$

where Q_f is the flow rate through the fracture (L^3/T), b is the external radius of the fracture (L), and p_a and p_b are the hydraulic fluid pressures (M/LT^2) at the inner and outer boundaries of the fracture. (Considering the overflow locations shown in Fig. 6.2, the fluid pressure at the outer boundary of the fracture will be maintained at nearly atmospheric pressure.) As the axial stresses increase, the fracture aperture will be reduced and the flow through the intact matrix can influence the flow process, although the flow rate measured in the steady-state flow experiment is the combined flow rate Q occurring through both the intact region and the fracture. Because the pulse testing of the intact cylinder provides an estimate for the matrix permeability K_i, we obtain the following result for the permeability of the fracture (K_f^I) that also accounts for the fluid flow through the intact matrix:

$$K_f^I = \left(\frac{\eta Q \log_e(b/a)}{4\pi\sqrt{3}(p_a - p_b)} - \frac{K_i l}{2\sqrt{3}} \right)^{2/3} \quad [\text{Units(length)}^2]. \qquad (6.5)$$

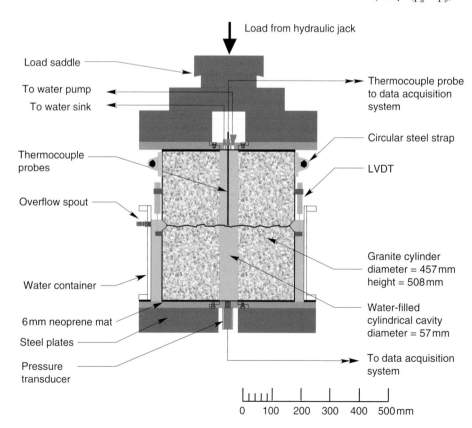

Load from hydraulic jack

Load saddle

To water pump
To water sink

Thermocouple probe to data acquisition system

Thermocouple probes

Circular steel strap

LVDT

Overflow spout

Granite cylinder diameter = 457 mm height = 508 mm

Water container

Water-filled cylindrical cavity diameter = 57 mm

6 mm neoprene mat

Steel plates

Pressure transducer

To data acquisition system

0 100 200 300 400 500 mm

Fig. 6.2. Details of the experimental arrangements for the application of external loads and for the pressurization of the central fluid-filled cavity.

These results can be used to examine the influence of matrix flow in accurately interpreting the permeability of a fracture that is subjected to a stress normal to its plane.

As the magnitude of $K_i l$ becomes small in comparison to the first term on the right-hand side of (6.5), the permeability of the fracture can be calculated using the result (6.4), where

$$\lambda = \left(\frac{3\eta Q \log_e(b/a)}{8\pi(p_a - p_b)} \right)^{1/3} \text{[Units(length)]}. \tag{6.6}$$

The permeability of the fracture can be represented by the parallel-plate model (Selvadurai 2000) where

$$K_f = \frac{(2\lambda)^2}{12} \text{[Units(length)}^2]. \tag{6.7}$$

The permeability of *relatively open fractures* (K_f^{II}) can be obtained from the result

$$K_f^{\text{II}} = \left(\frac{\eta Q \log_e(b/a)}{4\pi\sqrt{3}(p_a - p_b)} \right)^{2/3} \text{[Units(length)}^2], \tag{6.8}$$

which is the expression obtained when K_i is set to zero in (6.5).

EXPERIMENTAL PROCEDURES

Aspects of the experimental investigations dealing with hydraulic pulse testing of the Barre Granite are documented by Selvadurai *et al.* (2005). A petrographic analysis conducted by Hardy (1991) indicates that the Barre Granite, a blue-grey granodiorite, contains approximately 23% quartz, 61% feldspar and 11% mica. The large diameter cylindrical samples of the Barre Granite used in the experimental investigations were obtained from Vermont, USA. The mechanical and physical properties of the granite were determined from ASTM and ISRM Standard Tests. [These results are summarized in Table 1 of Selvadurai *et al.* (2005).] Diametral compression was then used to create the fracture. The V-notch ensured that the fracture plane was nominally perpendicular to the axis of the cylinder. The plane ends of the cylinder were machine polished to a mirror finish (Fig. 6.1) by the suppliers to ensure parallel end planes that could be mechanically sealed both during the transient pulse tests and the steady-state fracture permeability tests.

The details of the test frame used to apply stresses normal to the plane of the fracture are described by Selvadurai *et al.* (2005). Separate stainless steel plates of 25 mm thick were provided at the plane ends of the cylinder to prevent corrosion of the loading plates in the event of water leakage from the pressurized central cavity. The stainless steel plates were also fitted with O-rings to provide an adequate seal during pressurization of the central cavity. The effectiveness of this sealing technique was verified by pressurization of a hollow stainless steel cylinder, which maintained the pressure without loss (except for fluctuations of ambient temperatures) for a period of 8 days. Neoprene mats (6 mm thick) were also provided at the upper and lower contact surfaces to minimize any stress

concentrations that could occur due to uneven contact between the test cylinder and the stainless steel plates. The stainless steel plates contained specially designed couplings to provide access to sensors and instrumentation within the fluid-filled cavity. The stainless steel base plate incorporated a pressure transducer (maximum pressure of 2.07 MPa) to measure fluid pressure. The upper stainless steel plate contained a specially designed coupling unit that provided access for the inlet and outlet water supply ports. A thermocouple located in this coupling was used to measure water temperature in the cavity during the tests. All inlet and outlet leads for water supply were made of stainless steel, and Swagelok valves (maximum pressure rating of 20.7 MPa) were used to seal the fluid pressure in the cavity during hydraulic pulse tests on the intact sample. The compression of the intact sample during application of sealing stresses was found to be negligible. During permeability testing of fractures, however, the lower stainless steel plate was incorporated with a Plexiglas reservoir to collect any fluid migrating through the fracture (Fig. 6.2). Testing of fracture permeability also required measurement of the closure of the fracture during application of axial stresses. Three linear variable differential transformers (LVDTs) (Fig. 6.3), located at orientations of 120° around the circumference of the cylinder, were used to measure closure of the fracture. The three values were averaged to arrive at the closure of the fracture during the application of axial stress. A schematic detail of the sample and test arrangement used when performing steady radial flow tests through the fracture is shown in Figure 6.4. The experimental facilities also included a Shimadzu precision pump for saturating the granite cylinder, for applying the pressure pulse required to conduct the hydraulic pulse tests and for maintaining a steady flow when conducting permeability tests on the fractured cylinder. The data sets from all the experiments were recorded with a computerized data acquisition system that uses Lab View Data Processing.

Prior to fracturing the granite cylinder, four markers (Demec Gauges) were installed on the surface of the cylinder to allow reassembly of the split specimen at the correct orientation (Fig. 6.5) and for the measurement of the fracture aperture

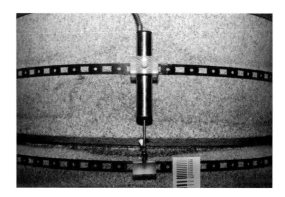

Fig. 6.3. The LVDT arrangement for movements of the fracture.

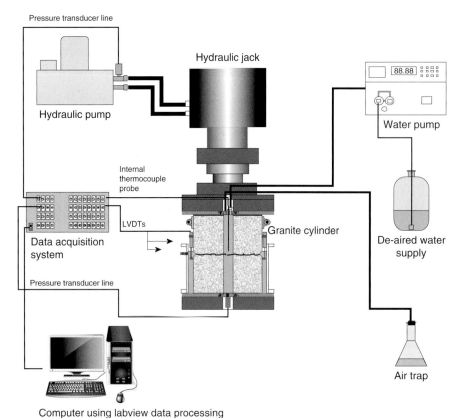

Fig. 6.4. Schematic view of the radial flow permeability testing arrangement.

prior to the application of normal stresses. In the central cavity of the granite cylinder, a steel bar was secured with end plates and rubber pads to allow axial expansion during splitting about a plane normal to the axis of the cylinder and to prevent relative movements between the separate parts of the fractured cylinder, which could lead to unwanted asperity shear. Splitting action was initiated by applying point loads through hardened steel wedge-shaped regions. The plane of the fracture can exhibit an uneven configuration depending on the microstructural fabric of the granite. The provision of the midplane groove localized the plane of failure, but where the groove was absent, the fracture plane could be nonplanar, which presents an obstacle for inducing pure compression on the fracture plane. The splitting test generally produced a relatively planar fracture with some evidence of compression failure near the point of applications of the loads. This result is, however, not the norm, and on occasions, the fracture surfaces can deviate from the position of the groove; this can result in a warped fracture surface and can also lead to edge fractures. The surfaces of the fractures of the sample tested were cleared of debris and loose particles by air blowing prior to laser scanning of the surfaces. Laser scanning of the surfaces was conducted at the *Center for Intelligent Machines* at McGill University. Laser scanning was used as a visual guide to establish the topography of the fracture and for assessing any damage to the surface during axial compression.

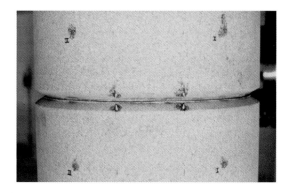

Fig. 6.5. Demec markers used for mated alignment of the fractured segments of the cylinder and measurement of the initial aperture width.

The scanning accuracy of the laser device was approximately 0.1 mm. The surface topography is shown in Figure 6.6. The fracture does not display any dominant features that would negate the use of the parallel-plate model.

The reassembled sample was placed in the testing machine, and LVDTs were installed to measure the reduction in the fracture aperture width during the application of compressive stresses. A reservoir collected the fluid migrating through the fracture during attainment of a steady flow rate. The outlet of the reservoir was maintained at approximately the level of the fracture plane.

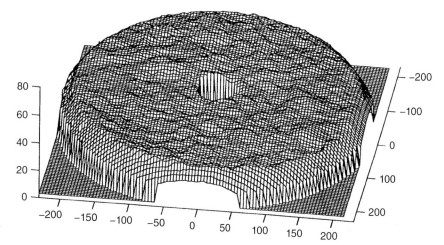

Fig. 6.6. The fracture surface topography (all dimensions are in mm).

EXPERIMENTAL PROCEDURES AND RESULTS

The fracture was subjected to axial stresses, and constant flow rates varying from 0.1 to 10 ml min^{-1} were maintained to achieve steady flow conditions depending on the level of axial stress and the resulting aperture closure. Altogether, three loading–unloading cycles were performed on the fracture, and the permeability of the fracture was calculated using Eq. 6.5. Figure 6.7 illustrates the closure of the fracture during application of the axial normal stress σ_n. The initial width of the fracture

aperture $2\lambda_0 \simeq 0.0707$ mm. The fracture exhibits significant closure during the first cycle of loading up to 2.5 MPa. At the peak load of 7.5 MPa, applied during the third cycle of loading, the fracture aperture reduces to $(2\lambda)_{min} \simeq 0.0022$ mm. Upon complete unloading of the fracture in the third cycle, the residual fracture aperture of $(2\lambda)_{res} \simeq 0.0321$ mm. Figure 6.8 illustrates the variation in the permeability of the fracture during three loading–unloading cycles of applied axial stress σ_n. The first cycle of loading was performed by subjecting the fracture to a maximum normal stress of 2.5 MPa. At this

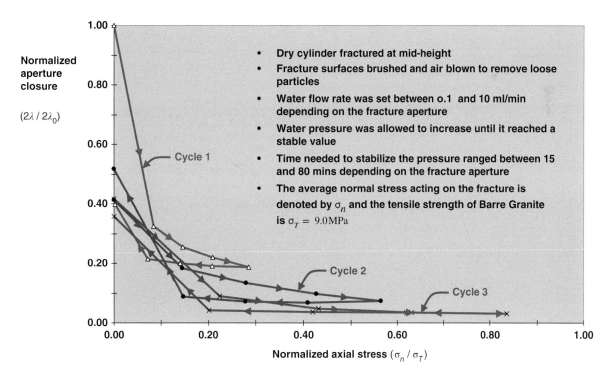

Fig. 6.7. Fracture closure during application of normal stresses. The starting position of the fracture during cycle 3 does not coincide with the end position of cycle 2. This is most likely due to unloading effects and the release of elastic stain energy in the system. The discrepancy, however, does not affect the third cycle of loading and the aperture closure trend with increasing axial stress is consistent with that observed in cycles 1 and 2. (*See color plate section for the color representation of this figure.*)

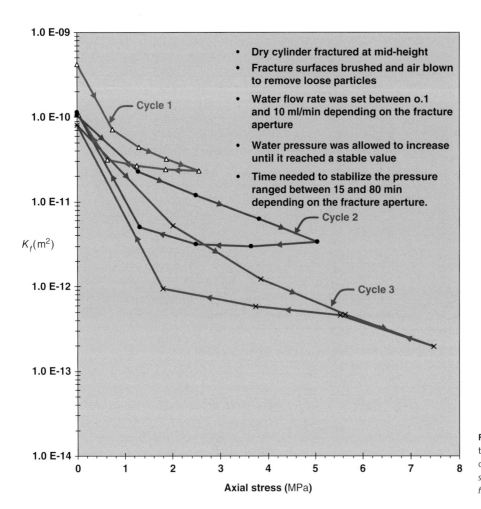

Fig. 6.8. Evolution of the permeability of the fracture during quasi-static application of loading–unloading cycles. (*See color plate section for the color representation of this figure.*)

level of axial stress, the flow rates need to be increased to approximately $10\,ml\,min^{-1}$ to ensure an accurate measurement of fluid pressure within the cylindrical cavity. The experiments were repeated for two other cycles of loading and unloading where the peak stresses were 5.0 and 7.5 MPa. At larger axial stresses, the lower flow rate of $0.1\,ml\,min^{-1}$ was sufficient to develop accurately measurable pressures within the central cavity. The first cycle of axial loading establishes a reduction in permeability that persists for subsequent load cycles despite the apparent absence of mechanical alteration of the fracture during the loading–unloading cycles. The permeability of the fracture reduces from approximately 1.0×10^{-10} to $1.2 \times 10^{-13}\,m^2$ over the axial stress range 0–7.5 MPa. In theory, axial stresses could be further increased, but the maximum stresses achievable are governed by both the capacity of the loading device and the tensile strength of the granite. With increasing axial stress on half of the sample a tensile fracture developed, similar to those that can occur in a Brazilian test.

In the experiment, the parameters that are measured are the pressure in the cylindrical cavity and the flow rate. The permeability of the fracture can be calculated by either
(1) including the flow through both the fracture and the intact matrix (Eq. 6.5) or

(2) by assuming that the flow takes place only in the fracture (Eq. 6.8). Because an estimate of permeability of the intact matrix is available from hydraulic pulse tests conducted previously (Selvadurai *et al.* 2005), it is possible to examine the combined influences of axial stresses on the fracture and permeability of the matrix on the interpretation of the fracture permeability.

Table 6.1 presents results that consider the two theoretical estimates outlined by (1) and (2) and indicates the relative error associated with the two estimates. As the normal stress acting on the fracture increases, omission of the component of flow through the intact matrix can lead to errors in the estimation of the permeability of the fracture.

The assessment of heterogeneity of fracture permeability is an important topic (Boulon 1995; Armand 2000; Giacomini *et al.* 2007), but requires a detailed sensing of fluid flow patterns both within the fracture and at the outer boundaries. After the series of axial compression cycles, the surfaces of the fracture were scanned and superposed to assess asperity degradation during pure axial stressing. The results of the laser scanning are shown in Figure 6.9 with colors indicating the level of mismatch. Red and purple colors indicate the level of strong mismatch that is expected at the boundary of the V-groove, and the points at

Table 6.1 Estimates for the permeability of a fracture in a Barre Granite cylinder during normal stress-induced fracture aperture reduction

Loading cycle	Maximum normal stress (σ_n) (MPa)	Permeability estimate (Eq. 6.5) (K_f^I) (m²)	Permeability estimate (Eq. 6.8) (K_f^{II}) (m²)	Overestimation of fracture permeability $\left(\frac{K_f^{II}-K_f^I}{K_f^I}\right)$
1	2.5	2.33×10^{-11}	2.33×10^{-11}	0
2	5.0	3.37×10^{-12}	3.43×10^{-12}	0.0178
3	7.5	1.95×10^{-13}	4.10×10^{-13}	1.103

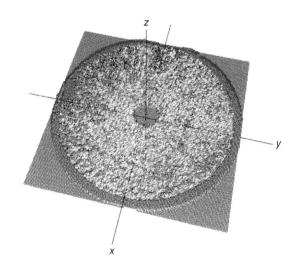

Fig. 6.9. Fracture topography mismatch after three cycles of axial compression normal to the plane of the fracture to a maximum stress of 7.5 MPa. (*See color plate section for the color representation of this figure.*)

which loads were applied to fracture the specimen. The blue color indicates a perfect match between the two surfaces, and the yellow color indicates a mismatch that is accurate to within 0.1 mm. There is no major deterioration of the mated fracture during axial compression.

CONCLUDING REMARKS

The chapter presents the results of a series of permeability tests conducted on cylinders of intact and fractured Barre Granite. The intact permeability measured via pulse tests provides a basis to accurately interpret the permeability of a cylindrical specimen containing a plane fracture subjected to normal stress. The range of axial stresses applied to the fracture is indicative of geostatic stresses that can be present in shallow earth environments (<100 m), and as such, the results are of particular interest to geomechanical and groundwater resources applications. Reduction of the *in situ* normal stress acting on a fracture to zero, for example by either excavation or stress relief, can result in a permeability increase of around three orders of magnitude. The tests were conducted on a mated set of fracture surfaces, which

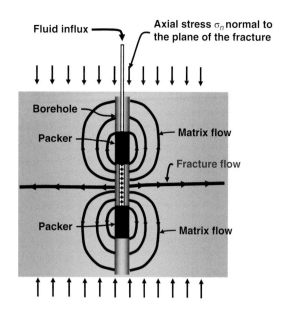

Fig. 6.10. Leakage through the matrix region during fracture permeability testing using a double-packer system.

is representative of fractures unaffected by shearing action. The relatively large scale of the experiments is relevant to interpreting the hydromechanical performance of near-surface fractures. For *in situ* permeability measurements of fractures at significant crustal depths, the conventional procedure is to employ packer systems to estimate the permeability. At significant depths, it is entirely possible that the fractures can remain tight owing to the *in situ* stress conditions, so that flow through the tight fracture can be comparable to leakage from the intact zone (Fig. 6.10). If the flow through the matrix is not accounted for, this will lead to an erroneous interpretation of the permeability of the fracture (Table 6.1). This effect will also be important for more highly permeable materials, such as sandstone and limestone where the intact permeability can approach that of the fracture at normal stress levels well within those required to induce diametral splitting of a fractured specimen. The aperture closure during the application of axial stresses exhibits hysteresis during the three loading–unloading cycles. The hysteretic behaviour is characteristic of interfaces that can consist of mated surfaces but with the possibility of displaying frictional phenomena at the scale of the asperities (Selvadurai & Boulon 1995; Selvadurai & Nguyen, 1997; Nguyen & Selvadurai 1998; Massart & Selvadurai, 2012, 2014). This is a complex interface process where asperities can deform at contact points but release the elastic energy to create frictional unloading that can lead to a persistent hysteresis until the asperities experience damage and failure.

ACKNOWLEDGEMENTS

The work described in this chapter was supported by a research contract awarded by the Canadian Nuclear Safety Commission, Ottawa, Canada and through a Discovery Grant awarded by the Natural Sciences and Engineering Research Council of Canada. The author is grateful to the reviewers for their highly constructive comments that led to significant improvements in the presentation. The work was carried out in the Environmental Geomechanics Laboratory of the Department of Civil Engineering and Applied Mechanics at McGill University. The author acknowledges the support of the researchers at CNSC, in particular Dr T. S. Nguyen, and the technical assistance of Mr Nick Vannelli, Engineer and Mr John Bartczak, Departmental Technician is greatly appreciated.

CHAPTER 7

Linking microearthquakes to fracture permeability evolution

TAKUYA ISHIBASHI[1,2], NORIAKI WATANABE[3], HIROSHI ASANUMA[1] AND NORIYOSHI TSUCHIYA[3]

[1] *Fukushima Renewable Energy Institute, National Institute of Advanced Industrial Science and Technology, Koriyama, Japan;*
[2] *Department of Energy and Mineral Engineering, EMS Energy Institute, Pennsylvania State University, University Park, PA, USA;*
[3] *Graduate School of Environmental Studies, Tohoku University, Sendai, Japan*

ABSTRACT

This study evaluates aperture distributions and fluid flow characteristics for variously sized laboratory-scale granite fractures under confining stress. The contact area in fracture plane was found to be virtually independent of scale for the investigated samples (from $0.05\,m \times 0.075\,m$ to $0.2\,m \times 0.3\,m$). By combining this characteristic with the self-affine fractal nature of fracture surfaces, a novel method for predicting fracture aperture distributions beyond laboratory scale is developed. This method enables us to predict scale dependencies of fluid flows through both joints – fractures without shear displacement – and faults – fractures with shear displacement of δ (m). Both joint and fault aperture distributions are characterized by a scale-independent contact area, a scale-dependent geometric mean, and a scale-independent geometric standard deviation of aperture. The contact areas for joints and faults are approximately 60% and 40%, respectively. Changes in the geometric means of joint and fault apertures (mm), $e_{m,\text{joint}}$ and $e_{m,\text{fault}}$, with fracture length (m), l, are approximated by $e_{m,\text{joint}} = 1.3 \times 10^{-1} l^{0.10}$ and $e_{m,\text{fault}} = 1.3 \times 10(\delta/l)^{0.59} l^{0.71}$, whereas the geometric standard deviations of both joint and fault apertures are approximately 3. Fluid flows through both joints and faults are characterized by formations of preferential flow paths (i.e., channeling flows) with scale-independent flow areas of approximately 10%, whereas the joint and fault permeabilities (m²), k_{joint} and k_{fault}, are scale dependent and are approximated as $k_{\text{joint}} = 9.8 \times 10^{-13} l^{0.16}$ and $k_{\text{fault}} = 2.3 \times 10^{-6}(\delta/l)^{1.18} l^{1.08}$. These scale dependencies suggest linkages between quantitative change in mean aperture ($e_{m,\text{fault}}/e_{m,\text{joint}}$) or fracture permeability ($k_{\text{fault}}/k_{\text{joint}}$) and moment magnitude of microearthquakes (M_w) during hydraulic stimulation for a fractured reservoir. These linkages are expressed as $e_{m,\text{fault}}/e_{m,\text{joint}} = 1.0 \times 10^{0.35 M_w}$ and $k_{\text{fault}}/k_{\text{joint}} = 116.4 \times 10^{0.46 M_w}$, and are useful in evaluating hydraulic stimulation for a fractured reservoir.

Key words: fracture flow, fracture aperture, permeability, scaling, preferential flow

INTRODUCTION

Fluid flow in the Earth's crust is of critical interest for a wide range of natural and engineering phenomena, including effective recovery of target fluids (e.g., hydrocarbons, geothermal fluids, and subsurface water), geological disposal of nuclear waste, and the earthquake cycle. Crustal permeability is the key parameter that controls crustal fluid flow and is generally believed to decrease with depth due to mechanical processes, and water–rock interactions. To capture the reality of fluid flow in the Earth's crust, it is essential to correctly understand how the crustal permeability is distributed in the Earth's crust (Manning & Ingebritsen 1999; Ingebritsen & Manning 2010).

Specifically, the permeability of crystalline rocks in the shallow crust (<5–10 km) is largely dominated by geometrical properties of rock fractures (fracture density, aperture, scale, location, and orientation) and the presence of zones of highly fractured rocks. In general, the permeability of rock fracture is much higher than the matrix permeability (e.g., see references in Ranjram *et al.* 2015; Preisig *et al.* 2015). Therefore, to predict overall fluid flow and permeability in the shallow crust, we generally use the concept of a rock fracture network. In analyzing fluid flows in rock fracture networks, discrete fracture network (DFN) model simulations have broad applicability because they can naturally incorporate the prescribed geometrical properties in fractures, and as a result can explicitly account for contributions of individual fractures to fluid flow and crustal permeability (Long *et al.* 1982a; Berkowitz 2002; Neuman 2005). In conventional DFN models, individual fractures are represented by pairs of parallel smooth plates having single aperture values (Jing *et al.* 2000), although real rock fractures have heterogeneous aperture distributions. Fractures with heterogeneous aperture distributions are sometimes classified based on the value of the joint roughness coefficient, which was

Crustal Permeability, First Edition. Edited by Tom Gleeson and Steven E. Ingebritsen.
© 2017 John Wiley & Sons, Ltd. Published 2017 by John Wiley & Sons, Ltd.
Companion Websites: www.wiley.com/go/gleeson/crustalpermeability/
http://crustalpermeability.weebly.com/

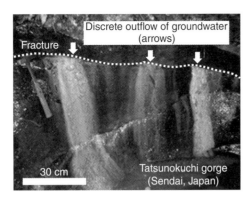

Fig. 7.1. Channeling flow seen as the discrete outflow of groundwater from a continuous fracture at the Tatsunokuchi gorge in Sendai city, Japan.

introduced by Barton *et al.* (1985). Fractures having a low joint roughness coefficient (e.g., <5) and a high joint roughness coefficient (e.g., >5) are generally accepted as smooth and rough fractures, respectively. Such heterogeneous aperture distributions can be precisely visualized using X-ray CT at atmospheric pressure or under confining stress (Watanabe *et al.* 2011a,b, 2012). Fluid flows through heterogeneous fractures are characterized by the formation of preferential flow paths (i.e., channeling flow) (Brown 1987b, 1989; Pyrak-Nolte *et al.* 1988; Watanabe *et al.* 2008; Nemoto *et al.* 2009). Channeling flows within fractures were identified through field observations as discrete outflows of groundwater from continuous fractures (Fig. 7.1) and have been examined through field investigations

at the Stripa mine in Sweden (Abelin *et al.* 1985; Tsang & Neretnieks 1998). We generally expect the occurrence of 3D preferential flow paths (i.e., 3D channeling flow) through rock fracture networks in nature.

We recently developed a novel 3D fracture network model simulator, in which fractures are characterized by 2D heterogeneous aperture distributions (Ishibashi *et al.* 2012). The ability of this simulator to analyze 3D channeling flow in a fracture network has been demonstrated in a laboratory multiple-fracture flow experiment (Fig. 7.2). In order to apply such simulators to practical field-scale problems, it is essential to understand aperture distributions and resulting fluid flow characteristics for fractures beyond the laboratory scale. Recent studies have discussed the scale problems (Selvadurai 2015; Rutqvist 2015) and highlighted their importance.

The fracture aperture distribution and resulting fluid flow characteristics, such as permeability and flow paths, are known to be constrained by fracture-surface topography, shear displacement, and confining stress (Goodman 1976; Gangi 1978; Durham & Bonner 1994; Durham 1997; Yeo *et al.* 1998; Esaki *et al.* 1999; Chen *et al.* 2000; Plouraboué *et al.* 2000; Pyrak-Nolte & Morris 2000). For instance, the hydromechanical response of a rock fracture during the application of normal stress should be considered in analyzing fluid flows through fractured rock networks (Rutqvist 2015) and has been studied experimentally (Selvadurai 2015). Moreover, fracture aperture and flow characteristics are known to vary with time as a result of mechanical and chemical influences, such as pressure solution at contacting asperities or dissolution at noncontacting

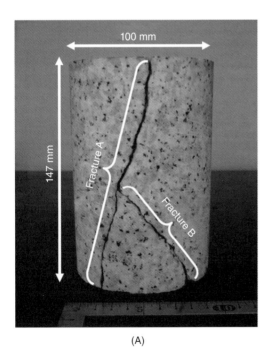

(A)

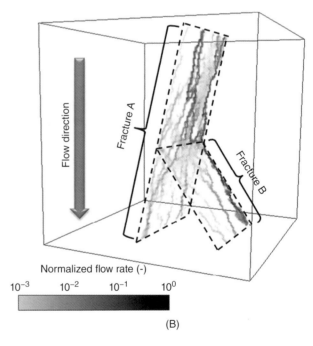

(B)

Fig. 7.2. (A) Cylindrical granite samples containing fracture network and (B) 3D channeling flow within the fracture network analyzed by GeoFlow. (see Ishibashi *et al.* [2012] for details)

asperities (Durham *et al.* 2001; Yasuhara *et al.* 2006; Ishibashi *et al.* 2013), and it is desirable to include these processes in analyses of fluid flows within rock fracture networks (Rutqvist 2015; Taron *et al.* 2009).

It has remained slightly unclear how the fracture aperture distribution and resulting fluid flow characteristics are constrained by the fracture scale. Based on the fluid flow experiments and field investigations, Witherspoon *et al.* (1979) and Raven & Gale (1985) reported that the scale of the rock fracture may influence permeability, although their results are inconsistent due mainly to the limited numbers of experiments. Matsuki *et al.* (2006) suggested that experimentally determining the scale-dependent permeability of a fracture is difficult because individual samples have unique aperture layouts. Therefore, they investigated scale dependencies in the aperture and permeability of fractures, with and without shear displacement, using numerically created fractures. Their investigations and results were very systematic, and this is the significant advantage of their study. However, the effect of confining stress on fracture flow characteristics was not strictly introduced in their evaluation because they assumed constant mean apertures. Furthermore, they never compared the absolute values of the aperture and the permeability with those of actual rock fractures. Thus, predicting the flow characteristics of rock fractures that include natural and artificial fractures remains difficult.

The objective of this study is to obtain insights into scale dependencies of aperture distribution and resulting fluid flow characteristics within a rock fracture beyond laboratory scale. The scale dependencies suggested in this study have advantages relative to those suggested in Ishibashi *et al.* (2015), because this study explores a wider variability of shear offset within the experimental fractures, thus allowing more detailed parameterizations. This treatment enables us to link microearthquake to fracture permeability evolution during hydraulic stimulation for a fractured reservoir. We first evaluate aperture distributions and resulting fluid flow characteristics for granite fractures at confining stresses. On this basis, we develop a novel method by which to predict fracture aperture distributions beyond the laboratory scale. The validity of the prediction method is confirmed by reproduction of the results obtained in the laboratory investigation and of the maximum aperture–fracture length relations for natural fractures. Subsequently, we predict extensively used scale dependencies of aperture distributions and fluid flow characteristics of subsurface fractures. Here, the classification of a fracture is defined in terms of shear displacement, and thus differs from Barton *et al.* (1985), where the joint roughness coefficient is used for fracture classification. With the scale dependencies in mind, we suggest a linkage between permeability enhancement of a rock fractures and the moment magnitude of microearthquakes.

CHANNELING FLOWS THROUGH HETEROGENEOUS FRACTURES AT LABORATORY SCALE

Surface measurement and fluid flow experiments for various scales of rock fractures

We determined the fracture aperture distribution under confining stress and the fluid flow through the aperture distributions at various scales according to the method developed by Watanabe *et al.* (2008, 2009). In their method, the fracture-surface topography data and fracture permeability data were used. Therefore, we measured the fracture-surface topography and permeability for single fractures at scales of $0.05\,\text{m} \times 0.075\,\text{m}$, $0.1\,\text{m} \times 0.15\,\text{m}$, and $0.2\,\text{m} \times 0.3\,\text{m}$ in this study. These fractures are contained in cylindrical samples of Inada medium-grained granite (Ibaraki Prefecture, Japan), which is characterized by uniaxial compressive strength between 160 and 180 MPa, uniaxial tensile strength between 4 and 8 MPa, and negligibly small matrix permeability between 10^{-19} and $10^{-18}\,\text{m}^2$ (Takahashi *et al.* 1990; Lin *et al.* 2008). As shown in Figure 7.3A, the short and long side lengths of the rectangular fractures correspond to the diameters and lengths of these cylindrical samples. For convenience, the radial and axial directions are defined along the short and long side lengths of the rectangular fractures.

We prepared these cylindrical fractured samples as follows. First, a tensile fracture of a larger section than the specified scale is induced in a cubic granite block using a wedge. Fractured samples are placed so that the fracture has either no shear displacement or a shear displacement of 5 mm in the radial direction, and the fracture is stabilized by surrounding and fixing the fractured block with mortar (Fig. 7.3B). Subsequently, we core the cylindrical fractured sample, and adjust the height of the sample by cutting both end planes for the axial direction. Since fractures with and without shear displacement are prepared for each fracture scale, the total number of rock samples is six.

Before the permeability measurements, we measured the two-dimensional distributions of surface height for each fracture surface in a 0.25-mm square grid system with a laser profilometer having a height resolution of 1 μm and a positioning accuracy of 20 μm. We use these data for surface topographies as input data for numerical determination of fracture aperture distributions and resulting flow-rate distributions. Then, we measured the fracture permeability, respectively, at confining stress of 10 and 30 MPa through a unidirectional (i.e., axial direction which is parallel to the direction of anisotropy of the aperture distribution) fluid flow experiment at room temperature (Fig. 7.3C). In a series of experiments, distilled water flows from a plunger pump to the bottom of the sample, and then flows from the bottom to the top of the sample via a stainless pipe. Confining stress is monitored by an analog pressure gage with an uncertainty of 0.5 MPa. Differential pressure between the inlet and the outlet is monitored by a digital pressure gage with an uncertainty of 0.001 MPa or 0.01 kPa depending on whether the fracture has shear displacement. We

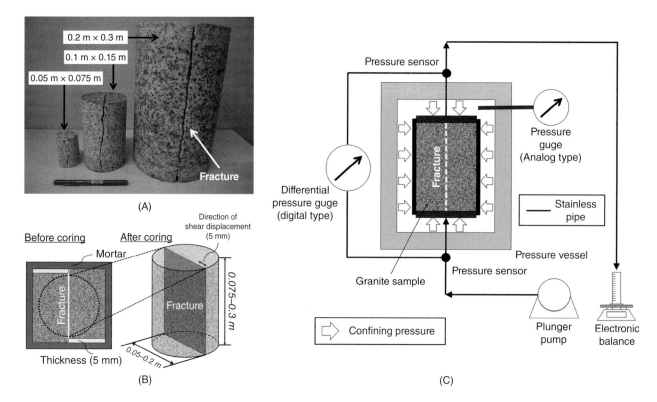

Fig. 7.3. (A) Cylindrical granite samples containing single tensile fractures of different sizes having (A) no shear displacement or (B) a shear displacement of 5 mm in the radial direction, and (C) experimental system for permeability measurement under confining stress.

consistently maintained very small pore pressure, less than 5% of the confining stress. Flow rate is monitored by measuring the weight of the effluent with an electronic balance with an uncertainty of 0.001 g.

Linear relationships between the flow rate and the differential pressure are observed for all given conditions (i.e., Darcy's law), and given the negligibly small matrix permeability of the rock samples, fracture permeability is determined based on the cubic law (Witherspoon *et al.* 1980; Tsang & Witherspoon 1981):

$$k = \frac{e_h^2}{12},\tag{7.1}$$

where k is the fracture permeability and e_h is the hydraulic aperture, which is described as follows:

$$e_h = \left(-\frac{12\mu L Q}{W \Delta P} \right)^{1/3},\tag{7.2}$$

where Q is the flow rate, ΔP is the differential pressure, μ is the viscosity of the fluid, and L and W are the long and short side lengths, respectively, of the fracture. It should be noted that the hydraulic characteristics of fracture are generally characterized by anisotropy between directions orthogonal and parallel to the fluid flow (Nemoto *et al.* 2009). Fracture permeabilities determined via our methodology correspond to the direction of maximum fracture permeability, since the imposed direction of shear displacement is orthogonal to the direction of fluid flow

and, as a result, this is the most favorable case for permeability gain.

Numerical modeling of fracture aperture distribution and resulting fluid flow

Based on the measurements of fracture-surface topography and permeability, we numerically determined 2D aperture distributions of the fractures under the confining stresses through a permeability matching method, where the pairs of fracture surfaces are in contact with each other so that the aperture distributions have the experimentally determined fracture permeabilities. In this permeability matching, we first create an aperture distribution with at least a single contact point in the 0.25-mm square grid system. Subsequently, the apertures are uniformly decreased to simulate fracture closure due to confining stress. The apertures are local separations between two opposite fracture surfaces in the direction normal to their mean planes. When we decrease the apertures, the two fracture surfaces come close together and some asperities overlap. These overlapping asperities are treated as contacting asperities (i.e., regions of zero aperture), since we do not take into account fracture surface deformations (Zimmerman *et al.* 1992; Matsuki *et al.* 2006).

In order to evaluate the fracture permeability of the aperture distribution, we simulate a unidirectional fluid flow by solving the Reynolds equation for a steady-state laminar flow of a viscous and incompressible fluid (Brown 1987b; Ge 1997; Yeo *et al.* 1998; Brush & Thomson 2003):

$$\frac{\partial}{\partial x}\left(e^3\frac{\partial P}{\partial x}\right) + \frac{\partial}{\partial y}\left(e^3\frac{\partial P}{\partial y}\right) = 0, \qquad (7.3)$$

where e is the aperture and P is the fluid pressure. The Reynolds equation generally overestimates flow rates when the ratio of contacting asperities is significantly large (Brown *et al.* 1995). Previous studies reported that the Reynolds equation could overestimate the flow rate by a factor of 1.5–2 (Hakami & Larsson 1996; Yeo *et al.* 1998); this implies that the permeabilities calculated in this study are also overestimated by a factor of 1.3–1.6. However, because such overestimation of fracture permeability is common regardless of fracture scale, the use of the Reynolds equation has little influence on the trend for permeability scaling described later, and we consider the use of the Reynolds equation acceptable for evaluating fluid flow characteristics. We solve the Reynolds equation with a finite-difference method, where boundary conditions corresponding to the fluid flow experiment are prescribed. In solving the Reynolds equation, contacting asperities are replaced with a negligibly small nonzero aperture of 1 μm for computational convenience. Since fluid flow simulation mimics the fluid flow experiment, permeabilities for the aperture distributions are also determined using Eqs 7.1 and 7.2.

Because aperture distributions obtained in this study are characterized by a significant number of contacting asperities (regions of zero aperture) and skewed distributions of nonzero apertures with long tails (i.e., lognormal-like distributions), the aperture distributions are evaluated based on the percentages of contacting asperities for all data points (i.e., contact area), the geometric mean, and the geometric standard deviation of the apertures (nonzero values). In evaluating the resulting fluid flows, the area of preferential flow paths in the fracture plane, referred to herein as the flow area, is evaluated because the flow of fluid is channelized along these preferential flow paths.

Channeling flows through heterogeneous aperture distributions of laboratory fractures

Figure 7.4A shows changes in permeability (or corresponding hydraulic aperture) with fracture scale (length) for the tested fractures, with and without shear displacement. Note that we represent the scales for the rectangular laboratory-scale fractures by the shorter side lengths. These permeabilities and corresponding hydraulic apertures are summarized in Table 7.1. For the case of a fracture without shear displacement, the permeability is lower at a higher confining stresses, increases with the fracture length from 0.05 to 0.1 m, and decreases with the fracture length from 0.1 to 0.2 m. This result suggests that the permeability of the fracture without shear displacement has no clear scale dependency. In contrast, for the case of a fracture with shear displacement, permeability increases with the fracture length, and the influence of confining stress on permeability is minor.

Subsequently, we determine the heterogeneous aperture distribution and the corresponding fluid flow of the laboratory-scale fractures at each given condition, and show representative results in Figure 7.5. In this figure, the flow-rate distributions, shown in color scale, are superimposed on the corresponding aperture distributions, shown in gray scale. In each panel of Figure 7.5, contacting asperities are represented by minimum apertures of 1 μm for convenience, and the flow rates are normalized by dividing by the maximum value of the flow rate, so that the maximum flow rate is equal to one. Considering that colorless points have negligibly small flow rates, we infer that fluid flow through a rock fracture is always characterized by preferential flow paths (i.e., channeling flow).

For each of the determined aperture distributions, the geometric mean and standard deviation of nonzero apertures and the percentage of zero apertures (contact area) are evaluated. Moreover, for each of the determined flow-rate distributions, flow area is evaluated. These values are summarized in Table 7.1. Figure 7.4B,C summarizes changes in the geometric mean of apertures (referred to hereinafter as the mean aperture) and in the contact area with fracture length. Fractures with shear displacement have a large mean aperture due to shear dilation. In contrast, fractures without shear displacement have a large contact area due to the high degree of matedness, defined as how well-matched opposite fracture surfaces are (Olsson & Barton 2001). For fractures with and without shear displacement, mean aperture does not change significantly with confining stress from 10 to 30 MPa. The mean aperture of the fracture without shear displacement increases with increasing fracture length, whereas the mean aperture of the fracture with shear displacement increases with the fracture length from 0.05 to 0.1 m and decreases with the fracture length from 0.1 to 0.2 m. Because the mean aperture of the largest fracture with shear displacement remains greater than that of the smallest fracture with shear displacement, we can argue that the mean aperture increases with increasing fracture length. The contact area of the fracture without shear displacement, slightly larger at higher confining stress, decreases slightly with increasing fracture length. In contrast, the contact area of the fracture with shear displacement, which does not change significantly with confining stress, increases slightly with increasing fracture length. The geometric standard deviations are essentially constant at approximately three for all given conditions, and flow areas range from 4% to 22% (see Table 7.1). Considering that the flow area is ≪100% of the fracture surface, we find that our analyzed fractures, which are considered to be relatively rough, are far from parallel smooth plates.

Based on these results, we now briefly discuss the scale dependency of the heterogeneous fracture aperture distribution and the channeling flow. Mean aperture is likely to increase with fracture length for fractures with and without shear displacement. This dependency may be caused by the increase in fracture surface roughness with increasing fracture length. On the other hand, contact area is likely to be scale

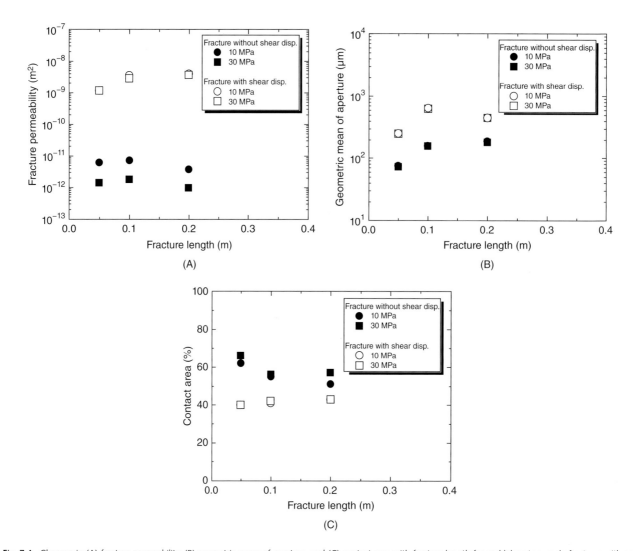

Fig. 7.4. Changes in (A) fracture permeability, (B) geometric mean of aperture, and (C) contact area with fracture length for real laboratory-scale fractures with no shear displacement and with a shear displacement of 5 mm at confining stresses of 10 and 30 MPa.

independent for fractures with and without shear displacement. Although the scale dependency of contact area is opposite in sense between the fractures with and without shear displacement, the changes in the absolute contact area are negligible in both instances. We interpret the scale-independent contact area as the physical balance between confining stress and reactive force at the contacting asperities. Moreover, we suggest that the scale-independent contact area is related to the scale-independent vertical stiffness of the fracture; this idea may be supported by the experimental results showing similar trends of fracture permeability-versus-confining stress regardless of fracture length. Finally, the scale dependencies in the fracture permeability are constrained by the combination of scale-dependent mean aperture and scale-independent contact area. For a fracture without shear displacement, permeability is nearly scale independent; the increase in mean aperture has little influence on permeability, due to the large contact area.

In contrast, for a fracture with shear displacement, permeability may have scale dependency, because the permeability can be influenced by the change in mean aperture given a relatively small contact area.

CHANNELING FLOWS THROUGH HETEROGENEOUS FRACTURES BEYOND LABORATORY SCALE

Prediction of fracture aperture distributions and fluid flows beyond laboratory scale

The laboratory investigation provides insight into the scale dependencies of the aperture distributions and channeling flows for subsurface fractures (i.e., fractures under confining stress). However, the results of laboratory investigation are limited to a relatively small range of fracture scale, and to a limited number

Table 7.1 Parameter values characterizing channeling flow through the heterogeneous aperture distribution of the real fractures

Shear displace-ment (mm)	Confining stress (MPa)	Fracture length		Fracture permeability (m²)	Hydraulic aperture (μm)	Flow area (%)	Aperture		Contact area (%)
		Short side (m)	Long side (m)				Geometric mean (μm)	Geometric std. (–)	
0	10	0.05	0.075	6.0×10^{-12}	8.5	8	75	3.7	62
0	10	0.1	0.15	7.0×10^{-12}	9.1	4	158	3.2	55
0	10	0.2	0.3	3.6×10^{-12}	6.6	10	188	3.2	51
0	30	0.05	0.075	1.4×10^{-12}	4.1	9	72	3.8	66
0	30	0.1	0.15	1.7×10^{-12}	4.6	5	157	3.2	56
0	30	0.2	0.3	9.6×10^{-13}	3.4	10	177	3.1	57
5	10	0.05	0.075	1.1×10^{-9}	116	22	246	2.8	40
5	10	0.1	0.15	3.4×10^{-9}	203	14	630	2.8	41
5	10	0.2	0.3	3.8×10^{-9}	213	8	443	2.9	43
5	30	0.05	0.075	1.1×10^{-9}	116	22	246	2.8	40
5	30	0.1	0.15	2.8×10^{-9}	184	14	619	2.9	42
5	30	0.2	0.3	3.6×10^{-9}	207	8	441	2.9	43

Reproduced from Ishibashi *et al.* (2015) by permission of John Wiley and Sons Ltd.

Fig. 7.5. Representative results for channeling flow within the heterogeneous aperture distribution of real laboratory-scale fractures (A) with no shear displacement and (B) with a shear displacement of 5 mm. (*See color plate section for the color representation of this figure.*)

of fracture types and experiments. To overcome this problem, we need a novel method that enables us to predict aperture distributions of subsurface fractures beyond laboratory scale.

As explained earlier, aperture distribution of a rock fracture is determined by using fracture-surface topography data. From such surface topography data, we can numerically create fractures beyond the laboratory scale. This is because the fracture-surface topography is generally characterized by a self-affine fractal nature (Brown 1987a; Power *et al.* 1987;

Kumar & Bodvarsson 1990; Power & Durham 1997), such that the surface topography is scaled by different amounts in the planar and height directions, and this nature is confirmed to be valid for the natural fault roughness from the microscale to the continental scale (Candela *et al.* 2012). In contrast, we are not sure how closely the two opposite fracture surfaces are situated under confining stress, and knowledge of this is also essential to predict aperture distributions of subsurface rock fractures beyond the laboratory scale.

Fracture permeability, mean aperture, or contact area may be used to predict the aperture distribution under confining stress, as long as their scale dependencies are clear and formulated. The scale dependencies of fracture permeability and mean aperture are not clear in the laboratory investigation, which precludes the formulation of their scale dependencies. On the other hand, the scale independency of the contact area has been revealed, and on this basis a novel method to predict the aperture distributions of fractures from laboratory to field scale has been developed, based on the fractal nature of the fracture surface and the scale-independent contact area. The aperture distribution of a fracture of any size can be predicted by simply placing the two fractal fracture surfaces in contact such that the fracture has the scale-independent contact area.

In this study, we create a pair of fractal fracture surfaces (i.e., hanging wall surface and foot wall surface) by using the method developed by Matsuki *et al.* (2006). Their method has the advantage of being able to take into account the desired degree of matedness (see Matsuki *et al.* 2006 or Ishibashi *et al.* 2015 for detail). For creating the hanging wall synthetic fracture surface, the fractal dimension of the fracture surface, D, and the standard deviation of the fracture surface height, σ_{h0}, along a linear profile of reference length, L_0, are required. In contrast, the mismatch length scale, λ_c, and the ratio of the power spectral density of the initial aperture to the power spectral density of the surface height as a function of spatial frequency, $R(f)$, are required to create the foot wall synthetic fracture surface. In this study, representative parameter values are computed for the fracture-surface topographies of the $0.2\,\text{m} \times 0.3\,\text{m}$ sample (Fig. 7.3A), providing a fractal dimension of 2.3, a standard deviation of 1.3 mm for a reference length of 0.2 m, and a mismatch length scale of 0.7 mm. Fractal dimension for this study is in good agreement with previous studies (Power & Durham 1997; Matsuki *et al.* 2006). In addition, the ratio of the power spectral density of the initial aperture to the power spectral density of the surface height, which is also measured for the actual rock fracture, is approximated by

$$R(f) = e^{\{-6.5\times10^{-3}\cdot(\ln f)^3 - 2.9\times10^{-1}\cdot(\ln f)^2 + 2.2\times10^{-1}\cdot(\ln f) + 5.5\times10^{-1}\}}. \quad (7.4)$$

With these parameter values, we numerically create paired surface topographies of square fractures on a 0.25-mm square grid system with a scale of $0.05\,\text{m} \times 0.05\,\text{m}$ to $0.6\,\text{m} \times 0.6\,\text{m}$. Aperture distributions having a scale-independent contact area are subsequently determined and evaluated as described earlier herein. In this study, the contact areas of synthetic fractures with and without shear displacement are set to 42% and 59%, respectively. These values can be obtained by calculating the average of contact area with respect to the fracture type (Table 7.1). Then, fluid flows through the predicted aperture distributions are determined and evaluated, with fluid flows limited to the orthogonal direction relative to shear displacement. As noted previously, this represents the most favorable case for shear dilation impact and fracture permeability enhancement.

Because the contact area and other characteristic parameters do not change significantly with confining stress, we can consider the predicted aperture distributions and fluid flows as representative of synthetic fractures with and without shear displacement under confining stress. Note that this study essentially provides results for factures beyond the laboratory scale, since even the present maximum fracture size (i.e., 0.6 m) is too large to allow experimental determination of its aperture distribution and corresponding fluid flow under confining stress.

Validity of the method for predicting fracture aperture distributions beyond laboratory scale

To evaluate the validity of the prediction method for fractures created in the laboratory, we first confirm the occurrence of channeling flows within the predicted synthetic aperture distributions. Figure 7.6 shows representative results for the aperture distribution and fluid flow, displayed in the same manner as in Figure 7.5. By comparing Figures 7.5 and 7.6, it is qualitatively evident that our suggested prediction method can reproduce channeling flow within the heterogeneous fracture aperture distribution. Channeling flow for the predicted synthetic fractures is observed regardless of fracture scale, implying that fluid flow through rock fractures is generally characterized by channeling flow. Next, we quantitatively evaluate the validity of the prediction method through comparison of two key parameters (mean aperture and fracture permeability) between the predicted synthetic fractures and the real laboratory-scale fractures. As shown in Figure 7.7, the absolute values of mean aperture and permeability for the predicted synthetic fractures are generally in agreement with those of the real laboratory-scale fractures. Although relatively large differences are observed between the predicted and real permeabilities for the fracture without shear displacement, the values are consistent within an order of magnitude. The parameter values that characterize the predicted synthetic aperture distribution and fluid flow are listed in Table 7.2, and are generally consistent with those for the real laboratory-scale fractures (see Table 7.1). Thus, we consider that the prediction method is valid for fractures created in the laboratory.

It is also desirable to assess the validity of the prediction method for natural fractures. To achieve this, we evaluate the maximum aperture–fracture length relations for the predicted synthetic fractures, and compare them with the relations for natural fractures (joints and faults) reported in the literature (Cowie & Scholz 1992; Vermilye & Scholz 1995; Schlische *et al.* 1996; Schultz *et al.* 2008). According to Vermilye & Scholz (1995), the maximum aperture (m), $e_{\max}$, and fracture length (m), l, are approximately related by the following formula:

$$e_{\max} = \alpha \cdot l^n, \quad (7.5)$$

where α is a pre-exponential factor related to rock properties or tectonic environments and n is an exponent. The representative fit curve is given as $e_{\max} = 2.5 \times 10^{-3} l^{0.48}$ for a joint (Schultz

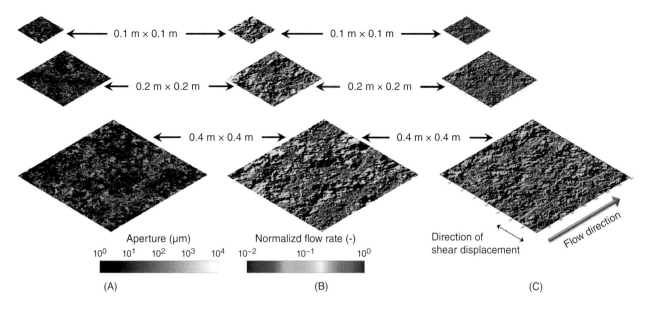

Fig. 7.6. Representative results for channeling flow within the heterogeneous aperture distribution of synthetic fractures (A) with no shear displacement, (B) with a shear displacement of 5 mm, and (C) with a shear displacement of constant δ/l ($\delta/l=0.01$). (*See color plate section for the color representation of this figure.*)

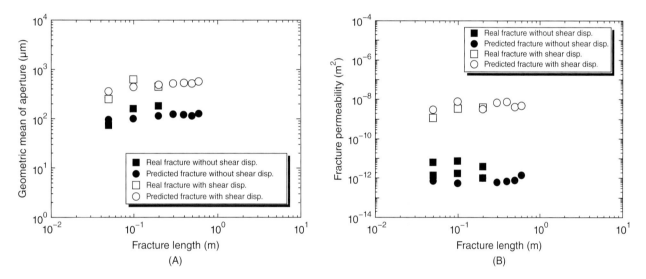

Fig. 7.7. Comparisons of (A) the geometric mean of aperture and (B) permeability between synthetic and real laboratory-scale fractures with no shear displacement and with a shear displacement of 5 mm. Values for the real fractures are determined at confining stresses of 10 and 30 MPa.

et al. 2008), and ranges from $e_{max}=0.001 \cdot l$ to $e_{max}=0.1 \cdot l$ for a fault. We compare the maximum aperture–fracture length relation for the synthetic fracture without shear displacement to that for joints and the relation for the synthetic fracture with shear displacement to that for faults. For the latter comparison, to imitate the natural faults, we assume a linear increase in shear displacement (m), δ, with fracture length (i.e., constant value of δ/l) for the predicted synthetic fractures (Kanamori & Anderson 1975). Since δ/l inherently varies depending on rock properties and tectonic environments, and most faults are expected to experience multiple slips, we set δ/l of the predicted fractures

to relatively large values from 0.0025 to 0.02. Note that fluid flows through the synthetic fractures are concurrently determined and evaluated in all cases. Representative results for the synthetic fractures having constant δ/l ($\delta/l=0.01$) are shown in Figure 7.6C, where the occurrence of channeling flow is clear.

Figure 7.8 compares the maximum aperture–fracture length relations between the numerically predicted synthetic fractures and natural fractures (i.e., joints and faults). The maximum aperture values for numerically predicted synthetic fractures are listed in Table 7.3. As shown in Figure 7.8, the maximum aperture–fracture length relation for the synthetic fracture

Table 7.2 Parameter values characterizing channeling flow through the heterogeneous aperture distribution of the predicted fractures under confining stress

Shear displace-ment (mm)	Fracture length		Fracture permeability (m^2)	Hydraulic aperture (µm)	Flow area (%)	Aperture		Contact area (%)	Maximum aperture (m)
	Short side (m)	Long side (m)				Geometric mean (µm)	Geometric std. (−)		
0	0.05	0.05	7.4×10^{-13}	3.0	14	95	3.0	59	7.6×10^{-4}
0	0.1	0.1	5.4×10^{-13}	2.5	12	100	3.0	59	8.9×10^{-4}
0	0.2	0.2	9.8×10^{-13}	3.4	13	112	3.0	59	1.0×10^{-3}
0	0.3	0.3	6.2×10^{-13}	2.7	10	121	3.0	59	1.1×10^{-3}
0	0.4	0.4	6.9×10^{-13}	2.9	8	118	3.0	59	1.1×10^{-3}
0	0.5	0.5	7.6×10^{-13}	3.0	11	113	3.0	59	1.2×10^{-3}
0	0.6	0.6	1.4×10^{-12}	4.1	10	123	3.0	59	1.3×10^{-3}
5	0.05	0.05	2.9×10^{-9}	186	14	350	2.9	42	2.2×10^{-3}
5	0.1	0.1	7.6×10^{-9}	301	13	423	2.9	42	3.3×10^{-3}
5	0.2	0.2	3.2×10^{-9}	196	13	471	2.9	42	3.3×10^{-3}
5	0.3	0.3	6.6×10^{-9}	281	12	503	2.9	42	3.5×10^{-3}
5	0.4	0.4	7.3×10^{-9}	296	13	521	2.9	42	3.9×10^{-3}
5	0.5	0.5	4.2×10^{-9}	224	12	509	2.9	42	4.4×10^{-3}
5	0.6	0.6	4.5×10^{-9}	232	12	555	2.9	42	4.3×10^{-3}

Reproduced from Ishibashi *et al.* (2015) by permission of John Wiley and Sons Ltd.

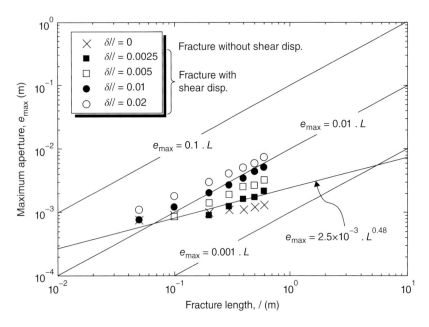

Fig. 7.8. Comparison of maximum aperture–fracture length relations between synthetic and natural fractures. The results are for synthetic fractures with constant ratios of shear displacement to fracture length ($\delta/l = 0.0025$, 0.005, 0.01, and 0.02). One of the linear curves, $e_{\max} = 2.5 \times 10^{-3} l^{0.48}$, corresponds to the relation for a joint, whereas the other linear curves correspond to the relation for a fault (Schlische *et al.* 1996; Schultz *et al.* 2008).

without shear displacement quantitatively coincided with that for a joint ($e_{\max} = 2.5 \times 10^{-3} l^{0.48}$). On the other hand, the maximum aperture–fracture length relations for the synthetic fractures with shear displacement ($\delta/l = 0.0025$, 0.005, 0.01, and 0.02) distribute approximately parallel to each other, and the respective relations are arranged adjacent to the middle relation for a fault ($e_{\max} = 0.01l$). These results support the validity of the prediction method for natural fractures.

Scale dependency of a channeling flow through a heterogeneous fracture aperture distribution

Now that the validity of the prediction method for natural fractures is demonstrated, we can now present the scale dependencies of channeling flows through the heterogeneous aperture distributions of natural fractures. The parameter values used for

the following discussion are listed in Table 7.3. According to the experimental results of Chen *et al.* (2000), there is little change in fracture permeability for shear displacement in the range of 0–0.4 mm, implying little shear dilation when shear slip is less than ~0.4 mm. Indeed, Tanikawa *et al.* (2014), studying granite fractures without large asperities, reported that fracture permeability may decrease rather than increase for small shear slip. These results may indicate a transition in fracture characteristics (i.e., from joint to fault) over the aforementioned shear displacement range, since it is unlikely to expect that the contact area abruptly changes from 59% to 42% at a specific value of shear displacement. Therefore, in this study, predicted aperture distributions of rock fractures are limited to those with shear displacement of more than ~0.4 mm. It is also difficult to apply the prediction method directly for fractures that were

Table 7.3 Parameter values characterizing channeling flow through the heterogeneous aperture distribution of predicted fractures with constant ratios of shear displacement to fracture length (i.e., $\delta/l = 0.0025, 0.005, 0.01$, and 0.02)

Shear displacement (m)	Fracture length		δ/l (−)	Fracture permeability (m²)	Hydraulic aperture (μm)	Flow area (%)	Aperture		Contact area (%)	Maximum aperture (m)
	Short side (m)	Long side (m)					Geometric mean (μm)	Geometric std. (−)		
5.0×10^{-4}	0.05	0.05	0.01	4.1×10^{-10}	70	19	109	2.8	42	7.5×10^{-4}
1.0×10^{-3}	0.05	0.05	0.02	1.0×10^{-9}	110	16	166	2.8	42	1.1×10^{-3}
5.0×10^{-4}	0.1	0.1	0.005	3.2×10^{-10}	62	15	108	2.8	42	8.4×10^{-4}
1.0×10^{-3}	0.1	0.1	0.01	7.7×10^{-10}	96	11	167	2.8	42	1.2×10^{-3}
2.5×10^{-4}	0.1	0.1	0.02	1.9×10^{-9}	151	7	260	2.8	42	1.8×10^{-3}
5.0×10^{-4}	0.2	0.2	0.0025	2.9×10^{-10}	59	18	115	2.8	42	8.9×10^{-4}
1.0×10^{-3}	0.2	0.2	0.005	6.7×10^{-10}	90	15	178	2.8	42	1.4×10^{-3}
2.0×10^{-3}	0.2	0.2	0.01	1.9×10^{-9}	152	13	277	2.8	42	2.0×10^{-3}
4.0×10^{-3}	0.2	0.2	0.02	3.5×10^{-9}	205	14	417	2.8	42	3.0×10^{-3}
7.5×10^{-4}	0.3	0.3	0.0025	5.0×10^{-10}	77	15	154	2.8	42	1.2×10^{-3}
1.5×10^{-3}	0.3	0.3	0.005	1.2×10^{-9}	118	14	243	2.8	42	1.9×10^{-3}
3.0×10^{-3}	0.3	0.3	0.01	3.2×10^{-9}	197	13	372	2.9	42	2.7×10^{-3}
6.0×10^{-3}	0.3	0.3	0.02	8.0×10^{-9}	311	12	555	2.9	42	4.0×10^{-3}
1.0×10^{-3}	0.4	0.4	0.0025	9.2×10^{-10}	105	13	189	2.8	42	1.6×10^{-3}
2.0×10^{-3}	0.4	0.4	0.005	2.4×10^{-9}	171	12	296	2.8	42	2.5×10^{-3}
4.0×10^{-3}	0.4	0.4	0.01	5.9×10^{-9}	266	13	456	2.9	42	3.4×10^{-3}
8.0×10^{-3}	0.4	0.4	0.02	1.5×10^{-8}	425	16	685	2.9	42	5.0×10^{-3}
1.25×10^{-3}	0.5	0.5	0.0025	7.2×10^{-10}	93	13	211	2.8	42	1.7×10^{-3}
2.5×10^{-3}	0.5	0.5	0.005	1.9×10^{-9}	149	12	330	2.9	42	2.6×10^{-3}
5.0×10^{-3}	0.5	0.5	0.01	4.0×10^{-9}	219	13	509	2.9	42	4.4×10^{-3}
1.0×10^{-2}	0.5	0.5	0.02	1.2×10^{-8}	380	13	762	2.9	42	5.8×10^{-3}
1.5×10^{-3}	0.6	0.6	0.0025	1.1×10^{-9}	114	13	258	2.8	42	2.1×10^{-3}
3.0×10^{-3}	0.6	0.6	0.005	2.6×10^{-9}	176	12	403	2.9	42	3.2×10^{-3}
6.0×10^{-3}	0.6	0.6	0.01	4.7×10^{-9}	236	12	617	2.9	42	5.0×10^{-3}
1.2×10^{-2}	0.6	0.6	0.02	7.3×10^{-9}	296	11	924	2.9	42	7.3×10^{-3}

sheared under a high confining pressure or at high temperature, since the prediction method does not consider gouge formation, surface melting, and so on (Tanikawa *et al.* 2014). As has been described previously, the prediction results for fractures without shear displacement ($\delta/l = 0$) are relevant to the typical scale dependencies of joints, whereas the prediction results for the fractures with shear displacement are relevant to the typical scale dependencies of faults. By treating δ/l as a variable number, we can consider the diverse characteristics of natural faults, which owe to differences in rock properties and tectonic environments.

Figure 7.9A shows changes in the mean aperture with fracture length for fractures with and without shear displacement (faults and joints, respectively). The mean aperture increases with fracture length for both joints and faults. For faults, the mean aperture increases with δ/l when the fracture length is constant. This increase in mean aperture is caused by the enhancement of shear dilation. The standard deviation is almost independent of fracture scale or δ/l, and is approximately three. The increase in mean aperture for joints is approximated by

$$\varepsilon_{m,\text{joint}} = 1.3 \times 10^{-1} l^{0.10}, \tag{7.6}$$

where $\varepsilon_{m,\text{joint}}$ and l are the mean aperture of joints (mm) and the fracture length (m). In contrast, for the faults with constant δ/l, the increase in mean aperture is approximated by

$$\varepsilon_{m,\text{fault}} = \alpha_1 l^{\beta_1}, \tag{7.7}$$

where $\varepsilon_{m,\text{fault}}$ is the mean aperture of faults (mm) and α_1 and β_1 are pre-exponential factors and exponents. Curves approximated by Eq. 7.7 are represented in Figure 7.9A, and α_1 and β_1 are calculated for fractures having different values of δ/l and summarized in Table 7.4. The values in Table 7.4 show that α_1 can be formulated as $\alpha_1 = 1.3 \times 10(\delta/l)^{0.59}$, whereas β_1 is virtually independent of δ/l (i.e., $\beta_1 = 0.71$). By substituting these values into Eq. 7.7, $\varepsilon_{m,\text{fault}}$ can be rearranged to yield

$$\varepsilon_{m,\text{fault}} = 1.3 \times 10\left(\frac{\delta}{l}\right)^{0.59} \cdot l^{0.71}. \tag{7.8}$$

Equations 7.6 and 7.8 suggest the scale dependency of $\varepsilon_{m,\text{joint}}$ is weak, whereas the scale dependency of $\varepsilon_{m,\text{fault}}$ is strong. This discrepancy is mainly caused by the difference in degree of matedness for the two opposite fracture surfaces caused by shear slip.

Figure 7.9B shows the change in permeability with fracture length for joints and faults. Here, we emphasize again that the evaluated fracture permeabilities represent the most favorable case of shear slip direction with respect to the anisotropy of aperture distribution. The permeabilities of both joints and faults increase with the increase in the fracture length. Furthermore, the permeability of faults increases with δ/l when the fracture length is constant. The increase in permeability for joints is approximated by

$$k_{\text{joint}} = 9.8 \times 10^{-13} l^{0.16}, \tag{7.9}$$

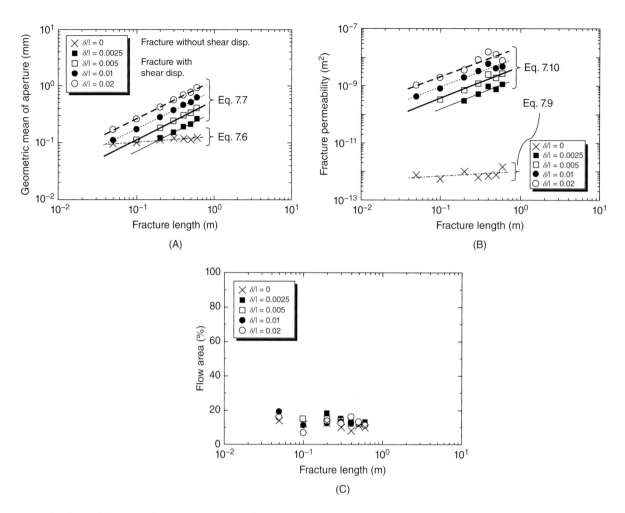

Fig. 7.9. Predicted typical changes in the (A) geometric mean of aperture, (B) permeability, and (C) flow area with fracture length, for fractures with no shear displacement (joints) and fractures with constant ratios of shear displacement to fracture length (faults). Note that the curves approximated by Eqs 7.6, 7.7, 7.9, and 7.10 are represented.

Table 7.4 Parameter values of α_1, α_2, β_1, and β_2 for the formulation of mean aperture and permeability of rock fracture

Shear displace-ment (δ/l)	α_1	β_1	α_2	β_2
0.0025	0.37	0.74	1.93×10^{-9}	1.09
0.005	0.57	0.71	4.73×10^{-9}	1.12
0.01	0.85	0.70	1.03×10^{-8}	1.07
0.02	1.27	0.68	2.23×10^{-8}	1.03

where k_{joint} is the permeability of joints (m²). For faults with constant δ/l, the increase in permeability is approximated by

$$k_{fault} = \alpha_2 l^{\beta_2}, \tag{7.10}$$

where k_{fault} is the fault permeability (m²) and α_2 and β_2 are pre-exponential factors and exponents. Curves approximated by Eq. 7.10 are represented in Figure 7.9B, and α_2 and β_2 are calculated for fractures having different values of δ/l and summarized in Table 7.4. Considering that α_2 can be formulated as $\alpha_2 = 2.3 \times 10^{-6}(\delta/l)^{1.18}$ and that β_2 is virtually

independent of δ/l (i.e., $\beta_2 = 1.08$), Eq. 7.10 for k_{fault} can be rearranged to yield

$$k_{fault} = 2.3 \times 10^{-6}\left(\frac{\delta}{l}\right)^{1.18} \cdot l^{1.08}. \tag{7.11}$$

For both joints and faults, permeabilities predicted by Eqs 7.9 and 7.11 are consistently ~1000 and ~10 times lower than those derived from Eqs 7.6 and 7.8 under the assumption of the parallel smooth plate model (i.e., cubic law). This discrepancy owes to the fact that the flow area within a fracture is channelized. Nevertheless, we can still conclude that scale dependencies of the permeability are stronger than those of the mean apertures. Moreover, as shown in Figure 7.9C, fluid flow through rock fractures is characterized by the limited flow area (~10% of the fracture plane) regardless of fracture scale or δ/l. This result is in good agreement with the flow area evaluated in field investigations (Abelin *et al.* 1985). Considering that the contact area is scale-independent (~60% for joints and ~40% for faults), noncontact area with stagnant fluid is estimated to occupy 30–50% of the fracture plane.

We suggest that the scale dependencies of channeling flows (e.g., Eqs 7.6, 7.8, 7.9, and 7.11) can be applied to predict fracture flow characteristics for confining stresses of up to ~100 MPa, because the change in the fracture flow characteristics with increasing confining stress of from 30 to 100 MPa is relatively small (Watanabe *et al.* 2008). An effective confining stress of 100 MPa, which corresponds to ~6–7 km depth in the continental crust, is sufficiently high to encompass the targeted depth of deep geothermal reservoirs.

Implications for permeability enhancement by microearthquakes during hydraulic stimulation

On the basis of these scale dependencies, we now discuss fracture permeability enhancement by microearthquakes during hydraulic stimulation of a fractured reservoir. *In situ* microearthquakes data are recorded during hydraulic stimulation to capture the underlying active processes and permeability evolution within the reservoir, and it is in general recognized that microearthquakes somehow represent permeability change in a fractured reservoir. More specifically, microearthquakes triggered by (hydro-)shear slip reactivation of fractures are considered to be linked to enhancement of fracture/reservoir permeability, whereas microearthquakes triggered by reorganization of the stress field may enhance fracture/reservoir permeability only when the slipped fractures are connected to hydraulically active fractures/fracture zones (Miller 2015; Preisig *et al.* 2015; Rutqvist 2015). However, it remains ambiguous how much the fracture permeability is enhanced by microearthquake. Therefore, if we can gain an insight into the quantitative linkage between the permeability enhancement and microearthquakes, we will be able to improve the reliability of models for *in situ* permeability evolution in a fractured reservoir.

To provide such a linkage, this study assumes a very simple model as follows: a fracture slip occurs on a square fracture, which is characterized by its fracture length (m), l, as a result of (hydro-)shear failure (microearthquakes), and two key parameters (mean aperture and fracture permeability) evolve. The use of a circular fracture where shear slip and resulting permeability enhancement are heterogeneously distributed within the fracture plane may be more generally appropriate, but such a model is difficult to treat and beyond the scope of this study. When (hydro-)shear failure occurs, a seismic moment is defined as (Aki & Richards 2002)

$$M_0 = \mu_0 A \delta, \tag{7.12}$$

where M_0 is the seismic moment (N m), μ_0 is the shear modulus of the host rock, A is the surface area of the fracture (rupture area) (m^2), and δ is the shear displacement averaged over the entire fracture surface (m). This seismic moment can be converted to an equivalent moment magnitude by

$$\log M_0 = 1.5 M_w + 9.1, \tag{7.13}$$

where M_w is the moment magnitude (Hanks & Kanamori 1979). As long as the value of stress drop is determined uniquely, A and δ can also be estimated uniquely as a function of M_w by using empirical scaling relations. This treatment of static stress drop is a key point and should be kept in mind in the following discussion, since the empirical scaling equations listed here (e.g., Eqs 7.14–7.17) are specific to a certain stress drop and can vary depending on values of stress drop. According to Leonard (2010), for strike-slip earthquakes with relatively small magnitude, fracture surface area is given by

$$\log M_0 = 1.5 \log A + 6.09 \tag{7.14}$$

or, in terms of moment magnitude

$$M_w = \log A - 2.01. \tag{7.15}$$

Because the fracture surface area is equivalent to the square of its fracture length (i.e., $A = l^2$), Eq. 7.15 can be rearranged to yield

$$l = 10^{0.5 M_w + 1.00}. \tag{7.16}$$

In contrast, shear displacement is given by

$$\log \delta = 0.5 \log A - 4.43 = \log l - 4.43, \tag{7.17}$$

which indicates that δ/l is constant ($\delta/l = 3.72 \times 10^{-5}$). The small value of δ/l is due to the fact that a fault experiences a single slip in using the aforementioned scaling. In this manner, moment magnitude of a microearthquake (M_w), fracture length (l), and shear displacement (δ) can be quantitatively correlated. In other words, for a fracture that generates a specific microearthquake, we can now determine the characteristic values of l and δ.

For a fracture that has not experienced shear failure (i.e., a pre-hydraulic stimulation, joint), we assume zero shear displacement. This assumption enables us to estimate initial values of mean aperture ($e_{m,joint}$) and fracture permeability (k_{joint}) for a fracture with a specified fracture length using Eqs 7.6 and 7.9. In contrast, for a fracture that has experienced a specific microearthquake (i.e., post-hydraulic stimulation, fault), subsequent values of mean aperture ($e_{m,fault}$) and fracture permeability (k_{fault}) are estimated by Eqs 7.8 and 7.11. Here, the range of application is limited to fractures with shear displacements smaller than ~0.4 mm (M_w <0 by Eqs 7.16 and 7.17). In this manner, quantitative changes in mean aperture and fracture permeability are successfully linked with the M_w of a microearthquake. More specifically, by combining Eqs 7.6, 7.8, 7.16, and 7.17, change in mean aperture caused by a specific M_w of microearthquake can be formulated as

$$e_{m,fault}/e_{m,joint} = 1.0 \times 10^{0.35 M_w}, \tag{7.18}$$

whereas by combining Eqs 7.9, 7.11, 7.16, and 7.17, the corresponding change in fracture permeability can be formulated as

$$k_{fault}/k_{joint} = 116.4 \times 10^{0.46 M_w}. \tag{7.19}$$

Table 7.5 Parameter values estimated for a fracture that has experienced a microearthquake with a specific moment magnitude

Moment magnitude (–)	Fracture length (m)	Shear displacement (m)	Mean aperture before slip (mm)	Mean aperture after slip (mm)	Fracture permeability before slip (m²)	Fracture permeability after slip (m²)
0.0	10.1	3.75×10^{-4}	0.16	0.16	1.4×10^{-12}	1.7×10^{-10}
0.5	17.9	6.65×10^{-4}	0.17	0.25	1.6×10^{-12}	3.1×10^{-10}
1.0	31.9	1.19×10^{-3}	0.18	0.37	1.7×10^{-12}	5.7×10^{-10}
1.5	56.7	2.11×10^{-3}	0.20	0.56	1.9×10^{-12}	1.1×10^{-9}
2.0	100.8	3.75×10^{-3}	0.21	0.84	2.1×10^{-12}	2.0×10^{-9}

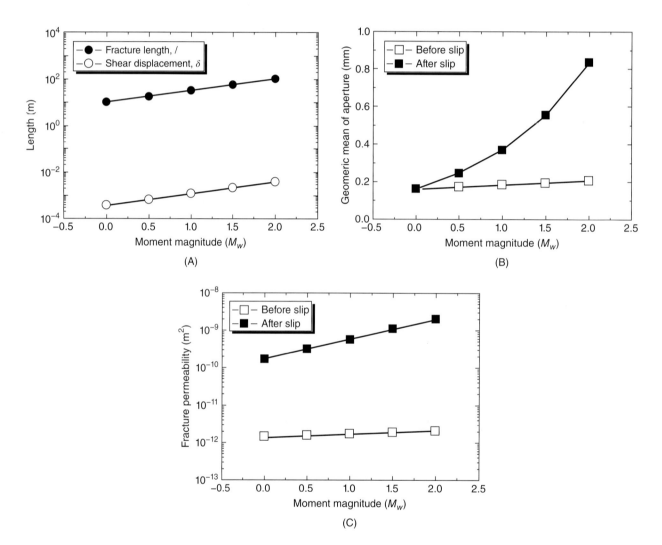

Fig. 7.10. (A) Estimated relations between fracture length and moment magnitude and shear displacement and moment magnitude, and changes in (B) mean aperture and (C) fracture permeability as a function of moment magnitude.

We note that the generality of these linkages is obviously limited, because the relations between $e_{m,\text{fault}}/e_{m,\text{joint}}$, $k_{\text{fault}}/k_{\text{joint}}$, and M_w can be influenced by rock type, roughness of fracture surface, fault gouge formation, and the absolute value of static stress drop.

With the aforementioned steps in mind, characteristic values of l and δ are first estimated for a microearthquake with a specific M_w of 0–2. For the respective microearthquake, we subsequently evaluate mean aperture and fracture permeability both pre- and post-microearthquake, and these parameter values are summarized in Table 7.5. These parameters are calculated assuming uniform distribution of shear slip along the fracture surface. Figure 7.10A shows the fracture length–moment magnitude relation and the shear displacement–moment magnitude relation. Figure 7.10B,C shows changes in mean aperture and fracture permeability caused by microearthquakes of a specific M_w. Microearthquakes of $M_w \geq 0$ cause an increase in mean aperture of a rock fracture due to shear dilation. For $M_w \geq 0$, degree of fracture permeability enhancement increases with M_w. On the other hand, for a microearthquake of $M_w < 0$, it is expected that the impact of shear dilation is very small or absent. As a result, permeability enhancement caused by a microearthquake of $M_w < 0$ is also estimated to be very small or zero, which may suggest that, during hydraulic stimulation of a fractured reservoir, permeability enhancement is limited to zones where seismic events with $M_w \geq 0$ are observed. Actual phenomena for fracture shear are much more complicated than those explained earlier (e.g., heterogeneous slip distribution along fracture), but we think that this approach is appropriate as a first step for linking microearthquake to fracture permeability evolution.

Finally, we compare absolute amounts of permeability gain evaluated by Eq. 7.19 (i.e., $k_{\text{fault}}/k_{\text{joint}}$) with some data from field experiments. According to Häring *et al.* (2008), hydraulic stimulation for creating an enhanced geothermal system in Basel, Switzerland, caused several microseismic events with local magnitudes ≥ 2.5, and the corresponding gain in fracture transmissivity was estimated to be approximately 400-fold. Similar values of transmissivity gain (200-fold) were reported in the enhanced geothermal system at Soultz-sous-Fôret (Evans *et al.* 2005). In contrast, Eq. 7.19 entails permeability gain of approximately 1000-fold for a microearthquake of $M_w = 2.0$, 2.5–5 times higher than that reported for field experiments. However, the results are consistent within an order of magnitude, and we consider such a degree of overestimation is acceptable. The difference in permeability gain may be attributed to three main causes. First, the permeability gain evaluated by Eq. 7.19 represents the most favorable case of shear slip direction, which is orthogonal to the anisotropy of aperture distribution. This is because empirical Eq. 7.19 is primarily developed from Eq. 7.11. Second, there is the uncertainty regarding values of stress drop. The scaling equations (e.g., Eqs 7.14–7.17) depend on values of stress drop, as do the pre-exponential factors and exponents in Eq. 7.19. A third and final consideration is the difference in characteristics between a single fracture and a fracture network. The actual geothermal system, where hydraulic stimulation was conducted, is a rock fracture network, whereas Eq. 7.19 was derived for a single rock fracture. These topics will be considered in the future work.

This study proposes a quantitative linkage between permeability enhancement and microearthquake, enabling us to roughly map evolving fracture permeabilities using *in situ* microearthquake data. To our knowledge, we have few insights linking microearthquake to fracture permeability evolution quantitatively and this study is the first step for that. In summary, this study suggests that the relationship between earthquake dynamics and crustal fluids is likely influenced by *in situ* characteristics of Earth's crust, including stress field and fault geometry (size and morphologic characteristics), as well as by the characteristics of crustal fluid. It may be formulated or constrained by considering both scale-independent parameters (ratio of asperity contact and flow area) and scale-dependent parameters (mean aperture, fracture permeability, and shear displacement).

CONCLUSIONS

We evaluated heterogeneous aperture distributions and the resulting channelized flows for granite fractures of various sizes under confining stress. The scale independency of contact area was revealed, which, in combination with the well-known self-affine fractal nature of the fracture surfaces, provides a novel method by which to predict fracture aperture distributions beyond the laboratory scale. The validity of the proposed method was tested by laboratory investigations and maximum aperture–fracture length relations, which have been reported in the literature for natural fractures (joints and faults).

Subsequently, representative aperture distributions and resulting channeling flows were numerically determined under confining stress (up to ~100 MPa) for joints and faults of various sizes. Aperture distributions and fluid flow were evaluated, and the changes in the mean aperture and permeability with fracture scale were formulated. We can now predict plausible scale dependencies of channelized flows through heterogeneous aperture distributions from laboratory scale to field scale for various combinations of fracture scale and shear displacement. These results allow us to suggest linkages between quantitative changes in mean aperture or fracture permeability and the moment magnitude of microearthquake during hydraulic stimulation for a fractured reservoir. Such linkages may enable rough inverse-mapping of evolving fracture permeabilities using *in situ* microearthquake data. This mapping will facilitate new insights into transport phenomenon within the Earth's crust and is relevant to engineering and scientific applications such as the development of hydrocarbon or geothermal reservoirs and clarification of earthquake mechanisms.

ACKNOWLEDGMENTS

Insightful reviews by Giona Preisig and Joshua Taron are greatly appreciated. The authors also thank Derek Elsworth for his constructive comments and English correction, and Atsushi Okamoto and Nobuo Hirano for their valuable suggestions. This work also benefitted from discussions within Geothermal Energy Team, Fukushima Renewable Energy Institute, AIST (FREA). This study was supported in part by JSPS Postdoctoral Fellowships for Research Abroad, No. 26-709 (to T.I.), and a Grant-in-Aid for Specially Promoted Research, No. 25000009 (to N.T.). The data in this paper are available from the corresponding author.

CHAPTER 8

Fractured rock stress–permeability relationships from *in situ* data and effects of temperature and chemical–mechanical couplings

JONNY RUTQVIST

Lawrence Berkeley National Laboratory (LBNL), Berkeley, CA, USA

ABSTRACT

The purpose of this chapter is to (i) review field data on stress-induced permeability changes in fractured rock; (ii) describe the estimation of fractured rock stress–permeability relationships through model calibration against such field data; and (iii) discuss the observations of temperature and chemically mediated fracture closure and its effect on fractured rock permeability. The field data that are reviewed include *in situ* block experiments, excavation-induced changes in permeability around tunnels, borehole injection experiments, depth- (and stress-) dependent permeability, and permeability changes associated with a large-scale rock-mass heating experiment. Data show how the stress–permeability relationship of fractured rock very much depends on local *in situ* conditions, such as fracture shear offset and fracture infilling by mineral precipitation. Field and laboratory experiments involving temperature have shown significant temperature-driven fracture closure even under constant stress. Such temperature-driven fracture closure has been described as thermal overclosure and relates to better fitting of opposing fracture surfaces at high temperatures, or is attributed to chemically mediated fracture closure related to pressure solution (and compaction) of stressed fracture surface asperities. Back-calculated stress–permeability relationships from field data may implicitly account for such effects, but the relative contribution of purely thermal–mechanical and chemically mediated changes is difficult to isolate. Therefore, it is concluded that further laboratory and *in situ* experiments are needed to increase the knowledge of the true mechanisms behind thermally driven fracture closure, and to further assess the importance of chemical–mechanical coupling for the long-term evolution of fractured rock permeability.

Key words: chemical and mechanical coupling, fractured rock, *in situ* experiments, permeability and stress, temperature

INTRODUCTION

Coupled thermal–hydrological–mechanical–chemical processes are critically important for many geological engineering practices, including geological disposal of nuclear waste, geothermal energy extraction, and underground carbon sequestration (Tsang 1991, 1999; Rutqvist & Stephansson 2003; Rutqvist 2012; Rutqvist & Tsang 2012). Results from the international DECOVALEX project on modeling of coupled processes associated with nuclear waste disposal have shown that some processes, such as thermal and thermal–mechanical processes, can be predicted with a relatively high confidence level, whereas other processes are much more difficult to predict (Rutqvist *et al.* 2005, 2009a). In particular, coupled hydromechanical processes, including the effects of stress and deformation on rock-mass permeability, are notoriously difficult to predict in complex fractured rock masses (Rutqvist & Stephansson 2003; Rutqvist *et al.* 2013a). This difficulty arises because mechanically induced changes in rock-mass permeability depend on site-specific hydraulic and mechanical interactions within a heterogeneous fracture network – which are very sensitive to small changes in fracture aperture and fracture connectivity (Min *et al.* 2004; Baghbanan & Jing 2008; Koh *et al.* 2011; Zhao *et al.* 2013). As a result, for a given rock mass, it is not easy to confidently predict how permeability might change under different mechanical forcing without observations from actual site-specific field experiments. Thus, as is discussed in this chapter, it is crucial that site-specific *in situ* experiments are undertaken at the appropriate scale (or across multiple scales), to avoid misleading results.

In theory, a relationship between stress and permeability may be derived using laboratory testing on single fractures

Crustal Permeability, First Edition. Edited by Tom Gleeson and Steven E. Ingebritsen.

Companion Websites: www.wiley.com/go/gleeson/crustalpermeability/
http://crustalpermeability.weebly.com/

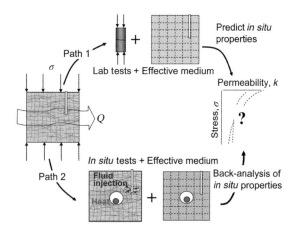

Fig. 8.1. Two alternative ways (Path 1 and Path 2) for deriving a stress–permeability relationship of a fractured rock unit. Path 1 involves laboratory testing on single fractures and an effective medium theory, whereas Path 2 involves back analysis by model calibration against field data. (*See color plate section for the color representation of this figure.*)

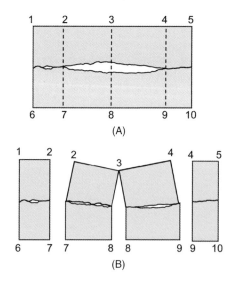

Fig. 8.2. Schematic showing the effects of unrepresentative sampling through a fracture. (A) The *in situ* fracture and (B) potential core samples if drilled through the fracture. The large void under *in situ* conditions may not be captured in laboratory experiment on core samples that would indicate a smaller aperture and higher stiffness than for the original larger scale fracture. (Adapted from Wei & Hudson 1988.)

and an effective medium theory (see Path 1 in Fig. 8.1). However, owing to issues related to unrepresentative sampling and sampling disturbances, it is difficult to upscale reliable stress–permeability relationships (Rutqvist & Stephansson 2003). An alternative approach described in this chapter is the back analysis of stress–permeability relationships by model calibration against field data of stress-induced changes in rock-mass permeability (illustrated by Path 2 in Fig. 8.1). Unrepresentative sampling of a rock fracture is illustrated in Figure 8.2. It shows how rock fractures in the field might contain large voids as a result of large shear offset and waviness along fracture surfaces. Using data from laboratory experiments on rock cores sampled from such a fracture could likely result in underestimation of aperture and permeability, whereas fracture normal stiffness would likely be overestimated.

Model calibration according to Path 2 in Figure 8.1 eliminates difficulties related to unrepresentative sampling and sample disturbances. However, model calibration against field data may be less controlled in terms of boundary conditions, and involves a number of underlying processes not directly related to stress that affect the overall permeability evolution. For example, field experiments involving elevated temperature may be strongly affected by additional temperature-dependent fracture closure. Such a temperature-driven fracture closure has been observed both in laboratory experiments and field tests and has been described as thermal overclosure related to better fitting of opposing fracture surfaces at high temperatures (Barton 2007). Similar phenomenon has also been described as chemically mediated fracture closure attributed to pressure solution (and compaction) of stressed fracture surface asperities (Yasuhara *et al.* 2004, 2011; Min *et al.* 2009). Such chemical–mechanical coupling and its impact on permeability has received renewed attention, especially related to long-term evolution of rock-mass permeability associated with nuclear

waste disposal, geothermal energy extraction, and even geologic carbon sequestration (Taron & Elsworth 2009; Rutqvist 2012; Rutqvist & Tsang 2012).

The objective of this chapter is to (i) review field data on stress-induced permeability changes in fractured rock; (ii) describe the back analysis of rock-mass stress–permeability relationships through model calibration against such field data; and (iii) discuss the observations of temperature and chemically mediated fracture closure and its effect on fractured rock permeability. Reviewed field data include *in situ* block experiments, borehole injection experiments, excavation-induced changes in permeability around tunnels, depth- (and stress-) dependent permeability, and permeability changes associated with a large-scale rock-mass heating experiment. Back analyses of stress–permeability relationships against such field data are presented, including one case involving a large-scale rock-mass heating experiment. Finally, observations of additional temperature-dependent fracture closure as a result of the so-called thermal overclosure or chemically mediated mechanical changes are discussed.

FRACTURED ROCK STRESS–PERMEABILITY RELATION AND SAMPLE SIZE EFFECT

Figure 8.3, from a review by Rutqvist & Stephansson (2003) on hydromechanical coupling in fractured rocks, presents "typical" hydromechanical behavior of single rock fractures as known from several decades of experimental and theoretical studies. In Figure 8.3A, a size dependency on fracture normal closure is indicated as an increased maximum closure, δ_{max}, with increased sample size. Such a sample size effect on fracture normal stiffness was observed by Yoshinaka *et al.* (1993) and through systematic experiments on concrete replica of rock fractures by Fardin (2003), as well as in a recent comprehensive review of joint stiffness data by Zangerl *et al.* (2008). Similar reduction in fracture normal stiffness with sample size has also been observed from seismic wave propagation through fractured rocks (Worthington & Lubbe 2007; Hobday & Worthington 2012). Figure 8.3B shows a corresponding increase in hydraulic conducting fracture aperture with sample size, which has been observed by comparing the hydromechanical behavior of different-sized rock fractures, including *in situ* block and ultra-large core experiments that is discussed in the following section. Such a correlation between fracture stiffness (or fracture compliance) and permeability is also being used to develop scaling relationships aiming at new methods for seismic (nonintrusive) evaluation of fractured rock permeability (Petrovitch *et al.* 2013). The effect of sample size on fracture shear behavior (Fig. 8.3C,D) was quantified in the late 1970s by Barton & Choubey (1977), with conclusive experiments by Bandis *et al.* (1983). For a larger fractured sample, the peak shear stress is smaller and takes place after a larger shear displacement magnitude and the onset of shear dilation is delayed.

In fractured rock, a change in the stress field may mechanically reactivate fractures, either as a result of changes in effective normal stress (normal stress reactivation) or shear stress (shear stress reactivation). Such reactivation may be triggered by fluid injection and associated increase in fluid pressure that reduces the effective normal stress and thereby also reduces the shear resistance (shear strength) of the fracture. Permeability change by effective normal stress reactivation is related to changes in the local stress field across fractures and occurs to some degree even for small and elastic mechanical responses. Substantial permeability changes by shear reactivation generally require shear failure and a shear displacement of several millimeters (Fig. 8.3D). However, the magnitude of shear displacement along a fracture is limited by its length and the elastic properties of the surrounding rock mass (Dieterich 1992; Rutqvist *et al.* 2013a). To allow for substantial shear-induced permeability changes, fractures must be sufficiently long or located near a free surface, such as the ground surface or a tunnel wall, or within a fracture zone of highly fractured rocks.

Figure 8.4 shows an example of calculated stress-induced permeability changes for a fractured rock mass as a combination of elastic fracture normal closure and shear dilation (Min *et al.* 2004). For the assumed Mohr–Coulomb mechanical properties, shear dilations were initiated at a stress ratio of 3, that is, a maximum compressive stress magnitude three times the minimum compressive stress magnitude. A substantial shear dilation occurred because the rock mass was assumed to be intensely fractured, and shearing was localized along multiple aligned fractures. Moreover, it appears that the substantial shearing observed in Figure 8.4 might also depend on the

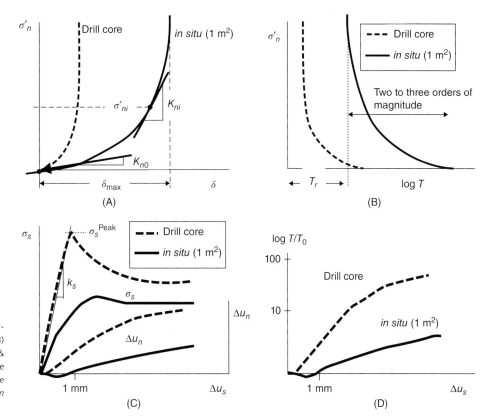

Fig. 8.3. Typical hydromechanical behavior of rock fractures during normal (A, B) and shear (C, D) deformation (Rutqvist & Stephansson 2003). Effects of sample size is indicated with the laboratory sample response (dashed lines) compared with *in situ* fracture response (1-m² size).

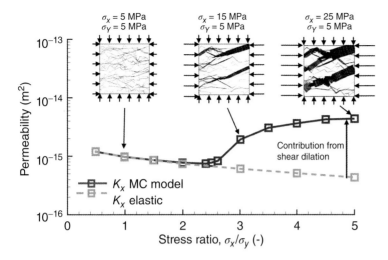

Fig. 8.4. Simulation of the horizontal permeability evolution of an intensely fractured rock mass subject to increasing shear stress (σ_x/σ_y) ratio (results from Min *et al.* 2004). Elastic simulation results are compared with that of elastoplastic using Mohr–Coulomb failure along fractures to investigate the effect of shear dilation on horizontal permeability.

model size, with free moving boundaries around the 5×5 m model domain. Indeed, the same fracture network was used in a different study, but considering a larger 20×20 m model, and as a result, much less shear-induced permeability increase occurred (Rutqvist *et al.* 2013a).

IN SITU BLOCK AND ULTRA-LARGE CORE EXPERIMENTS

In situ block and ultra-large core experiments enable controlled studies of coupled hydraulic and mechanical behavior of fractured rocks and rock fractures at meter scale. Figure 8.5 presents the results of a number of such *in situ* block and ultra-large core experiments conducted in the 1970s and 1980s. *In situ* block experiments are typically loaded with flat jacks – a thin envelope-like bladder that may be pressurized with hydraulic oil – in slots around the block, whereas ultra-large core experiments involve collecting a large core (including fractures) and then loading the core in a laboratory. Figure 8.5A compares results from several experiments from the 1970s (Iwai 1976; Pratt *et al.* 1977; Witherspoon *et al.* 1979). From these data, Witherspoon *et al.* (1979) identified an apparent dependency of fracture permeability on sample size. According to their findings, for the maximum stress level that could be attained, the minimum values of fracture hydraulic conductivity increased with specimen size. Similar behavior for a high minimum fracture hydraulic conductivity is observed in Figure 8.5B,C for two other large-scale experiments (Hardin *et al.* 1982; Sundaram *et al.* 1987). In these experiments, aperture and permeability change rapidly at low stress, but for stress increases above 5 MPa, aperture and permeability do not change as rapidly and reach irreducible (or residual) values. Figure 8.5D presents results from an *in situ* block experiment at the Stripa Mine in Sweden exhibiting different behaviors (Makurat *et al.* 1990a). In this *in situ* block experiment, the fracture permeability goes to zero when the normal stress exceeds about 5 MPa. This was attributed to soft

fracture infilling of precipitated minerals that completely clogs the fracture at high stress (Makurat *et al.* 1990a).

Figure 8.6 presents another unique experimental result of a uniaxial compression test conducted at the Lawrence Berkeley National Laboratory on a large core sample (about 1 m high and 1 m diameter) of fractured granite from the Stripa Mine (Thorpe *et al.* 1982). The fractured sample was loaded in unconfined compression (axial load) until shear failure was initiated, with shear in some existing subvertical fractures and creation of some new fractures. Fluid flow from an internal borehole was used to evaluate changes in transmissivity with axial loading. Before the actual compression test, a seepage test was conducted showing that the fluid flow was dominated by flow through the subhorizontal fracture B and a steeply dipping fracture D (Fig. 8.6B). Fracture B was inadvertently opened during the sample preparation, which may explain its dominance. At the intersection between fractures D and B, an offset of 1–2 cm was observed, indicating previous shear in fracture D.

Figure 8.6E shows the flow rate versus axial load, with initial decreasing flow until shear failure is initiated at an axial load of about 6 MPa. Thereafter, the flow increases with the additional axial load. In Figure 8.6C,D, the shear and normal stresses have been estimated (with great uncertainty) by projection from the axial load normal and along the steeply dipping fracture plane. The initial flow decrease is caused by fracture normal compression under increasing normal stress, whereas the subsequent flow increase is caused by shear dilation once shear failure occurs. Figure 8.6E shows measured total outflow as well as calculated outflow from fractures B and D using measurements of aperture changes (from strain measurements). There are discrepancies between calculated and observed flows, possibly related to differences between hydraulically conducting and mechanical apertures. Nevertheless, the overall response resembles that of the numerical simulations in Figure 8.4. Moreover, the mechanical behavior of fracture D follows in general that of generic curves in Figure 8.3 for 1 m² sample size, that is, an initial (elastic) shear displacement of about 1 mm, followed by

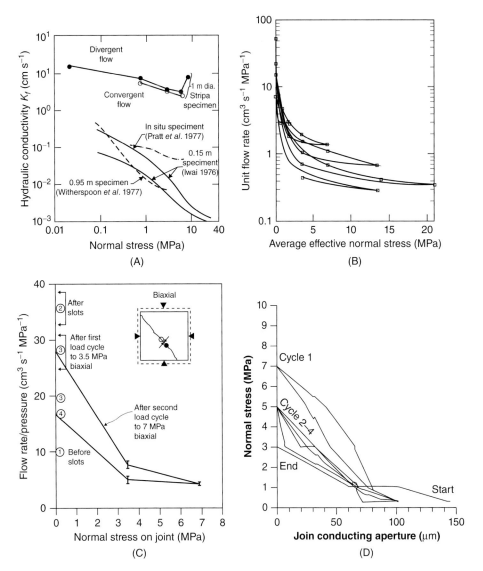

Fig. 8.5. Hydromechanical behavior of fractures from *in situ* block and ultra-large core experiments: (A) fracture conductivity (in cm s⁻¹) as a function of normal stress for several experiments of different sample sizes compared in Witherspoon *et al.* (1979), (B) unit flow rate (in cm³ s⁻¹ MPa⁻¹) as a function of normal stress for a fractured ultra-large core specimen (Sundaram *et al.* 1987), (C) unit flow rate (in cm³ s⁻¹ MPa⁻¹) as function of normal stress for an *in situ* block experiment (Hardin *et al.* 1982), (D) stress-versus-hydraulic conducting aperture from an *in situ* block experiment, including a mineral-filled fracture (Makurat *et al.* 1990).

a shear failure without a pronounced peak shear stress. Then, during an additional 1 mm shearing, a shear dilation of about 0.3 mm was observed, resulting in an increased permeability when the axial load was increased up to 7.5 MPa. Further shear-induced permeability increases would be expected if the loading had been continued.

In summary, the large-scale *in situ* block and ultra-large core experiments in Figures 8.5 and 8.6 show the important effects of sample size and fracture filling, effects that can lead to very different permeability evolution with stress – which would be difficult to predict from tests on small-scale core samples of the same fractures. Moreover, the large core experiment in Figure 8.6 shows the heterogeneous nature of the flow, with flow focused in a few open fractures and their intersections, and heterogeneously distributed along fracture planes. In this large core experiment, the flow response to loading was complex, but demonstrated theoretically predicted behavior, with compression and shear dilation according to the loading on

the dominant fractures. Nevertheless, at a fractured rock site, it would be difficult to predict the magnitude of these changes, as well as the initial permeability, without actual site-specific *in situ* testing. In the following section, one type of such *in situ* tests is described.

BOREHOLE INJECTION TESTS

Hydraulic jacking test (or step-rate test) is a type of borehole injection test that has been used to investigate pressure-sensitive permeability within dam foundations (Louis *et al.* 1977) and also used in association with hydraulic fracturing stress measurements (Doe & Korbin 1987; Rutqvist & Stephansson 1996). Rutqvist (1995) and Rutqvist *et al.* (1997) applied hydraulic jacking tests combined with coupled numerical modeling to determine the *in situ* hydromechanical properties of fractures at two crystalline rock sites in Sweden (Fig. 8.7). Hydraulic jacking tests were conducted by a step-wise increase in fluid pressure.

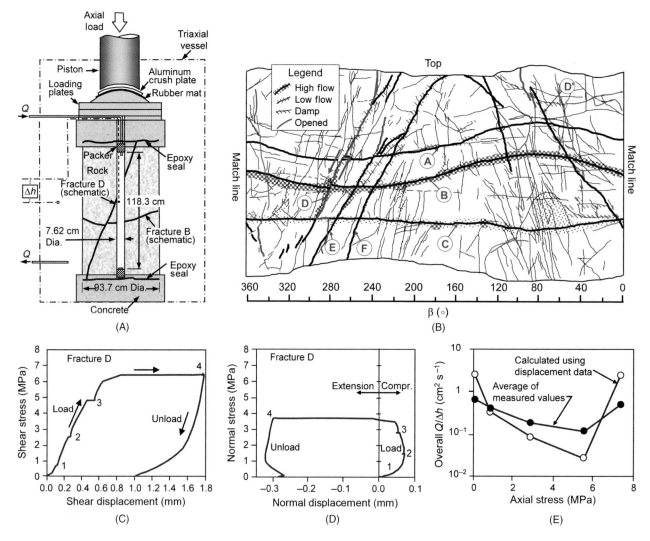

Fig. 8.6. Uniaxial compression on an ultra-large core of fractured Stripa Granite conducted at the Lawrence Berkeley National Laboratory (Thorpe *et al.* 1982). (A) Schematic of test arrangement, (B) fracture mapping and seepage during a hydraulic test conducted before the unixial compression test, (C) shear stress versus shear displacement, (D) normal stress versus normal displacement, and (E) flow versus axial stress. (Adapted from Thorpe *et al.* 1982.) (*See color plate section for the color representation of this figure.*)

At each step, the well pressure was kept constant for a few minutes until the flow was steady (Fig. 8.7A). Numerical analysis of these injection tests showed that the flow rate at each pressure step is strongly dependent on the aperture and normal stiffness of the fracture in the vicinity of the borehole, where the flow resistance and pressure gradient are the highest (Fig. 8.7B). Such numerical modeling of injection tests are also presented in Preisig *et al.* (2015), using a discrete fracture model, and indicating the potential for more complex responses, including shear slip.

Figure 8.7C shows one example of a pressure-versus-flow-rate response from a hydraulic jacking test at 267 m depth in a borehole at the Laxemar crystalline rock site in southeastern Sweden (Rutqvist *et al.* 1997). At the first cycle of step-wise increasing pressures, the flow rate increased as a nonlinear function of pressure. A temporal peak pressure was obtained at a flow rate

of $1.3 \, l \, min^{-1}$, and thereafter the pressure began to decrease with increasing flow rate. A shear-slip analysis of the particular fracture, which was inclined to the principal *in situ* stresses, indicated that these irreversible fracture responses could be caused by shear slip, because the increasing fluid pressure reduces fracture shear strength. The subsequent step pressure cycle took a different path because of the change in hydromechanical properties, possibly as a result of shearing or break-up of fracture filling.

Figure 8.8A presents the back-calculated relationships between fracture transmissivity and effective normal stress for several fractures at depths ranging from 266 to 338 m. Results showed that the transmissivity of the most conductive fractures is relatively insensitive to stress. From borehole images, these fractures appeared to be open fractures that were incompletely cemented, suggesting flow channels in a fracture that appears

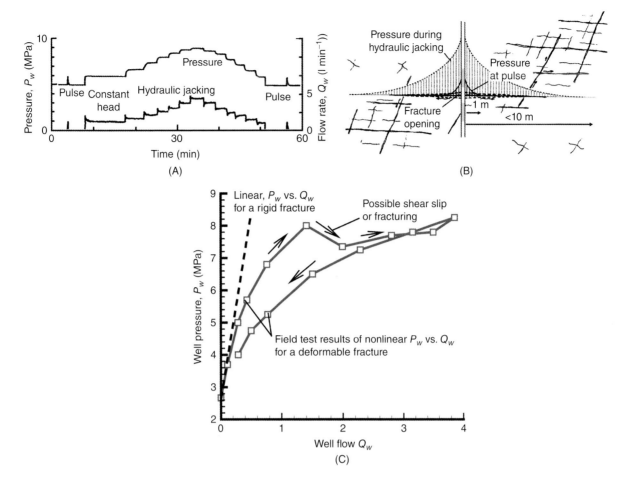

Fig. 8.7. *In situ* determination of stress–transmissivity relationships, using a combination of pulse, constant head, and hydraulic jacking tests. (A) Schematic representation of pressure and flow versus time, (B) the radius of influence in each test, and (C) results of a hydraulic jacking test at 267 m depth. (Rutqvist *et al.* 1997.)

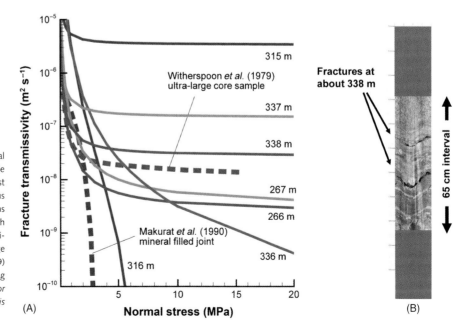

Fig. 8.8. Back-calculated hydromechanical properties of fractures intersecting a borehole at a crystalline rock site in Sweden (Rutqvist *et al.* 1997): (A) Fracture transmissivity versus effective normal stress for fractures at depths between 266 and 338 m (solid lines) with comparison to results from *in situ* block experiment by Makurat *et al.* (1990a) and ultra-large core experiment by Witherspoon *et al.* (1979) (dashed lines), (B) borehole image indicating open fractures at about 338 m depth. (*See color plate section for the color representation of this figure.*)

to be "locked open" (e.g., fractures at about 338 m shown in Fig. 8.8B). The stress–permeability behavior of these fractures resembles that of the large fracture-scale experimental data in Figure 8.5, such as that of Witherspoon *et al.* (1979). That is, permeability changes rapidly at low stress, but for stress increases above 5 MPa aperture and permeability do not change as rapidly and reach irreducible (or residual) values. The back-calculated stress-versus-transmissivity relationships for fractures at 316 and 336 m resemble that of the Makurat *et al.* (1990a) *in situ* block test shown in Figure 8.5D. That is, soft mineral filling is postulated to have a strong effect, making transmissivity of these fractures extremely stress dependent, clogging the fractures completely for fluid flow when the stress normal to the fracture exceeds a certain threshold.

Recently, specialized borehole testing tools enabling simultaneous measurements of hydraulic and mechanical fracture responses have been developed, both at Clemson University in South Carolina (Schweisinger *et al.* 2007; Svenson *et al.* 2008) and at Geosciences Azur at University of Nice and Marseille, France (Cappa *et al.* 2006a,b; Guglielmi *et al.* 2014). This kind of specialized equipment has been used to evaluate *in situ* properties such as fracture normal stiffness and storativity during various types of injection tests, including step-rate and pulse tests. As described in Schweisinger *et al.* (2009), and further developed in Schweisinger *et al.* (2011), Murdoch & Germanovich (2012), and Slack *et al.* (2013), the use of simultaneous pressure and deformation measurements improves the estimates of storativity and reduces nonuniqueness during the evaluation of hydraulic well tests. The most recent development of this concept involves a three-component borehole deformation sensor that enables simultaneous measurements of fracture normal and shear displacement during a borehole injection test (Guglielmi *et al.*, 2014). The equipment has been tested for measurements of injection-induced shear deformation in fractured carbonate rock (Derode *et al.* 2013). Moreover, the use of this equipment has been qualified as an International Society of Rock Mechanics "Suggested Method," denoted "Step-Rate Injection Method for Fracture In-Situ Properties" (Guglielmi *et al.* 2014). Future applications of this equipment will include *in situ* measurements of coupled hydromechanical responses during reactivation of faults in shale. These measurements will help to constrain model parameters for permeability change during reactivation of minor faults in shales, which is relevant for assessing caprock sealing performance at geologic carbon sequestration sites and for assessing the potential for fault activation and leakage during shale-gas hydraulic fracturing operations (Rutqvist 2012; Rutqvist *et al.* 2013b).

MODEL CALIBRATION AGAINST EXCAVATION-INDUCED PERMEABILITY CHANGES

Stress-induced changes in permeability around excavations, including the excavation disturbed zone, have been studied since the early 1980s at many sites with various rock types (Bäckblom & Martin 1999; Tsang *et al.* 2005; Blümling *et al.* 2007). The excavation disturbed zone includes a damage zone of induced rock failure and fracturing by the excavation process, a zone with altered stress distribution around an excavation, and a zone of reduced fluid pressure. Permeability measurements in the zone of altered stress around an excavation may be utilized for model calibration of stress–permeability relationships. Here, two examples are presented: one from a sparsely fractured rock site at the underground rock laboratory in Manitoba, Canada, and another from an intensely fractured rock site at Yucca Mountain, Nevada.

The Manitoba underground rock laboratory TSX tunnel experiment

The TSX tunnel (Room 425) at the Manitoba underground rock laboratory (Martino & Chandler 2004) was one of several experimental tunnels at the underground rock laboratory dedicated to studying the excavation disturbed zone. To minimize the excavation disturbed zone, the TSX tunnel was excavated using smooth drill-and-blast techniques in an elliptical cross-section 3.5 m high and 4.375 m wide (a horizontal to vertical aspect ratio of 1.25). At the depth of the TSX tunnel (420 m), the principal stresses were estimated to be 60 MPa (maximum stress), 45 MPa (intermediate stress), and 11 MPa (minimum stress), with the maximum principal stress parallel with the tunnel axis and the minimum principal stress subvertical. During excavation, the occurrence and location of microseismic events were monitored. After excavation, the resulting excavation disturbed zone was characterized by a variety of methods, including the microvelocity probe method for measuring changes in sonic velocities, and the SEPPI, which is a borehole pressure pulse probe for measuring changes in permeability. The results of the SEPPI permeability measurements were used for model calibration of a stress–permeability relationship (Rutqvist *et al.* 2009b).

The excavation-induced permeability changes at the TSX tunnel were simulated using a simplified but practical model that could be both implemented in the numerical simulator at hand and capture reasonably well the observed damage and permeability changes at the underground rock laboratory field experiments (Rutqvist *et al.* 2009b). Using recommended *in situ* rock-strength parameters (corresponding to an *in situ* strength about 50–60% of short-term laboratory strength) the simulation resulted in a limited mechanical failure at the crown of the tunnel, in agreement with observed increased macroscopic fracturing. This is the region where the highest shear stress occurs and most microseismic events were clustered.

The permeability around the tunnel was simulated using an empirical stress–permeability relationship in which permeability is a function of effective mean stress, σ'_m, and deviatoric stress, σ_d, according to (Rutqvist *et al.* 2009b):

$$k[k_r + \Delta k_{\max} \exp(\beta_1 \sigma'_m)] \cdot \exp(\gamma \Delta \sigma_d), \qquad (8.1)$$

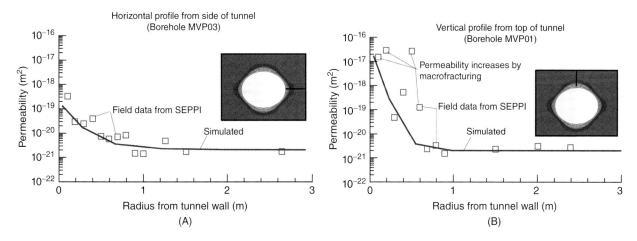

Fig. 8.9. Calculated and measured permeability changes around the TSX tunnel (Rutqvist *et al.* 2009b). Permeability versus radius along (A) a horizontal profile from the side of the tunnel and (B) a vertical profile from the top of the tunnel. (*See color plate section for the color representation of this figure.*)

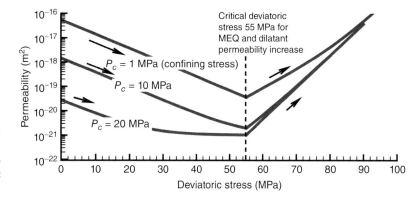

Fig. 8.10. Calibrated stress-versus-permeability relationship according to Eq. 8.1, with $\beta_1 = 4 \times 10^{-7}$ Pa^{-1}, $k_r = 2 \times 10^{-21}$ m^2, $\Delta k_{max} = 8 \times 10^{-17}$ m^2, $\gamma = 3 \times 10^{-7}$ Pa^{-1}, and the critical deviatoric stress for onset of shear-induced permeability is set to 55 MPa. MEQ refers to micro-earthqaukes, which were also triggered in areas around the tunnel where deviatoric stress exceeded 55 MPa. (Rutqvist *et al.* 2009b.)

where k_r is residual (or irreducible) permeability at high compressive mean stress, Δk_{max}, β_1, and γ are fitting constants, and $\Delta \sigma_d$ is the change in deviatoric stress relative to a critical deviatoric stress for onset of shear-induced permeability.

Figure 8.9 compares simulated and measured permeability changes for $\beta_1 = 4 \times 10^{-7}$ Pa^{-1}, $k_r = 2 \times 10^{-21}$ m^2, $\Delta k_{max} = 8 \times 10^{-17}$ m^2, $\gamma = 3 \times 10^{-7}$ Pa^{-1}, and with the critical deviatoric stress for onset of shear-induced permeability set to 55 MPa. The 55 MPa critical deviatoric stress roughly coincides with the extent of the observed cluster of microseismic events at the top of the tunnel and to about 0.3 of the instantaneous uniaxial compressive stress of small-scale core samples, which is consistent with the stress level at which crack initiation has been observed in studies of Lac du Bonnet granitic samples Martin & Chandler (1994). Thus, at least part of the observed permeability increase above the tunnel is caused by microfracturing under high compression, whereas permeability increase off to the side of the tunnel is caused by opening of existing microfractures as a result of decreased mean stress.

The back-calculated stress–permeability relationship is presented in Figure 8.10. Having determined this relationship by *in situ* calibration at the scale of the tunnel, one can implicitly take into account scale- and time-dependent strength degradation. Studies at the Manitoba underground rock laboratory have indicated that strength reduction of the granitic rock around open tunnels reaches a steady state at 50–60% of the short-term strength within a few weeks. It is possible that the several-orders-of-magnitude increase in permeability measured at the top of the tunnel was caused by macroscopic fracturing that was indeed observed in the boreholes. The macrofracturing implies that a simple relationship between mean and deviatoric stress, as defined in Eq. 8.1, may no longer be valid. Instead, the permeability may be governed by fracture permeability as a function of stress normal to the created fracture planes.

The Yucca Mountain niche excavations

Figure 8.11 presents the measurements of excavation-induced changes in permeability above a niche excavated off a tunnel in unsaturated fractured tuff at Yucca Mountain, Nevada (Wang *et al.* 2001; Rutqvist & Stephansson 2003; Rutqvist & Tsang 2012). The complete data set involved measurements of air permeability at four niches excavated by a mechanical (alpine mining) method. Permeability was measured before and after excavation in 30 cm packed-off sections along 10 m long boreholes located about 0.65–1 m above the niche,

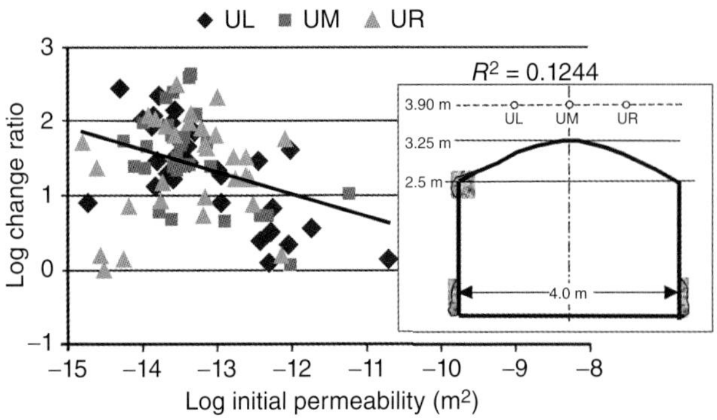

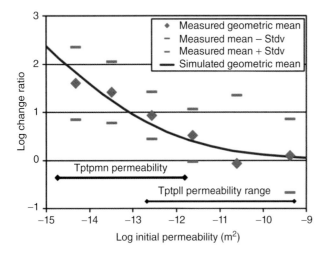

Fig. 8.11. Results of pre- to postexcavation air-permeability tests above a niche in fractured unsaturated tuff. The results shown are for a niche with three boreholes (UL, UM, and UR) located above niche 3560 (Wang *et al.* 2001). (*See color plate section for the color representation of this figure.*)

Fig. 8.12. Measured and calculated mean values of pre- to postpermeability change ratio at three niches (Niche 3107, 3560, and 4788). (Rutqvist 2004.)

supposedly outside the damaged zone (Fig. 8.11). The results in Figure 8.11 for one niche show that the pre-/postpermeability ratio ranges between a factor of 1 and 400, with a trend of relatively stronger permeability increase for those sections where the initial permeability is smaller.

Figure 8.12 presents the best match of calculated and measured geometric mean permeability change ratio for data from two welded tuff units, denoted Tptpmn and Tptpll. The calculation was conducted assuming three orthogonal fracture sets (consistent with fracture mapping at the site), with permeability depending on the stress normal to each fracture set (Rutqvist 2004; Rutqvist & Tsang 2012). The calculated geometric mean permeability matches the measured geometric mean change (Fig. 8.12). However, the wide range of permeability changes (from 1 to 400) observed in the field (Fig. 8.11) may be related to opening of fractures with different orientation relative to the niche. It is likely that pressure–flow response in each section is determined by flow into a few dominant fractures and that

their mechanical response upon excavation depends on their orientation.

Figure 8.13 presents the stress–aperture relationship used for the best match solution. In this relationship, the aperture b depends on the effective normal stress σ'_n according to (Rutqvist & Tsang 2003):

$$b = b_r + b_{max} \exp(\alpha \sigma'_n), \tag{8.2}$$

where b_r is the irreducible aperture, and b_{max} and α are fitting parameters used to match measured and calculated permeability responses. In this case, $b_{max} = 200\ \mu m$ and $\alpha = 0.8\ MPa^{-1}$. The irreducible aperture, b_r, varies with initial permeability, which explains the trend of stronger permeability increase at lower initial permeability. The stress-versus-aperture function defined by Eq. 8.2 was developed and applied as an empirical function for matching the permeability changes observed in the intensely fractured tuff units. However, such an exponential function was later derived by a closed-form solution and verified using a number of data sets in the literature for fracture closure versus stress (Liu *et al.* 2013).

The functions in Figure 8.13 display some resemblance to the results of the *in situ* block and ultra-large core tests shown in Figure 8.5. In essence, permeability is very stress sensitive at low stress, but for normal stresses above 5 MPa, the permeability is relatively insensitive to stress and approaches an irreducible permeability. Moreover, the tendency for stronger permeability change in fractures with smaller initial aperture is consistent with observations at the borehole injection tests at the Swedish crystalline rock site (Fig. 8.8). This behavior could be related to the notion that fractures with high initial aperture might have been locked open by shearing offset between rough fracture surfaces (Rutqvist & Stephansson 2003). However, Eq. 8.2 would be expected to be valid at relatively low stress, since ultimately, at very high stresses, fractures are not expected to stay open at such a high residual aperture. The exponential function in Liu *et al.* (2013) contains additional terms that allow for gradual fracture closure at very high stress.

DEPTH-DEPENDENT PERMEABILITY OF SHALLOW BEDROCK

Measurements of permeability in vertical boreholes and their variations with depth (and stress) might be used to calibrate *in situ* stress–permeability relationships. However, one has to consider that rock-mass permeability is strongly affected by other parameters or factors, such as fracture frequency, connectivity, infilling by mineral precipitation, and history. For example, Figure 8.14 presents a schematic by Rutqvist & Stephansson (2003) based on a vertical permeability profile of short-interval (3-m packer separation) well tests performed in crystalline rock at Gideå, Sweden. The figure shows a general decrease in permeability with depth; however, the depth dependency is obscured by a very large permeability variation at any given

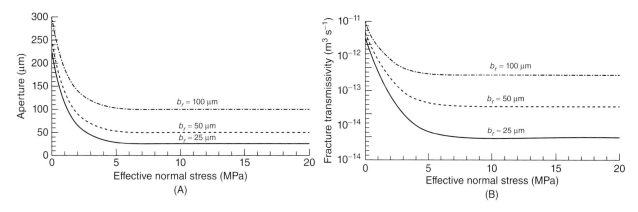

Fig. 8.13. Fracture stress–transmissivity behavior of fractures back-calculated from permeability measurements during excavation of niches: (A) exponential stress–aperture relationship and (B) corresponding stress–transmissivity relationship.

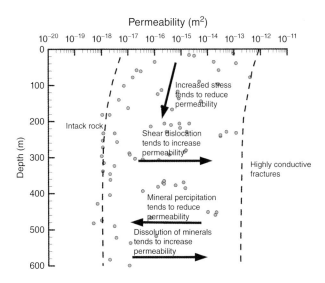

Fig. 8.14. Depth-dependent permeability in fractured rock obscured by chemical and mechanical processes (Rutqvist & Stephansson 2003). Permeability was measured in short-interval well tests in fractured crystalline rocks at Gideå, Sweden (data points from Wladis *et al.* 1997). Schematic of effects of shear dislocation and mineral precipitation/dissolution processes that obscure the dependency of permeability on depth (stress). The permeability values on the left-hand side represent intact rock granite, whereas the permeability values on the right-hand side represent highly conductive fractures.

depth (up to six orders of magnitude). This large variability may also be affected by difficulties in evaluating appropriate permeability from transmissivity values obtained from well tests (Stober & Bucher 2015). One general observation from Figure 8.14, and from many fractured crystalline rock sites, is that a more pronounced depth dependency can be found in the upper 100–300 m of the bedrock. The more pronounced depth dependency in shallow areas can be explained by the nonlinear normal stress–aperture relationship of single extension joints. Such a decrease in hydraulic aperture with depth was observed in the 1960s by Snow (1968a), who used detailed pumping tests and fracture statistics to determine fracture apertures at numerous dam foundations.

The minimum values of permeability on the left-hand side of Figure 8.14 represent the permeability of the rock matrix, implying that either no fracture intersected the 3-m-long test interval, or if fractures were intersecting the interval, they are either completely cemented with minerals or isolated from a conducting fracture network. The maximum permeability values on the right-hand side of Figure 8.14 represent the transmissivity of at least one intersecting fracture, which is highly conductive and connected to a larger network of conducting fractures. It is likely that these highly conductive fractures are "locked open" either by bridges of hard mineral filling or by large shear dislocation. In crystalline rock, the "locked open" fracture category could comprise mineral-filled or shear-dislocated fractures throughout the rock mass; however, large channels are more likely to occur in fault zones where large movements allow substantial shear dislocation and extension of oblique fractures (National Research Council 1996). Interestingly, in Figure 8.14, the maximum permeability values of about $k = 1 \times 10^{-13}\,\mathrm{m^2}$ below a few 100 m depth correspond quite accurately to the transmissivity value for of the most transmissive fracture in Figure 8.8. That is, for a packer spacing of $b = 3\,\mathrm{m}$, $k = 1 \times 10^{-13}\,\mathrm{m^2}$ corresponds to a transmissivity of about $T = (k \times b \times \rho \times g)/\mu \approx (1 \times 10^{-13} \times 3 \times 1000 \times 10)/1 \times 10^{-3} = 3 \times 10^{-6}\,\mathrm{m^2\,s^{-1}}$, which is very similar to the transmissivity of the fracture at 315 m in Figure 8.8. The "average" permeability at depth in Figure 8.14 is about 1×10^{-16} to $1 \times 10^{-15}\,\mathrm{m^2}$ and this corresponds to a transmissivity of about 3×10^{-9} to $3 \times 10^{-8}\,\mathrm{m^2\,s^{-1}}$, which is similar to the transmissivity of the fractures at 266 and 267 m in Figure 8.8.

When properly considering the many processes that could impact the depth-dependent permeability, such data could be useful for bounding and/or validating stress–permeability relationships at a fractured rock site. An example of such development applied to fractured sandstone is presented in Jiang *et al.* (2009) and for granite and shale in Jiang *et al.* (2010b). Here, a practical example from Yucca Mountain is presented, in which the fracture permeability and fracture frequency have been carefully characterized for different layers of a sequence

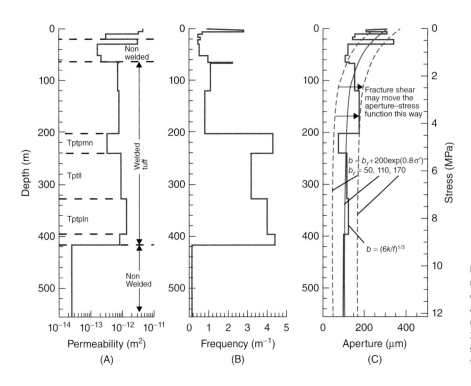

Fig. 8.15. Field data of (A) fractured rock permeability, (B) fracture frequency, and (C) calculated hydraulic conducting aperture through a vertical sequence of fractured volcanic rock units at Yucca Mountain, Nevada (Rutqvist 2004). Subparallel curved lines in (C) correspond to the exponential stress–aperture function for various residual apertures.

of welded and unwelded tuff units, from the ground surface of the mountain down to the groundwater table at 600 m depth (Fig. 8.15). Figure 8.15A shows that the permeability varies by about two orders of magnitude, that is, much less than the six-order variation for the crystalline rock site in Figure 8.14. The permeability was evaluated from air-injection tests in packed-off sections of vertical boreholes (with 3 m packer separation) and with one average permeability value assigned to the entire thickness of each geological unit. Figure 8.15B shows that the fracture frequency ranges from 0.1 to 5 fractures per meter, with some tendency toward higher permeability in layers with higher fracture frequency. Finally, Figure 8.15C presents the equivalent hydraulic conducting aperture, b_h, calculated from the permeability, k, and fracture frequency, f, using the cubic relation between equivalent fractured rock permeability and aperture according to Rutqvist & Stephansson (2003)

$$b_h = \sqrt[3]{6k/f} \qquad (8.3)$$

assuming a fractured medium with three orthogonal sets of fractures and with uniform fracture spacing, $s = 1/f$. The uniform fracture spacing assumption in Eq. 8.3 is certainly a simplification that results in an average b_h for the particular rock unit. At the Yucca Mountain, the rock units are intensely fractured with three orthogonal fracture sets and, therefore, Eq. 8.3 is relatively accurate. At other sparsely fractured rock sites, such as the aforementioned Gideå crystalline rock sites, Eq. 8.3 is likely not appropriate.

The subparallel curved lines in Figure 8.15C represent the stress–aperture function that was back-calculated from the aforementioned niche-excavation experiments, that is,

the exponential function of stress according to Eq. 8.2 with $b_{max} = 200\,\mu m$ and $\alpha = 0.8\,MPa^{-1}$. Both permeability and frequency vary substantially between different rock units, whereas the back-calculated hydraulically conducting aperture is quite consistent around the average stress–aperture function. Note that the rock mass at Yucca Mountain has a dense, well-connected fracture network, which also explains why permeability does not vary as much as in a sparsely fractured rock site such as the Gideå crystalline rock site shown in Figure 8.14. Again, the aperture is relatively constant for stresses above 5 MPa, indicating that the fracture aperture approaches an irreducible value.

The stress–aperture functions matched to field data in Figure 8.15C, with a residual aperture, could be expected to be valid for the shallow Earth crust, where stresses are much lower than the uniaxial compressive rock strength. Going much deeper, large voids created by fracture shear dislocation may not be kept open, but would be expected to collapse under ambient stress. As is further discussed in Section "Application to geoengineering activities and potential implications for crustal permeability," such collapse of larger voids at depth may have important implications for crustal permeability–depth distribution.

MODEL CALIBRATION AGAINST THE YUCCA MOUNTAIN DRIFT SCALE TEST

The Yucca Mountain drift scale test provides perhaps the most comprehensive data set of thermally induced changes in fractured rock permeability during rock-mass heating and

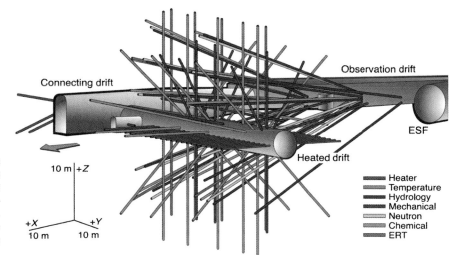

Fig. 8.16. Three-dimensional view of the Yucca Mountain drift scale test. The color-coded lines indicate boreholes for various measurements of thermally driven thermal–hydrological–mechanical–chemical responses. (Tsang *et al.* 2009; Rutqvist & Tsang 2012.) (*See color plate section for the color representation of this figure.*)

cooling (Rutqvist *et al.* 2008). During the experiment, a volume of over 100,000 m³ of intensely fractured volcanic tuff was heated for 4 years, including several tens of thousands of cubic meters heated to above-boiling temperature (Fig. 8.16). This massive heating induced strongly coupled thermal–hydrological–mechanical–chemical changes that were continuously monitored by thousands of sensors embedded in the fractured rock mass. Among the monitoring data were periodic active pneumatic (air-injection) measurements that were used to track changes in air permeability within the variably saturated fracture system around the heated drift. Air-injection testing was conducted in 44 several-meter-long packed-off sections in hydrological boreholes, located in three clusters forming vertical fans that bracketed the heated drift (Fig. 8.16).

Thermally induced permeability changes at the Yucca Mountain drift scale test were analyzed using coupled thermal–hydrological–mechanical modeling (Rutqvist *et al.* 2008). The analysis showed that observed changes in air permeability at the site can be explained by a combination of thermal–mechanically induced changes in fracture aperture and moisture-induced changes (Fig. 8.17). The modeling reproduced the average permeability evolution using the stress–permeability function shown in Figure 8.18A, which is based on the exponential stress–aperture function in Eq. 8.2. Moreover, a better match between calculated and measured responses was achieved when assuming that the measured permeability was dominated by air flow into the dominant vertical fracture set oriented normal to the subhorizontal boreholes. However, the analysis also indicated irreversible permeability changes that significantly deviated from the reversible thermohydroelastic solution (Fig. 8.17). The identified irreversible permeability changes may be attributed to inelastic thermal–mechanical processes consistent with either inelastic fracture shear dilation (where permeability increased) or inelastic fracture surface asperity shortening (where permeability decreased).

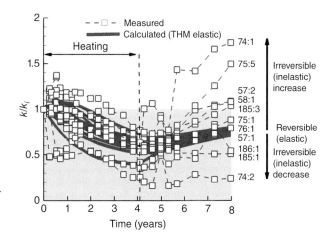

Fig. 8.17. Schematic of the range of measurements of air permeability at the Yucca Mountain drift scale test. Details of calculated and measured responses in all 44 measurement intervals can be found in Rutqvist *et al.* (2008) (*See color plate section for the color representation of this figure.*)

Whereas the stress–permeability relationship shown in Figure 8.18A provides a good match to the measured average permeability responses, the underlying aperture-versus-stress relationship depends on the spacing of active (or dominant) fractures. Therefore, two alternative functions are presented in Figure 8.18B. Note that the aperture-versus-stress relationships shown in Figure 8.18B are quite different from those that were back-calculated from the niche-excavation experiments (Fig. 8.13A). A possible explanation for this apparent inconsistency in the stress–aperture relationships derived from the two different field experiments in the same rock unit may be related to thermal overclosure and/or chemically mediated changes that resulted in additional fracture closure during heating at the drift scale test. This possibility is further discussed in the following section.

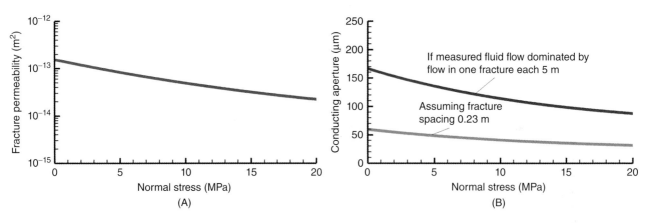

Fig. 8.18. Fracture stress-versus-permeability behavior back-calculated from air-permeability measurements at the Yucca Mountain drift scale test: (A) fractured rock permeability versus stress along one set of fractures and (B) corresponding aperture-versus-stress relationship for two assumptions regarding the spacing of dominant fractures.

THERMAL AND CHEMICALLY MEDIATED MECHANICAL CHANGES

Chemically mediated changes in fracture permeability received renewed attention after a series of laboratory experiments conducted at the Penn State University (Polak *et al.* 2003; Yasuhara *et al.* 2004). For example, experiments conducted on single fractures in Arkansas Novaculite (essentially pure quartz) show that fracture permeability decreases with increasing temperature up to 150 °C, even under a constant stress. During these experiments, the chemical composition of the fluid flowing through the fracture was carefully monitored. Minerals were dissolved as the permeability decreased, indicating fracture closure as a result of pressure solution of fracture asperities. Similar behavior has been observed in laboratory experiments on fractures in samples of volcanic tuff (Lin & Daily 1990) and in granite (Yasuhara *et al.* 2011).

The effect of temperature on fracture permeability under constant normal stress has also been observed during previously reported *in situ* block heater tests. For example, during the Terra Tek *in situ* heated block tests in gneiss (Hardin *et al.* 1982), aperture magnitude was reduced by more than three times as the temperature increased from 12 to 74 °C under constant stress. These results and other field observations were reviewed by Barton (2007), who emphasized that these effects, although known from experiments for almost 40 years, appear to have been ignored in the analysis of coupled thermal–hydrological–mechanical processes of fractured rock. The explanation for this phenomenon, which Barton (2007) described as thermal overclosure, is that the joint in question, and perhaps the huge majority of joints developed in the crust, was formed at variously elevated temperatures. They were thereby given a primeval "fingerprint" of three-dimensional roughness influenced by all the minerals (or grains) forming the joint walls. When cooled, various subtle changes would occur, causing reduced fit (Barton 2007) as a result of

thermal–mechanical changes in the fracture surface geometry, rather than chemically mediated mechanical changes.

Min *et al.* (2009) developed a thermal–hydrological–mechanical–chemical model of fracture permeability, including dissolution-like compaction of the fracture surfaces, based on the experimental results for Arkansas Novaculite. They applied the model to explain the effects of temperature on permeability observed during the Terra Tek *in situ* heated block experiment (Fig. 8.19). The model did not explicitly include the time-dependent pressure solution process, but instead considered the steady-state aperture value achieved for a given load and temperature (Min *et al.* 2009). The model explains the additional temperature-dependent permeability reduction as being caused by pressure solution of highly stressed surface asperities, causing additional irreversible fracture closure. Figure 8.19B shows that the agreement between modeling (lines) and observations (filled circles) is excellent for all loading stages except for the end point, that is, point 6. The modeling shows irreversible fracture closure as a result of shortening of fracture surface asperities by pressure solution, while the experiments indicate that the fracture rebounds to the preheating conditions. This might be a result of the fact that the *in situ* block was fixed to underlying rock at the bottom, which was not considered in the modeling.

The G-tunnel *in situ* heated block experiment conducted in welded fractured tuff close to Yucca Mountain, Nevada, provides another example of temperature-dependent fracture closure (Fig. 8.20) (Zimmerman *et al.* 1985). Fracture flow testing was first conducted under a number of isothermal stress cycles, and then the block was exposed to heat. Figure 8.20B presents the interpreted hydraulic conducting aperture (estimated from the flow rate) versus effective normal stress from the flat-jack loading of the block. For loading under isothermal conditions, the stress–aperture function resembles that back-calculated from the niche experiments (Fig. 8.13A) and other large-scale fracture experiments on clean open fractures (Fig. 8.5A–C). That is, the fracture appears to reach an

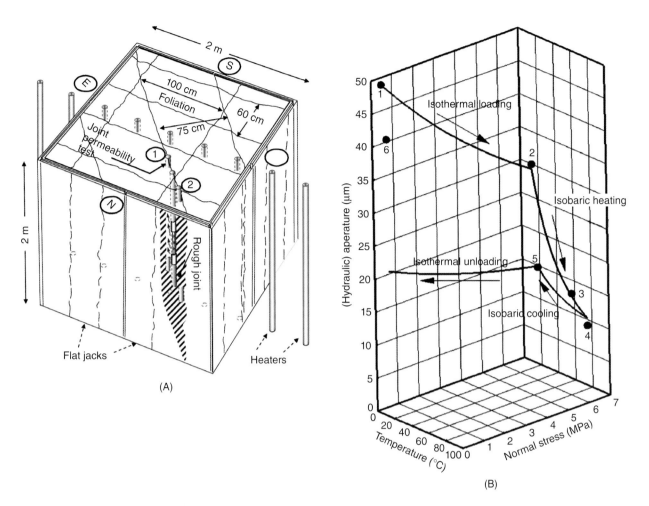

Fig. 8.19. (A) Perspective of the Terra Tek *in situ* block experiment (Adapted from Hardin *et al*. 1982 and (B) comparison of calculated (lines) and measured (filled circles) evolution of hydraulic conducting fracture aperture with stress and temperature at the Terra Tek *in situ* heated block experiment. (Min *et al*. 2009.) The calculated evolution shown by solid lines considers pressure solution of fracture surface asperities to be the mechanism causing fracture closure during temperature increase under constant stress.

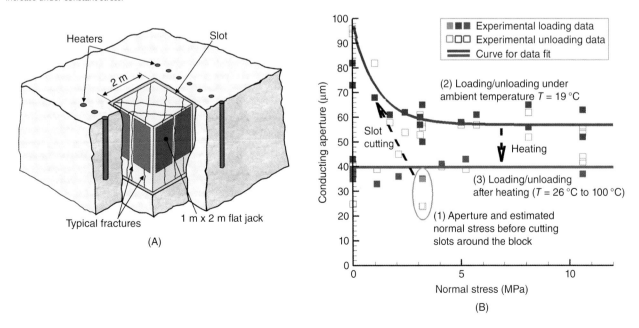

Fig. 8.20. (A) Perspective of G-tunnel heated block experiment (Adapted from Zimmerman *et al*. 1985) and (B) hydraulic conducting aperture as a function of normal stress evaluated from the G-tunnel *in situ* block experiment. The data from Zimmerman *et al*. (1985) are separated in the sequential steps showing additional fracture closure as a result of heating. (Rutqvist & Tsang 2012.) (*See color plate section for the color representation of this figure*.)

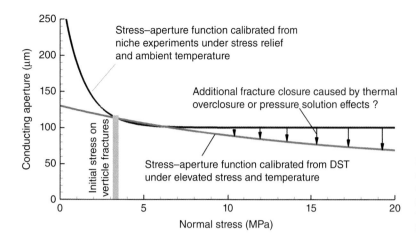

Fig. 8.21. Comparison of stress-versus-aperture relationships derived by model calibration from Yucca Mountain niche-excavation experiments (dashed curve) and the drift scale test (solid curve) and possible chemically mediated fracture closure as a result of rock-mass heating during the drift scale test.

irreducible aperture at stresses above 5 MPa. Figure 8.20B also shows a marked reduction in aperture after heating and a relatively modest temperature increase, from the ambient 19 °C to about 40 °C. This heating appears to have caused an irreversible decrease in hydraulic conducting aperture.

Considering the observations at the G-tunnel *in situ* block experiment, it is very likely that similar permeability reduction took place during the 4-year rock-mass heating at the Yucca Mountain drift scale test. This may explain the differences in the back-calculated stress–aperture relationships in Figures 8.13A and 8.18B: At the drift scale test, temperature and stress increased simultaneously, and it is difficult to separate the pure thermal–mechanical responses from chemically mediated thermal–mechanical responses. Therefore, the back-calculated stress–aperture function in Figure 8.18B may implicitly include additional fracture closure caused by thermally and chemically mediated changes. This possibility is illustrated in Figure 8.21, which presents aperture-versus-stress functions back-calculated from both field experiments. In Figure 8.21, the functions are plotted for vertical fractures at an equivalent initial normal stress and aperture, which illustrates how the temperature increase at the drift scale test might have caused additional fracture closure. Such additional fracture closures may have been the result of pressure solution and compaction of stressed fracture surface asperities, or so-called thermal overclosure.

APPLICATION TO GEOENGINEERING ACTIVITIES AND POTENTIAL IMPLICATIONS FOR CRUSTAL PERMEABILITY

One important use of stress–permeability functions derived from field tests is to provide reliable input data to numerical models for predictions of thermal–hydrological–mechanical behavior associated with geoengineering activities, such as geologic nuclear waste disposal, geothermal energy extraction, hydrocarbon production, and geologic carbon sequestration. For example, in the Yucca Mountain project for geologic nuclear waste disposal (Rutqvist & Tsang 2012),

stress–permeability relationships for different rock units were determined by back analysis from underground Yucca Mountain experiments described in Sections "The Manitoba underground rock laboratory TSX tunnel experiment" and "Model calibration against the Yucca Mountain drift scale test." These stress–permeability relations were then part of the input to a forward analysis for predicting the long-term behavior of a heat releasing nuclear waste repository (Rutqvist & Tsang 2003, 2012; Rutqvist 2004). Similarly, the stress–permeability function back-calculated from the TSX experiment described in Section "The Manitoba underground rock laboratory TSX tunnel experiment" was applied in a forward analysis related to the excavation disturbed zone evolution around a nuclear waste repository deposition tunnel (Nguyen *et al.* 2009).

This approach could also be used for other applications, such as geologic sequestration of carbon and geothermal energy extraction. *In situ* data from experiments on faults would be particularly useful for predicting the potential for leakage along faults, which is an important issue related to the attempts of sequestering carbon in deep sedimentary formations (Rutqvist 2012; Rinaldi *et al.* 2013). Related to geothermal energy extraction, enhanced geothermal systems involve stimulation of a rock volume to enhance permeability for an economic energy production (Tester 2006). The approach favored in current enhanced geothermal system projects is the so-called hydroshearing involving the injection of water to cause existing fractures to shear and dilate. In such a case, the injection pressure is kept below the fracturing pressure but sufficiently high to cause fractures to fail in shear. Hydroshearing can permanently enhance the permeability of natural fractures that, in theory, should remain open because of fracture self-propping (due to surface roughness) even after the stimulation period ends and fluid pressure is reduced. In this context, the borehole hydraulic jacking described in Section "Borehole injection tests" can be used to test the rock mass for potential permeability increase that can be achieved on individual fractures. During the actual stimulation an increase in permeability might be achieved similar to that shown in Figure 8.4 and will depend on the initial permeability as observed in Figure 8.12. The

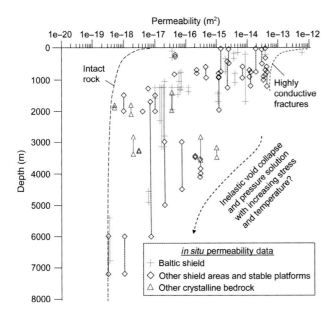

Fig. 8.22. Compilation of permeability measurements in boreholes in crystalline bedrock (from Juhlin *et al.* 1998) with added schematic of upper and lower limits of permeability related to mechanical and chemomechanical behavior. (*See color plate section for the color representation of this figure.*)

potential permeability increase that can be achieved can also be inferred from Figure 8.14, where the permeability values to the left represent intact rock or a rock mass with completely closed fractures, whereas the highest permeabilities to the right represent highly conductive fractures that might have been propped open by shear dilation. If the initial average permeability is somewhere around 1×10^{-15} m^2, and if the maximum permeability is about 1×10^{-13} m^2 (as shown in Figure 8.14), then at best a two orders of magnitude increase is feasible.

Finally, the review presented in this chapter on *in situ* stress–permeability relations as well as effects of thermal and chemical coupling may be relevant to crustal permeability distribution with depth. Figure 8.22 shows a compilation of permeability data from crystalline bedrock to a depth of 7000 m (Juhlin *et al.* 1998). At shallow depth, the variability of permeability shown in Figure 8.22 is similar to that of Figure 8.14, that is, about six orders of magnitude. Such variability in permeability is also observed in a recent comprehensive compilation of permeability data by Ranjram *et al.* (2015). Again, in Figure 8.22, the minimum permeability values to the left are similar to stress–permeability functions for intact rock (e.g., Rutqvist & Stephansson 2003), whereas maximum permeability values to the right correspond to highly transmissive fractures that could be propped open by shear dilation. Figure 8.22 displays few data at the greatest depths, but those available show decreasing permeability with depth toward intact rock permeability. At 7000 m depth, the lithostatic stress would be about 200 MPa, that is, on the order of the uniaxial compressive strength of granite. Large voids may not stay open; that is, an open void in a fracture as shown in Figure 8.2 is likely

to collapse under high ambient stress. However, laboratory experiments by Durham & Bonner (1994) on a fracture in granite propped open by slight shear offset showed that the fracture remained open and conductive even at a very high normal stress of 160 MPa. This indicates that fractures could stay open even at great depths, leading to permeability higher than that of the intact rock. However, at these great depths, high temperatures and stress may affect the long-term behavior, and chemically mediated mechanical processes such as pressure solution and associated creep could further compact fractures. Still, significant permeability can persist at very great depths, as inferred from hydrothermal modeling and the progress of metamorphic reactions (Ingebritsen & Manning 2010), and from seismicity (Townend & Zoback 2000).

CONCLUDING REMARKS

A number of field observations and experiments on stress-induced permeability changes in fractured rock are reviewed here to summarize our general understanding of the relationship between stress and permeability in fractured rock, and to describe the use of such field data and model calibration to back-calculate the stress–permeability functions for fractured rock units. Data presented show how the stress–permeability relationship depends strongly on local *in situ* conditions such as shear offset or fracture fillings. The various model calibrations presented here suggest that while coupled hydromechanical behavior in fractured rock is very complex and cannot be predicted in detail at every location of the rock mass, the overall rock-mass stress–permeability relationship can be estimated and bounded. Model calibration of the stress–permeability relationship against field data of stress-induced permeability changes has the advantage of eliminating issues related to unrepresentative sampling and sample disturbance, issues usually associated with laboratory testing on drill cores.

However, as highlighted in this chapter, model calibration of stress–permeability relationships against field experiments involving simultaneous stress and temperature changes may be affected by additional fracture closure, a phenomenon that has been described as thermal overclosure, due to thermal–mechanical changes in the fracture surface geometry and/or to chemically mediated fracture closure as a result of the pressure solution of stressed fracture surface asperities. Model calibration against field experiments may implicitly account for such effects, but the relative contribution of purely mechanical versus thermal or chemically mediated mechanical changes is difficult to isolate. It may well be that the observed temperature-dependent fracture closure results from a combination of the two proposed mechanisms. Therefore, further laboratory and *in situ* experiments are needed to increase the knowledge of the true mechanisms behind thermally driven fracture closure, and to further assess the importance of chemically mediated mechanical changes on the long-term evolution of fractured rock permeability. Such thermal–hydrological–mechanical–chemical effects on permeability are important to consider when estimating the long-term permeability evolution associated with geoengineering activities, such as geologic nuclear waste disposal, and can also explain crustal permeability distribution with depth.

ACKNOWLEDGMENTS

This chapter was prepared with funds from the Swedish Radiation Safety Authority (SSM) to the Lawrence Berkeley National Laboratory through the US Department of Energy Contract No. DE-AC02-05CH11231. Technical review comments of the initial manuscript by Victor Vilarrasa and editorial review by Dan Hawkes of the Lawrence Berkeley National Laboratory are greatly appreciated. Technical and editorial reviews for the journal publication by Steven Micklethwaite, University of Western Australia, and by Steven Ingebritsen, U.S. Geological Survey, substantially improved the manuscript and are greatly appreciated by the author.

Static permeability

CHAPTER 9

Static permeability

TOM GLEESON AND STEVEN E. INGEBRITSEN

[1] Department of Civil Engineering, University of Victoria, Victoria, BC, Canada; [2] U.S. Geological Survey, Menlo Park, CA, USA

Key words: static permeability, sedimentary rocks, igneous rocks, metamorphic rocks, brittle-ductile transition

The chapters broadly categorized as dealing with static permeability include contributions related to sediments and sedimentary rocks (two chapters) and igneous and metamorphic rocks (four chapters). Volcanic, sedimentary, plutonic, and metamorphic rocks represent about 9%, 73%, 7%, and 11%, respectively, of the exposed continental crust (Wilkinson *et al.* 2009). This book includes representative publications concerning each of these broad categories. Extensive studies and compilations of permeability in sedimentary rock exist in the petroleum geology literature (e.g., Ehrenberg & Nadeau 2005) and industry archives, but this is not a large emphasis here. Much of the data in petroleum industry archives is not publically disclosed, which limits its utility for study, analysis, and modeling.

In contrast to the near-surface crust, the bulk of the continental crust is dominated by metamorphic rocks (approximately 90% of crustal volume despite only approximately 11% of surface exposures) rather than sedimentary rock (only a few percent of crustal volume despite approximately 73% of surface exposures) (Fig. 9.1). Much less is known about the permeability of the predominantly metamorphic deeper crust compared with the near-surface crust, as echoed in this book by the lack of well-test data from the deeper crust and differing perspectives on the likelihood of Darcian flow below the brittle–ductile transition (BDT) (Yardley 2009; Stober & Bucher, this book; Connolly & Podladchikov, this book).

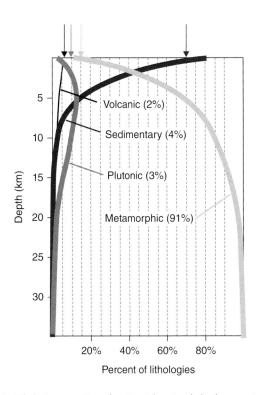

Fig. 9.1. Lithologic composition of continental crust with depth; arrows indicate proportions of outcrop at the Earth's surface from Food and Agricultural Organization (FAO) global maps. (From Wilkinson *et al.* 2009.)

SEDIMENTS AND SEDIMENTARY ROCKS

Luijendijk & Gleeson (this book) explore porosity–permeability relations for clastic sediments (sand–clay mixtures) to a depth of 2 km, a depth restriction that largely avoids complications such as pressure solution and hydrothermal alteration. They find that the laboratory-scale k of natural sand–clay mixtures is best predicted as the geometric mean of the permeabilities of the sand and clay components and suggest that their algorithm could be applied to well-log data by using neutron and density logs to estimate clay content and porosity (Fig. 10.9). Daigle & Screaton (this book) explore a large laboratory-scale sediment-permeability data set (530 samples) from seafloor drill sites worldwide; the emphasis is on the relationship between

permeability (10^{-21} to 10^{-14} m^2) and porosity (0.2–0.8) for various subsets of the data. They find that anisotropy is modest (Fig. 11.8), that field and laboratory k measurements are quite consistent (Fig. 11.9), and that porosity–permeability trends seem to be maintained through burial and diagenesis to porosity <0.10, suggesting that extrapolation to significant depth is reasonable.

IGNEOUS AND METAMORPHIC ROCKS

Ranjram *et al.* (this book) explore large *in situ* data sets ($n = 977$ from Sweden, Germany, and Switzerland) to assess the depth dependence of the permeability of crystalline rocks in the shallow ($\leq$2.5 km) crust. The complete data set (Fig. 12.2) does not support a general k–z relation; however, some specific lithologies and tectonic settings display a statistically significant decrease of permeability with depth. Burns *et al.* (this book) demonstrate that regional groundwater flow can explain lower-than-expected heat flow in a thick sequence of highly anisotropic ($k_x/k_z \sim 10^4$) continental flood basalts (Columbia River Basalt Group). The limited *in situ k* data are compatible with a steep permeability decrease (approximately 3.5 orders of magnitude) at 0.6–0.9 km depth (Fig. 13.4) and approximately 40 °C (Fig. 13.5), possibly a result of low-temperature hydrothermal alteration; these authors note that substantial k decreases at similar temperatures have also been observed in the volcanic rocks of the Cascade Range and at Kilauea Volcano, Hawaii. Geochemical analyses and numerical modeling by Pepin *et al.* (this book) suggest that the Truth or Consequences, New Mexico, hot-spring system is supported by deep (2–8 km) circulation in permeable (10^{-12} m^3) crystalline basement rocks. Circulation of meteoric fluids to these depths may occur through "hydrologic windows" in overlying lower permeability units. Their Figure 14.1 compares basement permeabilities inferred for Rio Grande Rift hydrothermal systems with published k–z curves. Finally, Stober & Bucher (this book) provide a conceptual, qualitative discussion of fluid flow in crystalline rocks above the BDT; their contribution includes observations of decreasing *in situ k* with depth in a single, deep borehole (Fig. 15.1).

PART II(A): Sediments and sedimentary rocks

How well can we predict permeability in sedimentary basins? Deriving and evaluating porosity–permeability equations for noncemented sand and clay mixtures

ELCO LUIJENDIJK[1] AND TOM GLEESON[2]

[1] *Geoscience Centre, Georg-August-Universität Göttingen, Göttingen, Germany;* [2] *Department of Civil Engineering, University of Victoria, Victoria, BC, Canada*

ABSTRACT

The permeability of sediments is a major control on groundwater flow and the associated redistribution of heat and solutes in sedimentary basins. While porosity–permeability relationships of pure clays and pure sands have been relatively well established at the laboratory scale, the permeability of natural sediments remains highly uncertain. Here, we quantify how well existing and new porosity–permeability equations can explain the permeability of noncemented siliciclastic sediments. We have compiled grain size, clay mineralogy, porosity, and permeability data on pure sand and silt ($n = 126$), pure clay ($n = 148$), and natural mixtures of sand, silt, and clay ($n = 92$). The permeability of pure sand and clay can be predicted with high confidence ($R^2 \geq 0.9$) using the Kozeny–Carman equation and empirical power law equations, respectively. The permeability of natural sediments is much higher than predicted by experimental binary mixtures and ideal packing models. Permeability can be predicted with moderate confidence ($R^2 = 0.26$–0.48) and a mean error of 0.6 orders of magnitude as either the geometric mean or arithmetic mean of the permeability of the pure clay and sand components, with the geometric mean providing the best measure of the variability of permeability. We test the new set of equations on detailed well-log and permeability data from deltaic sediments in the southern Netherlands, showing that permeability can be predicted with a mean error of 0.7 orders of magnitude using clay content and porosity derived from neutron and density logs.

Key words: permeability, sediments

INTRODUCTION

Fluid flow in sedimentary basins and the associated redistribution of heat and solutes depends strongly on the permeability of sediments. However, data on the permeability of sediments are scarce and tend to be restricted to permeable units that form shallow aquifers or deeper geothermal or hydrocarbon reservoir units (Neuzil 1994; Ehrenberg & Nadeau 2005; Gleeson *et al.* 2011). Permeability of pure granular material or clays can be relatively well approximated using porosity–permeability equations that have been calibrated to experimental data (Mesri & Olson 1971; Bourbie & Zinszner 1985; Revil & Cathles 1999). However, the permeability of mixed sand, silt, and clay materials that form the bulk of sediments in most sedimentary basins remains difficult to predict.

The high variability of permeability of natural sand and clay sediments is illustrated by the permeability data shown in Figure 10.1. The relatively well-constrained porosity–

permeability trends for pure quartz sand and the clay minerals kaolinite, illite, and smectite contrast with the high variability of permeability of sand–clay mixtures based on shallow (<2 km deep) samples from the Roer Valley Graben in the Netherlands and the Beaufort-Mackenzie Basin in Canada (Heederik 1988; Hu & Issler 2009; Luijendijk 2012).

A number of previous studies that predominantly focus on clay-rich lithologies have found a linear correlation between log-transformed permeability and clay content (Yang & Aplin 1998; Dewhurst *et al.* 1999a; Schneider *et al.* 2011). In contrast, Koltermann & Gorelick (1995) and Revil & Cathles (1999) derive equations for the porosity and permeability of ideal mixtures of sand and clay that predict a rapid decrease of permeability with increasing clay content, with a minimum at clay contents of approximately 40%. These two models create very different predictions of permeability. However, they have each only been tested on a limited range of natural sediments.

Crustal Permeability, First Edition. Edited by Tom Gleeson and Steven E. Ingebritsen.
© 2017 John Wiley & Sons, Ltd. Published 2017 by John Wiley & Sons, Ltd.
Companion Websites: www.wiley.com/go/gleeson/crustalpermeability/
http://crustalpermeability.weebly.com/

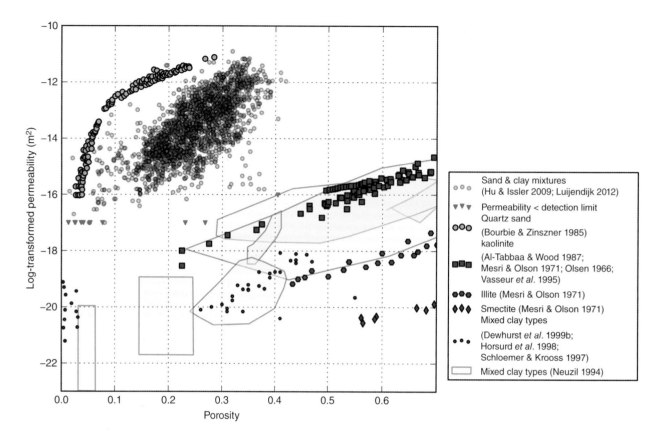

Fig. 10.1. Porosity and permeability data of sand–clay mixtures and pure sands and clays. (*See color plate section for the color representation of this figure.*)

The Koltermann & Gorelick (1995) and Revil & Cathles (1999) models are based on mainly laboratory-scale binary mixtures of sand and clay (Marion 1990; Knoll & Knight 1994), while the log-linear relation between clay content and permeability is mainly based on clay-rich sediments. Therefore, the extent to which porosity–permeability equations can be used to predict permeability of natural sediments at larger scales remains uncertain.

A number of studies have demonstrated that permeability can be successfully predicted using data on pore-size distributions (Marshall 1958; Yang & Aplin 1998; Dewhurst *et al.* 1999b; Schneider *et al.* 2011). However, such data are rarely available, and thus, pore-size distributions presently offer little opportunity to characterize sediment permeability at larger scales (Walderhaug *et al.* 2012).

Our objective was to quantify how well permeability of sand–clay mixtures can be predicted using simple mixing models and information on porosity, grain size, and clay content that are frequently available in sedimentary basins or can be inferred from sample descriptions or well-log data. We first evaluate how well a number of existing porosity–permeability equations such as the Kozeny–Carman equation fit a compilation of data on the permeability of pure sands and clays. We then use existing and new data sets of mixed siliciclastic sediments to evaluate how permeability relates to the permeability of the pure sand

and clay end members. We also use these data sets to evaluate existing permeability equations and develop and test a new approach based on the power mean. Previous studies have used power mean equation to explore the effective permeability of heterogeneous distributions of permeability at reservoir scales (Gómez-Hernández & Gorelick, 1989; McCarthy 1991; de Dreuzy *et al.* 2010), but this approach has to our knowledge not yet been combined with permeability equations of pure sands and clays to study the permeability of sediment mixtures at sample scale. We focus most of our analysis on core-plug (0.1 m) scale, which is the scale of most of the permeability data available in the literature. In the last section, we combine the power mean porosity–permeability equation with well-log data to scale-up permeability estimates from core plug to formation (50 m) scale.

Our analysis focuses exclusively on noncemented sediments. Note that throughout this chapter, the term sand is used to denote any granular siliciclastic material, that is, sands and silts. Clay refers to clay minerals. For data sets where there was no direct information on the percentage of clay minerals in sediment mixtures, we used a grain size cutoff of 2 μm to estimate clay content, which follows the cutoff values used for the main data sets that were included in our analysis (Heederik 1988; Dewhurst *et al.* 1999a).

DATA AND METHODS

We apply existing and new equations for the permeability of pure sand and clay and sand–clay mixtures using several permeability data sets. The data sets consist of a compilation of published experimental and field data on the permeability of pure sands and clays and a combination of published and newly compiled data on the permeability of sand–clay mixtures from sedimentary basins.

Permeability data sets

Pure sands and clays

Permeability data for pure quartz sand were obtained from Bourbie & Zinszner (1985), who report permeability data for the Oligocene Fontainebleau sandstone in the Paris Basin. Permeability was measured using a falling-head permeameter. The porosity varies from 2% to 30% as a function of burial depth. The median grain size is constant for all samples at 250 μm.

Data on the permeability of pure clays were obtained from several experimental studies in which porosity and permeability were measured during compaction experiments (Olsen 1966; Mesri & Olson 1971; Al-Tabbaa & Wood 1987; Vasseur *et al.* 1995). Permeability was measured in these studies using either consolidation (Olsen 1966; Mesri & Olson 1971), steady-state flow (Vasseur *et al.* 1995), or falling-head tests (Al-Tabbaa & Wood 1987). Al-Tabbaa & Wood (1987) measured permeability normal and perpendicular to the normal stress, while for the remainder of studies the anisotropy of permeability was not discussed. Descriptions of experimental procedures suggest that the measured permeability likely represents permeability parallel to the applied stress for the Olsen (1966) and Vasseur *et al.* (1995) data sets, whereas for the Mesri & Olson (1971) data set, the exact test setup is unknown.

Sand–clay mixtures

The relation of the permeability of sand–clay mixtures to porosity and clay content was examined using two data sets: one consisting of Cenozoic shallow marine sands in the Roer Valley Graben in the southern Netherlands and the second consisting of unconsolidated marine clays and silts of the Eocene London Clay formation in southeast England. While large compilations of permeability data have been published (Neuzil 1994; Nelson & Kibler 2001; Ehrenberg & Nadeau 2005; Wilson *et al.* 2008; Yang & Aplin 2010), the Roer Valley Graben and London Clay data sets are to our knowledge the only available data sets that combine detailed porosity, permeability, grain size, and clay content data, as well as some constraints on the mineralogy of the clay fraction. Both data sets consist of sediments that were buried less than 2 km deep and, therefore, have not been affected significantly by diagenesis.

The first data set from the Roer Valley Graben consists of 67 core samples from the geothermal exploration well AST-02 (Heederik 1988). The samples were derived from the Paleocene Reusel Member and the Eocene/Oligocene Vessem Member, which both consist of shallow marine (deltaic) fine sand and silt deposits with low clay contents. Detailed permeability and grain size data were reported separately in industry reports (Jones 1987; Anonymous 1988) that are available on the website of the Dutch Geological Survey (http://www.nlog .nl). Porosity and permeability were measured on 0.1-m-long core plugs with a diameter of 0.025 m. Porosity was measured by helium porosimetry. Both horizontal and vertical permeabilities were measured using nitrogen as a flowing fluid, with a detection limit of 1.0×10^{-17} m^2. The grain size data are shown in Figure 10.2A. Clay mineralogy data were available for 24 samples in these members and an adjacent stratigraphic unit; see Table 10.1. Note that while the permeability data were derived from a geothermal exploration well, geothermal gradients in the area are moderate, approximately 35 °C km^{-1} (Luijendijk *et al.* 2011). The grain size, porosity, and permeability for this data set are available as supplemental information (Table SI), from the authors' web page (http://wwwuser.gwdg.de/ eluijen) and on http://www.figshare.com.

A second data set consists of four samples from the Eocene London Clay deposit (Dewhurst *et al.* 1999a). The London Clay contains sizeable fractions of silt and fine sand, with clay contents ranging from 56% to 67%. Compared with Dewhurst *et al.* 1999b, we did not use a number of samples with low clay contents due to insufficient grain size distribution data. The samples were compacted experimentally with pressures up to 33×10^6 Pa. Permeability was measured parallel to the applied stress using steady-state flow tests. Porosity and permeability were measured at various stages of experimental compaction of the four samples, resulting in a total of 25 porosity and permeability data. Information on the clay mineralogy of the London Clay was derived from Kemp & Wagner (2006); see Table 10.1 for a summary of the clay mineralogy data and Figure 10.2B for the grain size data.

We compare both natural sediment data sets to a third data set of experimental binary mixtures of kaolinite clay and quartz sand published by Knoll (1996), who measured porosity and permeability in seven samples consisting of homogeneous mixtures with a uniform grain size of 7×10^{-4} to 8×10^{-4} m. Permeability was measured using steady-state flow tests. The specific surface of the sand component was reported as 39 m^2 kg^{-1}.

In addition to the three main data sets, we use an additional data set of siliciclastic sediments from the Beaufort-Mackenzie Basin in Canada (Hu & Issler 2009) to explore the variation of permeability anisotropy in natural sediments. This data set contains $n = 2112$ porosity and permeability data of noncemented siliciclastic sediments from Cenozoic formations that were already shown in Figure 10.1. For $n = 224$ samples, both horizontal (bedding-parallel) and vertical permeability data were available. Detailed clay content and grain size data were not available for this data set. Sample descriptions show that lithology ranges from clay to coarse sand and predominantly consists of very-fine to medium-sized sand.

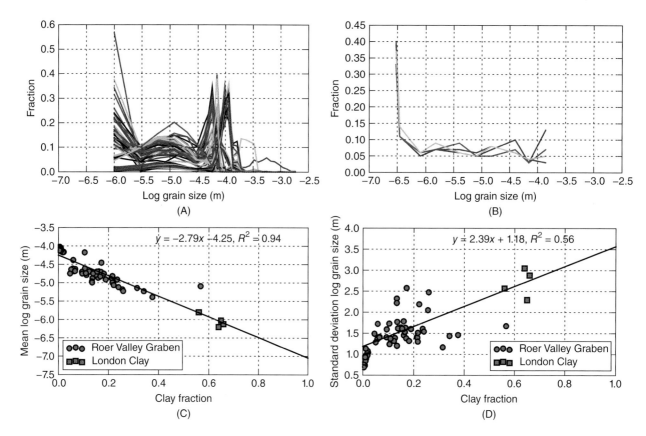

Fig. 10.2. Grain size distribution data for samples from (A) well AST-02 in the Roer Valley Graben (Heederik 1988) and (B) the London Clay data set (Dewhurst *et al*. 1999a). (C) Good correlation ($R^2 = 0.94$) between mean grain size and clay content suggests that independent information on clay content can be used to provide rough estimates of mean grain size. In general, the standard deviation of the distribution of grain size increases with increasing clay content, although clay content only explains 56% of the variation of the grain size distribution (D).

Table 10.1 Clay mineralogy data for Cenozoic sediments in the Roer Valley Graben (Heederik 1988) and the London Clay formation (Kemp & Wagner 2006).

Data set	Sample id.	Depth	Unit	Kaolinite	Illite	Smectite	Mixed layer illite–Smectite	Chlorite
Roer Valley Graben	1	875	Breda Formation	0.20	0.50	0.20	0.10	0.00
	2	875	Breda Formation	0.20	0.50	0.15	0.15	0.00
	3	875	Breda Formation	0.20	0.60	0.10	0.10	0.00
	4	876	Breda Formation	0.15	0.70	0.10	0.05	0.00
	5	876	Breda Formation	0.25	0.45	0.20	0.10	0.00
	6	876	Breda Formation	0.20	0.50	0.20	0.10	0.00
	7	1204	Voort Sand Member	0.00	0.50	0.40	0.10	0.00
	8	1201	Voort Sand Member	0.00	0.45	0.45	0.10	0.00
	9	1217	Voort Sand Member	0.00	0.65	0.25	0.10	0.00
	10	1213	Voort Sand Member	0.00	0.40	0.45	0.15	0.00
	11	1210	Voort Sand Member	0.00	0.60	0.30	0.10	0.00
	12	1458	Rupel Clay Member	0.40	0.50	0.05	0.05	0.00
	13	1453	Rupel Clay Member	0.35	0.45	0.10	0.10	0.00
	14	1462	Rupel Clay Member	0.30	0.50	0.10	0.10	0.00
	15	1466	Vessem Member	0.35	0.35	0.15	0.15	0.00
	16	1479	Vessem Member	0.15	0.35	0.30	0.20	0.00
	17	1476	Vessem Member	0.40	0.45	0.10	0.05	0.00
	18	1474	Vessem Member	0.45	0.45	0.05	0.00	0.05
	19	1484	Vessem Member	0.05	0.40	0.40	0.15	0.00
	20	1488	Vessem Member	0.05	0.20	0.65	0.10	0.00
	21	1491	Vessem Member	0.00	0.20	0.65	0.15	0.00
	22	1496	Vessem Member	0.00	0.30	0.60	0.10	0.00
	23	1492	Vessem Member	0.00	0.25	0.65	0.10	0.00
	24	1502	Vessem Member	0.00	0.30	0.65	0.05	0.00
London Clay	Arithmetic mean			0.15	0.44	0.30	0.10	0.00
	Min. k			0.00	0.20	0.65	0.15	0.00
	Max. k			0.45	0.45	0.05	0.00	0.05
	Arithmetic mean			0.25	0.26	0.38	n/a	0.12
	Min. k			0.09	0.17	0.65	n/a	0.09
	Max k			0.33	0.34	0.15	n/a	0.19

Permeability equations

Kozeny–Carman equation

Permeability of granular material such as sand and silt was calculated using the Kozeny–Carman equation (Kozeny 1927; Carman 1937, 1956):

$$k = \frac{1}{\rho_w \rho_s C S_s^2} \frac{n^3}{(1-n)^2}, \qquad (10.1)$$

where C is a constant, ρ_w and ρ_s are the density of the fluid and solid phase ($kg\,m^{-3}$), S_s is the specific surface ($m^2\,kg^{-1}$) of the solid phase, and n is porosity. The Kozeny–Carman equation was derived from the Hagen–Poiseuille equation (Poiseuille 1844) and calculates flow through a series of cylindrical pipes that represent the connected pore space. The empirical Kozeny–Carman constant C was introduced by Carman (1937) to account for the tortuosity of flow paths and was reported to equal five for uniform spheres (Carman 1956). Previous authors have shown that, while successful at high porosities, the Kozeny–Carman equation overestimates permeability at lower values of porosity (Bourbie & Zinszner 1985), perhaps due to the disproportional closure of pore throats at low porosities (Doyen 1988). Mavko & Nur (1997) demonstrated that permeability can be more successfully predicted by replacing total porosity (n) in the Kozeny–Carman equation with the effective or connected porosity $n_e = n - n_t$, where n_t is the percolation threshold.

For granular material, the specific surface (S_s) in Eq. 10.1 was calculated as a function of the grain size distribution (Holdich 2002; Chapuis & Aubertin 2003):

$$S_s = \frac{6.0}{\rho_s} \sum \frac{f}{D}, \qquad (10.2)$$

where f is the mass fraction of grain size D (m). Previous research has shown that this equation can estimate the specific surface of a range of sediments with an error of 20% or less and that using the specific surface provides better predictions of permeability than modified formulations of the Kozeny–Carman equation that use a representative grain size instead of specific surface (Chapuis & Aubertin 2003; Chapuis 2012).

For cases where detailed grain size distribution data were absent, but median grain size was known, the specific surface was calculated assuming a log-normal distribution of grain size. A log-normal distribution is a good first estimate for the grain size distribution of sediments (Tanner 1964; Folk 1966; Weltje & Prins 2003).

No data were available on the specific surface or the grain size distribution for the Fontainebleau sand data set. Instead, the value of specific surface S_s of the Fontainebleau sandstone was calibrated. The median grain size of the Fontainebleau sandstone is known (250 μm, Bourbie & Zinszner 1985), and therefore, the corresponding grain size distribution could be calculated to ensure that the S_s value was realistic. Typical values of specific surface for kaolinite, illite, and smectite used for

estimating permeability using the Kozeny–Carman equation were estimated as 14×10^3, 116×10^3, and $600 \times 10^3\,m^2\,kg^{-1}$ (Mesri & Olson 1971; Ames et al. 1983).

Empirical power law equations

As noted by previous authors (Taylor 1948; Michaels & Lin 1954; Freeze & Cherry, 1979), the Kozeny–Carman equation is less successful in predicting the permeability of clays than sands. As an alternative, previous studies have proposed empirical porosity–permeability relationships, in which permeability is calculated as a log-linear or power law function of porosity or void ratio (Mesri & Olson 1971; Tavenas et al. 1983; Al-Tabbaa & Wood 1987; Tokunaga et al. 1998; Yang & Aplin 1998; Revil & Cathles 1999; Schneider Reece et al. 2012). Here, we test two proposed empirical power law equations.

Revil & Cathles (1999) and Tokunaga et al. (1998) suggest that clay permeability can be calculated as a power law function of porosity:

$$k = k_0 \left(\frac{n}{n_0} \right)^m, \qquad (10.3)$$

where k_0 is the permeability at a reference porosity n_0 (m^2) and m is an empirical coefficient. Following Revil & Cathles (1999), we choose a reference porosity of 0.5. Al-Tabbaa & Wood (1987), Mesri & Olson (1971), Tavenas et al. (1983), and Vasseur et al. (1995) suggest that permeability can be approximated as a power law function of the void ratio:

$$k = k_0 v^m, \qquad (10.4)$$

where k_0 is permeability at a void ratio of 1 (m^2), v is the void ratio, and m is an empirically determined coefficient. The void ratio (v) is the ratio of the volume of the void space to the volume of solids and is related to porosity (n) by $v = n/(1-n)$.

Ideal packing model

Given the strong contrast between the permeability behaviors of granular material (sands and silts) and clays, previous workers have developed methods to calculate sediment permeability by treating sediments as binary mixtures of sand or silt and clay and estimating permeability from the permeability of the sand and clay components. Revil & Cathles (1999) developed a model based on ideal packing of sand–clay mixtures in which clays are dispersed homogeneously in the sand pores. Following Revil & Cathles (1999) and Revil (2002), permeability of sand–clay mixtures is calculated as

$$k = k_{sd*}^{1-w/n_{sd}} k_{cfs}^{w/n_{sd}} \quad 0 \leq w \leq n_{sd}, \qquad (10.5)$$

$$k = k_{cl*} w^{3/2} \quad w > n_{sd}, \qquad (10.6)$$

$$k_{cfs} = k_{cl*} n_{sd}^{3/2}, \qquad (10.7)$$

where w is the fraction of clay, k_{cl*} and k_{sd*} are the permeability of the sand and clay fraction of the sediment (m^2), k_{cfs} is the permeability of sand of which the pore space is completely filled by clay (m^2), and n_{sd} is the porosity of the sand fraction, that is, the theoretical porosity if one would remove all the clay from the sediment. Revil & Cathles (1999) and Revil (2002) used a modified Kozeny–Carman equation to calculate the permeability of the sand fraction (k_{sd*}) and a power law equation similar to Eq. 10.3 to calculate the permeability of the clay fraction (k_{cl*}). In both cases, the permeability is not calculated using the observed porosity of the sample (n), but using a theoretical porosity of the sand fraction or the clay fraction only. We refer to the porosity of the clay fraction as n_{cl}.

To evaluate the ideal packing model, we calculated permeability for two of the three mixed sediment data sets, the binary sand–kaolinite and the Roer Valley data sets. For the kaolinite–sand data set by Knoll (1996), we follow Revil & Cathles (1999) and use a value of $k_{sd*} = 4.4 \times 10^{-10}\,m^2$ and $k_{cl*} = 1.5 \times 10^{-15}\,m^2$ for the permeability of the sand and clay fraction. For the Roer Valley Graben data set, k_{sd*} and k_{cl*} were calculated using the Kozeny–Carman equation (Eq. 10.1) and power law–void ratio equation (Eq. 10.4), respectively. The porosity of the sand fraction (n_{sd}) was estimated to be 0.4 for the Roer Valley Graben data set based on observed porosity of clay-free sands in this data set. The porosity of the clay fraction (n_{cl}) was set to 0.2, given the 1500 m burial depth of this data set and clay compaction curves by Revil (2002). For the London Clay data set, we could not estimate the porosity of the sand and clay end members with sufficient confidence, as this data set contained no data close to the pure clay or sand end members. In addition, permeability was measured on experimentally compacted samples. For each new compacted permeability measurement, one would need to recalculate n_{sd} and n_{cl}, which cannot be easily derived from the observed porosity. Therefore, this data set was not used to evaluate the ideal packing model.

Power mean model

As an alternative to the ideal packing model, we estimate the permeability of sediment mixtures using the geometric, arithmetic, and harmonic mean of the clay and sand or silt components. Warren & Price (1961) have shown that the effective permeability of randomly distributed components is equal to the geometric mean of the components, which for a random mixture of sand and clay yields:

$$\log(k) = w \log(k_{cl}) + (1 - w) \log(k_{sd}), \tag{10.8}$$

where w is the fraction of clay, and k_{cl} and k_{sd} are the permeability of the sand and clay fraction of the sediment (m^2). Note that, in contrast to k_{sd*} and k_{cl*} in the ideal packing model (Eqs 10.5–10.7), the permeability of the clay and sand fractions are based on the observed porosity (n) of each sample, instead of the porosity of the sand and clay end members. If the clay is distributed in a laminar manner in a sand matrix, the effective permeability for flow parallel to the layers is given by the arithmetic mean, and the effective permeability for flow perpendicular to

a layered sequence is given by the harmonic mean (Cardwell & Parsons 1945). These three different means describe different relations between clay content and permeability, which can be generalized using the power mean or Holder mean of the sand and clay fractions:

$$k = (w k_{cl}^p + (1 - w) k_{sd}^p)^{(1/p)}, \tag{10.9}$$

where P is the power mean exponent, which can vary between -1 and 1. For $P = 1$, Eq. 10.9 is equal to the arithmetic mean, and for $P = -1$, Eq. 10.9 reduces to the harmonic mean. For $\lim_{P \to 0}$, Eq. 10.9 equals the geometric mean (Eq. 10.8).

We quantify which values of P are able to describe the permeability of sand–clay mixtures for two data sets with detailed porosity, permeability, clay content, and grain size distribution data (see "Sand–clay mixtures"). The range between the harmonic and arithmetic means ($-1 < P < 1$) is expected to capture the full variation of permeability in natural sand–clay mixtures. The harmonic mean results in a permeability that is close to that of the clay component and represents samples in which flow is dominated by the clay fraction. Conversely, the arithmetic mean represents samples in which flow is dominated by the most permeable fraction.

Evaluating permeability equations

We analyze the relation between the observed permeability of mixed siliciclastic sediments and the permeability of pure sand and clay end members using a new metric, the normalized permeability difference:

$$\Delta \log(k) = \frac{\log k - \log k_{cl}}{\log k_{sd} - \log k_{cl}}, \tag{10.10}$$

where $\Delta \log(k)$ is the normalized permeability difference, the difference between the observed permeability and the theoretical permeability of the pure clay component, normalized by the difference between the pure sand and clay components. Here, k denotes the observed permeability, and k_{sd} and k_{cl} are the permeability of sand and clay components, respectively (m^2). The permeability of pure sand and clay was calculated using Eqs 10.1 and 10.4, respectively, using the available porosity and grain size data for each sample. Specific surface (S_s) of the sand and silt fraction of each sample was calculated from the grain size distribution using Eq. 10.2.

The permeability of the clay component was calculated as the geometric mean of the permeability of each clay mineral, which is justified by experimental results from Mondol et al. (2008). Direct information on the clay mineralogy for each sample was not available. However, Heederik (1988) and Kemp & Wagner (2006) report clay mineralogy for the formations that were sampled to obtain the Roer Valley Graben and London Clay data sets, respectively (Table 10.1). The uncertainty of the permeability of the clay component (k_{cl}) was taken into account by calculating minimum and maximum estimates of the permeability using the clay samples with the highest and lowest kaolinite contents. A best estimate was calculated using the average clay mineral content.

The performance of the permeability equations was evaluated by calculating the coefficient of determination (R^2) and the mean absolute error (MAE) of the predicted permeability. The coefficient of determination was calculated as (Anderson-Sprecher 1994):

$$R^2 = 1 - \frac{\sum (k_{obs} - k_{pred})^2}{\sum (k_{obs} - \overline{k_{obs}})^2}, \qquad (10.11)$$

where k_{obs} is the observed and k_{pred} is the predicted permeability (m^2), respectively. Note that for nonlinear models such as those used in this chapter, R^2 can be negative if the variance of the prediction error is greater than the variance of the data set.

Estimating porosity and clay content using well-log data

Detailed core-plug measurements are typically only available for relatively permeable formations that are of interest for hydrocarbon or geothermal energy exploration. An alternative way to estimate permeability on larger scales is to utilize information from well logs. We explore how well core-scale permeability can be estimated from well-log data using estimates of porosity and clay content derived from well logs to calculate permeability for the Roer Valley Graben data set. We subsequently compare the calculated values and their uncertainty to the measured permeability. Porosity was calculated from the bulk density log using

$$n = \frac{\rho_m - \rho_b}{\rho_m - \rho_f}, \qquad (10.12)$$

where ρ_m and ρ_f are the density of the sediment matrix and pore fluid, respectively ($kg\,m^{-3}$), and ρ_b is the bulk density as measured using the gamma–gamma ray log tool. The matrix density and fluid density in the analyzed section of well AST-02 are $2660\,kg\,m^{-3}$ and $1025\,kg\,m^{-3}$ (Heederik 1988).

The clay content of sediments was estimated by comparing the porosity calculated from bulk density logs with neutron logs. The neutron log detects the presence of water in the formation. Water is located in the pore space but also occurs as part of the mineral formula of clay minerals and as water bound to the mineral surface. If the porosity is known, the percentage of clay

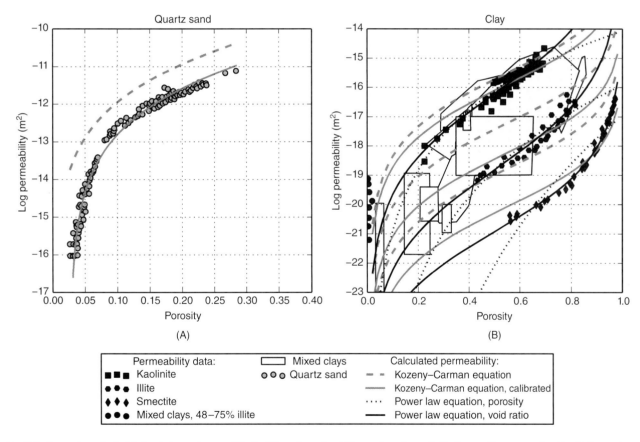

Fig. 10.3. Comparison of calculated and observed permeability for (A) quartz sand and (B) pure clays. For sands, the Kozeny–Carman equation (Eq. 10.1) reproduces the data well, but only when the equation includes a percolation threshold and the value of the specific surface is calibrated. For clays, the permeability data are closely matched when permeability is calculated as a power law function of the void ratio (Eq. 10.4). Data for sands were reported by Bourbie & Zinszner (1985). Permeability data for pure clays were obtained from Al-Tabbaa & Wood (1987), Mesri & Olson (1971), Olsen (1966), and Vasseur *et al.* (1995). The figure also shows data on mixed clay types from Schloemer & Kroos (1997) and Neuzil (1994) that were not used to calibrate the porosity–permeability equations. See Table 10.2 for the fit statistics of the permeability equations and Table 10.3 for calibrated parameter values. (*See color plate section for the color representation of this figure.*)

minerals, w, can be determined as

$$w = \frac{\text{NPHI} - n}{\text{NPHI}_{\text{clay}}}, \qquad (10.13)$$

where NPHI is porosity measured by a neutron log and NPHI$_{\text{clay}}$ is neutron porosity of a pure clay. Typical values of neutron porosity for kaolinite, illite, and smectite are 0.37, 0.30, and 0.44 neutron porosity units, respectively (Serra 1982; Rider 2002).

The permeability of the sand and clay components was calculated using Eqs 10.1 and 10.4, respectively. For the sand component, the specific surface was calculated from the grain size distribution, which was estimated using an empirical correlation between grain size and observed clay content shown in Figure 10.2C,D. For both the Roer Valley Graben and London Clay data sets, the median grain size decreases and the standard deviation of the log-transformed grain size distribution increases with increasing clay content. We use the linear correlation as best estimates for grain size distribution. These relations are likely to be slightly specific to these formations, although consistent with general trends in the grain size literature. For basins where such correlations are not available, several sources in the literature provide rough estimates of grain size distribution (Spencer 1963).

Given the high uncertainty of the correlation between clay fraction and grain size distribution (Fig. 10.2), all further calculations were performed using the lowest and highest values of the standard deviation of the log-transformed grain size in Figure 10.2 (0.7 and 3.0 m) as an uncertainty range.

RESULTS

Comparison predicted and observed permeability pure sands and clays

The comparison of permeability data in Figure 10.3 confirms that while the Kozeny–Carman equation can successfully predict the permeability of sands (Fig. 10.3A), it fails to predict the permeability of clays (Fig. 10.3B). The Kozeny–Carman equation is reasonably close for kaolinite ($R^2 = 0.51$, mean absolute error $\log(k) = 0.39$), but overpredicts permeability by an order of magnitude for the clay minerals illite and smectite. When the value of the Kozeny–Carman constant (C) is calibrated to the clay permeability data, the predicted values of permeability are much closer to the observed values. However, the equation still overestimates permeability at low porosity. In contrast, the empirical power law relation of permeability to the void ratio is able to closely match the observed values of permeability with a mean absolute error of log-transformed permeability ranging between 0.1 and 0.2 and a coefficient of determination (R^2) between 0.90 and 0.99 for kaolinite, illite, and smectite. The difference between observed and calculated permeability values and the calibrated model parameters is listed in Tables 10.2 and 10.3.

Table 10.2 Performance of permeability equations for pure clays and sands

Material	n	Equation	Equation number	Mean absolute error $\log(k)$	R^2
Sand	126	Kozeny–Carman	1	1.16	0.26
		Kozeny–Carman, calibrated	1	0.19	0.97
Kaolinite	79	Kozeny–Carman	1	0.39	0.51
		Kozeny–Carman, calibrated	1	0.22	0.82
		Power law, porosity	4	0.17	0.90
		Power law, void ratio	5	0.17	0.90
Illite	33	Kozeny–Carman	1	1.01	−0.65
		Kozeny–Carman, calibrated	1	0.30	0.81
		Power law, porosity	4	0.25	0.87
		Power law, void ratio	5	0.20	0.92
Smectite	36	Kozeny–Carman	1	1.29	−0.04
		Kozeny–Carman, calibrated	1	0.37	0.91
		Power law, porosity	4	0.43	0.86
		Power law, void ratio	5	0.10	0.99

As discussed by Tokunaga *et al.* (1998), experimental permeability data for pure clays for porosities lower than 0.2 are scarce. Comparing calculated permeabilities to data on natural mixed clay types (Neuzil 1994) and mudstones predominantly composed of illite (Schloemer & Krooss 1997) shows that neither the power law permeability–void ratio equation nor the Kozeny–Carman equation can match the data at low porosities. The porosity–permeability equations still underestimate permeability. However, the natural clays included in the Neuzil (1994) and Schloemer & Krooss (1997) data sets include a sizeable silt fraction and may therefore have a higher permeability than pure clays.

The permeability of the Fontainebleau sands shown in Figure 10.3A can be calculated with a mean absolute error of $\log(k)$ of 0.19 provided that both the value of specific surface (S_s) and the percolation threshold (n_t) are calibrated. When specific surface is estimated using a uniform grain size (i.e., $\sigma = 0$) and the median grain size of 250 μm reported by Bourbie & Zinszner (1985), permeability is overestimated by up to 1 order of magnitude. The calibrated value of the specific surface is 14.8 m^2 kg^{-1}. Following Eq. 10.2, this value of specific surface corresponds to a standard deviation of log-transformed grain size of 1.0, which conforms to the literature values for well-sorted sands. The misfit of the calculated permeability when using a uniform grain size shows the importance of taking into account grain size distributions for calculating permeability using the Kozeny–Carman equation (Chapuis & Aubertin 2003). The calibrated value of the percolation threshold n_t is 0.027.

Table 10.3 Calibrated parameter values for the permeability equations of pure sands and clays

Equation	Eq. number	Parameter	Units	Calibrated parameter values			
				Sand	Kaolinite	Illite	Smectite
Kozeny–Carman	10.1	S_s	$m^2\,kg^{-1}$	14.80			
		C	Dimensionless		12	51	100
		n_t	Dimensionless	0.027	0.0016	0.0025	0.0025
Power law, porosity	10.4	k_0	m^2		7.65×10^{-17}	1.53×10^{-19}	8.44×10^{-23}
		m	Dimensionless		6.82	9.65	17.02
Power law, void ratio	10.5	k_0	m^2		6.16×10^{-17}	1.54×10^{-19}	1.18×10^{-21}
		m	Dimensionless		3.61	3.58	3.01

Predicting the permeability of sand–clay mixtures

The permeabilities of natural sand–clay mixtures from the London Clay and the Roer Valley Graben data sets show strong correlations with porosity, clay content, and grain size (Fig. 10.4A–C). The permeabilities of the sand and clay fractions of each sample as calculated using Eqs 10.1, 10.2, and 10.4 are shown in Figure 10.4A. The difference between the permeability of the sand and clay fraction is six orders of magnitude, while the internal variation for the sand and clay components due to grain size distribution and clay mineralogy is two orders of magnitude. This illustrates that clay content is the dominant factor determining the permeability of noncarbonate sediments (Dewhurst *et al.* 1999b).

Figure 10.5 shows how the three data sets compare with the permeability of pure sand and clay at the same porosity. The three data sets show markedly different relations between clay content and permeability. The experimental homogeneous sand–clay mix by Knoll (1996) shows a rapid decline of permeability with increasing clay content. The London Clay shows similar low permeability values at clay contents of 50–70%. In contrast, deltaic sands from the Roer Valley Graben retain high values of permeability at clay contents up to 60%. Even at moderate clay contents, the permeability remains several orders of magnitude higher than the estimated permeability of the clay fraction (see also Fig. 10.4A).

Comparisons between the data sets and the ideal packing model and the power mean permeability equation are also shown in Figure 10.5 and model error statistics are shown in Table 10.4. The ideal packing model underestimates the permeability of the Roer Valley Graben data set by up to 2 orders of magnitude. The negative value of R^2 indicates that the variance of the model error is greater than the variance of the observed permeability data. The ideal packing model is much more successful in predicting the permeability of the London Clay data set. Note that due to the difficulty of estimating n_{sd} and n_{cl}, we could not calculate the model error of the ideal packing model for this data set.

The permeability of the Roer Valley Graben data set is much better predicted by the power mean model than by the ideal packing model. The modeled permeability is close to the observed permeability for either a power mean exponent (p)

of 0 or 1, which corresponds to the geometric and arithmetic mean, respectively (see Fig. 10.6B,C). The predictive power is moderate; coefficients of determination (R^2) of 0.26–0.48 show that approximately a quarter to half of the variance of the data set can be explained by the power mean equation with a fixed exponent of 0 or 1. The calculated value of the power mean exponent p for each sample is shown in Figure 10.6D. The mean value of p for the Roer Valley Graben data set is 0.01 and ranges from -0.25 to 0.8 (Fig. 10.6D).

In contrast, the power mean exponents for the London Clay data set all fall between the values for harmonic mean and geometric mean, with a mean of -0.39. Permeability is well predicted by the harmonic mean equation, with R^2 of 0.39 and a mean absolute error of $0.4\,m^2$ (see Fig. 10.6A and Table 10.4).

The lower permeability of the London Clay samples may be attributed in part to the fact that this data set represents permeability perpendicular to the normal stress, while permeability for the Roer Valley Graben data set was measured parallel to the subhorizontal bedding. The vertical permeability is likely to be lower. A compilation of anisotropy for siliciclastic sediments of the Beaufort-Mackenzie data set shows that the anisotropy (the ratio of horizontal to vertical permeability, k_h/k_v) in 90% of the core-plug samples lies between 0 and 10 (Fig. 10.7). Assuming that this database is representative of natural sand–clay mixtures, the vertical permeability of the Roer Valley Graben data set could be up to 1 order of magnitude lower than the horizontal permeability shown in Figure 10.4. A decrease of 1 order of magnitude would shift the normalized permeability difference values (Fig. 10.5) down by approximately 20%, which results in values that are still much higher than the experimental sand–kaolinite mixture or the London Clay data sets.

Predicting permeability using well logs

We used only well-log-derived estimates of porosity, grain size distribution, and clay content to calculate permeability for a section of well AST-02, from which the Roer Valley Graben data set was derived. We first derived clay content from neutron and density log data as explained in estimating porosity and clay content using well-log data. Figure 10.8 shows that the observed clay content for well AST-02 is best matched using an

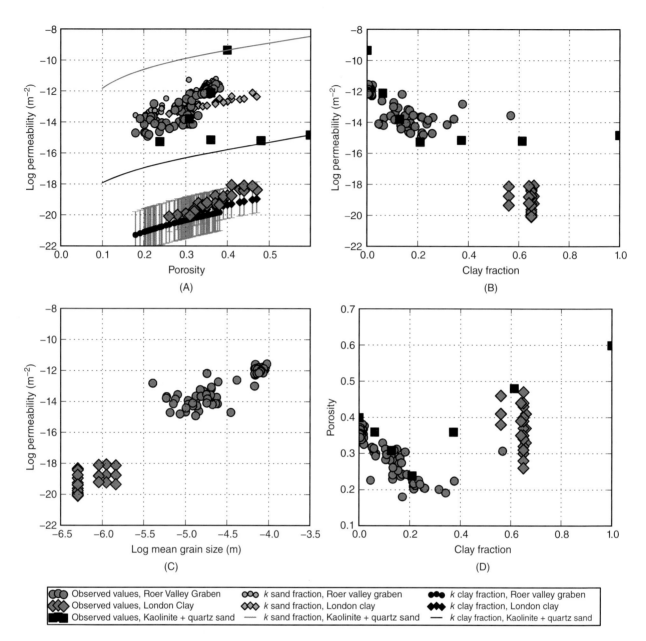

Fig. 10.4. Relation of permeability to (A) porosity, (B) clay content, and (C) mean grain size and (D) the relation between clay content and porosity for two data sets of natural sand–clay mixtures and one experimental data set that consists of a mixture of kaolinite and quartz sand with a uniform grain size. The data for natural sediments were derived from unconsolidated shallow marine sands in the Roer Valley Graben (Heederik 1988) and the London Clay in southeast England (Dewhurst *et al.* 1999a). The experimental data were reported by Knoll (1996). The calculated permeabilities of the clay and sand fraction of each sample of the Roer Valley Graben and London Clay data sets are also shown in (A). The permeabilities of the Roer Valley Graben and the London Clay data sets are relatively close to the calculated permeabilities of their sand and clay fractions, respectively. The error bars for the clay fraction reflect the uncertainty in the mineral composition. (*See color plate section for the color representation of this figure.*)

apparent neutron porosity of 0.42. The correlation coefficient is relatively low ($R^2 = 0.52$), possibly due to lithological variation, such as minor carbonate and organic matter contents or minor offsets between the depths of core samples and well logs. In addition, samples may contain a minor portion of nonclay minerals smaller than 2 μm, and, conversely, some clay particles may be larger than 2 μm.

A comparison between the well-log and core data and calculated and observed permeability is shown in Figure 10.9. The clay content and grain size calculated from well-log data match the observed data from core-plug samples and correctly show the transition between moderate clay content and small grain sizes of the Reusel Member to the clay-free sediments of the overlying Vessem Member. The calculated permeability curves

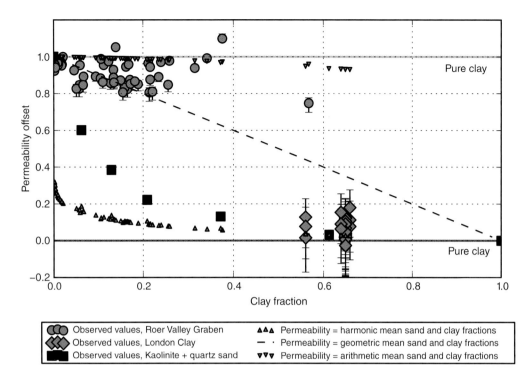

Fig. 10.5. Normalized permeability difference for two natural and one experimental data sets of sand–clay mixtures. The normalized permeability difference is calculated as the difference between the log-transformed permeability and the calculated permeability of the pure clay fraction, normalized by the difference in permeability between pure sand and clay. The three data sets show markedly different behavior. The fine-sand and silt-dominated samples of the Roer Valley Graben data set maintain relatively high permeability over the entire range of clay contents of 0–60%. In contrast, the experimental data set shows permeability decreasing rapidly with increasing clay content. The London Clay data set shows relatively low values of permeability that are close to the predicted permeability of the clay fraction, even though the samples also contain a silt fraction of 35–45%. The permeabilities calculated as the harmonic, geometric, or arithmetic mean of the sand and clay components are also shown for reference. Note that, due to the normalized log scale on the y-axis, the harmonic mean and arithmetic mean permeability cannot be shown as a single line.

show that the permeability calculated as the geometric mean of the sand and clay components is close to the observed values and shows a similar sensitivity to clay content.

A comparison of the error between the observed and calculated values of permeability is shown in Figure 10.6 and the model error statistics are shown in Table 10.5. Using observed data on porosity, clay content, and grain size, the permeability can be estimated with a mean absolute error of log k of 0.57–0.61 m^2 for the geometric and arithmetic mean equations. When only density and neutron log data are used, the permeability can still be predicted with a mean absolute error of 0.72–0.75 and an R^2 value of 0.23 and 0.33 for the geometric mean and arithmetic mean permeability, respectively (see Table 10.5).

The higher value of the coefficient of determination for the arithmetic mean permeability ($R^2 = 0.33$) compared with the geometric mean ($R^2 = 0.23$) is caused by a number of outliers (e.g., see Fig. 10.6A–C) and a higher variance of the model error for the geometric mean equation. However, the arithmetic mean model results in a much lower sensitivity of permeability to clay content than is observed in the data. The range of

Table 10.4 Performance of permeability equations for mixed sand and clay sediments

Permeability equation	Parameter	Roer Valley Graben data set	London Clay data set
Ideal packing	R^2	−4.8	n/a
Ideal packing	MAE	1.7	n/a
Harmonic mean	R^2	−33.5	0.39
Harmonic mean	MAE	6.0	0.40
Geometric mean	R^2	0.26	−8.4
Geometric mean	MAE	0.57	1.9
Arithmetic mean	R^2	0.48	−84.1
Arithmetic mean	MAE	0.61	5.8

log-transformed permeability predicted by the arithmetic mean equation is −13.5 to −12.4 m^2, while the range of the observed values of log(k) is −14.9 to −11.6 m^2 (see Fig. 10.9). While overall underpredicting permeability, the geometric mean predicts a similar variation in permeability as observed in the data, with a range of −15.6 to −12.4 m^2. Therefore, in this case, the geometric mean is a better measure for the variability of permeability in a siliciclastic formation.

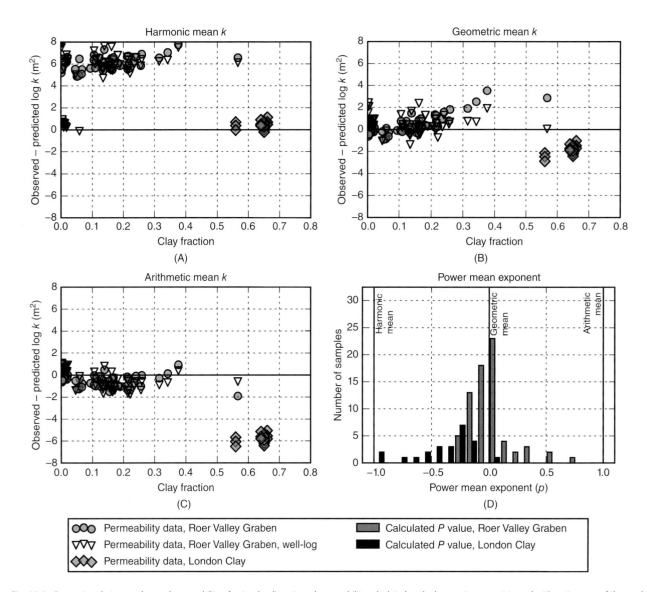

Fig. 10.6. Comparison between observed permeability of natural sediments and permeability calculated as the harmonic, geometric, and arithmetic mean of the sand and clay fractions (A–C) and calculated values of the power mean exponent (*P*) for each sediment sample (D). Calculated values of the power mean exponent (*P*) for each sediment sample. For the Roer Valley Graben data set, the calculated power mean exponent clusters around a value of 0 (D). For the London Clay data set, permeability is best predicted using a value of *P* that lies approximately halfway between the geometric and harmonic means, with a mean calculated value of *P* of −0.4.

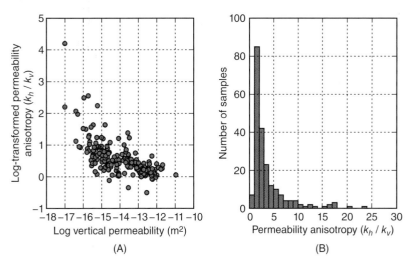

Fig. 10.7. Permeability anisotropy in $n = 224$ sediment samples from the Beaufort-Mackenzie Basin (Hu & Issler 2009); k_h and k_v denote horizontal and vertical permeability, respectively.

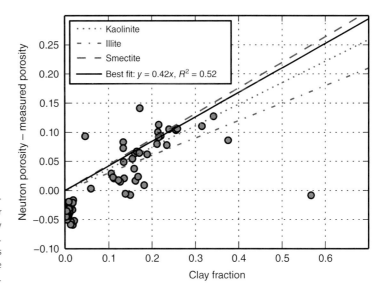

Fig. 10.8. Clay content measured in core samples versus the difference between neutron porosity and observed porosity for the Roer Valley Graben data set. Neutron porosity includes water bound to clay minerals and is higher than the actual porosity in clay-rich sediments. The theoretical neutron porosity of kaolinite, illite, and smectite is shown for comparison (Rider 2002). The clay content of the core samples was calculated as the fraction of grain sizes smaller than 2 μm.

CONCLUSIONS

We have compiled $n = 148$ data on the permeability of pure clays (kaolinite, illite, and smectite) and $n = 126$ data on clay-free sand from published data sets, as well as detailed data on porosity, permeability, and grain size distribution of shallow (<2 km) sediments from an existing data set of London Clay ($n = 29$) and a newly compiled data set of deltaic silts and fine sands ($n = 67$). In addition, we compare permeability of the natural sediments with an experimental data set consisting of homogeneous mixture of kaolinite and quartz sand with a uniform grain size.

The Kozeny–Carman equation was able to predict permeability of quartz sands with a mean absolute error of $\log(k)$ of 0.19 and an R^2 value of 0.97, but only if a percolation threshold was introduced that accounts for the difference between connected and total pore space at low porosity. The Kozeny–Carman equation was less successful in predicting permeability of pure clays. However, an empirical function that calculates permeability as a power law function of void ratio was able to match the observed permeability closely with a mean absolute error of 0.1–0.2 orders of magnitude and R^2 values exceeding 0.90.

The permeability of sand–clay mixtures shows a strong contrast between the behavior of natural sediments and experimental homogeneous sand–clay mixtures. The permeability of the experimental binary sediment mixture showed a rapid decrease with increasing clay content, with permeability decreasing to minimum values at clay contents of approximately 20%. However, the permeability of a data set consisting of natural silts and fine sands retained relatively high values of permeability at clay contents ranging from 0% to 60%. The comparison between these data sets suggests that permeability equations developed for ideally packed sediment mixtures have limited applicability to natural sediments.

For the deltaic sand and silt data set, log-transformed permeability can be estimated with mean absolute errors of 0.57 and 0.61 and moderate predictive power ($R^2 = 0.26$ to $R^2 = 0.48$) using either the geometric mean or arithmetic mean value for the power mean exponent, the Kozeny–Carman equation for the permeability of the pure sand component, and a power law equation for the clay component. In contrast, a second data set consisting of shallow marine clays and silts showed much lower values of permeability, which fall in between the geometric and harmonic means of permeability of their clay fraction and the sand or silt fraction.

The contrast in permeability trends of the two data sets may be related to the internal structure of the core-plug-sized (0.1 m) samples. Clay particles are not likely to be distributed homogeneously in deltaic sediments and are therefore not able to block all pore throats throughout the sample, even at high clay contents. The comparatively low permeability of the shallow marine clays and silts of the London Clay may be due to a more homogeneous distribution of clays in these sediments. The contrast between the two data sets points to a nonlinear relation between permeability and clay content that has been suggested by several previous studies (Dewhurst *et al.* 1999b).

The comparison of anisotropy and permeability of siliciclastic sediments shown in Figure 10.7A points to the importance of sediment structure in core-plug samples. The permeability anisotropy increases with decreasing vertical permeability, which presumably correlates with increasing clay contents. This may reflect a laminar distribution of clays, which would decrease vertical permeability, while the horizontal permeability is maintained by relatively clay-free intervals in the sample. An additional explanation for the correlation between anisotropy and permeability could be that compaction and the realignment of clay minerals increases anisotropy and reduces porosity and permeability for clay-rich samples.

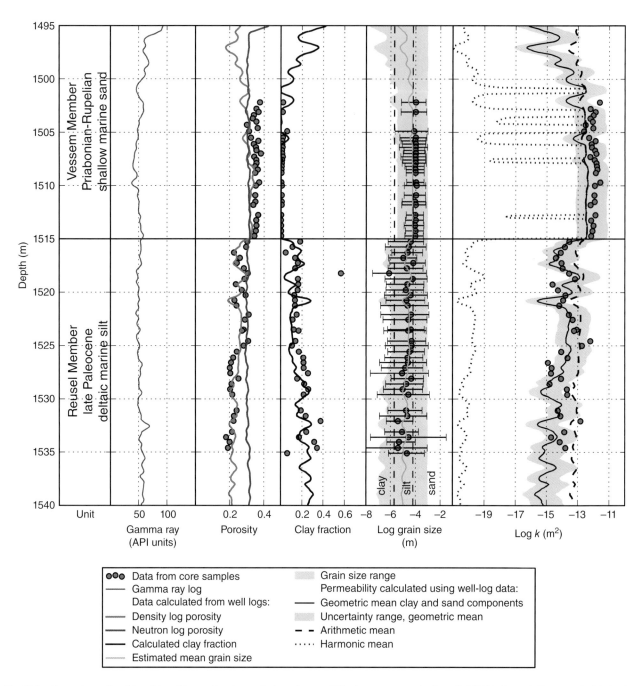

Fig. 10.9. Comparison of well-log data with porosity, clay content, and permeability from core samples in well AST-02. Permeability was calculated using well-log-derived estimates of porosity, clay content, and grain size distribution. Grain size distribution was calculated using the empirical correlations between clay content and grain size distributions shown in Figure 10.2C and D. The calculated permeability shows a relatively good match with observed permeability and estimates permeability within 1 order of magnitude for 80% of the samples. The uncertainty range of the calculated permeability averages ±1.0 orders of magnitude and was calculated using minimum and maximum estimates of clay mineralogy and grain size distribution. (*See color plate section for the color representation of this figure.*)

For the deltaic sediment data set, the model error only increases in minor amounts if neutron and density log data are used to estimate porosity, clay content, and grain size distribution, instead of core-plug data. The mean error increases to 0.72 and 0.75 and the R^2 decreases to 0.23 and 0.33 for power mean exponents of 0 and 1, respectively. The relatively accurate prediction of permeability from widely available neutron and density log data provides new opportunities for estimating permeability for formations where no core samples are available and for determining the variation of permeability at larger scales.

Table 10.5 Performance of permeability equations for the Roer Valley Graben data set using well-log data as an input

Permeability equation	R^2	MAE
Harmonic mean	−26.0	4.7
Geometric mean	0.23	0.72
Arithmetic mean	0.33	0.75

A comparison of well-log-derived permeability shows that, while the model error is slightly higher, the geometric mean equation replicates the variability of permeability much better than the arithmetic mean. Thus, the geometric mean equation would be the best first estimate for the variability of permeability in heterogeneous siliciclastic sediments.

ACKNOWLEDGMENTS

We would like to thank Michael Cardiff and three anonymous reviewers for their excellent and very thorough reviews that helped us improve the manuscript considerably.

SUPPORTING INFORMATION

Additional supporting information may be found in the online version of this chapter, *Geofluids* (2015) **15**, 67–83:

Table SI. Grain size, porosity, and permeability data for core-plug samples of Cenozoic sediments from geothermal well AST-02 in the Roer Valley Graben, southern Netherlands.

CHAPTER 11

Evolution of sediment permeability during burial and subduction

HUGH DAIGLE[1] AND ELIZABETH J. SCREATON[2]

[1] *Department of Petroleum and Geosystems Engineering, University of Texas at Austin, Austin, TX, USA;* [2] *Department of Geological Sciences, University of Florida, Gainesville, FL, USA*

ABSTRACT

We assembled a data set of permeability measurements from 317 subduction zone and reference site samples worldwide made over nearly 25 years of scientific drilling. This data set allowed us to examine the influence of grain size, structural domain, and measurement type on permeabilities ranging from 10^{-21} to 10^{-14} m^2. We found that porosity–permeability behavior is a function of clay-size fraction, which is consistent with previous works. Sediments within the slope, accretionary prism, and fault-zone structural domains are strongly affected by shearing, which alters the permeability behavior with burial. Consolidation, flow-through, and transient pulse-decay measurements all provide comparable results. Measurements of horizontal and vertical permeability show significant centimeter-scale permeability anisotropy (ratio of horizontal to vertical permeability >10) in the slope and accretionary prism structural domains, further indicating shear deformation in these domains. Laboratory consolidation trends match large-scale (10^2 m) field trends in structural domains with negligible shear, but tend to underestimate the rate of permeability reduction with porosity loss where shear is significant. Comparison with downhole measurements shows that permeability is controlled by higher-permeability ($>10^{-15}$ m^2) layers at the meter to tens of meters scale, while wireline formation tester measurements closely match laboratory results. Sediments from the underthrust and reference structural domains exhibit similar porosity–permeability trends, which suggests that shallow subduction (total burial <1 km) does not significantly alter the porosity–permeability behavior of incoming sediments. Comparison with measurements of deeper analog data from 14 passive-margin samples show that porosity–permeability trends are maintained through burial and diagenesis to porosities <10%, suggesting that the behavior observed in shallow samples is informative for predicting the behavior at depth following subduction.

Key words: Integrated Ocean Drilling Program, Ocean Drilling Program, permeability

INTRODUCTION

In subduction zones, fluid flow and pore pressure generation are closely interrelated with deformation. Excess pore pressures reflect a balance between fluid escape and pore pressure increase due to forcing mechanisms such as burial, tectonic stresses, and volume increases during chemical reactions. Excess pore pressures build where sediment permeability is low enough to prevent pore fluid escape at a rate comparable to the rate of forcing (Neuzil 1995). These excess pore pressures can affect the location and strength of the plate boundary fault, the amount of sediment accreted or subducted, and the taper angle of material on the overriding plate (e.g., Davis *et al.* 1983). Fluid escape and pore pressures may also play a role in the generation of earthquakes and propagation of seismic slip (e.g., Moore & Saffer 2001).

An intense effort has been made to characterize subduction zone permeabilities through collection of core samples from the Ocean Drilling Program (ODP) and Integrated Ocean Drilling Program (IODP). Permeabilities in marine sediments may generally be modeled as a function of porosity (e.g., Neuzil 1994), void ratio (e.g., Taylor & Leonard 1990), or effective stress (e.g., Shi & Wang 1988). Porosity–permeability relationships are widely used because porosity is a routine, readily available measurement.

Results from individual transects have been incorporated into numerical modeling through development of porosity–permeability relationships (e.g., Bekins *et al.* 1995; Gamage & Screaton 2006; Matmon & Bekins 2006; Skarbek & Saffer 2009). Numerical modeling consistently indicates that simulated pore pressures are very sensitive to the assumed permeability–porosity relationship of the subducting and accreting sediments (e.g., Gamage & Screaton 2006; Skarbek & Saffer 2009; Screaton 2010; Rowe et al. 2012; Screaton & Ge 2012).

The intense sampling and testing effort has built a substantial data set to address questions concerning the evolution of permeability during early burial and subduction. On the other hand, current data sets are limited to a handful of subduction zone transects, and samples are primarily from depths of approximately 1 km or less. This depth range examines the transition from unconsolidated near-seafloor sediment to those with porosities <0.3. Because an important aim of subduction zone research is to understand seismogenic processes, permeability–porosity relationships have been extended by testing of samples during consolidation (e.g., Skarbek & Saffer 2009). These results have then been used to simulate processes at depths up to 10 km. An important question remains as to how realistically consolidation tests replicate *in situ* permeability. In particular, the progress of diagenetic reactions may alter the permeability–porosity relationship by causing recrystallization of grains within the pore space (e.g., Day-Stirrat et al. 2008, 2010).

In this chapter, the existing data from subduction zone sediments are analyzed. We build on previous work by Gamage et al. (2011) by adding data from >250 samples. In addition, existing data from passive margins are used to fill important gaps in the available subduction zone data. Our analysis considers a total of 317 samples, including 303 samples from subduction zones and 14 samples from passive, pelagic sites, which allows us to revisit previous conclusions presented in Gamage et al. (2011). The larger data set also enables the comparison of permeability test methods, evaluation of anisotropy, and consideration of differences between structural domains. Furthermore, the inclusion of existing data from passive margins allows insight on extrapolating results to greater depths during subduction.

SUBDUCTION ZONE SEDIMENTS

Sampled subduction zones include the Kumano, Muroto, and Ashizuri transects offshore southwest Japan, the Tohoku region near the Japan Trench, the northern Barbados accretionary prism, the Cascadia accretionary prism offshore Oregon (United States), the Peru Margin, and the Nicoya and Osa Peninsulas offshore Costa Rica (Table 11.1, Fig. 11.1A). Although not immediately seaward of a subduction zone, samples collected from the equatorial Pacific and from the South Pacific Gyre are included in this study to provide additional data on carbonate-rich sediments and pelagic clay, respectively. Planned drilling as part of IODP aims to recover data and samples from depths of up to 5 km below the seafloor (Tobin & Kinoshita 2006). However, currently available samples generally are from <1 km below the seafloor. Fine-grained lithologies are generally well represented; however, there is significantly less availability of silt- and sand-rich turbidite deposits due to issues with drilling and recovery of coarse-grained, unconsolidated sediments.

The thickness and lithology of sediments entering subduction zones vary depending on proximity to continental sediment sources, water column productivity, and seafloor bathymetry during deposition (Kennett 1982). Seafloor depth increases as the ocean crust cools and becomes denser, roughly following a linear trend of depth increasing with the square root of crustal age (Parsons & Sclater 1977). Superimposed on this general trend are variations due to basement highs such as seamounts and ridges (Fig. 11.1B). Crustal topography has an important impact on sedimentation patterns related to the carbonate compensation depth. Above the carbonate compensation depth, carbonate in sediments can be preserved, whereas dissolution removes carbonate material below the carbonate compensation depth. Seafloor bathymetry can also affect the distribution of terrigenous sediments through funneling of sediments in submarine canyons or blockage of sediment pathways by basement highs.

Due to variations in sediment supply, the total thickness of sediments arriving at subduction zones can range from several hundred meters to 8 km (Clift & Vannucchi 2004). At subduction zones with >1 km (±0.5 km) of sediment, only the lower portion is subducted, and the remainder is offscraped and accreted to form a sedimentary prism. Through time, this accretionary prism builds seaward through imbrication of frontal thrusts. Mass is also added to the prism from below through underplating of initially subducted sediments beneath the wedge as inferred from seismic data, mass balance considerations, and detrital zircon fission track ages (e.g., Platt et al. 1985; Clowes et al. 1987; Brandon & Vance 1992; Kopp et al. 2000; Fuis et al. 2008). In contrast, the entire sediment column is generally subducted at margins with <1 km (±0.5 km) of incoming sediment, and the base of the upper plate may be eroded and subducted. This subduction erosion can result in large-scale subsidence of the seafloor (Clift & Vannucchi 2004).

Diagenesis is an important control on sediment fabric, and thus may be an important factor in permeability (e.g., Milliken et al. 2012). As a result, the thermal state of sediments is important. Differences in incoming crustal ages, sediment thicknesses and deposition history, and vigor of hydrothermal circulation yield a wide range of thermal histories of incoming sediments. *In situ* measurements of conductive heat flow show that shallow thermal gradients of sediments approaching the subduction zones represented in this study range from $6\,°C\,km^{-1}$ offshore the Nicoya Peninsula of Costa Rica to $183\,°C\,km^{-1}$ offshore the Muroto Peninsula of Japan. During subduction, heat conduction to the seafloor is slowed by the overlying plate, resulting in warming of the downgoing sediments. Observational evidence suggests that the onset of seismogenic behavior occurs at temperatures between 100 and 150 °C, corresponding to subseafloor depths of 5–10 km (Dixon & Moore 2007).

Table 11.1 Summary of permeability sample locations

Location	Sediment types	ODP/IODP leg or expedition	Depth range (mbsf)	Temperature range (°C)	Porosity-depth (m) relationship	References
Barbados	Hemipelagic clay	110, 156	10.64–530.4	3–41	$n = 0.715e^{-0.000786z}$	Taylor & Leonard (1990), Brückmann et al. (1997), Zwart et al. (1997), and Screaton et al. (1990)
Cascadia	Silty clay and sand-rich turbidites	146	30.50–447.00	6–28	NA	Brown (1995) and Moran et al. (1995)
Costa Rica: Nicoya	Hemipelagic clay and pelagic oozes	170, 205	80.85–598.00	3–9	Clay: $n = 0.8e^{-0.001z}$ Ooze: $n = 0.8e^{-0.0007z}$	Saffer et al. (2000), Screaton et al. (2006), and McKiernan & Saffer (2006)
Costa Rica: Osa	Hemipelagic clay and pelagic oozes	344	88.40–779.04	N/A	NA	Daigle & Screaton (2015b)
Nankai: Ashizuri	Hemipelagic silty clay, silt- and sand-rich turbidites	190	482.23–732.54	53–79	$n = 0.79e^{-0.00087z}$	Gamage et al. (2011) and Saffer (2010)
Nankai: Kumano	Hemipelagic silty clay, silt- and sand-rich turbidites	315, 316, 319, 322, 333	2.39–920.63	2–84	NA	Dugan & Daigle (2011), Ekinci et al. (2011), Reuschle (2011), Rowe et al. (2011), Saffer et al. (2011), Hüpers & Kopf (2012), Yue et al. (2012), Screaton et al. (2013), Dugan & Zhao (2013), Daigle & Dugan (2014), Guo & Underwood (in review), and Song et al. (in review)
Nankai: Muroto	Hemipelagic silty clay	190	107.14–842.75	21–157	$n = 0.77e^{-0.0011z}$	Gamage & Screaton (2003, 2006) and Skarbek & Saffer (2009)
Peru Margin	Sand-rich turbidites and biogenic oozes	204	17.10–409.40	2–25	$n = 0.73e^{-0.0009z}$	Gamage et al. (2005) and Matmon & Bekins (2006)
Tohoku	Pelagic clay and hemipelagic silty clay	343	697.60–826.70	NA	NA	Tanikawa et al. (2013)
Equatorial Pacific	Biogenic oozes and chalks; silty clay	320–321	57.28–333.81	2–13	NA	Screaton et al. (2014)
South Pacific Gyre	Pelagic clay	329	6.30–37.50	3–5	NA	Daigle & Screaton (2015b)
North Sea and Gulf of Mexico	Mudstone	NA	2047.9–6884.3	NA		Yang & Aplin (2007)
North Sea	Chalks	NA	2085–3935	NA		Mallon et al. (2005)
Belgium, Switzerland, offshore Norway	Claystones and siltstones	NA	250–1580	NA		Hildenbrand et al. (2002)
Belgium, Germany, Switzerland, offshore Norway	Mudrocks and sandstone	NA	220–4979	NA		Hildenbrand et al. (2004)

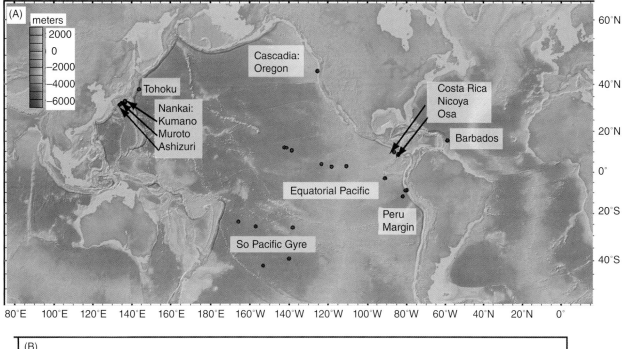

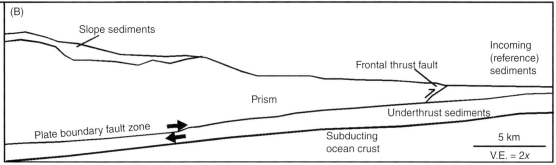

Fig. 11.1. (A) Locations of samples used in this investigation. (B) Schematic of a shallow subduction zone showing domains. This figure was based on an accretionary system. For simplicity, only the plate boundary and frontal thrust are shown. In reality, the prism is extensively faulted. In an erosive system, only the seawardmost portion of the upper plate would be a sediment prism, typically consisting of reworked slope sediments. Scale bar gives example dimensions and vertical exaggeration (V.E.). (*See color plate section for the color representation of this figure.*)

Many hydrogeologic modeling studies focus on the evolution of pore pressures from the deformation front to these depths. For clastic sediments, important diagenetic reactions in this depth range include the transformations of opal to quartz and of smectite to illite (Underwood 2007). For carbonate sediments, mechanical compaction is thought to dominate above 50–200 m below seafloor (mbsf), below which dissolution and reprecipitation control compaction trends (Mallon & Swarbrick 2002).

METHODS

Permeability measurement

Throughout the history of scientific ocean drilling in subduction zones, sediment permeabilities have been measured in the laboratory mainly by steady-state flow-through tests and uniaxial consolidation tests. Less commonly, transient pulse-decay tests are used. *In situ* permeability values have been determined by wireline formation testing tools and drill stem tests. To compare the permeabilities determined by these different techniques, it is important to understand the scales of measurements and the assumptions implicit in the measurement and data analysis.

Laboratory methods

Steady-state flow-through tests (Bernabe 1987; ASTM International 2004) measure permeability by applying either a constant pressure gradient ΔP or a constant flow rate Q axially along a saturated, cylindrical sample. Permeating fluids may be deionized water or simulated seawater. Samples are typically 2.5–3.8 cm in diameter with length equal to or greater than the diameter, although some investigators perform measurements on intact whole-round (approximately 6 cm diameter) samples (e.g., Screaton *et al.* 2013). Tests are run until the pressures and flow rate reach steady state. Several permeability tests may be performed on the same sample by reversing the flow direction in successive tests, and multiple tests may be run at different stress states by applying different isotropic confining pressures or at various porosities during uniaxial consolidation tests (e.g., McKiernan & Saffer 2006).

Uniaxial consolidation tests (Craig 2004; ASTM International 2006) allow permeability determination by the analysis of pore pressure data, as a sample is consolidated and drained at only one end. Samples typically have initial dimensions of 2.5–5 cm diameter and 1–2 cm height. Samples are trimmed into a fixed ring and then placed in a consolidation cell. The consolidation cell is filled with water (deionized, simulated seawater, or tap water) and pressurized to a few tens to hundreds of kPa. During testing, the sample is consolidated axially by applying force to a piston that presses on the porous stone at the top of the sample. The fixed ring ensures no radial deformation and true uniaxial consolidation. In constant rate of strain consolidation tests, the sample is deformed at a constant rate, and the stress required to maintain the rate of deformation is monitored. In oedometer tests, a constant force is applied to the sample, and the resulting deformation is monitored. Permeability is computed continuously during the test from the pressure gradient across the sample. The permeability at the *in situ* porosity of the sample is determined by isolating the virgin consolidation portion of the data in a constant rate of strain test or the final permeabilities from each load increment in virgin consolidation in an oedometer test.

Transient pulse-decay permeability measurements (Brace *et al.* 1968; Hsieh *et al.* 1981) are useful in low-permeability sediments in which steady-state flow-through methods are not feasible. During a transient pulse-decay measurement, a cylindrical sample roughly 2 cm in diameter and 5 cm in length is loaded into a pressure vessel similar to that used for steady-state flow-through experiments. The sample is then saturated and placed under confining pressure until the pore pressure within the sample equilibrates, taken as the time when the rate of change in pore pressure in the sample falls below a certain threshold (e.g., equivalent volume strain $<10^{-7}\,\mathrm{s}^{-1}$; Kitajima *et al.* 2012). Once the pressure has equilibrated, the pressure at one end of the sample is suddenly increased to a new value, and the pressure at the opposite end of the sample is recorded as the transient pressure pulse equilibrates. Permeability is computed from numerical fitting of the pressure–time data.

Downhole methods

Wireline formation testing tools may be used to determine *in situ* permeability by single-probe drawdown tests, by dual-packer drawdown or injection tests, or by single-packer pulse or flow tests. During a single-probe drawdown test (Schlumberger Limited 1981), a rubber packer with a probe in the middle is pressed against the formation. This hydraulically isolates the area around the probe from the wellbore. The pore fluid in the isolated volume of 5–10 cm³ is withdrawn by a pump in the tool, and the pressure response is recorded by the probe as fluid moves from the formation to replace the withdrawn volume. The permeability obtained by this method is an equivalent spherical permeability, that is, it is assumed that fluid flows at an equal rate from every direction to replenish the drawdown volume (Dussan & Sharma 1992).

During dual-packer drawdown or injection tests, a section of the wellbore, a few meters in length, is isolated hydraulically by inflatable packers on the formation testing tool. Fluid is removed from or injected into the isolated section of borehole, and the pressure response is recorded as fluid flows from or into the formation. Single-packer tests follow a similar procedure to dual-packer tests, except that the isolated interval of wellbore extends from the packer to the bottom of the hole. In practice, single-packer tests generally test much longer intervals of wellbore than single-probe or dual-packer tests.

The permeability may be determined from the pressure–time data by graphical curve matching, analytical solution, or by numerical solution. The solution will vary depending on whether the withdrawal/injection is a pulse or a sustained, constant rate (Papadopulos *et al.* 1973; Bredehoeft & Papadopulos 1980; Becker 1990; Boutt *et al.* 2012). The geometry of the testing affects whether the permeability obtained is the equivalent spherical or radial permeability (Matthews & Russell 1967; Boutt *et al.* 2012). Additional *in situ* tests have been conducted at sealed boreholes, either through pulse and flow tests or passive monitoring of perturbations due to nearby drilling (e.g., Bekins *et al.* 2011).

Analysis of data

Sample locations, depth ranges, and temperature ranges are listed in Table 11.1. The depth range corresponds to the deepest and shallowest sample from each location in our data set. Temperature ranges were approximated from the depth range of the samples and the measured thermal gradients. Permeability–porosity behavior was analyzed in the data set by the determination of relationships of the form $\log k = \gamma n + k_0$, where k is permeability, n is porosity, and γ and k_0 are constants. Other relationships, such as the Kozeny–Carman equation (e.g., Yang & Aplin 2007; Luijendijk & Gleeson 2015) or generalizations thereof (e.g., Revil and Cathles 1999; Revil and Florsch 2010), may be used to predict permeability in cases where specific surface area of grain-size distribution is known. As these data were not available for most of the samples in our data set, the log-linear porosity–permeability relationship was selected to examine broad trends.

In many cases, values were provided at more than one porosity value. For the plots and analysis, only one value was used per sample unless two different methods were applied to the same sample. The value used is generally the reported value

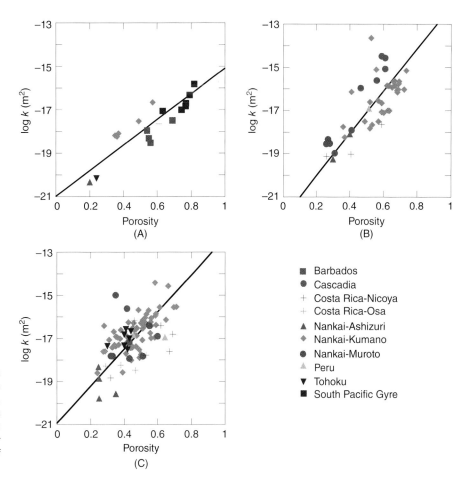

Fig. 11.2. Permeability–porosity relationships for clays. (A) Group 1 (>80% clay). (B) Group 2 (60–80% clay). (C) Group 3 (<60% clay). Best-fit lines are shown. Regression coefficients are given in Table 11.2. Tohoku data shown were not used for the best-fit lines. (*See color plate section for the color representation of this figure.*)

closest to the expected *in situ* porosity value. Regression was performed using the reduced major axis (RMA) technique due to the assumption of error in both porosity and permeability measurements (e.g., Williams & Troutman 1987).

Where grain size and carbonate content data were available, we divided samples into five categories following the categorization of Gamage *et al.* (2011). Carbonate samples contained more than 50% $CaCO_3$ mass percent in the solid matrix. Of the remaining samples, those with >5% sand-sized (>63 μm) particles by mass were examined separately. Although we term these sediments as sands/silts, percentage of clay-sized material is significant, ranging up to 55%. The remaining samples are referred to as fine grained. Where grain size and carbonate content data were not available, we classified samples based on shipboard sedimentologic descriptions.

Fine-grained sediments comprise the bulk of the samples and exhibit a wide range of porosity (0.20–0.89) and permeability (4.6×10^{-21} to 2.3×10^{-14} m²). To assess specific trends within the data, we analyzed permeability–porosity trends by further subdividing the fine-grained samples with associated grain-size data into three groups (Bryant 2002; Gamage *et al.* 2011): Group 1, with >80% of the matrix composed of clay-sized grains by mass; Group 2, with between 60% and 80% of the matrix composed of clay-sized grains by mass; and Group 3,

with <60% of the matrix composed of clay-sized grains by mass (Fig. 11.2). Clay-sized particles were defined as smaller than 4 μm for consistency with other works that use the Wentworth grain-size classification (Steurer & Underwood 2003; Gamage *et al.* 2011). Available data from Dugan & Daigle (2011), Kopf *et al.* (2011), and Daigle & Dugan (2014) suggest that this yields clay percentages 1.3–1.6 times greater than if a 2 μm threshold is applied, as is commonly performed in the engineering community (e.g., Dewhurst *et al.* 1999a).

RESULTS AND DISCUSSION

The constants γ and k_0 for all trends shown in figures, along with standard errors and coefficient of determination (R^2) values, are given in Table 11.2. The standard errors are used as a basis for determining whether values are statistically different.

Variation with measurement type

Permeabilities obtained by flow-through, consolidation, and transient pulse-decay measurements all fall within the same range (Fig. 11.3). Previous subduction zone work noted possible discrepancies of up to an order of magnitude between results from flow-through and consolidation tests (e.g., Zwart

Table 11.2 Regression lines for figures

Figures	Description	γ	k_0	R^2
11.2	Group 1 clays	5.75 ± 0.79	-20.924 ± 0.410	0.658
	Group 2 clays	9.78 ± 0.99	-21.971 ± 0.360	0.527
	Group 3 clays	8.39 ± 0.70	-20.862 ± 0.235	0.351
11.3	Ashizuri	11.64 ± 3.52	-22.489 ± 0.612	0.268
	Barbados	9.36 ± 2.54	-23.574 ± 1.449	0.337
	Cascadia	5.86 ± 1.29	-18.280 ± 0.615	0.272
	Equatorial Pacific	4.43 ± 1.63	-18.249 ± 1.390	0.057
	Kumano	9.45 ± 0.61	-21.467 ± 0.209	0.161
	Muroto	7.19 ± 0.90	-20.665 ± 0.307	0.718
	Nicoya	11.15 ± 2.27	-22.882 ± 0.745	0.410
	Osa	8.20 ± 2.28	-20.964 ± 0.708	0.458
	Peru	8.20 ± 1.92	-21.492 ± 1.088	0.009
	South Pacific Gyre	7.66 ± 2.33	-22.448 ± 2.186	0.354
	Tohoku	15.95 ± 3.86	-23.476 ± 0.781	0.532
11.4	Fault zone clays	10.94 ± 4.43	-23.191 ± 1.327	0.343
	Prism clays	6.79 ± 0.92	-20.465 ± 0.338	0.078
	Reference clays	8.37 ± 0.51	-21.656 ± 0.198	0.536
	Slope clays	7.32 ± 0.67	-20.010 ± 0.266	0.545
	Underthrust clays	8.18 ± 2.53	-21.303 ± 0.758	0.042
	Prism Group 3	9.20 ± 1.53	-21.240 ± 0.471	0.112
	Reference Group 3	8.86 ± 1.14	-21.582 ± 0.410	0.638
	Slope Group 3	6.80 ± 0.86	-19.779 ± 0.329	0.417
	Underthrust Group 3	9.43 ± 4.71	-21.548 ± 1.460	0.001
11.5	Constant rate of strain	9.66 ± 0.66	-21.517 ± 0.223	0.331
	Flow-through	8.10 ± 0.56	-21.034 ± 0.212	0.199
	Transient pulse decay	7.94 ± 2.53	-21.843 ± 1.078	0.392
11.7	Prism clays, experimental	6.60 ± 1.05	-19.855 ± 0.811	
	Prism clays, field	6.79 ± 0.92	-20.465 ± 0.338	0.078
	Prism silts/sands, experimental	5.28 ± 1.46	-17.733 ± 1.175	
	Prism silts/sands, field	11.75 ± 2.60	-20.222 ± 0.599	0.413
	Reference clays, experimental	11.98 ± 3.55	-23.635 ± 2.456	
	Reference clays, field	8.37 ± 0.51	-21.656 ± 0.198	0.536
	Reference silts/sands, experimental	7.86 ± 1.74	-19.508 ± 1.198	
	Reference silts/sands, field	5.56 ± 1.42	-18.833 ± 0.675	0.149
	Slope clays, experimental	8.97 ± 2.47	-20.628 ± 1.523	
	Slope clays, field	7.32 ± 0.67	-20.010 ± 0.266	0.545
	Slope silts/sands, experimental	6.86 ± 2.34	-18.759 ± 1.599	
	Slope silts/sands, field	7.69 ± 2.14	-19.606 ± 0.830	0.069
11.9	Silts/sands	6.53 ± 0.89	-18.916 ± 0.338	0.152
11.10	Carbonate >200 m plus deep analogs	8.40 ± 0.49	-20.598 ± 0.108	0.876
11.11	All fault zone samples	14.49 ± 3.68	-23.406 ± 1.034	0.226
	All prism samples	8.75 ± 1.09	-20.929 ± 0.339	0.004
11.13	Groups 2 and 3	8.73 ± 0.56	-21.148 ± 0.194	0.425
	Groups 2 and 3 plus analogs	9.49 ± 0.44	-21.554 ± 0.128	0.660
	Silts/sands plus analogs	11.36 ± 0.78	-21.434 ± 0.169	0.642

et al. 1997), but these differences are not apparent in this compilation. Although γ and k_0 for flow-through and consolidation tests are statistically different, the relationships suggest similar values at porosities between 0.4 and 0.5 with increasing differences at higher or lower porosities.

Large confidence intervals on the transient pulse-decay trend make it difficult to distinguish from the flow-through results. In general, the $k-n$ trends for flow-through and transient pulse-decay measurements have similar slopes, but the transient pulse-decay measurement trend is lower than the flow-through trend by approximately an order of magnitude.

We do not observe a large discrepancy of up to three orders of magnitude such as was found by Mallon & Swarbrick (2008) when comparing gas permeability measurements and transient pulse-decay measurements on nonreservoir chalk samples from the North Sea.

Variation with grain size

Fine-grained sediments

The overall trends obtained from the data match those reported by Gamage *et al.* (2011) with greater clay-sized fraction yielding

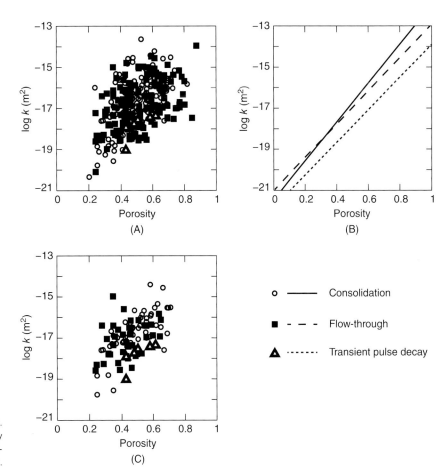

Fig. 11.3. Comparison by measurement method. (A) Data for all clay samples. (B) Trends for all clay samples. (C) Comparison restricted to Group 3 samples. Regression coefficients are given in Table 11.2.

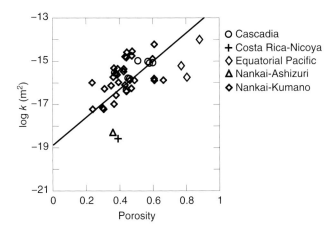

Fig. 11.4. Permeability of silts/sands sediments. Regression coefficients are given in Table 11.2.

lower permeability at a given porosity. This is slightly consistent with the results of Luijendijk & Gleeson (2015) who showed that permeability of sand–clay mixtures can be represented by the geometric mean of the end-member permeabilities. However, the observed differences are greater between Groups 1 and 2 than between Groups 2 and 3. Although Groups 2 and 3 have statistically different $k–n$ relationships, the resulting k values are

similar at high porosity $(n = 0.8;$ Fig. 11.2B,C). Permeability values differ by an order of magnitude at the lower end of porosity $(n = 0.2)$. Schneider *et al.* (2011) and Reece *et al.* (2013) reported that, in mixtures of silt and clay, the silt grains shield the clay grains from porosity loss, with most of the permeability reduction occurring in the silt fraction. This causes the bulk $k–n$ behavior of the sediment to be controlled by the behavior of the silt fraction when the silt fraction exceeds some threshold. This threshold varies depending on the clay mineralogy, but the data of Reece *et al.* (2013) suggest that this threshold is around 20% silt fraction for Nankai Trough sediments, which corresponds to 80% clay fraction. The global data set suggests that this threshold is typical of marine sediments in convergent margins and should be considered in permeability predictions based on grain-size distribution.

Silts/sands

Samples containing more than 5% sand-sized (>63 μm) particles by mass, and <60% clay (Group 4 of Bryant 2002 and Gamage *et al.* 2011) were recovered from the Kumano Basin transect offshore of Nankai (Fig. 11.4). Permeability is consistently higher for a given porosity than in the clay Groups 1–3 $(\log k = 6.53n - 18.92)$. The scatter likely relates to the variability in clay content.

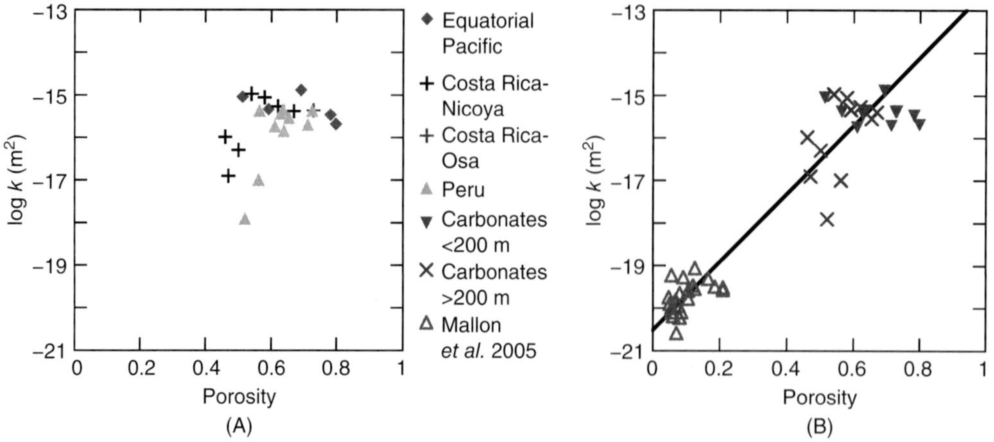

Fig. 11.5. (A) Permeability of carbonate oozes and chalks by location. (B) Results separated by burial depth and combined with deep sediment data from Mallon *et al.* (2005). Regression coefficients are given in Table 11.2. (*See color plate section for the color representation of this figure.*)

Carbonate oozes and chalks

Carbonate-rich samples were obtained from Costa Rica, Peru, and the Equatorial Pacific (Fig. 11.5). Permeabilities from samples above 200 mbsf range between 1×10^{-16} and 1×10^{-15} m^2, with no clear trend (Fig. 11.5). In contrast, samples obtained from below 200 mbsf combined with nonreservoir carbonate chalks tested by Mallon *et al.* (2005) yield a k–n relationship of log $k = 8.40n$–20.60. Mallon & Swarbrick (2002) suggest that 50–200 mbsf marks the onset of chemical compaction due to dissolution and reprecipitation. Although the compaction method of carbonate-rich samples differs from clastic sediments, the resulting k–n relationship is similar to that of Group 3 clays (log $k = 8.39n$–20.86) with overlapping confidence intervals.

Fine-grained sediment variation with geographic region

Little systematic variation in permeability trends with geographic region is observed in the fine-grained sediment data (Fig. 11.6). The samples from Ashizuri, Kumano, Osa, and Tohoku exhibit high sensitivity to porosity with $\gamma > 9.7$ and have the lowest k_0 values ($<10^{-22}$ m^2). The high γ value for samples from Tohoku ($\gamma = 16.5$) could be due to the fact that these samples are all close to the plate boundary fault and have probably undergone more shearing than the rest of the samples, altering the k–n trends. Alternatively, separation by grain-size groups (Fig. 11.2) suggests that grain-size variations could be the major factor in the apparent geographic differences.

Fine-grained sediment variation with structural domain

The k–n trends among the fault zone, prism, and slope structural domains are similar (Fig. 11.7A). The sediments from the reference domain have a larger γ value and smaller k_0 value than the other domains, but this could be due to the large spread of the data; the reference domain data have the largest range of *in situ* porosity (0.20–0.82) and permeability (4.6×10^{-21} to 2.3×10^{-14} m^2). Confidence intervals of γ and k_0 overlap

for the fault zone and reference domains. Sediments from the slope domain exhibit higher permeabilities than those in other domains, while sediments from the fault-zone domain have generally lower permeabilities. The higher permeability of the slope sediments is likely due to their more proximal location to sediment sources, resulting in coarser-grained sediments. The low permeability of the fault-zone sediments is likely related to the lithologic controls on the location of plate boundary faults in convergent margins, in which low-permeability, clay-rich sediments allow buildup of excess pore pressure, weakening the rock and allowing slip (e.g., Brückmann *et al.* 1997; Tobin & Saffer 2009; Nakajima *et al.* 2013; Tanikawa *et al.* 2013) but could also be affected by shearing. Permeabilities within fault zones may achieve transient values as large as 10^{-11} m^2 as suggested by numerical modeling, but these values would not be captured in the laboratory because sampling preferentially recovers the low-permeability fault-zone cores and not the surrounding brecciated zones (Saffer 2015). Although the underthrust domain is represented by relatively few samples (10), the trend is similar to that of the reference domain within the confidence intervals (log $k = 8.18n$–21.3 and log $k = 8.37n$–21.7, respectively). This suggests that incoming sediments from the reference domain maintain similar k–n trends as they are underthrust. Permeability–porosity trends determined at reference sites are, therefore, useful for predicting future down-dip behavior.

Restricting the comparison with Group 3 sediments reduces the impact of grain-size variation and also shows similar behavior in the different structural domains (Fig. 11.7B). The permeability–porosity fit is much better in the underthrust sediments than in the prism sediments. This may reflect the variable effects of deformation in the prism.

Permeability anisotropy

Vertical and horizontal permeabilities were measured on adjacent samples from the Kumano area of the Nankai Trough

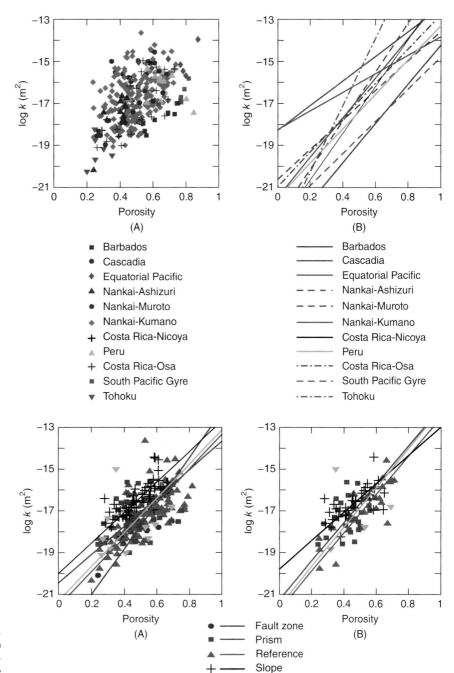

Fig. 11.6. Comparison by location. (A) Data. (B) Trends. Regression coefficients are given in Table 11.2. (*See color plate section for the color representation of this figure.*)

Fig. 11.7. Comparison by structural domain. (A) Data for all clay samples. (B) Comparison restricted to Group 3 samples. Regression coefficients are given in Table 11.2. (*See color plate section for the color representation of this figure.*)

in several studies (Dugan & Daigle 2011; Ekinci *et al.* 2011; Ekinci 2012; Yue *et al.* 2012; Dugan & Zhao 2013). In these studies, vertical permeability was defined as parallel to the borehole axis, and horizontal permeability was defined as perpendicular to the borehole axis. Horizontal permeabilities were not aligned in any specific azimuth in the horizontal plane. These measurements were performed by both constant rate of strain and flow-through experiments. The permeability anisotropy ratio k_h/k_v is the ratio of horizontal to vertical permeability. For samples from reference sites, this quantity ranges

from 1.01 to 2.62 and generally increases with decreasing porosity (Fig. 11.8). For samples from prism sites, k_h/k_v varies greatly, ranging from 0.02 to 83.3.

Uniaxial consolidation experienced during burial is generally expected to cause an increase in k_h/k_v as initially randomly oriented clay grains are rotated to produce a preferential alignment in the horizontal plane (e.g., Leroueil *et al.* 1990; Dewhurst *et al.* 1998, 1999). Permeability anisotropy development may be predicted by considering clay grain rotation as a function of uniaxial strain following the theory of March (1932). Daigle &

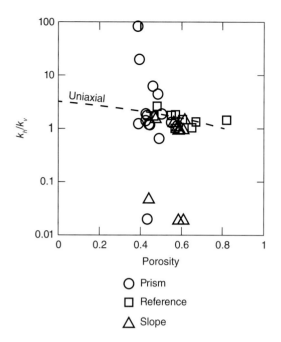

Fig. 11.8. Anisotropy as a function of porosity for the prism, reference, and slope structural domains. Prediction from uniaxial consolidation and grain rotation (Eq. 11.1) is shown for reference. The parameters used for the uniaxial prediction were $n_0 = 0.8$ and $\theta_0 = 45°$.

Dugan (2011) derived an equation for k_h/k_v as a function of grain orientation and porosity in sediments composed of flat, cylindrical grains:

$$\frac{k_h}{k_v} = \left[\frac{1 + \left[\frac{8}{9} m \cos\theta + \frac{2}{\pi} \sin\theta \right] / \left[\frac{3\pi}{8(1-n)} - \frac{1}{2} \right]}{1 + \left[\frac{8}{9} m \sin\theta + \frac{2}{\pi} \cos\theta \right] / \left[\frac{3\pi}{8(1-n)} - \frac{1}{2} \right]} \right]^2 \quad (11.1)$$

where m is the grain aspect ratio (ratio of diameter to thickness) and θ is the average angle of the grains with respect to the horizontal plane. Under conditions of uniaxial strain, θ can be computed as $\tan^{-1}[(1-\varepsilon_v)\tan\theta_0]$, where θ_0 is the initial grain orientation at deposition and ε_v is the vertical strain. Vertical strain may be computed as a function of porosity as $n_0 - n(1 - n_0)/(1 - n)$ where n_0 is the initial porosity at deposition. Grains are usually deposited at the seafloor with a random orientation (Bennett et al. 1989) so θ_0 may be assumed equal to 45°, and marine sediments typically have $n_0 = 0.8$ (e.g., Bryant et al. 1974; Bennett et al. 1989; Long et al. 2011).

The k_h/k_v versus porosity trend for the samples from reference sites is represented well by Eq. 11.1 with $m = 10$ and $n_0 = 0.8$, but the samples from prism and slope sites exhibit extreme departures from the trend (Fig. 11.8). This is likely attributable to significant shear strains and rotation experienced by these samples. Daigle & Dugan (2011) showed that k_h/k_v greater than roughly 10 is only possible by shearing, and this is consistent with experimental and field observations (e.g., Wilkinson & Shipley 1972; Al-Tabbaa & Wood 1987; Arch

& Maltman 1990; Leroueil et al. 1990; Dewhurst et al.1996; Crawford et al. 2008; Day-Stirrat et al. 2008; Haines et al. 2009; Solum & van der Pluijm 2009; Adams et al. 2013). The k_h/k_v values <1 are probably due to shearing followed by rotation such that the present-day vertical direction aligns with the original horizontal direction. Milliken & Reed (2010) observed subtle, localized deformation bands in slope and prism sediments from the Kumano region of the Nankai Trough, which provide permeability heterogeneity at the millimeter to centimeter scale and may also contribute to some of the larger anisotropy ratios (>2) observed in the laboratory.

Comparison with laboratory consolidation tests

Data from laboratory consolidation tests allow comparisons between $k-n$ behavior during virgin consolidation in the laboratory and trends observed in the field. During burial, shallow (<2 km depth) sediments are expected to consolidate uniaxially in the absence of extreme surface topography or shear stresses (Jaeger et al. 2007). Changes in deviatoric stress and diagenesis can alter the $k-n$ relationships in sedimentary basins (e.g., Jones & Addis 1985; Bjørlykke 1999), while stress history, over- or under-consolidation, cements, and the type of permanent fluid can influence $k-n$ relationships determined in the laboratory (Mesri & Olson 1971; Santagata & Germaine 2002, 2005; Santagata & Kang 2007; Morgan et al. 2008). However, trends observed across the data set are indicative of the degree to which laboratory conditions may recreate the subsurface.

We determined or compiled γ and k_0 for 101 individual consolidation tests. These tests were mainly from the Kumano region (89), with six tests from Muroto, five from Ashizuri, and one from the South Pacific Gyre. Tests included samples from the prism, reference, and slope structural domains. Samples consisted of fine-grained sediments (grain size Groups 1–3) and silts/sands. Data were not analyzed for carbonate-rich sediments. Because carbonate-rich sediments compact through dissolution and reprecipitation (Mallon & Swarbrick 2002), laboratory consolidation tests are not expected to reproduce field consolidation.

Within the reference and slope domains, the laboratory and field trends are similar for both fine-grained sediments and silts/sands based on overlap of confidence intervals (Fig. 11.9B,C). The trend for silts/sands in the slope domain is offset to lower permeability in the field than in the laboratory, which may be attributable to shearing during sedimentary processes on the slope (e.g., erosion and mass transport; Sassa et al. 2012; Strasser et al. 2012), although the confidence interval of k_0 for the experimental data is large ($\pm$1.599). Within the prism structural domain, the laboratory-determined consolidation trend for clays is offset by 0.5–1 order of magnitude to higher permeabilities, while the field trend for silts/sands is much steeper than the laboratory trend and has a much lower k_0 (−20.2 vs. −17.7; Fig. 11.9A). This may be indicative of a greater degree of shearing experienced by sediments in the prism, consistent with the anisotropy results.

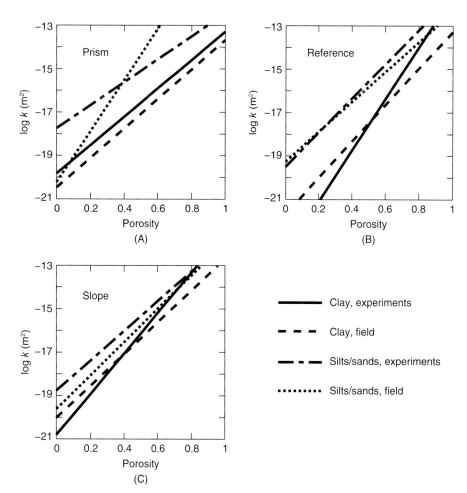

Fig. 11.9. Comparison of permeability–porosity relationships from field data and laboratory consolidation tests. (A) Prism structural domain. (B) Reference structural domain. (C) Slope structural domain. Regression coefficients are given in Table 11.2.

Overall, in the laboratory, the clays had higher average γ than in the field, while silts/sands had lower γ. In both lithology types, the k_0 values were similar between the laboratory and the field (Fig. 11.10A). Comparing all samples separated by structural domain, the prism domain samples had similar laboratory and field γ, while the reference and slope domains had higher laboratory γ. Again, k_0 between the field and laboratory was similar in all structural domains (Fig. 11.10B). The differences in γ may be caused by natural processes not reproduced in the laboratory such as diagenesis and shear or could be due to sample disturbance altering the microstructure of the samples and changing the virgin consolidation behavior. This can be caused by the type of coring tool used. For example, clays cored by piston coring (the HPC or APC tool) have similar γ and k_0 values between the laboratory and the field. In contrast, clays recovered by the extended core barrel (XCB) have much higher γ in the laboratory (Fig. 11.10C,D). This may reflect much greater disturbance in the collection of XCB cores, which rotates the samples and may cause vertical cracks.

Formation tester and dual-packer measurements

Permeabilities were determined from downhole tests at Site 892 in the Cascadia region, Sites 948 and 949 in the Barbados region, and Site C0009 in the Kumano region. At Cascadia, a single packer was used to isolate an approximately 85 m section of Hole 892B across an out-of-sequence thrust fault (Screaton *et al.* 1995). Pulse tests and constant-rate injection and recovery tests were performed after drilling and partially casing the borehole during ODP Leg 146 (Shipboard Scientific Party 1994) while pulse, constant-pressure, and constant-flow aquifer tests were performed roughly 1 year later using the submersible *Alvin*. We assigned porosity values to the tested interval based on the average shipboard-derived moisture and density porosity value over the interval (Shipboard Scientific Party 1994). The permeability values are higher than the values obtained in discrete samples from fault zones (Fig. 11.11A), although the lower end of the data overlaps with permeabilities determined on silts/sands.

At Barbados, a single packer was used to isolate 60–70 m sections of two boreholes spanning the plate boundary fault zone. Both pulse and constant-rate aquifer tests were performed with injection and withdrawal. Porosity values assigned to the tests were determined from the average density porosity over the isolated intervals using the data presented by Fisher & Zwart (1997). As in the case of Cascadia, the permeabilities obtained are larger than the values obtained on discrete samples

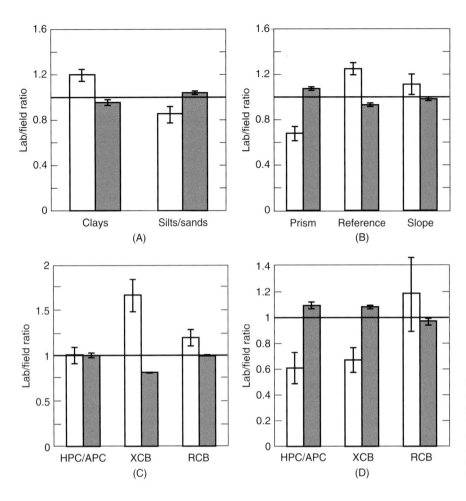

Fig. 11.10. Comparison of γ (white) and k_0 (gray) determined from field and laboratory data. (A) By lithology. (B) By structural domain. (C) Only clays by coring system. (D) Only silts/sands by coring system. HPC/APC = hydraulic piston core; XCB = extended core barrel; and RCB = rotary core barrel. Error bars are given for all laboratory/field ratios.

in the laboratory (Fig. 11.11A). This is probably due to the much larger scale of the measurement, and the fact that single packer measurements yield an equivalent radial permeability, which is orthogonal to the permeability measured in the laboratory. The scale dependence is likely the main cause of the difference, driven mainly by the presence of fractures near the décollement that would not be sampled by discrete laboratory samples (Shipboard Scientific Party 1995a,b). One important but often unknown parameter in interpreting single-packer tests is the condition of the borehole during the test. The results from Bekins *et al.* (2011), which we present for Barbados, are reinterpretations of data from Fisher & Zwart (1997) and Screaton *et al.* (1997). The reinterpretation is based on the observation that some of the downhole test conditions caused failure of the borehole wall, partially filling the borehole with sediment. The reinterpreted permeability values are up to three orders of magnitude higher than the original values, illustrating the sensitivity of the analysis to borehole conditions.

At Site C0009 in the Kumano region, permeabilities were determined from single-probe measurements and dual draw-down packer measurements. The single-probe measurements investigate a spherical volume with radius approximately equal to three times the probe radius (Dussan & Sharma 1992),

yielding a spherical volume approximately 4 cm in radius for the large-diameter probe used for these tests. The single-probe tests, therefore, investigated a volume comparable to discrete laboratory samples, but the resulting permeability is an equivalent spherical permeability rather than a vertical permeability. The dual-packer measurements isolated a 1-m section of borehole, and the resulting permeability was approximately a spherical permeability (Boutt *et al.* 2012). The results of these tests are shown in Figure 11.11B along with discrete laboratory measurements from the prism domain at Kumano. Porosities were determined from the average shipboard moisture and density values over the measured intervals. Both types of tests yield similar values to the laboratory measurements, although there is a bias toward higher permeability values. We interpret this difference as caused by the scale of the measurements (slightly larger than laboratory measurements) and the fact that the spherical permeability contains some component of horizontal permeability, while the laboratory measurements determine only vertical permeability.

Overall, single-probe and dual-packer measurements yield results that are comparable to laboratory measurements, while larger-scale single-packer measurements yield permeabilities that are larger than those obtained in the laboratory. The

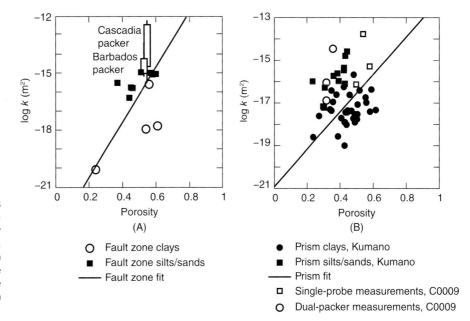

Fig. 11.11. (A) Single-packer test results from Barbados (Holes 948D, 949C) and Cascadia (Hole 892B) shown as box-and-whisker plots with fault-zone permeability data. (B) Permeability results from prism domain in the Kumano area with single-probe and dual-packer test results from Hole C0009A. Regression coefficients are given in Table 11.2.

downhole test results at Cascadia and Barbados overlap the higher end of the silt/sand permeability values, which we interpret as being caused by preferential flow through higher permeability silty, sandy, or fractured units within the tested interval. In general, these zones of higher permeability will control fluid flow at the meter scale and greater.

General comments

While covering diverse geographical locations and a wide range of porosity and permeability, our large data set allows us to make some general observations about permeability in marine sediments at convergent margins, and how sediment permeability evolves during burial and subduction.

Controls on permeability–porosity behavior in marine sediments at convergent margins

Permeability measurements on samples from convergent margins parallel a large body of work on permeability of mudstones in passive sedimentary basins. Much of this work has focused on understanding the controls on permeability in fine-grained, clay-rich sediments in the context of hydrocarbon exploration and geohazard mitigation in passive sedimentary basins (e.g., Aplin *et al.* 1999; Dugan & Sheahan 2012). In the absence of diagenesis, the primary control on k–n behavior during consolidation and burial is the grain size. Luijendijk & Gleeson (2015) showed that detailed knowledge of grain size distribution is essential to accurate permeability prediction in sediments containing mixtures of clay and larger grains. Kwon *et al.* (2004) and Yang & Aplin (2007, 2010) found that permeability variation of up to three orders of magnitude at the same porosity in passive basin sediments could be explained by differences in clay-sized fraction and that samples with similar clay-sized fraction follow the same k–n trend. Comparing the data and

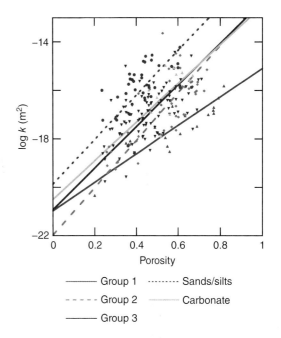

Fig. 11.12. Compilation of permeability data and best-fit lines for each group. Carbonate best-fit line only for samples >200 mbsf and includes deeper data from Mallon *et al.* (2005), which are shown in Figure 11.10B. (*See color plate section for the color representation of this figure.*)

trends from all the grain-size groups (Fig. 11.12) shows the distinct behavior of Group 1 (>80% clays) and the sediments with >5% sand (labeled sand/silt). In contrast, Groups 2 and 3 (60–80% clay and <60% clay, respectively) and the carbonate samples buried >200 m show similar trends. This suggests that clay-sized grains affect permeability the most when clay content is large.

Detailed analysis of resedimented natural samples and laboratory mixtures has provided insight into the mechanism of permeability reduction during consolidation. When sediment is consolidated, the larger pores collapse first (Griffiths & Joshi 1989; Dewhurst *et al.* 1998, 1999). This causes sediments with larger initial pore size to experience greater reductions in permeability with loss of porosity. While the log-linear $k-n$ relationship implies that permeability reduction is generally more rapid at higher porosities, sediments with larger pores should be characterized by higher γ values as the large pores additionally increase the rate of porosity loss. In our data set, we observed higher γ values for clay Groups 2 and 3 and silts/sands compared with clay Group 1, indicating greater reduction in permeability with porosity in these samples. This is consistent with

experimental observations. The similarity in $k-n$ trends between clay Groups 2 and 3 reflects the control that silt content exerts on the relationship between permeability and porosity. In sediments containing more than 20% silt by weight, the collapse of larger, intergranular silt pores causes a more rapid reduction in permeability than samples with more clay. Experimental and microstructural observations by Curtis *et al.* (1980), Milliken & Reed (2010), Schneider *et al.* (2011), and Reece *et al.* (2013) have shown that the clay fraction does not undergo significant consolidation in samples with greater silt content and that most of the porosity reduction occurs by collapse of pores associated with silt grains. Combining clay Groups 2 and 3 shows that the $k-n$ behavior of both groups is similar and may be fit well with a single trend (Fig. 11.13A). The samples from Group 2 fall

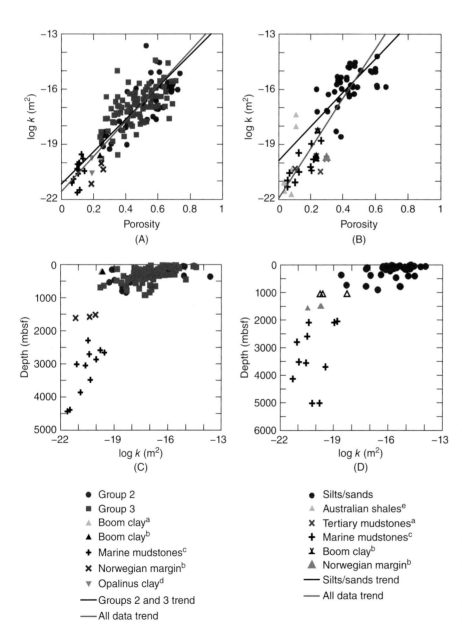

Fig. 11.13. (A) Comparison of Groups 2 and 3 with data from deep sediments belonging to Groups 2 and 3. Best-fit lines shown for subduction zone sediments (black) and all data (green). (B) Comparison of silt/sand data with deep silt/sand sediment data. Best-fit lines shown for subduction zone sediments (black) and all data (green). All regression coefficients are given in Table 11.2. (C) Permeability plotted against depth for Groups 2 and 3 and deep sediments showing depth range of deep sediments. (D) Permeability plotted against depth for silt/sand data and deep silt/sand samples. Note that the Boom Clay depth reported is the present-day depth; this formation is presumed to have been uplifted (Hildenbrand *et al.* 2002). Data sources: [a]Hildenbrand *et al.* (2002), [b]Hildenbrand *et al.* (2004), [c]Yang & Aplin (2007), [d]Marschall *et al.* (2005), [e]Amann-Hildenbrand *et al.* (2013). (*See color plate section for the color representation of this figure.*)

slightly below this trend, while those from Group 3 are slightly above, which is consistent with clay content causing a slight overall reduction in permeability at a given porosity.

In addition to mechanical controls, diagenesis has the potential to alter $k–n$ relationships although a previous synthesis suggested consistent behavior between mud and shale (Neuzil 1994). In clastic marine sediments in convergent margins, the most important diagenetic reactions are the breakdown of opal-A and opal-CT to quartz, transformation of smectite to illite, and alteration of volcanic ash to zeolites and clays (Kastner et al. 1991). Breakdown of opal and alteration of volcanic ash occur at lower temperatures (30–100 °C for opal breakdown; <35° for volcanic ash alteration to zeolite (Kastner et al. 1991; Morad et al. 2000)), while smectite transforms into illite at temperatures of 60–160 °C (Kastner et al. 1991). With the exception of samples from Sites 1173 and 1174 in the Muroto area, all the samples in our data set had in situ temperatures below 60 °C and are not expected to have undergone significant illitization. The main diagenetic reactions that are expected to have occurred in the bulk of the data set are, therefore, opal breakdown and volcanic ash alteration. Both these reactions are hypothesized to have contributed to anomalous porosity–depth trends in the Ashizuri and Muroto areas (Spinelli et al. 2007; White et al. 2010). While the collapse of anomalous porosity appears to alter the pore structure of these sediments (e.g., Daigle et al. 2014), we do not see any evidence in the $k–n$ relationships of abrupt changes associated with these diagenetic reactions. We therefore conclude that the influence of diagenesis on $k–n$ relationships within our data set is minimal.

Finally, there is evidence from numerical simulation of fluid flow at accretionary margins and downhole measurements that permeability within fault zones is transient and strongly affected by fluctuations in pore pressure (Saffer 2015). This behavior is very difficult to capture in the laboratory because sampling tends to be biased toward more competent intervals. Bolton et al. (1999) found that permeability of fault-zone samples from Costa Rica varied over three orders of magnitude during triaxial consolidation and unloading. This behavior has been captured by downhole measurements at Cascadia and Barbados, where permeability increased by two to three orders of magnitude when the wellbore pressure was increased to 50–60% of the overburden stress (Screaton et al. 1995, 1997; Fisher & Zwart 1997). Laboratory measurements therefore represent minimum values on permeability, especially within fault zones.

Projecting observed properties down-dip through the subduction zone

The challenges associated with predicting sediment behavior within the subduction zone involve changes in stress state and diagenesis as sediments are either accreted in the prism or underthrust with the subducting plate. These processes can have competing effects on permeability and porosity. Sediments that remain in the subducting plate may experience significant

overpressures at depth (e.g., Tobin & Saffer 2009), which may reduce the in situ effective stress. Any elastic permeability increase that occurs as a result of unloading by overpressuring is expected to be very small as the sediments occupy a confined volume, and the amount of permeability increase during elastic decompression is generally small because of the permanent alteration of sediment fabric caused by virgin consolidation (e.g., Kwon et al. 2004). Furthermore, subducting sediments are subjected to increasing total stress. As a result, excess pore pressures can increase without decreasing effective stress.

Sediments that are accreted in the prism may experience a decrease in effective stress if they are uplifted. However, sediments within the prism often undergo shear and rotation (e.g., Bray & Karig 1985; Karig & Lundberg 1990; Taira et al. 1992). In the case of shear, the permeability normal to the shearing plane (i.e., vertical permeability during horizontal shearing) will be greatly reduced (e.g., Arch & Maltman 1990; Haines et al. 2009; Ikari & Saffer 2012), while in the case of rotation, the vertical permeability may be increased by the rotation of the bedding planes. We believe these effects are responsible for the permeability anisotropy ratios >10 observed in our data set, particularly the samples from the prism domain. Permeability evolution due to mechanical processes within the prism is therefore difficult to predict.

Sediments that are subducted to depths at which temperatures exceed the threshold for onset of the smectite–illite transition will experience changes in pore structure and pore water chemistry. Illite has a smaller surface area than smectite (Santamarina et al. 2002), and the conversion of smectite to illite releases some interlayer water (Kastner et al. 1991), reducing the volume of the resulting clay grains. These effects alone will result in a slightly higher porosity and permeability of sediments that undergo a transformation of some smectite to illite. However, recrystallization of illite during this process can result in strongly preferential mineral fabric, with grains oriented perpendicular to the maximum principal stress (Day-Stirrat et al. 2008). Depending on the stress orientation in the subducting sediments, this could result in either enhanced or reduced vertical permeability and either a large or very small permeability anisotropy ratio.

A secondary effect related to dewatering is release of interlayer water. This may alter the $k–n$ relationship because most porosity determinations include interlayer water. Porosity correction to remove the interlayer water volume can yield a shifted $k–n$ relationship (Gamage et al. 2011). In addition, the pore fluid is freshened as the interlayer water from smectite is expelled into the surrounding pore space (Bekins et al. 1994; Kastner et al. 1991; Brown et al. 2001). Permeability of sediments containing significant amounts of clay has been observed experimentally to depend on pore water salinity (e.g., Mesri & Olson 1971; Moore et al. 1982). Revil et al. (2005) showed that the permeability of clay-rich sediments decreases with decreasing salinity due to an electroosmotic pressure that opposes flow through clay grains with negative surface charge. This is related to the thickening of the electrical double layer (Stern & Guoy-Chapman layers) at low salinities (Clavier *et al.* 1984; Leroy & Revil 2004). The result of this process may be a slight reduction in permeability as the pore water freshens. Although the relative changes in permeability due to the physical alteration of the clay minerals and pore water freshening are difficult to constrain, the processes are important sources of fluids contributing to overpressures in the subducting sediments (e.g., Bekins *et al.* 1995; Spinelli & Saffer 2004).

Despite the possible diagenetic effects, deep marine mudstones and uplifted Mesozoic and Paleogene mudstones yield results consistent with $k–n$ relationships from shallow Groups 2 and 3 samples (Fig. 11.13A). This conclusion is consistent with the finding of Neuzil (1994) for permeabilities of clays and shales spanning nearly eight orders of magnitude. Similarly, deep and old, uplifted silts/sand-bearing sediments from around the world show $k–n$ behavior that is similar to the results from shallow samples in our data set (Fig. 11.13B). It may be that the effects of diagenesis are not discernable due to the already very low permeability of these sediments. In the case of both fine-grained sediments and silts/sands, most of the permeability loss occurs during the first approximately 1 km of burial (Fig. 11.13C,D). Presumably the additional permeability reduction due to diagenesis during deeper burial has a comparatively smaller influence.

The tentative conclusion that pelagic carbonate sediments may follow a $k–n$ relationship similar to Group 3 clays would allow the projection of down-dip permeability distribution if porosity has been measured or inferred from seismic reflection data. On the other hand, porosity prediction may be challenging for chalks. Effective-stress-dependent porosity is commonly assumed for clays, allowing predictive simulations of behavior during subduction (e.g., Rowe *et al.* 2012). Although similar behavior is often assumed for chalks (e.g., Mallon & Swarbrick 2002), temperature-dependent porosity loss has also been suggested (Nordgård Bolås *et al.* 2008).

CONCLUSIONS

We examined a data set of 317 permeability measurements on sediments from subduction zones and reference sites worldwide. This large data set allowed us to make general observations on porosity–permeability trends in these sediments. Overall, the porosity–permeability relationship is controlled by grain size, as has been observed by previous authors (e.g., Gamage *et al.* 2011). We found that samples with >80% clay-sized grains and samples with >5% sand-sized grains exhibit different porosity–permeability behavior from other sediments. This suggests that samples with little sand but silt fractions >20% behave similarly, which is consistent with laboratory observations of Schneider *et al.* (2011) and Reece *et al.* (2013). For silt contents larger than this threshold, the silt grains shield the clay matrix from deformation.

Sediments recovered from the fault-zone domain generally have the lowest permeabilities at a given porosity, while sediments in the slope domain have the highest. We interpret this as related to proximity to sediment sources in the case of slope sediments and physical processes such as shearing in the case of fault-zone sediments. Sediments from the reference and underthrust domains exhibit similar porosity–permeability trends, which suggest that permeability trends observed at reference sites are useful for predicting behavior down-dip in subducting sediments.

Different permeability measurement techniques yield comparable results. Consolidation and flow-through test results are generally quite close (within less than half an order of magnitude), while transient pulse-decay measurements generally yield permeabilities that are lower by roughly half an order of magnitude at a given porosity. This may be due to sampling bias toward lower-permeability samples for transient pulse-decay measurements.

Measurements of permeability parallel and perpendicular to the borehole axis yield permeability anisotropy ratios that range from 0.02 to 83. For samples from the reference domain, the trend of anisotropy ratio with porosity is described by the uniaxial consolidation grain rotation model of Daigle & Dugan (2011). For samples from the prism and slope domains, the anisotropy ratios deviate significantly from this trend. We attribute this to shearing and bed rotation in these domains, which can enhance permeability anisotropy or rotate the principal axes of the permeability tensor.

Comparison of porosity–permeability trends for individual consolidation experiments with the trends observed in the field shows that laboratory conditions reproduce the field conditions in the reference and slope domains. However, there is serious mismatch in the trends in the prism domain. This is attributable to shearing in the prism that is not recreated in uniaxial consolidation tests. An additional source of mismatch is coring disturbance. We note that clays cored by XCB display the greatest mismatch between laboratory and field.

Silt- and sand-rich sediments have uniformly higher permeability than clays at a given porosity but more scatter. This is probably related to greater variability in clay content displayed by these sediments. Carbonate chalks and oozes only show an organized porosity–permeability trend below 200 mbsf, where they start to behave like Group 3 clays.

Downhole measurements are scale dependent. Single-packer tests sample larger borehole intervals, and the results are controlled by layers with highest permeability. Results from Barbados and Cascadia overlap the higher end of the discrete silts/sands because of this effect. Dual-packer and single-probe measurement sample volumes that approach those investigated in the laboratory, but the results are not directly comparable because of the spherical or cylindrical flow that occurs during the tests. Care must, therefore, be taken when comparing downhole measurements with laboratory data.

We did not observe an evidence of diagenetic effects on the porosity–permeability trends in this data set. Comparison with deeper data sets shows that similar k–n trends are followed to porosities <5%. Any effects of diagenesis are therefore small enough not to alter the matrix permeability trends significantly.

ACKNOWLEDGMENTS

This research used samples and data provided by the ODP and the IODP. Funding for this research was provided by the Consortium for Ocean Leadership US Scientist Support Program postcruise grants to E. Screaton, H. Daigle, and K. Gamage, National Science Foundation grant OCE-0751497 and EAR-0819769, and the University of Texas at Austin. The authors thank Kusali Gamage, Rob Harris, Stephanie James, and Michael Nole for help, assembling the data set, and valuable discussions.

SUPPORTING INFORMATION

Additional Supporting Information may be found in the online version of this chapter, *Geofluids* (2015) 15, 84–105.

Table SI. Data used for this study organized in four worksheets.

CHAPTER 12

Is the permeability of crystalline rock in the shallow crust related to depth, lithology, or tectonic setting?

MARK RANJRAM[1], TOM GLEESON[2] AND ELCO LUIJENDIJK[3]

[1] Civil Engineering Department, McGill University, Montreal, QC, Canada; [2] Department of Civil Engineering, University of Victoria, Victoria, BC, Canada; [3] Department of Structural Geology and Geodynamics, Georg-August-Universität Göttingen, Göttingen, Germany

ABSTRACT

The permeability of crystalline rocks is generally assumed to decrease with depth due to increasing overburden stress. While experiments have confirmed the dependence of permeability on stress, field measurements of crystalline permeability have not previously yielded an unambiguous and universal relation between permeability and depth in the shallow crust (<2.5 km). Large data sets from Sweden, Germany, and Switzerland provide new opportunities to characterize the permeability of crystalline rocks in the shallow crust. Here, we compile *in situ* permeability measurements ($n = 973$) and quantitatively test potential relationships between permeability, depth (0–2.5 km), lithology (intrusive and metamorphic), and tectonic setting (active and inactive). Higher permeabilities are more common at shallow depths (<1 km), but trend analysis does not support a consistently applicable and generalizable relationship between permeability and depth in crystalline rock in the shallow crust. Results suggest that lithology has a weak control on permeability–depth relations in the near surface (<0.1 km), regardless of tectonic setting, but may be a more important control at depth. Tectonic setting appears to be a stronger control on permeability–depth relations in the near surface. Permeability values in the tectonically active Molasse basin are scattered with a very weak relationship between permeability and depth. Although results indicate that there is no consistently applicable relationship between permeability and depth for crystalline rock in the shallow crust, some specific lithologies and tectonic settings display a statistically significant decrease of permeability with depth, with greater predictive power than a generalized relationship, that could be useful for hydrologic and earth system models.

Key words: crystalline rock, data mining, data synthesis, hydraulic conductivity, permeability

INTRODUCTION

The relationship between permeability and depth is critical in the study of groundwater in the shallow crust (<2.5 km). It is often assumed or suggested that the permeability of crystalline rock decreases with depth (Snow 1968; Anderson *et al* 1985; Morrow & Lockner 1997; Ingebritsen & Manning 1999; Shmonov *et al*. 2003; Saar & Manga 2004; Stober & Bucher 2007a; Jiang *et al*. 2010c; Stober 2011), although several studies identify anomalies and uncertainties in this expected relationship (Brace 1980, 1984; Huenges *et al*. 1997). Where a relationship is accepted, it is often estimated as an exponential/logarithmic relationship fit to highly variable data (Snow 1968; Anderson *et al*. 1985; Wladis *et al*. 1997; Shmonov *et al*. 2003; Saar & Manga 2004). These relationships typically explain only a small percentage of the variation in the data.

In the shallow crust, lithology may be an important control on permeability. A recent compilation of near-surface (<0.1 km) data clearly indicates that regional-scale permeability values are controlled by lithology (classified in the compilation as unconsolidated, sedimentary, crystalline, volcanic, or carbonate) (Gleeson *et al*. 2011). Similarly, the permeability of crystalline rock has been shown to depend on whether the lithology is gneissic or granitic in the Black Forest region of Germany (Stober 1996). At depths where contact metamorphism can occur (<5 km), the permeability of metamorphosed rocks is lithology dependent, whereas at depths of regional metamorphism (>5–10 km), permeability is not controlled by lithology (Manning & Ingebritsen 1999).

Permeability in crystalline rock is predominantly secondary fracture permeability, which is controlled by fracture density, aperture, and connectivity (Berkowitz 2002; Neuman 2005;

Crustal Permeability, First Edition. Edited by Tom Gleeson and Steven E. Ingebritsen.
© 2017 John Wiley & Sons, Ltd. Published 2017 by John Wiley & Sons, Ltd.
Companion Websites: www.wiley.com/go/gleeson/crustalpermeability/
http://crustalpermeability.weebly.com/

Ingebritsen *et al.* 2006), as well as hydromechanical coupling (Earnest & Boutt 2014) and fracture in-filling (Rutqvist 2015). Fracture density, aperture, and connectivity are a function of lithology, deformation history, and current tectonic setting. The deformation history of crystalline regions is typically long lasting and complex with multiple events that can reactivate previous structures. For example, Viola *et al.* (2009) suggest that the crystalline bedrock in Sweden is effectively "saturated" for fractures such that fracture reactivation is more common than fracture generation (Munier & Talbot 1993). Fracture permeability can also be affected by temperature-dependent fluid–rock interactions and fracture in-filling (Rutqvist 2015) that are a function of geochemistry, temperature history, and fluid flux. A recent study by Earnest & Boutt (2014) suggests that hydromechanical coupling also plays a role in controlling fractured rock permeability in the upper crust, with fracture normal stiffness being more important than shear dilation. Horizontal stresses are typically much greater than vertical stresses at shallow depths, but the ratio of horizontal to vertical stress decreases significantly in the upper 1 km of the crust as overburden stress increases (Brown & Hoek 1978; Maloney *et al.* 2006; Earnest & Boutt 2014). For example, Maloney *et al.* (2006) show that for crystalline rock in the Canadian Shield, the near surface (<300–600 m in their study) is dominated by local horizontal stresses, while stresses at greater depths are smaller and controlled by distant boundary conditions.

Our objective was to quantitatively evaluate the relationship between the permeability of crystalline rock and depth, lithology, and tectonic setting. We compiled a data set of 973 *in situ* permeability measurements in crystalline rock from the surface to depths of 2.5 km, from metamorphic and intrusive lithologies and from three different locations representing inactive and active tectonic settings. We focus on permeability–depth relations in the upper 2.5 km of the crust for two reasons. First, this is the depth of "traditional data" such as core samples, pumping tests, and drill stem tests, rather than inferential data on permeability such as metamorphic fluid fluxes. Second, this depth is crucial for hydrologic research and examining the role of groundwater in earth processes at the earth surface and in the shallow crust. We do not explicitly examine the potential role of topography and climatic conditions as most of our data are derived from low-to-moderate topographic settings with humid climates. We significantly expand and update previous permeability compilations and quantitatively assess trends of permeability with depth, lithology, and tectonic setting for the first time.

DATA SOURCES, SYNTHESIS, AND ANALYSIS

Crystalline rock permeability has been measured *in situ* at various depths in metamorphic and intrusive lithologies, as well as in active and inactive tectonic settings. Laboratory permeability tests are excluded from this compilation because of the

well-described discrepancy between laboratory and field estimates of permeability (Brace 1980). Focusing on *in situ* values allows this study to make conclusions about permeability values in the field rather than in the laboratory. We significantly expand on previous permeability compilations that have presented data only as synthesized ranges (Brace 1980) ($n = 21$, 21 sources) or a combination of synthesized ranges and individual data points: Clauser (1992) ($n = 67$, 48 sources); Ingebritsen & Manning (1999) ($n = 201$, 25 sources); Shmonov *et al.* (2003) ($n = 35$, 4 sources); and Juhlin & Sandstedt (1989) ($n = 18$, 7 sources). Note that we use "compilation" to describe a collection of permeability values from different sources. In this study, permeability–depth data ($n = 973$) were synthesized from 16 data sources, primarily from research projects for nuclear waste repositories or geothermal resource exploration in Sweden, Germany, and Switzerland, with additional small amounts of data from the United States and Canada (Table 12.1). Herein, we focus our analysis on data from Sweden, Germany, and Switzerland, as this is where the majority of the data are from (94% of total data set). Previous compilations have used specific lithologic categories such as granite and gneiss. A more generalized but consistent lithologic categorization is used herein (intrusive and metamorphic) as some rocks categorized as "granites" or "gneiss" are not technically granites or gneiss, respectively.

The Fennoscandian Shield in Sweden, the Black Forest region in southern Germany, and the Molasse basin in Switzerland represent three distinct tectonic settings. The data locations are presented in Figure 12.1 along with indicators of current tectonic activity (seismicity) and long-term tectonic history (apatite fission track ages). The Fennoscandian Shield has a low density of seismic events, and fission track data around the sample locations in Sweden show that these rocks have exhumed extremely slowly from depths of 4 to 5 km over more than 250 million years (Hendriks *et al.* 2007). The rocks sampled from the Black Forest region and the basement underlying the Molasse basin are relatively close and consist of similar crystalline lithologies, but are derived from different tectonic settings. The Black Forest developed as the eastern rift shoulder of the Upper Rhine Graben following the onset of rifting in the Eocene (Illies 1972). The Black Forest region has a moderate density of seismic events and has experienced exhumation of 1–2 km since the late Eocene, with vertical motion predominantly taking place in the Miocene (Timar-Geng *et al.* 2006; Meyer *et al.* 2010). The Molasse basin has experienced more recent exhumation, with up to 1.2 km of exhumation since the Pliocene (Mazurek *et al.* 2006; Cederbom *et al.* 2011). Sample locations in the Molasse basin are all located within 5 km of seismic events that exceed the magnitude of 3 on a Richter scale (Fig. 12.1). The Black Forest region and the Molasse basin are influenced by similar maximum horizontal stress directions (Hinzen 2003; Reinecker *et al.* 2010). Earthquake fault plane solutions show a normal faulting regime in the Upper Rhine Graben and surrounding areas (Hinzen 2003),

Table 12.1 Summary of data sources.

References	N	Depth (m)	Reported units	Location	Test method	Lithology	Length of tested intervals (m)
Snow (1968)	25	1.9–89	m²	Colorado, USA	Injection	Metamorphic	<31
Brace (1980)	14	0–2015	Darcys	Manitoba, Canada; Cornwall, England, Nevada, New Mexico, South Carolina, Colorado, Wyoming USA	Various	Metamorphic and intrusive	0–30
Gale et al. (1982)	147	51–287	m²	Stripa Mine, Lindesberg, Sweden	Packer	Intrusive	2
Belanger et al. (1989)	76[a]	238–1610	m s⁻¹	Leuggern, Switzerland	Packer	Metamorphic	1–60, 924
Butler et al. (1989)	10	2007–2472	m s⁻¹	Weiach, Switzerland	Packer; slug; pulse; drill stem	Metamorphic	7–39, 416
Juhlin & Sandstedt (1989)	14	310–2240	m²	Cornwall, England; Siljan, Sweden; Bottstein, Switzerland; Cajon Pass, USA	Various	Metamorphic and intrusive	N/A
Ostrowski & Kloska (1989)	27	405–1480	m s⁻¹	Siblingen, Switzerland	Packer; slug; pulse; drill stem	Intrusive	5–359
McCord & Moe (1990)	40[a]	299–1240	m s⁻¹	Kaisten, Switzerland	Packer; slug; pulse; drill stem	Metamorphic	7–68
Moe et al. (1990)	23[a]	1510–2000	m s⁻¹	Schafisheim, Switzerland	Packer; slug; pulse; drill stem	Intrusive	9–326
Ahlbom et al. (1991)	164[a]	10–695	m s⁻¹	Baven, Sweden	Packer	Metamorphic	25
Stober(1995)	149	12–661	m s⁻¹	Black Forest, Germany	Open-hole	Intrusive and metamorphic	5–358
Huenges et al. (1997)	8	208–2130	m²	Windischeschenbach, Germany	Drill stem	Metamorphic	30–317
Morrow & Lockner (1997)	15	679–1610	m²	Illinois, USA	Pulse; injection	Intrusive	76–1470
Walker et al. (1997)	125	0–1390	m s⁻¹	Oskarshamn, Sweden	Packer	Intrusive	26–389
Wladis et al. (1997)	78[a]	0–625	m s⁻¹	Gidea, Sweden	Injection	Metamorphic	25
SKB (2008)	58[a]	0–985	m s⁻¹	Forsmark, Sweden	Packer	Metamorphic	20

[a] These data sets have a detection limit, which establishes an artificial minimum permeability.

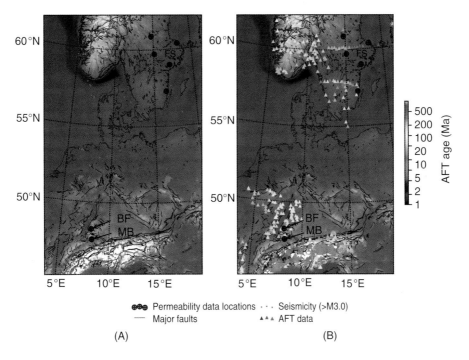

Fig. 12.1. Locations of permeability data and indicators of (A) short-term (years) and (B) long-term (million years) tectonic activity. Permeability data are derived from southern Germany and the Black Forest (BF), the Molasse basin (MB) in Switzerland and the Fennoscandian Shield (FS) in Sweden. Seismic events in (A) denote events since the year 2000 that exceed the magnitude of 3 on the Richter scale from the National Earthquake Information Center (http://earthquake.usgs.gov/regional/neic/). (B) AFT denotes apatite fission track data obtained from Herman et al. (2013). Apatite fission track data are a proxy for long-term tectonic activity. The apatite fission track age is approximately equal to the last time the rock outcrop was at a temperature of 120 °C (Wagner & Reimer 1972), which at normal geothermal gradients corresponds to a depth of approximately 3–5 km. (*See color plate section for the color representation of this figure.*)

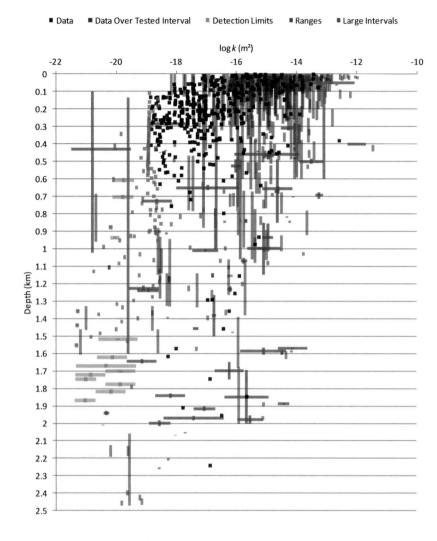

■ Data ■ Data Over Tested Interval ▪ Detection Limits ▪ Ranges ▪ Large Intervals

Fig. 12.2. Full data set of permeability data for crystalline rock (*n* = 973). Black points are singular or average permeability values (*n* = 422). Red lines are permeability values reported over a tested interval (*n* = 426). Gray points are data with reported detection limits (*n* = 80). Green points are the midpoints of permeability values reported as ranges, with the error bar showing the range (*n* = 37). Purple lines are data with tested intervals >500 m (*n* = 8). The vertical extent of a point indicates the extent of the tested interval. (*See color plate section for the color representation of this figure.*)

while the Molasse basin is currently under a thrust or strike-slip faulting regime (Reinecker *et al.* 2010).

Six different *in situ* permeability measurement methods were used in the synthesized studies: open-hole tests, drill stem tests, packer tests, injection tests, pulse tests, and slug tests. Test intervals range from 2 to 1400 m. To be included in the database, data points had to be *in situ* values at depths shallower than 2.5 km. To be included in statistical analysis, data had to be collected from tested intervals smaller than 500 m. The values of hydrogeological parameters are known to change with the scale of observation (Neuman 1994); this 500-m limit reduces the potential for permeability values in the database to be grossly affected by the scale of measurement. An earnest effort was made to include information regarding fracture and fault zone control on permeability in the synthesis. Unfortunately, this information was rarely provided in our compiled data sources and thus could not be included in our analysis. An important assumption in our analysis is that the *in situ* tests represent the permeability over the reported depths. In reality, testing is often controlled by more permeable features such as fractures or fault zones. However, we exclude the potential

impact of specific fractures or fault zones as we do not have data on their location, size, and hydraulic importance, while also acknowledging the importance of permeable features and the inherent difficulties in determining representative elementary volumes for hydraulic tests (Stober & Bucher 2015). The results of the data synthesis are presented in Figure 12.2, and summaries of the 16 data sources are provided in Table 12.1. In Table 12.1, note that studies in Switzerland and Sweden provide site-specific permeability for one distinct location each, while data from the Black Forest in Germany are a regional synthesis wherein each permeability value represents a different location.

All data were converted to permeability values (m²) where necessary to ensure consistency in the data set. Permeability (*k*) data measured in darcys were converted to m² through unit conversion (1 darcy = 9.87×10^{-13} m²). Converting hydraulic conductivity (*K*) data measured in m s^{-1} is more complex, requiring values of fluid viscosity and density at depth. We estimated values of viscosity and density by gathering location-specific salinity and temperature data and employing known viscosity and density functions dependent on total

Table 12.2 Summary of salinity and temperature values

References	Location	Salinity range (kg kg⁻¹)	Temperature gradient	Salinity source	Temperature source	Salinity source location	Temperature source location
Ahlbom et al. (1991)	Båven, Sweden	4.6E-5 to 5.2E-4	$5.25°C + 12°C\,km^{-1}$	Ahlbom et al. (1991)	SKB (2008)	Båven, Sweden	Forsmark, Sweden
Walker et al. (1997)	Oskarshamn, Sweden	0 to 7.4E-2	$5.25°C + 12°C\,km^{-1}$	Walker et al. (1997)	SKB (2008)	Oskarshamn, Sweden	Forsmark, Sweden
Wladis et al. (1997)	Gidea, Sweden	1.3E-4 to 5.0E-4	$5.25°C + 12°C\,km^{-1}$	Ahlbom et al. (1991) & Gale et al. (1982)	SKB (2008)	Båven & Lindesberg, Sweden	Forsmark, Sweden
SKB (2008)	Forsmark, Sweden	9.9E-5 to 1.4E-2	$5.25°C + 12°C\,km^{-1}$	SKB (2008)	SKB (2008)	Forsmark, Sweden	Forsmark, Sweden
Belanger et al. (1989)	Leuggern, Switzerland	8.1E-4 to 4.8E-3	$11.3°C + 32.9°C\,km^{-1}$	Wittwer (1986)	Wittwer (1986)	Leuggern, Switzerland	Leuggern, Switzerland
Butler et al. (1989)	Weiach, Switzerland	9E-3 to 3.1E-2	$7.8°C + 46.8°C\,km^{-1}$	Wittwer (1986)	Butler et al. (1989)	Weiach, Switzerland	Weiach, Switzerland
Ostrowski & Kloska (1989)	Siblingen, Switzerland	8.8E-4 to 8.9E-2	$9.2°C + 41.1°C\,km^{-1}$	Wittwer (1986)	Wittwer (1986)	Siblingen, Switzerland	Siblingen, Switzerland
McCord & Moe (1990)	Kaisten, Switzerland	1.2E-3 to 1.3E-3	$11.2°C + 36.6°C\,km^{-1}$	Wittwer (1986)	Wittwer (1986)	Kaisten, Switzerland	Kaisten, Switzerland
Moe et al. (1990)	Schafisheim, Switzerland	8E-3 to 1.4E-2	$5.4°C + 39.3°C\,km^{-1}$	Wittwer (1986)	Moe et al. (1990)	Schafisheim, Switzerland	Schafisheim, Switzerland
Stober (1995)	Black Forest, Germany	1.2E-4 to 7.4E-3	$11.2°C + 33.2°C\,km^{-1}$	Stober (1995)	Stober (1995)	Black Forest, Germany	Black Forest, Germany

dissolved solids (TDS) and temperature. Due to the nonlinear relationship of salinity with depth, depth- and location-specific salinity values are determined through linear interpolation of known salinity–depth values from literature. Point-specific temperatures are determined using location-specific geothermal gradients. Summaries of the salinity ranges, temperature functions, and data sources for each reference that measured conductivity in $m\,s^{-1}$ are provided in Table 12.2. Stuyfzand (1989) specifies the change in density with changes in salinity and temperature as follows:

$$\rho(T, TDS) = 1000 + 805(TDS)$$
$$- 6.5 \times 10^{-3}(T - 4 + 220(TDS))$$

where ρ represents density in $kg\,m^{-3}$, T represents temperature in °C, and TDS represents salinity in $kg\,kg^{-1}$. Batzle & Wang (1992) specify the change in viscosity with changes in salinity and temperature as follows:

$$\mu(T, TDS) = (0.1 + 0.333(TDS)$$
$$+ (1.65 + 91.9(TDS)^3) \times \exp(-a))$$
$$a = (0.42((TDS)^{0.8} - 0.17) + 0.045) \times T^{0.8}$$

where μ represents viscosity in centipoises (this value is converted to Pa s by dividing by 10^4). After viscosity and density have been estimated, permeability values are calculated from conductivity values as follows:

$$k = K\left(\frac{\mu}{\rho g}\right)$$

where k represents permeability in m^2, K represents hydraulic conductivity in ms^{-1}, and g represents the gravitational constant $9.81\,m\,s^{-2}$.

Logarithmic functions are fit to the data using simple linear regression. Logarithmic functions are used due to their prevalence as a fitting function in the literature (Snow 1968; Anderson et al. 1985; Stober 1995; Wladis et al. 1997; Ingebritsen & Manning 1999; Shmonov et al. 2003; Saar & Manga 2004; Stober & Bucher 2007a). Note that the use of a logarithmic function implies an assumption of a lower limit on permeability due to the asymptotic nature of logarithmic functions. Permeability values reported as a range are included in the regression by selecting the midpoint of the range. Permeability values that are reported as a methodological cutoff ($n = 80$ points of the total data set), which are herein referred to as "detection limits," are not included in the regressions. We note that excluding detection-limit data may impact the statistical analysis by eliminating a number of low permeability data from the regressions. However, we choose to exclude these values as they are objectively lower quality data that do not describe an actual permeability value. We tested the importance of excluding these cutoff values by artificially assigning them a permeability value one and two orders of magnitude lower than their reported values, and found that this did not significantly change any of the statistical results. The R^2 value of the regression is used to quantify the quality of the derived fit. A t test on the slope parameter was performed for each regression. The t test evaluates the discrepancy between the derived slope and a slope of zero, indicating no relationship, and requires an assumption of normality in the regression error. Passing the t test implies that there is a statistically significant relationship of permeability with depth; low R^2 values imply that the derived function is a poor predictor of permeability with depth.

The importance of different variables (depth, lithology, and tectonic setting) was examined by dividing the permeability data into different categories and comparing these categories using the nonparametric Kolmogorov–Smirnov (KS) test (Lilliefors 1967). The KS test is a statistical method that identifies whether two distributions are derived from the same distribution. Failing the KS test indicates that the two distributions are not similar enough to be derived from the same distribution. We use the KS test to quantify the difference between permeability distributions at different depth intervals, as well as to test relationships between lithologies (intrusive or metamorphic) and tectonic setting (Fennoscandian Shield, southern Germany, and Molasse basin).

RESULTS AND DISCUSSION

In Figure 12.2, data points with tested intervals greater than 500 m ($n = 8$) and data points representing detection limits ($n = 80$) are presented for context, but excluded from the following statistical analyses. A summary of the regression analyses and KS tests are provided in Tables 12.3 and 12.4, respectively.

All data

The average permeability of the entire data set excluding tested intervals >0.5 km and detection limits (Fig. 12.3, $n = 885$) is -16.3 ± 1.81 m^2 ($\mu_{\log k} \pm \sigma_{\log k}$, where $\mu_{\log k}$ is the arithmetic mean and $\sigma_{\log k}$ is the standard deviation. Note that all reported "averages" refer to the arithmetic mean). The frequency of permeability data decreases with depth (Fig. 12.4). Although an ideal statistical analysis would have data randomly distributed over the 2.5-km range examined in this analysis, the realities of *in situ* data acquisition create a shallow data bias in the synthesized data set.

A statistically significant logarithmic fit exists through the data at $<1\%$ significance ($P = 1.32\text{E-}9$), although this function has a low predictive power ($R^2 = 0.230$). The logarithmic fit shows minimal qualitative agreement with both the Shmonov *et al.* (2003) fit and the Manning–Ingebritsen fit (Ingebritsen & Manning 1999) in the entire 2.5-km range. The lack of agreement with the Manning–Ingebritsen fit is not unexpected, as this was derived to describe much deeper permeability data than examined in this analysis. Although the data support the assumption of a decrease of permeability with depth, the low

predictive power of the derived logarithmic fit illustrates the ineffectiveness of a general logarithmic permeability–depth relationship as a tool to predict permeability values. Stober & Bucher (2007a) also reached this conclusion in the analysis of a smaller crystalline rock data set.

Multiple KS tests were performed to determine an appropriate cutoff between "deeper" and "near-surface" data (Table 12.4). KS tests examining cutoffs from 0.1 to 1.0 km display P values at least two orders of magnitude below the 5% significance cutoff in all cases, indicating that P values are not useful for assigning a depth cutoff. Therefore, we use the arbitrary depth cutoff of 0.1 km, which (i) maintains a reasonable

Table 12.3 Summary of regression analyses

Data set	t Test P value	R^2	N
All	1.32E-09	2.30E-01	885
Intrusive	2.49E-03	1.29E-01	390
Metamorphic	1.99E-07	3.00E-01	495
Southern Germany	3.91 E-03	3.91 E-01	152
Southern Germany metamorphic	5.05E-04	5.43E-01	107
Southern Germany intrusive	**9.46E-01**	**4.98E-03**	**45**
Fennoscandian Shield	1.25E-02	1.53E-01	515
Fennoscandian Shield metamorphic	**1.54E-01**	**1.91E-01**	**236**
Fennoscandian Shield intrusive	**1.03E-01**	**9.11E-02**	**279**
Molasse basin	3.82E-03	5.21E-02	159
Molasse basin metamorphic	1.33E-03	8.80E-02	119
Molasse basin intrusive	1.78E-02	1.26E-01	40

Bold indicates data sets that show no statistically significant decrease of permeability with depth at 5% significance.

Table 12.4 Summary of Kolmogorov–Smirnov tests

Data set a (km)	Data set b (km)	n_a	n_b	P value
All <0.1	All >0.1	265	620	1.66E-31
All <0.2	All > 0.2	425	460	1.60E-22
All <0.3	All >0.3	557	328	3.07E-15
All <0.4	All >0.4	622	263	3.00E-15
All <0.5	All >0.5	676	209	1.17E-14
All <0.6	All >0.6	698	187	2.44E-14
All <0.7	All >0.7	719	166	2.15E-15
All <0.8	All >0.8	735	150	5.82E-13
All <0.9	All >0.9	757	128	1.08E-14
All <1.0	All >1.0	776	109	3.24E-13
Intrusive <0.1	**Metamorphic <0.1**	**137**	**128**	**4.83E-01**
Intrusive >0.1	Metamorphic >0.1	253	367	4.20E-08
Fennoscandian <0.1	S. Germany <0.1	156	81	1.20E-10
Fennoscandian >0.1	S. Germany >0.1	359	71	3.00E-13
Fennoscandian intrusive <0.1	S. Germany intrusive <0.1	106	29	2.32E-05
Fennoscandian metamorphic <0.1	S. Germany metamorphic <0.1	50	52	2.49E-04
Fennoscandian intrusive <0.1	**Fennoscandian metamorphi 0.1**	**106**	**50**	**7.59E-01**
S. Germany metamorphic <0.1	S. Germany metamorphic <0.1	29	52	4.93E-02
Fennoscandian intrusive 0.4–2	**Molasse intrusive 0.4–2**	**25**	**40**	**1.23E-01**
Fennoscandian 0.4–2	Molasse 0.4–2	72	140	1.58E-04
Fennoscandian >0.3	Molasse >0.3	129	155	8.79E-07

Bold indicates data sets that show statistical similarity at 5% significance.

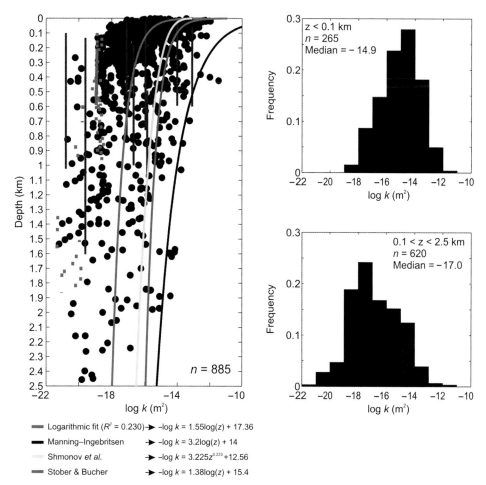

Fig. 12.3. The relationship between permeability and depth for the full data set, with error bars removed for clarity. Ranges are plotted as the midpoint. Gray rectangles indicate measurements at a detection limit. Purple lines indicate data points from tested intervals greater than 500 m. Red line indicates logarithmic fit through data ($R^2 = 0.230$). Black line indicates Manning–Ingebritsen fit (Ingebritsen & Manning 1999). Blue line indicates Shmonov *et al.* (2003) fit. Green line indicates Stober & Bucher (2007a) fit. Histograms display distribution of permeability data above and below 0.1 km. (*See color plate section for the color representation of this figure.*)

Logarithmic fit ($R^2 = 0.230$) ➤ $-\log k = 1.55\log(z) + 17.36$
Manning–Ingebritsen ➤ $-\log k = 3.2\log(z) + 14$
Shmonov *et al.* ➤ $-\log k = 3.225z^{0.223} + 12.56$
Stober & Bucher ➤ $-\log k = 1.38\log(z) + 15.4$

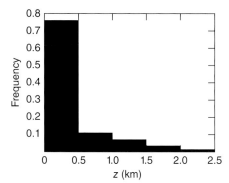

Fig. 12.4. The distribution of permeability values in the full data set.

statistical size above and below the cutoff, (ii) is consistent with previous near-surface permeability compilations (Gleeson *et al.* 2011), and (iii) allows calculation of permeability values, which could be useful for near-surface hydrologic modeling. Hereafter, "near-surface" permeability refers to <0.1 km depth and "deeper" permeability refers to >0.1 km depth. The average permeability in the near-surface data is $-15.0 \pm 1.36\,\text{m}^2$ ($n = 265$), approximately two orders of magnitude higher than

the average permeability in the deep data ($-16.8 \pm 1.71\,\text{m}^2$, $n = 620$). Higher permeabilities at shallow depths could be due to larger fracture apertures, greater connectivity, or higher fracture density due to low overburden stress, unloading following glacial isostatic rebound and/or the development of sheeting fractures. Rutqvist (2015) describes how large stresses can create highly conducive "locked-open" fractures that do not close in response to large overburden stresses, potentially introducing large permeability values at depth. Rutqvist (2015) also notes that mineral precipitation and dissolution may play a role in creating "locked-open" fractures. Earnest & Boutt (2014) describe an even more explicit relationship between permeability and stress in fractured rock, describing how stress magnitude, shear stiffness and normal stiffness are dominant controls on fracture aperture, and thus permeability, in the upper 1 km of the subsurface.

Lithology

Both intrusive and metamorphic lithologies display a statistically significant logarithmic decrease of permeability with depth, although again with a low predictive power (Fig. 12.5, Table 12.3). The average permeability of the

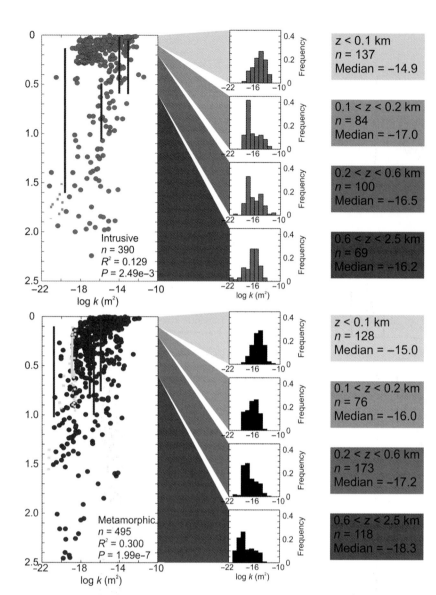

Fig. 12.5. The relationship between permeability and lithology for metamorphic (blue) and intrusive (red) rocks. All data points are midpoints of tested intervals. Pink rectangles indicate intrusive detection limits. Cyan rectangles indicate metamorphic detection limits. Purple lines indicate data points from tested intervals >500 m. Reported R^2 and P values are for logarithmic fits through data. Histograms identify the permeability distribution in four depth ranges. From top to bottom: <100, 100–200, 200–600, and >600 m. (*See color plate section for the color representation of this figure.*)

intrusive data set is almost one order of magnitude larger than the metamorphic average (intrusive $= -15.9 \pm 1.69 \, \mathrm{m}^2$; metamorphic $= -16.6 \pm 1.83 \, \mathrm{m}^2$) although this difference is within one standard deviation. The metamorphic data display a fit with more predictive power than the all-data case, although the R^2 value is still low ($R^2 = 0.300$). A KS test on data in the near surface (<0.1 km) in each lithology shows that intrusive and metamorphic data are statistically similar at 5% significance ($P = 0.483$), indicating that lithology may be a weak control on crystalline rock permeability in the near surface. A KS test on deeper data shows that intrusive and metamorphic data are statistically dissimilar at 5% significance ($P = 7.41 \times 10^{-3}$). The histograms for metamorphic data in the four arbitrary depth intervals in Figure 12.5 display a smoother transition to low permeability values with depth (a steady decrease in permeability) as compared with the intrusive data, which display a much more discontinuous transition toward deeper

depth intervals. Both data sets include large values of permeability at depth (e.g., $10^{-14} \, \mathrm{m}^2$ values below 1.5 km), although large permeability values are less frequent in the metamorphic data.

This analysis suggests that lithology (classified broadly as either "metamorphic" or "intrusive") might not be a critical control on crystalline rock permeability at near-surface depths. Metamorphic data display better agreement with a logarithmically declining permeability–depth function as compared with intrusive data. Intrusive rocks display a higher average permeability than metamorphic rocks over the entire 2.5-km-depth range (Fig. 12.5). Both intrusive and metamorphic data sets show a statistically significant logarithmic decrease in permeability with depth. This conclusion agrees with Stober (1996) who found that granitic rocks had higher conductivities than gneissic rocks and that gneissic rocks display a decrease in permeability with depth. Note, however, that in the Stober

(1996) analysis, granitic rocks display no decrease with depth, which is not the case with the intrusive data in this analysis.

Tectonic setting

Each tectonic setting displays a statistically significant logarithmic decrease of permeability with depth, although with low predictive power (Fig. 12.6). The fit derived from the southern Germany data displays the highest predictive power ($R^2 = 0.391$), while the fit from the Molasse basin displays almost no predictive power ($R^2 = 0.052$), although the lack of near-surface data in the Molasse basin and the deeper data in the Fennoscandian Shield and southern Germany limits the veracity and application of these statistics. Permeabilities in the Molasse basin ($\sigma_{\log k} = 2.10 \, \text{m}^2$) display the largest amount of scatter as compared with the Fennoscandian Shield Basin ($\sigma_{\log k} = 1.53 \, \text{m}^2$) and southern Germany ($\sigma_{\log k} = 1.36 \, \text{m}^2$). The scatter in permeability correlates with tectonic activity, with low scatter in the tectonically inactive Fennoscandian Shield and

higher scatter in the Molasse basin, which has undergone high rates of vertical motion in the Pliocene and Pleistocene (Genser *et al.* 2007; Cederbom *et al.* 2011). The large scatter and poor permeability–depth fit in the Molasse basin are also reflected in the bimodal distribution of the Molasse basin histogram in Figure 12.6.

A KS test on near-surface data in the Fennoscandian Shield (average $= -15.3 \pm 1.38 \, \text{m}^2$, $n = 156$) and southern Germany (average $= -14.2 \pm 0.937 \, \text{m}^2$, $n = 81$) shows that the two data sets are statistically dissimilar at 5% significance ($P = 1.5 \times 10^{-7}$). The deeper data in these regions show the same result ($P = 3.0 \times 10^{-13}$). In the light of the statistically similar near-surface result from the lithology analysis, this suggests that tectonic setting may be a stronger control on permeability in the near surface. This is consistent with the observation of Maloney *et al.* (2006) who noted a similar relationship in the Canadian Shield between stresses and depth. In their study, the near surface (<300–600 m) was much more influenced by local horizontal stresses, while stresses at depth

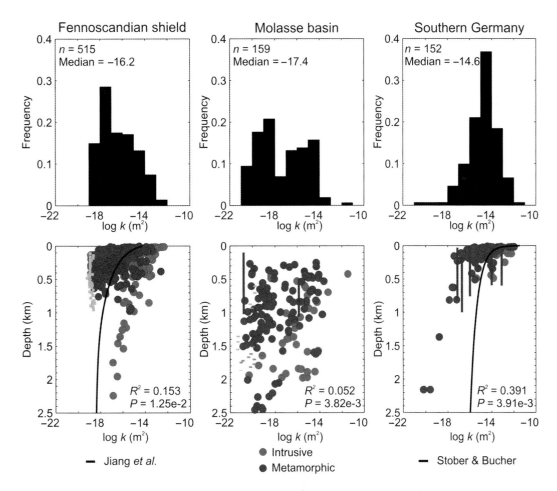

Fig. 12.6. The relationship between permeability and tectonic setting. Red points indicate intrusive rocks. Blue points indicate metamorphic rocks. Pink rectangles indicate intrusive detection limits. Cyan rectangles indicate metamorphic detection limits. Purple lines indicate data points from tested intervals >500 m. All data points are midpoints. Reported R^2 and P values are for logarithmic fits through the combination of intrusive and metamorphic data. Gray lines are functions from the literature (Stober & Bucher 2007; Jiang *et al.* 2010c). (*See color plate section for the color representation of this figure.*)

reflected a stress regime determined by some distant boundary. Thus, local tectonics may be more important in the near surface but less important at depth. We exclude the Molasse basin from this comparison due to the lack of near-surface data. In the 0.3- to 2.5-km-depth range where both Molasse basin and Fennoscandian Shield data are available, a KS test shows that the data sets are statistically dissimilar at 5% significance ($P = 8.8 \times 10^{-7}$). Considering tectonic setting provides useful insight into the applicability of a generalized logarithmic permeability–depth relationship. For example, applying a more general permeability–depth function to the data in the Molasse basin would be nonsensical due to the large amount of scatter inherent in the data.

Tectonic setting and lithology

Three tectonic setting–lithology combinations display no statistically significant permeability–depth relationship at 5% significance: Fennoscandian Shield intrusive ($P = 0.103$, $n = 279$); Fennoscandian Shield metamorphic ($P = 0.154$, $n = 236$); and southern Germany intrusive ($P = 0.946$, $n = 45$). An important caveat to this observation is that the Fennoscandian metamorphic and southern Germany intrusive data sets have no data below 1 and 0.5 km, respectively; furthermore, the Molasse intrusive data include no data above 0.4 km (Fig. 12.7 and Table 12.2). KS tests on near-surface intrusive and metamorphic data in the Fennoscandian Shield and

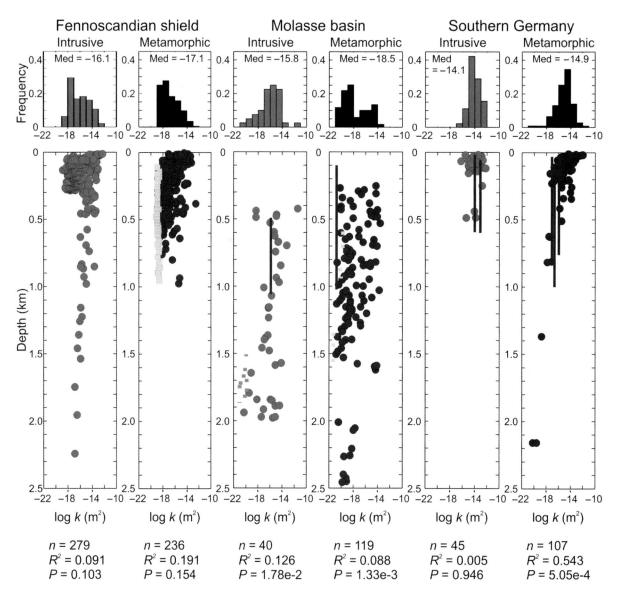

Fig. 12.7. The relationship between permeability and lithologies in different tectonic settings. Red indicates intrusive rocks. Blue indicates metamorphic rocks. Pink rectangles indicate intrusive detection limits, while cyan rectangles indicate metamorphic detection limits. Purple lines indicate data points from tested intervals >500 m. All data points are midpoints. Reported R^2 and P values are for logarithmic fits through the data. P values in boldface indicate data sets that fail the t test at 5% significance. Histograms include text that indicates the median value of the distribution. (*See color plate section for the color representation of this figure.*)

southern Germany indicate that these data are statistically dissimilar at 5% significance ($P = 2.3 \times 10^{-5}$ and $P = 2.5 \times 10^{-4}$). KS tests indicate that near-surface metamorphic and intrusive data in the Fennoscandian Shield are statistically similar at 5% significance, while near-surface metamorphic and intrusive data in southern Germany are dissimilar at just under 5% significance ($P = 4.9 \times 10^{-2}$). The similarity of near-surface data for multiple lithologies in a single tectonic setting relative to the dissimilarity between tectonic settings provides additional evidence that lithology may be a weaker control than tectonic setting. A KS test on intrusive data from Fennoscandian Shield and Molasse basin in the 0.4- to 2-km interval ($n = 25$ and $n = 40$, respectively) indicates that these data are statistically similar at 5% significance ($P = 0.123$), suggesting that lithology may be a more important control on permeability for deeper data. Accounting for both tectonic setting and lithology defines stronger and more credible permeability–depth relationships, although categorization of data in this way decreases the number of points in each statistical analysis.

CONCLUSIONS

We compiled a large data set ($n = 973$) of permeability data from metamorphic and intrusive crystalline rocks in the shallow crust to depths of 2.5 km. The data were obtained mainly from three tectonic settings as follows: the Molasse basin in Switzerland, the Fennoscandian Shield in Sweden and southern Germany. We used trend analyses and KS tests to quantify relationships between permeability and depth for the entire data set (excluding data measured under a detection limit and data from tested intervals greater than 500 m, $n = 885$) and subsets that distinguish tectonic settings and intrusive or metamorphic lithologies.

(1) The trend analysis does not support a consistently applicable and generalizable relationship between permeability and depth in crystalline rock in the shallow crust ($z < 2.5$ km), in agreement with conclusions drawn previously by Brace (1980, 1984), Huenges et al. (1997), and Stober & Bucher (2007a). A logarithmic fit to the entire data provides a very low R^2 value of 0.230 (Fig. 12.3). Although a t test indicates a statistically significant decrease in permeability with depth at 5% significance, the low predictive power of the fitted function suggests that a generalized permeability–depth function should not be used in hydrologic and earth system models of the shallow crust without further justification.

(2) Higher permeabilities are more common at shallow depths in crystalline rock (Fig. 12.3). The KS test shows that near-surface permeabilities are statistically dissimilar (at 5% significance) from deeper permeabilities regardless of the depth cutoff (100–1000 m). The average near-surface (<0.1 km) permeability ($\mu_{\log k} = -15.0 \pm 1.36$ m^2, $n = 265$) is almost two orders of magnitude higher than the average of deeper permeability values ($\mu_{\log k} = -16.8 \pm 1.71$ m^2,

$n = 624$). Higher permeabilities at shallow depths could be due to fracture aperture, density, or connectivity, and hydromechanical responses due to lower vertical stresses and/or minimal fracture in-filling.

(3) Lithology has a weak control on crystalline rock permeability at near-surface depths: the KS test shows no statistical difference between metamorphic and intrusive rocks in the near surface at 5% significance. Intrusive rock permeabilities with depth are poorly described using a logarithmic function ($R^2 = 0.129$). Metamorphic rock permeabilities show a better agreement, but the predictive power of the function is still low ($R^2 = 0.300$). In both cases, a statistically significant decrease in permeability is apparent at 5% significance.

(4) Tectonic setting has a stronger control than lithology on crystalline rock permeability in the near surface and may be a weaker control than lithology on crystalline rock permeability in the deeper subsurface (Fig. 12.6). A KS test on near-surface data in the Fennoscandian Shield and southern Germany (where near-surface data are available) indicates that these data are statistically dissimilar at 5% significance. On the contrary, a KS test indicates that near-surface metamorphic and intrusive data in the Fennoscandian Shield are statistically similar, while near-surface metamorphic and intrusive data in southern Germany are dissimilar at just under 5% significance ($P = 4.9 \times 10^{-2}$). Thus, tectonic setting appears to have more of an influence on permeability than lithology in the near surface. In the deeper subsurface, however, a KS test on Fennoscandian intrusive data and Molasse basin intrusive data in the 0.4- to 2-km interval indicates that these data are statistically similar at 5% significance ($P = 0.123$), suggesting that lithology may have more influence on permeability in the deeper subsurface.

(5) Tectonic activity may be a strong control on the variation in permeability with depth in crystalline rocks. Larger stress magnitudes in tectonically active regions may produce larger than expected fracture apertures at depth (Earnest & Boutt 2014; Rutqvist 2015), confounding a logarithmically decreasing permeability–depth relationship. The Molasse basin is an active tectonic region, as indicated by high rates of vertical motion since the Pliocene (Genser et al. 2007; Cederbom et al. 2011) (Fig. 12.1). Permeabilities in the Molasse basin are very scattered at depth, with the corresponding logarithmic function displaying an R^2 of just 0.052. While we did not explicitly explore the physical processes causing the higher values of permeability, the compiled data suggest that active tectonics may lead to higher permeabilities in the shallow crust, a hypothesis that may focus future research efforts.

(6) The clearest permeability–depth relationships in crystalline rock are defined when lithology and tectonic setting are both accounted for (Fig. 12.7), although the smaller data sets available at this level of categorization limit the efficacy of the derived logarithmic fits. Three of six data sets that

distinguish both tectonic setting and lithology demonstrate no statistically significant decrease in permeability with depth (Fennoscandian intrusive, Fennoscandian metamorphic and southern Germany intrusive). Of the remaining three, the Molasse metamorphic and Molasse intrusive data display very low predictive power ($R^2 = 0.088$ and $R^2 = 0.126$, respectively), while the southern Germany metamorphic data display the largest predictive power of any data set analyzed ($R^2 = 0.543$).

DATA AVAILABILITY

The full data set is available from the research web page of the corresponding author and also on figshare.

ACKNOWLEDGMENTS

We thank I. Stober for providing guidance on translating publications on the Black Forest that helped in our data collection. We also thank I. Stober, K Bucher, S. Ingebritsen, M. Person, and an anonymous reviewer for insightful and useful suggestions which significantly improved this manuscript.

CHAPTER 13

Understanding heat and groundwater flow through continental flood basalt provinces: insights gained from alternative models of permeability/depth relationships for the Columbia Plateau, United States

ERICK R. BURNS[1], COLIN F. WILLIAMS[2], STEVEN E. INGEBRITSEN[2], CLIFFORD I. VOSS[2], FRANK A. SPANE[3] AND JACOB DEANGELO[2]

[1] U.S. Geological Survey, Portland, OR, USA; [2] U.S. Geological Survey, Menlo Park, CA, USA; [3] Pacific Northwest National Laboratory, Richland, WA, USA

ABSTRACT

Heat-flow mapping of the western United States has identified an apparent low-heat-flow anomaly coincident with the Columbia Plateau Regional Aquifer System, a thick sequence of basalt aquifers within the Columbia River Basalt Group. A heat and mass transport model was used to evaluate the potential impact of groundwater flow on heat flow along two different regional groundwater flow paths. Limited *in situ* permeability (k) data from the Columbia River Basalt Group are compatible with a steep permeability decrease (approximately 3.5 orders of magnitude) at 600–900 m depth and approximately 40 °C. Numerical simulations incorporating this permeability decrease demonstrate that regional groundwater flow can explain lower-than-expected heat flow in these highly anisotropic ($k_x/k_z \sim 10^4$) continental flood basalts. Simulation results indicate that the abrupt reduction in permeability at approximately 600 m depth results in an equivalently abrupt transition from a shallow region where heat flow is affected by groundwater flow to a deeper region of conduction-dominated heat flow. Most existing heat-flow measurements within the Columbia River Basalt Group are from shallower than 600 m depth or near regional groundwater discharge zones, so that heat-flow maps generated using these data are likely influenced by groundwater flow. Substantial k decreases at similar temperatures have also been observed in the volcanic rocks of the adjacent Cascade Range volcanic arc and at Kilauea Volcano, Hawaii, where they result from low-temperature hydrothermal alteration.

Key words: advection, advective heat transport, anisotropy, conduction-dominated, flood basalts, heat flow, hydrothermal alteration, permeability, regional groundwater flow

INTRODUCTION

Within the Earth's upper crust, heat flow occurs via two primary mechanisms, conduction and advection (commonly by groundwater). Heat-flow maps (e.g., Fig. 13.1) are derived from the vertical component of heat flow measured in a borehole and are typically intended to represent the conductive component of heat flow from the Earth's upper crust. In areas where there is significant advection of heat, shallow heat-flow measurements may not reflect thermal conditions deeper in the crust.

Advective heat transport can be driven by forced convection and free convection (Saar 2011). Free convection refers to density-driven flow, and forced convection refers to groundwater flow driven by hydraulic gradients resulting from topography and other factors. Free convection results from unstable fluid-density configurations and depends on the ease with which the fluid can move through the geologic material, the rate of heat addition, and the rate at which heat is conducted away from the heat source. The Rayleigh number is a measure of the relation between these factors, and free convection occurs when the Rayleigh number exceeds a critical value. In natural systems, both free convection and forced convection can influence groundwater flow patterns.

The hydrogeologic relation between heat flow and fluid flow has been the topic of considerable research, which is summarized in review articles by Anderson (2005) and Saar (2011). Cross-sectional numerical models have been used to demonstrate thermal anomalies associated with active groundwater

Crustal Permeability, First Edition. Edited by Tom Gleeson and Steven E. Ingebritsen.
© 2017 John Wiley & Sons, Ltd. Published 2017 by John Wiley & Sons, Ltd.
Companion Websites: www.wiley.com/go/gleeson/crustalpermeability/
http://crustalpermeability.weebly.com/

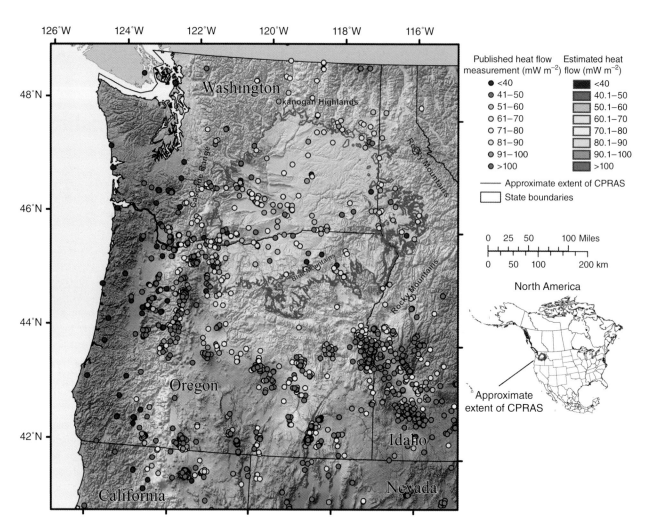

Fig. 13.1. Heat-flow map of the northwestern United States showing apparent low heat flow in the vicinity of the Columbia Plateau Regional Aquifer System. Contours are based on published heat-flow data for the Cascade Range and adjacent regions assembled for the USGS geothermal database. The map was constructed using the methods of Williams & DeAngelo (2008). The *Approximate Extent of Columbia Plateau Regional Aquifer System* is the extent of Columbia River Basalts from the geologic model of Burns *et al.* (2011). (*See color plate section for the color representation of this figure.*)

flow (e.g., Smith & Chapman 1983; Forster & Smith 1988a,b, 1989; Nunn & Deming 1991; Deming 1993). Early work by Smith & Chapman (1983) demonstrated most of the salient features and dependencies for systems with no free convection. Compared to pure heat conduction through an aquifer system, advective transport of heat results in a decrease in temperature and near-surface heat flow in upland recharge areas and an increase in temperature and near-surface heat flow in lowland discharge areas. Using reasonable hydrogeologic parameters, temperature and heat-flow distributions have been shown to be sensitive to variations in recharge and discharge magnitude and location, depth and length of the aquifer system, and permeability.

Recent regional mapping of geothermal resources of the western United States (e.g., Williams & DeAngelo 2008, 2011) identified an apparent low-heat-flow anomaly associated with the Columbia Plateau Regional Aquifer System

(Fig. 13.1). Although heat flow in most areas immediately to the east of the Cascade Range is above the global average for Cenozoic tectonic provinces (approximately >65 mW m^{-2}; Pollack *et al.* 1993), heat-flow measurements from most of the Columbia Plateau are significantly lower. Low heat flow through the Columbia Plateau is particularly enigmatic because this province is seismically active and lies adjacent to active volcanoes of the Cascade Range. However, throughout the Columbia Plateau, most temperature profile data used to measure heat flow have been collected from depth intervals that are within the active regional groundwater flow system (typically <500 m deep, Fig. 13.2).

In order to understand geothermal heat flow at the province scale, it is necessary to understand the permeability structure of the upper crust and the resulting coupled heat and groundwater flow. The Columbia Plateau Regional Aquifer System is comprised almost entirely of continental flood basalts, a type

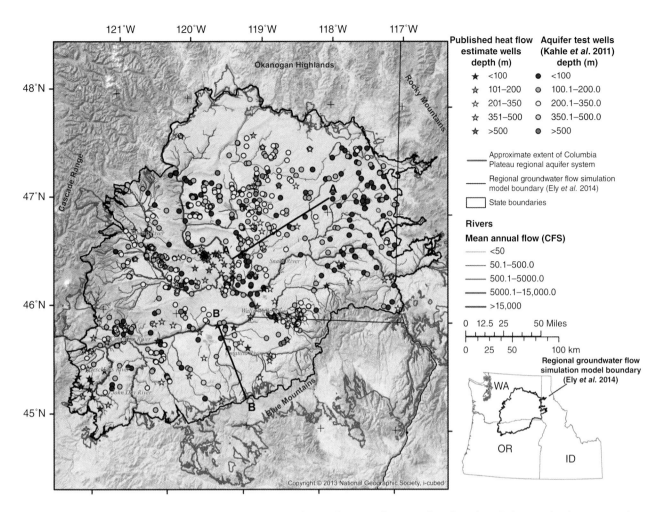

Fig. 13.2. Map of compiled borehole temperature logs (stars) and aquifer tests from groundwater supply wells (circles). The bottom of each temperature log is frequently at or above the depth of nearby aquifer tests, indicating that most temperature data were collected within the active groundwater system. To estimate intrinsic permeability from aquifer tests, the temperature of water pumped was estimated using the nearest temperature log. Representative cross sections A–A′ and B–B′ were used for coupled groundwater and heat-flow simulations. (*See color plate section for the color representation of this figure.*)

example of large igneous provinces (Jerram & Widdowson 2005) typified by thick and laterally extensive mafic lava flows. The massive sheet flows of the Columbia Plateau result in confined aquifers over much of the region, but individual lobe-shaped flows of smaller areal extent or frequent joints and fractures can result in unconfined aquifers. This study and previous studies agree that the permeability of the Columbia River Basalt Group – and presumably other continental flood basalts – is highly anisotropic ($k_x/k_z \sim 10^4$). The thinner, less laterally extensive basalt flows of the adjacent Cascade Range (Manga 1996, 1997) and the Hawaiian Islands (e.g., Gingerich & Voss 2005) exhibit much less anisotropy ($k_x/k_z \leq 10^2$).

This manuscript summarizes two-dimensional (2D) cross-sectional coupled groundwater and heat-flow models that demonstrate that, for the most likely distributions of permeability and aquifer tests (Kahle *et al.* 2011; Spane 2013), the apparent low-heat-flow anomaly in the Columbia Plateau Regional Aquifer System can be explained by accounting for

the heat transported to large regional rivers by the groundwater system. The heat-flow pattern is shown to be spatially complex both laterally and vertically, illustrating the difficulties in using the current borehole temperature data set to estimate regional heat flow.

BACKGROUND

Location and geology of study area

The Columbia Plateau Regional Aquifer System lies in a structural and topographic basin within the Columbia River drainage. It is bounded on the west by the Cascade Range, on the east by the Rocky Mountains, on the north by the Okanogan Highlands, and on the south by the Blue Mountains (Fig. 13.1). The Columbia Plateau is underlain by the Miocene Columbia River Basalt Group, a series of more than 300 sheet and intracanyon lava flows and sedimentary interbeds. Sheet flows are areally extensive, sometimes covering tens of thousands of square kilometers. Flows that separate into lobes or channels controlled by paleotopography are called intracanyon flows. Sheet flows commonly take on intracanyon character near their margins. Individual flows range in thickness from 3 m to more than 100 m (Tolan *et al.* 1989; Drost *et al.* 1990). Near the center of the Columbia Plateau Regional Aquifer System, the total thickness of the Columbia River Basalt Group is estimated to exceed 5 km (Reidel *et al.* 2002; Burns *et al.* 2011). Thick sedimentary deposits overlie the basalt in only a few basins, with basalt units occurring at or near land surface over most of the Plateau (Burns *et al.* 2011). Columbia River Basalt Group units are variably folded and faulted. Generally, structural intensity is greater near the uplifted Blue Mountains and Cascade Range. An expanded description of the Columbia River Basalt Group is provided by Reidel *et al.* (2013).

Groundwater occurrence and movement

Groundwater moves through the regional aquifer system along preferential pathways developed during lava deposition. Most Columbia River Basalt Group lava flows consist of a dense flow interior and irregular flow tops and flow bottoms with a variety of textures (Reidel *et al.* 2002). Although flow interiors have joints and fractures, they typically do not transmit water easily. Flow tops and bottoms are commonly vesicular or brecciated, and may or may not be permeable. Local permeability of flow tops and bottoms can be highly variable over short distances as a result of depositional processes, but tends to be high over long distances, resulting in highly transmissive aquifers at the regional scale. The Columbia River Basalt Group thus comprises a stack of laterally extensive lava flows with relatively thin permeable productive zones at flow tops and flow bottoms separated by relatively thick dense flow interiors of low permeability. Thin permeable aquifers are estimated to occupy about 10% of the total flow thickness (Burns *et al.* 2012a; Ely *et al.* 2014). Flow interiors have low permeability and low storage characteristics, so that they form effective confining units between permeable flow tops. As a result, the aquifer system is highly anisotropic, with effective horizontal hydraulic conductivity controlled by the fraction of the thickness occupied by thin aquifers and the effective vertical hydraulic conductivity controlled by the dense flow interiors. Effective bulk horizontal permeability is frequently more than 10^4 times greater than bulk vertical permeability (summary Tables 2–4 of Kahle *et al.* 2011). Flow interiors also commonly have negligible porosity compared to an approximate value of 0.25 for thin aquifers (Reidel *et al.* 2002). The bulk porosity of the Columbia River Basalt Group is about 0.025.

The limited data available suggest that horizontal hydraulic conductivity may decrease with increasing depth. Hansen *et al.* (1994) simulated groundwater flow through the entire thickness of the Columbia Plateau Regional Aquifer System, but during model calibration, it was found that horizontal

hydraulic conductivity (K_h) values decreased with increasing depth. To correct for the reduction in permeability with depth, the equations of Weiss (1982) were used, under the assumption that K_h would decrease as a function of overburden pressure only, resulting in a typical reduction in permeability by a factor <2 over the entire thickness. Recent simulation of the Columbia Plateau Regional Aquifer System (Ely *et al.* 2014) tested the hypothesis that horizontal permeability decreases with increasing depth using computer-assisted parameter estimation techniques, but found that the calibration data were insufficient to positively identify a persistent vertical trend in permeability. This was attributed to the fact that almost all of the calibration data were collected from very shallow depths compared to the total thickness of the Columbia Plateau Regional Aquifer System, so that model calibration was only sensitive to the upper part of the groundwater flow system. It was postulated that the Hansen and Vaccaro model may have required a vertical trend in permeability to compensate for model structural error. In particular, the Hansen and Vaccaro model had very coarse vertical discretization (cells were hundreds to thousands of meters thick) compared to the Ely *et al.* model (cells were typically 30 m thick), possibly resulting in lowering of permeability during calibration to reduce erroneous overconnection of the deep aquifers with surface water boundaries. Hansen *et al.* (1994) estimated that Columbia River Basalt Group bulk K_h ranges from 3.0×10^{-7} to 3.0×10^{-5} m s^{-1}, and Ely *et al.* (2014) estimated that Columbia River Basalt Group bulk K_h ranges from 3.5×10^{-6} to 9.9×10^{-5} m s^{-1}. Both studies found that K_h varies laterally, with lower K_h occurring in areas with more intense geologic structure (more folds and faults) and higher K_h in relatively undeformed areas.

Published K_h values from hydraulic tests for Columbia River Basalt Group units range from 10^{-15} to 0.21 m s^{-1}, over 14 orders of magnitude (Kahle *et al.* 2011). These tests include both lava interflows (aquifers) and flow interiors (confining units). The K_h estimates from aquifer pump tests documented a narrower range of K_h of 2.8×10^{-7} to 0.21 m s^{-1}. Near the center of the Columbia Plateau Regional Aquifer System, Spane (1982) found that hydraulic conductivity in the deeper Grande Ronde Basalts tended to be 2–3 orders of magnitude lower than that in the overlying Saddle Mountains and Wanapum Basalts. However, the Grande Ronde Basalt does not have uniformly lower K_h values throughout the Columbia Plateau Regional Aquifer System (plate 6 of Hansen *et al.* 1994; Ely *et al.* 2014), indicating that the reduction in permeability observed by Spane (1982) may be related to depth, temperature, and the degree of secondary mineralization, which all increase along the deep groundwater flow paths near the center of the basin (Reidel *et al.* 2002).

Changes in lithology associated with folds and faults also affect flow paths through the Columbia Plateau Regional Aquifer System by forming flow barriers or preferential pathways for groundwater flow (Newcomb 1959; Porcello *et al.* 2009; Snyder & Haynes 2010; Kahle *et al.* 2011; Burns

EXPLANATION

Overburden

Dense basalt flow interior (confining unit)

Permeable basalt flow top (aquifer)

Accretion wedge

Rivers and creeks

Rocky Prairie thrust fault
—dashed where extinction
depth uncertain

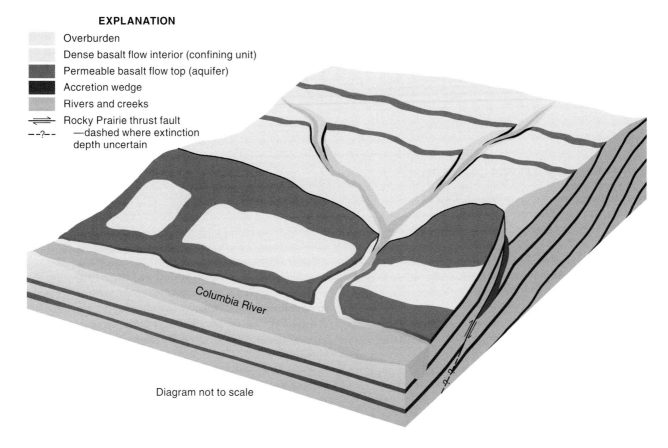

Columbia River

Diagram not to scale

Fig. 13.3. Conceptual model of aquifer system geometry (from Burns *et al.* 2012a). Upland recharge can enter thin aquifers at flow margins. Geologic structures can act as flow barriers that may or may not crosscut all aquifers. (*See color plate section for the color representation of this figure.*)

et al. 2012a,b). Faults create flow barriers by juxtaposing thin aquifers with flow interiors; furthermore, fault gouge consists largely of low-permeability clay-rich minerals. At the regional scale, geologic structure apparently has the net effect of reducing the effective horizontal hydraulic conductivity. As a result, regional modeling efforts suggest lower horizontal hydraulic conductivity than the values reported from aquifer tests (e.g., Hansen *et al.* 1994; Kahle *et al.* 2011). Similarly, modeling efforts result in higher predicted vertical hydraulic conductivity (e.g., Hansen *et al.* 1994) than the values suggested by aquifer tests on basalt flow interiors (e.g., Long *et al.* 1982b) because vertical connectivity at the regional scale is also controlled by the geometry of individual lava flows, allowing water to flow through the vertical stack of lavas via preferred pathways.

Groundwater flows through thin aquifers from the uplands toward large regional rivers, entering aquifers preferentially in the uplands where precipitation is higher and lava flow margins are exposed (typical aquifer geometries shown in Fig. 13.3). The older lava flows tend to be more voluminous and areally extensive and have also undergone more deformation, forming structural and erosional troughs into which younger, less-voluminous flows were deposited (Fig. 13.3). Some deeper aquifers may be completely covered by younger lava flows (Fig. 13.3), restricting recharge.

Groundwater exits the aquifers into rivers and streams. The strongest control on hydraulic head in any Columbia River Basalt Group aquifer is often the lowest elevation at which the aquifer intersects the land surface (Burns *et al.* 2012a). Thus, if an aquifer intersects a large regional river, hydraulic head in that aquifer may be similar to hydraulic head in that river. If there is a flow barrier between an aquifer and the large regional rivers, then the aquifer may discharge at higher elevations where upland streams intersect the aquifer (Fig. 13.3). Groundwater that bypasses the stream–aquifer intersection has poor connection with discharge locations and may be very old, while water that seasonally fills and drains from the upper portions of the aquifer may be much younger. Horizontal head gradients within individual aquifers can be very small in deep, almost stagnant parts of the system, with much steeper gradients in generally shallower, active parts of the system. The steepest gradients (approaching the magnitude and direction of the gradient of the land surface) occur in structurally complex, high-recharge upland areas. This conceptual model explains why water in deep aquifers near the center of the Columbia Plateau is more geochemically evolved (Kahle *et al.* 2011), and why there is significant groundwater contribution year-round to springs and streams near the crest of the Blue Mountains (the anticlinal ridge covered by Columbia River

Basalt Group lavas near the southeastern boundary of the study area).

More detailed descriptions of the Columbia Plateau Regional Aquifer System are provided in reports by Kahle *et al.* (2009), who discuss the geologic framework; Burns *et al.* (2011), who describe the 3D characteristics of the geology; and Kahle *et al.* (2011), who discuss the hydrogeologic framework and the hydrologic budget.

METHODS OF ANALYSIS AND RESULTS

The SUTRA model (Voss & Provost 2002) was used to perform coupled groundwater and heat-flow simulations. The hypothesis that advective transport of heat can account for the apparently low heat flow under the Columbia Plateau Regional Aquifer System was tested by simulating a constant $80\,\mathrm{mW\,m^{-2}}$ heat flow at the bottom of the domain and quantifying the effect of advection on the apparent near-surface heat flow. An analysis of existing permeability data was conducted, and three different permeability/depth relations, constrained by these data, were evaluated. Prior to performing SUTRA simulations, a Rayleigh number analysis was performed to evaluate the likelihood that density-driven convective flow occurs in the Columbia Plateau Regional Aquifer System.

Rayleigh analysis

Free convection can occur if the Rayleigh number (Ra) exceeds the critical Rayleigh number (Ra_{crit}). For a system with anisotropic permeability and isotropic heat conduction, Phillips (1991, pp. 144–145) derives the following special form of the Raleigh number (Ra) and the associated Ra_{crit}.

$$Ra_h = \frac{\alpha_w \rho_w^2 c_m g k_h L \Delta T}{\mu_w \lambda_{\mathrm{bulk}}} \tag{13.1}$$

and

$$(Ra_h)_{\mathrm{crit}} = \left(1 + \sqrt{\frac{k_h}{k_v}}\right)^2 \pi^2 \tag{13.2}$$

where the relation of the form of the Rayleigh number to the horizontal permeability (k_h) has been made explicit and the critical Rayleigh number depends on the ratio of the horizontal-to-vertical permeability $\frac{k_h}{k_v}$. The characteristic length scale L is the thickness across which a temperature difference (ΔT) is experienced, α_w is the thermal expansivity of the fluid phase, ρ_w is the density of the fluid phase, c_w is the heat capacity of the fluid phase, μ_w is the dynamic viscosity of the fluid, λ_{bulk} is the bulk thermal conductivity of the medium, and g is the gravitational constant.

Ra_h and $(Ra_h)_{\mathrm{crit}}$ were estimated using conservative values for both the upper permeable zone and the thickest part of the Columbia River Basalt Group (approximately $5\,\mathrm{km}$). Because of the large horizontal-to-vertical anisotropy in permeability, in both cases Ra_h was approximately 100 times less than $(Ra_h)_{\mathrm{crit}}$,

indicating that free convection can be ignored during the analysis.

Permeability/depth relations

Rapid reductions in permeability over a narrow depth range have been identified in studies of several volcanic terrains and have generally been attributed to hydrothermal alteration at temperatures above $30\,°\mathrm{C}$. Investigations at geothermal sites throughout the Cascade Range identified a significant decrease in permeability with depth in the range from approximately 200 to $1000\,\mathrm{m}$ (e.g., Swanberg *et al.* 1988; Blackwell 1994; Hulen & Lutz 1999) associated with low-temperature hydrothermal alteration of Cascade Range rocks at temperatures in the range of 30–$50\,°\mathrm{C}$ (e.g., Bargar & Keith 1999). At the Kilauea Volcano, Hawaii, Keller *et al.* (1979) attributed a significant reduction in permeability to hydrothermal alteration below $488\,\mathrm{m}$ where temperatures increased from 30 to $60\,°\mathrm{C}$ over a short interval. Recently published detailed borehole test data (Spane 2013) shows that for several deep boreholes located near the center of the Columbia Plateau Regional Aquifer System, rapid reduction in permeability starts at a similar depth of approximately $600\,\mathrm{m}$. Ely *et al.* (2014) tested the hypothesis that permeability is significantly less in Columbia River Basalt Group units deeper than $600\,\mathrm{m}$ by lowering K_h below this depth by a factor of 1000 and recalibrating their regional model. Model fit was not significantly changed, and shallow groundwater parameters remained within the range of expected values. It was concluded that model calibration was not sensitive to the deep system permeability because most groundwater flows through the upper $600\,\mathrm{m}$, where aquifers are directly connected to recharge and discharge boundaries. In other words, the available hydrogeologic data were insufficient to allow unique estimation of the deep permeability structure of the Columbia Plateau Regional Aquifer System.

Ely *et al.* (2014) simulated the Columbia Plateau Regional Aquifer System using model cells that were approximately $30\,\mathrm{m}$ thick, representing a typical lava flow thickness including the permeable flow top and bottom, and using bulk horizontal and vertical permeability. Locally, effective horizontal permeability is controlled by thin aquifers, while vertical permeability is controlled by the thick flow interiors (confining units). Because individual aquifers are not continuously mapped across the study area and because hundreds of hydraulically distinct aquifers can exist in areas where the Columbia River Basalt Group is thickest, individual aquifers and confining units were not subdivided. Instead, bulk horizontal permeability represents the net effect of the horizontal connection, and bulk vertical permeability represents the permeability of the confining units. This implementation reduces the number of model cells (reducing computational burden) and reduces numerical instability that would occur as a result of simulating thin high-contrast aquifers adjacent to relatively thick confining units every $30\,\mathrm{m}$ vertically. The SUTRA modeling described

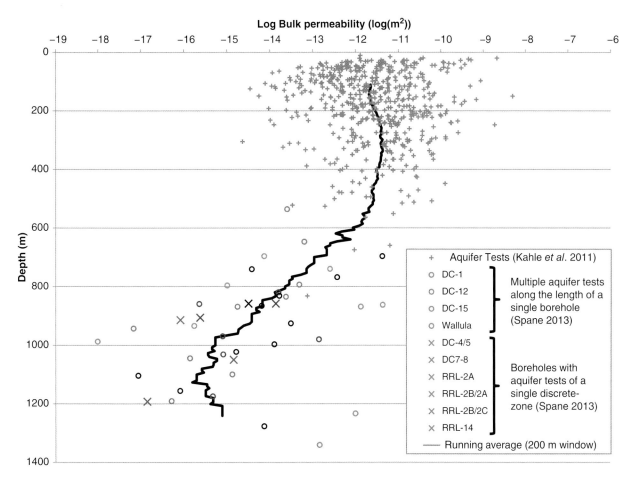

Fig. 13.4. Estimates of bulk horizontal permeability from Kahle *et al.* (2011) and Spane (2013). Using a 200 m window to compute a running average identifies a depth range of 600–900 m where permeability rapidly decreases. Above 600 m depth, average permeability is approximately constant. The deep data are sparse, but below 900 m depth, the rate of decrease in permeability apparently slows. Circles of the same color denote packer tests at different elevations within the same borehole. All other symbols represent single tests within a discrete borehole. (*See color plate section for the color representation of this figure.*)

herein (in the next section) also utilizes this numerical scheme and is appropriate for steady-state simulations of the regional aquifer system where the precise location of individual aquifers is not known.

Bulk permeability/depth relations have been estimated from published estimates of hydraulic conductivity and transmissivity (Fig. 13.4). Hydraulic conductivity estimates for Columbia River Basalt Group aquifers were taken from the USGS National Water Information System (NWIS) database (Kahle *et al.* 2011) and converted to permeability estimates using temperature estimates from the nearest temperature profile log (Fig. 13.2). Temperature was estimated by using a best linear fit to the 100 m of borehole temperature log closest to the depth of the aquifer test. Viscosity and density were estimated as a function of temperature, and hydraulic conductivity was converted to permeability by multiplying by viscosity and dividing by the product of density and the gravitational acceleration constant (Fetter 1994, p. 96). Bulk permeability was computed using the layered system approximation (the

arithmetic mean), assuming that the aquifer occupies 10% of the total thickness and that the permeability of the lava flow interiors is negligibly small.

The detailed Columbia River Basalt Group aquifer transmissivity estimates of Spane (2013) were also converted to estimated permeability (Fig. 13.4). First, estimated transmissivity was converted to bulk hydraulic conductivity by dividing by 30 m or the open length of the borehole, whichever was longer. Short open boreholes were assumed to test individual aquifers, so a typical lava flow thickness of 30 m was assumed to convert the aquifer test to a bulk permeability by using the layered system approximation (Fetter 1994, pp. 123–124). Long open borehole tests (>30 m) potentially intersected multiple aquifers and a representative length of confining unit. Temperature measured in these boreholes (Schroder & Strait 1987; McGrail *et al.* 2009; Spane *et al.* 2012) was used to correct viscosity and density, and permeability was computed and assigned to the middle of the test interval. Four boreholes have multiple packer tests vertically along the borehole length.

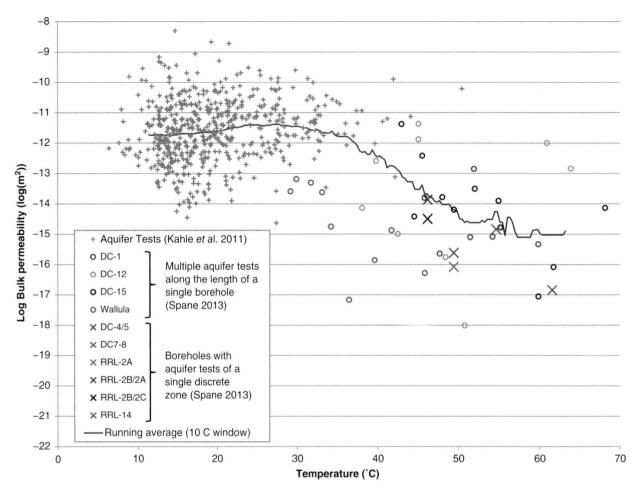

Fig. 13.5. Plot showing that permeability transitions rapidly in the temperature range 35–50 °C. For the Kahle data, temperatures were estimated from the nearest temperature log (Fig. 13.2). For the remaining aquifer tests, temperatures were estimated from temperature logs collected within the borehole during a battery of geophysical tests. (*See color plate section for the color representation of this figure.*)

The permeability estimates exhibit considerable variability, but no persistent regional trends, so a running average value was computed using a 200 m window (line on Fig. 13.4). Available data define a substantial decline in permeability starting at a depth of about 600 m and slowing below 900 m. Similar rapid reductions in permeability with depth are documented for the volcanic rocks of the nearby Cascade Range by Swanberg *et al.* (1988), Blackwell (1994), Hulen & Lutz (1999), and Saar & Manga (2004). Because this rapid reduction in permeability has been suggested to correspond to hydrothermal alteration of volcanic rocks to pore-clogging clay minerals in the temperature range of 30–50 °C, the estimated aquifer-test temperatures were plotted against permeability (Fig. 13.5). The rapid reduction in aquifer-test permeability corresponds to a similar temperature range of 35–50 °C, suggesting that hydrothermal alteration may also cause the reduction in permeability with depth in the Columbia Plateau Regional Aquifer System.

Although there are no well-test data below 1400 m depth, permeability must be assigned to the full thickness of the

Columbia River Basalt Group layers, which is almost 5000 m thick near the center of the Columbia Plateau Regional Aquifer System. Manning & Ingebritsen (1999) observed that for geologic materials in general, the rate of reduction in permeability is rapid near the land surface but slows with depth. Manning & Ingebritsen proposed a power-law fit to the data to define the permeability/depth relation. For the Cascade Range, Saar & Manga (2004) proposed altering the Manning & Ingebritsen relation by using piece-wise continuous exponential relations near the land surface to capture three distinct zones: an upper high-permeability zone, a middle zone of rapid reduction in permeability, and the deep Manning & Ingebritsen power-law relation.

For the SUTRA simulations of the Columbia Plateau Regional Aquifer System, three different permeability/depth relations were considered (Fig. 13.6): (i) Manning & Ingebritsen (hereafter called the "power-law depth relation"), (ii) Saar & Manga (hereafter called the "piece-wise-600 m depth relation"), and (iii) constant permeability with depth (hereafter

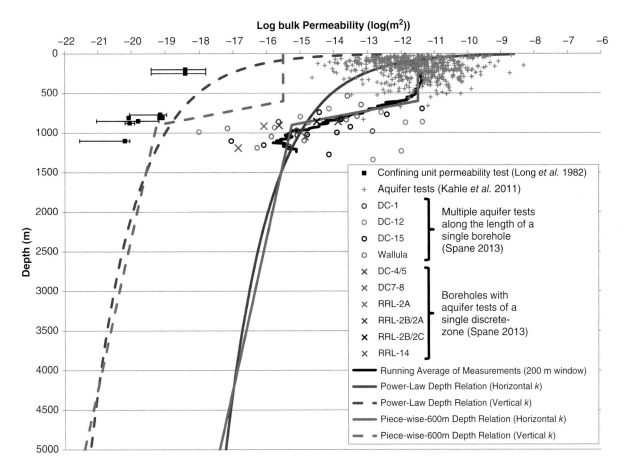

Fig. 13.6. Estimated permeability vs. depth relations for use in the SUTRA models. The solid lines are the maximum (subhorizontal) permeability, and the dashed line is minimum (subvertical) permeability. The red lines are the power-law depth relation. The blue lines are the piece-wise-600 m depth relation. The constant-permeability depth relation model is not shown, but has a log permeability of −11.55, indistinguishable from the shallow exponential model value of −11.5. (*See color plate section for the color representation of this figure.*)

called the "constant-permeability depth relation"). The original Manning & Ingebritsen power-law depth relation was shifted to the left by one order of magnitude to better fit the horizontal permeability data (in other words, permeability is 10 times less for the Columbia Plateau Regional Aquifer System), so that the curve passes through the running average of deep permeability observations. The power-law decay rate was unaltered. The piece-wise-600 m depth relation was fit to the horizontal permeability data such that above 900 m, the exponential relations are fit to the data, and below 900 m, the power-law depth relation is approximated.

Because the deep borehole tests are from a very limited areal extent, near a major fault zone, it is possible that they are not representative across the Columbia Plateau Regional Aquifer System. Consistent with recent regional modeling (Ely *et al.* 2014), the other end-member to consider is the constant-permeability depth relation. The permeability value used for the constant-permeability depth relation model was adjusted slightly to ensure a similar distribution of groundwater fluxes and shallow hydraulic heads between scenarios, as discussed in more detail later.

Consistent with previous regional groundwater flow simulations (Ely *et al.* 2014), bulk vertical permeability is assumed to be four orders of magnitude lower than horizontal permeability for each permeability–depth relation (Fig. 13.6). With increasing depth, the bulk vertical permeability should approach the permeability of individual flow interiors (Confining Unit Permeability Tests on Fig. 13.6), which are typically >4 orders of magnitude less than the aquifer-test data.

SUTRA simulations

Coupled groundwater and heat flow was simulated for two cross sections (Figs 13.2 and 13.7) using SUTRA (Voss & Provost 2002). The cross sections represent two typical groundwater flow paths from recharge areas to the Columbia River. Cross section A–A′ represents the gently sloping Palouse Slope, which receives relatively little recharge and drains toward both the Snake and Columbia Rivers. Cross section B–B′ represents the flow from one of the highest recharge areas, the Blue Mountains, toward the Columbia River. To the northwest of the section lines (Fig. 13.2), the dominant direction of geologic

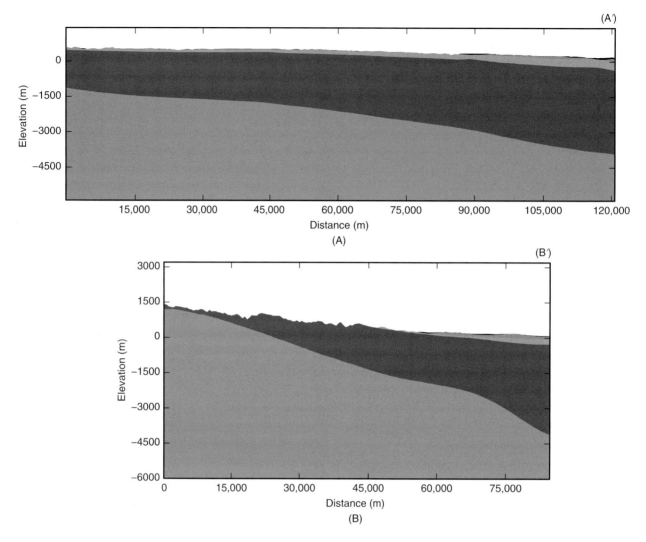

Fig. 13.7. Typical cross sections through the Columbia Plateau Regional Aquifer System (see Fig. 13.2). For both cross sections (units in m), the Columbia River is at the right-hand boundary, incised 250–300 m into the geologic units. (A) Cross section through the relatively gently dipping, low-recharge Palouse Slope. (B) Cross section from the crest of the higher recharge Blue Mountains through the Umatilla River basin. (Blue, pre-Miocene rocks; brown, Grande Ronde Basalts; green, Wanapum Basalts; red, Saddle Mountains Basalts; black, thin deposits of sedimentary overburden). (*See color plate section for the color representation of this figure.*)

structure prevents accurate simulation of conditions using a 2D model. The two selected cross sections represent the range of conditions encountered in the remainder of the Columbia Plateau Regional Aquifer System.

Modifications to SUTRA

Version 2.2 of SUTRA was modified to improve the simulation capabilities over a wider range of temperatures (Alden Provost, U.S. Geological Survey, written communication, 2013). In general, water density and viscosity are functions of both pressure and temperature, though up to about 300 °C and 500 bars (about 5000 m of water pressure) the dependence on pressure is weak compared to the dependence on temperature (see Fig. 4.1 of Ingebritsen *et al.* 2006). SUTRA does not account for

pressure dependence, but the nonlinear temperature-dependent water density and viscosity relations were improved to provide reasonably good fit up to approximately 300 °C (McKenzie &Voss 2013).

SUTRA was also altered to allow for spatially distributed thermal conductivity. SUTRA 2.2 only allows for a single homogeneous, possibly anisotropic thermal conductivity. Because the Columbia River Basalt Group has lower thermal conductivity than the underlying rocks (approximately 1.6 W m^{-1} per °C compared to approximately 2.5 W m^{-1} per °C), the thermal contrast can result in heat refraction at an inclined interface.

The last alteration to SUTRA 2.2 was to add a new type of boundary condition. For boundary conditions of the fluid, SUTRA 2.2 only allows for prescribed flux and prescribed pressure conditions. The new boundary condition is analogous to the MODFLOW drain boundary condition (McDonald & Harbaugh 1984), where water never flows into the model domain, but is allowed to leave the domain when pressure exceeds a fixed reference pressure. The rate of flow is proportional to the difference in these two pressures.

Model formulation, boundary conditions, and discretization

For the case where density and viscosity depend on temperature, the steady-state solution is achieved by running the simulation with steady boundary conditions for a very long time. Deming (1993) notes that the characteristic time of response for the fluid-flow problem is short compared to the heat-flow problem, so it is only necessary to estimate the characteristic timescale of response ($\hat{t}$) for heat flow:

$$\hat{t} = \frac{L^2}{4\alpha} \tag{13.3}$$

with

$$\alpha = \frac{\lambda_{\text{bulk}}}{(1-\varepsilon)\rho_s c_s + \varepsilon\rho_w c_w} \tag{13.4}$$

where α is the thermal diffusivity, λ_{bulk} is the bulk thermal conductivity of the Columbia River Basalt Group (approximately 1.6 W m^{-1} per °C), ε is the estimated bulk porosity (estimated as 0.025), ρ_s is the density of the solid (rock) phase (approximately 3000 kg m^{-3}), c_s is the heat capacity of the solid (estimated as 840 J kg^{-1} per °C), ρ_w is the density of the fluid phase (estimated for this computation as 1000 kg m^{-3}), and c_w is the heat capacity of the fluid phase. The thickest part of the Columbia Plateau Regional Aquifer System is approximately 5 km and the simulated thickness below the Columbia Plateau Regional Aquifer System is 20 km, so the simulated domain is up to 25 km thick, providing an upper estimate for the characteristic length scale L. While the bulk thermal conductivity of the underlying rocks is higher, using the lower value for the Columbia River Basalt Group provides a conservative estimate of $\hat{t}$ of 2.52×10^{14} s (approximately 8 Ma). All variable density or variable viscosity simulations were run for a period greater than five times the characteristic timescale (1.26×10^{15} s) to ensure that the steady-state solution is approximated well.

A 2D regular mesh (Voss & Provost 2002) was used to represent each cross-sectional model, with approximately 30 m vertical discretization of the Columbia River Basalt Group on the right-hand side of the model (Fig. 13.8). Model cells thin from right to left, with the same number of vertical nodes representing the Columbia River Basalt Group across the domain. Horizontal node spacing is the same for all

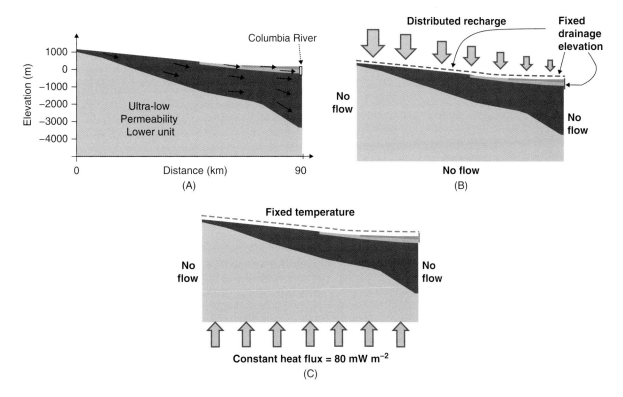

Fig. 13.8. Model formulation and boundary conditions: (A) hydrogeologic units with arrows showing that the preferential direction of permeability is rotated to align with the plane of basalt deposition (blue, pre-Miocene rocks; brown, Grande Ronde Basalts; green, Wanapum Basalts; red, Saddle Mountains Basalts); (B) hydrologic boundary conditions showing that recharge and discharge occur near the upper boundary and into the river on the right-hand side; (C) heat-flow boundary conditions showing the constant heat flux at the lower boundary and the prescribed temperature at land surface, with cooler temperatures in the uplands to the left. The thickness of the underlying ultralow-permeability unit was varied during model testing, with the final thickness being 20 km (figure not to scale for this unit). (*See color plate section for the color representation of this figure.*)

layers and is about 3000 m, except for the two vertical strings of nodes closest to the river, which are 300 m apart. The 20-km-thick ultralow-permeability unit below the Columbia Plateau Regional Aquifer System is simulated using 37 additional nodes (in the vertical direction) with a mesh that increases with depth from 30 m near the Columbia River Basalt Group units to a maximum of 1235 m at depth.

Within each hydrogeologic unit, the anisotropic permeability tensor is rotated so that the preferential flow direction is along the depositional horizon for each lava flow (Fig. 13.8A). The maximum rotation (6.5°) occurs at the contact between the Columbia River Basalt Group and the underlying unit near the right-hand model boundary. At the upper model boundary, dip parallels land surface. Permeability within the Columbia River Basalt Group is represented as a function of depth. Permeability in the ultralow-permeability region is set to 10^6 less than the overlying Columbia River Basalt Group permeability, ensuring that advective heat transport is negligible in the lower domain (Smith & Chapman 1983). Porosity is assumed to be 0.025 for all model cells. The isotropic bulk thermal conductivity of Columbia River Basalt Group units is set to 1.59 W m^{-1} per °C, and the underlying pre-Miocene rocks are set to 2.5 W m^{-1} per °C (typical values taken from Lachenbruch & Sass 1977).

The lower heat-flow boundary condition (Fig. 13.8C) is a constant uniform heat flux at a rate of 80 mW m^{-2}. Upper-boundary and above-river-elevation nodes are set to the approximate annual average temperature for the period 1895–2007 extracted from the spatially distributed PRISM maps (PRISM Climate Group 2004). Except for the Columbia River, lateral boundaries are no-flow for fluid and fully insulated for heat.

All fluid flow ultimately enters and leaves the Columbia Plateau Regional Aquifer System at the land surface (Fig. 13.8B), and virtually every 3 km model cell contains a river or stream channel that can receive groundwater discharge (Ely *et al.* 2014). However, because the model is used to simulate the temperature of groundwater leaving the Columbia Plateau Regional Aquifer System, water cannot be allowed to flow out through the fixed-temperature boundary nodes. Instead, water is allowed to drain from the system at all nodes in the approximate depth range of 15–60 m, except at the Columbia River boundary. This depth range is consistent with the typical amount of stream incision shown schematically in the conceptual model (Fig. 13.3). At the Columbia River (right) boundary, boundary conditions transition at 30 m elevation (the river surface elevation), a depth of 150 m from the model upper boundary. Above 30 m elevation, in the depth range of 0–150 m, temperature is fixed on a vertical string of nodes on the right-hand boundary, and water is allowed to drain from the string of nodes of 300 m to the left in the depth range of 15–150 m. These conditions represent the incised "canyon" associated with the Columbia River. Below 30 m elevation, in the depth range of approximately 150–350 m,

water is allowed to drain from the right-hand boundary nodes with a fixed pressure corresponding to 30 m of elevation. Extending the drain nodes as much as 200 m below the river elevation at this location is consistent with the 3D geometry of the Columbia River, which intersects the Grande Ronde unit at various locations along its length (controlled by faults, folds, and inclined beds).

Groundwater recharge from precipitation (Fig. 13.8B) enters the model into the same nodes where drainage is simulated (between 15 and 60 m depth) for all locations except the Columbia River, where no precipitation recharge is simulated. Temperature of recharge water is set to the same temperature as the fixed-temperature nodes. Recharge rates were estimated using a regression equation for the Columbia Plateau that relates average annual precipitation to recharge (Bauer & Vaccaro 1990). Annual average precipitation for the period 1895–2007 was estimated from the PRISM model (PRISM Climate Group 2004). There are no persistent trends in annual precipitation during this period though there are decadal scale variations corresponding to wet and dry periods.

Calibration

The remainder of this section and all figures pertain to the Blue Mountain–Umatilla cross section (Section B–B′ on Figs 13.2 and 13.7B), where groundwater flow is relatively vigorous. The Palouse cross section (Section A–A′ on Figs 13.2 and 13.7B) is considered only in "Discussion" Section.

Prior to examination of temperature simulation results, the groundwater simulation was "calibrated" for each of the permeability/depth relations. For the piece-wise-600 m depth relation, drain conductance was adjusted in two groups: all land surface drains and the Columbia River drains. These two parameters were adjusted such that the resulting hydraulic heads in the uplands near the crest of the Blue Mountains (left-hand model boundary) were near land surface elevation, hydraulic heads midslope are above land surface (confined conditions with upward gradient), and 38% of the total groundwater is simulated as exiting the system into the Columbia River boundary. These values are consistent with measured conditions.

The other two permeability/depth relations were calibrated to these same conditions, with the most important consideration being a similar distribution of groundwater discharge (which strongly influences advective transport of heat). Failure to incorporate hydrologic constraints can lead to erroneous conclusions regarding the importance of certain parameters. For example, applying the shallow permeability value from the piece-wise-600 m depth relation (Fig. 13.6) to the entire model thickness causes more than half of the total groundwater to exit the system at the Columbia River, transporting significantly more heat. The constant-permeability depth relation model was calibrated using a modest reduction in permeability (reduced from $10^{-11.5}$ to $10^{-11.55}$ m^2) along with adjustment of drain conductance parameters. The power-law depth relation model was calibrated using only drain conductances.

RESULTS

We first consider the effects of the piece-wise-600 m depth relation. Advective transport of heat by groundwater depresses temperatures near the higher recharge areas in the uplands and raises temperatures near the Columbia River. If only conduction of heat is considered (Fig. 13.9A), temperatures are slightly influenced by refraction as heat flows around the thicker parts of the relatively low-thermal-conductivity Columbia River Basalt Group units. When advective heat transport is added (Fig. 13.9B), cool groundwater recharged in the uplands cools much of the domain and raises temperatures below the Columbia River. The difference between the conduction-only simulation (Fig. 13.9A) and the combined groundwater and heat-flow simulation (Fig. 13.9B) shows that groundwater flow depresses temperatures by as much as 15 °C in some areas and elevates temperatures by up to 3 °C under the Columbia River (Fig. 13.9C).

The power-law depth relation is characterized by very high-permeability near land surface that decreases very rapidly

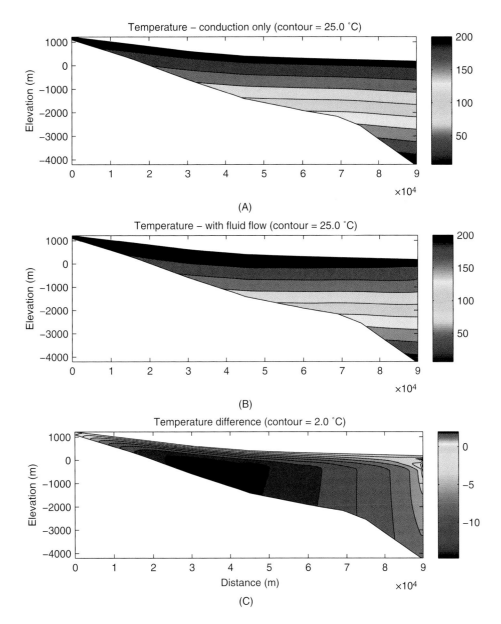

Fig. 13.9. Comparison of simulated temperatures for cross section B–B′ (Umatilla) using the piece-wise-600 m depth relation. (A) shows the simulated subsurface temperature distribution assuming heat conduction only (no fluid flow). (B) shows the simulated temperature distribution for combined groundwater and heat flow. (C) shows the temperature difference between the two simulations; warmer colors show areas where temperatures are elevated by advective transport of heat, and cooler colors show areas where subsurface temperatures are lowered by heat advection. Plots are only colored in the part of the domain occupied by Columbia River Basalt Group units. (*See color plate section for the color representation of this figure.*)

with depth (Fig. 13.6). Because permeability is approximately 1000 times greater than either of the other permeability relations very near land surface, virtually all of the recharge moves along the thin skin of the system to discharge in rivers and streams. Below this very thin upper region, there is very little groundwater flow, and therefore little advective perturbation of the heat-flow field. As a result, the temperature distribution is almost indistinguishable from the conduction-only model (Fig. 13.9A).

The constant permeability/depth relation model has a very similar distribution of groundwater discharge, with discharge to the Columbia River within 0.5% of the 38% simulated using the piece-wise-600 m depth relation model. The constant-permeability depth relation allows for deeper penetration of groundwater flow along the cross section. A comparatively narrow range of acceptable combinations of permeability and drain conductance can be used to calibrate

this model, all of which support the same general pattern of temperature seen in the piece-wise-600 m depth relation model (Fig. 13.9B), albeit with significantly amplified temperature perturbations relative to the heat-conduction-only model. The maximum depression of temperature (centered at approximately 3500 m on Fig. 13.9C) shifts right to approximately 5000 m and is about 20 °C lower. The constant-permeability depth relation model is approximately 35 °C warmer than the piece-wise-600 m depth relation model at a depth of approximately 500–2000 m in a very narrow zone directly under the regional groundwater discharge location (right-hand side).

Local vertical conductive heat flow was estimated by computing the vertical temperature gradient between nodes and multiplying by the thermal conductivity. For the piece-wise-600 m depth relation, vertical heat-flow estimates are strongly affected within the higher permeability zone above 600 m, with deeper effects most notably below the Columbia River (Fig. 13.10A).

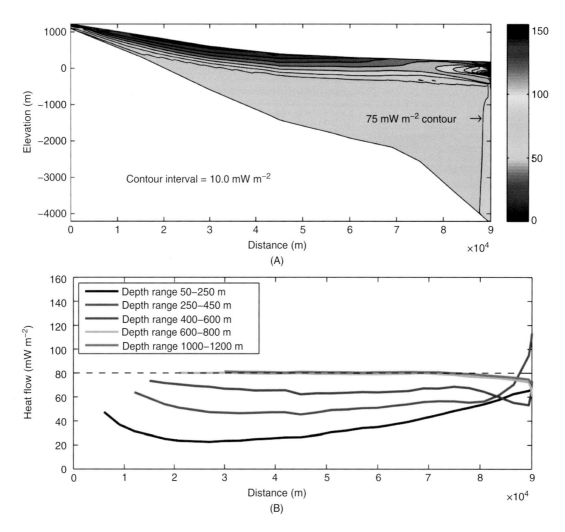

Fig. 13.10. Illustrations showing estimated vertical conductive heat flow for cross section B–B' (Umatilla) for the piece-wise-600 m depth relation: (A) local estimated vertical conductive heat flow and (B) estimated vertical heat flow for five depth ranges, computed as the bulk thermal conductivity times the representative gradient (slope of the best-fit line) of all temperatures in each depth range. Below approximately 600 m depth, heat flow is dominated by conduction. Plot A is only colored in the part of the domain occupied by Columbia River Basalt Group units. (*See color plate section for the color representation of this figure.*)

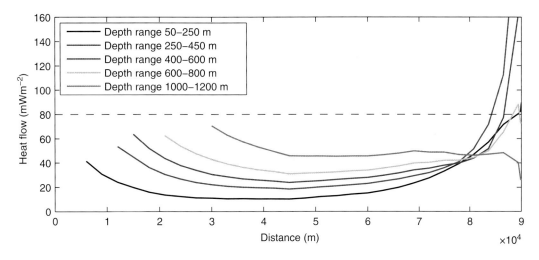

Fig. 13.11. Illustrations showing estimated vertical heat flow for cross section B–B′ (Umatilla) for the constant-permeability depth relation. Estimated vertical heat flow for five depth ranges, computed as the bulk thermal conductivity times the representative gradient (slope of the best-fit line), of all temperatures in each depth range (compare with Fig. 13.10B). (*See color plate section for the color representation of this figure.*)

To allow comparison with heat-flow measurements (estimated from thermal gradients measured in discrete intervals of boreholes), best linear fits to simulated profiles at different depth intervals were used to estimate heat flow (Fig. 13.10B). Vertical conductive heat flow is broadly reduced in the uplands and more locally increased near the Columbia River. Except below the Columbia River, simulated vertical conductive heat flow is approximately 80 mW m^{-2} at most depths below 600 m (Fig. 13.10B).

Because groundwater only flows in a thin skin near land surface for the power-law depth relation model, the entire model domain is very nearly constant at approximately 80 mW m^{-2}. Heat flow is slightly higher to the left and slightly lower to the right, as heat is preferentially conducted around the thickest part of the low-thermal-conductivity Columbia River Basalt Group units, but when plotted on the same color scale as Figure 13.10A, virtually the entire domain falls within the 75–85 mW m^{-2} heat-flow contours.

For the constant-permeability depth relation, the rapid change in heat flow at a depth of approximately 600 m does not occur (Fig. 13.11). Simulated vertical conductive heat flow varies more smoothly with depth, asymptotically approaching 80 mW m^{-2}. Otherwise, the heat-flow pattern is similar to Figure 13.10A though variability is amplified, with higher heat flow directly under the river, and lower heat flow in the shallow subsurface in the recharge areas.

Despite the overall similarity in the groundwater discharge pattern at land surface, and groundwater discharge at the Columbia River being 0.5% lower, the heat removed advectively was significantly larger for the constant-permeability depth relation model. Groundwater carried 71.2% of the total heat exiting the system, compared to 58.2% of the heat exiting the system for the piece-wise-600 m depth relation model.

Comparison between model results and measured heat flow

In geophysical practice, vertical heat flow is determined from field measurements using a selected interval or intervals from a borehole temperature log. In order to compare estimates of vertical conductive heat flow from the simulations with field data, a best linear fit to the simulated vertical temperature profile at fixed depth intervals was used to produce transects of simulated vertical heat flow (Fig. 13.10B shows the estimates for the piece-wise-600 m depth relation).

Two depth intervals (50–250 and 250–450 m) were selected to represent the typical range of available heat-flow data (Fig. 13.2). To evaluate the role of the simulated rapid decline in permeability at 600–900 m depth, intervals of 400–600 and 600–800 m were also selected. Finally, a depth range below the permeability transition (1000–1200 m) was selected for comparison with the heat flow prescribed at the bottom boundary of the model.

For the piece-wise-600 m depth relation model, vertical heat flow below 600 m depth is very similar to the lower-boundary heat flow of 80 mW m^{-2} over most of the simulated cross section (Fig. 13.10B). Above 600 m depth, simulated conductive heat flow is significantly less than the deep conductive heat flow over much of the transect. The obvious exception is the high simulated heat flows near the Columbia River.

Heat-flow estimates from measurements (Fig. 13.12) are generally consistent with both the simulated values for the piece-wise-600 m depth relation model (Fig. 13.10B) and the constant-permeability depth relation model (Fig. 13.11). Several near-surface heat flow measurements near the Columbia River are markedly higher than 80 mW m^{-2}, and upslope estimates are significantly lower. Spatial patterns in the measured data generally match the simulated profiles from <600 m depth

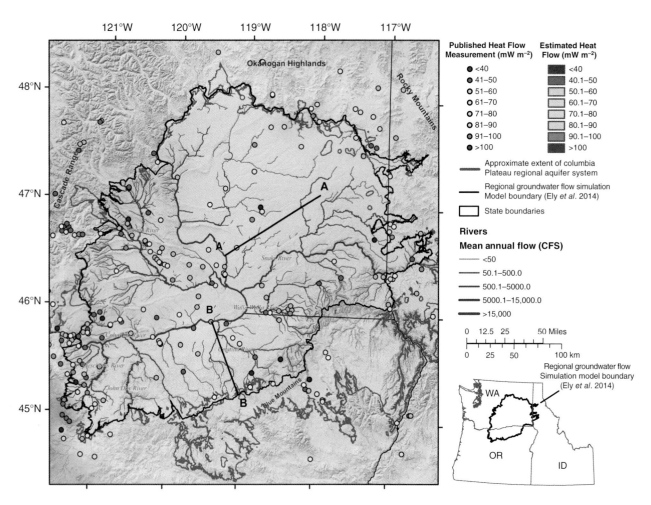

Fig. 13.12. Heat-flow map with rivers; river line width is proportional to mean annual flow rate. Larger rivers are at the lowest elevations and receive water from long, regional groundwater flow paths. (*See color plate section for the color representation of this figure.*)

(Figs 13.10B and 13.11), with lowest heat flows predicted midslope and higher heat flows near the crest of the Blue Mountains and the Columbia River. Because the power-law depth relation model predicts a constant value of 80 mW m^{-2} at all locations and depths, this model can be rejected. Because both the piece-wise-600 m and constant-permeability depth relations are deemed reasonable, the 2D cross-sectional model gives no indication of whether or not the shifted Manning and Ingebritsen relation is an appropriate representation of the deep (approximately >1 km) Columbia Plateau Regional Aquifer System. Future regional 3D modeling may provide some insight, but it is likely that additional deep thermal profile data would also be required to evaluate any deep permeability model.

Given the same permeability distribution, the Palouse cross section (Section A–A′ on Figs 13.2 and 13.7A) exhibits less reduction in simulated vertical heat flow because the rates of groundwater recharge are lower and the groundwater flow system is less vigorous. Simulated heat flow under the Palouse was about 55 mW m^{-2} over most of the transect at a depth range of

50–250 m (analogous to Fig. 13.10B). At 250–450 m depth, simulated vertical heat flow was about 70 mW m^{-2}, and below 600 m simulated heat flow was approximately 80 mW m^{-2}, the lower-boundary value. Simulated heat flow near the Columbia River was elevated, but to a lesser extent than depicted in Figure 13.10.

DISCUSSION

Simulation results using best estimates of hydrogeologic properties show that groundwater can remove sufficient heat to account for the apparent low-heat-flow anomaly without requiring anomalously low heat flow at the base of the Columbia River Basalt Group. Except for the power-law depth relation model, which resulted in most of the groundwater flowing through a very shallow zone, all simulation results show that, for a given depth, the highest temperatures will occur near major groundwater discharge zones. Previous heat-flow measurement has generally not focused on major groundwater discharge zones (Fig. 13.12), and most of the temperature measurements from wells are within depths perturbed by groundwater flow (Fig. 13.2). Deep boreholes far from regional groundwater discharge zones are likely the best estimators of regional heat flow.

Hansen *et al.* (1994) simulated the Columbia Plateau Regional Aquifer System using a range of hydraulic conductivities, with most of the system having bulk horizontal hydraulic conductivities in the range 3.5×10^{-6} to 3.5×10^{-5} m s^{-1} (1–10 ft d^{-1}). These values correspond approximately to horizontal intrinsic permeabilities of $10^{-12.5}$ to $10^{-11.5}$ m^2. The latter value was employed in this study (Fig. 13.6), and the former value may account for the role of geologic variability in restricting groundwater flow at a length scale longer than that represented by aquifer-test data. Lower permeability would require more water to drain in the uplands, which would result in less advective transport of heat. Our comparison of permeability/depth distributions assumed that the distribution of groundwater discharge was known, but in practice, groundwater discharge distribution is uncertain. An improved understanding of the distribution of groundwater discharge from the system would allow an improved understanding of both permeability and heat flow. Conversely, an improved understanding of the heat flow field can be used to constrain the groundwater flow rate.

Smith & Chapman (1983) showed that systems with much less anisotropy may exhibit much larger thermal effects. In our case, higher vertical permeability would allow water to flow more easily to the Columbia River, resulting in a larger thermal anomaly. A larger thermal anomaly would also result from increasing the depth of the permeability transition (600–900 m), with the end-member case being the constant-permeability depth relation model.

For most of the rivers receiving a large fraction of the deep regional groundwater flow, groundwater inflow is a relatively small fraction of total stream flow, so the thermal influence of groundwater is difficult to detect. The Umatilla cross-sectional model predicts that approximately 5.4 m^3 s^{-1} of approximately 30°C groundwater flows directly into the Columbia River (median annual flow approximately 5500 m^3 s^{-1}) along a 85-km reach. Although this simulated groundwater influx carries hundreds of megawatts of geothermal heat, the heat input will likely be thermally detectable only if focused monitoring is conducted at preferential groundwater discharge locations (subaqueous springs).

A limited amount of deep (>600 m) temperature data are available (Fig. 13.2), but these data are very near regional groundwater discharge zones and areas of significant structural complexity, indicating that heat flow in these areas may still have significant influence from groundwater flow.

Temperature measurements for 18 deep wells (>600 m) were compiled and, assuming a bulk thermal conductivity of 1.59 W m^{-1} per °C, estimated vertical heat flow below 600 m ranged from 55 to 71 mW m^{-2}, with a median value of 63.5 mW m^{-2}. Simulation results from the piece-wise-600 m

depth relation model (Fig. 13.10) indicate that heat flow in discharge areas is on the order of 10% lower than the deep regional heat flow (Fig. 13.10). The constant-permeability depth relation model predicts larger reductions in conductive heat flow, so that either model supports an average vertical heat flow of ≥70 mW m^{-2}, at or above the global average for tertiary tectonic provinces.

One confounding element in understanding the effects of permeability distribution is the hypothesis that a controlling factor is temperature, with an important transition at approximately 35–45°C. In the uplands, these temperatures occur at greater depth, so our lowland-based estimate of the depth of transition (600–900 m) may be too shallow. Furthermore, elevated paleotemperatures (e.g., on the flanks of the Cascades) could have controlled hydrothermal alteration in some areas. Allowing a deeper transition to the low-permeability zone would result in behavior intermediate between the piece-wise-600 m and constant-permeability depth relation models (compare Figs 13.10B and 13.11). Preliminary exploratory simulations combining lower permeability values (similar to the lower permeabilities used in calibrated regional groundwater flow models [$10^{-12.5}$ m^2]) with deeper permeability transition depths (approximately 1000 m) yield simulated near-surface heat flow patterns that are also consistent with the existing heat flow estimates.

We have little information about the distribution of hydrothermal alteration minerals in the continental flood basalts of the Columbia Plateau Regional Aquifer System with respect to depth and temperature, although pore-filling alteration minerals (dominated by Fe-rich smectite clays, celadonite, zeolites, and silica) are commonly observed in flow tops and filling the joints and fractures of flow interiors (Reidel 1983; Horton 1991; Reidel *et al.* 2002; Zakharova *et al.* 2012). Detailed paragenetic studies of hydrothermal mineralization in basaltic rocks of the nearby Cascade Range and in the Hawaiian Islands do explicitly link hydrothermal alteration assemblages to depth and temperature. In Cascade Range, a significant decrease in permeability with depth in the range of approximately 200–1000 m is associated with temperatures in the range of 30–50°C (e.g., Swanberg *et al.* 1988; Blackwell 1994; Hulen & Lutz 1999) and the appearance of a suite of pore-filling alteration minerals including clays and zeolites (e.g., Bargar 1988; Bargar & Keith 1999). At Kilauea Volcano, Hawaii, a significant reduction in permeability below 488 m is associated with temperatures of 30–60°C and a suite of alteration minerals including calcite, Fe–Ti oxides, and Ca–Mg smectites (Keller *et al.* 1979).

For this study, heat generation within the model domain, either by decay of radionuclides (e.g., Deming 1993) or through viscous dissipation from fluid flow (Manga & Kirchner 2004), was neglected. Radiogenic heat production in basalts averages <0.4 μW m^{-3} (Vila *et al.* 2010), so the resulting difference in heat flow from the base to the surface of the Columbia River Basalt Group due to radiogenic sources will be

<2 mW m^{-2} or approximately 2.5% of the background value. Manga & Kirchner (2004) estimate that the temperature of water increases through viscous dissipation by approximately 2.3 °C per 1000 m of elevation change from recharge to discharge location, indicating that the maximum heating of water for either cross-sectional model (Fig. 13.5) will be approximately 2 °C. This is potentially a non-negligible component of the temperature field (Fig. 13.9C) and the heat budget, but the general patterns of temperature and heat flow will be unaffected. The maximum error in estimated heat flow (Fig. 13.10A) resulting from neglecting this term is approximately 2% and occurs where the vertical temperature gradient is lowest (between 20 and 50 km; Fig. 13.9C).

CONCLUSIONS

Average heat flow at depth under the Columbia Plateau Regional Aquifer System is likely closer to the value of approximately 70–80 mW m^{-2} expected on the basis of its tectonic context than previous estimates from mapping of near-surface heat flow measurements (Williams & DeAngelo 2008, 2011) have indicated. Patterns and rates of groundwater flow computed using best-available estimates of hydrologic properties and boundary conditions can explain the apparent low-heat-flow anomaly. Near-surface heat flow is predicted to be relatively low in high-recharge upland areas and relatively high near major groundwater discharge locations (e.g., where deep aquifers intersect lowland rivers). Highest subsurface temperatures at a given depth are predicted to occur under the major rivers.

Compiled hydraulic test data from near the center of the Columbia Plateau Regional Aquifer System indicate that permeability decreases significantly in the depth range 600–900 m, corresponding approximately to a subsurface temperature range of 35–50 °C. Thus, *in situ* heat flow estimated below such depths will more reliably predict the deep, background heat flow.

The permeability decrease in the Columbia Plateau Regional Aquifer System occurs at temperatures similar to those associated with sharp reductions in permeability with depth in basalts of the adjacent Cascade Range and in the Hawaiian Islands, two localities where loss of permeability is known to be associated with hydrothermal alteration. Thus, we surmise that the permeability decrease at 600–900 m depth in the Columbia Plateau Regional Aquifer System also owes to hydrothermal alteration and, based on the similarity of pore-filling alteration minerals across these volcanic terrains, postulate that rapid permeability reductions with depth will likely occur in other continental flood basalt provinces.

The semiarid Columbia Plateau Regional Aquifer System provides a good example of modest groundwater flow rates significantly altering crustal heat flow. Because continental flood basalt provinces exhibit high degrees of lateral hydraulic connectivity over considerable distances (at least in the upper permeable zone), advective disturbances to crustal heat flow are expected in the active groundwater flow domain, with a return to predominantly conductive heat flow below the active groundwater system. Given similar thermal and hydrologic boundary conditions, temperatures would be affected to greater depths in systems with less extreme permeability anisotropy.

ACKNOWLEDGMENTS

Dave Norman, Jeff Bowman, and Jessica Czajkowski (all from the Washington State Department of Natural Resources) provided new data to augment the USGS geothermal database. Alden Provost (USGS Reston) provided the new version of SUTRA for use in this study, altering the code to ensure that matrix singularities did not occur in the deep (heat conduction only) part of the domain. He was also a wonderful resource when considering how to best apply the new general head boundary conditions. Jonathan Haynes (USGS Oregon Water Science Center) provided geologic model GIS support and figure preparation. Many thanks to reviewers, Ingrid Stober, Ying Fan, and one anonymous reviewer, and editors Mark Person and Tom Gleeson, for the obvious care and many excellent comments and suggestions that resulted in significant improvement of the original manuscript. Funding for this project was provided by the US Department of Energy – Geothermal Technologies Program and the USGS Energy Resources Program.

CHAPTER 14

Deep fluid circulation within crystalline basement rocks and the role of hydrologic windows in the formation of the Truth or Consequences, New Mexico low-temperature geothermal system

JEFF D. PEPIN[1], MARK PERSON[1], FRED PHILLIPS[1], SHARI KELLEY[2], STACY TIMMONS[3], LARA OWENS[4], JAMES WITCHER[5] AND CARL W. GABLE[6]

[1] *Department of Earth & Environmental Science, New Mexico Institute of Mining and Technology, Socorro, NM, USA;* [2] *Geologic Mapping, New Mexico Bureau of Geology and Mineral Resources, Socorro, NM, USA;* [3] *Aquifer Mapping Program, New Mexico Bureau of Geology and Mineral Resources, Socorro, NM, USA;* [4] *Ormat Technologies, Inc., Reno, NV, USA;* [5] *Witcher and Associates, Las Cruces, NM, USA;* [6] *Los Alamos National Labs, Los Alamos, NM, USA*

ABSTRACT

Hot springs can occur in amagmatic settings, but the mechanisms of heating are often obscure. We have investigated the origin of the Truth or Consequences, New Mexico low-temperature (approximately 41 °C) hot springs in the southern Rio Grande Rift. We tested two hypotheses that could account for this amagmatic geothermal anomaly: lateral forced convection in a gently dipping carbonate aquifer and circulation through high-permeability crystalline basement rocks to depths of 8 km that is then focused through an overlying faulted hydrologic window. These hypotheses were tested using a regional two-dimensional hydrothermal model. Model parameters were constrained by calibrating to measured temperatures, specific discharge rates, and groundwater residence times. We collected 16 temperature profiles, 11 geochemistry samples, and 6 carbon-14 samples within the study area. The geothermal waters are Na^+/Cl^- dominated and have apparent groundwater ages ranging from 5500 to 11,500 years. Hot-spring geochemistry is consistent with water/rock interaction in a silicate geothermal reservoir rather than a carbonate system. Peclet number analysis of temperature profiles suggests that specific discharge rates beneath Truth or Consequences range from 2 to 4 m year^{-1}. Geothermometry indicates that maximum reservoir temperatures are around 170 °C. Observed measurements were reasonably reproduced using the deep-circulation permeable-basement modeling scenario (10^{-12} m^2) but not the lateral forced-convection carbonate-aquifer scenario. Focused geothermal discharge is the result of localized faulting, which has created a hydrologic window through a regional confining unit. In tectonically active areas, such as the Rio Grande Rift, deep groundwater circulation within fractured crystalline basement may play a more prominent role in the formation of geothermal systems than has generally been acknowledged.

Key words: geochemistry, geothermal, hydrologic windows, numerical modeling, permeability

INTRODUCTION

Interest in constraining crustal permeability has grown over the last 15 years (Ingebritsen & Manning 1999, 2010; Manning & Ingebritsen 1999; Shmonov *et al.* 2003; Stober & Bucher 2007). This parameter plays a key role in understanding the genesis of ore deposits (Raffensperger & Garven 1995), contact metamorphism (Gerdes *et al.* 1995), and metasomatism (Fritz *et al.* 2006). Crustal permeability is widely believed to

decay with depth due to mechanical loading and fluid-rock geochemical reactions. Manning & Ingebritsen (1999) used geothermal and metamorphic data to suggest that crustal permeability decreases from 10^{-12} to 10^{-16} m^2 at 1 km depth to 10^{-15} to 10^{-18} m^2 at 10 km depth (Fig. 14.1) (Ingebritsen & Manning 1999, 2010; Manning & Ingebritsen 1999; Shmonov *et al.* 2003; Stober & Bucher 2007). Independent results from various studies throughout the world provide a range for crustal permeabilities of $10^{-16.6}$ m^2 all the way up to $10^{-7.3}$ m^2 for

Crustal Permeability, First Edition. Edited by Tom Gleeson and Steven E. Ingebritsen.
© 2017 John Wiley & Sons, Ltd. Published 2017 by John Wiley & Sons, Ltd.
Companion Websites: www.wiley.com/go/gleeson/crustalpermeability/
http://crustalpermeability.weebly.com/

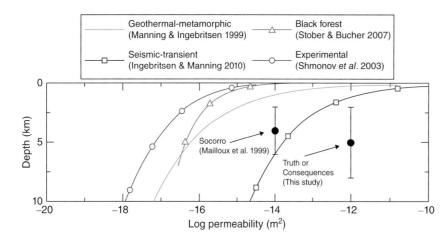

Fig. 14.1. Crustal permeability versus depth relationships from previous investigations that provide a range of crustal permeabilities of 10^{-18} to 10^{-10} for depths <10 km. Crystalline basement permeabilities within the Rio Grande Rift are also indicated (black circles) and are high in comparison. This suggests that large fault structures and significant fracture networks have substantially increased the permeability of crystalline basement rocks in this region of New Mexico above global averages. (Adapted from Ingebritsen & Manning 2010.)

depths <10 km (see studies in Ingebritsen & Manning 2010). These studies indicate that crustal permeabilities may vary by orders of magnitude across various geological settings at any given depth. In 2010, Ingebritsen and Manning argued that permeability has a relatively high variance while maintaining clear depth dependence for depths <10 km (Fig. 14.1). Deviation from this depth dependence may be due to time-dependent permeability changes as a result of tectonism and seismicity (Rojstaczer & Wolf 1992; Ingebritsen & Manning 2010).

Previous numerical modeling efforts focusing on the Rio Grande Rift suggest that crystalline basement permeability is relatively elevated (approximately 10^{-14} m^2), allowing for deep groundwater circulation (2–6 km) (Barroll & Reiter 1990; Mailloux *et al.* 1999). This permeability value is within the crustal scale permeability range determined by the aforementioned synthesis studies, although it marked a new upper limit for crystalline basement permeability along the Rio Grande Rift. In many instances, fault-block rotation, erosion, lithological variations, and emplacement of Cenozoic dikes can provide gaps or "hydrologic windows" through overlying low-permeability Paleozoic to Cenozoic confining units, allowing pore water from the crystalline basement to discharge. In areas where crustal permeability is relatively high, the rapid rise of groundwater through hydrologic windows is probably responsible for the development of several low-temperature geothermal systems in the Rio Grande Rift (Witcher 1988; Barroll & Reiter 1990; Mailloux *et al.* 1999).

Between October 2012 and 2013, we collected temperature, geochemical, and isotopic data to assess circulation patterns within the Truth or Consequences (T or C) hot-springs district (HSD). We also developed a cross-sectional hydrothermal model to investigate groundwater circulation patterns that formed the Truth or Consequences heat-flow anomaly (approximately 41 °C at depths <54 m). Approximately 13 MW of heat are discharged by 0.1 m^3 s^{-1} of geothermal water at Truth or Consequences (Theis *et al.* 1941). The majority of the geothermal groundwater ascends under the town's historic HSD and then flows laterally through a shallow alluvial aquifer before discharging into the Rio Grande.

The goal of this study is to determine the groundwater circulation patterns and subsurface permeability structure responsible for generating the geothermal conditions within the HSD. Our analysis is intended to be broadly relevant to regions with relatively permeable fractured crystalline basement rocks and low-temperature geothermal anomalies. We tested two hypotheses using a two-dimensional numerical model (Fig. 14.2). The first hypothesis considers relatively deep (>5 km) groundwater circulation within permeable crystalline basement rocks. Highly fractured limestone crops out locally in Truth or Consequences, indicating that large fault networks are present and could foster deep groundwater circulation by greatly increasing the permeability of the subsurface. The second hypothesis is based on inferences presented in previous studies investigating the Truth or Consequences geothermal anomaly and assumes that transmission of geothermally heated water is primarily through the permeable carbonate aquifers that overlie the crystalline basement rocks (Powell 1929; Theis *et al.* 1941; Wells & Granzow 1981). These carbonate layers are laterally continuous and occur at depths of about 2.7 km approximately 15 km north of Truth or Consequences.

We constrained sediment and crystalline basement permeability by reproducing HSD temperature profiles and ^{14}C residence time data. A Peclet number analysis provided estimates of vertical specific discharge rates constrained by temperature profiles logged as part of this study. We compared the discharge rates from the Peclet number analysis to simulated vertical specific discharge rates from the two-dimensional model. We also conducted a geochemical analysis of the geothermal waters, comparing geochemistry results to those of other geothermal systems of similar temperature, including New Mexico's Socorro geothermal system and the carbonate-hosted Etruscan Swell of Italy. We compared the temperatures determined by silica and cation geothermometry to simulated maximum temperatures experienced by water particles discharging in the HSD. Finally, we compared our model-derived permeability estimates to those found elsewhere in the Rio Grande Rift and internationally.

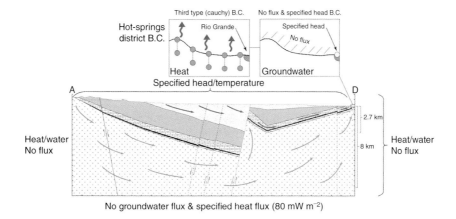

Fig. 14.2. Boundary conditions used in our two-dimensional hydrothermal model and a schematic diagram showing the two groundwater flow hypotheses evaluated as part of this study. Boundary conditions are shown for both heat transport and groundwater flow. The inset portrays a close-up of the boundary conditions applied to the hot-springs district. The basement-circulation hypothesis (blue arrows) involves deep circulation of groundwater within highly fractured crystalline basement rocks. Groundwater discharges where hydrologic windows exist in overlying confining units, such as the Percha Shale (black). The shallow-circulation hypothesis (red arrows) considers shallow groundwater circulation through the carbonate Magdalena Group. (*See color plate section for the color representation of this figure.*)

Hydrogeologic setting

The Rio Grande Rift is an active 100–300 km wide, north-trending zone of crustal extension that cuts through central New Mexico (Seager & Morgan 1979; Morgan *et al.* 1986; Baldridge *et al.* 1995). The town of Truth or Consequences (formerly known as Hot Springs) is located on the banks of the Rio Grande in south-central New Mexico at the southern terminus of the Engle sub-Basin within the Rio Grande Rift (Fig. 14.3). The location of the geothermal system at Truth or Consequences is controlled, in part, by geologic structures that formed during three major tectonic events that affected New Mexico and the southwestern United States.

First, folds and thrust faults with a northwest-striking trend and strike-slip faults with a north-northeast-striking trend formed in the vicinity of Truth or Consequences during northeast-southwest-directed compressional Laramide deformation between 75 and 45 million years ago (Seager & Mack 2003; Harrison & Cather 2004). A northwest-trending overturned syncline and at least one low-angle fault are preserved in a limestone outcrop at the northern edge of the HSD. This faulted, overturned syncline is an important structure facilitating discharge of groundwater migrating southward out of the Engle Basin toward the surface (Fig. 14.4). Local faulting associated with this overturned syncline has created a hydrologic

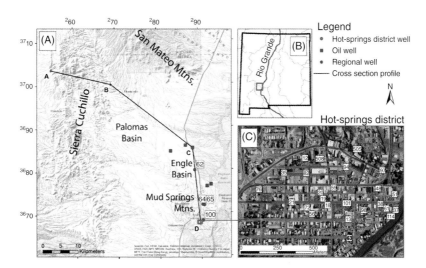

Fig. 14.3. Basemap (A) showing surface water drainages (light blue lines) and the location of the study area in south-central New Mexico (B). The presumed recharge area of the Sierra Cuchillo and San Mateo Mountains north of Truth or Consequences is also indicated for reference. The orientation of the geologic cross section and two-dimensional hydrothermal model is shown in black (A–B–C–D) in addition to the locations of wells discussed in this paper. The inset of the hot-springs district (C) shows geothermal well locations. The Rio Grande can also be seen in the lower right corner of this inset. The delineated areas on the New Mexico state map (B) are major drainage basins. Regional map coordinate datum is UTM NAD83 Zone 13. (*See color plate section for the color representation of this figure.*)

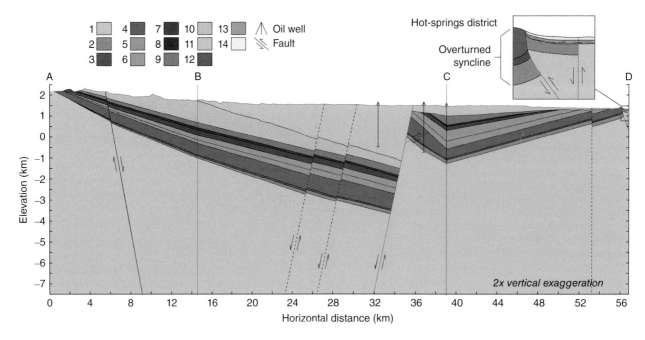

Fig. 14.4. Geologic cross section depicting the stratigraphic units used in our two-dimensional hydrothermal model. Additional information about model parameters is provided in Table 14.1. The color legend of this figure corresponds to the hydrostratigraphic units and descriptions in Table 14.3. The cross section was constructed by utilizing oil-well data, an east–west regional cross section (Lozinsky 1987), gravity data (Gilmer *et al.* 1986), and surface geologic maps. (Harrison *et al.* 1993; Harrison & Cather 2004.) (*See color plate section for the color representation of this figure.*)

window in the Percha Shale that otherwise confines flow in the crystalline basement. This gap allows geothermal waters to surface in Truth or Consequences. Laramide compression was followed by voluminous eruptions in the Mogollon-Datil volcanic field starting about 37 million years ago (Harrison *et al.* 1993). Volcanic units from eruptions in the San Mateo Mountains cover the highest elevations of the Sierra Cuchillo and are buried in the Engle Basin. The third major tectonic event that shaped the landscape is Rio Grande Rift extension that began about 36 million years ago and peaked between 16 and 5 million years ago (Kelley & Chapin 1997; Seager & Mack 2003). This event caused the uplift of the Caballo Mountains to the east and the Mud Springs Mountains, the Sierra Cuchillo, and the San Mateo Mountains to the west and northwest (Fig. 14.3A). Material eroded from the rising rift flank uplifts filled the Engle Basin with as much as 2.7 km of sediment (Lozinsky 1987).

Mean annual air temperature and annual precipitation are 16.5 °C and 25.2 cm year^{-1}, respectively. Potential evapotranspiration is 84 cm year^{-1} as estimated by the temperature-based Thornthwaite equation and greatly exceeds precipitation (Dingman 2002). Regional precipitation and water-table patterns indicate that the principal source of recharge to the geothermal system is probably from the Sierra Cuchillo and San Mateo Mountains to the northwest (Fig. 14.5). Enhanced fracture permeability in the recharge area is suggested by collapse structures associated with seven calderas identified in the San Mateo Mountains (Ferguson *et al.* 2012). We estimated recharge for

the watershed using the Maxey-Eakin method, which assumes that the fraction of precipitation available for subsurface recharge increases directly with precipitation. The empirical scaling coefficients used in this study are those utilized by Anderholm (2001) to estimate mountain-front recharge from the Sandia Mountains east of Albuquerque, NM. We estimate the total recharge available as between 0.1 and 0.9 m^3 s^{-1}. The higher estimate reflects the recharge over the entire watershed (Fig. 14.5). The lower estimate only considers the recharge where Paleozoic sediments and crystalline basement rocks are exposed at the surface. The HSD geothermal discharge estimate of 0.1 m^3 s^{-1} by Theis *et al.* (1941) is consistent with the lower end of our recharge estimates. Theis *et al.* (1941) calculated this value by coupling average HSD geothermal water chloride content with measurements of Rio Grande streamflow and chloride concentration above and below the HSD.

Early studies of HSD geothermal waters proposed magmatic heat sources (Powell 1929; Theis *et al.* 1941; Wells & Granzow 1981). However, the youngest magmatism documented near Truth or Consequences is approximately 5 million years old in the Mud Springs Mountains to the northwest (Dunbar 2005). Conceptual numerical models presented by Furlong *et al.* (1991) suggest that heat from this magmatism dissipated within <1 million years. Furthermore, swarms of microearthquakes, characteristic of magmatic activity, have not been detected beneath the study area (NMT/IRIS-PASSCAL Data Center). It is therefore unlikely that magmatic heating is involved in this geothermal system.

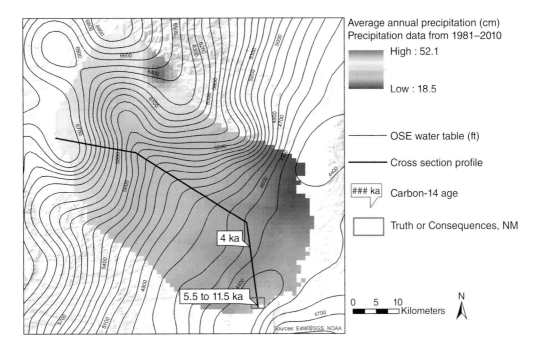

Fig. 14.5. Groundwater contributing area (color-shaded contours) to the hot-springs district (HSD) and water-table contours in relation to our two-dimensional hydrothermal model transect (bold black line). The color contours denote the spatial distribution of annual precipitation across the watershed. The black contour lines are estimated water-table elevations from the New Mexico Office of the State Engineer. Precipitation and water-table patterns suggest recharge to the HSD primarily occurring in the Sierra Cuchillo and San Mateo Mountains northwest of Truth or Consequences. Apparent carbon-14 groundwater ages collected as part of this project are displayed on the basemap as well (annotated circles). Oldest groundwater ages are within the HSD (precipitation data from PRISM Climate Group, Oregon State University 2012). (*See color plate section for the color representation of this figure.*)

Previous studies (e.g., Theis *et al.* 1941) also hypothesized that circulation of the geothermal waters was primarily confined to the Magdalena Group carbonates (Fig. 14.2). Highly fractured limestone outcrops in the vicinity of the HSD suggest that permeability may be relatively high in the subsurface. The Magdalena Group carbonates dip to the north and are buried to a maximum depth of about 2.7 km, where groundwater could be heated to approximately 110 °C, assuming a conductive geothermal gradient of 40 °C km^{-1} (Reiter *et al.* 1986). In order for groundwater flow within the Magdalena Group carbonates to carry heat into the HSD, significant lateral convective heat transfer must occur. An alternative hypothesis involves deep circulation through crystalline basement rocks (Fig. 14.2). A recently drilled 74-m-deep well completed in fractured crystalline basement rocks yielded a permeability estimate of 2.6×10^{-10} m^2 from a preliminary specific capacity test in June 2014; this is a very high permeability for crystalline basement rocks (this well is TC-114 in Appendix A and Fig. 14.3; specific capacity test method outlined by Theis 1963). In the basement-circulation hypothesis, elevated temperatures within the HSD are primarily the result of vertical convective heat transfer. Geothermal water discharges due to the localized absence of the confining Percha Shale, consistent with the hydrologic window hypothesis first proposed by Witcher in 1988.

FIELD MEASUREMENTS

Temperature profiles

We measured 16 temperature profiles to depths of 54 m. Ten of these were capped, cased wells that had not been pumped for at least 1 year (Type 1). Five profiles were from wells drilled within 2 weeks of measurement or that had their pumps removed just prior to the time of measurement but were recently active (Type 2). Data from these wells are less reliable, as the wells had recently been disturbed. One flowing artesian well was logged (Type 3).

Additional details about the wells logged are included in Appendix A, and their locations are shown in Figure 14.3. Most wells displayed temperature profiles indicative of an upward flow regime (Fig. 14.6), as signified by negative-slope concave downward temperature profiles (Bredehoeft & Papadopulos 1965). In some instances, such curvature can be the result of horizontal flow at near steady-state conditions (Ziagos & Blackwell 1986). However, the nearly isothermal conditions at depth argue for vertical flow (Bredehoeft & Papadopulos 1965).

Groundwater residence times

We collected five carbon-14 (^{14}C) samples within the HSD and one sample north of town to assess groundwater residence times. Uncorrected age precision of the samples collected is

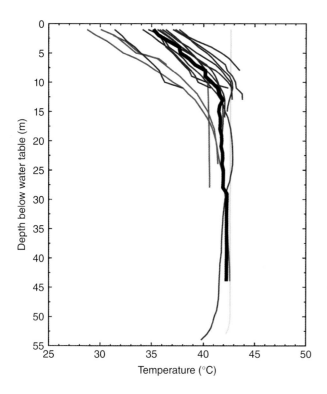

Fig. 14.6. Temperature–depth profiles measured within the hot-springs district during October 2012–2013. The type of well the profile was measured in is indicated by its color (blue = type 1, red = type 2, green = type 3; see methods section for details). A calculated average profile for depths 0–44 m is also plotted (bold black line). The average profile was only calculated to this depth due to lack of abundant data at greater depths and was used when interpreting hydrothermal model results. (*See color plate section for the color representation of this figure.*)

±50 years. All ^{14}C ages reported in this study are uncorrected, meaning their ages have not been adjusted for reservoir effects; the chemical and isotopic data necessary to confidently make corrections are not available. Plummer *et al.* (2004) used ^{14}C, ^{3}H, and CFC data to assess groundwater ages in the Middle Rio Grande Basin within the Rio Grande Rift about 250 km north of our study area. Their results suggest that geochemical corrections to radiocarbon groundwater ages may not be necessary in this region. Nonetheless, these uncorrected ages should be viewed as estimated maximum groundwater residence times. The ages acquired from our samples were used to calibrate our two-dimensional hydrothermal model.

Geochemistry

We collected 11 water samples from within the HSD for geochemical analysis (Appendix B). All wells were purged with multiple borehole volumes prior to sample collection. Field parameters such as discharge temperature, pH, and specific conductance were recorded for each sample. Trace metal samples were filtered and acidified on site using a 0.45-μm filter and 10 drops of nitric acid per sample. The New Mexico Bureau of

Geology and Mineral Resources chemistry laboratory analyzed the samples. Geochemical results were used in conjunction with those reported in previous studies to investigate the geochemistry and origin of the geothermal waters and to estimate geothermal reservoir temperatures using geothermometry.

THERMAL PECLET NUMBER ANALYSIS METHODS

Curvature in well temperature profiles can be used to estimate vertical flow velocities. We conducted a curve-matching exercise using the analytical solution of Bredehoeft & Papadopulos (1965) while assuming a fluid density (ρ_f) of 1000 kg m^{-3}, a fluid heat capacity (c_f) of 4180 J kg^{-1}, and bulk thermal conductivity of the sediments (λ) of 2 W (m °C)$^{-1}$. We used the bottom-hole temperature in each well bore for T_2, while T_1 was prescribed as the water-table temperature at the top of the well; L was assigned a value of 50 m, as this is the typical depth to bedrock in the HSD.

The vertical specific discharge rates from the Peclet number analysis were compared to those calculated by our two-dimensional hydrothermal model and to the geothermal discharge rate estimated by Theis *et al.* (1941).

GEOTHERMOMETRY METHODS

The chemistry of geothermal groundwater is commonly used to estimate reservoir temperatures. Constituents such as dissolved silica, calcium, potassium, and sodium are ideal for geothermometry due to their slow retrograde mineral-solute re-equilibration timescales. As geothermal groundwater ascends and cools, the temperature signature of the geothermal reservoir is preserved in the concentrations of these constituents due their slow reaction times. However, this is only true when several assumptions about the waters and their flow path are satisfied. One key assumption is that dissolution of minerals at depth is controlled by temperature-dependent chemical reactions that equilibrate at the reservoir's maximum temperature. It is also assumed that composition changes due to mixing are insignificant. Additional assumptions are described in detail by Fournier (1989) and Karingithi (2009).

Geothermometry analysis of 13 chemistry samples collected within the HSD was carried out using a Microsoft Excel spreadsheet program published by Powell & Cumming (2010) entitled "liquid_analysis_v3_powell-cumming_2010_stanfordgw .xlsx." Of the 13 samples analyzed, 8 were collected as part of this study, 4 were from Theis *et al.* (1941), and 1 was from Summers (1976). The composition of the samples collected as part of this study is presented in Appendix B. The Powell & Cumming (2010) spreadsheet program calculates many common geothermometers. Silica geothermometers typically represent minimum reservoir temperatures, whereas cation geothermometers provide information about maximum reservoir temperatures. This is due to the more rapid reaction rate of silica (Fournier 1989; Karingithi 2009; Powell & Cumming 2010). Some geothermometers have proven to be more consistently representative of geothermal reservoir temperatures than others; we used those suggested by Karingithi (2009) and Powell & Cumming (2010).

HYDROTHERMAL MODELING METHODS

We employed a finite element method of characteristics hydrothermal model to characterize hydrology and paleohydrology within the study area (Person *et al.* 2008). The numerical approximations to the fluid flow and heat transport equations implemented in this code were originally validated, in part, by reproducing simulations by Smith & Chapman (1983).

Groundwater flow

We solve for variable-density groundwater flow using the following stream-function-based groundwater equation (Senger & Fogg 1990):

$$\nabla_x \cdot \left[\frac{K}{|K|} \nabla_x \psi \right] = -\frac{\partial \rho_r}{\partial x} \qquad (14.1)$$

where ∇_x is the gradient operator $[L^{-1}]$, K is the hydraulic conductivity tensor $[L^1 t^{-1}]$, $|K|$ is the determinant of K, ψ is the stream function $[L^2 t^{-1}]$, and ρ_r is the relative density $[-]$ (defined below). The right-hand side of this equation accounts for variable-density-driven groundwater flow. Although this is a steady-state groundwater flow equation, time-dependent changes in density can result in transient flow conditions.

Relative density (ρ_r) used in Eq. 14.1 is defined as

$$\rho_r = \frac{\rho_f - \rho_o}{\rho_o} \qquad (14.2)$$

where ρ_f is the density of groundwater at its elevated temperature and pressure $[M^1 L^{-3}]$, and ρ_o is the density of water at standard conditions (10 °C and atmospheric pressure) $[M^1 L^{-3}]$.

Specific discharge is related to the stream function through the Cauchy–Riemann equations:

$$\frac{\partial \psi}{\partial z} = q_x \qquad (14.3)$$

$$\frac{\partial \psi}{\partial x} = q_z \qquad (14.4)$$

where q_x and q_z are specific discharge $[L^1 t^{-1}]$ in the x and z directions $[L^1]$, respectively.

Heat transport

Temperature can affect groundwater density in our model. The model solves a conductive and convective–dispersive heat transfer equation given by

$$\left[c_f \rho_f \varphi + c_s \rho_s (1 - \varphi) \right] \frac{\partial T}{\partial t} = \nabla_x [\lambda \nabla_x T] - \overline{q} \rho_f c_f \nabla_x T, \qquad (14.5)$$

where λ is the thermal dispersion-conduction tensor $[M^1 L^1 t^{-3} T^{-1}]$, t is the time $[t^1]$, φ is the porosity $[-]$, T is the temperature $[T^1]$, c_s is the specific heat capacity of the solid phase $[L^2 t^{-2} T^{-1}]$, c_f is the specific heat capacity of the liquid phase $[L^2 t^{-2} T^{-1}]$, $\overline{q}$ is the specific discharge vector $[L^2 t^{-1}]$, ρ_f is the density of the liquid phase $[M^1 L^{-3}]$, and ρ_s is the density of the solid phase $[M^1 L^{-3}]$. Thermal conductivities are assumed to be isotropic, as we have used scalar quantities.

Groundwater residence time

Reverse particle tracking was used to determine simulated groundwater residence times. The particle tracking algorithm utilizes the final seepage velocity field generated by the model to advect particles throughout the finite element domain (Person *et al.* 1998). Particle tracking along flow paths permitted the determination of maximum temperatures experienced by groundwater discharging in the HSD.

Equations of state

Thermodynamic equations of state are used to compute the density and viscosity of groundwater at elevated temperature and pressure conditions. The model uses the polynomial expressions of Batzle & Wang (1992), valid for temperatures between 10 and 350 °C.

Boundary conditions

The boundary conditions assigned in our model are consistent with field evidence. For groundwater flow, the lateral regional water-table gradient published by the New Mexico Office of the State Engineer was used to assign specified head values across the top of the model domain (Fig. 14.2). The model uses this head gradient to determine a surface flux using the following steady-state stream function equation:

$$\frac{\partial \psi}{\partial z} = -K \frac{\partial h}{\partial x} \qquad (14.6)$$

where K *is* the hydraulic conductivity $[L^1 t^{-1}]$, h is the hydraulic head $[L^1]$, and ψ is the stream function $[L^2 t^{-1}]$. Assigned water-table elevations varied from 2143 m in the Sierra Cuchillo to 1331 m in the HSD (Fig. 14.5). A no-flux boundary condition was assigned locally near the southern edge of the model domain within the HSD. This was performed to emulate shallow lateral geothermal water flow toward the Rio Grande. At the southern terminus of the model domain, the Rio Grande was represented as a specified head node. The sides and base of the solution domain were represented as no-flux boundaries. The no-flux boundary condition on the northern end of the domain represents the watershed divide in the Sierra Cuchillo. The no-flux boundary condition on the southern end, at the Rio Grande, forces waters to rise and discharge. Water-table temperature measurements in wells on the opposite side of the Rio Grande near Truth or Consequences are cold relative to the HSD. This suggests that the Rio Grande serves as a no-flux boundary for the geothermal waters; it is also the regional topographic low point in the area.

For heat transport, a specified temperature boundary condition was assigned across most of the top of the solution domain at the water table (Fig. 14.2). Assigned water-table temperatures ranged from 15 °C in the uplands to 24 °C near the HSD. A third-type or Cauchy boundary condition was assigned along the surface locally in the HSD. This was performed so that near-surface temperatures were free to vary with the magnitude of convective heat transfer. This boundary condition utilizes simulated temperature gradients just below the water table and the effective thermal conductivity of the vadose zone to compute the vertical conductive heat flux across the surface. Along the base of the solution domain, we assigned a basal heat flux of 80 mW m^{-2}, which closely resembles the measured heat flux values near Truth or Consequences (Sass *et al.* 1971; Sanford *et al.* 1979; Reiter *et al.* 1986). No-flux boundaries were imposed on the sides of the domain.

Initial conditions

Initial subsurface temperatures were assumed to increase linearly with depth using a 40 °C km^{-1} temperature gradient that is representative of the region (Reiter *et al.* 1986). All model simulations were run for 300,000 years using a time step of 20 years to reach steady-state conditions in order to eliminate the influence of initial conditions.

Mesh configuration

The cross-sectional finite element mesh included 3904 nodes and 7493 triangular elements. We used 123 nodal columns to discretize 14 hydrostratigraphic units present in the study area. The mesh was vertically and laterally refined within the HSD.

The horizontal dimensions of the grid varied from 1000 m to the north to 50 m within the HSD. Except for very thin hydrostratigraphic units of 10–20 m in thickness, there are two to three nodes per stratigraphic layer in the vertical direction. Within the unconsolidated deposits of the HSD, vertical grid dimensions average about 10 m. The vertical discretization of the crystalline basement varies from 100 to 800 m beneath the HSD.

Hydrostratigraphy and simulation parameters

We constructed our northwest-southeast cross-sectional model perpendicular to the New Mexico State Engineer's published regional water-table contours (Figs 14.4 and 14.5). Although no one cross section fully captures three-dimensional flow patterns, this cross section provides a reasonable estimate of groundwater flow patterns associated with the Truth or Consequences geothermal system. The stratigraphy of the cross section was constrained using oil-well data, an east–west regional cross section (Lozinsky 1987), gravity data (Gilmer *et al.* 1986), and surface geologic maps (Harrison *et al.* 1993; Harrison & Cather 2004). The hydrologic parameters assigned to each of the 14 stratigraphic units are presented in Table 14.1. Insufficient well-test data exist to assign locally derived hydrologic parameters. Therefore, we used representative permeability and porosity values consistent with those reported by Freeze & Cherry (1979). Thermal transport and petrophysical parameters that were assigned to all hydrostratigraphic units and held constant are presented in Table 14.2. Brief geologic descriptions of the hydrostratigraphic units are given in Table 14.3. Stratigraphic offsets caused by faults were included in the model, but discrete faults were not modeled as conduits or barriers to groundwater flow.

Table 14.1 Hydrologic parameters assigned to different stratigraphic units.

Unit	Unit name	$\log(k_x)$ (m^2)	Anisotropy (k_x/k_z)	Porosity
1	Precambrian Granitic and Metamorphic Rocks	−19 to −11	1	0.05
2	Additional Lower Paleozoic Formations	−14	1	0.15
3	Lower Paleozoic Percha Shale, Lake Valley Formation, Kelley Limestone	−19	1	0.15
4	Pennsylvanian Magdalena Group	−13 to −11	1	0.20
5	Permian Abo Formation	−15	1	0.25
6	Permian Yeso Formation	−17	1	0.25
7	Permian San Andres Formation	−17	1	0.25
8	Cretaceous Mancos Shale & Dakota Sandstone	−17	1	0.25
9	Cretaceous Sediments	−17	1	0.25
10	Tertiary Volcanics	−16	1	0.15
11	Tertiary Palomas Formation	−12	100	0.30
12	Quaternary Clay	−17	1	0.30
13	Quaternary Fine Sand	−12	100	0.30
14	Quaternary Sand and Gravel	−12	100	0.30

Crystalline basement (Unit 1) and Magdalena Group (Unit 4) permeabilities are variable.

Table 14.2 Thermal, solute transport, and physical parameters held constant for all simulations and all hydrostratigraphic units.

Symbol	Variable name	Magnitude
α_L	Longitudinal dispersivity	10 m
α_T	Transverse dispersivity	1 m
λ_f	Fluid thermal conductivity	$0.58\,\mathrm{W\,m^{-1}\,°C^{-1}}$
λ_r	Solid thermal conductivity	$2.5\,\mathrm{W\,m^{-1}\,°C^{-1}}$
ρ_s	Rock density	$2600\,\mathrm{kg\,m^{-3}}$
S_s	Specific storage	$10^{-7}\,\mathrm{m^{-1}}$

Numerical implementation

We solved the governing equations sequentially using the finite element method. The groundwater flow equation was solved using Galerkin's method with triangular elements and linear trial solutions. The heat transport equation was solved using the modified method of characteristics. Reverse particle tracking to obtain groundwater residence times and advective ages was performed using the seepage velocity field from the last time step in each simulation.

Permeability can vary by several orders of magnitude for a given lithology. We conducted a sensitivity study and calibration procedure in which we varied the permeability of the Magdalena Group carbonates (unit 4) between 10^{-13} and $10^{-11}\,\mathrm{m^2}$ and that of the crystalline basement (unit 1) from 10^{-19} to $10^{-11}\,\mathrm{m^2}$.

These permeability values are within the range noted by Freeze & Cherry (1979) and Gleeson *et al.* (2011) for these lithologies. Forty-one simulations were run to complete the trial and error calibration process. In the following, we present model output from seven of these simulations.

RESULTS

Peclet number analysis results

Curvature in nearly all of the temperature profiles collected within the HSD is evidence of upward groundwater migration (Fig. 14.6). Specific discharge was determined for 10 of these wells using Peclet number analysis. Our best-fit Peclet number results indicate vertical specific discharge rates ranging from 2 to $4\,\mathrm{m\,year^{-1}}$ beneath Truth or Consequences (Fig. 14.7). Theis *et al.* (1941) estimated geothermal discharge from the HSD to be $0.1\,\mathrm{m^3\,s^{-1}}$. Most of the geothermal waters discharge in the HSD, which comprises an area of 750 by 750 m. Dividing the $0.1\,\mathrm{m^3\,s^{-1}}$ discharge estimate by the approximate area of the HSD yields a vertical specific discharge rate of around $6\,\mathrm{m\,year^{-1}}$. Considering the typical accuracy of Peclet number analysis and the uncertainty associated with the discharge approximation by Theis *et al.* (1941), the calculated specific discharge estimates are in reasonable agreement.

Table 14.3 Brief lithologic description of each of the 14 stratigraphic units.

ID	Name	Thickness (m)	Description
1	Precambrian Granitic and Metamorphic Rocks		Metamorphosed volcanic rocks, sandstone, and shale deposited in an extensional basin about 1.60–1.65 billion years ago, later intruded by 1.4 billion year old granite
2	Additional Lower Paleozoic Formations	0–175	Cambrian to Silurian shallow marine limestone, dolomite, shale, and sandstone; includes, from oldest to youngest, Bliss Sandstone, El Paso Formation, Montoya Formation, and Fusselman Dolomite
3	Lower Paleozoic Percha Shale, Lake Valley Formation, Kelley Limestone	0–50	The Devonian Percha Shale includes shale intercalated with thin siltstone and limestone beds. The Percha Shale grades up into the carbonates of the Mississippian Lake Valley Formation. The Percha Shale is a confining unit
4	Pennsylvanian Magdalena Group	200–816	Fossiliferous limestone, cherty limestone, shale, dolomite, and conglomerate deposited in shallow ocean water that grade up into Abo Formation
5	Permian Abo Formation	10–397	River floodplain mudstone and siltstone, sinuous channel sandstones, and rare carbonate lake deposits. This formation is generally a confining unit
6	Permian Yeso Formation	10–533	Predominantly sandstone and gypsum with layers of limestone, siltstone, and shale
7	Permian San Andres Formation	20–231	Fossiliferous marine limestone, shale, and fine-grained sandstone
8	Cretaceous Mancos Shale and Dakota Sandstone	0–50	Fluvial sandstone, shale, and conglomerate of the Dakota grades up into two tongues of Mancos Shale. The Mancos Shale tongues act as confining units
9	Cretaceous Sediments	0–435	Sandstone, shale, and conglomerate of the marginal marine Gallup Sandstone and fluvial Crevasse Canyon Formation
10	Tertiary Volcanics	375–1020	Lava flows, ash flow tuffs, debris, and stream deposits that formed during voluminous eruptions in the Mogollon-Datil volcanic field. Fracturing has increased this unit's permeability
11	Tertiary Palomas Formation	10–2000	Weakly to moderately cemented sandstones, conglomerates, and siltstones deposited in the Engle Basin to the north of Truth or Consequences during Rio Grande Rift extension
12	Quaternary Clay	0–7	Clay layers within the hot-springs district's Quaternary fine sands
13	Quaternary Fine Sand	0–30	Quaternary fluvial sediments that contain clay lenses. Units 13 and 12 combine to form a leaky confining unit in the hot-springs district
14	Quaternary Sand and Gravel	0–20	These Quaternary alluvial sediments deposited locally in Truth or Consequences are a mixture of alluvial fan and fluvial sediments

Thickness estimates are based on oil-well logs and surface outcrops (see Lozinsky 1987).

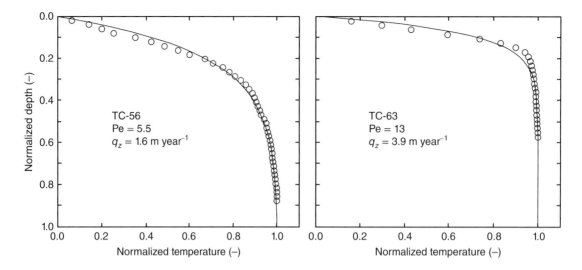

Fig. 14.7. Peclet number analysis results from two representative wells. Our best-fit curve-matched results indicate vertical specific discharge rates ranging from 2 to 4 m year^{-1} beneath the hot-springs district (see Figure 14.3C for well locations).

Groundwater residence time results

Uncorrected carbon-14 groundwater ages within the HSD ranged between 5490 and 11,480 years. A ^{14}C sample collected approximately 14 km north of town provided a younger uncorrected age of 4040 years. Shallow groundwater within alluvial deposits will typically yield ages on the order of hundreds of years (Weissmann *et al.* 2002). The ages collected in Truth or Consequences support a deep flow path for the geothermal waters.

Geothermometry results

Geothermometry results are summarized in Table 14.4. Silica geothermometers and the K/Mg geothermometer provide information about near-discharge area conditions, as they react more rapidly than other geothermometers (Fournier 1989; Karingithi 2009; Powell & Cumming 2010). Na/K and Na–K–Ca cation geothermometers preserve a longer record of the flow history due to slower retrograde equilibration rates.

Consequently, they typically reflect more distant and deeper reservoir temperatures.

The silica geothermometers and K/Mg geothermometer suggest temperatures ranging from 63 to 102 °C. The slower-to-equilibrate cation geothermometers indicate temperatures ranging from 158 to 207 °C. When Na–K–Ca geothermometry indicates that reservoir temperatures are below 180 °C, a magnesium correction is sometimes applied. The magnesium-corrected Na–K–Ca geothermometer yields reservoir temperatures ranging from 100 to 125 °C. The value of $[\log_{10}(\sqrt{CaNa^{-1}}) + 2.06]$ is slightly positive at 0.3 for all of the samples analyzed. This suggests that the most appropriate cation geothermometer for determining deeper reservoir temperatures may be the Na–K–Ca geothermometer, which yields temperatures ranging from 122 to 183 °C, or the Na–K–Ca magnesium-corrected geothermometer, which ranges from 100 to 125 °C (Karingithi 2009). Averaging the slower-to-equilibrate cation geothermometers indicates maximum reservoir temperatures around 167 °C. Dividing this

Table 14.4 Summary of silica and cation geothermometry results.

Type	Geothermometer	Source	Mean (°C)	Min. (°C)	Max. (°C)	Range (°C)	Est. depth (km)
Silica	Chalcedony cond.	Fournier & Potter (1982)	63	51	77	26	1.6
	Quartz cond.	Fournier & Potter (1982)	94	83	107	24	2.4
	Quartz adiabatic	Fournier (1981)	96	85	107	21	2.4
Cation	Na–K–Ca	Fournier (1981)	158	122	183	61	4.0
	Na–K–Ca Mg corr	Fournier (1981)	113	100	125	25	2.8
	Na/K	Fournier (1979)	190	168	208	39	4.7
	Na/K	Giggenbach (1988)	207	186	223	37	5.2
	K/Mg	Giggenbach (1986)	102	93	111	18	2.6

Silica geothermometers represent a minimum reservoir temperature while cation geothermometers provide information about maximum reservoir temperatures. Estimated depths were calculated by dividing the mean geothermometer temperature by the geothermal gradient of 40 °C km^{-1} (Reiter *et al.* 1986).

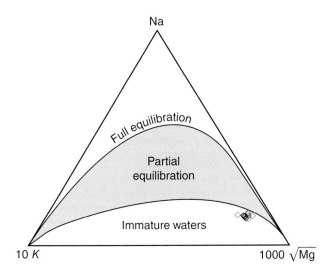

Fig. 14.8. Giggenbach (1991) plot of the geothermal waters collected in the hot-springs district. This plot classifies them as "immature waters," implying that cation geothermometry results may not be representative of geothermal reservoir temperatures at depth due to mixing or lack of equilibration.

value by the local geothermal gradient ($40\,°C\,km^{-1}$; Reiter *et al.* 1986) implies geothermal groundwater circulation to depths around 4 km.

Geothermometry temperatures should be applied cautiously. Groundwater at temperatures in excess of $180\,°C$ is generally in equilibrium with silica (Karingithi 2009). Cation geothermometers, including the Na–K–Ca magnesium-corrected geothermometer, average $167\,°C$ and imply that silica may have not reached equilibrium in the reservoir.

Furthermore, a Giggenbach plot classifies the HSD waters as immature, signifying they may not be the representative of geothermal reservoir conditions (Fig. 14.8). In general, cation geothermometry temperature uncertainties are typically about $20\,°C$ (Fournier 1989).

Groundwater geochemistry results

Geochemical data from this study, Theis *et al.* (1941) and Summers (1976), are summarized on a Piper diagram in Figure 14.9. The HSD waters have total dissolved solid (TDS) contents averaging about 2600 ppm, which is fairly high for the region. Shallow nonthermal waters typically have TDS concentrations of <500 ppm. The analyses of the geothermal waters are very similar, such that most of them overlie each other in Figure 14.9. The Na^+/Cl^- composition of the HSD geothermal waters is the characteristic of high-to-moderate temperature water in crystalline basement rocks (Kühn 2004; Stober & Bucher 2007a; Bucher & Stober 2010). The Na^+ is known to be derived from fluid–rock alteration reactions at relatively high temperatures, and the Cl^- is released from fluid inclusions as the rock is altered (Ellis & Mahon 1964, 1967).

Two analyses of Rio Grande streamflow are plotted in Figure 14.9 for comparison. The samples were collected 7

months apart from the Rio Grande just north of Truth or Consequences, below Elephant Butte Dam. The Rio Grande has a mixed Na^+/Ca^{2+} cation composition, anions dominated by HCO_3^- and SO_4^{2-}, and is relatively dilute. Several nearby (within 16 km), mildly geothermal waters (20 to $30\,°C$) are similar to the HSD geothermal waters, but have higher concentrations of Ca^{2+} and SO_4^{2-}. Some of these waters have a composition intermediate between the HSD groundwater and the Rio Grande, suggesting mixing of dilute shallow water with geothermally derived water.

We further explore the possibility of mixing by plotting sodium and trace element concentrations closely associated with typical geothermal waters against the concentration of chloride. Throughout the Great Basin, geothermal waters tend to show a close correlation between reservoir temperature and chloride concentrations. Geothermal waters from the Basin and Range typically have elevated boron and lithium concentrations that also correlate with chloride (Arehart *et al.* 2003). Covariation with the chloride concentration thus supports a mixing hypothesis (Fig. 14.10). The mildly geothermal waters in some cases do not follow the general trend; temperatures fall below the trend and one sample has notably high boron content (indicated with an arrow in Fig. 14.10). The elevated boron content of marine shales makes them a common source of boron in groundwater (Harder 1959; Walker 1975), so a likely source is the marine Percha Shale, thought to have been removed from the HSD by faulting. The covariation of the remaining samples with chloride strongly supports the hypothesis that Truth or Consequences is supplied by a geothermal aquifer of quite homogeneous composition that is locally diluted by cooler shallow waters. It also provides further evidence that the geothermometry results may not accurately represent geothermal reservoir conditions due to mixing during ascent.

Groundwater residing in carbonate rocks at high temperatures would be expected to exhibit a $Ca^{2+}/HCO_3^-/SO_4^{2+}$ composition (Chiodini *et al.* 1995), similar to those from the "Etruscan Swell" area of Italy. However, the geochemical composition of groundwater from the carbonate system is quite distinct from the Truth or Consequences waters (Fig. 14.9).

The Socorro, New Mexico geothermal system (Owens 2013), is thought to result from deep circulation within highly fractured crystalline rocks (Barroll & Reiter 1990; Mailloux *et al.* 1999). Average Socorro water temperatures are slightly lower than the HSD waters, averaging about $36\,°C$, and their TDS content is a little lower (approximately 2100 ppm), but their overall chemistry is similar to the HSD samples. This suggests the Truth or Consequences and Socorro groundwaters circulate primarily in rocks of similar composition, most likely igneous and metamorphic basement rocks.

Two-dimensional hydrothermal modeling results

Here, we present seven representative hydrothermal model results that serve to evaluate the groundwater circulation

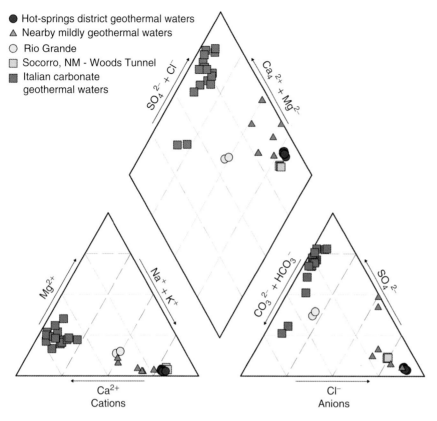

● Hot-springs district geothermal waters
△ Nearby mildly geothermal waters
○ Rio Grande
□ Socorro, NM - Woods Tunnel
■ Italian carbonate
  geothermal waters

Fig. 14.9. Piper diagram summarizing geochemical analyses discussed in this paper. The geothermal waters from the hot-springs district (HSD) are shown with red circles. Mildly geothermal waters in the vicinity (within 16 km) of the HSD are plotted as orange triangles. Two samples of Rio Grande surface waters collected upstream of Truth or Consequences are shown using yellow circles. Data from a low-temperature carbonate geothermal reservoir in the Etruscan Swell of Italy (Chiodini *et al.* 1995) are shown with green squares. Data from the Woods Tunnel slim hole from the Socorro, New Mexico, geothermal system are shown by blue squares (Owens 2013). Truth or Consequences waters have a Na^+/Cl^- signature characteristic of geothermal waters derived from igneous and metamorphic rocks. (*See color plate section for the color representation of this figure.*)

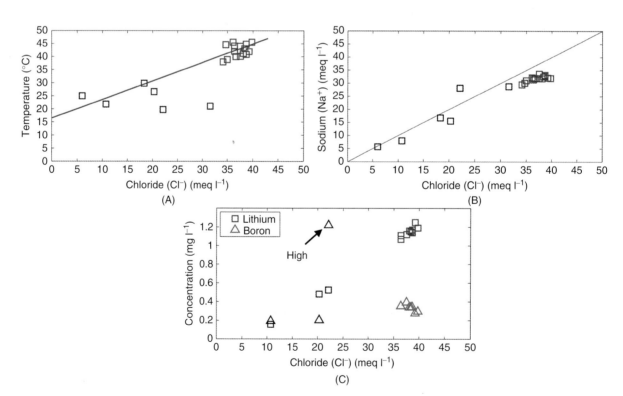

Fig. 14.10. Plots of groundwater temperature (A), Na^+ (B), and B and Li^+ (C) against Cl^-. Mildly geothermal groundwaters near the hot-springs district (HSD) (within 16 km) are shown using black squares. The line in (A) connects mean annual temperature at Truth or Consequences (16.5 °C) to the temperature of the geothermal waters. The line in (B) is a 1:1 line. The arrow in (C) highlights a well located about 5 km north of the HSD that has elevated concentrations of boron. All plots show covariation, suggesting that geothermal Na^+/Cl^- waters have undergone mixing with shallow non-geothermal groundwater.

hypotheses mentioned in the introduction and depicted in Figure 14.2. The first hypothesis involves highly permeable crystalline basement rocks and the relatively rapid ascent of geothermal waters through a hydrologic window in a regional confining unit (i.e., the Percha Shale) and will be referred to as the basement-circulation hypothesis. The second hypothesis considers shallow circulation through the permeable Magdalena Group carbonates and will be referred to as the shallow-circulation hypothesis. The goal of each scenario was to try to reproduce the temperature profiles, vertical specific discharge rates, and carbon-14 groundwater ages measured or estimated.

We tested the deep-circulation hypothesis by varying the permeability of crystalline basement rocks from 10^{-14} to 10^{-11} m^2 (Fig. 14.11D–G). This allowed circulation to depths of around 8 km. Geothermal waters discharge in the HSD and to a lesser degree in the center of the model domain, near the Mud Springs Fault. In both of these areas, hydrologic windows are present in overlying confining units due to faulting. The regional temperature distribution for all deep-circulation scenarios is the result of a forced-convection heat-flow regime, with the exception of the lowest basement-permeability case (10^{-14} m^2), which is conduction-dominated (Fig. 14.11D–G; Smith & Chapman 1983). Regional forced convection results in the redistribution of heat toward the topographically low southern side of the model domain. The warmest simulated temperatures are on the south side of the HSD (Fig. 14.12D–G), which is somewhat inconsistent with field observations, as hottest

measured temperatures occur in the center of the district, approximately 300 m north of our warmest simulated temperatures. We extracted temperature profiles from the warm region in the model for comparison to locally measured temperature profiles. Comparing a representative average of these simulated profiles to the average measured profile from Figure 14.6 shows that assigned basement permeabilities of 10^{-13} and 10^{-12} m^2 produce the best agreement with observed HSD temperatures (Fig. 14.13D, Table 14.5). Basement permeabilities outside the 10^{-13} to 10^{-12} m^2 range resulted in groundwater temperatures that were too low. The net cooling effect with high permeabilities, as evidenced in our 10^{-11} m^2 scenario, is consistent with patterns observed in many prior studies (e.g., Smith & Chapman 1983). Specific discharge rates calculated in our Peclet number analysis and inferred from Theis *et al.* (1941) most closely matched our basement-permeability scenario of 10^{-12} m^2 (33% and −29% errors, respectively). Simulated HSD residence times for the basement-circulation scenarios had a wide range of 7770 years to 491,012 years (Table 14.5). The simulated residence times closest to carbon-14 ages were obtained when employing basement permeabilities of 10^{-11} m^2 (7% error) and 10^{-12} m^2 (167% error). Finally, comparing the maximum estimated reservoir temperatures from particle tracking to the average cation geothermometry results shows best agreement with basement-permeability scenarios of 10^{-12} m^2 (28% error) and 10^{-11} m^2 (5% error). In addition to the constant permeability-with-depth scenarios presented here, we also considered a permeability-decay-with-depth scenario for

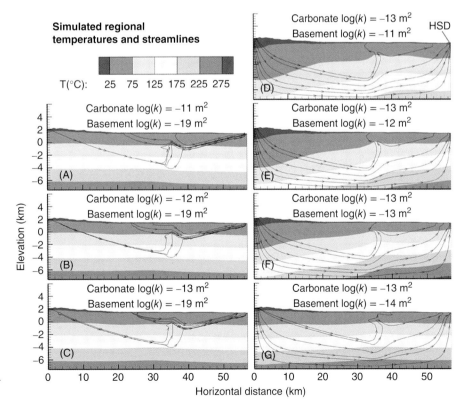

Fig. 14.11. Comparison of computed regional groundwater flow patterns (black lines with arrows) and temperatures for three shallow-circulation scenarios (left, A–C) and four basement-circulation scenarios (right, D–G). The base-10 logarithm of permeabilities used for the Magdalena Group and the crystalline basement are listed above each plot. Refer to Table 14.5 for simulation parameters and goodness of fit for subplots A–G. The location of the hot-springs district (HSD) is shown in graphic D. Groundwater flow directions are parallel to streamlines. Groundwater reaching the HSD in the shallow-circulation scenarios flows primarily through the shallow Magdalena Group. In contrast, deep-circulation scenarios are characterized by geothermal waters derived predominately from crystalline basement rocks. Shallow-circulation scenarios yield thermal patterns typical of a conductive thermal regime. (*See color plate section for the color representation of this figure.*)

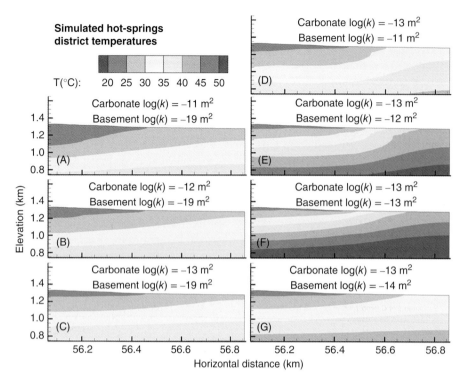

Fig. 14.12. Comparison of computed hot-springs district (HSD) temperature patterns for three shallow-circulation scenarios (left) and four basement-circulation scenarios (right). Refer to Table 14.5 for simulation parameters and goodness of fit for subplots A–G. The relative regional location of these cross sections is shown in Figure 14.11D denoted by "HSD". Only two (simulations E and F) of the presented simulations reproduced average measured HSD temperatures of 41°C. Both of these simulations required highly permeable crystalline basement rocks (10^{-13} and 10^{-12} m²). Increasing or decreasing basement permeabilities beyond this range resulted in reduced temperatures in the HSD. (*See color plate section for the color representation of this figure.*)

the crystalline basement rocks (not shown) using the relation presented by Manning & Ingebritsen (1999). However, this scenario resulted in near-conductive conditions due to the rapid decay of permeability and resulting shallow groundwater circulation patterns.

To test the shallow-circulation hypothesis, we set the crystalline basement permeability to 10^{-19} m² and varied the Magdalena Group carbonates' permeability from 10^{-13} to 10^{-11} m². This effectively restricts groundwater circulation to the shallow carbonate aquifer, as shown in Figure 14.11A–C. Water discharging in the HSD enters the system from both the primary recharge area in the north and a zone near the Mud Springs Fault in the center of our model domain. All computed shallow-circulation temperature distributions are indicative of conduction-dominated heat transport (Fig. 14.11A–C). Simulated HSD temperature patterns are similar for these scenarios and are warmer toward the south end of the district (Fig. 14.12A–C). Temperature profiles extracted from this zone do not match the observed average temperature profile from the HSD (Fig. 14.13C). Increasing the permeability of the carbonate aquifer did not result in higher regional or HSD temperatures. Specific discharge calculations for these shallow-circulation scenarios were smaller than both the Theis *et al.* (1941) estimate and the Peclet number analysis results by more than 95%. Also computed HSD residence times poorly matched field measurements. Finally, cation geothermometry results were not in agreement with the simulated maximum temperatures.

For both the shallow- and deep-circulation simulations, computed regional temperature profiles agree fairly well with

bottom-hole temperature data measured in two oil wells located approximately 15 km north of Truth or Consequences (Fig. 14.13A,B). This is likely due to the fact that groundwater flow is largely horizontal in this region, and temperatures are close to conductive conditions. The reasonable agreement between simulated and observed temperatures near these wells suggests that the basal heat flux and thermal conductivities we assigned in our model are representative.

DISCUSSION AND CONCLUSIONS

Numerical modeling results and geochemical interpretation strongly suggest that the Truth or Consequences geothermal system is the result of deep groundwater circulation within fractured permeable crystalline basement rocks. Faulting has made regional confining units locally discontinuous. This hydrologic window allows waters circulating deep within the crystalline basement to ascend relatively rapidly and discharge in the HSD.

Two-dimensional hydrothermal modeling does not support a shallow groundwater circulation hypothesis in which flow is primarily confined to the Magdalena Group carbonates, as argued by most previous studies. In this scenario, convective heat transport was negligible and observed HSD temperatures could not be reproduced. Allowing groundwater to circulate deeply within the system through permeable crystalline basement rocks and ascend directly under the HSD best-matched observations. Consequently, we conclude that groundwater circulation to depths ranging from 2 to 8 km within fractured crystalline basement rocks of effective permeability on the

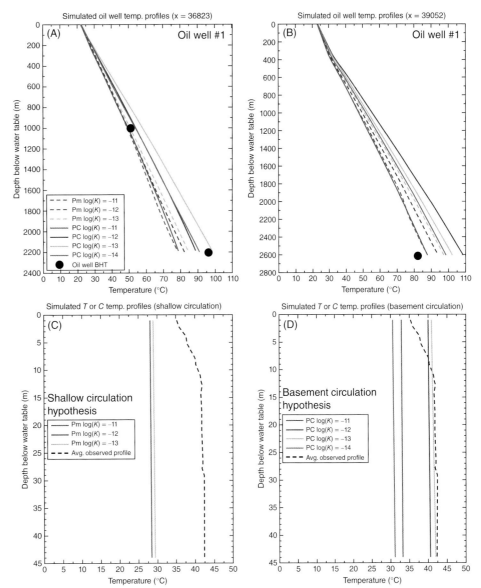

Fig. 14.13. Comparison of simulated and observed temperatures along the model domain. The assigned base-10 logarithm permeabilities for the crystalline basement rocks (PC) and Magdalena Group (Pm) are shown in the legends (graphs A and B share the same legend). Simulated temperature profiles are compared to bottom-hole temperature data collected in oil wells approximately 15 km north of Truth or Consequences (Top, A and B). They agree fairly well with bottom-hole temperature data, suggesting that our assigned thermal properties, such as basal heat flux and thermal conductivities, represent those of the study area. Average simulated and measured temperature profiles from the hot-springs district (HSD) are compared in C and D. Only simulations having high crystalline basement permeability (10^{-13} and 10^{-12} m²) were able to reproduce observed HSD temperatures. (*See color plate section for the color representation of this figure.*)

order of 10^{-12} m² is required to account for the geothermal anomaly. Although geothermal water is transmitted to the surface through a carbonate aquifer, our results indicate that the geothermal water enters the carbonate aquifer from the fractured crystalline basement at relatively shallow depths and that the carbonate aquifer is not the route for long-distance flow.

Our hydrothermal model was able to reproduce the temperature range of measured temperature profiles, but not their shallow curvature patterns. The temperature curvature typically takes place in the upper 10 m of the subsurface (Fig. 14.6). Capturing this level of detail would require a more refined finite element mesh not practical for our regional-scale application. To some degree, these shallow, cooler temperatures may be the result of transient hydrologic conditions during periods of time when the Rio Grande river stage is higher than the alluvial aquifer. Our models were quasi-steady-state.

Down valley transport of cool water coming into the HSD perpendicular to our cross-sectional model may also explain, in part, this curvature in the shallow temperature profiles. The warmest HSD temperatures in our model results are about 300 m south of where observed hottest temperatures were measured in the HSD. This discrepancy may simply be the result of focusing on a localized area of a regional model; it is also possible that lithologic heterogeneities or faults not represented in our model are influencing local geothermal water flow. For example, south-dipping faults may be acting as conduits. In our simulations, faults are not modeled as conduits or barriers and would therefore fail to capture this effect. Considering the regional scale of our modeling effort, these localized divergences are not unexpected.

There remains some nonuniqueness in our model results due to parameter uncertainty. Like many other studies, we found

Table 14.5 Comparison of average simulated hot-springs district (HSD) temperatures and ages to those observed.

Scenario	ID	Pm log(k) (m²)	PC log(k) (m²)	HSD temperatures		Particle temperatures	
				Average (°C)	% Error	Maximum (°C)	% Error
Shallow circulation (varying Pm)	A	−11	−19	28.2	−31	85.5	−49
	B	−12	−19	29.1	−29	85.0	−49
	C	−13	−19	29.1	−29	62.5	−63
Basement circulation (varying PC)	D	−13	−11	33.0	−20	175.5	5
	E	−13	−12	40.3	−2	213.0	28
	F	−13	−13	41.3	0	241.2	44
	G	−13	−14	30.8	−25	129.1	−23

Scenario	ID	Pm log(k) (m²)	PC log(k) (m²)	Specific discharge			Age	
				Average (m year⁻¹)	% Error (1941)	% Error (Peclet)	Average (years)	% Error
Shallow circulation (varying Pm)	A	−11	−19	0.1	−99	−97	502,802	6,795
	B	−12	−19	0.1	−99	−97	63,667	773
	C	−13	−19	0.1	−99	−97	498,818	6,741
Basement circulation (varying PC)	D	−13	−11	10.1	81	237	7,770	7
	E	−13	−12	4.0	−29	33	19,465	167
	F	−13	−13	0.6	−90	−80	128,000	1,655
	G	−13	−14	0.1	−99	−97	491,012	6,634

Average observed (41.1°C) and simulated temperatures have been calculated by averaging temperature profiles spanning 0–44 m. Observed carbon-14 dates collected within the hot-springs district are interpreted as representative groundwater residence times and have been averaged (7292 years) for comparison with advective particle travel times using the steady-state flow field simulated by our cross-sectional model. All shallow-circulation scenarios fail to reproduce hot-springs district temperatures. Scenarios E and F accurately simulate temperatures. However, all other calibration parameters are better fit by scenario E, suggesting a crystalline basement permeability of 10^{-12} m² is most likely. The permeability assigned to the crystalline basement (PC) and Magdalena Group carbonates (Pm) are listed. Observed data: hot-springs district temp. = 41.1°C; geothermometry = 167°C. Observed data: Spec. discharge (1941) = 5.6 m year⁻¹; Spec. discharge (Peclet) = 3.0 m year⁻¹; Age (Carbon-14) = 7292 years.

that permeability had the largest effect on model results (e.g., Smith & Chapman 1983). This is why our calibration procedure focused on varying permeability. However, other parameters such as porosity can also have some effect on model results. Due to lack of data, it was necessary to assume average representative porosities for lithologies in our model consistent with Freeze & Cherry (1979). To assess the implications of this assumption, we re-ran one of our deep-circulation scenarios (basement permeability 10^{-13} m²) but doubled the porosity of the modeled crystalline basement rocks to 0.10. This increased simulated HSD temperatures by 0.8%, groundwater ages by 3.8%, and maximum reservoir temperatures by 2.2% and decreased average specific discharge by 0.2%. Groundwater age was most sensitive to changes in porosity. However, increasing crystalline basement porosity by 100% led to changes in all parameters of <4%, some <1%. This nonlinear relationship to porosity is due to the coupled nature of fluid flow and heat transport. Doubling porosity in our model decreased seepage velocities, but not by a factor of 2. It also resulted in a lower bulk thermal conductivity and a slight temperature increase. However, this temperature increase was accompanied by a decrease in viscosity, which increased seepage velocities and partially compensated for changing porosity. This led us to conclude that the model sensitivity to porosity is small relative to permeability.

Geochemistry results also support the basement-circulation hypothesis. The Na⁺/Cl⁻-dominated composition of typical HSD geothermal waters is characteristic of geothermal water derived from igneous and metamorphic rocks (Kühn 2004;

Stober & Bucher 2007a; Bucher & Stober 2010). In addition, the spatial distribution of water composition and the covariation of several geothermally associated parameters suggest that geothermal waters mix with shallow groundwater. The distinct chemical differences between Italian carbonate-reservoir geothermal waters and the Truth or Consequences waters are strong evidence against long-term residence in carbonate rocks in our study area (Chiodini *et al.* 1995). In contrast, the chemistry of geothermal waters from the Socorro, New Mexico, and the Truth or Consequences HSD is nearly identical (Owens 2013). Studies investigating groundwater flow patterns of the Socorro geothermal system have found that its geothermal anomaly is likely the result of deep groundwater circulation within crystalline basement rocks (Barroll & Reiter 1990; Mailloux *et al.* 1999), thus supporting the basement-circulation hypothesis for Truth or Consequences.

We hypothesize that the geothermal waters enter the system in the Sierra Cuchillo and San Mateo Mountains and flow downward until they contact crystalline basement rocks. High permeability enables the water to move through, and geochemically react with, the granitic/metamorphic rocks at relatively high temperatures (probably >150 °C) for thousands of years at depths ranging from 2 to 8 km. Geothermometry indicates maximum geothermal reservoir temperatures around 170 °C. This suggests that geothermal groundwater circulation is focused toward the shallower end of the 2–8 km depth range (around 4 km assuming a temperature gradient of 40 °C km⁻¹; Reiter *et al.* 1986). The rock/water chemical reactions in this environment produce a characteristic geothermal Na^+/Cl^- composition. The geothermal waters eventually ascend under the HSD where a hydrologic window exists due to the absence of overlying confining units. In transit toward the shallow Quaternary alluvial sediments in which most HSD wells are currently completed, the waters traverse a Paleozoic limestone unit and in some instances mix with shallow groundwater. The geothermal water moves through the limestone rapidly enough that it does not have time to re-equilibrate with the carbonate rocks. Once in the shallow alluvial aquifer, geothermal waters flow laterally until they discharge to the Rio Grande or are extracted by wells.

This study provides additional evidence that the crystalline basement rocks beneath the Rio Grande Rift can be remarkably permeable. Mailloux *et al.* (1999) and Barroll & Reiter (1990) studied the Socorro geothermal system located approximately 115 km north of Truth or Consequences and estimated crystalline basement permeability of 10^{-14} m^2 (Mailloux *et al.* 1999). Figure 14.1 compares these Rio Grande Rift permeabilities to the crustal permeability trends determined by previous studies. Permeabilities within the Rio Grande Rift are high in comparison. This suggests that large fault structures and significant fracture networks have substantially increased the permeability of the crystalline basement, potentially during the Laramide orogeny or due to the ongoing extension of the rift.

It is noteworthy that we were able to reproduce temperature anomalies in the HSD when using a constant permeability-with-depth modeling scheme for each lithology. This is consistent with the results of Burns *et al.* (2015). Burns *et al.* (2015) tested additional schemes involving step-function permeability transitions at depth based on available well data.

The results from both step-function and constant permeability schemes were similar in that they reasonably reproduced observed temperature anomalies. Like Burns *et al.* (2015), our modeling results indicated that allowing permeability to decay with depth in accordance with the Manning & Ingebritsen (1999) power law lead to conduction-dominated heat transfer due to the rapid decline of permeability. A constant permeability scheme for regional hydrologic models of this scale appears to be a reasonable simplification, and the resulting permeabilities should be interpreted as effective regional permeabilities.

Future modeling efforts within continental rifts should consider the possibility of highly permeable crystalline basement rocks that promote deep circulation of groundwater, in conjunction with hydrologic windows in confining units, as key factors in the generation of low-temperature geothermal systems. Modeling efforts that capture the three-dimensional geometry of systems could better interrogate the effects of water-table configuration and geologic heterogeneity. The possibility that discrete high-permeability faults serve as conduits for geothermal waters also warrants consideration.

ACKNOWLEDGMENTS

We would like to extend our gratitude to the City of Truth or Consequences and its residents for their assistance and cooperation throughout our study. This research was funded by the City of Truth or Consequences, New Mexico Bureau of Geology and Mineral Resources (Aquifer Mapping Program), Department of Energy award DE-EE0002850 (funded part of the temperature logging effort), and the National Science Foundation (EPSCoR) under Grant No. IIA-1301346.

APPENDIX A – WELL INVENTORY, INDICATING WHICH WELL DATA PROVIDED THE BASIS FOR VARIOUS ANALYSES AND FIGURES

ID #	UTM X	UTM Y	Type	Depth (m)	Piper	Chem. mixing	Temp. profile	Geothermometry	Carbon-14
TC-12	289524	3667924	Pump	34	x	x		x	x
TC-13	289772	3667577	Artesian	53	x	x	x	x	x
TC-18	289651	3667673	Pump	12	x	x		x	
TC-19	289627	3667745	Open	12			x		
TC-20	289609	3667730	Open	12			x		
TC-21	289439	3667679	Open	12			x		
TC-35	289398	3667845	Pump	63	x	x	x	x	
TC-37	290053	3667623	Open	17			x		
TC-38	289989	3667658	Open	18			x		
TC-46	289889	3667656	Open	14			x		
TC-48	289951	3667745	Open	12			x		
TC-51	290048	3667712	Open	15			x		
TC-56	289558	3667652	Pump	44			x		
TC-59	289688	3667756	Open	15	x	x	x	x	x
TC-61	289996	3667633	Open	22	x	x	x	x	x
TC-62	288638	3679885	Pump	183					x
TC-63	289775	3667472	Pump	30	x	x	x	x	
TC-64	290436	3671828	Pump	30	x	x			
TC-65	290782	3671749	Pump	61	x	x			
TC-72	289530	3667647	Pump	67			x		
TC-76	289253	3667755	Pump	65	x	x		x	x
TC-82	289528	3667839	Pump	32		x			
TC-94	289765	3667869	Pump	8	x	x		x	
TC-97	289977	3667846	Pump	30		x			
TC-98	289863	3667626	Artesian	56	x	x		x	
TC-100	290458	3668222	Artesian	37	x	x			
TC-101	289756	3667667	Pump	4	x	x		x	
TC-114	290016	3667630	Open	74			x		
TC-505	289825	3667969	Spring	–	x	x		x	
TC-508	289602	3667919	Spring	–	x	x		x	
Datum: NAD83, UTM Zone 13				Count	16	18	16	13	6

APPENDIX B – CHEMICAL ANALYSES FROM THIS STUDY

Parameter	D. Limit	Units	TC-12	TC-13	TC-18	TC-35	TC-59	TC-61	TC-62	TC-63	TC-64	TC-65	TC-76
Date Collected	-	-	1/9/2013	9/5/2013	1/9/2013	1/8/2013	1/8/2013	1/8/2013	1/8/2013	1/8/2013	1/8/2013	1/8/2013	9/5/2013
Alkalinity as CaCO$_3$	5	mg l^{-1}	176	173	175	173	177	181	111	182	137	23	173
Aluminum	0.0005	mg l^{-1}	ND	ND	ND	ND	ND	ND	0.0035	ND	ND	0.006	ND
Anions total		meq l^{-1}	44.28	44.67	43.88	43.77	42.56	43.66	23.86	43.2	16.18	43.87	45.05
Antimony 121	0.0005	meq l^{-1}	ND	ND	ND	ND	ND	ND	ND	ND	ND	ND	ND
Arsenic	0.0005	meq l^{-1}	ND	ND	ND	ND	ND	ND	0.0025	ND	ND	ND	ND
Barium	0.001	meq l^{-1}	0.197	0.190	0.206	0.216	0.206	0.182	0.197	0.170	0.102	0.017	0.197
Beryllium	0.0005	meq l^{-1}	ND	ND	ND	ND	ND	ND	ND	ND	ND	ND	ND
Bicarbonate (HCO$_3$)	5	meq l^{-1}	215	211	214	211	216	221	136	222	168	28	211
Boron 11	0.005	meq l^{-1}	0.345	0.303	0.340	0.345	0.354	0.342	0.203	0.394	0.194	1.220	0.296
Bromide	0.01	meq l^{-1}	0.126	ND	0.134	0.122	0.129	0.180	ND	0.121	0.041	0.210	ND
Cadmium 111	0.0005	meq l^{-1}	ND	ND	ND	ND	ND	ND	ND	ND	ND	ND	ND
Calcium	0.05	mg l^{-1}	163	165	160	160	157	163	139	140	128	246	164
C13/C12 ratio		-	-21.1	-19.8			-10.0	-18.5	-17.3				-21.3
Carbon-14		years	5490	6870			11480	7110	4040				5510
Carbon-14 Fraction		fmdn	-0.211	-0.198			-0.100	-0.185	-0.173				-0.213
Carbonate (CO$_3$)	5	meq l^{-1}			ND	ND	ND	ND	ND	ND	ND		ND
Cations total		meq l^{-1}	43.85	43.07	43.41	44.11	42.37	42.87	23.55	41.81	15.80	41.79	42.92
Chloride	1	mg l^{-1}	1370	1390	1360	1370	1290	1350	720	1330	380	784	1410
Chromium	0.0005	mg l^{-1}	ND	ND	ND	ND	ND	ND	0.0033	ND	0.0031	ND	ND
Cobalt	0.0005	mg l^{-1}	ND	ND	ND	ND	ND	ND	ND	ND	ND	ND	ND
Copper 65	0.0005	mg l^{-1}	0.0069	0.0062	ND	ND	ND	0.0053	0.0081	ND	ND	0.0093	0.0520
Fluoride	0.1	mg l^{-1}	3.10	3.14	3.11	3.12	3.18	3.09	2.87	3.42	1.43	4.26	3.37
Hardness (CaCO$_3$)		mg L^{-1}	483	479	474	474	479	482	382	411	380	667	477
Iron	0.02	mg l^{-1}	ND	ND	ND	ND	0.611	ND	ND	ND	ND	ND	ND
Lead	0.0005	mg l^{-1}	ND	ND	ND	ND	ND	ND	ND	ND	ND	ND	ND
Lithium	0.001	mg l^{-1}	1.140	1.200	1.150	1.170	1.110	1.160	0.482	1.120	0.158	0.526	1.190
Magnesium	0.05	mg l^{-1}	18.3	16.7	18.3	18.0	21.2	18.4	8.4	14.8	14.9	13.1	16.4
Manganese	0.001	mg l^{-1}	ND	ND	ND	ND	0.034	ND	ND	ND	ND	0.057	0.016
Molybdenum 95	0.001	mg l^{-1}	ND	ND	ND	ND	ND	ND	ND	ND	ND	0.032	ND
Nickel	0.0005	mg l^{-1}	ND	ND	ND	ND	ND	ND	0.0045	ND	0.0053	0.0092	0.0059
Nitrate	0.1	mg l^{-1}	2.71	2.40	2.34	2.35	2.04	2.40	1.35	1.86	1.81	ND	2.49
Nitrite	0.1	mg l^{-1}	ND	ND	ND	ND	ND	ND	ND	ND	ND	ND	ND
Orthophosphate	0.5	mg l^{-1}	ND	ND	ND	ND	ND	ND	ND	ND	ND	ND	ND
Percent difference		%	-0.48	-1.83	-0.55	0.38	-0.23	-0.91	-0.64	-1.64	-1.19	-2.43	-2.42
pH		ph Units	7.3	7.1	7.4	7.2	7.2	7.1	7.5	7.2	7.6	8.0	7.6
Potassium	0.05	mg l^{-1}	57.80	56.00	56.60	58.60	55.20	55.50	13.10	65.60	7.17	11.60	56.20
Selenium	0.001	mg l^{-1}	ND	ND	ND	ND	ND	ND	ND	ND	ND	ND	ND
Silicon	0.02	mg l^{-1}	20.3	22.2	20.3	20.7	19.0	20.9	32.0	25.9	15.1	5.8	21.4
Silver 107	0.0005	mg l^{-1}	ND	ND	ND	ND	ND	ND	ND	ND	ND	ND	ND
SiO$_2$	0.05	mg l^{-1}	43.4	47.5	43.5	44.2	40.6	44.6	68.5	55.4	32.3	12.5	45.8
Sodium	0.05	mg l^{-1}	752	737	747	762	721	731	358	734	184	647	734
Specific Conductance		µS cm^{-1}	4680	4660	4650	4650	4560	4650	2570	4530	1710	4170	4700
Strontium	0.001	mg l^{-1}	3.96	3.89	3.90	3.96	4.03	3.92	2.48	3.47	1.56	8.48	3.90
Sulfate	1	mg l^{-1}	85.2	80.2	81.8	75.2	112.0	87.5	54.3	85.9	125.0	1010.0	77.6
TDS calc		mg l^{-1}	2610	2610	2590	2600	2520	2570	1440	2550	961	2750	2620
Field temperature		°C	40.85	42.70	43.09	44.80	44.10	41.20	26.62	40.12	21.83	19.77	46.10
Thallium	0.0005	mg l^{-1}	ND	ND	ND	ND	ND	ND	ND	ND	ND	ND	ND
Thorium	0.0005	mg l^{-1}	ND	ND	ND	ND	ND	ND	ND	ND	ND	ND	ND
Tin	0.0005	mg L^{-1}	ND	ND	ND	ND	ND	ND	ND	ND	ND	ND	ND
Titanium	0.001	mg l^{-1}	ND	ND	ND	ND	0.10	0.18	0.006	ND	ND	ND	ND
Tritium		mg l^{-1}	ND	ND	ND	ND	ND	ND	-0.02	ND	ND	ND	ND
Uranium	0.0005	mg l^{-1}	ND	ND	ND	ND	ND	ND	0.0076	ND	ND	ND	ND
Vanadium	0.0005	mg l^{-1}	0.0083	0.0101	0.0088	0.0090	ND	0.0089	0.0077	0.0111	0.0066	ND	0.0079
Zinc 66	0.0005	mg l^{-1}	0.0139	ND	0.0093	ND	ND	0.0068	0.0062	ND	ND	0.0168	0.0351

CHAPTER 15

Hydraulic conductivity of fractured upper crust: insights from hydraulic tests in boreholes and fluid–rock interaction in crystalline basement rocks

INGRID STOBER[1] AND KURT BUCHER[2]

[1] *Institute of Applied Geosciences, Karlsruhe Institute of Technology (KIT), Karlsruhe, Germany;* [2] *Mineralogy and Petrology, University of Freiburg, Freiburg, Germany*

ABSTRACT

Permeability (k (m^2)) of fractured crystalline basement of the upper continental crust is an intrinsic property of a complex system of rocks and fractures that characterizes the flow properties of a representative volume of that system. Permeability decreases with depth. Permeability can be derived from hydraulic well test data in deep boreholes. Only a handful of such deep wells exist on a worldwide basis. Consequently, few data from hydraulically tested wells in crystalline basement are available to the depth of 4–5 km. Permeability of upper crust varies over a very large range depending on the predominant rock type at the studied site and the geological history of the drilled crystalline basement. Hydraulic tests in deep boreholes in the continental crystalline basement revealed permeability (k) values ranging over nine log-units from 10^{-21} to 10^{-12} m^2. This large variance also decreases with depth, and at 4 km depth a characteristic value for the permeability k is 10^{-15} m^2. The permeability varies with time due to deformation-related changes of fracture aperture and fracture geometry and as a result of chemical reaction of flowing fluids with the solids exposed along the fractures. Dissolution and precipitation of minerals contribute to the variation of the permeability with time. The time dependence of k is difficult to measure directly, and it has not been observed in hydraulic well tests. At depths below the deepest wells down to the brittle–ductile transition zone, evidence of permeability variation with time can be found in surface exposures of rocks originally from this depth. Exposed hydrothermal reaction veins are very common in continental crustal rocks and witness fossil permeability and its variation with time. The transient evolution of permeability can be predicted from models using fictive and simple starting conditions. However, a geologically meaningful quantitative description of permeability variation with time in the deeper parts of the brittle continental crust resulting from combined fracturing and chemical reaction appears very difficult.

Key words: brittle–ductile transition, deep well, hydrothermal, permeability, veins, well test

INTRODUCTION

Features of brittle deformation, such as fractures, faults, joints and veins, are the principal water-(fluid-)conducting structures in crystalline basement rocks and provide the dominant conduits for fluid flow in the brittle upper continental crust. Fluid flow in this environment can be described by the flow law of Darcy. Flow is driven by a hydraulic gradient and controlled by the prime parameters of advective fluid, solute, and heat transport in fractured crystalline rocks, namely permeability and porosity. Depending on the scale of interest, different methods are used to determine these two parameters. These include laboratory measurements (e.g., Berckhemer *et al.* 1997), fracture analysis (Committee on Fracture Characterization and Fluid Flow 1996; Mazurek *et al.* 2003; Jakob *et al.* 2003; Caine *et al.* 1996; Konzuk & Kueper 2004), geophysical modeling of heat-flow data (e.g., Hayba & Ingebritsen 1994), and *in situ* testing of boreholes (e.g., Cooper & Jacob 1946; Ferris 1951; Butler 1998; Nielsen 2007; Peters 2012). Well tests provided permeability data from the crystalline basement to depths of 5 km and offered insights into the permeability structure of the crust and its variation with depth and lithology.

It is important, however, to critically analyze and evaluate permeability data derived from well tests reported in the literature before they can be used (or discarded) in large-scale models of the permeability structure of the crust. In this paper, we address several aspects and difficulties in deriving permeability data from hydraulic well tests.

Crustal Permeability, First Edition. Edited by Tom Gleeson and Steven E. Ingebritsen.
© 2017 John Wiley & Sons, Ltd. Published 2017 by John Wiley & Sons, Ltd.
Companion Websites: www.wiley.com/go/gleeson/crustalpermeability/
http://crustalpermeability.weebly.com/

Porosity and permeability are subject to unceasing changes in the brittle upper continental crust. Reactive fluid flow along the fractures continuously modifies the aperture and surface structure of the fractures mostly by chemical dissolution and precipitation reactions (Weisenberger & Bucher 2011; Bucher *et al.* 2012; Alt-Epping *et al.* 2013). After initial porosity and permeability generation by creation of a fracture, the permeability tends to increase during a certain period. Later, permeability and porosity decrease due to precipitation of secondary minerals along the fractures until they become completely sealed. This fracture–reaction–seal process of reaction vein formation may occur in several cycles and cause incremental vein growth. Although there are ample field observations that imply the significance of the described permeability–time relationship in fractured upper crust (e.g., Bucher-Nurminen 1981), the time-dependent permeability evolution cannot usually be recorded by hydraulic well tests because of incompatible timescales of the fracture–reaction–seal process (Bucher-Nurminen 1989) and the well test methods. Nevertheless, in this paper, we try to illustrate structural characteristics of transient permeability in fractured upper crustal rocks.

Fluid flow in fractured upper crustal aquifers is generally driven by hydraulic gradients, which may result from a number of different feasible causes and imbalances including topography, thermal, and chemical disequilibrium. Pumping tests or injection tests carried out in boreholes are artificially induced hydraulic gradients as forcing for fluid flow. Hydraulic tests provide data on the hydraulic properties and the nature of aquifers, the permeability structure of the upper crust, and the depth variation of permeability. Hydraulic well tests also have the potential to identify and locate faults or fault systems in the crust. The tests can quantify the hydraulic conductivity of the faults if it is different from that of the fractured rock matrix. Fault conductivity can be lower than, equal to, or higher than that of the rock matrix. Examples of low-permeability faults have been presented in Stober & Bucher (2005a) and Stober *et al.* (1999). Important and frequently overlooked driving forces for fluid flow are tidal forces that prevent fluids in fractured systems from ever becoming stagnant and chemically inactive. We also illuminate the effect of tidal forces on flow and reaction of fluids in fractured basement rocks.

PERMEABILITY – SIGNIFICANCE IN FRACTURED BASEMENT ROCKS

The permeability k (m^2) of a volume of fractured crystalline rocks forming the upper continental crust relates to the structure and connectivity of the fluid-filled pore space. The permeability controls, together with the viscosity and density of the fluid, the ability of the volume of rock to conduct a fluid phase. Permeability is, together with the fluid properties, a decisive parameter controlling fluid flow in brittle rocks. It can be retrieved from the transmissivity T (m^2 s^{-1}) calculated

from hydraulic test data. From the experimentally derived transmissivity T, the hydraulic conductivity K (m s^{-1}) can be computed. Quantifying T and K from pressure–time data obtained in well tests requires geological expertise and ideas about the geological and hydraulic structures of the underground. The parameter K is not measured data; it is ultimately the result of a modeling process.

The Darcy flow law describing fluid flow through porous media can be written as

$$\vec{q} = -[K]\frac{1}{g\rho}\nabla P \tag{15.1a}$$

where $\vec{q}$ is the specific discharge vector per cross-sectional area (m^3 s^{-1} m^{-2}) with the components q_x, q_y, q_z; (K) (m s^{-1}), the tensor of hydraulic conductivity; ρ (kg m^{-3}), the density of the fluid; g (m s^{-2}), the acceleration due to gravity; and ∇P (kg m^{-2} s^{-2} or Pa m^{-1}), the vector of the pressure gradient.

The one-dimensional form of Darcy flow is

$$q = -K\frac{1}{g\rho}\frac{\Delta P}{\Delta l} \tag{15.1b}$$

with ΔP (Pa) the pressure difference along the baseline Δl (m).

In hydraulic tests, Q (m^3 s^{-1}) corresponds to the rate of pumping fluid from or injecting fluid into a wellbore. It is a parameter that can be varied during the test. Q is related to q in Eq. 15.1b by $Q/A = q$ with A (m^2) being the cross section perpendicular to flow. The imposed input signal Q results in a pressure response ΔP by the conductive system that can be measured with instruments.

Data analysis of hydraulic test data requires a hydraulic aquifer model. For instance, during the radial flow period in a homogeneous isotropic aquifer, the hydraulic conductivity K is proportional to the pumping or injection rate and is the inverse of the drawdown and the thickness of the tested formation in a simple and ideal case. This means that the derived hydraulic conductivity K depends on the chosen concept for aquifer characterization. The hydraulic conductivity K, and hence also the permeability k, is not a parameter that can be measured directly and unequivocally (such as the length with a measuring tape or a folding ruler).

The permeability k (m^2) can then be derived from hydraulic conductivity K:

$$k = K\frac{\mu}{g\rho} \tag{15.2}$$

with the viscosity of the fluid μ (kg m^{-1} s^{-1} or Pa s) and K, g, k, and ρ from above.

In order to better understand the significance and meaning of both the hydraulic conductivity K and the derived permeability k, it can be helpful to look behind the routine of hydraulic well testing. How are such hydraulic tests performed, how is the wanted parameter derived from the measured pressure data, and what kind of compulsory assumptions need to be made during data processing? This analysis may help evaluating the strength and significance of the parameter "hydraulic conductivity," but also disclose its weaknesses.

Because the subject of interest in this book is the permeability structure of brittle continental crust, hydraulic tests in deep boreholes are of prime significance. At deep drilling sites, additional monitoring wells are typically not available. The deep well is simultaneously used as testing and monitoring well.

There is a generous repertoire of hydraulic tests tailored and fitted to the specific application, environment, and problem definition (and budget) (Stober 1986; Krusemann & de Ridder 1991; Horne 1996). Some continuously record the pressure directly in the tested section (at depth); in other tests, pressure is measured relatively close to the surface. In many tests, the water level in the wellbore is measured as a proxy for pressure rather than the pressure itself. If no pressure data are available from the depth of interest, convoluted corrections must be applied to near-surface pressure or water-level data. The flow rate Q can be varied in hydraulic tests in many ways. Fluid can be pumped or injected at a constant rate during the entire test period; however, pumping rates can also be varied stepwise according to a detailed protocol. The flow rate may also be continuously adjusted to maintain a constant drawdown or recovery of the water level in the wellbore. In the following discussion, we presuppose that all tests are performed according to the state of the art, that all data corrections are precise and correct, and that data evaluation and analysis are based on a hydraulic model that correctly reflects the geological situation underground.

Test duration varies from a few minutes to several weeks. The test produces information on the hydraulic properties of the subsurface to a certain distance from the wellbore. This distance may be named radius of influence. The radius depends strongly on the test duration, the type of test method, and the hydraulic properties of the rocks. Moreover, the radius of influence also depends on the technical details of the special testing tools and the testing procedure.

The hydraulic properties of the rocks in the vicinity of the wellbore are commonly altered by the drilling process itself or by drilling mud and added chemicals. Short-term hydraulic tests will not be able to "see" beyond this alteration zone near the wellbore.

An example of the consequences of hydraulic tests with small radii of influence is the 4444-m-deep wellbore Urach 3 (SW Germany). Urach 3 was drilled as a research borehole for deep geothermal energy utilization. From 1604 to 4444 m below surface, the bore is in granite and gneiss of the Variscan basement. The research borehole provided an unusually long series of water-table measurements over a period of 13 years. These Urach 3 data have been used to decipher the hydraulic properties of the basement, the chemical composition of the deep fluid, and the interaction of this fluid with the rock matrix (Stober & Bucher 2004). Urach 3 has also been used for a large number of diverse hydraulic tests studying the influence of test methods and test duration on derived hydraulic properties. For example, long-term injection tests have been

repeatedly performed, along with slug tests of short duration. A comparison of data from the two different types of tests (slug test and long-term injection test) shows that the hydraulic conductivity around the wellbore is higher than of the pristine formation by a factor of 7 (Stober 2011).

Flow rate and pressure variations during injection tests can be delimited to the capacity of the formation. Derived parameters will then reflect the intrinsic properties of the formation. If tests are run outside these conditions, fracture apertures may artificially increase and elastic reactions of the formation alter the conductivity of the system during the test. The derived parameters reflect the artificial properties created by the test itself. For example, injection tests in the Urach 3 research well induced elastic reactions by the gneissic basement at well head pressures above 170 bar, resulting in a significantly increased conductivity (Stober 2011). In tectonic active areas, injection tests may reduce effective stress and frictional strength of fractures. The induced stress reduction by shearing can result in an irreversible and permanent increase in the hydraulic conductivity (Evans 2005). Consequently, injection tests performed at an overpressure that is too high for the tested bedrock alter the natural hydraulic properties of the tested rock volume. Hydraulic parameters derived from such tests do not reflect the properties of the natural system.

Hydraulic tests may test the entire open hole or segments of it, or the section with perforated casing. Often packers or other technical tools are used to isolate certain sections of the wellbore for testing. The tested sections typically vary from several hundred to only a few meters.

The prime result of processing and analyzing the data of a hydraulic well test is the so-called transmissivity of the tested section and not directly the hydraulic conductivity or the permeability. The transmissivity $T\,(\mathrm{m^2\,s^{-1}})$ is an integral quantity characterizing the hydraulic properties of the entire tested section. If the tested section $H\,(\mathrm{m})$ is homogeneous and isotropic, the hydraulic conductivity can be computed from the determined transmissivity:

$$K = T/H \tag{15.3}$$

The transmissivity of tested heterogeneous and anisotropic stratified formations can be written as an integral over all contributions:

$$T = \int_0^H K\,\mathrm{d}h \tag{15.4}$$

The hydraulic conductivity $K\,(\mathrm{m\,s^{-1}})$ can be calculated from the experimentally derived transmissivity T from Eqs 15.3 and 15.4.

Fluid flow in brittle crystalline basement rocks occurs along fractures and faults, and along lithological contacts and other discrete 2D structures. The water-conducting features form an interconnected network with characteristic hydraulic properties (Mazurek 2000; Mazurek *et al.* 2003; Stober & Bucher

2007a). Water-conducting structures can be regularly or unevenly distributed. Consequently, at some drilling sites, the data can be interpreted as hydraulic response of a homogeneous isotropic aquifer. At other drilling sites, other concepts may appear more appropriate. The choice of the concept depends on the amount of available geological information. If not much is known about the geological structure, the derived parameter can be correct if the structure is simple or incorrect if the structure is more complex than has been recognized.

Such considerations are not made during planning of hydraulic tests, and the hydraulic conductivity is computed from Eq. 15.3. This practice may lead to serious misjudgments of the hydraulic conductivity and permeability of fractured rocks. The following considerations may illustrate the problem: A hydraulic test performed in a 600-m-thick open hole of a deep wellbore resulted in a measured transmissivity $T = 2 \times 10^{-4}\,\mathrm{m^2\,s^{-1}}$. The calculated hydraulic conductivity (Eq. 15.3) $T/H = K = 3.3 \times 10^{-7}\,\mathrm{m\,s^{-1}}$ appears feasible for fractured basement rocks. However, the main water inflow to the open hole can be restricted to a 10-m-thick very strongly fractured fault zone. The resulting hydraulic conductivity for the fault zone is $K = 2 \times 10^{-5}\,\mathrm{m\,s^{-1}}$, 60 times higher than the first value. Strongly channeled fluid flow is not uncommon in fractured basement. Such zones can be identified and localized by geophysical well logging. The example shows that errors and misinterpretations can increase with increasing length of the tested section. On the other hand, the example also shows that the selection of an invariant test length can be very difficult and the result questionable. The high K-value may characterize the conductive fracture zone. The lower K-value, however, may not characterize the entire system of fractured rocks and highly conductive fault zones.

Geophysical well logging methods such as flow meter logs, conductivity logs, and temperature logs help to better characterize heterogeneous hydraulic properties of tested formations. Prominent inflow zones can be identified and separately tested using packer systems. The methods may produce a quantitative vertical permeability profile. However, such elaborate and costly test designs are rarely used (typically in the context of development of repositories for nuclear waste, e.g., Nagra 1985; Nagra 1992). Usually one relies on T/H data to derive hydraulic conductivity K and permeability k. It is thus evident that conductivity K and permeability k in the end do not represent true and absolute values. The derived permeability k of the continental crust, for instance, at a given point of interest, strongly depends on the details of the test design of the hydraulic test used for deriving the parameter. It may be useful and representative, but the contrary may be true as well if the basement is hydraulically complex and heterogeneous and the design of the test is unable to adequately resolve the heterogeneity, or the tests are inappropriate to derive a meaningful permeability at a point in the bedrock. However, we do not propose to give up measuring properties by experiments, but rather to highlight possible pitfalls of the methods and of uncritically using published permeability data for modeling projects.

The hydraulic conductivity K characterizing the hydraulic properties of the bedrock has two components. Firstly, it describes the properties of the fracture network expressed by the permeability of the rock volume; secondly, it depends on the properties of the fluid itself. This follows from Eq. 15.2, which may be rewritten to highlight this aspect of K and k:

$$K = k\frac{g\rho}{\mu}. \tag{15.5}$$

It can be seen from Eq. 15.5 that hydraulic conductivity data derived from well tests depend on the properties of the fluid residing in the fracture porosity. If one is interested in the permeability of the tested rocks at a point of interest, the density ρ and the viscosity μ of the fluid at that point must be known (see Eqs 15.1a and 15.2 and definition of the parameters there).

The fluid parameters ρ and μ depend on temperature, pressure, mineralization, and the gas content of the fluid. This means that hydrochemical analyses must be available from sampling points at depth, in addition to depth and temperature data. The factor $\mu/g\rho$ in Eq. 15.2 converts hydraulic conductivity to permeability. It has the value $1.0 \times 10^{-7}\,\mathrm{m\,s}$ at $20\,^\circ\mathrm{C}$ for pure water, $1.7 \times 10^{-8}\,\mathrm{m\,s}$ at $180\,^\circ\mathrm{C}$ for pure water, and $3.1 \times 10^{-8}\,\mathrm{m\,s}$ at $180\,^\circ\mathrm{C}$ for $100\,\mathrm{g\,kg^{-1}}$ NaCl solution (using μ and ρ data from Wagner & Kretschmar 2008; Sun *et al.* 2008; Häfner *et al.* 1985). Particularly the viscosity of the fluid in the example above is markedly different in hot and highly mineralized fluid from that of pure water. Pumping tests in deep thermal wells can produce hydraulic conductivity values up to eight times higher than would be obtained under cool low-salinity conditions, independent of the permeability structure of the formation itself, just as a result of fluid properties. Derived permeability may be erroneously high if fluid properties are unknown or not adequately considered. The opposite effect can be recognized in injection tests, which normally use cool low-mineralized surface water as injection fluid.

The correct conversion of K to k data requires fluid data from samples representative of the tested section. Collecting and analyzing fluid samples from deep wells, particularly if the fluid contains high amounts of dissolved gases, are technically challenging and expensive and therefore fluid data are often not available.

The permeability of rocks can also be measured on drill cores in the laboratory. It is difficult or impossible to correctly represent fractures, faults, and larger cavities in core samples. The lab-measured permeability typically characterizes the property of the unfractured rock matrix. In general, it is significantly lower than the permeability of large volumes of fractured and faulted basement derived from well tests.

The above discussion of the parameters hydraulic conductivity and permeability relates to well tests in deep boreholes that are the prime source of K and k data. The technical details of such tests and their cross-links to the geology at the drilling site have specific consequences for the derived data. The user of these data must be aware of these aspects, evaluate data carefully, and not adopt permeability data uncritically.

Long-term pumping tests in deep boreholes may be well suited to detect the hydraulic properties of fractured basement. Under favorable circumstances, long-term tests can identify, locate, and characterize large-scale hydraulic structures (the so-called hydraulic boundaries). An excellent example is the highly developed pumping test in the 4-km-deep pilot hole of the continental deep drilling project (KTB) in Germany (Stober & Bucher 2005a). The test produced a reliable value for the hydraulic conductivity of homogeneous fractured continental basement at 4 km depth $K = 4.07 \times 10^{-8}\,\mathrm{m\,s^{-1}}$, which results in a characteristic permeability $k = 1.18 \times 10^{-15}\,\mathrm{m^2}$ given the fluid properties at depth (120 °C, TDS = 62 g l^{-1}, $\mu = 0.27 \times 10^{-3}\,\mathrm{Pa\,s}$, $\rho = 952.8\,\mathrm{kg\,m^{-3}}$) (molecular weight of NaCl = 58.44 g mol^{-1}). The fluid temperature and the salinity have been taken from Stober & Bucher (2005a), the dynamic viscosity of the fluid at 120 °C and 1 molal salinity from Kestin *et al.* (1981), and the density of a 1 molal salt solution at 120 °C from Rogers & Pitzer (1982). This value of the permeability $k = 1.176 \times 10^{-15}\,\mathrm{m^2}$ may be considered a representative value for fractured continental crust at about one-third of the distance from the surface to the brittle–ductile transition zone in moderately warm crust (MOHO temperature ~650–700 °C). The value is in excellent agreement with permeability data from continental crust published by Ingebritsen & Manning (2010). Their permeability data have been compiled from continental crust worldwide and have been derived by various geophysical tools and methods but not from hydraulic well tests. Our $k = 1.18 \times 10^{-15}\,\mathrm{m^2}$ value is closer to the new k versus depth fit of Ingebritsen & Manning (2010) than to their previous fit (Ingebritsen and Manning 1999).

Hydraulic testing of the KTB wellbore discovered a hydraulic boundary with a lower conductivity than the fractured basement. The seal appears to be located at a distance of several hundred meters from the wellbore at 4 km depth. The hydraulic boundary was identified as a fault and correlated with a major fault zone known from surface geology. Similar "water-tight" hydraulically prominent faults have been identified at other localities in the central European crystalline basement (Stober *et al.* 1999). Both examples show that prominent fault systems in the basement are not necessarily characterized by a higher conductivity than the surrounding fractured basement.

PERMEABILITY AND FLUID FLOW IN THE CRUST

Hydraulic tests in wellbores to 5000 m depth worldwide reveal a remarkable variation of hydraulic conductivity of the crystalline basement from 10^{-14} to $10^{-4}\,\mathrm{m\,s^{-1}}$. The upper 1000 m are generally characterized by the higher values and also by a greater variance (Stober 1996; Stober & Bucher 2007a). The mean variation of the hydraulic conductivity decreases rapidly with increasing depth. The decrease in the permeability k with depth has been derived from well tests in the basement of southwest Germany, northeast France, and north Switzerland (Stober & Bucher 2007a,b) and can be described by

$$\log k = -1.38 \log z - 15.4 \qquad (15.6)$$

with z being the depth in kilometers and k in square meters. The described decrease in the permeability with depth (Eq. 15.6) is mainly controlled by the properties of biotite gneiss and other metamorphic rocks because test data for greater depths were mostly available from wellbores in gneiss. We suspect that the decrease in the hydraulic conductivity in granitic basement would be slower with depth, because of contrasting mechanical properties of mica and feldspar (see above and in Stober & Bucher 2007a). At our studied sites, granitic basement has a higher permeability at a given depth and tectonic environment than gneissic basement (given that granite is dominated by feldspar and quartz and gneiss typically contains significantly more mica than granite). Ingebritsen and Manning (1999) in their geophysical study of terrestrial heat flow derived a surprisingly similar power-law function for the permeability decrease with depth. At 4 km depth in a crystalline basement, their function predicts that $\log k = -15.9\,\mathrm{m^2}$. Ingebritsen and Manning (2010) compared two k–z relations: (i) for "tectonically active" crust, that is, their 1999 curve and (ii) another curve for "disturbed crust" that included hypocenter migration, enhance geothermal systems and other sources of data. The new function predicts a permeability for "disturbed crust" at 4 km depth (KTB pilot hole) of $\log k = -13.5\,\mathrm{m^2}$ (Eq. 12 Ingebritsen & Manning 2010) in contrast to our derived value of $\log k = -14.9\,\mathrm{m^2}$ at the KTB site.

The decrease of the hydraulic conductivity and permeability with depth has, to our knowledge, never been reported and demonstrated for a single discrete deep borehole (>1000 m) in the crystalline basement. In deep wellbores adequate hydraulic tests at different depths over sufficiently long test intervals consistent with the requirements of a correctly chosen representative elementary volume are not normally available. Urach 3 is an exceptionally deep research bore in fractured crystalline basement of SW Germany. Several hydraulic tests at different depths and a total test interval of more than 2000 m produced unique hydraulic data. The basement belongs to the so-called Moldanubic zone of the central European continental crust. The basement consists of mafic biotite-hornblende gneiss (amphibolite), migmatitic gneiss, quartz-diorite, biotite-cordierite gneiss and other high-grade metamorphic rocks. The unaltered high-grade rocks contain zones of hydrothermal alteration. The permeability of the basement is related to a fracture network of variable orientation. There are open fractures functioning as water conducting structures and sealed impervious fractures. The fracture density is about two to three open fractures per meter (Dietrich 1982).

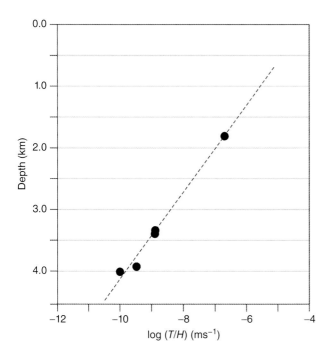

Fig. 15.1. log (T/H) data from well tests at five different depths in the Urach 3 borehole in gneiss of the Black Forest crystalline basement.

In the years 1978–1979 the wellbore was drilled to 3334 m (bottom hole at that time) with a 14 m thick open-hole section. The casing has been perforated at different sections and the accessible rock has been hydraulically tested. The wellbore has been deepened from time to time (for different reasons) in the following years. At each increment the newly accessible rocks have been hydraulically tested. In 1992, the wellbore reached its present depth of 4444 m (Stober 2011). The incremental drilling history of the Urach 3 bore offers a unique opportunity to evaluate the variation of the hydraulic conductivity with depth (Fig. 15.1).

Transmissivity data have been derived from pumping and injection tests at 1.77, 3.30, 3.35, 3.88, and 3.97 km depth, respectively. The hydraulic conductivity (T/H) of fractured basement gneiss decreases systematically with depth z. At 3.97 km, the hydraulic conductivity is about 2000 times lower than at 1.77 km. The depth z–log (T/H) data follow a linear trend (Fig. 15.1). The log (T/H) versus log z data can be regressed to a linear equation:

$$\log(T/H) = -8.748\,z - 4.465\ (R^2 = 0.976) \quad (15.7)$$

The log-linear Eq. 15.7 can be converted to Eq. 15.8 expressing the depth dependence of the permeability in crystalline basement rocks (here using data for a fixed salinity of 1 molal NaCl, 120 °C, and 300 bar):

$$\log k = -8.748z - 12.0\,(R^2 = 0.976), \quad (15.8)$$

where log k ranges from −14.24 at 1.77 km depth to −17.56 at 4 km depth. Note that the permeability of gneiss basement

at the Urach 3 site is about 400 times lower than that of the basement at the KTB site at the same depth of 4 km.

In general, the hydraulic conductivity of granite is higher than that of gneissic basement in areas of strong and young deformation (e.g., the three-country corner of Germany–Switzerland–France). In tectonically inactive areas, fracture density and, consequently, the hydraulic conductivity of large volumes of granite can be very low (Stober & Bucher 2007a).

With increasing depth, the vertical stress component usually exceeds the horizontal stress components (Brown & Hoek 1978) with the result that the open water-conducting fracture system changes from predominantly horizontal to mostly vertical orientation. Vertical fluid flow and fluid exchange become dominant at depth, particularly in areas with significant topography (Fig. 15.2). The steep fluid-conducting fractures support deep fluid circulation systems (Stober & Bucher 2004; Stober et al. 1999; Bucher et al. 2012). Hydrothermal alteration zones along vertical fractures document the existence of significant vertical permeability (Lee et al. 2011) (see also "Reactive fluid flow in the crust and its effect on permeability" Section).

One notable exception has been described from the Blancket well (Australia), where the vertical pressure component becomes lower than the horizontal component at depths below 1800 m. Consequently, fractures and faults have a preferred horizontal orientation in the 4000-m-deep stimulated geothermal Habanero 4 well (Bendall et al. 2014).

Hydrochemical and isotope data suggest that thermal spring waters represent upwelling deep waters (Stober 1995; Chebotarev 1955; Tóth 1962; Allen et al. 2006; Bucher et al. 2009). The conclusions from chemical data are supported by the evidence from numerical modeling (Kukkonen 1995; Forster & Smith 1989). Topography-driven hot water flow from several thousand meter deep sources has been reported (Tóth 1978; Bucher et al. 2009).

The circulation systems depend on the existence of steep fault structures that must merge into high-conductivity fault zones allowing for rapid upward flow of deep water (e.g., Baden–Baden) (Sanner 2000; Stober 1995) (Fig. 15.2). The final temperature of the deep fluid reaching the surface depends, among other factors, on the flow velocity, which is proportional to the hydraulic conductivity for a given hydraulic gradient. Typically, major fault zones with vertical displacements of hundreds or thousands of meters tend to have an inner, low-permeability central zone mantled by high-permeability rims on both sides (Mazurek et al. 2003).

Topographical-driven deep water circulation in the Black Forest region (Germany) results in the outflow of warm saline deep water into the freshwater aquifer of the Quaternary fill of the drainage valley (Kinzig valley). The natural contamination of the freshwater by deep salty fluids can be traced as a NaCl plume (Ohlsbach Plume) for a long distance into the Quaternary gravel aquifers of the river Rhine plain (Stober et al. 1999). The deep saline warm water flows upward along a steep fault system forming the boundary of the Black Forest

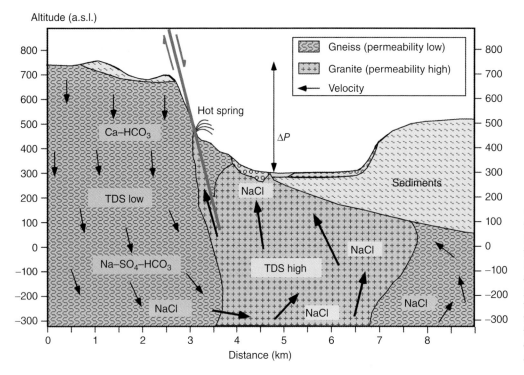

Fig. 15.2. Schematic water flow systems in a crystalline basement area with moderate topography (e.g., Black Forest basement). The permeability contrast between granite and gneiss has a strong control on the flow paths, in addition to topography and fault structures. The figure also illustrates changing water composition along flow.

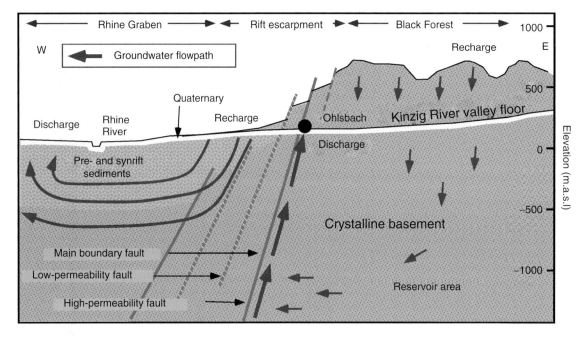

Fig. 15.3. Section from the Rhine River rift valley to the Black Forest basement along the Kinzig River valley. Horizontal scale of section about 20 km. Recharge water from the Black Forest is channeled along a fault system to the surface and does not reach the Rhine graben structure. For more details see Stober et al. 1999.

topographic high and the rift system of the river Rhine valley (Fig. 15.3). Although the rift escarpment represents a generally extensional environment, strike-slip movements along the faults form fine-grained fault gauge, which clogs permeability, thus forming hydraulic barriers. However, many of the Rhine rift boundary faults may be highly transmissive and could channel fluid flow (e.g., Bruchsal and Landau) (Aquilina et al. 2000; Bächler 2003).

In the case of the Ohlsbach plume (Fig. 15.3), deep water is forced to flow upward by the hydraulic boundary fault through fractured granitic rocks that have similar conductive properties as the area of recharge. The flow velocity is slow so that the water reaches the surface with only 27 °C although the reservoir at >3000 m depth has a temperature close to 100 °C (Stober et al. 1999). If upwelling deep water from similar reservoirs follows a highly permeable fault zone such as in Baden–Baden, the outflow temperature is as high as 68 °C although the reservoir temperature may be as high as 150 °C. The water is characterized by a discharge temperature that reflects the permeability structure of the faults and the basement as described by theoretical models of fluid flow in fault zones (López & Smith 1995).

Hydrothermal circulation systems, ascent and descent channels alike, require open highly permeable steep fracture systems. Ascent channels must be extremely conducting structures permitting high flow rates in order for hot water to reach the surface environment at high temperature and appear as a hot spring. The ascent flow may follow zones of strongly fractured granite (Fig. 15.2) or low-permeability fault zone (Fig. 15.3). Driving force for the circulation is topography in both examples. In areas lacking topography, or in the absence of suitable steep high-permeability structures, deep-reaching vertical circulation systems may not develop, and natural hot springs will not be present.

REACTIVE FLUID FLOW IN THE CRUST AND ITS EFFECT ON PERMEABILITY

At depth below the deepest borehole that has been hydraulically tested, the permeability of the continental crystalline crust can be derived from other data and observations including geothermal data (Ingebritsen & Manning 2010).

An important source of information for understanding permeability and permeability evolution with time in deeper parts of the brittle crust is the preserved effects of fossil fluid–rock interaction in surface outcrops of fractured rocks from the zone between the deepest wells to the brittle–ductile transition zone. The temperature range where these structures formed is about 200–400 °C. Fluid flow along fractures in crystalline basement rocks is generally accompanied by chemical reactions between the aqueous fluid and the rock exposed along the fracture. The reactions occur because the fluid is rarely in chemical equilibrium with the mineral assemblages of granites and gneisses predominantly present in the continental crust (Garrels & Howard 1959). At the generally relatively low temperature above the ductile–brittle boundary (<400 °C), reactions involving primary silicates are generally slow (Browne *et al.* 1989; Yadav & Chakrapani 2006; Brantley 2004) and most high-grade minerals are not stable in the presence of H_2O at the prevailing conditions. Fluid–rock reaction (retrograde metamorphism) occurs from the ductile–brittle boundary to the surface in fractured basement that ranges in temperature from about 400 °C down to ambient temperature. Hydration of primary high-grade assemblages is the major chemical process. Newly formed clay, zeolite, Fe-hydroxide, and other alteration products are common on fractures studied at drill cores from deep boreholes and from tunnels (Iwatsuki & Yoshida 1999; Borchardt *et al.* 1990; Borchardt & Emmermann 1993; Bauer 1987; Stenger 1982; Lindberg & Siltari-Kaupi 1998; Weisenberger & Bucher 2010). This shows that fluid–rock interaction and associated mineral dissolution and precipitation along the fractures contribute to the variation of the permeability with time (Moore *et al.* 1983; Polak *et al.* 2003).

Mineral coats along the fractures develop from the instant of fracture formation and its saturation with aqueous fluid until the fracture becomes sealed by reaction products. It is difficult to relate veins of observed secondary minerals in drill cores to the flow-active fracture porosity and to the quantitative permeability history of the crust. However, these structures relate to fossil flow and reaction systems, and the observations can be regarded as analogs for present-day processes progressing in the deeper parts of the upper crust.

There is abundant evidence for permeability-relevant reactions in deep fractured rocks. The effects of the reactions can be easily recognized in the so-called reaction veins (Fig. 15.4). The dissolution and precipitation processes may proceed at different time-varying rates. Reaction veins have a structural component caused by rock deformation and a chemical reaction component

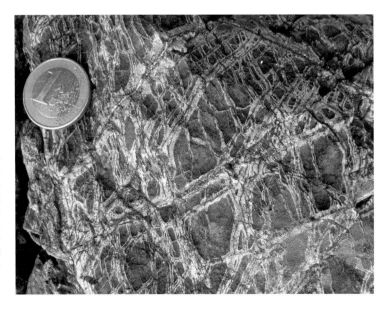

Fig. 15.4. Serpentine reaction veins in peridotite (Erro Tobbio mantle, Italy). The serpentinization of peridotite occurred along symmetrical reaction veins along brittle fractures. During the active period of serpentinization, the permeability of the fractured rock permitted flow and reaction of aqueous fluid. Flow and reaction ceased because of self-sealing of the fractures, preventing the peridotite (brown crust) from being completely serpentinized. Later, the fractures became completely sealed, and the vein system does not contribute to the near-surface permeability of the partly serpentinized peridotite. (*See color plate section for the color representation of this figure.*)

Fig. 15.5. Talc veins in peridotite (Vlisarvatnet, Norway). Low-temperature hydration of peridotite (coarse-grained spinel harzburgite with brownish weathering rind) produced white silvery zones of talc along open brittle fractures that represent fluid-conducting structures. Note the vertical orientation of the fractures and the large aperture of these young structures. High content of dissolved silica in the hydrothermal fluid resulted in talc formation rather than serpentinization (Fig.15. 4). The open fracture system contributes to the present-day permeability of the rocks (and could be measured with well tests). (*See color plate section for the color representation of this figure.*)

that follows from irreversible reaction of the advective fluid with the exposed rock. Both components have consequences for the permeability of the fractured system and its variation with time.

Simple monomineralic veins may form by a crack–seal mechanism (Durney & Ramsay 1973; Ramsay 1980; Ankit *et al.* 2013). These veins have a predominantly deformational component and include quartz veins in quartzite (or granite) and calcite veins in limestone and marble. Reactive flow may dissolve components from the exposed rocks, import dissolved components from external sources, and precipitate new minerals in the fractures that were not initially present. These veins result from a fracture–reaction–seal mechanism (Bucher-Nurminen 1989). Examples include serpentine veins in peridotite (Fig. 15.4), talc veins in peridotite (Fig. 15.5), chlorite veins in biotite gneiss and in garnet amphibolite (Fig. 15.6), and tremolite veins in marble (to name a few examples). The structures shown in Figures 15.4–15.6 contain a wealth of qualitative information on the development and time dependence of permeability in the brittle crust. Some of the field-deduced conclusions are summarized in the figure captions.

These examples suggest that vein growth starts with an initial brittle fracture, followed by a reaction period during which permeability increases, then a period of decreasing permeability until the vein becomes impervious and fluid flow stops. Thus, permeability follows a time evolution similar to the porosity wave proposed for the lower crust (Connolly & Podladchikov 2007). The schematic permeability–time relationship shown in Figure 15.7 characterizes a single vein or a single fluid-conducting structure (e.g., shown in Fig. 15.6). The permeability evolution path (Fig. 15.7) shows the transient local effect of a single fracture on top of the original background permeability. After sealing the fracture, the system returns to the background permeability, which represents the sum of all effects of all flow-active fractures currently contributing to the

permeability of the considered volume of rock. If many veins are active simultaneously, transient permeability increases and then, with reaction progress, permeability decreases. However, if vein formation is not strictly synchronous in an area, the single fractures contribute statistically to the background permeability.

Outcrops on Vannøya (northern Norway) representatively show the sequence of a crack–reaction–seal cycle (Fig. 15.8). First, fluid-conducting fractures were formed in the green mafic igneous rock. The distinctive green–brown reaction front resulted from reaction of a CO_2-rich fluid with the green rock. These reactions destroyed all Fe-bearing minerals and carried away dissolved iron, leaving a brown albite–carbonate rock. It is not a redox reaction front (e.g., Yamamoto *et al.* 2013) but rather a carbonation front (Priyatkina *et al.* 2011).

Hydrothermal alteration reactions may, in their simplest form, be plain hydration reactions. No components dissolved in the fluid are lost or gained in such reactions. Hydration reactions consume H_2O and cease if water supply to the reactive fracture ends. Consumption of H_2O passively increases the total mineralization of the fluid (TDS). Because of hydrothermal alteration (retrograde metamorphism) of upper crustal rocks and the associated H_2O consumption, deep fluids are generally highly saline (Edmunds & Savage 1991; Frape & Fritz 1987; Stober & Bucher 2005b; Bucher & Stober 2010). The salinity is typically much higher than that of seawater (up to NaCl saturation). Hydration reactions also desiccate the ductile lower continental crust (Frost & Bucher 1994). Evidence for active desiccation of the lower crust is the presence of high-temperature halite in eclogites (Markl & Bucher 1998).

Plagioclase is the most abundant mineral of the continental crust. This feldspar mineral is a mixture of a Ca-component (anorthite) and a Na-component (albite). At low-temperature hydrothermal conditions, Ca-bearing versions of the mineral

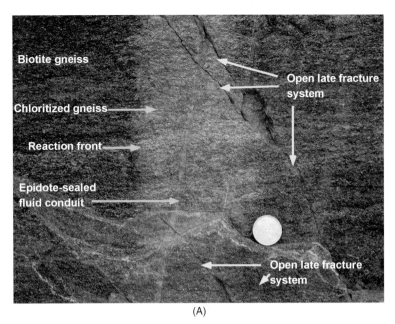

(A)

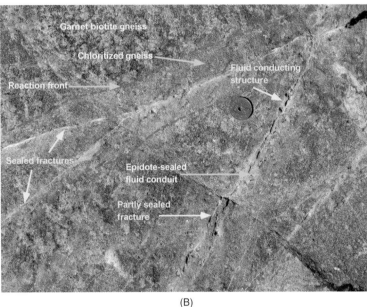

Fig. 15.6. Reaction veins in Caledonian amphibolite-grade gneiss: (A) Early chlorite vein in biotite gneiss (Hammerfest, Norway). Biotite gneiss has been chloritized along a brittle fracture that served as fluid conduit for hydrothermal fluid. The open fracture has been sealed later by epidote and became inactive. Young brittle fractures are open water-conducting structures and represent the present-day permeability of the rock. The later structures (blue arrows) control the hydraulic conductivity detected and measured by well tests. (B) Chlorite vein in garnet biotite gneiss (Torsnes, Kvaløya, Norway). Chloritization occurred along fractures by interaction of advecting hydrothermal fluid with the primary gneiss assemblage. Most of the fractures are completely sealed by solid reaction products. One of the fractures is only partly sealed. The visible fracture porosity contributes to the present-day permeability of the rocks (detectable by well tests). (*See color plate section for the color representation of this figure.*)

(B)

cannot be in equilibrium with H_2O, and it hydrates to clay or various zeolite minerals such as stilbite, laumontite, and others depending on the temperature (Nishimoto & Yoshida 2010; Weisenberger & Bucher 2010) and pressure (depth). Low-temperature hydration of the anorthite component of plagioclase can be written as

$$\text{Anorthite} + 5\,\text{quartz} + 7H_2O = \text{stilbite}(CaAl_2Si_7O_{18} \cdot 7H_2O).$$
$$(15.9)$$

The mineral name stilbite is used here for a Na-free end-member component stilbite-Ca. It forms at very low temperature ($<120\,°C$). Quartz on the reactant side of Eq. 15.9 can

be taken from the minerals exposed on the fracture surface or from dissolved SiO_{2aq} in the fluid. In the process, the albite component of the plagioclase recrystallizes as pure albite. Laumontite, another Ca-zeolite, forms from hydration of plagioclase at higher temperature ($\sim150–250\,°C$) corresponding to a depth range of 6–10 km. Its formation reduces hydraulic conductivity and may efficiently desiccate the fracture system (Fig. 15.9).

Many primary igneous minerals hydrate to zeolites and other hydrous minerals when the igneous high-temperature rocks pass through the brittle upper crust during exhumation before reaching the erosion surface. Their production on fractures typically involves a large volume increase relative to the original igneous rock. The process efficiently seals the water-conducting

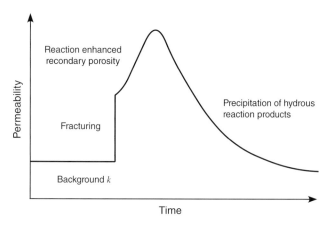

Fig. 15.7. Schematic graph showing the time-dependent permeability development related to a single fracture that serves as a fluid conduit for reactive fluids. This simple pattern can be modified in many ways, for example, by multiple fracturing, by time-dependent aperture variations due to complex extension and shearing, and by externally controlled changes of fluid composition. In addition, the permeability of the large-scale permeability is controlled by the combined result of k–t curves of a large number of fractures of different orientations.

fracture and the volume increase of the solids can accommodate significant fracture extension. The large volume increase of the solids in hydration reaction seals the fracture porosity and stops veins from growing. The lifetime of high-permeability conditions permitting fluid flow is difficult to quantify. However, the permeability of sealed veins may not be higher than that of the rock matrix.

Generally, zeolites are widespread and common in fractured continental basement. They have been reported to occur on fractures and fissures encountered during road and rail tunnel construction in the Alps (Armbruster *et al.* 1996; Weissenberger

& Bucher 2010), in deep drillholes in the continental basement (Stober & Bucher 2005b), and in drilled young granitic plutons (Yoshida *et al.* 2013). An example of zeolite-bearing veins is shown in Figure 15.9. Pink Ca-zeolite laumontite has formed along vertical fractures through massive coarse-grained gabbro of the Seiland igneous province in northern Norway. The exposed fractures are efficiently sealed by zeolite that formed according to a reaction analogous to Eq. 15.9. The presence of laumontite instead of stilbite as fracture filling (and other petrologic details) suggests that the hydration reaction occurred at a temperature of about 250 °C. Thus, the hydration veins in fractured basement displayed on Figure 15.9 represents a situation at about 10 km depth, assuming a geothermal gradient at the time of hydrothermal vein formation of about 25 K km^{-1}. Note also that the orientation of the fractures is nearly vertical as a result of predominance of vertical stresses over horizontal stresses at this depth (this is also the case shown in Fig. 15.5). The hydration reactions progress in two steps: The first step dissolves the reactive high-temperature minerals such as igneous plagioclase, pyroxene, and olivine. The dissolution step produces porosity and increases related permeability. Plagioclase conversion to albite and zeolite has produced initially up to 15 vol% porosity in fractured granites of the Alps (Weisenberger & Bucher 2010). In the second step, the created secondary porosity is replaced by solid reaction products that eventually seal the open fractures. Fluid–rock reactions in fractured granites thus first enhance permeability that originally has been created by mechanical fracturing of the granite. Increasing amounts of solid reaction products gradually decrease permeability of the porous, fractured wall rocks (Fig. 15.7).

Hydration reactions consume H$_2$O from the fluid and create a local pressure depression inducing of fluid flow toward the reaction site. In the Urach 3 borehole, vertical downward

Fig. 15.8. Vein system in mafic rocks (dolerite on Vannøya, Norway). The structures suggest that first a fracture system opened and then advecting fluid reacted with the mafic igneous rock (green) producing an albite–calcite rock (light brown) before the fluid conduit was finally sealed by brown (and white) carbonate in the central part of the veins. The structures support the proposed k–t curve shown in Figure 15.7. (*See color plate section for the color representation of this figure.*)

Fig. 15.9. Laumontite (Ca-zeolite) clogs vertical fluid-conducting fractures in gabbro (Langfjorden, Norway). The zeolite forms from hydrothermal alteration of primary labradorite (plagioclase) of the gabbro. The large volume increase of the solids in the reaction efficiently seals the vertical fracture system. The vein formation occurred at about 250 °C and 10 km depth during the late Caledonian formation of the Langfjorden–Repparfjorden fault system. (*See color plate section for the color representation of this figure.*)

In tectonically active areas, water-level measurements and other data collected after earthquakes have been used to show that active fracturing temporarily increased permeability. The healing of the created deformation structures occurred relatively rapidly, within the timescale of observation. Generally, the observed and documented permeability changes refer to mixed crust (basement and sediments) or ocean island crust, and not the inactive normal continental crust (Cappa *et al.* 2009; Xue *et al.* 2013; Kitagawa *et al.* 2007; Elkhoury *et al.* 2006).

Historical reports on the behavior of temperature and discharge of thermal waters in geothermal or mineral water spas indicate variations with time. One example is the spa Baden-weiler in SW Germany. This large and luxurious spa was built and used by the Romans about 1800 years ago. The buildings were once heated by hot water using a sophisticated thermal heating system. Today, discharge and temperature of the hot springs are insufficiently low for spa operation. This indicates that the permeability structure has changed on the timescale of hundreds of years (Filgris 2001).

FLUID FLOW AND PERMEABILITY STRUCTURE OF THE UPPER CRUST

The Darcy flow law (Eq. 15.1b) implies that fluid flow ceases if the pressure gradient disappears. The major forces driving fluid flow in the crust are thermal disequilibrium and topographic relief. If the driving force ΔP (Eq. 15.1b) approaches zero, fluid flow stops. The fluid becomes stagnant. The stagnant fluid resides in the fracture pore space and chemically interacts with the rock matrix at a very slow rate. The fluid interacts with solids and pores by diffusion. In low-permeability near-surface rocks such as clay and shale, with permeability lower than 10^{-20} m^2 (10 nD), fluids are considered stagnant, and the Darcy flow law cannot describe fluid transport because at very low permeability and feasible pressure gradients the flow–force relationship is not linear (Bear 1979).

In fractured crystalline basement rocks with a typical permeability of 10^{-15} m^2, fluid flow stops if the driving pressure gradient vanishes, for instance, because the topographical relief has been eroded and removed. Therefore, the fluid in the fracture pore space of basement of flat vast plains has been considered essentially stagnant.

This view ignores a very efficient driving force for fluid flow, the Earth tides. Two times every day a tidal wave moves through the Earth crust. The rise and fall of the Earth surface amounts to some tens of centimeters (Emter *et al.* 1999). This is because the Earth is not a rigid body; it reacts elastically to the gravitational forces of Moon and Sun. The moving layers of rocks exert compressional and extensional forces on each other. The pore space of the rocks also deforms elastically in reaction to tidal

migration of fluid has been associated with progressing hydration reactions at depth (Stober & Bucher 2004). H$_2$O consumption at depth may not be fully compensated by downward fluid flow because of low permeability. Consequently, a measurable vertical hydraulic potential gradient in the Urach 3 bore, with decreasing potential with increasing depth, is enhanced and supported by low and decreasing hydraulic conductivity with depth. Thus, downward fluid flow can be small despite of favorable steep fracture orientation at depth simply due to low conductivity of the fractures. Deep fluid circulation systems may not develop in such low-permeability environments.

There are few direct observations of changing permeability of crustal rocks with time. This is a consequence of the timescale of the fracture–reaction–seal processes in the crust. The effects cannot usually be detected in well tests. Even if well tests could be repeatedly performed in, for example, 5 km deep boreholes over periods of 20–30 years, well deterioration and other technical effects would probably obliterate the signals from the undisturbed ground.

forces, with the result that the fluid residing in the pore space experiences an everlasting alternation of compression and extension. The consequence of the ever-changing tidal forces can be observed as water-table fluctuations in boreholes. The magnitude of the fluctuations depends on the position of the Sun and the Moon (e.g., Bredehoeft 1967; Evans & Wyatt 1984; Hsieh *et al.* 1988). Therefore, the highest amplitudes occur at full moon and new moon. In the fractured crystalline basement at the deep drilling site Urach 3, water-table fluctuations related to tidal forces of up to 20 cm per day have been measured (Stober 2011) (Fig. 15.10).

It can be concluded that the observed fluctuations of the water table in the wellbore indicate the reaction of a very large volume of deep water. This conclusion is in accordance with the very low fracture porosity of about $\phi = 0.002$ and the very low compressibility of deep fluid $c_w = 5.3 \times 10^{-10}$ Pa^{-1}. A pressure change of $\Delta P = 1$ bar results in a relative change of the volume of water of $\Delta V_w/V_w = 5.3 \times 10^{-10}$ and relative to the crust of $\Delta V_w/V_{rock} = 1.1 \times 10^{-12}$. This implies that the fracture pore space is interconnected on a large scale and the crust reacts in a hydraulically coherent manner.

Long-term hydraulic tests in deep boreholes confirm that a large volume of fluid hydraulically reacts and that the fracture porosity of the basement is an interconnected network of water-conducting structures. The 1-year-long pumping test at the 4000 m deep research drillhole (KTB-VB: see above) in southwest Germany extracted a total of 22,300 m^3 saline thermal fluid (Erzinger & Stober 2005) from the open hole at 3850–4000 m at a rate of 0.5 l s^{-1} and later 1.0 l s^{-1} (Fig. 15.11). The composition of the thermal fluid remained constant during this time (Stober & Bucher 2005a). Given the

porosity of 0.5%, the homogeneous volume of fluid originates from a cylinder of fractured rocks with a radius of about 100 m around the wellbore. The hydraulic signal reaches far beyond this cylindrical volume around the bore. The extracted fluid must be replenished from a much larger volume around the borehole.

In regions with high topographic relief, deep flow systems develop in the crystalline basement. Temperature profiles in very deep boreholes can be used to estimate the water circulation depths. For instance, in the 5000 m deep geothermal well GPK-2 in Soultz-sous-Forêts near Strasbourg, France (Genter *et al.* 2010), efficient water circulation reaches to 3700 m depth, which is 2300 m into the crystalline basement. Below 3700 m, the geothermal gradient is identical to the regional average (~27 K km^{-1}). Above this depth, water temperatures indicate advective heat transfer by upwelling hot deep water. Finally, in the near-surface fluid, temperature decreases very rapidly, resulting in a very high geothermal gradient (>100 K km^{-1}). Thermal signatures of advective fluid flow may be absent in some areas because permeability of the basement is low or because pressure gradients gradually diminish (with the exception of the cyclic tidal forces).

Temperature profiles of kilometer-deep boreholes, data, and observations from hydraulic tests and tidal water-level fluctuations all consistently show that fluids in the continental crust occupy an interconnected communicating pore space that permits fluid flow on a large scale, provided that an appropriate driving force exists in addition to the tides. In the above-mentioned examples, topography drives the fluid flow at Soultz-sous-Forêts and pumping, an anthropogenic force, drove fluid flow at KTB. Hot springs discharging

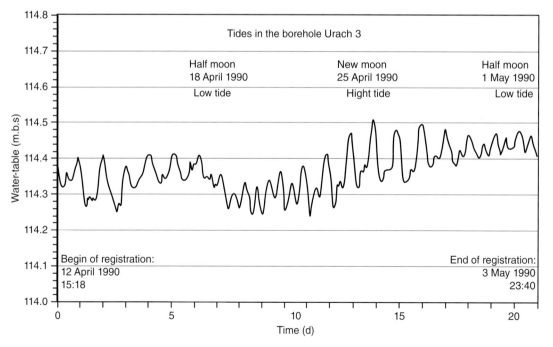

Fig. 15.10. Water-table fluctuations in the Urach 3 borehole caused by tidal forces.

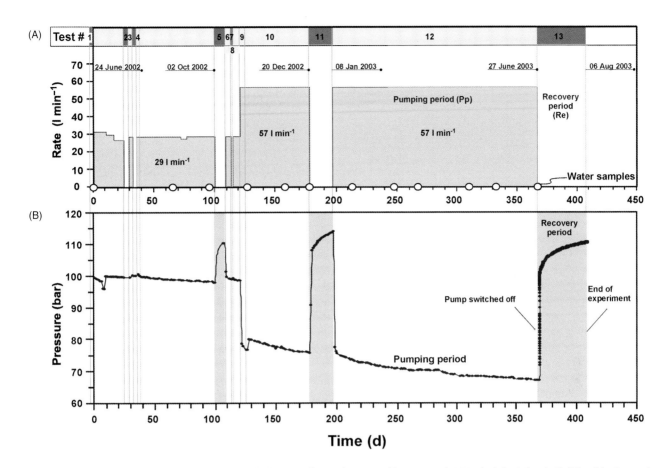

Fig. 15.11. Well test design and test-response data from hydraulic tests in fractured continental basement in the KTB pilot hole at 4 km depth (Oberpfalz, Germany).

highly mineralized waters from crystalline basement for 2000 and more years are clear evidence for deep fluid circulation and an interconnected fracture network providing sufficient permeability for fluid flow in the brittle crust.

Fluids in the crust are daily and perennially in flow motion driven and pumped by tidal forces. This motion mixes deep fluids and chemical reactions can be reactivated. The small-scale movement of aqueous solutions in and out of microcracks and the associated mixing processes have a significant effect on the kinetics of hydrothermal reactions. Deep waters in the brittle upper crust are never truly stagnant because tides keep them permanently in movement. The main effect of fluid movement caused by tidal forces is not its contribution to advection or bulk flow, but on reaction kinetics and local fluid composition, and thus on the timing of the permeability increase and decline.

The Earth tidal forces press the fluids into fine hairline cracks and dead-end pores and recover it from these pores during the decompression cycle. This cycling has significant geochemical consequences. The flow mixing is much more efficient than diffusional mixing. It maintains hydrogeochemical reactions going at significant rates. The tidal flow generates chemical reaction fronts along water-conducting fractures.

Extension of Darcy flow concepts to the lower crust is problematic. The lower crust is usually dry, with any free H_2O/CO_2

fluids removed by retrograde reactions (e.g., Frost and Bucher 1994). In an unusually active crust, prograde dehydration of H_2O-bearing minerals produces a free aqueous fluid, and these fluids must be removed from the lower crust. The exact transport mechanism is unknown. Nonetheless, Connolly and Podladchikov (this issue) and Weis (this issue) invoke a linear force–flux equation to quantitatively describe fluid removal from ductile lower crustal rocks.

SUMMARY AND CONCLUSIONS

Permeability k and hydraulic conductivity K of the brittle and fractured upper continental crust can be derived from pressure–time data obtained from hydraulic tests in deep boreholes. A handful of bores worldwide have produced permeability data for the crystalline basement to depths of about 4–5 km below surface. Below this depth accessible to direct measurement of transmissivity, k and K data can only be derived from indirect geophysical observations (e.g., seismicity propagation rates). However, it appears that permeability gradually decreases to the depth of the brittle–ductile transition zone at about 15 km depth and 350 °C in normal stable continental crust. As brittle water-conducting structures become fewer toward the transition zone, the representative elementary volume (Bear 1979) increases toward the ductile lower crust, where long-lived water-conducting structures do not exist. In the ductile lower crust, free fluid exists only temporarily (Frost & Bucher 1994; Yardley & Bodnar 2014).

For the uppermost 4–5 km of the brittle continental crust, where direct transmissivity (T) data can be measured in deep wells, the conversion of the T to k and K (=T/H) data is not simple. The conversion to permeability requires, for instance, the composition and properties of the fluid occupying the fracture pore space. These data require representative water samples. The sampling requires sophisticated and expensive sampling techniques. Derived T data from well tests are strongly influenced by the type of test, its duration (long, short), and the technical details of the test design and structure around the borehole. Injection tests may alter the undisturbed permeability structure of the formation, reversibly or irreversibly. A critical parameter for converting T to K (k) is the test length of the hydraulic test. It is often thought or assumed that permeability characterizes the rock formation at a certain spot in the crust. However, it may vary strongly with the test length (H). A meaningful permeability refers to a representative test length, correctly representing the entire system of rocks and fractures with all their properties including fracture density, aperture, connectivity, the number, distribution, and discharge of water inflow points and other properties and structures relevant to fluid flow. The representative test length may be very large in little-deformed basement. Thus, the characteristic permeability refers not to a spot in the basement but rather to a potentially large volume (e.g., several hundred meters). The representative test length may vary considerably along a depth profile (borehole). It tends to be smaller near the surface and generally increases with depth.

Deep fluid flow systems are controlled by steep or vertical fracture systems that develop as a result of increasing load with depth. Thermal water circulation in continental basement is typically topography driven and follows the vertical fracture systems to the surface where it may ultimately discharge as a hot spring. The permeability structure dictates the ascent velocity of thermal water and determines the final discharge temperature.

Faults or fault systems are prominent features of the crust. The hydraulic conductivity of faults can be lower than, equal to, or higher than that of the fractured rock matrix. Examples of low-permeability faults have been presented in Stober & Bucher (2005a) and Stober et al. (1999).

Permeability changes with time and the representative volume is also temporal and variable. The transient nature of k cannot usually be recorded by well test methods. The variation of the permeability structure of the upper continental crust is related to neotectonic processes and the chemical interaction of fluid with the rocks it comes in contact to along the flow path. Reactive fluid flow causes a number of chemical effects. Dissolution and precipitation of minerals on the fracture walls are relevant for the permeability of the system. These reactions tend to be kinetically slow at temperatures below 200 °C, corresponding to about 5 km depths from where well test T/H data are available. Thus, indications of permeability changes in normal continental crust from that depth range are scarce and mainly inferential (e.g., long-term temperature changes of hot springs). An analysis and discussion of permeability–time relationships related to coupled thermal–hydrological–mechanical–chemical processes can be found in Rutqvist (2015).

In the depth range of 5–15 km, brittle deformation and Darcy fluid flow still dominate. Evidence for transient permeability and its variability with time can be studied at outcrops from this depth range exposed at the present-day erosion surface. The evidence indicates that hydrothermal fluid–rock interaction tends to first increase permeability after initial fracture formation and then later reduce permeability by depositing solid reaction products on the fracture surfaces until the fracture is completely sealed and impervious to fluid flow. The typical time dependence of the conductivity of a single fracture is mostly related to the variation of the aperture with time. Flow at a given instant in time can be approximated by the cubic law for fluid flow, where the flow rate per hydraulic head difference is proportional to the cube of fracture aperture. The fracture aperture is a function of time due to mechanical aperture variations such as extension, compression, shearing, and other deformational effects, in addition to the progressing chemical reactions. The permeability of a representative volume in the lower part of the upper crust above the ductile transition zone and below the reach of deep boreholes, at a given instant in time, comprises the integral conductive property of all fractures that contribute to flow. Although each single fracture probably has its conductivity–time relationship, the permeability of the representative volume may not necessarily vary considerably with time. The time dependence of the permeability in this part of the upper continental crust is not realistically accessible from hydraulic data, geophysical data, or numerical models. This rather pessimistic view is well grounded in the extreme complexity of the processes controlling dissolution–precipitation reactions in hydrothermal environments. One often-neglected complexity is the effects of the Earth tidal forces, which keep even deep fluids in continuous motion despite the lack of obvious forces for flow such as topography in continuous motion. The tidal pumping of fluids into reactive microporosity increases reaction kinetics of fluid–rock interaction in a complex manner that is difficult to quantify and to study experimentally. The effect of tidal fluid motion in the upper crust is a feature that deserves more attention in the geoscience community.

ACKNOWLEDGMENTS

We acknowledge the very constructive reviews of all five reviewers, who helped to improve the manuscript substantially. Of course, all remaining errors remain in the responsibility of the authors.

Dynamic permeability

CHAPTER 16

Dynamic permeability

TOM GLEESON AND STEVEN E. INGEBRITSEN

[1] *Department of Civil Engineering, University of Victoria, Victoria, BC, Canada;* [2] *U.S. Geological Survey, Menlo Park, CA, USA*

Key words: static permeability, sedimentary rocks, igneous rocks, metamorphic rocks, brittle-ductile transition

The chapters broadly categorized as dealing with the dynamic variability of permeability include contributions related to oceanic crust (one chapter), fault zones (four chapters), crustal-scale phenomena (four chapters), and the effects of fluid injection at the reservoir or ore-deposit scale (four chapters).

OCEANIC CRUST

Cann *et al.* (this book) consider the crystalline crust near the mid-ocean ridge (MOR). They show that alteration of sheeted dikes to episodite "stripes" entails extensive dissolution of primary dike minerals and creates significant transient porosity (≤ 0.20) and permeability. Each dike was altered before being cut by a later dike, indicating a maximum timescale of 2000 years for hydrothermal alteration. This "reaction permeability" is rarely invoked in models of high-temperature MOR circulation, but may have complemented fracturing to provide the high k required for vigorous circulation. The stripe-like reaction k facilitated and channeled fluid flow while ongoing hydrothermal alteration conditioned fluid chemistry (metasomatism). The contribution of Cann *et al.* is unique in this book in considering the implications of outcrop- to thin section-scale mineralogy and paragenetic sequencing.

FAULT ZONES

Saffer (this book) considers active subduction-zone faults, which are important conduits for dewatering and transport of heat and solutes. He shows that fault-zone permeability values are consistent between margins, with time-averaged values of 10^{-15} to 10^{-14} m^2 and transient values of 10^{-13} to 10^{-11} m^2. Such values are approximately 10^6 times higher than estimated sediment permeabilities at (for instance) 5 km depth (Fig. 18.6). Permeable zones occupy a small fraction of the fault surface at any one time and migrate over time. Cox *et al.* (this book) show

that intermediate- to far-field earthquakes (generating seismic energy densities of approximately 0.1 J m^{-3}) cause a characteristic 5-day delayed, approximately 1 °C cooling of thermal springs in the Southern Alps, New Zealand. They attribute this behavior to permanent strains of 0.1–1 microstrain that open or cause fractures and allow greater mixing between thermal waters and cool groundwater. Micklethwaite *et al.* (this book) focus on underlapping fault stepovers, a type of stepover that is relatively rare but anomalously associated with gold deposits. They model the associated Coulomb failure stress changes and explain the association with gold in terms of localized fault damage. Supergiant gold deposits imply transiently high permeabilities on the order of 10^{-12} m^2, with healing on a timescale of 10^0 to 10^1 years (Fig. 20.9); a 5 Moz goldfield could perhaps form in 10–8000 years. Howald *et al.* (this book) use the age and isotopic composition of sinter in the (former) Beowawe geyser field to infer two long-lived (5×10^3 year) hydrothermal-discharge events following $M < 5$ earthquakes on the Malpais fault zone. Simulation of temperature and isotopic composition (Fig. 21.8) suggest that the earthquakes caused a $> 10^3$ increase in k (from $< 10^{-14}$ to $> 10^{-11}$ m^2).

CRUSTAL-SCALE BEHAVIOR

The continental crust is dominated by metamorphic rocks; approximately 90% of crustal volume despite only approximately 10% of surface exposures (Wilkinson *et al.* 2009). Yardley (this book) posits a fundamental difference between regions where the continental crust is being thickened and/or heated (prograde metamorphism, less common) and where the crust is stable and/or cooling (retrograde metamorphism, more common). Pervasively wet rocks and near-lithostatic fluid pressures seem to accompany prograde metamorphism, whereas localized hydration of dry rocks by fluid flow under near-hydrostatic fluid pressures is likely characteristic of retrograde metamorphism. In regions of prograde metamorphism, porosity waves are a mechanism by which fluids can be expelled from ductile rocks below the brittle–ductile transition. Connolly & Podladchikov (this book) present a general, analytical steady-state solution to the hydraulic equation that governs such flow and predict the dynamic variations in fluid pressure and permeability necessary to accommodate fluid production. Figure 23.8 is a conceptual model of flow regimes from the deep, ductile crust to the surface and assumes an approximately 10-km-thick transition zone above and below the brittle–ductile transition. Okada *et al.* (this book) show that, following the M_w 9.0 Tohoku earthquake in 2011, earthquake swarms at approximately 4–10 km depth exhibited temporal expansion that can be explained by fluid diffusion (e.g., Fig. 24.8; note the consistent rates), with inferred $k \sim 5 \times 10^{-16}$ m^2. The M_w 8.0 Wenchuan earthquake in 2008 caused coseismic groundwater-level changes across China. Shi *et al.* (this book) show that the sign and amplitude of water-level response was essentially random; thus, poroelastic response to coseismic static strain was not responsible for most water-level changes, even in the near field. Rather, hydrogeologic and tectonic settings were the dominant controlling factors, and permeability enhancement was the dominant mechanism.

EFFECTS OF FLUID INJECTION AT THE SCALE OF A RESERVOIR OR ORE DEPOSIT

Hydromechanical simulations of fluid injection into the very shallow crystalline crust by Preisig *et al.* (this book) show development of connected permeability to be almost exclusively orthogonal to the minimum principle stress, resulting in strongly anisotropic flow regardless of injection design. This is because tensile opening of a hydraulic fracture generates an increase in stress that limits the response of neighboring fractures in both tensile opening and shear. Most chapters in this book deal with an isotropic permeability (implying $k = k_x = k_y = k_z$) or two primary k orientations (e.g., k_x and k_z), whereas the full permeability tensor is relevant to the Preisig *et al.* analysis (e.g., Eq. 26.6). Miller (this book) simulates the well-known enhanced geothermal (EGS) experiment in Basel, Switzerland. He assumes a range of existing fault orientations subject to Mohr–Coulomb failure (Fig. 27.4) and effective-stress-dependent permeability that increases stepwise ($\times 1000$) when a failure criterion is satisfied; assumes a pre-existing permeability distribution; solves a nonlinear form of the diffusion equation (Eq. 27.5) subject to the observed pressure–time history; catalogs an "earthquake" when any gridpoint fails; and successfully simulates the time history of permeability evolution (Figs 27.5 and 27.6), hypocenter migration (Fig. 27.9), and earthquake rates (Fig. 27.10). Miller (this book) shows that dynamic permeability changes can be modeled using a set of empirically constrained rules, whereas Taron *et al.* (this book) attempt to simulate the actual highly nonlinear and interdependent physical processes that act upon fracture sets to enhance or degrade permeability during EGS experiments. They employ a new thermohydromechanical simulator capable of coupling the dominant physics of shear stimulation, allowing permeability to evolve under constitutive models that are tailored to fractured rock and represent the influence of thermohydromechanical stress and elastoplastic shear and dilation. Finally, Weis (this book), using a rule-based approach to permeability enhancement generally similar to that of Miller (this book), simulates multiphase H$_2$O–NaCl fluid flow in concert with a dynamic permeability model. Weis isolates, for instance, the influence of NaCl (Figs 29.7 vs. 29.9) and subaerial topography (Figs 29.10 and 29.11).

PART III(A): Oceanic crust

Rapid generation of reaction permeability in the roots of black smoker systems, Troodos ophiolite, Cyprus

JOHNSON R. CANN, ANDREW M. MCCAIG AND BRUCE W. D. YARDLEY

School of Earth and the Environment, University of Leeds, Leeds, UK

ABSTRACT

The deep levels of former black smoker hydrothermal systems are widespread in the Troodos ophiolite in Cyprus. They are marked by zones of hydrothermal reaction in the sheeted dike unit close to the underlying gabbros. These zones are characterized by the presence of epidosite (epidote–quartz rock). In the reaction zones, the dikes are altered to a range of greenschist facies mineral assemblages from a low degree of alteration with a five- to seven-phase metabasaltic assemblage to a high degree of alteration with a two- to three-phase epidosite assemblage. Individual dikes may contain the full range, with the epidosites forming yellow–green stripes within a darker background, often extending for more than several metres, parallel to the dike margins. Field relations show that the alteration took place on a dike-by-dike basis and was not a regional process. Scanning electron microscopy (SEM) petrography reveals that the epidosites contain millimetre-scale pores. The minerals surrounding the pores show euhedral overgrowths into the free pore space, indicating a former transient porosity of up to 20%. We conclude that the epidosites formed by reaction between newly intruded basaltic dikes and actively circulating black smoker fluid leading to extensive dissolution of primary dike minerals. This reaction generated the porosity in the stripes and transiently led to a much increased permeability, allowing the rapid penetration of the black smoker fluid into the dikes and flow along them in fingers. As the system evolved, the same flow regime allowed mineral precipitation and partial infilling of the porosity. This mechanism allows rapid recrystallization of the rock with the release of metals and other components into the fluid. This explains the depletion of these components in epidosites and their enrichment in black smoker vent fluids, and the relatively constant composition of vent fluids as fresh rock is continually mined.

Key words: epidosite, hydrothermal, permeability, sheeted dikes, Troodos

INTRODUCTION

The transformation of cold sea water to hot black smoker fluid at oceanic spreading centres is one of the most dramatic of hydrological processes. A major black smoker vent field may emit 0.5–1 GW of thermal energy, corresponding to about 0.5–1 m^3 s^{-1} of high-temperature fluid (Humphris & Cann 2000; Baker & German 2004; Di Iorio *et al.* 2012). Such flow may continue for long enough to generate major sulfide deposits containing up to several million tonnes of sulfides (Humphris & Cann 2000). Geophysical evidence from the East Pacific Rise and the Juan de Fuca Ridge is that the heat transfer to the circulating fluid takes place close to an underlying magma chamber at a depth of 1–2 km below the spreading axis (Detrick *et al.* 1987; Van Ark *et al.* 2007). Ocean drilling and submersible observations (Francheteau *et al.* 1992) and the structure of ophiolites (Baragar *et al.* 1990)

show that this region is composed of sheeted dikes formed by the repeated intrusion of dikes into one another as spreading takes place.

In the numerical modeling of black smoker systems, one critical problem has been the nature of a permeability structure of the crust that must consistently allow a high-heat output at a high temperature (Lowell & Germanovich 2004; Driesner 2010; Ingebritsen *et al.* 2010). Although modellers have used many approaches, developing over time, all models require a crustal permeability structure within the regions of hydrothermal circulation that is very closely defined. Their permeability estimates are significantly higher than the limited data set of *in situ* measurements in sheeted dikes from ODP Hole 504b (Becker & Davis 2004), suggesting that ridge crest permeability is transient. What processes control the permeability structure and its evolution so closely?

Almost all authors assume that permeability in the ocean crust is largely controlled by fracturing due to either cooling and thermal contraction (Lister 1974) or tectonic faulting and fracturing (Scott *et al.* 1974; German & Lin 2004). In this study, we document an additional mechanism, the creation of connected porosity by mineral dissolution in hydrothermal fluid. This mechanism for generating permeability is well known in karst and other diagenetic systems and has been suggested for skarns and marbles (Yardley & Lloyd 1989; Yardley *et al.* 1991b; Balashov & Yardley 1998), but textural evidence has not previously been documented in black smoker systems.

The key lies in the direct investigation of the regions within which the hot, metal-bearing hydrothermal fluid circulates through the crust and rises to the ocean floor. Active regions of this kind are not accessible to direct investigation in the oceans by drilling, and fossil regions have been elusive in drill holes that have penetrated deep into older ocean crust, but they are exposed in ophiolites such as the Troodos ophiolite in Cyprus and the Semailophiolite in Oman. In this study, we re-examine the processes that shape the zones of high-temperature hydrothermal alteration in the Troodos ophiolite through a combination of field evidence, petrography and chemistry, with the aim of understanding the links between alteration and fluid flow. In particular, we investigate the permeability structure that allowed the flow of fluid to take place through the reaction zone and its origin in the metasomatic reactions that conditioned the chemistry of the fluid. We show that mineral dissolution and creation of new porosity occurred during these metasomatic reactions, and occurred rapidly, between intrusion of one dike and the next.

THE TROODOS OPHIOLITE: GEOLOGICAL SETTING

A general introduction to the hydrothermal systems of the Troodos ophiolite of Cyprus can be found in Cann & Gillis (2004). Zones of high-temperature hydrothermal alteration are especially well exposed in the ophiolite, formed at a Cretaceous spreading centre above a subduction zone in a back arc environment (Pearce *et al.* 1984). This ophiolite extends for more than 100 km across the strike of the ancient spreading centre and contains remarkably complete units of the upper crust that have been affected only by low-temperature alteration since they formed (Staudigel *et al.* 1986; Gallahan & Duncan 1994). The ophiolite is capped by a unit of submarine lavas containing more than 30 sulfide deposits (Bear 1963; Adamides 1990) similar to those found associated with oceanic black smoker systems (Humphris & Cann 2000) and containing fossilized vent organisms (Little *et al.* 1999). Beneath the lavas is a unit of sheeted dikes that overlies a unit of gabbro formed by crystallization of the axial magma chamber (Fig. 17.1).

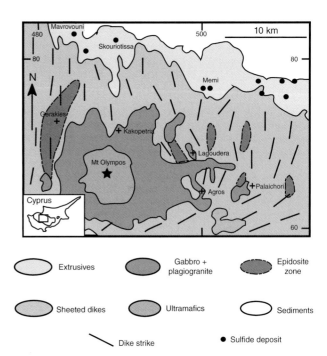

Fig. 17.1. Outline geological map of the central part of the Troodos ophiolite. The ophiolite is in the form of a dome. The deepest parts of the section are the ultramafics, gabbros and plagiogranites, which are overlain by sheeted dikes, and in turn by volcanics. The overlying sediments include deep ocean chalks and more recent deposits. The epidosite zones lie close to the base of the sheeted dike unit, except for the northern end of the big western zone, including the Gerakies localities, which reaches close to the volcanic unit, though there accompanied by outcrops of gabbro (Bettison-Varga *et al.* 1995). The variable dike strike reflects the complex processes of crustal generation at the ancient spreading centre. For a general introduction to the ophiolite, explaining this and other features, see Edwards *et al.* (2010). The grid numbers are UTM coordinates using the WGS84 grid, truncated by removing the first two digits and the last three digits of both eastings and northings. Place names are selected villages as well as the highest point, Mount Olympos. The sulfide deposits are all within the volcanic sequence. Three of them are named. (*See color plate section for the color representation of this figure.*)

Most of the thickness of the sheeted dike unit in the ophiolite is entirely made up of dikes intruded at the ancient spreading centre. The dikes intrude one another and are marked by well-defined chilled margins. They range in thickness from a few tens of centimetres up to a few metres. In every outcrop, there are some late dikes that are preserved complete with two chilled margins, but most of the outcrop is of dikes that have been intruded by other dikes, often in several stages. These earlier dikes are present as elongate lenses, sometimes with one chilled margin and at other times without a chilled margin visible. Careful field observations show that most lenses without chilled margins have a chilled margin on a segment of the dike on the other side of one of the intruding dikes (see Fig. 17.2) (Kidd & Cann 1974). Although dikes are generally nearly parallel to one another, in most outcrops, there are some dikes that cut obliquely across others.

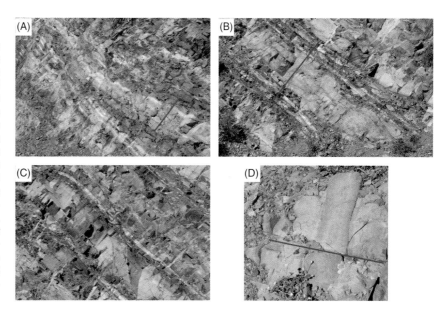

Fig. 17.2. Field photographs of epidosite zone exposures. Location references are UTM coordinates using the WGS84 grid, truncated by removing the first two digits of both eastings and northings. Red lines show the width of a prominent individual dike in each image. (A) Exposure of epidosite zone rocks showing the variety of colors in a single outcrop (exposure about 3 m high) (482160 68520). The paler lithologies are epidosites, and the darker colors are less altered dikes. (B) A 1.5-m-wide striped dike showing the symmetrical pattern of striping typical of wider dikes (481873 68871). (C) A 0.7-m-wide dike with a simpler alteration pattern. At the margins are narrow stripes of epidosite (481873 68871). The core of the dike is more altered than the outer zones inside the epidosite stripes. (D) Incipient alteration shown by a pod, 0.2 m wide, of more highly altered rock inside a less altered dike matrix (498774 67931). (*See color plate section for the color representation of this figure.*)

The dikes were intruded in a narrow zone at the ancient spreading centre. Modeling based on detailed field observations in Troodos (Kidd 1977) indicated that this zone might be as narrow as a few tens of metres. In the oceans, the zone of intrusion is marked by a chain of fissuring along the spreading axis that is about 200 m wide (Crane 1987). From this intrusion zone, the completed sheeted dike complex was rafted off laterally by spreading.

Throughout most of the thickness of the sheeted dike unit, the dikes have been metamorphosed to a typical greenschist facies assemblage: albite, actinolite, chlorite, quartz, epidote and titanite. This assemblage is referred to as the background alteration in what follows. Some parts of the sheeted dike unit have been rotated by up to tens of degrees. This rotation occurred after the metamorphism was complete (Varga *et al.* 1999).

EPIDOSITE ZONES: PREVIOUS WORK

Towards the lower part of the sheeted dike unit are complex zones marked by the presence of epidosite (epidote–quartz rock), formed by profound chemical and mineralogical transformation of the dikes, set within dikes showing less profound alteration.

These 'epidosite zones' are typically a few hundred metres wide perpendicular to the strike of the dikes and extend for a kilometre or more parallel to the dike strike. They have formed by the alteration of the sheeted dikes, as the chilled margins and the intrusive relations of the dikes are well preserved in outcrop. The mineral assemblages within the epidosite zones are much more variable than in the background sheeted dikes. Some of the dikes show the five- to seven-phase greenschist

facies metadiabase assemblage of the background, but many of the dikes contain fewer phases. At the most extreme degree of alteration, the assemblage is the classic epidosite assemblage of epidote, quartz and either titanite or Fe oxides. At a somewhat less extreme degree of alteration and more abundant than true epidosites are epidote, chlorite, quartz and titanite rocks. The degree of alteration varies on a small scale within and between dikes so that a 2-m-wide single dike may contain thin stripes of epidosite of at least several metres long, spaced 10–20 cm apart, associated with stripes of epidote, chlorite, quartz rock, and set in a variable matrix containing regions with a larger number of phases including epidote, chlorite and quartz assemblages (Jowitt *et al.* 2012).

The epidosites in Troodos were first identified during the earliest detailed geological survey of the ophiolite (Wilson 1959). Later work showed that the sulfide deposits of the ophiolite had been formed by the exhalation of hydrothermal fluid at the ocean floor (Constantinou & Govett 1972, 1973) pre-dating the discovery of active black smokers in 1979. Detailed work began on the epidosites in the 1980s (Richardson *et al.* 1987; Schiffman *et al.* 1987).

It was recognized (Richardson *et al.* 1987) that the extreme phase reduction of the epidosites might be the product of intense metasomatic alteration of the dikes under open-system conditions as envisaged by Korzhinskii (1959, 1965). After their discovery in Troodos, epidosites were identified and described in several other ophiolites, the Josephine ophiolite in California/Oregon (Harper *et al.* 1988 and related papers), the Semail ophiolite in Oman (Nehlig & Juteau 1988; Nehlig *et al.* 1994), the Pindos ophiolite in Greece (Valsami & Cann 1992) and the Solund ophiolite in Norway (Fonneland-Jorgensen *et al.* 2005). Epidosites have only rarely been found in the modern oceans, but have been dredged from the fore arc of the

Tonga Arc (Banerjee *et al.* 2000; Banerjee & Gillis 2001) and have been reported in rocks dredged from the Mid-Atlantic Ridge at 30°N (Shand 1949; Quon & Ehlers 1963).

Previous work on epidosites clarified a number of issues. The major element chemistry of rocks from epidosite zones shows the wide range in composition expected from the changing mineralogy (Richardson *et al.* 1987; Harper *et al.* 1988; Nehlig *et al.* 1994; Bettison-Varga *et al.* 1995). All of the rocks within epidosite zones, including the whole range from diabase to epidosite, are strongly depleted in Cu, Zn and Mn between 50% and 90%, with Cu the most depleted and Mn the least (Richardson *et al.* 1987; Harper *et al.* 1988; Nehlig *et al.* 1994; Bettison-Varga *et al.* 1995). Jowitt *et al.* (2012) demonstrate both the spatial scale of mineral assemblages and the extent of metal depletion, including Ni as well as Cu, Zn and Mn. In contrast, Sr is enriched in the epidosites and $^{87}Sr/^{86}Sr$ ratios are strongly altered in both epidosites and the surrounding dikes showing more typical greenschist facies assemblages, with values of ~0.705, intermediate between fresh basalt and Cretaceous sea water (Bickle & Teagle 1992; Bickle *et al.* 1998). Recent alteration of diabase in ODP Hole 504B is much less intense than that seen in Troodos (Teagle *et al.* 1998). Fluid inclusions within the epidosite zones, including all rock types, show homogenization temperatures broadly within the range 250–400°C, and salinities typically ranging from sea water to twice sea water concentrations (Richardson *et al.* 1987; Nehlig *et al.* 1994; Bettison-Varga *et al.* 1995; Juteau *et al.* 2000). Oxygen isotopes give evidence of extensive water–rock reaction at similar temperatures (Harper *et al.* 1988; Schiffman & Smith 1988; Nehlig *et al.* 1994; Fonneland-Jorgensen *et al.* 2005). All previous studies conclude from this evidence that epidosite zones have been the sites of intense hydrothermal interaction between black smoker fluids and dike rock, with the true epidosites within them representing the places where fluid–rock interaction has been most intense.

There have been two end-member models of the generation of the high permeability in the rocks that allowed the intense fluid flow to take place. One is that permeability has been related to generation of fractures in the rocks by tectonic processes at the spreading centre (Bettison-Varga *et al.* 1992; Nehlig *et al.* 1994). The other is that the replacement of lower density silicates by dense epidote has generated porosity within the rocks and hence higher permeability. This concept has been referred to in passing by a number of authors (Harper *et al.* 1988; Nehlig & Juteau 1988; Bettison-Varga *et al.* 1992; Bickle *et al.* 1998), but only Bettison-Varga *et al.* (1995) have developed the model. No previous authors have given textural evidence for reaction porosity generation in epidosites.

This study extends the previous work in Troodos and other ophiolites in two directions. First, we report a detailed study of field relationships in the epidosite zones of Troodos. Second, we discuss new petrographic studies of Troodos epidosites, using scanning electron microscopy and cathodoluminescence.

RESULTS

Field relationships of epidosites

The progressive alteration of the mineral assemblages in the sheeted dikes, from background metabasaltic greenschist assemblages to epidosites, can be seen in excellent roadside exposures through the varying color of the altered dikes (Fig. 17.2A). Precise field localities for examining epidosites may be found in Edwards *et al.* (2010), and localities are also given in the figure captions in this study. Epidosite, representing the highest degree of alteration, is a very pale greenish yellow, while rocks at a lower degree of alteration, with a greater content of chlorite and/or actinolite, are darker colored reaching a dark bluish green where the content of quartz and epidote is low. This contrast in color allows the spatial variation in mineral assemblage to be recorded in field photographs. The mineral assemblages vary from dike to dike and on a smaller scale within dikes. Individual dikes can be recognized by their chilled margins and the cross-cutting jointing typical of dikes in general, and these are preserved to the highest degrees of alteration.

Within individual dikes at higher degrees of alteration, the alteration is frequently striped parallel to the dike margins, with stripes at a higher degree of alteration alternating with stripes of a lower degree. The edges of the stripes are often sharp, within a few millimetres. The pattern of striping is commonly symmetrical about the centre of the dike, varying with the width of the dike (Fig. 17.2B). Narrow dikes, up to a metre wide, commonly have a broad higher degree stripe down the centre and may have a narrow stripe of higher alteration close to the margin (Fig. 17.2C). Wider dikes commonly contain several stripes of higher alteration close to the margin, which is separated by darker, wider stripes of lower alteration. The centres of dikes are commonly more altered than the regions close to the margins. Dikes at a lower degree of alteration may contain pods of a higher alteration elongate parallel to the dike margins. These are inferred to have finger-like geometries in three dimensions (Fig. 17.2D).

The joints cutting dikes within epidosite zones are only rarely filled with epidote and quartz. Most joints are empty of hydrothermal minerals. Epidote–quartz veins are commoner along dike margins but are still not abundant. The rarity of alteration associated with the cross-cutting joints indicates that most of the alteration predates the jointing.

Cross-cutting dikes show three important features (Fig. 17.3). (i) When one striped dike crosses another, the margin of the younger dike cuts across the stripes in the older dike. These terminate at the margin of the younger dike and reappear on the other side of it. The stripes in the younger dike follow the bends in the dike margin and remain parallel to each other. (ii) The degree of alteration of the older dike does not change as it approaches the margin of the younger dike. (iii) Dikes that are more altered frequently cut dikes that are less altered, which would not be expected if progressive alteration affected the

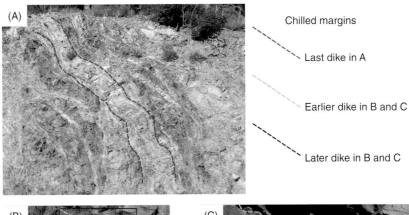

Chilled margins

Last dike in A

Earlier dike in B and C

Later dike in B and C

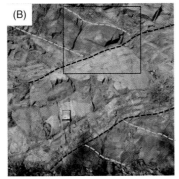

(B)

(C)

Fig. 17.3. Field photographs showing the relationships between dikes in an epidosite zone. (A) A single highly altered 1-m-wide dike cutting through a complex of earlier dikes altered to different degrees. (B) A 1.5-m-wide striped epidosite dike at a high degree of alteration, cutting a narrow 0.7-m-wide dike at a lower degree of alteration and displacing its outcrop. (C) Enlarged view of the area outlined in (B). Note the abrupt truncation of the earlier dike and its marginal and central stripes by the later dike and the apparent lack of any further alteration of the earlier dike by the higher degree of alteration of the later one. (*See color plate section for the color representation of this figure.*)

whole dike injection zone. All of these features indicate that the alteration occurred in one dike at a time, while the physical nature of each dike was distinct.

At a larger scale, epidosite zones can be distinguished from background sheeted dikes by the presence of striped dikes in outcrops. Individual epidosite zones are typically a few hundred metres wide and separated from each other by a few kilometres. They are best developed in the lower parts of the sheeted dike unit, but in the area west of the Solea Graben reach higher structural levels close to the major sulfide deposits at Mavrovouni and Skouriotissa (Fig. 17.1) although there they are associated with outcrops of gabbro (Bettison-Varga *et al.* 1995).

Petrography and mineral chemistry of epidosites

Samples from an epidosite zone can be put in a broad series of decreasing number of mineral phases, although this does not necessarily imply a sequence of development. The background dikes from outside the epidosite zones and many of the dikes within an epidosite zone show a typical metabasalt greenschist facies assemblage of albite (with some relict calcic plagioclase), actinolite (with some relict clinopyroxene), chlorite, quartz, epidote, iron oxides and titanite. Their texture is a typical ophitic to subophitic igneous texture dominated by albite laths. As the degree of alteration increases and the number of mineral phases decreases, minerals drop out regularly. Typically, albite disappears first, followed by actinolite. Chlorite is the next to go, although it survives to a high degree of alteration, and may

reappear as a late phase at high degrees of alteration. At the highest degree of alteration, in the epidosites, the assemblage is quartz, epidote and titanite or iron oxide, a true epidosite. At this stage, alteration is no longer pseudomorphic and no trace of igneous texture remains.

The modal compositions of the true epidosites established by point counting are close to 30% quartz, 55% epidote and smaller amounts of titanite and voids. Less highly altered rocks with a significant amount of chlorite contain about 30% quartz, 35% epidote, 25% chlorite and small amounts of other phases such as actinolite, Fe oxides, titanite and void space.

The texture of the epidosites is pervasively granoblastic (Fig. 17.4). The grain size varies with the position in the dikes, and hence with the original igneous grain size, suggesting some memory of the original igneous texture. The proportions of epidote and quartz may vary over a single polished block from areas where quartz is more abundant than epidote to areas in which epidote is dominant. In many cases, the eventual grain size of the quartz is much greater than that of the epidote, with small subprismatic crystals of epidote included within larger quartz crystals (Fig. 17.5).

Most of the epidosites contain voids, some more or less equant, and in other cases, a network of cracks on a scale of a few millimetres, each crack around 0.1 mm wide. These void spaces are commonly fringed by euhedral terminations of the bordering crystals, indicating that adjacent crystals were growing into open space until the filling of the voids was been interrupted (Fig. 17.6).

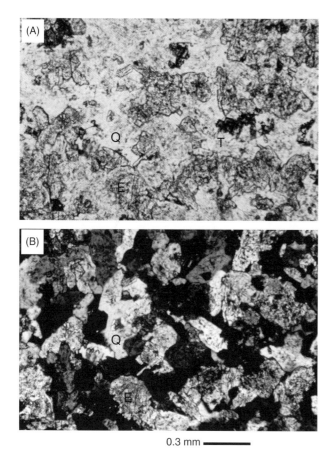

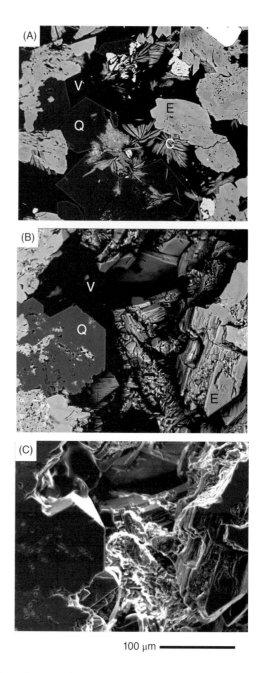

0.3 mm ▬▬▬

Fig. 17.4. Thin section photomicrograph of an epidosite in plane polarized light (A) and crossed polars (B), showing the absence of any relict igneous texture. The section shows epidote E, quartz Q and titanite T. The rock specimens in Figures 17.4–17.9 come from the road section west of Gerakies between (480830 73370) and 480940 73010). (*See color plate section for the color representation of this figure.*)

100 μm ▬▬▬

Fig. 17.5. Scanning electron microscope (SEM) images showing the relations between epidote (E), quartz (Q), chlorite (C) and the void spaces (V). (A) (58503) back-scattered electron image. (B) (58519) back-scattered electron image. (C) Secondary electron image of the same area as (B), showing euhedral grain facets within the pore. Epidote, quartz and chlorite are all clearly growing at least in part into porosity. Five-digit numbers in Figures 17.5–17.9 are University of Leeds specimen numbers.

Typically, there is a variation in the texture within the matrix. Some regions show traces of the original igneous texture, often seen best in the scattered presence of pseudomorphs in titanite or magnetite of original Fe–Ti oxides. In other regions, any sign of igneous texture is lacking, and instead, there are replacive or space-filling textures shown by optical zoning or trains of inclusions. This variation is typically on a scale of a few millimetres.

Back-scattered electron images of polished surfaces show voids and the euhedral terminations of epidote and quartz at the edges of the voids. They also show that the crystals forming the euhedral fringe to the voids are free of inclusions. Areas of similarly inclusion-free crystals within the samples are interpreted to indicate the former presence of voids, now filled (Fig. 17.7).

Cathodoluminescence images (Connell 2000) of coarse quartz close to voids show a clear growth zoning in cathodoluminescence, with bright zones and dark zones alternating parallel to crystal terminations (Fig. 17.8). The pattern of zoning is similar in samples from the same dike and often

different between dikes close together in the same outcrop. Some patterns show a cathodoluminescence bright-zoned core to crystals, a dark rim to crystals and late dark veins that cut across bright areas (Fig. 17.8). Other patterns show a dark zoned core, rimmed by a thin, bright zone, rimmed again by a dark outer zone (Fig. 17.9). Electron probe traverses (Connell

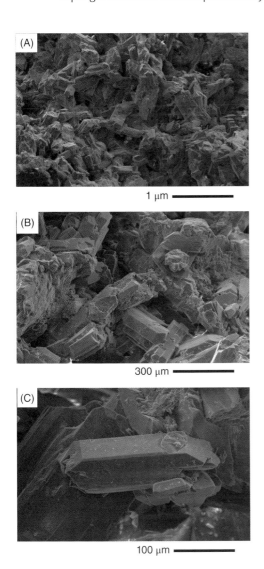

Fig. 17.6. SEM images of epidosite surfaces broken in the laboratory, showing euhedral crystals of quartz and epidote fringing the void spaces.

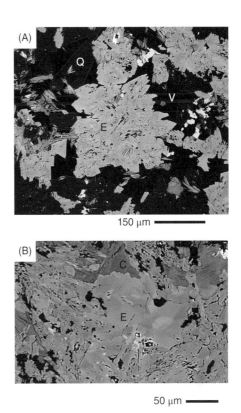

Fig. 17.7. SEM back-scattered electron images of polished specimens. (A) (58503) Cluster of epidote (E) and quartz (Q) crystals growing into void space (V). Note the euhedral terminations of the crystals towards the void. Note also the inclusion-free rims to the epidote crystals surrounding cores that are full of inclusions. Most of the inclusions are extremely small voids. (B) (58602) Coarse epidote crystals free of inclusions that have totally filled a void, growing from surrounding inclusion-rich crystals. The chlorite (C) appears to be late and to be corroding and replacing the epidote. The variation in brightness of the epidote crystals is the result of varying chemical composition ranging from Al/Al + Fe 0.85 to 0.80. Small euhedral cores can be seen within the inclusion-free epidote.

2000) across zoned quartz show that Al and Ti are enriched in the brighter zones (up to 1000 and 700 ppm, respectively) dropping to close to the detection limit in darker zones and late veins, while Fe is enriched from 1000 to 5000 ppm from the core to the margins of the crystals. Application of the Ti-in-quartz geothermometer (Wark & Watson 2006) gives meaningless temperatures that can exceed 1000°C, probably reflecting fast disequilibrium quartz precipitation (Huang & Audétat 2012). The associated epidote in areas close to voids is also zoned, as can be seen in the back-scattered electron images (Fig. 17.9). The epidote composition in such zoned crystals varies on a scale of a few tens of micrometres from Al/(Al + Fe) 0.7 to 0.85 (Connell 2000).

Samples taken from the epidosite stripes are especially rich both in existing voids and in the evidence that these voids were once larger than now. There is a continuum between clear evidence for void filling and the more subtle textures, suggesting that such void filling may have occurred in nearby coarser-grained areas. These include a common contrast between inclusion-rich cores to epidote-rich regions and wide inclusion-free rims adjacent to remaining voids. The voids would have been larger than any that now remain, perhaps reaching sizes of 5–10 mm and are inferred on the basis of the clear zones of epidote crystals and correlatable cathodoluminescence zones in quartz to have occupied volumes of up to 20% of the rock volume. Growth of euhedral crystal faces accompanied by fluctuations in mineral composition such as oscillatory zoning has been shown by Yardley et al. (1991b) to be indicative of infiltration metasomatism.

DISCUSSION

Significance of epidosite stripes

The association of the formation of epidosites with the circulation of black smoker fluid is inferred from a number of lines of

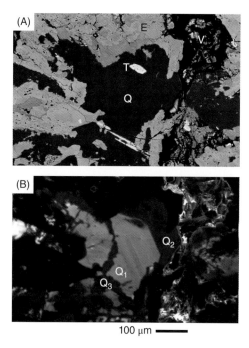

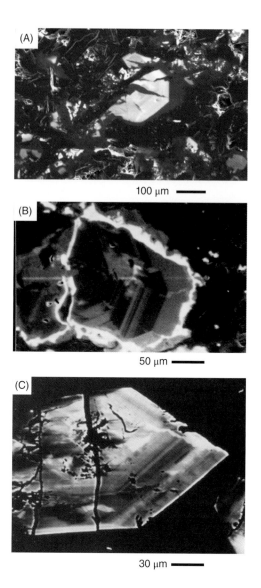

Fig. 17.8. Comparison of back-scattered electron and cathodoluminescence images of the same area of 59133. Much of the image is composed of void-filling epidote and quartz with a small euhedral titanite (T). In (A), the variable chemical composition of the epidote is clear. (B) shows the complexity of the crystallization of the quartz. The bright areas (Q1) show a growth zoning of the quartz and are rich in Al and Ti compared to the darker areas, which have higher Fe contents (Connell 2000). Dark Q2 quartz overgrows bright Q1 quartz and both Q1 and Q2 are cut by late dark veins (Q3) that seem to be replacing the earlier Q1 and Q2.

Fig. 17.9. Examples of growth zoning in quartz and epidote. (A) (58508) cathodoluminescence image showing quartz with a bright core surrounded by a dark rim and cut by late replacement quartz and an even later fracture now filled with dark quartz, a similar relation to that of Figure 17.8B. (B) (58596) cathodoluminescence image showing a very different sequence of quartz zoning. Here, the early, zoned growth is of dark quartz, surrounded by a thin rim of very bright quartz in turn surrounded by dark quartz again. (C) (59133) back-scattered electron image of epidote projecting into a void, showing oscillatory zoning of the composition of the crystal resulting from growth from the circulating fluid. Note that the contrast has been turned up to emphasize zoning in the epidote, so that both quartz and void space are black in this image.

evidence. The fluid inclusion data, the alteration mineral assemblages and the removal of Cu, Zn and Mn from the epidosite zones all coincide, as does the presence of VMS sulfide deposits within the overlying lava pile. The decreasing number of mineral phases in the series of alteration products between the background sheeted dikes and the epidosites can be linked to the open-system metasomatic model of Korzhinskii (1959, 1965), which predicts that increasing the proportion of infiltrated fluid leads to an increase in the number of mobile components in a system and hence to a decrease in the number of mineral phases present.

The field relations allow this line of evidence to be taken further. If the presence of stripes of epidosite within individual dikes reflects the variation in the local fluid/rock ratio or mass flux, then the elongation of the stripes parallel to the dike margins must indicate heterogeneous fluid flow confined at that time within a single dike. The flow must be linked to the margins of the dike. This conclusion is confirmed by the continuity of stripes parallel to dike margins, even when a dike curves quite sharply, and is consistent with the typically higher degree of alteration of dike centres relative to the margins. The consistent sharp truncation of the stripes in an older dike as it is cut by a younger dike is relevant too. It indicates that the stripes

were already present in the older dike when the younger dike was intruded and were not modified by the intrusion or by the fluid flow in the younger dike. Another indicator that the alteration is local rather than regional is the common occurrence of less altered dikes cut by more altered dikes. If the alteration had been regional and gradual, then newly intruded dikes would be less altered than older dikes.

Related to this is a further observation; only rarely is epidote seen along the joints that cut across a dike. In those rare cases, the joints are coated with a thin layer of epidote and quartz, as is often seen in metabasaltic dikes in other types of environment. Much more commonly in the epidosite zones the joints are empty, even when the core of the dike is clearly striped and must have seen intense fluid–rock interaction. This suggests that, in most cases, the alteration took place before the cross-cutting joints had formed.

Rapidity of epidotization

The combination of the field evidence summarized here indicates strongly that within an epidosite zone individual dikes are altered soon after they have been intruded, with the intensity of alteration variable both within a dike and between different dikes. An approximate maximum timescale for cross-cutting dikes can be calculated as follows. The Valu Fa spreading centre in the Tonga back arc has a similar magma chemistry to that in Troodos (Jenner *et al.* 1987) and has a half spreading rate of $25–30\,\mathrm{mm\,year^{-1}}$ (Taylor *et al.* 1996). Given the Valu Fa spreading centre as a model and a half-width of the dike injection zone at Troodos of <100 m (Kidd 1977), it would take about 3000 years for a dike intruded close to the centre of the injection zone to spread to the edge of the zone. Effectively, this represents the time taken for the construction of the sheeted dike complex at any one site. The interval between intrusions of two cross-cutting dikes would be expected to be much shorter than this.

This is the time frame of complete replacement of much of the earlier dike by new minerals and is extremely rapid by normal metamorphic standards. The rapidity of alteration indicates that the protolith for the epidosites was not the background dike material found outside the epidosite zones, but was the primary igneous mineralogy of a newly crystallized dike. The close association of intrusion and alteration indicates that, to form epidosites, intrusion must have taken place within an active black smoker system.

Flow of hydrothermal fluid through a newly intruded dike requires the generation of permeability within the dike, especially given the flow necessary to produce the intense alteration that yields the epidosites, The petrographic evidence summarized earlier and shown in Figures 17.4 and 17.5 demonstrates that the alteration has generated significant porosity within the epidosite stripes. Some of this porosity still remains, with some pores isolated, and others as a network of cracks. The petrography shows that there was originally much greater porosity (up to 20%) that was subsequently filled by precipitation of epidote, quartz and chlorite.

These substantial changes in mineral assemblage and porosity show that the volume of fluid flow along an epidosite stripe must have been considerable. This alone indicates that the permeability in the stripe must have been high, and thus that the porosity now visible was linked to a network.

Formation of stripes

How then were the stripes formed? A newly crystallized dike could have about 10% of grain-scale porosity from an approximate 10% increase in density on solidification of basaltic magma (Hooft & Detrick 1993). In addition, thermal contraction during extremely rapid cooling from 1000 to 400°C could introduce grain boundary cracks. This porosity would be likely to have given rise to a somewhat higher permeability in the new dike compared with the surrounding older dikes. Black smoker fluid, already flowing through the dike wall rocks, would enter the dike and start to flow along it. The fluid would immediately start to react with the igneous minerals, potentially increasing the porosity and permeability. Numerical models of this process show that the result would be the development of fingers of altered and permeable material propagating in the direction of fluid flow into unaltered rock (Steefel & Maher 2009). This results from the positive feedback between dissolution rate, fluid flow and permeability. In the case of a dike intruded into a black smoker upflow zone, this process of fingering will begin simultaneously over a large vertical extent of the dike. Once the fingers began to link together, the dike as a whole would become permeable and fluid flow would focus increasingly into it. However, mineral precipitation would be continuously and simultaneously occluding the porosity and reducing the permeability. For this reason, the highest local permeabilities would be at the metasomatic front between altered and fresh rock; this may provide a mechanism for the initial fingers to coalesce into the sheet-like stripes.

From a geochemical standpoint, the major problem is how a single fluid flow regime such as this could change from giving rise to net dissolution of silicate rock to create porosity to infilling the porosity by silicate precipitation. Creation of this porosity immediately after intrusion requires the rapid dissolution of primary minerals. A black smoker fluid at 350–400°C is likely to be 'far from equilibrium' with respect to igneous minerals formed at 1000°C. Typical behaviour is that rates of dissolution increase from zero at equilibrium and then plateau at affinities $>60\,\mathrm{kJ\,mol^{-1}}$ (Brantley *et al.* 2008). Under far from equilibrium conditions, silicate dissolution rates are extremely rapid, particularly at low pH. *In situ*, pH in basalt-hosted black smoker systems is estimated to be about 5 (Pester *et al.* 2012). For example, at 300°C and pH 5, orthopyroxene is predicted to dissolve at $4.3 \times 10^{-11}\,\mathrm{mol\,cm^{-2}\,s^{-1}}$ (Oelkers & Schott 2001), which corresponds to a rate of enlargement of a grain boundary crack or pore of $1.2\,\mathrm{\mu m\,day^{-1}}$. Rates at 500°C would be about an order of magnitude faster, and an alteration front could pass through the width of a dike in about 200 years. In practice, many sites of dissolution would operate simultaneously, and a dike could be fully altered in much less time than that. Rates for other minerals including clinopyroxene are similar although rarely measured at higher temperatures (Brantley & Chen 1995; Chen & Brantley 1998; Bandstra *et al.* 2008; Kaszuba *et al.* 2013).

Simple mineral dissolution is a good analogy to capture the general nature of the changes, but fails to account for the overall changes that have taken place. This is because the concentrations of the key rock-forming elements (Si and Al) in black smoker brines are much less than those of Ca, Fe and most other mineral-forming cations (Von Damm 1995), whereas simple mineral dissolution requires that they are dissolved stoichiometrically. The actual dissolution reactions likely involved simultaneous growth of secondary phases. The retention of the original dike geometry, combined with relict porosity, nevertheless suggests that there has been net removal of the material, and components such as Na and Mg have been quantitatively removed from the epidosites. In particular, Si was dissolved and then was returned to the rock as the void space was infilled, but there is no evidence for a significant change in fluid composition or flow regime to account for this change in behaviour. We can propose two alternative end-member hypotheses for the formation of the epidosite stripes. It is usually assumed that Si in black smoker systems is buffered at saturation by the presence of quartz (Wells & Ghiorso 1991). In fact, quartz is a rare phase in ocean floor alteration assemblages and recharging fluids may be below quartz saturation, but the abundance of quartz in epidosites suggests that saturation is reached or exceeded. Si content of black smoker systems is widely used as a geothermometer or geobarometer for the base of hydrothermal flow (Fontaine *et al.* 2009), showing that supersaturation of silica (and by inference other components) is normal during the rapid upflow of black smoker fluids to vents. In view of this, one hypothesis is that the epidosite stripes develop as primary igneous minerals rapidly dissolve while their secondary reaction products grow nearby. The reaction is driven by the instability of the primary assemblage under greenschist facies conditions, rather than by gradients in P, T or fluid composition.

The alternative hypothesis is that porosity in any one stripe is variable through time and the sequence of porosity generation followed by infilling reflects cooling of the flow system.

Over a wide range of P–T conditions, silica solubility in gas-poor aqueous fluids increases with pressure and temperature (prograde solubility), but retrograde behaviour occurs at temperatures above the critical point for sea water salinities, in a P–T region that is relevant to epidosite formation (Von Damm *et al.* 1991; Akinfiev & Diamond 2009). Steele-MacInnes *et al.* (2012a) showed that dissolution of quartz can occur in the deep upflow part of a black smoker system as the fluid cools in the retrograde solubility region. As the fluid cools further, silica precipitation will occur in the prograde solubility field.

In addition to cooling, silica solubility may also be influenced by pressure fluctuations. These may arise in black smoker systems from a number of causes, such as slip on spreading axis faults or by the precipitation and dissolution of anhydrite as fluid temperature rises and falls (Cann & Strens 1989; Tivey *et al.* 1995).

Importance of reaction permeability

It is doubtful whether the hydrothermal reaction zone close to the base of a black smoker system has ever been sampled in the modern oceans, but it is also possible that modern reaction zones have a different mineralogy from Cretaceous ones, perhaps due to changes in sea water chemistry. Modern ocean waters have a much higher Mg/Ca ratio than did the oceans of the Cretaceous (Lowenstein *et al.*, 2001; Holland 2005; Coogan 2009; Coggon *et al.* 2010). In a modern high-Mg/Ca ocean, the residual phases might be chlorite and quartz, rather than epidote and quartz. Porosity-generating processes similar to those documented here may well exist in systems with different mineralogy. The main driver for the process is the dissolution of primary minerals in black smoker fluid, rather than the precipitation of epidosite assemblages per se. This driver is present in any circumstance where fresh igneous minerals come into contact with rapidly flowing black smoker fluids, and the reactions between them are far from equilibrium. Depending on whether or not the existence of a specific range of P–T conditions in which silica has retrograde solubility is also essential (above-mentioned hypothesis 2), reaction permeability therefore may be a widespread feature of ocean floor alteration close to the spreading axis; it is probably particularly important in stock works beneath sulfide deposits, see, for example Steele-MacInnes *et al.* (2012b).

We are not suggesting here that reaction permeability replaces fracture permeability as the dominant control on fluid flow in the ocean crust, but that it is a major factor in the development of the most heavily metasomatized rocks in areas where fresh igneous rocks encounter actively flowing hydrothermal fluid. A single dike forms only a small part of the overall flow system, most of which will be less reactive. In most cases, the increase over fracture permeability caused by reactions would be transient and, while locally very high, might not be able to carry the volume of fluid seen in black smoker discharge. However, each dike might create a disturbance in the flow pattern and may be a reason for the shifting pattern of venting seen in some well-studied black smoker fields (Rona *et al.* 1993; Kelley *et al.* 2012).

One important aspect of reaction permeability is that it generates positive feedback and thereby allows further fluid access into the reacting parts of the dike and to previously unreacted minerals. The relatively constant chemistry and high metal content of black smoker fluids require continuous reaction with fresh hot rock (Pester *et al.* 2012); if fluid flowed only along a network of spaced fractures, this would not be possible, and at temperatures <400°C, lattice diffusion is too sluggish to deliver components into the fluid even on a grain scale (see, e.g. Ganguly 2002). The model documented here leads to both rapid and quantified access of fluid to primary minerals and also to continuous mining of remaining fresh rock by the fluid. A continuous access of hydrothermal fluid to fresh basalt is also required by the high ^{3}He/^{4}He ratios observed in both black smoker vent fluids and fresh mid-ocean ridge basalts (Craig & Lupton 1981; Graham *et al.* 1993). Reaction permeability is thus important both to the chemistry of the oceans and the formation of mineral deposits.

Regional implications

Within the epidosite, zones are many dikes metamorphosed to the same degree as the regional background alteration outside the epidosite zones. These dikes are interpreted here as the result of alteration at a lower flux of hydrothermal fluid than the intense flux that formed the epidosites. The low-degree dikes may be cut by later dikes containing epidosite stripes and hence altered at a higher fluid flux. On a wider scale, this is consistent with the model of Coogan (2008) that the regional metamorphism of the sheeted dikes is the result of widespread low-fluid-flux reactions with axial hydrothermal systems. The epidosites would thus be the markers of more intense hydrothermal alteration within a broad zone of axial hydrothermal circulation.

CONCLUSIONS

We have shown that the process of epidotization in the sheeted dike complex of the Troodos ophiolite was accompanied by the generation of significant porosity, probably at least 20% on the scale of a single thin section. We infer that the primary minerals were dissolved in hydrothermal fluid and then epidote, quartz and sometimes chlorite grew euhedrally into the resulting porosity. Epidosite is concentrated into stripes (i.e. bands) parallel to dike margins, separated by less altered (but still completely recrystallized) rock. Cross-cutting relationships show that each dike was epidotized before being cut by any later dike, indicating a maximum timescale of around 2000 years for epidotization and a more likely timescale as little as one-tenth of this. This requires a rapid self-organizing process. Numerical modeling of similar systems (Steefel & Maher 2009) shows that dissolution-induced permeability is a positive feedback process that leads to the development of porosity wormholes, which evolve into fingers of alteration.

We suggest that epidosite formed in the lower part of the upflow zone of a black smoker system, above the melt lens that would have supplied heat to the system. A newly intruded dike would cool very rapidly to the temperature of black smoker fluid, 350–400°C, and subsequently remain at that temperature throughout its vertical extent. The dike would have a significant initial permeability due to the density increase during crystallization and subsequent rapid cooling from 1000 to 400°C. The black smoker fluid would be far from equilibrium with the primary minerals in the dike and far in excess relative to the volume of the fluid in the dike. In far from equilibrium conditions, experimentally determined dissolution rates of minerals are extremely rapid, with data on clinopyroxene suggesting rates of up to a few microns per day in low-pH fluids at black smoker temperatures. Maintaining far from equilibrium conditions requires a rapidly flowing fluid and therefore a high permeability. Transport of major components in the fluid may have been aided by retrograde solubility in the vicinity of the critical point for H_2O and perhaps by pressure fluctuations due to the dynamic nature of porous medium convection. As soon as the porosity formed, it would begin to be occluded, and the highest permeability in the system would therefore be at the interface between fresh and altered rock. This gives a mechanism for initial pods of porosity within the dike to coalesce into layers, forming the characteristic stripes.

Both epidosites and less altered dikes in the Troodos ophiolite are strongly depleted in the metals that are enriched in massive sulfide deposits. If permeability is restricted to cracks as is widely assumed, it is extremely difficult for components in primary minerals to be released into the fluid at temperatures <400°C where volume diffusion is extremely sluggish. Reaction permeability provides a mechanism for rapid complete recrystallization, with access of fluid to all parts of the rock. In this respect, it may be a key part of the exchange of components between the lithosphere and the ocean–atmosphere system and the recycling of components back into the mantle at subduction zones.

The critical part of our model is the rapid dissolution of primary minerals in circulating hydrothermal fluid. Reaction permeability may be important in other parts of the ocean crust and in other environments where fresh igneous rocks are brought into contact with hydrothermal fluid. In such environments, the rock alteration may well proceed as rapidly as we demonstrate here, where the alteration takes place in the interval between the intrusion of one dike and another. This conclusion may shine a different light on many other examples of extreme metasomatism, often assumed to have happened at the slow speeds of continental geology.

ACKNOWLEDGEMENTS

We are indebted to the work of our students Jonathan Cowan (Newcastle upon Tyne PhD thesis, 1990), Jennifer Connell (Leeds MSc thesis 2000) and Graeme Penwright, who collected and documented large amounts of data on the epidosites of the Troodos ophiolite. This research was partly supported by NERC grant NE/I015035/1 to Andrew McCaig. The authors confirm that they have no conflict of interest.

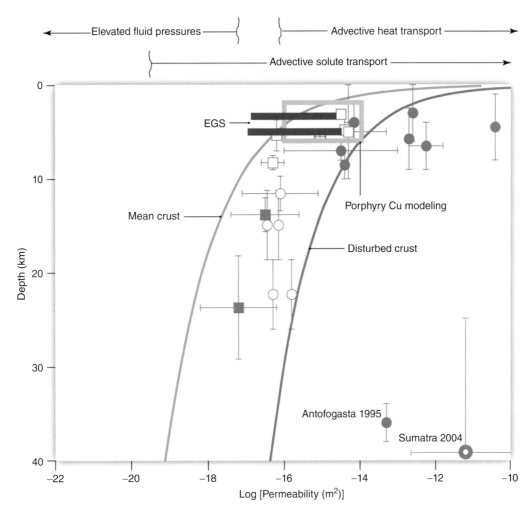

Fig. 1.1. Crustal-scale permeability (k) data. Arrows above the graph indicate approximate ranges of k over which certain geologically significant processes are likely. The "mean crust" k curve is based on k estimates from hydrothermal modeling and the progress of metamorphic reactions (Manning and Ingebritsen 1999). However, on geologically short timescales, k may reach values significantly in excess of these mean crust values (Ingebritsen and Manning 2010). The power-law fit to these high-k data–exclusive of the Sumatra datum (Waldhauser *et al.* 2012)–is labeled "disturbed crust." The evidence includes rapid migration of seismic hypocenters (solid circles), enhanced rates of metamorphic reaction in major fault or shear zones (open circles), recent studies suggesting much more rapid metamorphism than had been canonically assumed (solid squares), and anthropogenically induced seismicity (open squares); bars depict the full permissible range for a plotted locality and are not Gaussian errors. Red lines indicate k values before and after enhanced geothermal systems reservoir stimulation at Soultz (upper line) (Evans *et al.* 2005) and Basel (lower line) (Häring *et al.* 2008) and green rectangle is the k-depth range invoked in modeling the formation of porphyry-copper ores (Weis *et al.* 2012).

Crustal Permeability, First Edition. Edited by Tom Gleeson and Steven E. Ingebritsen.
© 2017 John Wiley & Sons, Ltd. Published 2017 by John Wiley & Sons, Ltd.
Companion Websites: www.wiley.com/go/gleeson/crustalpermeability/
http://crustalpermeability.weebly.com/

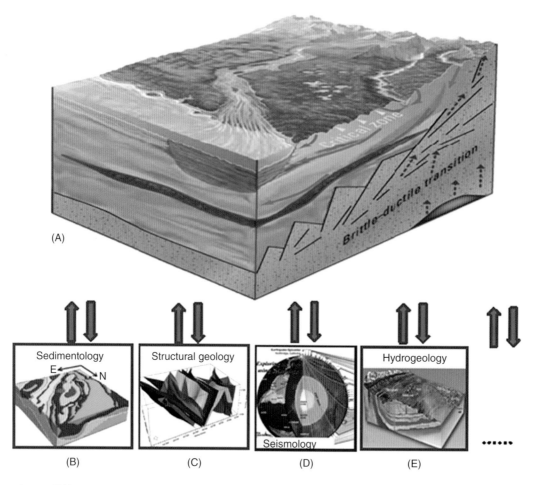

Fig. 2.1. The geologic scaffolding of DigitalCrust from the critical zone to the brittle–ductile transition (A), receiving contribution from and delivering service to a wide range of Earth science disciplines (B–E). (Image source: (A) Adapted from Winter *et al.* (1998), (B) McInerney *et al.* (2005), (C) Hinz *et al.* (2012), (D) IRIS (http://www.iris.edu/hq/), and (E) Paschke *et al.* (2011)).

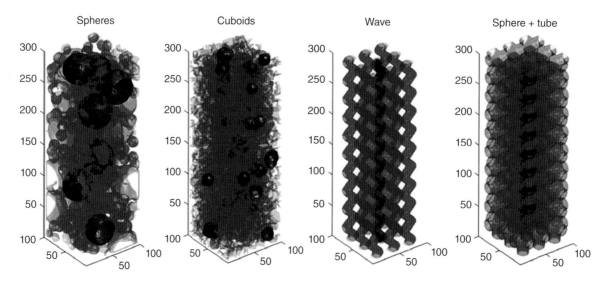

Fig. 4.1. Pore-scale representations of the four porous media. Warm colors (red) show larger pores. All textures are synthetically constructed; spheres and cuboids are constructed with a stochastic nucleation and growth algorithm following the procedure described in Hersum and Marsh (2006).

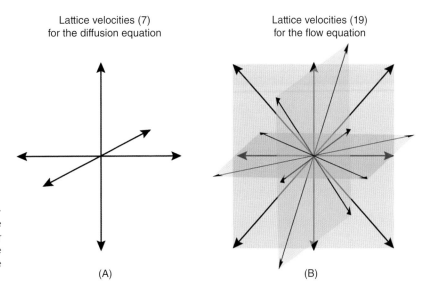

Lattice velocities (7)
for the diffusion equation

Lattice velocities (19)
for the flow equation

Fig. 4.2. Diagram showing the velocity discretization used for the lattice Boltzmann modeling. (A) The seven-velocity e_i (including a rest velocity e_0) model for the calculation of the formation factor is shown. (B) The 19-velocity model c, used for the flow calculations at the pore scale, is illustrated.

(A)   (B)

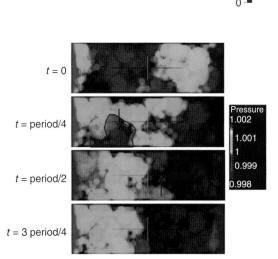

Fig. 4.4. Thermal field in the porous medium (cuboids) at steady state. Here, the solid fraction is assumed as a perfect insulator. The formation factor is computed from the effective thermal conductivity of the medium at steady state.

Temperature

1

0

$t = 0$

$t = $ period/4

$t = $ period/2

$t = 3$ period/4

Pressure
1.002

1.001

1

0.999

0.998

Fig. 4.11. 3D visualization of the pore-pressure field (normalized) at the forcing frequency corresponding to the maximum of the resonance peak in the spheres medium. The four images show the temporal evolution of the pressure field every quarter period. Note the regions with large lateral pore-pressure gradients (the imposed gradient is left to right) highlighted in red. It shows that flow pathways with different hydraulic connectivity have different response times to the forcing and that large pore-scale pore-pressure imbalance can emerge, violating the assumption of planar pressure wave propagating from the outlet.

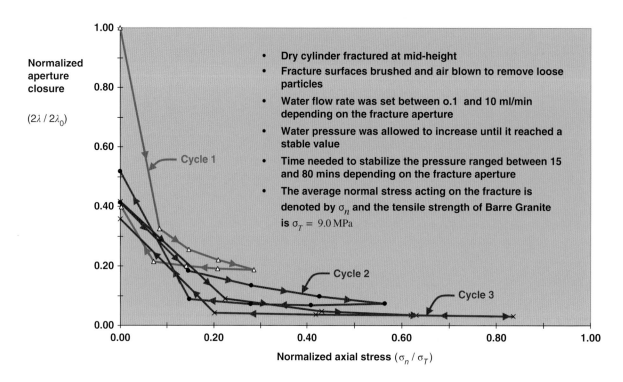

Fig. 6.7. Fracture closure during application of normal stresses. The starting position of the fracture during cycle 3 does not coincide with the end position of cycle 2. This is most likely due to unloading effects and the release of elastic stain energy in the system. The discrepancy, however, does not affect the third cycle of loading and the aperture closure trend with increasing axial stress is consistent with that observed in cycles 1 and 2.

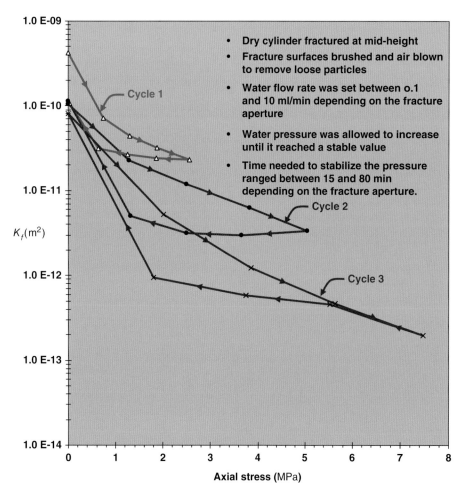

Fig. 6.8. Evolution of the permeability of the fracture during quasi-static application of loading–unloading cycles.

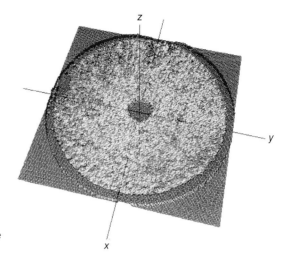

Fig. 6.9. Fracture topography mismatch after three cycles of axial compression normal to the plane of the fracture to a maximum stress of 7.5 MPa.

0.05 m × 0.075 m

0.1 m × 0.15 m

Flow direction

Direction of
shear displacement

0.2 m × 0.3 m

30 MPa confining stress

Aperture (μm)

10^0 10^1 10^2 10^3 10^4

Normalized flow rate (–)

10^{-2} 10^{-1} 10^0

(A)   (B)

Fig. 7.5. Representative results for channeling flow within the heterogeneous aperture distribution of real laboratory-scale fractures (A) with no shear displacement and (B) with a shear displacement of 5 mm. (Reproduced from Ishibashi *et al.* (2015) by permission of John Wiley and Sons Ltd.)

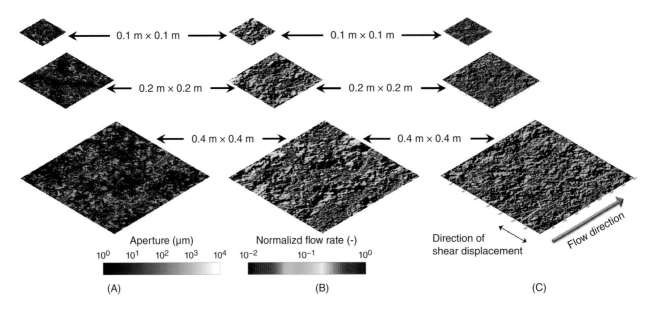

0.1 m × 0.1 m

0.1 m × 0.1 m

0.2 m × 0.2 m

0.2 m × 0.2 m

0.4 m × 0.4 m

0.4 m × 0.4 m

Aperture (μm)

10^0 10^1 10^2 10^3 10^4

Normalizd flow rate (-)

10^{-2} 10^{-1} 10^0

Direction of shear displacement

Flow direction

(A)

(B)

(C)

Fig. 7.6. Representative results for channeling flow within the heterogeneous aperture distribution of synthetic fractures (A) with no shear displacement, (B) with a shear displacement of 5 mm, and (C) with a shear displacement of constant $\delta/l (\delta/l = 0.01)$. (Reproduced from Ishibashi *et al.* (2015) by permission of John Wiley and Sons Ltd.)

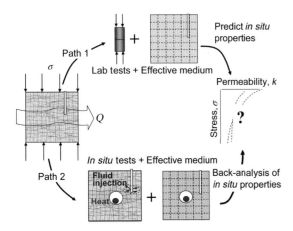

σ

Path 1

Lab tests + Effective medium

+

Predict *in situ* properties

Permeability, k

?

Stress, σ

Q

Path 2

In situ tests + Effective medium

Fluid injection

Heat

+

Back-analysis of *in situ* properties

Fig. 8.1. Two alternative ways (Path 1 and Path 2) for deriving a stress–permeability relationship of a fractured rock unit. Path 1 involves laboratory testing on single fractures and an effective medium theory, whereas Path 2 involves back analysis by model calibration against field data.

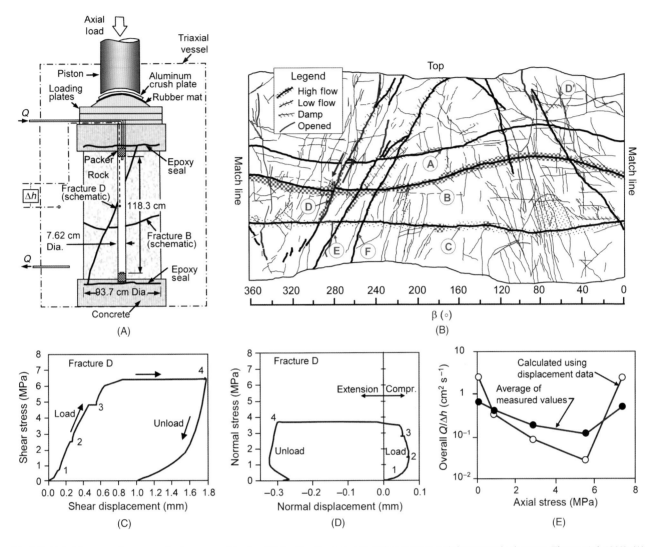

Fig. 8.6. Uniaxial compression on an ultra-large core of fractured Stripa Granite conducted at the Lawrence Berkeley National Laboratory (Thorpe *et al.* 1982). (A) Schematic of test arrangement, (B) fracture mapping and seepage during a hydraulic test conducted before the unixial compression test, (C) shear stress versus shear displacement, (D) normal stress versus normal displacement, and (E) flow versus axial stress. (Adapted from Thorpe *et al.* 1982.)

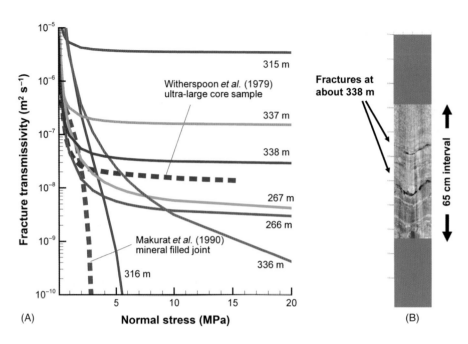

Fig. 8.8. Back-calculated hydromechanical properties of fractures intersecting a borehole at a crystalline rock site in Sweden (Rutqvist *et al.* 1997): (A) Fracture transmissivity versus effective normal stress for fractures at depths between 266 and 338 m (solid lines) with comparison to results from *in situ* block experiment by Makurat *et al.* (1990a) and ultra-large core experiment by Witherspoon *et al.* (1979) (dashed lines), (B) borehole image indicating open fractures at about 338 m depth.

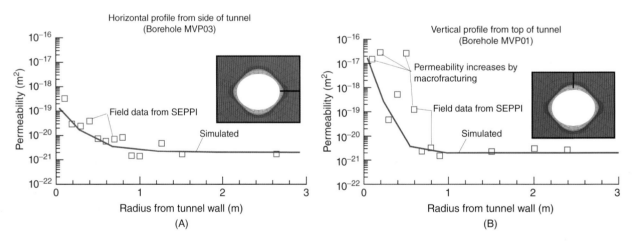

Fig. 8.9. Calculated and measured permeability changes around the TSX tunnel (Rutqvist *et al.* 2009b). Permeability versus radius along (A) a horizontal profile from the side of the tunnel and (B) a vertical profile from the top of the tunnel.

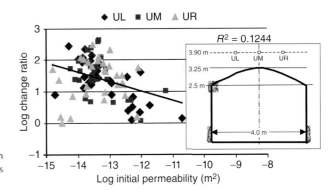

Fig. 8.11. Results of pre- to postexcavation air-permeability tests above a niche in fractured unsaturated tuff. The results shown are for a niche with three boreholes (UL, UM, and UR) located above niche 3560 (Wang *et al.* 2001).

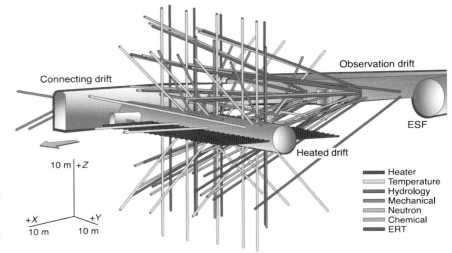

Fig. 8.16. Three-dimensional view of the Yucca Mountain drift scale test. The color-coded lines indicate boreholes for various measurements of thermally driven thermal–hydrological–mechanical–chemical responses. (Tsang *et al.* 2009; Rutqvist & Tsang 2012.)

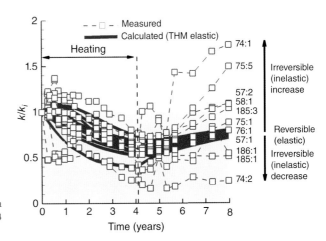

Fig. 8.17. Schematic of the range of measurements of air permeability at the Yucca Mountain drift scale test. Details of calculated and measured responses in all 44 measurement intervals can be found in Rutqvist *et al.* (2008)

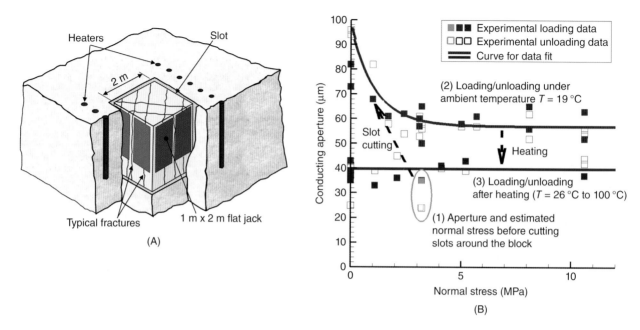

Fig. 8.20. (A) Perspective of G-tunnel heated block experiment (Adapted from Zimmerman *et al.* 1985) and (B) hydraulic conducting aperture as a function of normal stress evaluated from the G-tunnel *in situ* block experiment. The data from Zimmerman *et al.* (1985) are separated in the sequential steps showing additional fracture closure as a result of heating. (Rutqvist & Tsang 2012.)

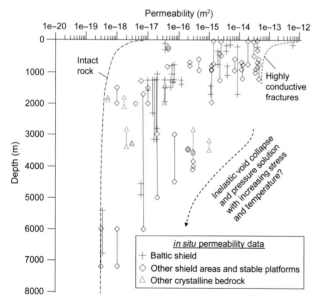

Fig. 8.22. Compilation of permeability measurements in boreholes in crystalline bedrock (from Juhlin *et al.* 1998) with added schematic of upper and lower limits of permeability related to mechanical and chemomechanical behavior.

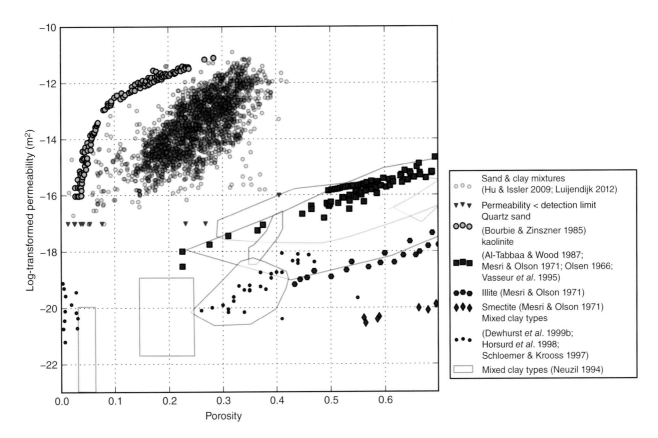

Fig. 10.1. Porosity and permeability data of sand–clay mixtures and pure sands and clays.

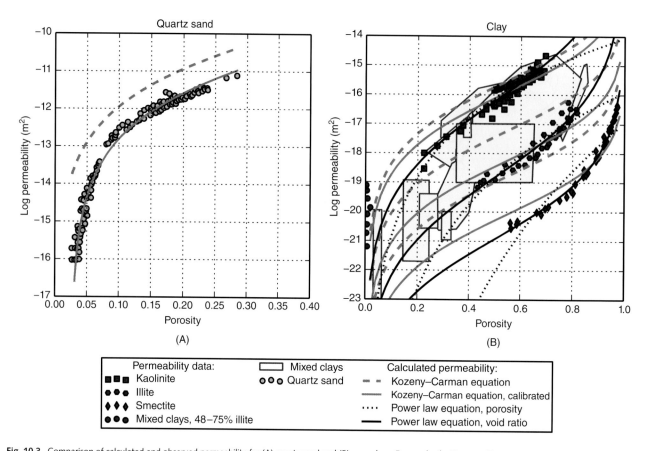

Fig. 10.3. Comparison of calculated and observed permeability for (A) quartz sand and (B) pure clays. For sands, the Kozeny–Carman equation (Eq. 10.1) reproduces the data well, but only when the equation includes a percolation threshold and the value of the specific surface is calibrated. For clays, the permeability data are closely matched when permeability is calculated as a power law function of the void ratio (Eq. 10.4). Data for sands were reported by Bourbie & Zinszner (1985). Permeability data for pure clays were obtained from Al-Tabbaa & Wood (1987), Mesri & Olson (1971), Olsen (1966), and Vasseur *et al.* (1995). The figure also shows data on mixed clay types from Schloemer & Krooss (1997) and Neuzil (1994) that were not used to calibrate the porosity–permeability equations. See Table 10.2 for the fit statistics of the permeability equations and Table 10.3 for calibrated parameter values.

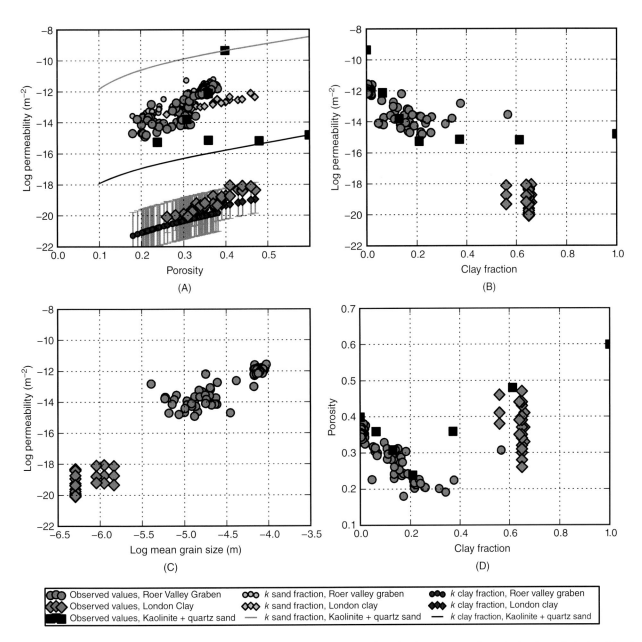

Fig. 10.4. Relation of permeability to (A) porosity, (B) clay content, and (C) mean grain size and (D) the relation between clay content and porosity for two data sets of natural sand–clay mixtures and one experimental data set that consists of a mixture of kaolinite and quartz sand with a uniform grain size. The data for natural sediments were derived from unconsolidated shallow marine sands in the Roer Valley Graben (Heederik 1988) and the London Clay in southeast England (Dewhurst *et al.* 1999a). The experimental data were reported by Knoll (1996). The calculated permeabilities of the clay and sand fraction of each sample of the Roer Valley Graben and London Clay data sets are also shown in (A). The permeabilities of the Roer Valley Graben and the London Clay data sets are relatively close to the calculated permeabilities of their sand and clay fractions, respectively. The error bars for the clay fraction reflect the uncertainty in the mineral composition.

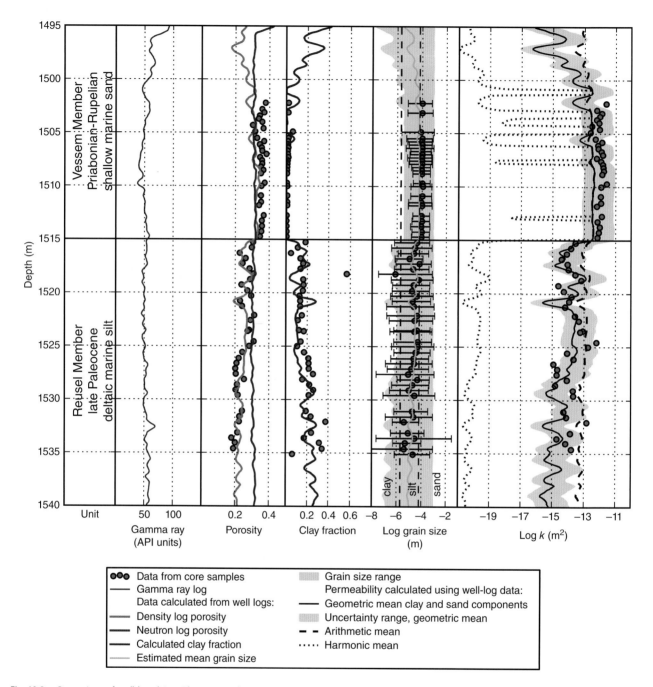

Fig. 10.9. Comparison of well-log data with porosity, clay content, and permeability from core samples in well AST-02. Permeability was calculated using well-log-derived estimates of porosity, clay content, and grain size distribution. Grain size distribution was calculated using the empirical correlations between clay content and grain size distributions shown in Figure 10.2C and D. The calculated permeability shows a relatively good match with observed permeability and estimates permeability within 1 order of magnitude for 80% of the samples. The uncertainty range of the calculated permeability averages ±1.0 orders of magnitude and was calculated using minimum and maximum estimates of clay mineralogy and grain size distribution.

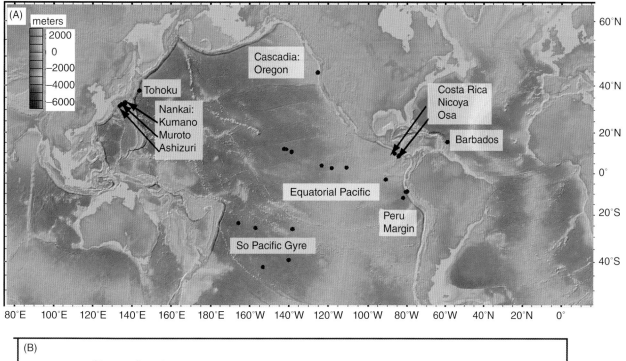

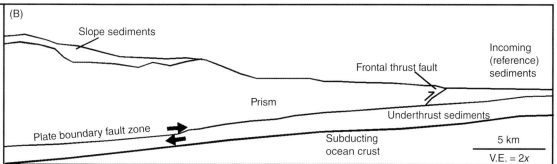

Fig. 11.1. (A) Locations of samples used in this investigation. (B) Schematic of a shallow subduction zone showing domains. This figure was based on an accretionary system. For simplicity, only the plate boundary and frontal thrust are shown. In reality, the prism is extensively faulted. In an erosive system, only the seawardmost portion of the upper plate would be a sediment prism, typically consisting of reworked slope sediments. Scale bar gives example dimensions and vertical exaggeration (V.E.).

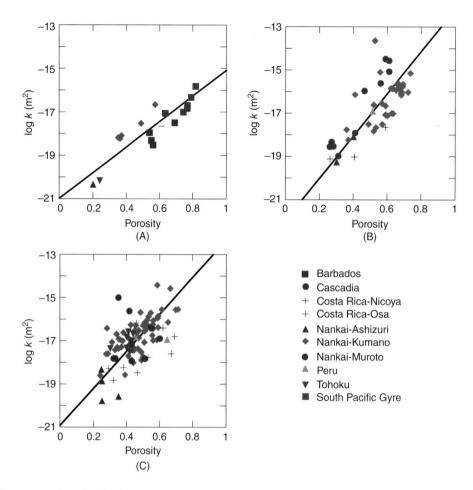

Fig. 11.2. Permeability–porosity relationships for clays. (A) Group 1 (> 80% clay). (B) Group 2 (60–80% clay). (C) Group 3 (<60% clay). Best-fit lines are shown. Regression coefficients are given in Table 11.2. Tohoku data shown were not used for the best-fit lines.

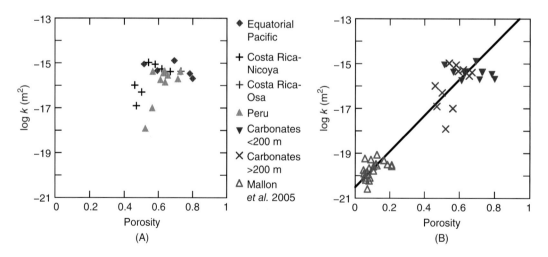

Fig. 11.5. (A) Permeability of carbonate oozes and chalks by location. (B) Results separated by burial depth and combined with deep sediment data from Mallon *et al.* (2005). Regression coefficients are given in Table 11.2.

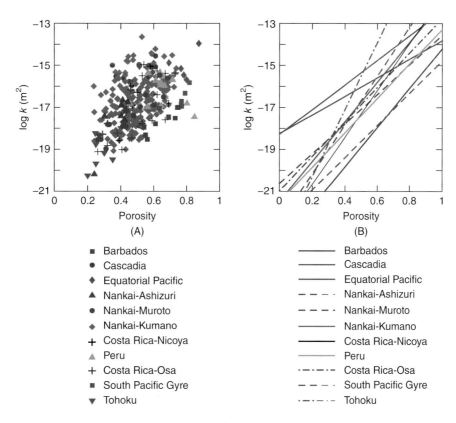

Fig. 11.6. Comparison by location. (A) Data. (B) Trends. Regression coefficients are given in Table 11.2.

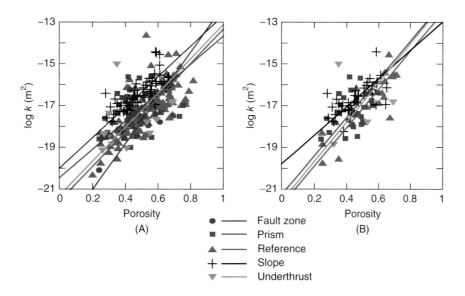

Fig. 11.7. Comparison by structural domain. (A) Data for all clay samples. (B) Comparison restricted to Group 3 samples. Regression coefficients are given in Table 11.2.

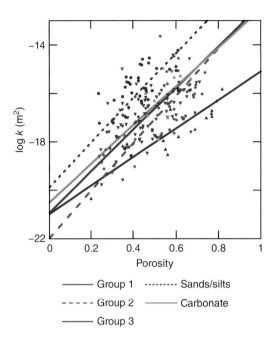

Fig. 11.12. Compilation of permeability data and best-fit lines for each group. Carbonate best-fit line only for samples > 200 mbsf and includes deeper data from Mallon *et al.* (2005), which are shown in Figure 11.10B.

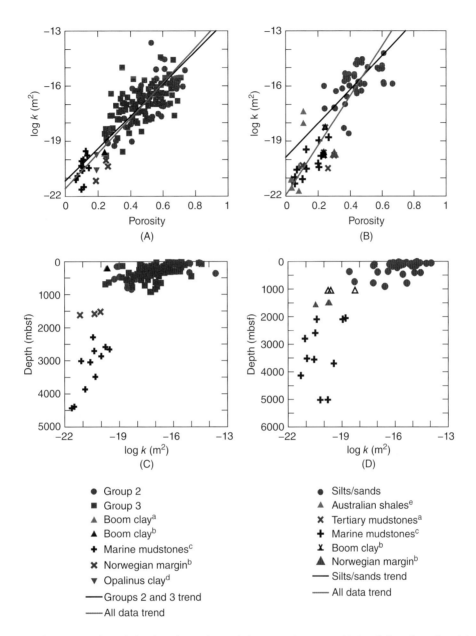

Fig. 11.13. (A) Comparison of Groups 2 and 3 with data from deep sediments belonging to Groups 2 and 3. Best-fit lines shown for subduction zone sediments (black) and all data (green). (B) Comparison of silt/sand data with deep silt/sand sediment data. Best-fit lines shown for subduction zone sediments (black) and all data (green). All regression coefficients are given in Table 11.2. (C) Permeability plotted against depth for Groups 2 and 3 and deep sediments showing depth range of deep sediments. (D) Permeability plotted against depth for silt/sand data and deep silt/sand samples. Note that the Boom Clay depth reported is the present-day depth; this formation is presumed to have been uplifted (Hildenbrand *et al.* 2002). Data sources: [a]Hildenbrand *et al.* (2002), [b]Hildenbrand *et al.* (2004), [c]Yang & Aplin (2007), [d]Marschall *et al.* (2005), [e]Amann-Hildenbrand *et al.* (2013).

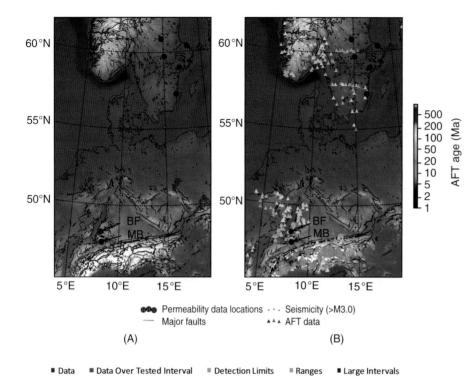

Fig. 12.1. Locations of permeability data and indicators of (A) short-term (years) and (B) long-term (million years) tectonic activity. Permeability data are derived from southern Germany and the Black Forest (BF), the Molasse basin (MB) in Switzerland and the Fennoscandian Shield (FS) in Sweden. Seismic events in (A) denote events since the year 2000 that exceed the magnitude of 3 on the Richter scale from the National Earthquake Information Center (http://earthquake.usgs.gov/regional/neic/). (B) AFT denotes apatite fission track data obtained from Herman *et al.* (2013). Apatite fission track data are a proxy for long-term tectonic activity. The apatite fission track age is approximately equal to the last time the rock outcrop was at a temperature of 120°C (Wagner & Reimer 1972), which at normal geothermal gradients corresponds to a depth of approximately 3–5 km.

Legend (A):
⬤⬤⬤ Permeability data locations   ··· Seismicity (>M3.0)
— Major faults   ▲▲▲ AFT data

(A)   (B)

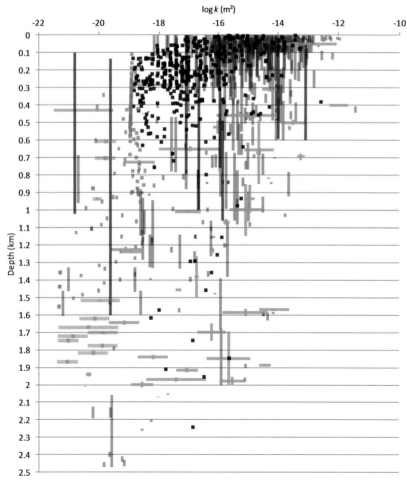

■ Data   ■ Data Over Tested Interval   ■ Detection Limits   ■ Ranges   ■ Large Intervals

Fig. 12.2. Full data set of permeability data for crystalline rock (n = 973). Black points are singular or average permeability values (n = 422). Red lines are permeability values reported over a tested interval (n = 426). Gray points are data with reported detection limits (n = 80). Green points are the midpoints of permeability values reported as ranges, with the error bar showing the range (n = 37). Purple lines are data with tested intervals > 500 m (n = 8). The vertical extent of a point indicates the extent of the tested interval.

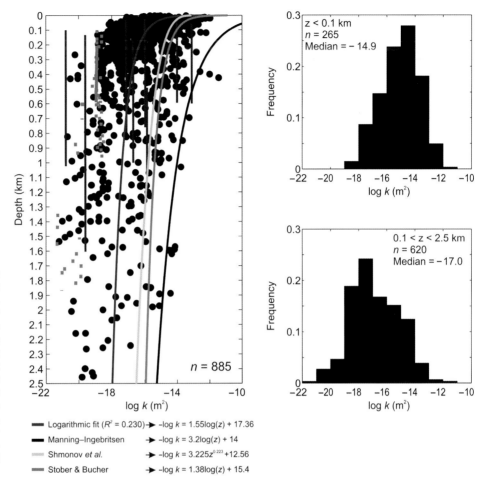

Fig. 12.3. The relationship between permeability and depth for the full data set, with error bars removed for clarity. Ranges are plotted as the midpoint. Gray rectangles indicate measurements at a detection limit. Purple lines indicate data points from tested intervals greater than 500 m. Red line indicates logarithmic fit through data ($R^2 = 0.230$). Black line indicates Manning–Ingebritsen fit (Ingebritsen & Manning 1999). Blue line indicates Shmonov *et al.* (2003) fit. Green line indicates Stober & Bucher (2007a) fit. Histograms display distribution of permeability data above and below 0.1 km.

Logarithmic fit ($R^2 = 0.230$) ➤ $-\log k = 1.55\log(z) + 17.36$

Manning–Ingebritsen ➤ $-\log k = 3.2\log(z) + 14$

Shmonov *et al.* ➤ $-\log k = 3.225z^{0.223} + 12.56$

Stober & Bucher ➤ $-\log k = 1.38\log(z) + 15.4$

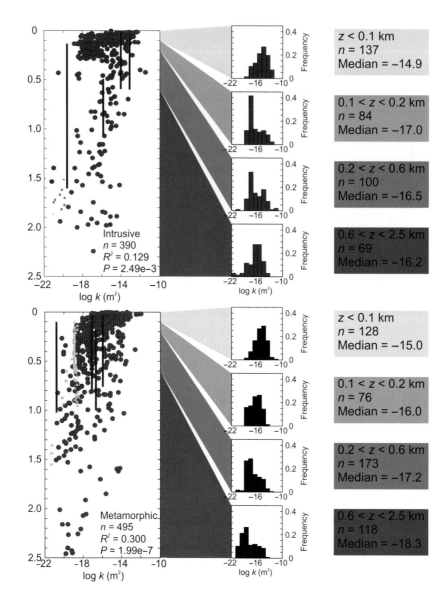

Fig. 12.5. The relationship between permeability and lithology for metamorphic (blue) and intrusive (red) rocks. All data points are midpoints of tested intervals. Pink rectangles indicate intrusive detection limits. Cyan rectangles indicate metamorphic detection limits. Purple lines indicate data points from tested intervals > 500 m. Reported R^2 and P values are for logarithmic fits through data. Histograms identify the permeability distribution in four depth ranges. From top to bottom: <100, 100–200, 200–600, and > 600 m.

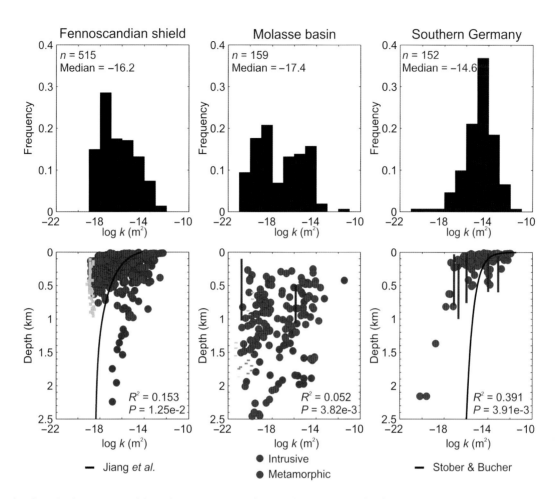

Fig. 12.6. The relationship between permeability and tectonic setting. Red points indicate intrusive rocks. Blue points indicate metamorphic rocks. Pink rectangles indicate intrusive detection limits. Cyan rectangles indicate metamorphic detection limits. Purple lines indicate data points from tested intervals > 500 m. All data points are midpoints. Reported R^2 and P values are for logarithmic fits through the combination of intrusive and metamorphic data. Gray lines are functions from the literature (Stober & Bucher 2007a; Jiang *et al.* 2010c).

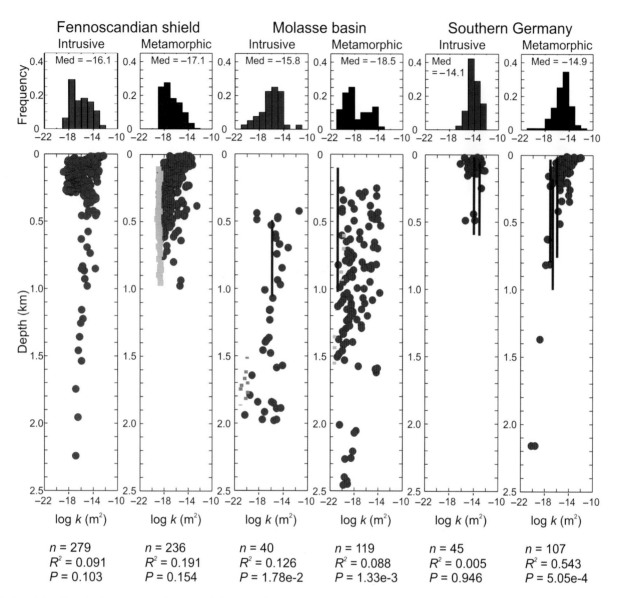

Fig. 12.7. The relationship between permeability and lithologies in different tectonic settings. Red indicates intrusive rocks. Blue indicates metamorphic rocks. Pink rectangles indicate intrusive detection limits, while cyan rectangles indicate metamorphic detection limits. Purple lines indicate data points from tested intervals > 500 m. All data points are midpoints. Reported R^2 and P values are for logarithmic fits through the data. P values in boldface indicate data sets that fail the t test at 5% significance. Histograms include text that indicates the median value of the distribution.

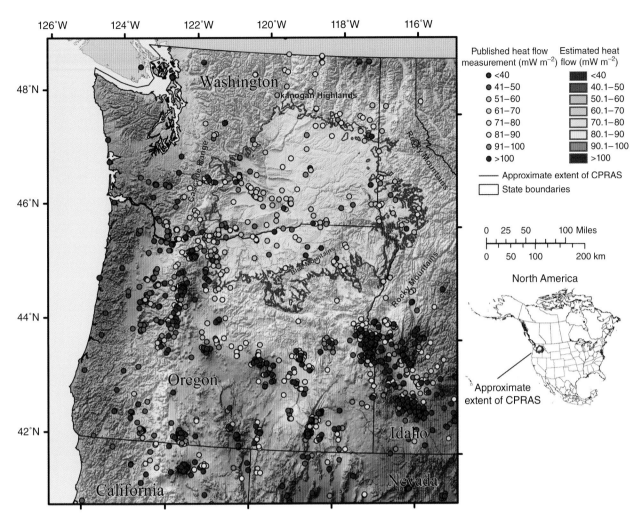

Fig. 13.1. Heat-flow map of the northwestern United States showing apparent low heat flow in the vicinity of the Columbia Plateau Regional Aquifer System. Contours are based on published heat-flow data for the Cascade Range and adjacent regions assembled for the USGS geothermal database. The map was constructed using the methods of Williams & DeAngelo (2008). The *Approximate Extent of Columbia Plateau Regional Aquifer System* is the extent of Columbia River Basalts from the geologic model of Burns *et al.* (2011).

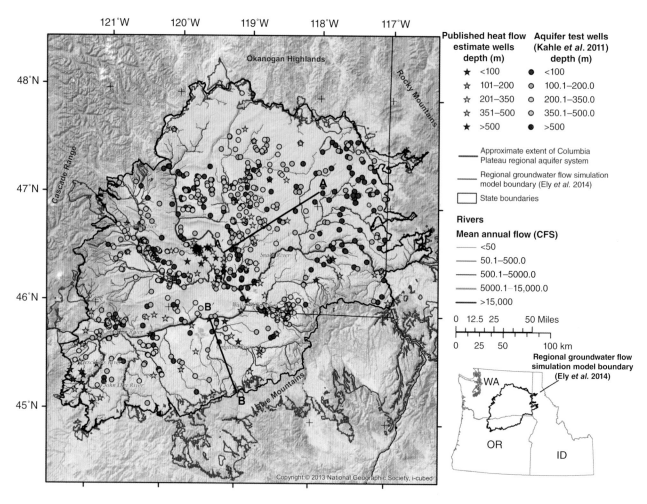

Fig. 13.2. Map of compiled borehole temperature logs (stars) and aquifer tests from groundwater supply wells (circles). The bottom of each temperature log is frequently at or above the depth of nearby aquifer tests, indicating that most temperature data were collected within the active groundwater system. To estimate intrinsic permeability from aquifer tests, the temperature of water pumped was estimated using the nearest temperature log. Representative cross sections A–A' and B–B' were used for coupled groundwater and heat-flow simulations.

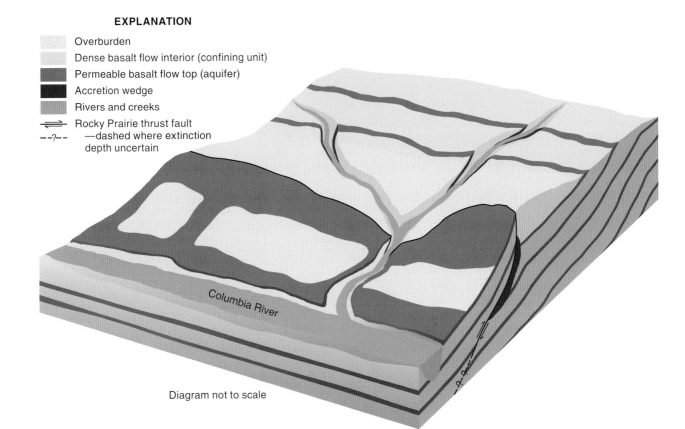

EXPLANATION

- Overburden
- Dense basalt flow interior (confining unit)
- Permeable basalt flow top (aquifer)
- Accretion wedge
- Rivers and creeks
- Rocky Prairie thrust fault
  —dashed where extinction depth uncertain

Columbia River

Diagram not to scale

Fig. 13.3. Conceptual model of aquifer system geometry (from Burns *et al.* 2012a). Upland recharge can enter thin aquifers at flow margins. Geologic structures can act as flow barriers that may or may not crosscut all aquifers.

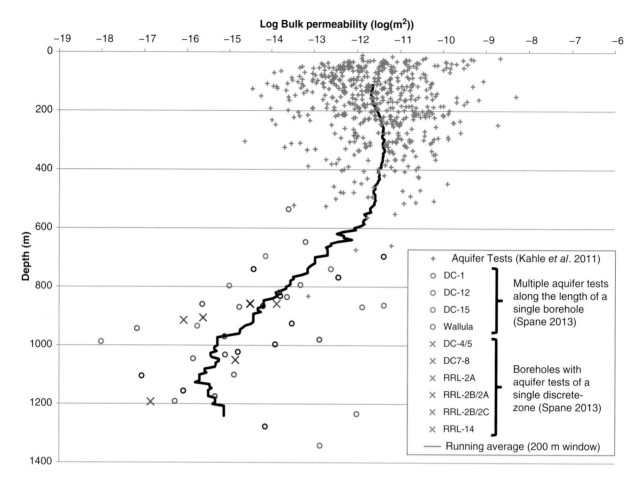

Fig. 13.4. Estimates of bulk horizontal permeability from Kahle *et al.* (2011) and Spane (2013). Using a 200 m window to compute a running average identifies a depth range of 600–900 m where permeability rapidly decreases. Above 600 m depth, average permeability is approximately constant. The deep data are sparse, but below 900 m depth, the rate of decrease in permeability apparently slows. Circles of the same color denote packer tests at different elevations within the same borehole. All other symbols represent single tests within a discrete borehole.

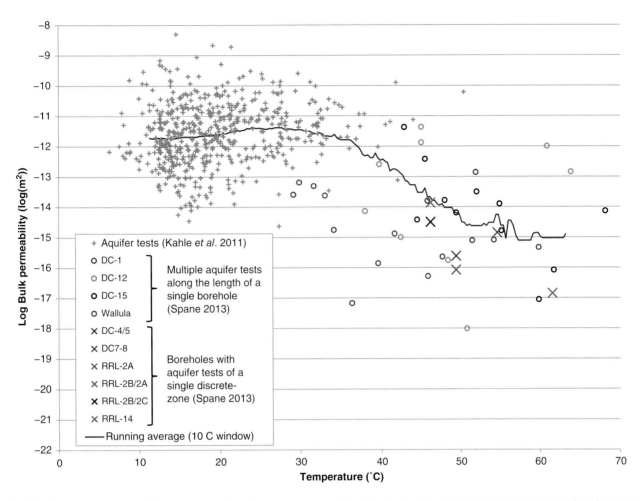

Fig. 13.5. Plot showing that permeability transitions rapidly in the temperature range 35 – 50°C. For the Kahle data, temperatures were estimated from the nearest temperature log (Fig. 13.2). For the remaining aquifer tests, temperatures were estimated from temperature logs collected within the borehole during a battery of geophysical tests.

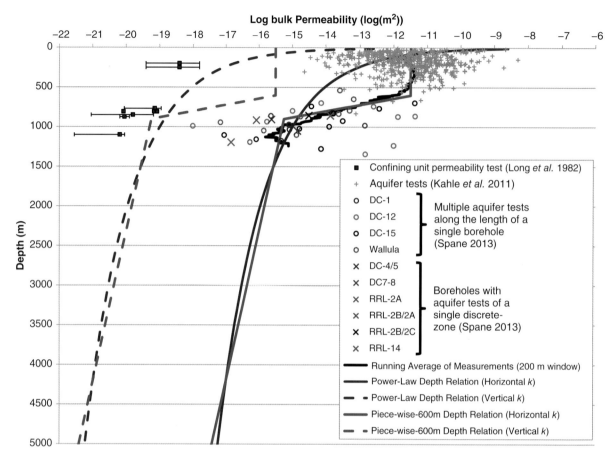

Fig. 13.6. Estimated permeability vs. depth relations for use in the SUTRA models. The solid lines are the maximum (subhorizontal) permeability, and the dashed line is minimum (subvertical) permeability. The red lines are the power-law depth relation. The blue lines are the piece-wise-600 m depth relation. The constant-permeability depth relation model is not shown, but has a log permeability of −11.55, indistinguishable from the shallow exponential model value of −11.5.

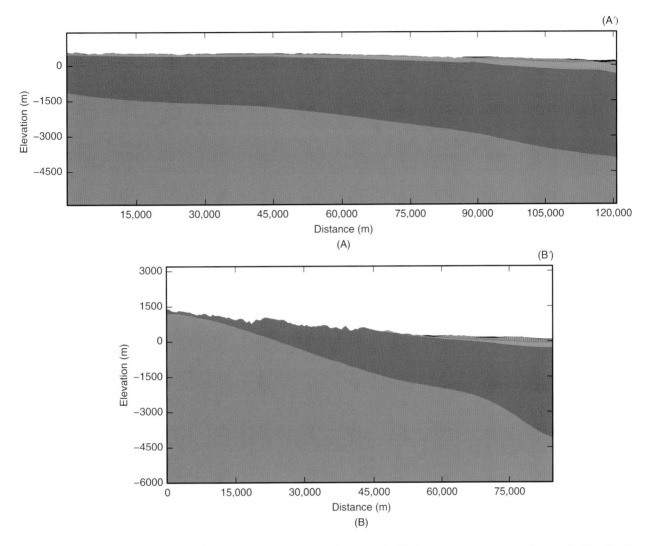

Fig. 13.7. Typical cross sections through the Columbia Plateau Regional Aquifer System (see Fig. 13.2). For both cross sections (units in m), the Columbia River is at the right-hand boundary, incised 250–300 m into the geologic units. (A) Cross section through the relatively gently dipping, low-recharge Palouse Slope. (B) Cross section from the crest of the higher recharge Blue Mountains through the Umatilla River basin. (Blue, pre-Miocene rocks; brown, Grande Ronde Basalts; green, Wanapum Basalts; red, Saddle Mountains Basalts; black, thin deposits of sedimentary overburden).

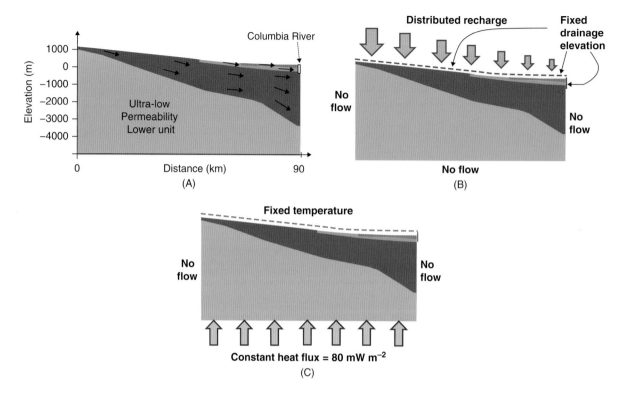

Fig. 13.8. Model formulation and boundary conditions: (A) hydrogeologic units with arrows showing that the preferential direction of permeability is rotated to align with the plane of basalt deposition (Blue, pre-Miocene rocks; brown, Grande Ronde Basalts; green, Wanapum Basalts; red, Saddle Mountains Basalts); (B) hydrologic boundary conditions showing that recharge and discharge occur near the upper boundary and into the river on the right-hand side; (C) heat-flow boundary conditions showing the constant heat flux at the lower boundary and the prescribed temperature at land surface, with cooler temperatures in the uplands to the left. The thickness of the underlying ultralow-permeability unit was varied during model testing, with the final thickness being 20 km (figure not to scale for this unit).

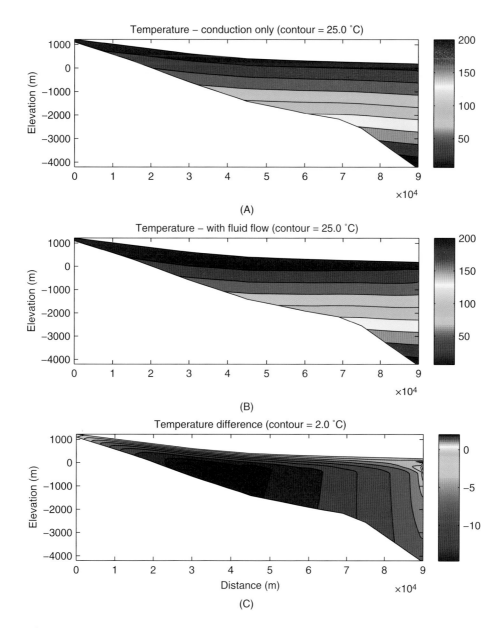

Fig. 13.9. Comparison of simulated temperatures for cross section B–B′ (Umatilla) using the piece-wise-600 m depth relation. (A) shows the simulated subsurface temperature distribution assuming heat conduction only (no fluid flow). (B) shows the simulated temperature distribution for combined groundwater and heat flow. (C) shows the temperature difference between the two simulations; warmer colors show areas where temperatures are elevated by advective transport of heat, and cooler colors show areas where subsurface temperatures are lowered by heat advection. Plots are only colored in the part of the domain occupied by Columbia River Basalt Group units.

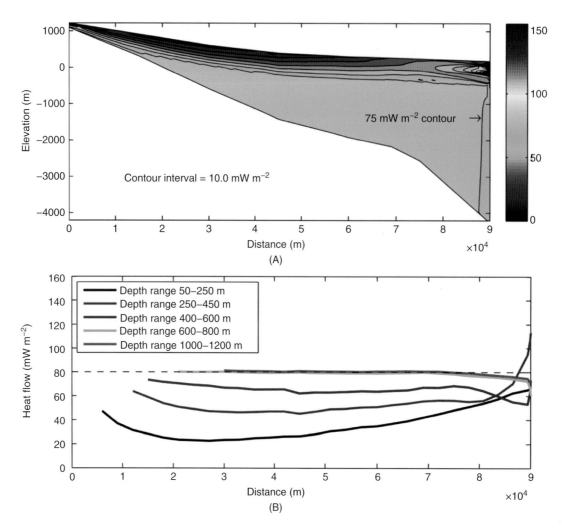

Fig. 13.10. Illustrations showing estimated vertical conductive heat flow for cross section B–B′ (Umatilla) for the piece-wise-600 m depth relation: (A) local estimated vertical conductive heat flow and (B) estimated vertical heat flow for five depth ranges, computed as the bulk thermal conductivity times the representative gradient (slope of the best-fit line) of all temperatures in each depth range. Below approximately 600 m depth, heat flow is dominated by conduction. Plot A is only colored in the part of the domain occupied by Columbia River Basalt Group units.

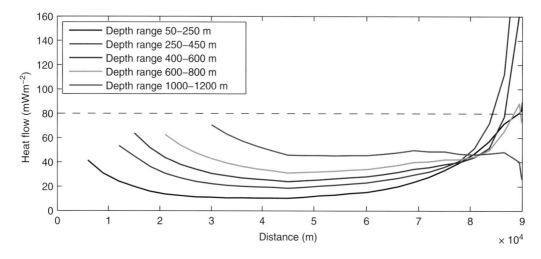

Fig. 13.11. Illustrations showing estimated vertical heat flow for cross section B–B′ (Umatilla) for the constant-permeability depth relation. Estimated vertical heat flow for five depth ranges, computed as the bulk thermal conductivity times the representative gradient (slope of the best-fit line), of all temperatures in each depth range (compare with Fig. 13.10B).

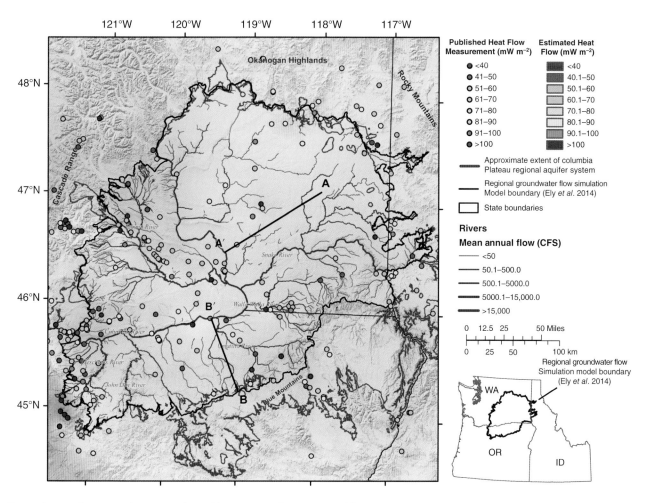

Fig. 13.12. Heat-flow map with rivers; river line width is proportional to mean annual flow rate. Larger rivers are at the lowest elevations and receive water from long, regional groundwater flow paths.

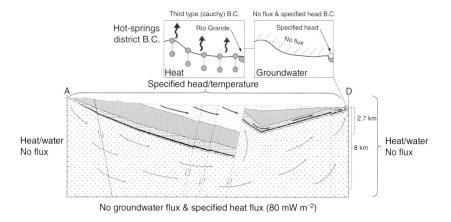

Fig. 14.2. Boundary conditions used in our two-dimensional hydrothermal model and a schematic diagram showing the two groundwater flow hypotheses evaluated as part of this study. Boundary conditions are shown for both heat transport and groundwater flow. The inset portrays a close-up of the boundary conditions applied to the hot-springs district. The basement-circulation hypothesis (blue arrows) involves deep circulation of groundwater within highly fractured crystalline basement rocks. Groundwater discharges where hydrologic windows exist in overlying confining units, such as the Percha Shale (black). The shallow-circulation hypothesis (red arrows) considers shallow groundwater circulation through the carbonate Magdalena Group.

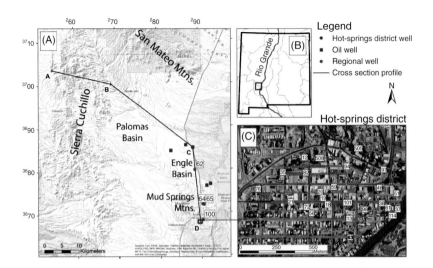

Fig. 14.3. Basemap (A) showing surface water drainages (light blue lines) and the location of the study area in south-central New Mexico (B). The presumed recharge area of the Sierra Cuchillo and San Mateo Mountains north of Truth or Consequences is also indicated for reference. The orientation of the geologic cross section and two-dimensional hydrothermal model is shown in black (A–B–C–D) in addition to the locations of wells discussed in this paper. The inset of the hot-springs district (C) shows geothermal well locations. The Rio Grande can also be seen in the lower right corner of this inset. The delineated areas on the New Mexico state map (B) are major drainage basins. Regional map coordinate datum is UTM NAD83 Zone 13.

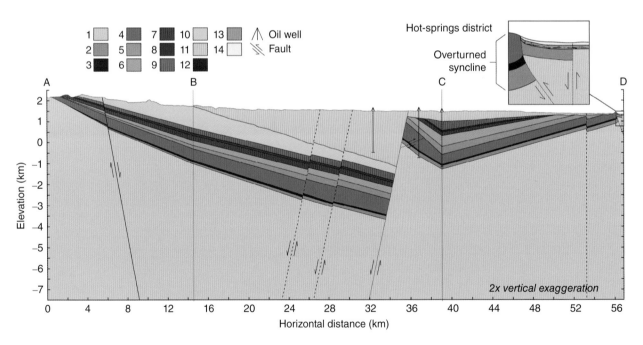

Fig. 14.4. Geologic cross section depicting the stratigraphic units used in our two-dimensional hydrothermal model. Additional information about model parameters is provided in Table 14.1. The color legend of this figure corresponds to the hydrostratigraphic units and descriptions in Table 14.3. The cross section was constructed by utilizing oil-well data, an east–west regional cross section (Lozinsky 1987), gravity data (Gilmer *et al.* 1986), and surface geologic maps.(Harrison *et al.* 1993; Harrison & Cather 2004.)

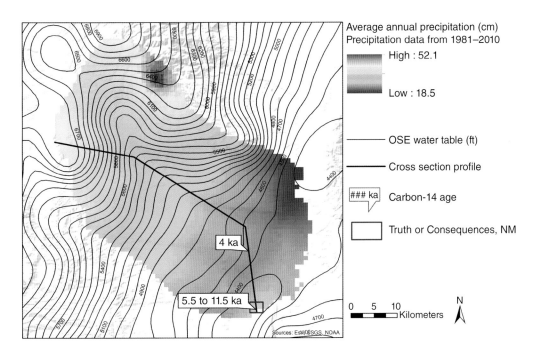

Fig. 14.5. Groundwater contributing area (color-shaded contours) to the hot-springs district (HSD) and water-table contours in relation to our two-dimensional hydrothermal model transect (bold black line). The color contours denote the spatial distribution of annual precipitation across the watershed. The black contour lines are estimated water-table elevations from the New Mexico Office of the State Engineer. Precipitation and water-table patterns suggest recharge to the HSD primarily occurring in the Sierra Cuchillo and San Mateo Mountains northwest of Truth or Consequences. Apparent carbon-14 groundwater ages collected as part of this project are displayed on the basemap as well (annotated circles). Oldest groundwater ages are within the HSD (precipitation data from PRISM Climate Group, Oregon State University 2012).

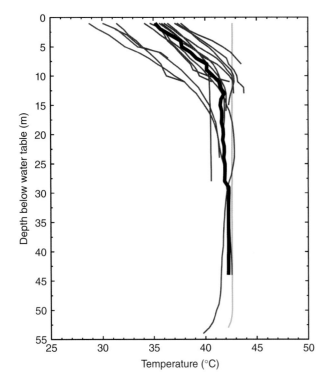

Fig. 14.6. Temperature–depth profiles measured within the hot-springs district during October 2012–2013. The type of well the profile was measured in is indicated by its color (blue = type 1, red = type 2, green = type 3; see methods section for details). A calculated average profile for depths 0–44 m is also plotted (bold black line). The average profile was only calculated to this depth due to lack of abundant data at greater depths and was used when interpreting hydrothermal model results.

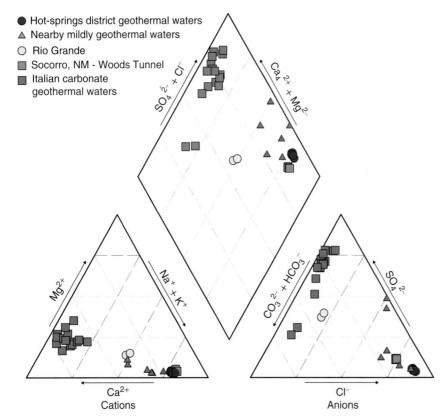

- ● Hot-springs district geothermal waters
- △ Nearby mildly geothermal waters
- ○ Rio Grande
- ▢ Socorro, NM - Woods Tunnel
- ▢ Italian carbonate geothermal waters

Fig. 14.9. Piper diagram summarizing geochemical analyses discussed in this paper. The geothermal waters from the hot-springs district (HSD) are shown with red circles. Mildly geothermal waters in the vicinity (within 16 km) of the HSD are plotted as orange triangles. Two samples of Rio Grande surface waters collected upstream of Truth or Consequences are shown using yellow circles. Data from a low-temperature carbonate geothermal reservoir in the Etruscan Swell of Italy (Chiodini *et al.* 1995) are shown with green squares. Data from the Woods Tunnel slim hole from the Socorro, New Mexico, geothermal system are shown by blue squares (Owens 2013). Truth or Consequences waters have a Na^+/Cl^- signature characteristic of geothermal waters derived from igneous and metamorphic rocks.

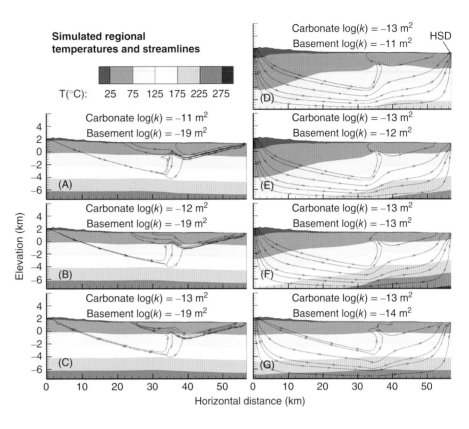

Fig. 14.11. Comparison of computed regional groundwater flow patterns (black lines with arrows) and temperatures for three shallow-circulation scenarios (left, A–C) and four basement-circulation scenarios (right, D–G). The base-10 logarithm of permeabilities used for the Magdalena Group and the crystalline basement are listed above each plot. Refer to Table 14.5 for simulation parameters and goodness of fit for subplots A–G. The location of the hot-springs district (HSD) is shown in graphic D. Groundwater flow directions are parallel to streamlines. Groundwater reaching the HSD in the shallow-circulation scenarios flows primarily through the shallow Magdalena Group. In contrast, deep-circulation scenarios are characterized by geothermal waters derived predominately from crystalline basement rocks. Shallow-circulation scenarios yield thermal patterns typical of a conductive thermal regime.

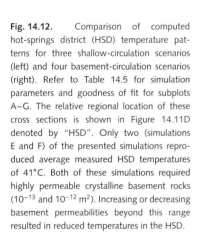

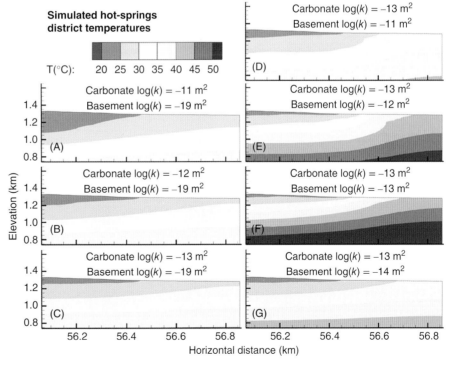

Fig. 14.12. Comparison of computed hot-springs district (HSD) temperature patterns for three shallow-circulation scenarios (left) and four basement-circulation scenarios (right). Refer to Table 14.5 for simulation parameters and goodness of fit for subplots A–G. The relative regional location of these cross sections is shown in Figure 14.11D denoted by "HSD". Only two (simulations E and F) of the presented simulations reproduced average measured HSD temperatures of 41°C. Both of these simulations required highly permeable crystalline basement rocks (10^{-13} and 10^{-12} m^2). Increasing or decreasing basement permeabilities beyond this range resulted in reduced temperatures in the HSD.

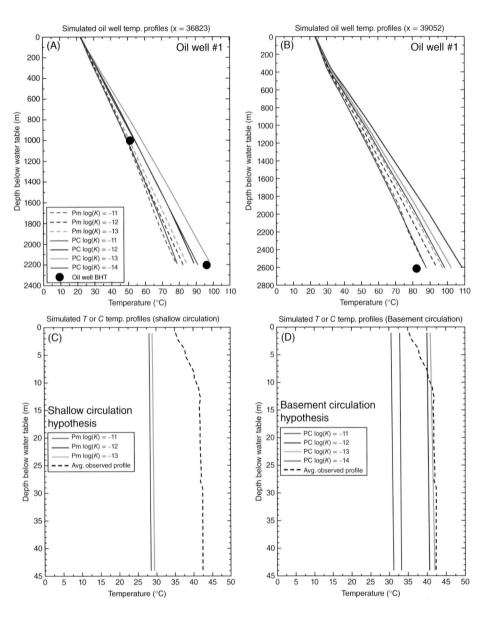

Fig. 14.13. Comparison of simulated and observed temperatures along the model domain. The assigned base-10 logarithm permeabilities for the crystalline basement rocks (PC) and Magdalena Group (Pm) are shown in the legends (graphs A and B share the same legend). Simulated temperature profiles are compared to bottom-hole temperature data collected in oil wells approximately 15 km north of Truth or Consequences (Top, A and B). They agree fairly well with bottom-hole temperature data, suggesting that our assigned thermal properties, such as basal heat flux and thermal conductivities, represent those of the study area. Average simulated and measured temperature profiles from the hot-springs district (HSD) are compared in C and D. Only simulations having high crystalline basement permeability (10^{-13} and 10^{-12} m^2) were able to reproduce observed HSD temperatures.

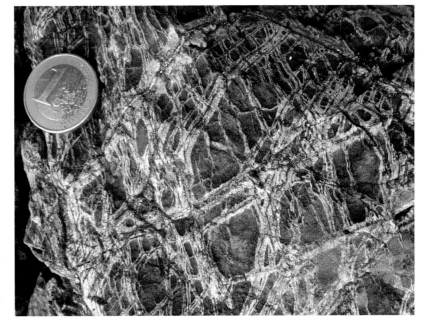

Fig. 15.4. Serpentine reaction veins in peridotite (Erro Tobbio mantle, Italy). The serpentinization of peridotite occurred along symmetrical reaction veins along brittle fractures. During the active period of serpentinization, the permeability of the fractured rock permitted flow and reaction of aqueous fluid. Flow and reaction ceased because of self-sealing of the fractures, preventing the peridotite (brown crust) from being completely serpentinized. Later, the fractures became completely sealed, and the vein system does not contribute to the near-surface permeability of the partly serpentinized peridotite.

Fig. 15.5. Talc veins in peridotite (Vlisarvatnet, Norway). Low-temperature hydration of peridotite (coarse-grained spinel harzburgite with brownish weathering rind) produced white silvery zones of talc along open brittle fractures that represent fluid-conducting structures. Note the vertical orientation of the fractures and the large aperture of these young structures. High content of dissolved silica in the hydrothermal fluid resulted in talc formation rather than serpentinization (Fig.15. 4). The open fracture system contributes to the present-day permeability of the rocks (and could be measured with well tests).

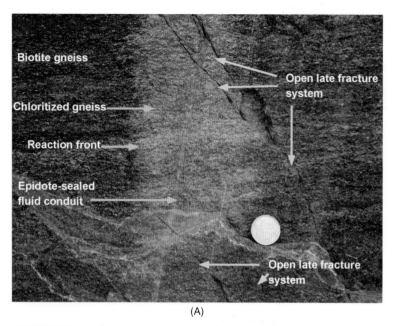

(A)

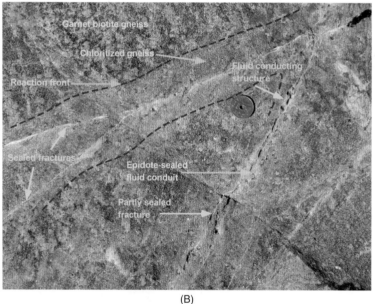

(B)

Fig. 15.6. Reaction veins in Caledonian amphibolite-grade gneiss: (A) Early chlorite vein in biotite gneiss (Hammerfest, Norway). Biotite gneiss has been chloritized along a brittle fracture that served as fluid conduit for hydrothermal fluid. The open fracture has been sealed later by epidote and became inactive. Young brittle fractures are open water-conducting structures and represent the present-day permeability of the rock. The later structures (blue arrows) control the hydraulic conductivity detected and measured by well tests. (B) Chlorite vein in garnet biotite gneiss (Torsnes, Kvaløya, Norway). Chloritization occurred along fractures by interaction of advecting hydrothermal fluid with the primary gneiss assemblage. Most of the fractures are completely sealed by solid reaction products. One of the fractures is only partly sealed. The visible fracture porosity contributes to the present-day permeability of the rocks (detectable by well tests).

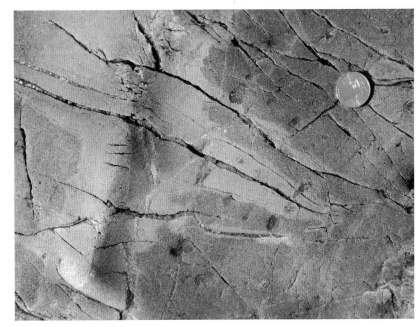

Fig. 15.8. Vein system in mafic rocks (dolerite on Vannøya, Norway). The structures suggest that first a fracture system opened and then advecting fluid reacted with the mafic igneous rock (green) producing an albite–calcite rock (light brown) before the fluid conduit was finally sealed by brown (and white) carbonate in the central part of the veins. The structures support the proposed k–t curve shown in Figure 15.7.

Fig. 15.9. Laumontite (Ca-zeolite) clogs vertical fluid-conducting fractures in gabbro (Langfjorden, Norway). The zeolite forms from hydrothermal alteration of primary labradorite (plagioclase) of the gabbro. The large volume increase of the solids in the reaction efficiently seals the vertical fracture system. The vein formation occurred at about 250°C and 10 km depth during the late Caledonian formation of the Langfjorden–Repparfjorden fault system.

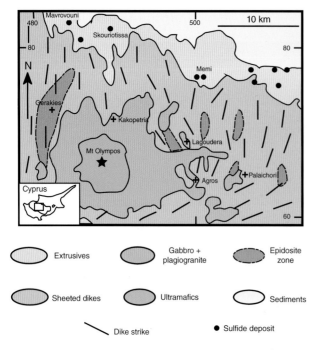

Fig. 17.1. Outline geological map of the central part of the Troodos ophiolite. The ophiolite is in the form of a dome. The deepest parts of the section are the ultramafics, gabbros and plagiogranites, which are overlain by sheeted dikes, and in turn by volcanics. The overlying sediments include deep ocean chalks and more recent deposits. The epidosite zones lie close to the base of the sheeted dike unit, except for the northern end of the big western zone, including the Gerakies localities, which reaches close to the volcanic unit, though there accompanied by outcrops of gabbro (Bettison-Varga *et al.* 1995). The variable dike strike reflects the complex processes of crustal generation at the ancient spreading centre. For a general introduction to the ophiolite, explaining this and other features, see Edwards *et al.* (2010). The grid numbers are UTM coordinates using the WGS84 grid, truncated by removing the first two digits and the last three digits of both eastings and northings. Place names are selected villages as well as the highest point, Mount Olympos. The sulfide deposits are all within the volcanic sequence. Three of them are named.

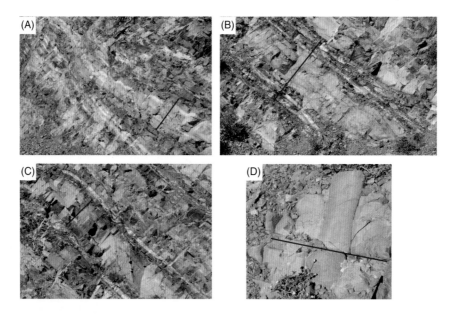

Fig. 17.2. Field photographs of epidosite zone exposures. Location references are UTM coordinates using the WGS84 grid, truncated by removing the first two digits of both eastings and northings. Red lines show the width of a prominent individual dike in each image. (A) Exposure of epidosite zone rocks showing the variety of colors in a single outcrop (exposure about 3 m high) (482160 68520). The paler lithologies are epidosites, and the darker colors are less altered dikes. (B) A 1.5-m-wide striped dike showing the symmetrical pattern of striping typical of wider dikes (481873 68871). (C) A 0.7-m-wide dike with a simpler alteration pattern. At the margins are narrow stripes of epidosite (481873 68871). The core of the dike is more altered than the outer zones inside the epidosite stripes. (D) Incipient alteration shown by a pod, 0.2 m wide, of more highly altered rock inside a less altered dike matrix (498774 67931).

Chilled margins

Last dike in A

Earlier dike in B and C

Later dike in B and C

Fig. 17.3. Field photographs showing the relationships between dikes in an epidosite zone. (A) A single highly altered 1-m-wide dike cutting through a complex of earlier dikes altered to different degrees. (B) A 1.5-m-wide striped epidosite dike at a high degree of alteration, cutting a narrow 0.7-m-wide dike at a lower degree of alteration and displacing its outcrop. (C) Enlarged view of the area outlined in (B). Note the abrupt truncation of the earlier dike and its marginal and central stripes by the later dike and the apparent lack of any further alteration of the earlier dike by the higher degree of alteration of the later one.

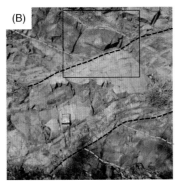

0.3 mm

Fig. 17.4. Thin section photomicrograph of an epidosite in plane polarized light (A) and crossed polars (B), showing the absence of any relict igneous texture. The section shows epidote E, quartz Q and titanite T. The rock specimens in Figures 17.4–17.9 come from the road section west of Gerakies between (480830 73370) and 480940 73010).

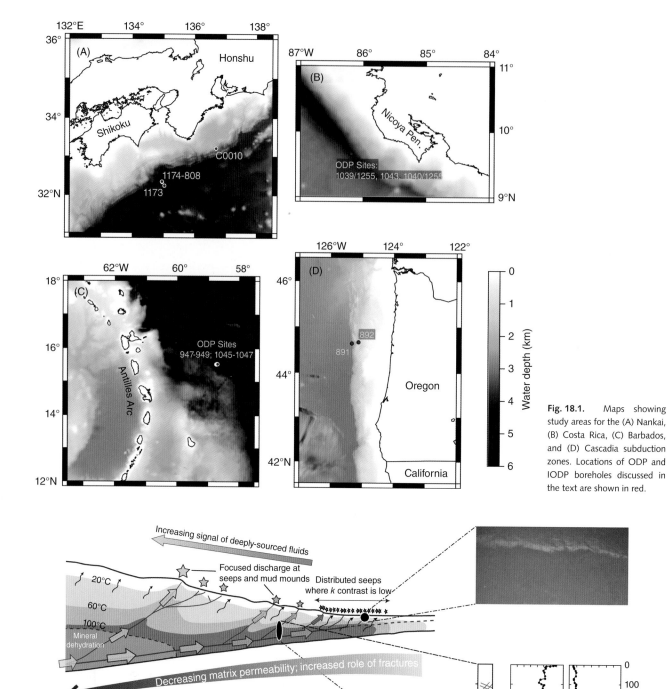

Fig. 18.1. Maps showing study areas for the (A) Nankai, (B) Costa Rica, (C) Barbados, and (D) Cascadia subduction zones. Locations of ODP and IODP boreholes discussed in the text are shown in red.

Fig. 18.2. Schematic subduction zone cross section synthesizing hydrologic, geologic, and geochemical observations indicating focused fluid flow along fault zones. Colors indicate temperature distribution and illustrate the role of faults in advecting heat (after Vrolijk *et al.* (1991)). Arrows indicate fluid flow; colors represent the relative contributions of deeply sourced fluids derived from diagenetic and metamorphic reactions (green) and initial pore water with seawater composition (blue). Stars at seafloor indicate sites of focused seepage, with the same color scheme as flow arrows. Inset at bottom right shows schematic structural column spanning the dècollement at ODP Site 1040 (Costa Rican margin) and pore-water geochemical data indicating active or recent flow in the dècollement and a splay fault in the hanging wall (Chan & Kastner 2000; Morris *et al.* 2003b). Photograph shows example of an antitaxial vein from the Barbados dècollement (Vrolijk & Sheppard 1987). Photograph at the top right shows approximately 50-cm-high authigenic carbonate crust formed at the outcrop of the dècollement at the Costa Rican margin.

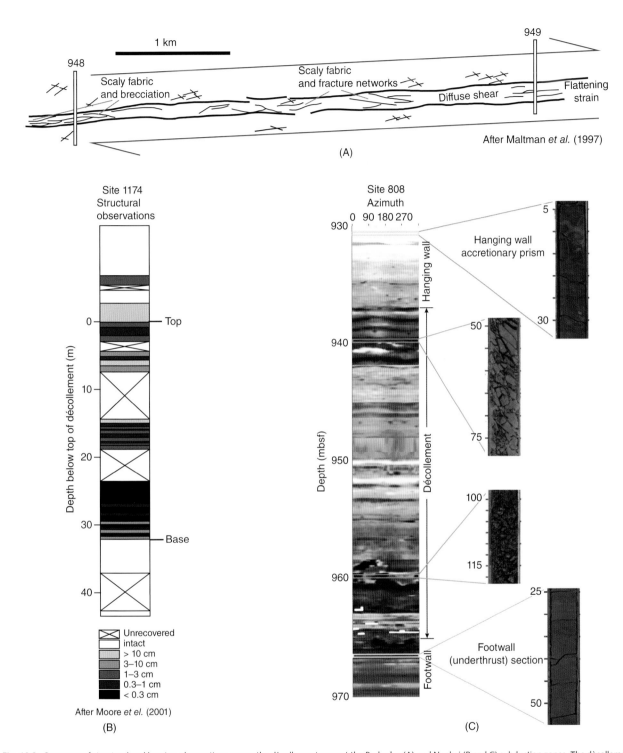

Fig. 18.3. Summary of structural and logging observations across the dècollement zone at the Barbados (A) and Nankai (B and C) subduction zones. The dècollement is commonly defined as a tens-of-meters-thick zone of brecciated and fractured scaly mudstone, with increasing intensity of brecciation downward, and a sharp basal contact with relatively undeformed underthrust sediment. (A) Synthesis of logging and structural observations from core at several ODP sites at the Barbados margin (after Maltman *et al.* (1997)). (B) Summary structural section across the dècollement at ODP Site 1174, located approximately 2 km landward of the trench in the Nankai Trough (after Moore *et al.* (2001)). (C) Logging while drilling resistivity image across the dècollement zone at ODP Site 808 in the Nankai Trough, located approximately 1 km landward of Site 1174. Core photographs are from ODP Site 1174 with equivalent locations at Site 808 estimated based on relative depth from the top of the dècollement (after Ienaga *et al.* (2006)).

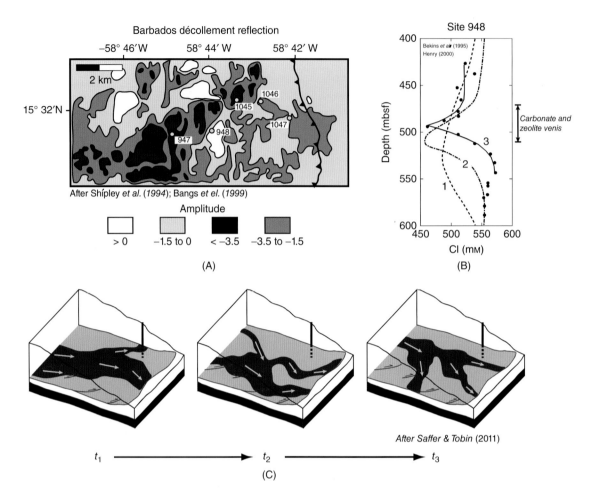

Fig. 18.5. (A) Seismic reflection amplitude of the Barbados dècollement zone (from Shipley *et al.* (1994)), showing locations of trench (barbed line; teeth facing subduction direction) and ODP drill sites (circles). (B) Pore-water chloride profile at Site 948, with freshening centered at the dècollement. Curves show model results from Bekins *et al.* (1995) for transiently increased dècollement permeability of 10^{-13} m^2 (1; duration 30 kyr) and 10^{-11} m^2 (2; duration 300 year) and transient diffusion model from Henry (2000) (3; duration approximately 11 kyr). Distribution of veining within the dècollement is shown at right. (C) Schematic showing conceptual model in which permeable channels along the dècollement shift over time, leading to transient behavior at a given location, and spatially distributed zones of high permeability and flow (after Saffer & Tobin (2011)).

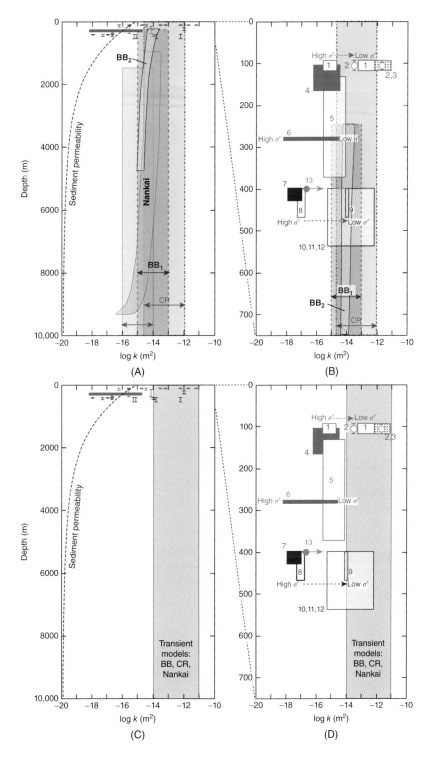

Fig. 18.6. Synthesis of published constraints on fault permeability reported in Table 18.1, shown as functions of depth below seafloor (mbsf). (A) and (B) show *in situ* permeability measurements or estimates described in the text and in Table 18.1, with permeability ranges derived from steady-state numerical modeling studies for the Barbados (BB_1: Bekins *et al.* (1995) and Screaton *et al.* (1990); BB_2: Henry & Le Pichon (1991)), Nankai (green; Saffer (2010); Skarbek & Saffer (2009)), Oregon (light brown; Saffer & Bekins (1998b)), and Costa Rican (blue; Spinelli *et al.* (2006)) margins. Mudstone sediment matrix permeability from Skarbek & Saffer (2009) (dashed line) is shown for comparison. (C) and (D) show the *in situ* permeabilities with transiently increased permeabilities from numerical models for the Nankai, Barbados, and Costa Rican margins (gray-shaded region). The *in situ* and laboratory measurements are blown up in panels (B) and (D). These include laboratory measurements (filled squares), *in situ* determinations from well testing (open squares), analyses of thermal anomalies and tidal responses in borehole observatories (open and closed circles, respectively), and analysis of vein chemistry (stippled square). These follow the same color scheme as for model results: brown = Oregon; maroon = Barbados; blue = Costa Rica; green = Nankai. Arrows show the range of permeabilities obtained over ranges of effective stress in well tests at the Oregon frontal thrust (*1*), Barbados dècollement (*8–9*), and in laboratory measurements of fault zone core from Costa Rica (*6*). *1* = Screaton *et al.* (1995); *2* = Davis *et al.* (1995); *3* = Sample (1996); *4* = Brown (1995); *5* = Bolton *et al.* (1999); *6* = Saffer & Screaton (2003); *7* = Zwart *et al.* (1997); *8* = Screaton *et al.* (1997); *9* = Screaton *et al.* (2000); *10* = Bekins *et al.* (2011); *11* = Fisher & Zwart (1997); *12* = Fisher & Hounslow (1990); *13* = Hammerschmidt *et al.* (2013).

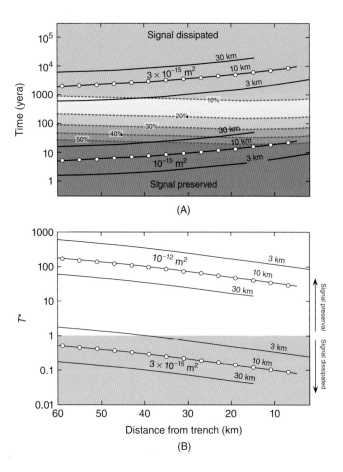

(A)

(B)

Fig. 18.8. (A) Time required for up-dip fluid migration over distances of 3, 10, or 30 km (black curves, as labeled) as a function of distance landward of the trench, for fault permeabilities of 10^{-12} m^2 (bottom, as labeled) and 3×10^{-15} m^2 (top, as labeled). Flow rates are computed assuming a fault zone porosity of 0.05, a hydraulic gradient equivalent to a pore pressure ratio ($\lambda = P_{fluid}/P_{lithostatic}$) of 0.80, and accounting for varying fluid viscosity as a function of temperature. Times required for the dissipation of a 1-m-wide initial solute concentration pulse are shown by color shading, computed assuming a molecular diffusivity (D) of 10^{-9} m^2 s^{-1}, and with effective diffusivity defined as a function of sediment matrix porosity by $D^* = D\eta/2$ and accounting for the effects of temperature and fluid viscosity by the Stokes–Einstein equation (Helfferich 1966). Dashed contours show the times at which the peak concentration of the pulse has decreased to a particular percentage of the initial peak value. (B) The ratio (T^*) of the times shown in panel (A) ($T_{diffusion}/T_{flow}$) as a function of distance from the trench, where $T_{diffusion}$ is defined as the time by which the pulse is reduced to 10% of its initial amplitude. For a fault zone permeability of 10^{-12} m^2, localized concentration pulses would dissipate by <50% over transport distances of <10 km, regardless of location. For fault permeability of 3×10^{-15} m^2, signals would dissipate by > 90% in most scenarios. Thermal structure, geometry, and sediment matrix porosity distribution are defined using the Nankai accretionary complex as an example (Skarbek & Saffer 2009).

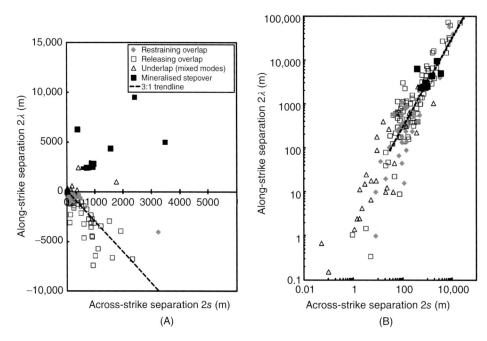

(A)

(B)

Fig. 20.4. Comparison of along-strike (2λ) versus across-strike ($2s$) separation between mineralised stepovers and examples from well-constrained strike-slip fault data (Aydin & Schultz 1990; Kim *et al.* 2004). (A) 2λ versus $2s$ demonstrating that mineralised stepovers investigated were all underlapping (positive R) in contrast to the majority of data (negative R). The mineralised stepovers for the most part had larger dimensions than other underlapping stepovers previously measured. (B) log 2λ versus log $2s$. When compared directly with one another, underlapping mineralised stepovers maintain an aspect ratio ~3:1, similar to their overlapping counterparts (for along-strike dimensions > 100 m). Restraining or releasing overstep data show no appreciable difference in geometry and scaling properties. For the sake of brevity, the releasing or restraining modes of underlapping stepovers are not distinguished.

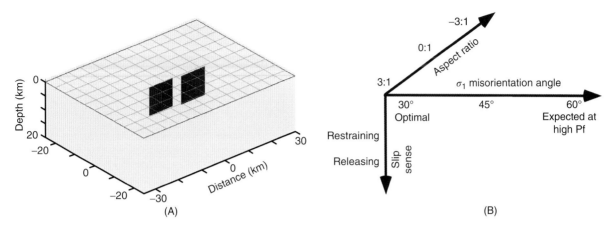

Fig. 20.5. (A) Example model of the three-dimensional boundary element modelling configuration. In this case, two underlapping fault segments are shown with aspect ratio $R = 3$ and a tapered slip distribution with maximum slip ~30% of the strike length from the stepover. (B) A 3-3-2 matrix of parameters was investigated to understand the impact of stepover geometry on $\Delta_{\sigma f}$, including the orientation of maximum principal stress (σ_1) relative to the fault strike. In some cases, we also ran additional models with aspect ratio $R = 1$.

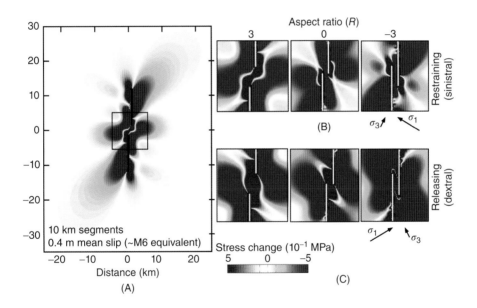

Fig. 20.6. Horizontal distributions of change in Coulomb failure stress ($\Delta\sigma_f$) for two models, with stepover geometries varying from underlapping, to neutral, to overlapping (modified aspect ratios of $R = 3, 0, -3$). The remote maximum and minimum principal stresses are σ_1 and σ_3, respectively, with $\sigma_1 = 60°$ relative to fault strike, representative of mineralised systems affected by high fluid pressures (e.g. Sibson *et al.* 1988; Cox *et al.* 2001; Micklethwaite 2008). In both cases, $\Delta\sigma_f$ is calculated for triggering of strike-slip faults. (A) Results for an underlapping, restraining stepover. Subsequent stresses are sampled from the more restricted domain shown around the stepover. (B) Model 1–restraining stepovers (sinistral fault slip). (C) Model 2–releasing stepovers (dextral fault slip). In Models 1 and 2, underlapping geometries have significantly larger areas of positive $\Delta_{\sigma f}$, relative to neutral and overlapping geometries.

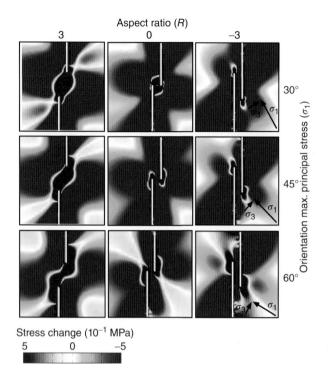

Aspect ratio (R)

3   0   −3

30°

45°

60°

Orientation max. principal stress (σ_1)

Stress change (10^{-1} MPa)

5   0   −5

Fig. 20.7. The impact of stepover geometry (modified aspect ratios of $R = 3, 0, -3$) on $\Delta_{\sigma f}$, under conditions of varying remote stress orientation. The remote maximum and minimum principal stresses are σ_1 and σ_3, respectively. In the examples shown, all models are for restraining stepovers.

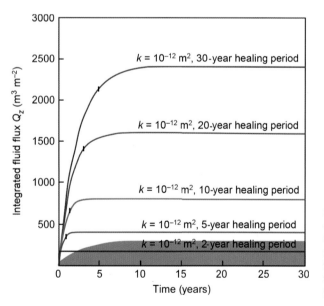

$k = 10^{-12}$ m^2, 30-year healing period

$k = 10^{-12}$ m^2, 20-year healing period

$k = 10^{-12}$ m^2, 10-year healing period

$k = 10^{-12}$ m^2, 5-year healing period

$k = 10^{-12}$ m^2, 2-year healing period

Fig. 20.9. Time-integrated fluid flux derived from initial coseismic permeabilities of 10^{-12} m^2 decaying to 10^{-18} m^2 over time periods of 2, 5, 10, 20 and 30 years (N). For comparison, the field in grey is the range of flux values if coseismic permeabilities were 10^{-13} m^2, illustrating the sensitivity of time-integrated fluid flux to transient coseismic permeability. The marked inflection points are 90% of the total time-integrated fluid flux for each curve–which occurs at ~15% of the healing period (N).

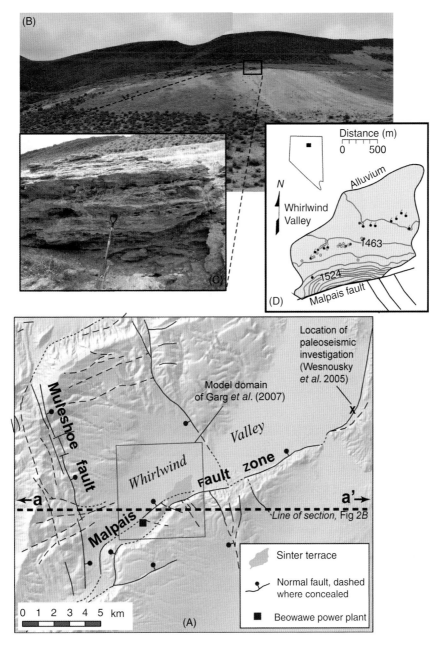

Fig. 21.1. (A) Structural map showing location of Beowawe sinter deposits, paleoseismic investigation of Wesnousky *et al.* (2005) along the Malpais fault zone, the Beowawe power plant (black square). (B) Photo mosaic of Beowawe sinter deposits, Whirlwind Valley, Nevada. The photo was taken looking southeast. (C) Outcrop scale photo of sinter deposits. (D) Sinter topographic map *with locations of historic and paleo hot springs (black circles, as well as locations of sinter samples (red and yellow circles). Multiple samples were collected at locations with yellow dots.*

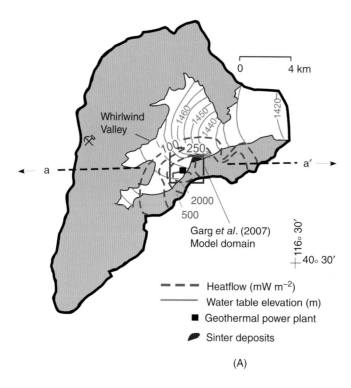

(A)

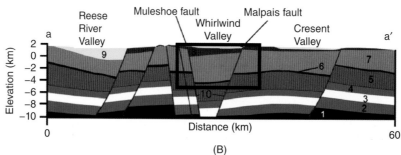

(B)

Fig. 21.2. (A) Heat flow (orange dashed contour lines; in mW m^{-2}) and water-table elevation (solid light blue lines; m) contour map across Whirlwind Valley, Nevada. Location of regional hydrothermal model cross section is depicted by a–a'. Dark blue box indicates the location of the Beowawe sinter terrace shown in Figure 21.1B. Heat-flow and water-table contour maps from Olmsted & Rush (1987) and Faulder et al. (1997). (B) Hydrogeologic cross section across Reese, Whirlwind, and Crescent Valleys. The inset (solid black rectangle) denotes the portion of the hydrothermal model domain presented in Figure 21.3A–F. Permeability data for the different numbered units in Figure 21.1C are presented in Table 21.1 and described in the supplemental materials section.

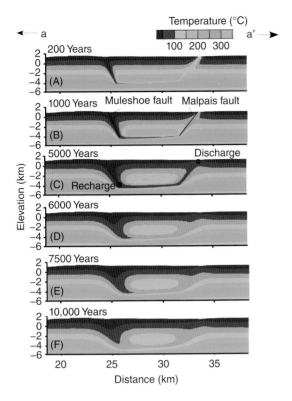

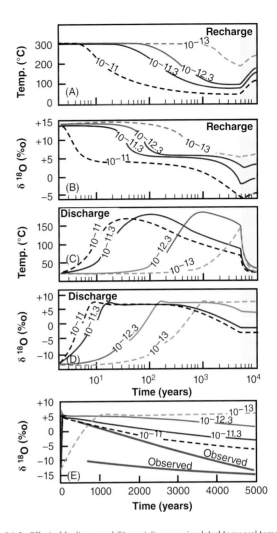

Fig. 21.7. Computed temperature contour maps illustrating thermal evolution along the Malpais fault, Muleshoe fault, and karst aquifer. Fault zones were assigned a uniform permeability of 10^{-12} m^2. Fluid flow is active for 5000 years (A-C) followed by 5000 years (6000–10,000 years; D-F) of conductive cooling. Shallow meteoric fluid infiltration occurs down the Muleshoe fault zone and then along a thin, permeable karst zone. Upflow occurs along the Malpais fault zone. Points labeled "recharge" and "discharge" are the locations within the Beowawe geothermal system where temperature and $\delta^{18}O_{fluid}$ were plotted through time in Figure 21.8A,C.

Fig. 21.8. Effect of fault permeability variations on simulated temporal temperature changes (A) and (B) oxygen isotopic composition along the Muleshoe fault zone at a depth of about 5 km. The numbers on the lines denote the permeability assigned to the Muleshoe and Malpais faults (in m^2). Effect of fault permeability variations on simulated temporal changes in temperature (C) and (D) oxygen isotopic composition along the Malpais fault zone at the water table. The shaded area of the graphs represents conductive recovery following 5000 years of convection and cooling of fluids moving through the system. Locations within the recharge and discharge area are shown in Figure 21.2c (red circles). (E) Comparison of observed trends (gray dashed lines) and computed changes in $\delta^{18}O_{H_2O}$ versus time for the Beowawe sinter terrace.

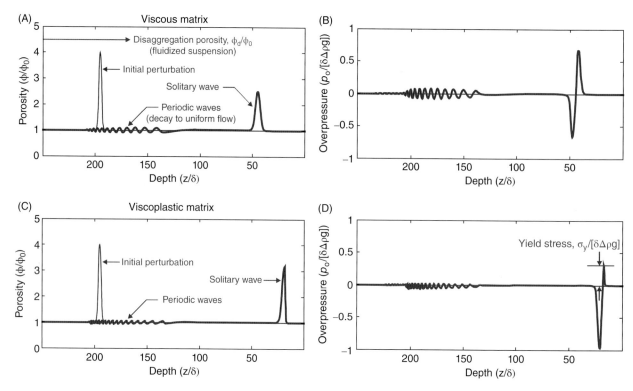

Fig. 23.1. Numerically simulated porosity wave evolution from a region of increased porosity within an otherwise uniform flow regime through viscous and viscoplastic porous media. (A) Porosity versus depth for the viscous case. The first wave to initiate from the flow perturbation (i.e., the region of elevated porosity) is a solitary wave that propagates above the background porosity in the same direction as fluid flow through the unperturbed matrix. The solitary wave is nondissipative in the steady-state limit; thus, it propagates infinite distance with essentially unchanging form. Due to transient effects, a periodic wave train initiates behind the solitary wave; the periodic wave train propagates both in and against the direction of fluid flow through the unperturbed matrix. The wave train corresponds to a periodic steady state in which the porosity oscillates about the background level. Porosity wave velocities are proportional to amplitude; thus, with time the solitary wave becomes isolated from the periodic wave train, which has no significant excess volume and degrades to the original uniform flow regime as it spreads. A narrow region of increased porosity was chosen for the initial conditions to emphasize the periodic solution. This choice leads to solitary waves that have lower porosity than the source region. Wider source regions tend to generate waves with porosities that are higher than in the source region; if these porosities exceed the disaggregation porosity (indicated schematically), the matrix disaggregates to a fluidized suspension. This range of porosity wave behavior can also be induced by perturbing the background fluid flux, as might occur as a consequence of devolatilization (melting). (B) Fluid overpressure (negative effective pressure) versus depth for the viscous case, the porosity dependence of the matrix permeability causes fluid pressure anomalies that are responsible for dilating and compacting the porosity during the passage of a wave. Although fluid flow from low to high overpressure may seem to contradict Darcy's law, if the overpressures are converted to hydraulic head, it is apparent that the direction of fluid flow in porosity waves is consistent with Darcy's law. (C) Porosity and (D) overpressure profiles for a viscoplastic scenario in which plasticity is manifest by hydrofracture when overpressure exceeds the brittle yield stress. This rheology affects wave symmetry, but does not fundamentally change porosity wave behavior because the rate of fluid expulsion remains limited by viscous compaction. In the numerical simulation, the effect of the viscoplastic rheology on the periodic waves with overpressures below the yield stress is an artifact of the reduced effective shear viscosity used to simulate brittle failure. Porosity (ϕ), depth (z), overpressure (p_o), and time are scaled relative to the background porosity (ϕ_0), viscous compaction length (δ), characteristic pressure ($\delta|\Delta\rho g|$), and the speed of fluid flow through the unperturbed matrix ($|v_0|$) discussed later in the text. The 1D volume of the initial perturbation is the same (8.862 $\delta\phi_0$) in both simulations, and the transient profiles (blue curves) are for the same model time ($50\delta/|v_0|$). The numerical simulations were obtained by finite difference methods for the small porosity formulation (Appendix) with $m_\sigma = 1$, $n_\phi = 3$, and $n_\sigma = 1$.

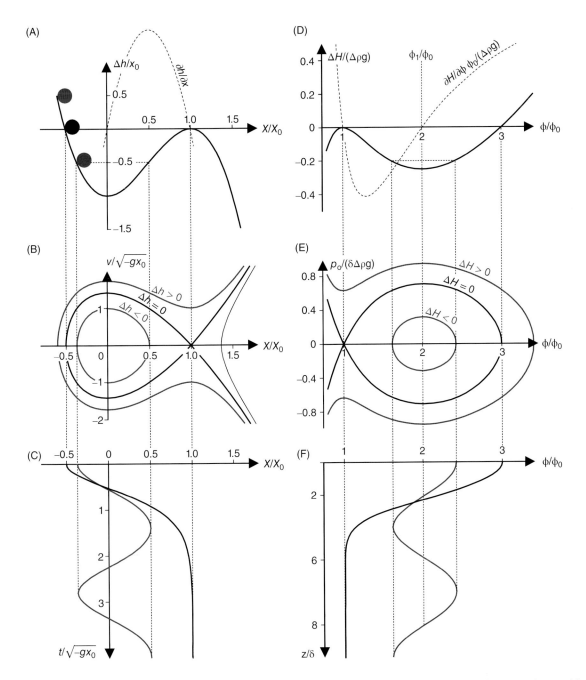

Fig. 23.2. Analogy of porosity waves with the oscillatory movement of a ball on a frictionless surface of variable height with a local maximum (i.e., a saddle point). In the analogy, the porosity wave properties H, ϕ, p_o, and z map to h, x, v, and t of the oscillating ball. If the ball is placed on the surface (e.g., the red ball in A) at a height below and to the left of the local maximum, the ball oscillates between points of equal height. In this case, the velocity–position trajectory of the ball defines a closed path around the focal point at $x = v = 0$ (e.g., the red curve in B) and its velocity–time (or position–time) trajectory defines a periodic wave train (e.g., the red curve in C). The location of the focal point ($x/x_0 = 0$) for the waves is dependent only on the shape of the surface, whereas the wave frequency is controlled by the gravitational acceleration. The solitary wave solution for the oscillating ball corresponds to the case that a ball (i.e., the black ball in A) is placed on the surface at the height of the saddle point. In this case, the ball would have no kinetic energy if it reached the saddle point; however, because the ball decelerates as it approaches the saddle point, the wave solution has an infinite period. If the ball is released from a height above the saddle point (i.e., the blue ball in A), it rolls continuously away from its initial position and there is no wave solution to the governing equations. The analogy of the oscillating ball to porosity wave solution is imperfect only in that the shape of function, or hydraulic potential, H of the porosity wave is dependent on wave velocity (Fig. 21.3), which is related to the intensity of the perturbation responsible for generating waves. For the velocity specified for H as illustrated in (D), the only wave solution capable of stably connecting the background porosity to a region of increased porosity is the solitary wave (i.e., the black trajectories in E and F); the periodic solutions (e.g., the red trajectories E and F) correspond to waves in which the porosity would oscillate about the focal point ($\phi/\phi_0 = 2$) between a porosity that is less than the maximum porosity of the solitary wave and greater than the background porosity. The solitary wave shape function illustrated in (D) is from the small porosity formulation of the compaction equations (Appendix) with $m_\sigma = 0$, $n_\phi = 3$, and $v_\phi/v_0 = 7$.

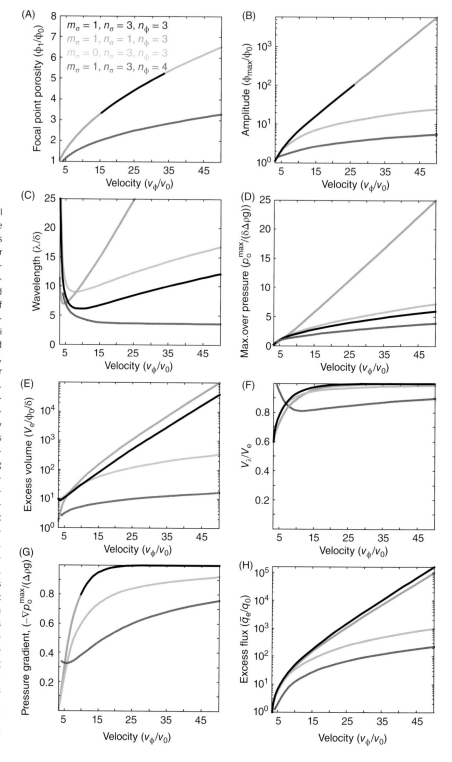

Fig. 23.4. Porosity wave properties in the small porosity limit ($\phi \ll \phi_d$) as a function of relative wave velocity for choices of the porosity exponents in the constitutive relations as indicated by color coding. Typical values for the exponents characterizing the porosity dependences of the permeability and effective bulk viscosity are $n_\phi = 3$ and $m_\sigma < 1$; a stress exponent $n_\sigma = 3$ is characteristic of dislocation creep, the viscous deformation mechanism commonly assumed for the lower crust (Ranalli 1995). The minimum relative velocity considered for each choice of exponents is $v_\phi/v_0 = n_\phi + 0.2$, slightly above the critical value, $vv_\phi/v_0 = n_\phi$, for the existence of the solitary solution (Eq. 23.44). Continuous curves drawn in different colors indicate that the properties are identical for the corresponding exponent choices. Focal point porosity (A) and amplitude (B). Focal point porosity, ϕ_1, is dependent only on n_ϕ, and the divergence of amplitude from ϕ_1 shows that the effect of increasing the nonlinearity (i.e., m_σ) of the effective bulk viscosity is to increase amplitude. In contrast, increasing nonlinearity of the porosity–permeability relationship decreases amplitude, which is independent of n_σ. (C) Wavelength, λ, is the distance separating the overpressure extrema within a wave (Fig. 23.3H). (D) Maximum fluid overpressure, the maximum underpressure (i.e., effective stress) is $-p_0^{max}$, both extrema occur at ϕ_1 (Fig. 23.3E). (E) Excess volume (Eq. 23.60) correlates with amplitude except at low wave speed, whereupon it decreases with speed in solutions for high n_σ/n_ϕ. That the excess volume for the nonlinear viscous cases ($n_\sigma = 3$) becomes larger than that for the linear viscous case ($n_\sigma = 1$) at low speeds ($v_\phi/v_0 \sim 6$) indicates a shifting of the porosity toward the tails of the nonlinear viscous solution. (F) Fraction of the volume that occurs within $\pm\lambda/2$ of ϕ_{max}. (G) Maximum $-p_0$ gradient, which occurs at ϕ_{max}, unity corresponds to a hydrostatic fluid pressure gradient. (H) Average excess fluid flux associated with wave passage.

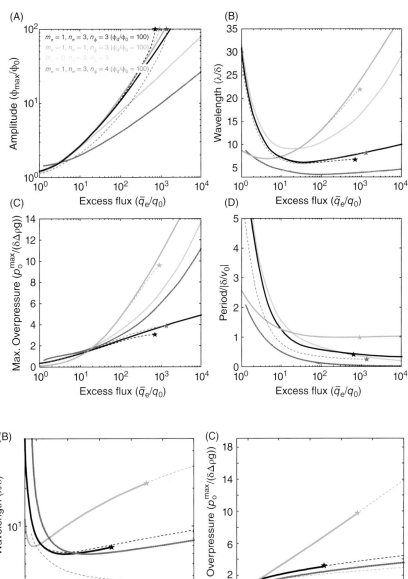

Fig. 23.5. Porosity wave properties as a function of the average excess fluid flux associated with wave passage for the exponent choices indicated by the legend and color coding. Solid curves are for the small porosity limit ($\phi \ll \phi_d$) as in Figure 23.4; dashed curves show the influence of a disaggregation in the intermediate porosity limit ($\phi_d \geqslant 1 - \phi$) for the specific case that $\phi_d/\phi_0 = 100$, discussed in the text (Fig. 23.36). Star symbol at the end of dashed curves indicates the disaggregation condition. Properties of waves likely to be generated in natural settings in response to fluid production can be estimated by equating vertically integrated fluid production to the average excess flux.

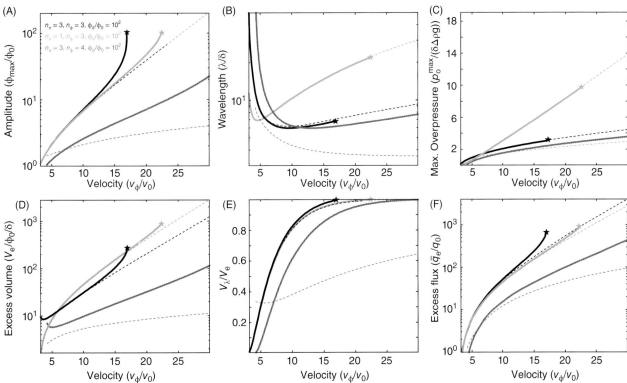

Fig. 23.6. Porosity wave properties in the intermediate porosity limit ($\phi_d \ll 1 - \phi$) as a function of velocity for a disaggregation porosity ϕ_d a hundred times greater than ϕ_0. All examples are for $m_\sigma = 1$ with other exponents of the formulation as indicated by the legend and color coding. The star symbol indicates the disaggregation condition. Dashed curves show the corresponding properties for the small porosity limit as in Figure 23.4. The solutions show that the effect of disaggregation is strongly dependent on both n_ϕ and n_σ; for $n_\phi = 3$, the disaggregation effect becomes significant if ϕ_{max} is within an order of magnitude of ϕ_d and is amplified by increasing n_σ. Properties for these solutions are shown as a function of $\bar{q}_e$ in Figure 23.5.

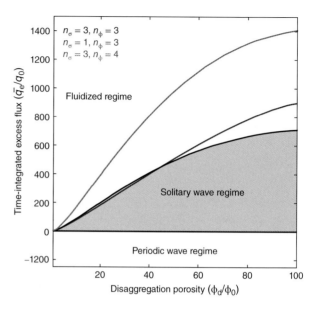

Fig. 23.7. Phase diagram depicting the hydrologic regimes predicted by the intermediate porosity limit solitary wave solution as a function of the disaggregation porosity and excess flux. The regime that develops in response to fluid production can be predicted by equating the magnitude of the excess flux carried by the waves to the vertically integrated fluid production $\bar{q}_s$. The boundary between the solitary wave and fluidized regime is dependent on the exponents m_σ, n_σ, and n_ϕ; it is shown for $m_\phi = 1$, with other exponents as indicated by the inset and color coding. Solitary waves become progressively more diffuse and indistinguishable from uniform flow as $\bar{q}_e/q_0 \to 0$. Periodic wave trains develop in response to negative fluid production, for example, the consumption of fluids by hydration reactions, or an obstruction to the background porosity (Spiegelman 1993). In nature, such periodic wave trains would decay to uniform flow.

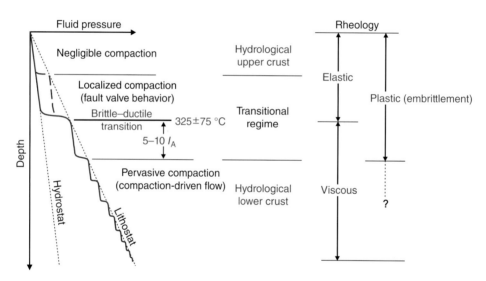

Fig. 23.8. Conceptual model of the hydrologic regimes that would result from superimposing thermally activated compaction on crustal column with heterogeneous permeability (Connolly & Podladchikov 2013). In the upper crustal regime, faulting maintains such high permeabilities that negligible deviation from hydrostatic fluid pressure is adequate to drive fluid circulation (Zoback & Townend 2001). This regime is limited at depth by the conditions at which localized compaction becomes an effective mechanism for sealing fault-generated permeability (Gratier et al. 2003; Tenthorey & Cox 2006). At greater depth, pervasive compaction and/or metamorphic fluid production may generate transient fluid overpressure that is periodically relieved by faulting (Sibson 1992). At the brittle–ductile transition (i.e., the base of the seismogenic zone), it is improbable that pervasive compaction can keep pace with metamorphic fluid production; thus, the transitional hydrologic regime is likely to persist over an interval that extends ~10l_A below the brittle–ductile transition, where l_A is the characteristic length scale for variation in the ductile rheology (typically ~1 km, Connolly & Podladchikov 2013). Beneath the transitional regime, pervasive compaction is capable of generating hydraulic seals, and fluid, if present, is at near-lithostatic pressure. Within this lowermost regime, fluid flow is truly compaction driven. In the absence of fluid production, the tendency of both time and depth is to decrease the wavelength of the fluid pressure compartments, resulting in a near-steady state. Barring the possibility of a subcrustal fluid source, the flux in this near-steady regime must decrease with depth. Thus, the magnitude of the perturbation caused by fluid production to the lower crustal regime is dependent on its depth. In the deepest portion of the crust, the rheology is viscous as assumed in the formulation presented here. Upward strengthening of the viscous rheology would cause porosity waves to provoke elastic and plastic deformation mechanisms at shallower levels. Because viscous porosity waves are associated with negative effective pressure anomalies, $\sim \lambda\Delta\rho g/2$, the first deviation from viscous behavior is likely to be viscoplastic. Viscoplastic rheology causes fluid flow to be focused into tube-like channels (Connolly & Podladchikov 2007; Connolly 2010). At still shallower depths, viscous compaction becomes entirely ineffective, leading to a viscoelastic transition. In numerical models, such a viscoelastic transition causes lower crustal solitary porosity waves to dissipate as porosity–pressure surges in the upper crust (Connolly & Podladchikov 1998).

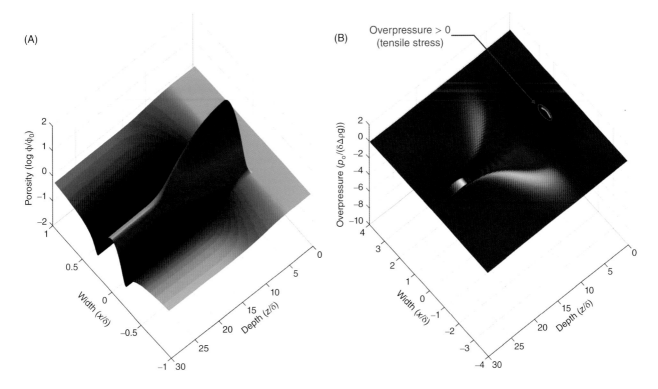

Fig. 23.9. Two-dimensional numerical simulation of a solitary porosity wave in a viscoplastic matrix. (A) Porosity; (B) fluid overpressure. The axial porosity and overpressure profiles of the wave are identical to the 1D case (Fig. 23.1C,D), but in the 2D case, the asymmetric overpressure distribution causes compaction of the background porosity on either side of the wave, leaving a tube-like channel that localizes subsequent fluid flow; the logarithmic scale for porosity emphasizes this effect. Numerical simulations (B. J. P. Kaus, personal communication 2005) have confirmed that 3D solitary waves in a viscoplastic matrix have radial symmetry orthogonal to the direction of propagation as in the viscous limit (Wiggins & Spiegelman 1995); thus, the 2D wave shown here corresponds to the axial section of a 3D wave.

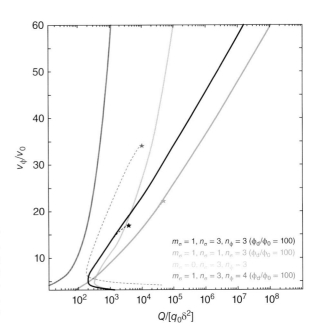

Fig. 23.10. Wave velocity as a function of volumetric fluid transport rate (Q) by spherically symmetrical 3D porosity waves (Wiggins & Spiegelman 1995) for exponent choices as indicated by the legend. For velocities at which the fluid pressure gradient (Fig. 23.4G) is nearly hydrostatic, the properties of the 1D and 3D solution are essentially identical; thus, at these conditions, the velocity can be used to predict 3D wave properties from the 1D solution (Fig. 23.4, and Examples #2 and #3). Where velocity is not a monotonic function of Q, it is probable that the high-velocity (short-wavelength) solution dominates.

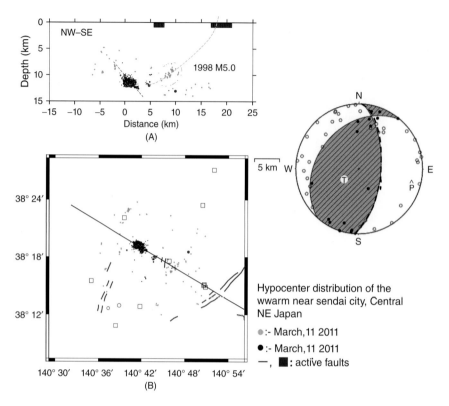

Fig. 24.2. Hypocenter distribution of the Sendai swarm in the central Miyagi (Region M). (A) NNW–SSE vertical cross section. (B) Epicenter distribution. Focal mechanism of the Mjma 3.2 event that occurred on April 30 is also shown. Thin broken lines in (A) and focal mechanism show the dip of possible fault plane. Red bold lines in (B) and red squares in (A) show the location of the surface trace of the active fault. White squares denote seismic stations.

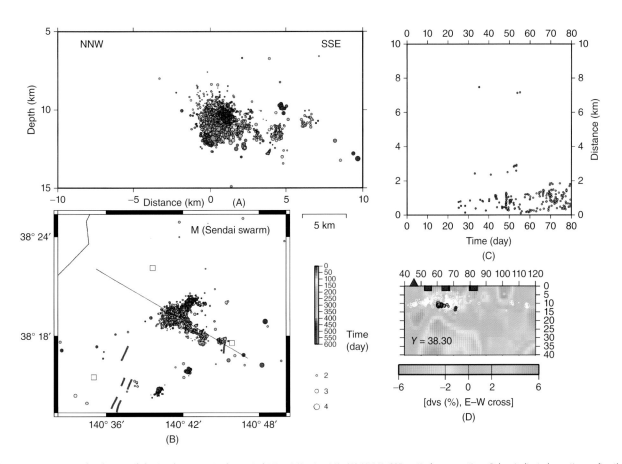

Fig. 24.3. Hypocenter distribution of the Sendai swarm in the central Miyagi (Region M). (A) NNW–SSE vertical cross section. Colors indicate lapse times after the 2011 Tohoku-Oki earthquake. (B) Epicenter distribution. (C) Time–distance plots. Distance from the first event of this cluster is shown. (D) E–W vertical cross section of Vs perturbation. White and colored circles show the hypocenters before and after the 2011 Tohoku-Oki earthquake. Red triangles show the volcano. Red bold lines in (B) and red squares in (D) show the location of the surface trace of the active fault. White squares in (B) denote locations of seismic stations.

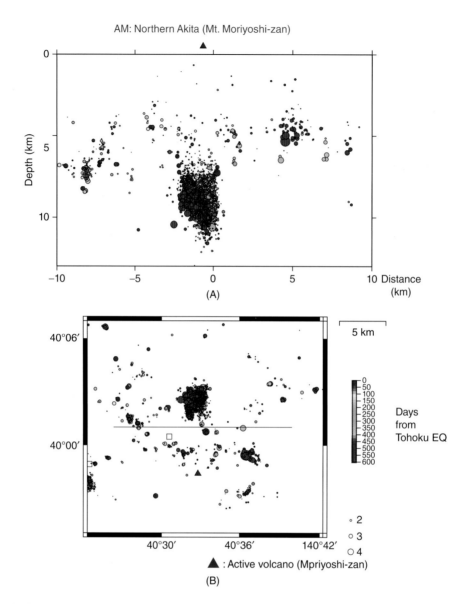

Fig. 24.4. Hypocenter distribution of the swarm near Mt. Moriyoshi in Northern Akita (Region AM). (A) E–W vertical cross section with a width of 10 km. Color of circle means the lapse time after the 2011 Tohoku-Oki earthquake. (B) Epicenter distribution. Line denotes the location of the cross section (A). Red triangle shows the volcano (Geological Survey of Japan 2013). White square in (B) denotes the location of a seismic station.

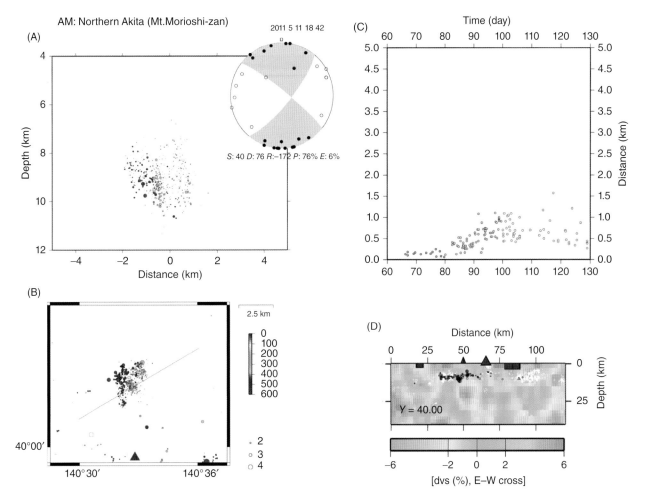

Fig. 24.5. Hypocenter distribution of the major swarm near Mt. Moriyoshi in Northern Akita (Region AM). Focal mechanism of the Mjma 2.4 event that occurred on May 11 2011 is also shown. (A) NE–SW vertical cross section. Color of circle means the lapse time after the 2011 Tohoku-Oki earthquake. (B) Epicenter distribution. Arrow shows the approximated front of hypocenters. (C) Time–distance plots. Distance from the first event of this cluster is shown. (D) E–W vertical cross section of Vp perturbation. Gray and colored dots show the hypocenters before and after the 2011 Tohoku-Oki earthquake. Red triangle shows the volcano. Red square in (D) shows the location of the surface trace of the active fault. White square in (B) denotes the location of a seismic station.

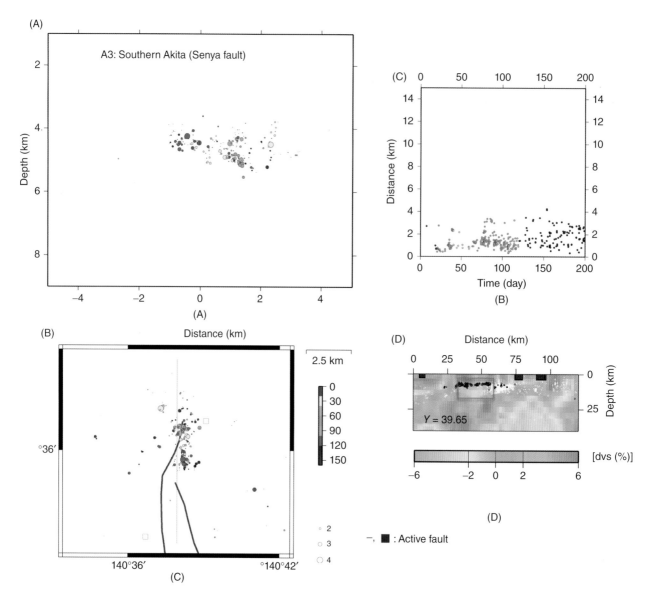

Fig. 24.6. Hypocenter distribution near Senya fault in Southern Akita (Region A3). (A) NNW–SSE vertical cross section. Color of circle means the lapse time after the 2011 Tohoku-Oki earthquake. (B) Epicenter distribution. (C) Time–distance plots. Distance from the first event of this cluster is shown. (D) E–W vertical cross section of Vs perturbation. Gray and colored dots show the hypocenters before and after the 2011 Tohoku-Oki earthquake. Red bold lines in (B) and red squares in (D) show the location of the surface trace of the active fault. White squares in (B) denote the locations of seismic stations.

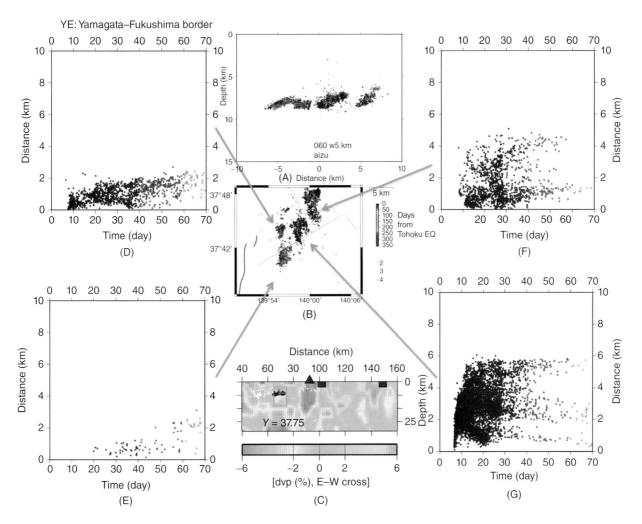

Fig. 24.7. Hypocenter distribution of the swarm on the Yamagata–Fukushima border. (A) NE–SW vertical cross section. Color of circle means the lapse time after the 2011 Tohoku-Oki earthquake. (B) Epicenter distribution. (C) E–W vertical cross section of Vs perturbation. Black and white dots show hypocenters before and after the 2011 Tohoku-Oki earthquake. Red triangle shows the volcano. Red bold lines in (B) and red squares in (C) show the location of the surface trace of the active fault. White square in (B) denotes the location of a seismic station. (D–G) Time–distance plots. Distance from the first event of each cluster is shown.

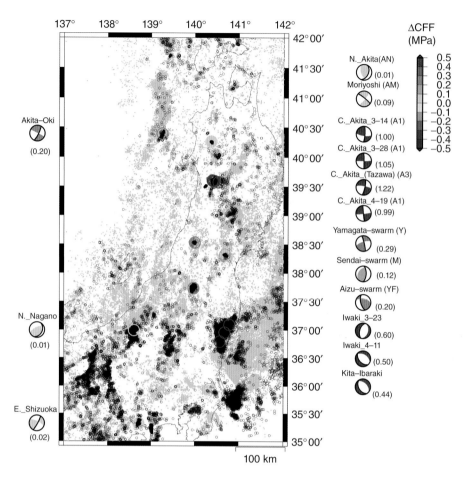

Fig. 24.9. Focal mechanisms with the plausible fault plane defined by bold great circles. Color of circle indicates the amount of Coulomb stress change (numerals in parenthesis) caused by the 2011 Tohok-Oki earthquake for each fault assumed in this study. Hypocenters of inland earthquakes that occurred in NE Japan based on the Japan Meteorological Agency (JMA) catalog are also shown. Inland earthquakes before and after the 2011 Tohoku earthquake are distinguished by gray and black circles, respectively. Stars denote earthquakes with magnitude > 5.

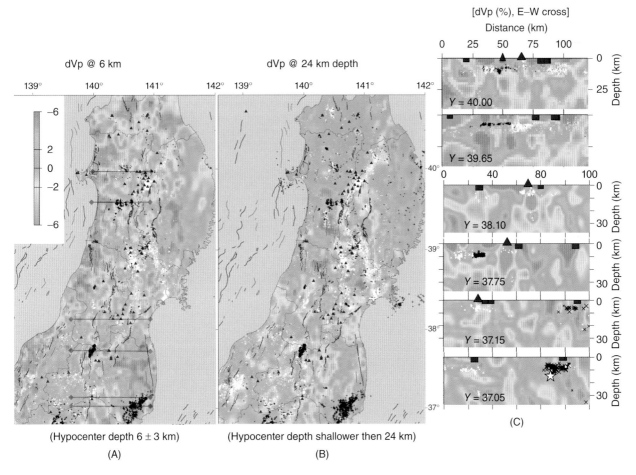

Fig. 24.10. Seismic velocity of the crust in NE Japan. (A) Vp perturbation at a depth of 6 km. Hypocenters of events at depths of 3–9 km are also shown. (B) Vs perturbation at a depth of 24 km. Hypocenters of events at depths <24 km are also shown. (C) E–W vertical cross section of Vp perturbation at various latitudes. Locations of cross sections are shown by gray lines in (A). White and black dots show hypocenters before and after the 2011 Tohoku-Oki earthquake. Large and small stars denote the hypocenters of a Mjma 7.0 earthquake on April 11, 2011 and a Mjma 6.4 earthquake on April 12, 2011 respectively. Red triangles show volcanoes (Geological Survey of Japan 2013). Red bold lines in (A) and (B), and red squares in (c) show the location of surface traces of active faults.

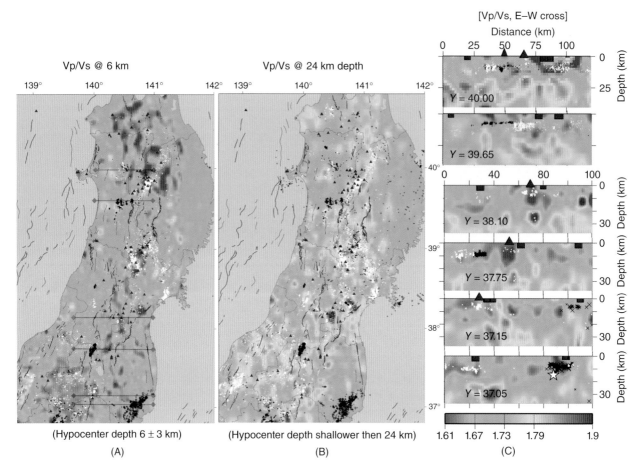

Fig. 24.11. Vp/Vs distribution of the crust in NE Japan. (A) At a depth of 6 km. Hypocenters of events at depths from 3 to 9 km are also shown. (B) At a depth of 24 km. Hypocenters of the events at depths <24 km are also shown. (C) E–W vertical cross section of Vp perturbation at various latitudes. Locations of cross sections are shown by gray lines in (A). White and black dots show hypocenters before and after the 2011 Tohoku-Oki earthquake. Large and small stars denote the hypocenters of a Mjma 7.0 earthquake on April 11, 2011 and a Mjma 6.4 earthquake on April 12, 2011, respectively. Red triangles show volcanoes. Red bold lines in (A) and (B), and red squares in (C) show the locations of surface traces of active faults.

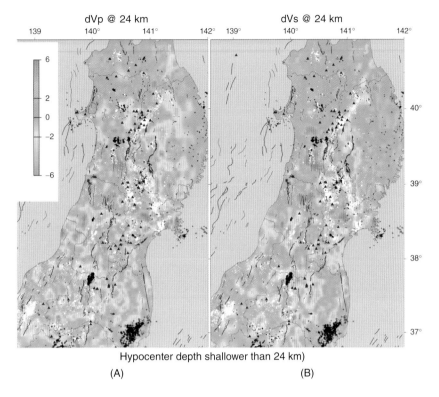

Fig. 24.12. Seismic velocity distribution of the crust in NE Japan. (A) dVp at a depth of 24 km. (B) dVs at a depth of 24 km. Hypocenters of events at depths <24 km are also shown.

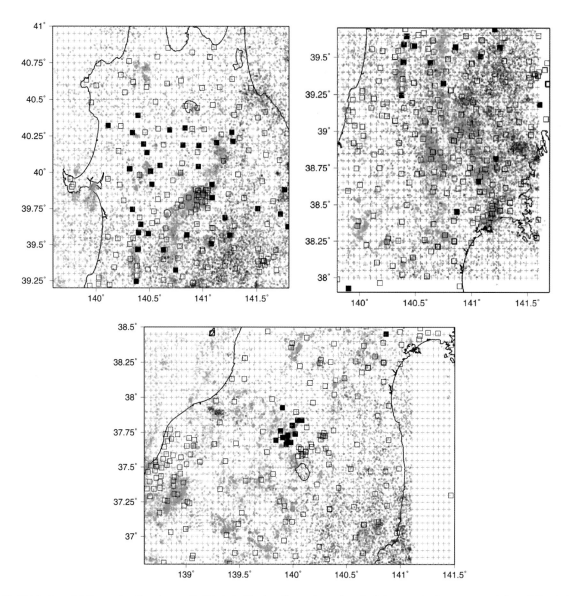

Fig. 24.A1. Distribution of stations and earthquakes used in this study. We show three subareas for seismic tomography. Square, green circle, and blue circle denote station, shallow earthquake, and deep earthquake, respectively. Black square denotes the station deployed by the Group for the aftershock observations of the 2011 off the Pacific coast of Tohoku earthquake. Grid used in this study is also shown by crosses.

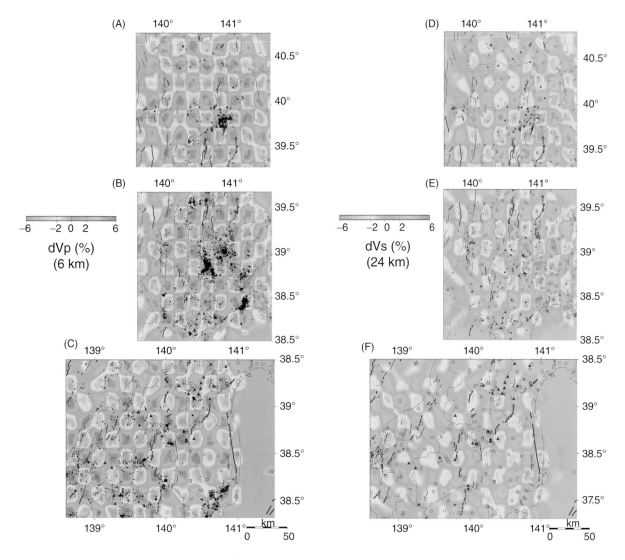

Fig. 24.B1. Results of the checkerboard resolution test. (A) dVp at a depth of 6 km for the northern part, (B) the central part, and (C) the southern part. (D) dVs at a depth of 24 km for the northern part, (E) the central part, and (F) the southern part. Black dots show the hypocenters before the 2011 Tohoku-Oki earthquake. Red triangles show the volcano. Red open square shows the location of the station used in this study. Red bold line denotes the surface trace of active fault.

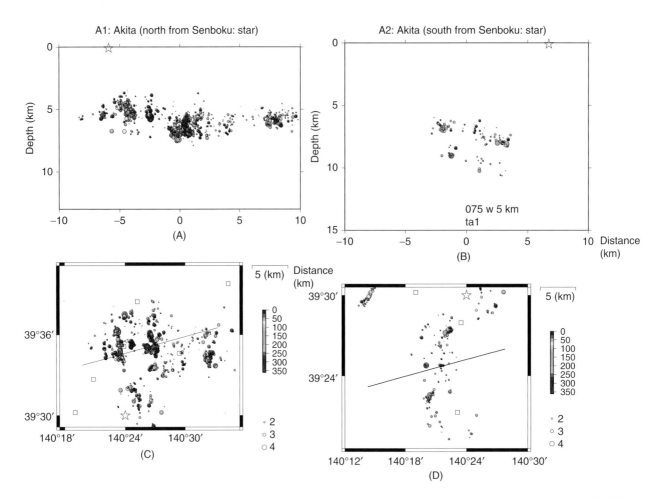

Fig. 24.C1. Hypocenter distribution near of the swarm in southern Akita (Region A1 and A2). (A) ENE−WSW vertical cross section for the swarm a few kilometers north of the hypocenter of the 1914 Akita-Senboku earthquake. Color of circle means the lapse time after the 2011 Tohoku-Oki earthquake. (B) Epicenter distribution for (A). (C) ENE−WSW vertical cross section for the swarm a few kilometers south of the hypocenter of the 1914 Akita-Senboku earthquake. Color of circle means the lapse time after the 2011 Tohoku-Oki earthquake. Gray circles are the earthquake within 1 year before the 2011 Tohoku-Oki earthquake. (D) Epicenter distribution for (C). White square denote the location of seismic station.

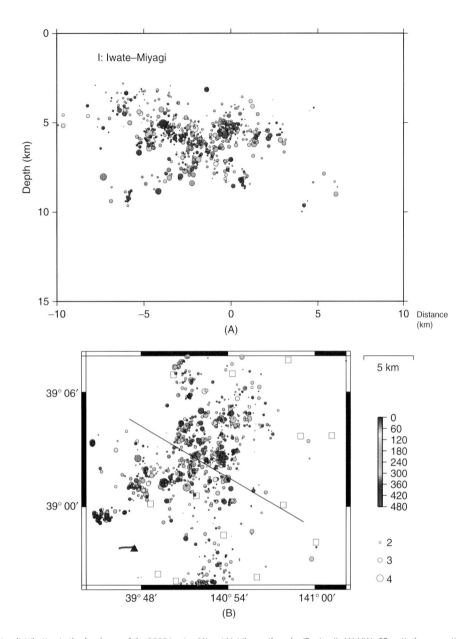

Fig. 24.C2. Hypocenter distribution in the focal area of the 2008 Iwate–Miyagi Nairiku earthquake (Region I). (A) NW–SE vertical cross section. Color of circle means the lapse time after the 2011 Tohoku-Oki earthquake. (B) Epicenter distribution. Red triangle and red bold line denote volcano and the surface trace of active fault, respectively. White square denote the location of seismic station.

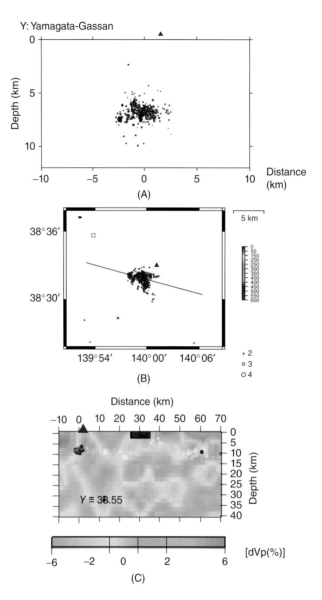

Fig. 24.C3. Hypocenter distribution near Mt. Gassan volcano in Yamagata (Region Y). (A) NNW–SSE vertical cross section. Color of circle means the lapse time after the 2011 Tohoku-Oki earthquake. (B) Epicenter distribution. (C) E–W vertical cross section of Vs perturbation. Black and white dots show the hypocenters before and after the 2011 Tohoku-Oki earthquake. Red triangles show the volcano (Gassan). Red square shows the location of the surface trace of the active fault. White square denote the location of seismic station.

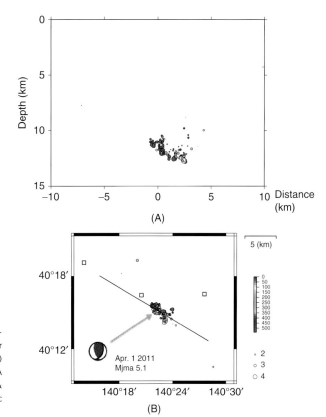

Fig. 24.C4. Hypocenter distribution in the focal area of the Mjma 5.1 earth-quake in northern Akita (Region AN). (A) NW–SE vertical cross section. Color of circle means the lapse time after the 2011 Tohoku-Oki earthquake. (B) Epicenter distribution. AQUA-CMT (2011-04-01 19:49:47/NORTHERN AKITA PREF on http://www.hinet.bosai.go.jp/AQUA/aqua_catalogue.php?y=2011& m=04&LANG=en) is also shown. Whites square denote the locations of seismic stations.

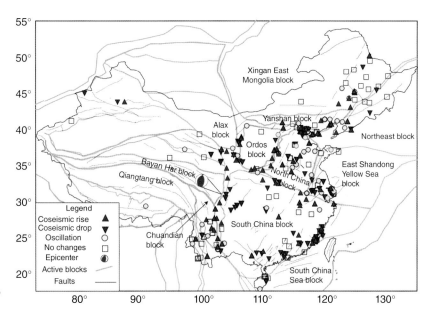

Fig. 25.1. Coseismic water-level changes in response to the Wenchuan earthquake.

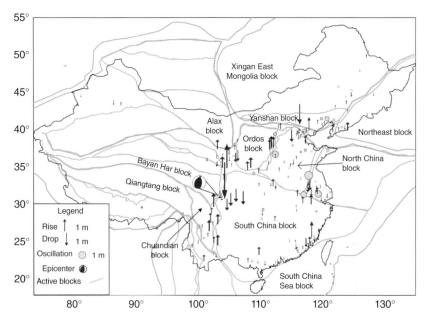

Fig. 25.2. Distribution of coseismic response amplitudes induced by the Wenchuan earthquake.

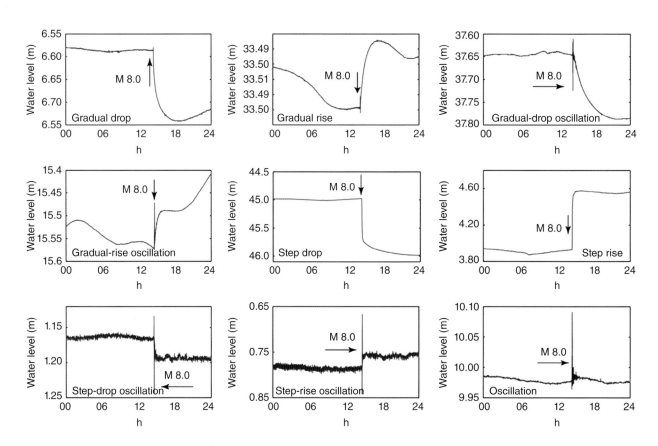

Fig. 25.3. Nine distinct response patterns of water-level changes induced by the Wenchuan earthquake. All the examples shown have sampling intervals of 1 min.

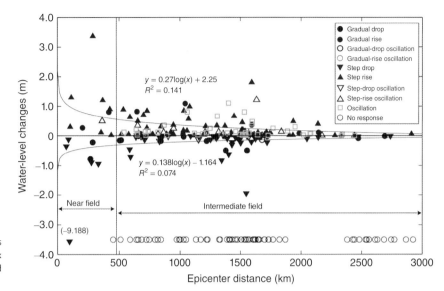

Fig. 25.4. Coseismic groundwater-level responses as a function of distance from the epicenter. The black circles indicate the locations of wells that recorded no coseismic response.

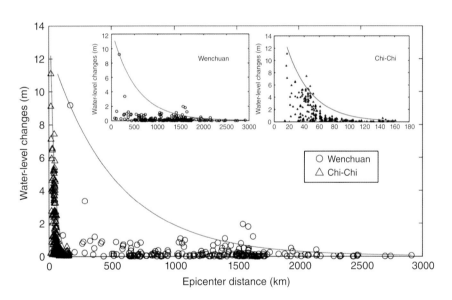

Fig. 25.5. Envelopes of the responses to the Wenchuan and 1999 Chi-Chi earthquakes. Triangles and circles show the coseismic responses following the 1999 Chi-Chi and the Wenchuan earthquakes, respectively. The red and blue curves are the envelopes for the Chi-Chi and the Wenchuan earthquakes, respectively.

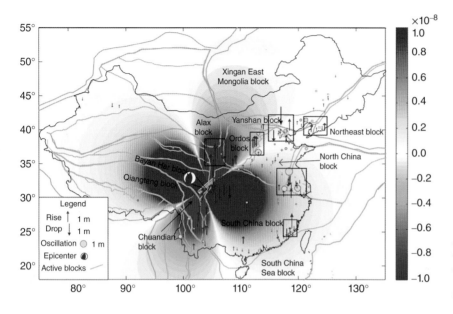

Fig. 25.6. Coseismic groundwater-level changes as a function of volumetric strain calculated from the dislocation model. Red colors indicate dilatation, and blue colors indicate contraction. The black rectangles highlight sensitive monitoring sites that have large coseismic responses to the earthquake.

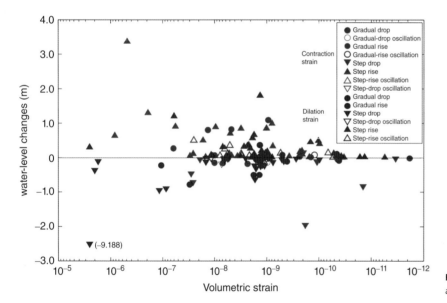

Fig. 25.7. Relationship between volumetric strain and the water-level response.

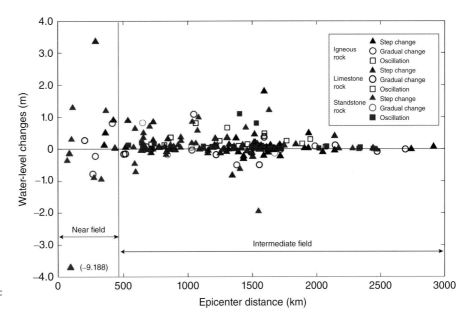

Fig. 25.9. Relationship between coseismic groundwater-level changes and lithology.

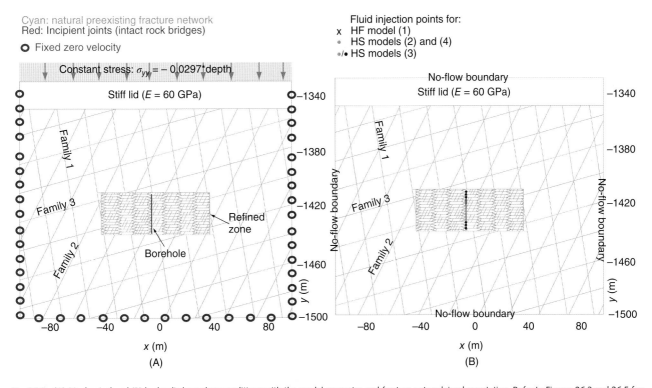

Fig. 26.2. (A) Mechanical and (B) hydraulic boundary conditions with the model geometry and fracture network implementation. Refer to Figures 26.3 and 26.5 for an enlargement of the refined zone.

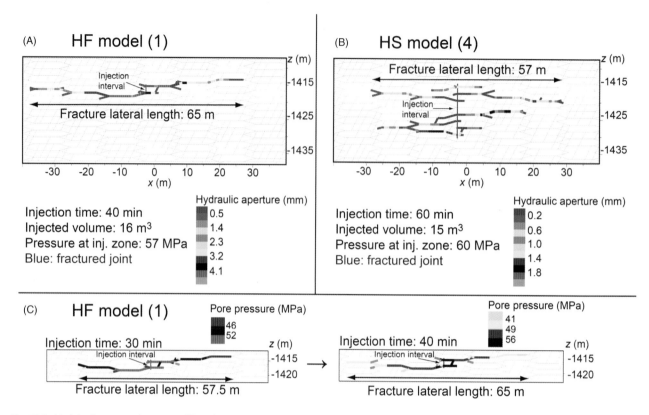

Fig. 26.3. Model enlargements showing total lateral extent and opening of hydraulic fractures for (A) hydraulic fracturing model (1) after 40 min of fluid injection at 400 l min⁻¹ into a 2-m packed interval, and (B) hydraulic shearing model (4) after 60 min of fluid injection at 250 l min⁻¹ into a 15-m packed interval. (C) Cyclical increase and decrease of pore pressures accompanying hydraulic fractures growth.

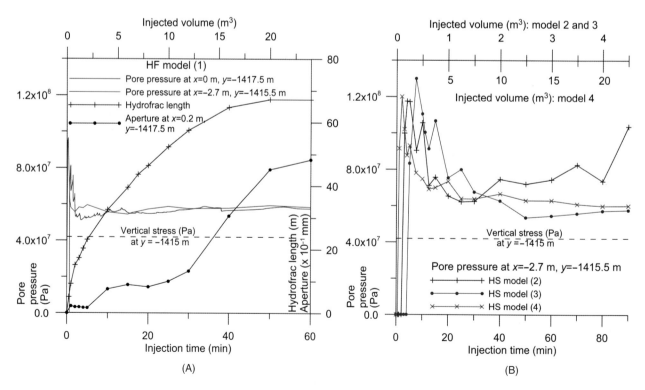

Fig. 26.4. (A) Pore pressure at and close to injection, hydraulic fracture length, and aperture as a function of injection time and volume (in m³) for hydraulic fracturing (HF) model (1): 400 l min⁻¹ into a 2-m interval. (B) Pressure–time response for hydraulic shearing (HS) models at injection point $x = -2.7$ m, $y = -1415.5$ m; HS model (2): 50 l min⁻¹ into a 15-m packed interval; HS model (3): 50 l min⁻¹ into a 30-m packed interval; HS model (4): 250 l min⁻¹ into a 15-m interval.

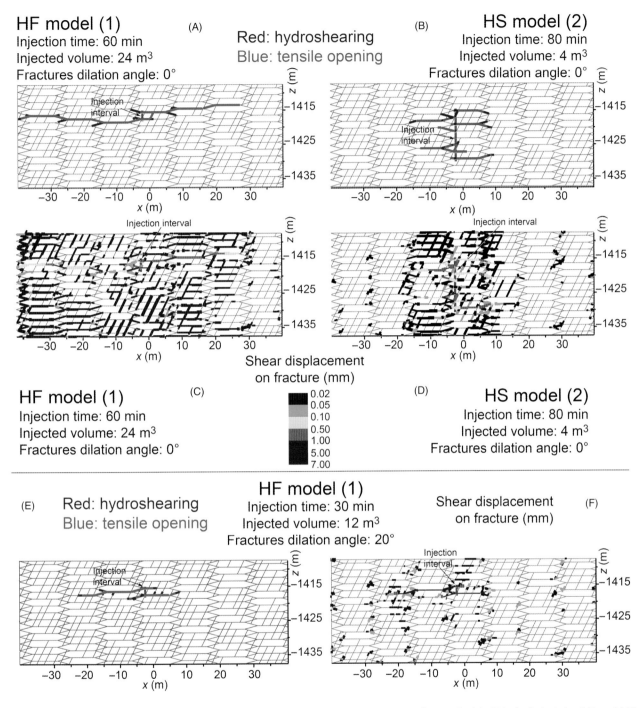

Fig. 26.5. Model enlargements illustrating hydroshear/slip and tensile opening and shear displacement on fractures for (A)–(C) hydraulic fracturing (HF) model (1) after 60 min of fluid injection at 400 l min⁻¹ into a 2-m packed interval, (B)–(D) hydraulic shearing model (2) after 80 min of fluid injection at 50 l min⁻¹ into a 15-m packed interval, and (E)–(F) HF model (1) after 30 min of fluid injection with fractures having nonzero dilation angle. Note that for shear displacement, fracture line thickness increases when its value is close to the upper boundary of its color grade.

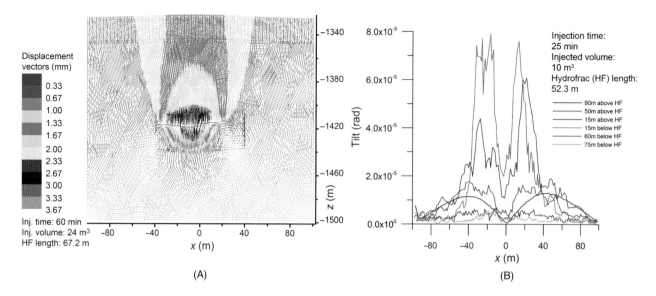

Fig. 26.6. Hydraulic fracturing model (1): induced (A) displacement field after 60 min of fluid injection at 400 l min^{-1} into a 2-m packed interval and (B) tilt after 25 min of fluid injection for six observation planes located above and below hydraulic fractures.

HF model (1): inj. time = 30 min, inj. volume = 12 m^3, HF length = 57.5 m

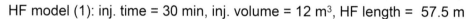

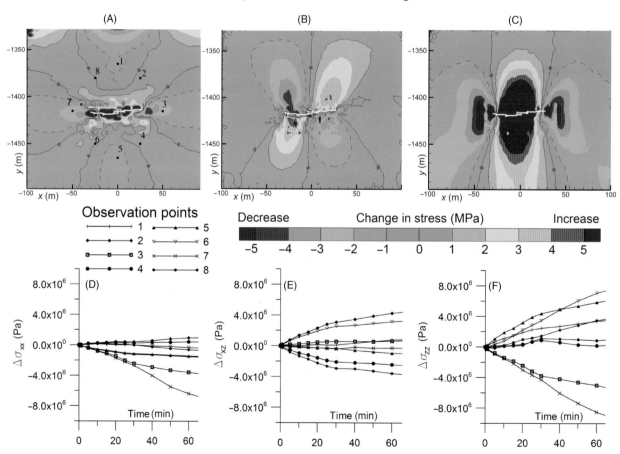

Fig. 26.7. Hydraulic fracturing model (1): change in (A) horizontal, (B) shear, and (C) vertical stress after 30 min of fluid injection at 400 l min^{-1} into a 2-m packed interval. Note that change in stress is computed via a kriging of data obtained by subtracting stress at a given injection time from the initial stress. (D)–(F) illustrate stress responses during fluid injection for eight observation points whose location is shown in (A).

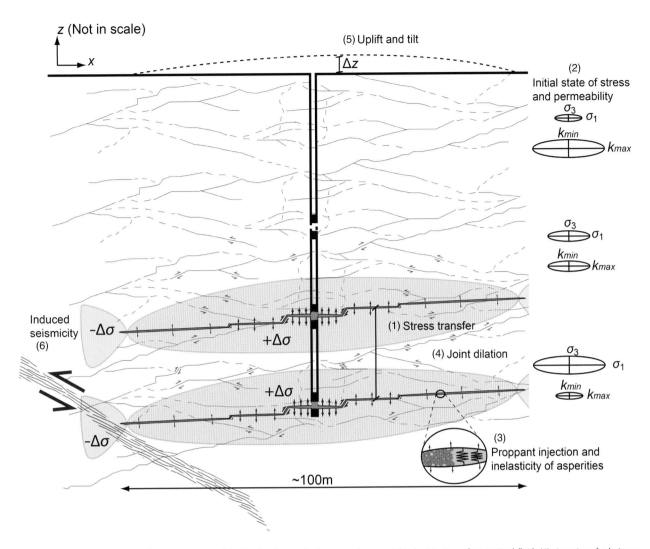

Fig. 26.8. Schematic cross section illustrating issues related to development of connected permeability by injection of pressurized fluid: (1) stress transfer between principal hydraulic fractures and its attenuation by intact rock blocks; (2) *in situ* stress state and initial permeability; (3) permanent aperture of tensile hydraulic fractures due to inelasticity of asperities or because of proppant injection; (4) permeability enhancement associated with fracture dilation under shearing; (5) generation of uplift and tilt, as well as (6) induced seismicity.

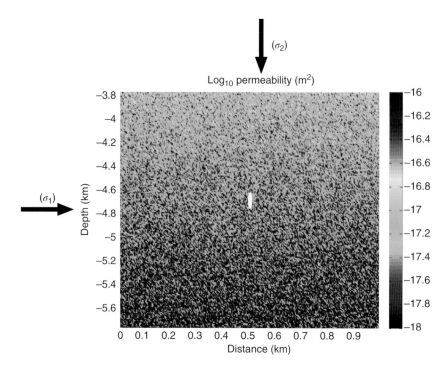

Fig. 27.2. Initial permeability field of the numerical model of a 2D cross section of the Basel experiment. The pressure–time history (Fig. 27.3) is applied as a boundary condition in the region approximated by the white ellipse in the center of the model. The average permeability of this configuration is about 10^{-17} m^2. No flow boundary conditions are applied on all external boundaries, and shear and normal effective normal stresses acting on virtual slip planes extending out of the page are calculated using Eqs 27.3 and 27.4.

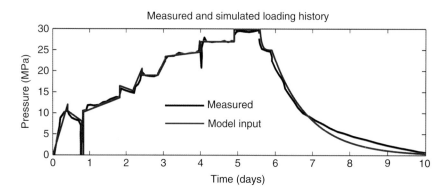

Fig. 27.3. Comparison of the recorded pressure–time history and the pressure–time history applied in the model.

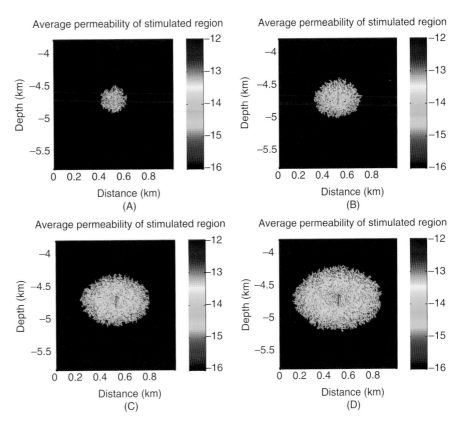

Fig. 27.5. Evolution of the permeability field in the model showing the stimulated region defined as those parts of the numerical domain that have reached the failure condition. Permeability field at (A) 1.4 day, (B) 2.8 days, (C) 4.2 days, and (D) 5.5 days of model stimulation. The scale is log(k), shown in the color bar.

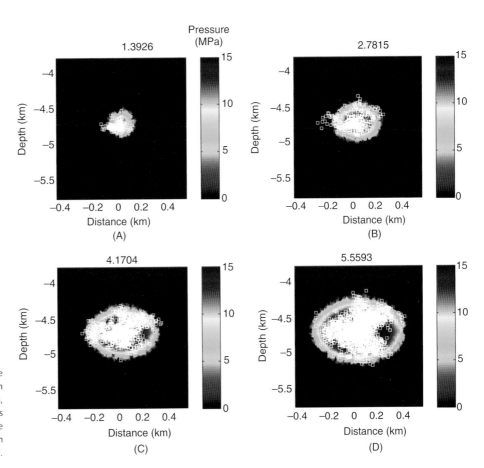

Fig. 27.7. Evolution of the fluid pressure field (above hydrostatic) and comparison with measured earthquake hypocenters, showing that the diffusing front compares well with the hypocenter migration. Note that the color bar is in MPa, showing high pressures propagating through the system.

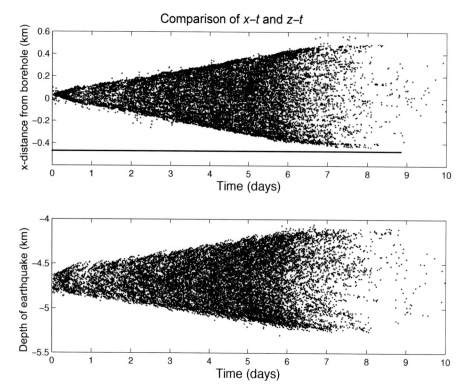

Fig. 27.9. Comparison between model (black) and observations (red) of hypocenter migration away from the borehole (top) and with depth (bottom).

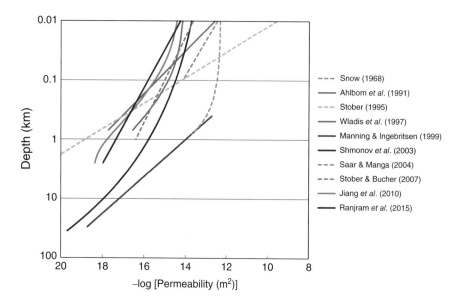

- - - - Snow (1968)
——— Ahlbom *et al.* (1991)
- - - - Stober (1995)
——— Wladis *et al.* (1997)
——— Manning & Ingebritsen (1999)
——— Shmonov *et al.* (2003)
- - - - Saar & Manga (2004)
- - - - Stober & Bucher (2007)
——— Jiang *et al.* (2010)
——— Ranjram *et al.* (2015)

Fig. 30.1. Proposed permeability–depth (*k–z*) relations. Of these, only Manning & Ingebritsen (1999), Shmonov *et al.* (2003), and Saar & Manga (2004) are conditioned on data from depths exceeding a few kilometers.

PART III(B): Fault zones

The permeability of active subduction plate boundary faults

DEMIAN M. SAFFER

Department of Geosciences, Center for Geomechanics, Geofluids, and Geohazards, The Pennsylvania State University, University Park, PA, USA

ABSTRACT

At subduction zones, continuous influx of fluids drives a dynamic system in which fault slip, fluid flow, and advective transport are tightly coupled. Field and numerical modeling studies have provided insight into the nature and rates of flow in these systems and illustrated that active subduction faults, including the master décollement and splay faults cutting the upper plate, are important conduits. Observations of *in situ* fracture dilation, modeling studies, and direct measurements documenting strong pressure dependence of fault permeability collectively suggest that permeability varies in time, perhaps due to pore pressure cycling. However, mechanical and fluid budget considerations dictate that increased fault permeability cannot be sustained, nor can it be present across the entire fault surface at a given time. The emerging conceptual model is that permeable patches or channels occupy only a fraction of the fault surface and shift transiently. Fault zone permeabilities obtained by several approaches are consistent between margins, with time-averaged values of approximately 10^{-15} to 10^{-14} m^2, several orders of magnitude higher than for the sediment matrix. Higher, transiently increased values of approximately 10^{-13} to 10^{-11} m^2 are required to explain geochemical and thermal signals and observed focused flow rates. Although faults accommodate significant fluid fluxes from dewatering of the surrounding sediment, they have little effect on pore pressures within the wall rock, where drainage is limited by low matrix permeability. However, fault permeability is a key control on the transport and preservation of localized geochemical and thermal anomalies from depths where temperatures are higher and low-temperature metamorphic reactions are underway. Despite significant recent progress, several key aspects of hydrologic behavior in these active faults remain incompletely understood, including the nature and timescale of transience, the causes of permeability enhancement and its relationship to fault slip and pore pressure fluctuations, and the depths and distances from which deeply sourced fluids are captured, mixed, and transported up-dip.

Key words: advective transport, dewatering, fault permeability, fluid flow, subduction zones

INTRODUCTION

Fault zone permeability, its spatial distribution, relationship to fault zone structural architecture, and variation in time are primary controls on the magnitude and distribution of fluid, heat, and solute fluxes in the Earth's crust (e.g., Faulkner *et al.* 2010; Bense *et al.* 2013). At subduction zones in particular, the combination of rapid fluid delivery and dewatering, tectonic deformation, and earthquakes leads to a dynamic environment in which fault zone processes and fluid flow are tightly linked. Fault permeability in these systems exerts a first-order control on pore-fluid pressure distribution, and thus the effective normal stress that governs the absolute strength of fault zones and immediately adjacent wall rock (e.g., Rice 1992; Townend & Zoback 2000), the overall geometry of the forearc (e.g., Davis

et al. 1983; Saffer & Bekins 2002), and the nature of fault slip (e.g., Scholz 1998; Kodaira *et al.* 2004; Liu & Rice 2007; Saffer & Tobin 2011; Kitajima & Saffer 2012). Subduction fault permeability is also a fundamental control on heat transport and elemental and volatile refluxes to the oceans (e.g., Chan & Kastner 2000; Hacker 2008).

Focused Ocean Drilling Program (ODP) and Integrated Ocean Drilling Program (IODP) efforts have enabled detailed investigations into fault zone structure, architecture, frictional properties, *in situ* pore pressure, and permeability at several subduction margins (Fig. 18.1) (e.g., Screaton *et al.* 1990, 1995, 2000; Tobin *et al.* 2001; Skarbek & Saffer 2009; Bekins *et al.* 2011; Saffer & Tobin 2011). These investigations have provided important new insights into fault zone hydraulic behavior (e.g., Mascle *et al.* 1988; Westbrook *et al.* 1994;

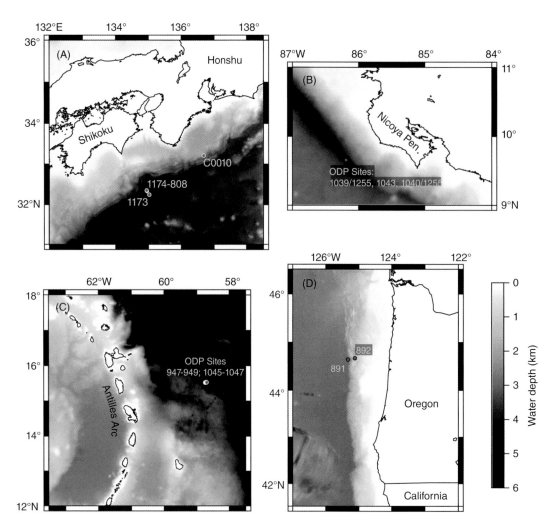

Fig. 18.1. Maps showing study areas for the (A) Nankai, (B) Costa Rica, (C) Barbados, and (D) Cascadia subduction zones. Locations of ODP and IODP boreholes discussed in the text are shown in red. (*See color plate section for the color representation of this figure.*)

Shipley *et al.* 1995; Moore *et al.* 2001; Kinoshita *et al.* 2009), based on diverse approaches that include laboratory experiments (e.g., Zwart *et al.* 1997); drilling, logging, and well testing (e.g., Screaton *et al.* 2000; Bourlange *et al.* 2003; Bekins *et al.* 2011; Conin *et al.* 2013); analyses of down-hole geochemical and thermal observations (e.g., Fisher & Hounslow 1990; Saffer & Screaton 2003); seafloor study of seep distributions, rates, and chemistry (e.g., Moore *et al.* 1990; Carson *et al.* 1994); imaging of gas hydrate bottom-simulating reflector geometry (Davis *et al.* 1995; Zwart *et al.* 1996); and regional-scale modeling studies constrained by field and laboratory data (e.g., Bekins *et al.* 1995; Saffer & Bekins 1998a; Spinelli *et al.* 2006) (Fig. 18.2).

Recent major reviews of fault zone permeability have summarized conceptual aspects of fault zone permeability, the state of knowledge of fault zone hydraulic architecture (Bense *et al.* 2013), and the interplay between fault structure and mechanical and hydrological behaviors (Faulkner *et al.* 2010) across a wide range of geologic and tectonic settings. Other recent work has synthesized data and modeling studies for particular well-studied localities (e.g., Bekins & Screaton 2007). To date, there has not been a comprehensive review of fault permeability that integrates information from multiple subduction zones needed to parameterize numerical models, provide a framework for comparison and interpretation of flow rate and geochemical data, and interpret data sets from ongoing borehole observatory deployments.

The goal of this paper is to synthesize a wide range of analyses and observations across several subduction margins and to summarize current knowledge about the permeability of active subduction zone faults. Discussion is restricted primarily to observations and constraints on flow rates, advective transport, and fault physical properties; although chemical processes associated with cementation and healing are likely to play a role in the evolution of fault permeability, there are few relevant observations at the shallow depths where active subduction faults have been accessed. I focus specifically on: (i) quantitative constraints on fault zone permeability, its links to observed fault zone architecture, structure, and seismic reflection character, and its variation with depth and effective normal stress; (ii) evidence for temporal and spatial variations in permeability and fluid fluxes; and (iii) the implications of existing observations for drainage of excess fluid pressure and the transport of heat and solutes.

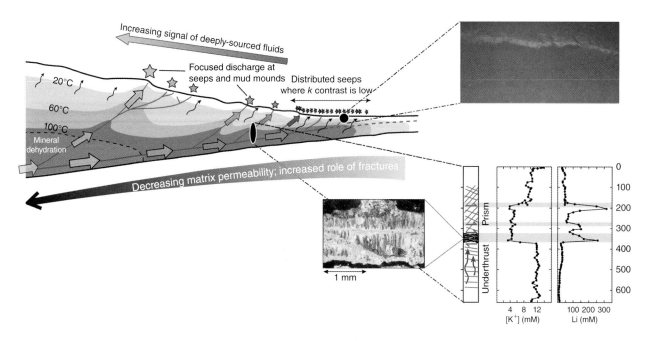

Fig. 18.2. Schematic subduction zone cross section synthesizing hydrologic, geologic, and geochemical observations indicating focused fluid flow along fault zones. Colors indicate temperature distribution and illustrate the role of faults in advecting heat (after Vrolijk *et al.* (1991)). Arrows indicate fluid flow; colors represent the relative contributions of deeply sourced fluids derived from diagenetic and metamorphic reactions (green) and initial pore water with seawater composition (blue). Stars at seafloor indicate sites of focused seepage, with the same color scheme as flow arrows. Inset at bottom right shows schematic structural column spanning the décollement at ODP Site 1040 (Costa Rican margin) and pore-water geochemical data indicating active or recent flow in the décollement and a splay fault in the hanging wall (Chan & Kastner 2000; Morris *et al.* 2003b). Photograph shows example of an antitaxial vein from the Barbados décollement (Vrolijk & Sheppard 1987). Photograph at the top right shows approximately 50-cm-high authigenic carbonate crust formed at the outcrop of the décollement at the Costa Rican margin. (*See color plate section for the color representation of this figure.*)

FAULT ZONE ARCHITECTURE: INFERENCES ABOUT HYDRAULIC PROPERTIES AND BEHAVIOR

A wide range of field observations suggest that tectonically active fault zones act as permeable conduits, with significant anisotropy associated with fault architecture that leads to higher permeability along structure than across it (e.g., Arch & Maltman 1990; Caine *et al.* 1996; Faulkner & Rutter 2001; Tobin *et al.* 2001; Bense *et al.* 2013). At subduction zones, observations are generally consistent with conceptual models of fault architecture that include one or more zones of fine-grained fault gouge where slip is localized, surrounded by damaged and fractured wall rock (e.g., Caine *et al.* 1996; Faulkner *et al.* 2010).

Structural descriptions of drill core, down-hole logging data, and regional-scale geophysical observations at shallow depths (less than approximately 2 km below sea floor, bsf) all indicate that major subduction zone thrust faults – including the master décollement, splay faults, and out-of-sequence thrusts – are characterized by tens-of-meters-thick zones of brecciated sediment and sedimentary rock with intense shear fabrics. These faults commonly include thin zones of localized shear bounded by deformed and fractured, brecciated, and possibly dilated fault rock (e.g., Bourlange *et al.* 2003; Ujiie *et al.* 2003; Vannucchi & Leoni 2007). Fracturing is primarily focused in the hanging

wall, with an increased intensity of brecciation downward, and a sharp basal contact with relatively less deformed underthrust sediments below (Fig. 18.3) (Shipboard Scientific Party 1994; Tobin *et al.* 2001; Ujiie *et al.* 2003; Vannucchi & Leoni 2007). Field observations also suggest that with increasing depth and decreasing porosity in the sedimentary and metasedimentary wall rock, fault zone permeability should become increasingly fracture dominated and localized (e.g., Moore 1989), whereas in high-porosity sediments at shallow depths, fracture networks are likely to be more distributed and complex (e.g., Maltman *et al.* 1997) (Figs 18.2 and 18.3).

For example, at the Costa Rica subduction zone, the active décollement was sampled by drilling at four ODP sites – two located approximately 500 m landward of the trench that sampled the fault at approximately 150 m below seafloor (mbsf) (Sites 1043 and 1255) and two located approximately 1.6 km from the trench where the fault depth is approximately 350 mbsf (Sites 1040 and 1254) (Fig. 18.1) (e.g., Tobin *et al.* 2001; Vannucchi & Leoni 2007). At these sites, the décollement zone comprises brecciated mudstone derived primarily from the hanging wall and ranges in thickness from 12 to 35 m. Brecciation generally increases in intensity with depth, with fragment sizes decreasing from 1 to 10 cm near the top of the fault zone to <0.3 mm at its base (Vannucchi & Leoni 2007). The base of the décollement is marked by an approximately

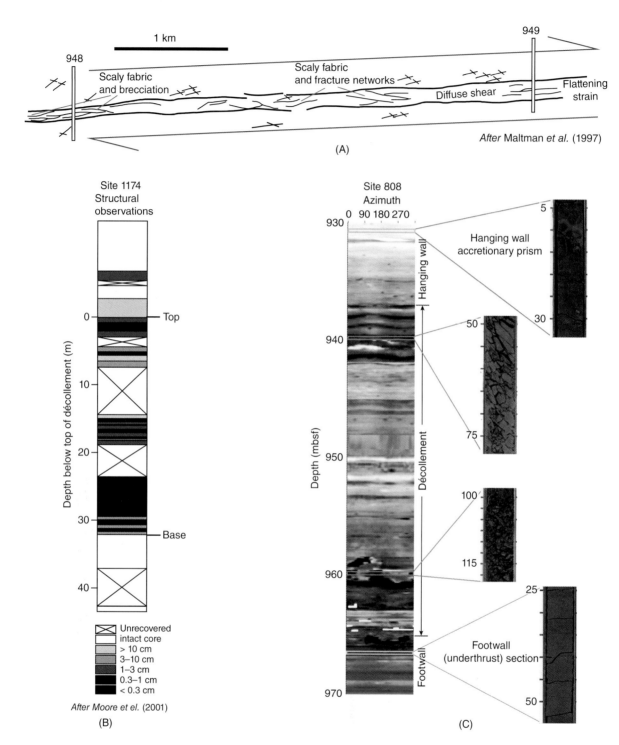

Fig. 18.3. Summary of structural and logging observations across the décollement zone at the Barbados (A) and Nankai (B and C) subduction zones. The décollement is commonly defined as a tens-of-meters-thick zone of brecciated and fractured scaly mudstone, with increasing intensity of brecciation downward, and a sharp basal contact with relatively undeformed underthrust sediment. (A) Synthesis of logging and structural observations from core at several ODP sites at the Barbados margin (after Maltman *et al.* (1997)). (B) Summary structural section across the décollement at ODP Site 1174, located approximately 2 km landward of the trench in the Nankai Trough (after Moore *et al.* (2001)). (C) Logging while drilling resistivity image across the décollement zone at ODP Site 808 in the Nankai Trough, located approximately 1 km landward of Site 1174. Core photographs are from ODP Site 1174 with equivalent locations at Site 808 estimated based on relative depth from the top of the décollement (after Ienaga *et al.* (2006)). (*See color plate section for the color representation of this figure.*)

10-cm-thick zone of interpreted highly localized shear that separates the damage zone above and the relatively undeformed underthrust sediment below.

The décollement at the Nankai Trough is characterized by a similar pattern of brecciation within an approximately 30- to 40-m-thick damage zone, composed of millimeter- to centimeter-scale overconsolidated mudstone fragments bounded by striations (Fig. 18.3) (e.g., Mikada *et al.* 2002; Ujiie *et al.* 2003). Coring across the frontal thrust of the Oregon accretionary prism sampled zones of intense shear fabrics, consisting of 1–5 mm polished angular chips of clay and siltstone that are more consolidated than the adjacent wall rock, and zones of well-developed scaly fabric, characterized by pervasive shear and highly polished parting surfaces (Shipboard Scientific Party 1994). Similar scaly fabrics are also a defining characteristic of the Barbados décollement zone (Fig. 18.3) (e.g., Maltman *et al.* 1997).

Textural and structural analyses, *in situ* measurements and images obtained by wireline logging or logging while drilling, and detailed study of veins suggest that these brecciated fault zones may be dilated by near-lithostatic pore pressures (Fig. 18.3) (e.g., Labaume *et al.* 1997a; Bourlange *et al.* 2003; Conin *et al.* 2013). For example, resistivity images and logging measurements through the décollement zone at the Nankai Trough document an approximately 30-m-thick zone containing abundant electrically conductive fractures. This, in combination with a contrast between the high bulk density of cored material and low bulk density measured by logging while drilling at a scale of tens of centimeters, has been interpreted to reflect a zone of dilated fractures bounding intact and densified (overconsolidated) breccia fragments (e.g., Bourlange *et al.* 2003; Ienaga *et al.* 2006). This is consistent with the observation that while fractures are abundant throughout the drilled and logged section, most conductive fractures are associated with major fault zones (e.g., McNeill *et al.* 2004; Ienaga *et al.* 2006; Conin *et al.* 2013). It is also consistent with waveform modeling of seismic reflections from active subduction zone faults, which suggest that they represent discrete zones of fractured and compliant rock with low *P*-wave velocity relative to the surrounding wall rock, and which are characterized by increased – although not necessarily lithostatic – pore-fluid pressures (e.g., Cloos 1984; Shipley *et al.* 1994; Tobin *et al.* 1994; Moore *et al.* 1995). At larger scales, the heterogeneous distribution of fault zone seismic reflection amplitude has been interpreted to represent fluid-enriched 'channels' within the fault where pore pressures are highest and permeability is enhanced due to dilation (e.g., Shipley *et al.* 1994).

Carbonate and mud-filled vein textures record growth and filling synchronous with deformation, providing additional evidence for both increased pore pressure and permeability, in both active and exhumed fault zones (e.g., Vrolijk & Sheppard 1987; Sample 1996; Labaume *et al.* 1997a,b; Kondo *et al.* 2005; Vannucchi & Leoni 2007; Rowe *et al.* 2009). Textural evidence and vein structures in the décollement at both Barbados and Nankai

further suggest that increased pore pressure within the faults is episodic, leading to overconsolidation relative to surrounding sediments at times when pore pressure is decreased, and post-consolidation embrittlement, fracturing, and vein formation at times when pore pressure is increased (e.g., Karig 1990; Ujiie *et al.* 2003; Vannucchi & Leoni 2007).

OBSERVATIONS OF FLUID FLOW, ADVECTIVE TRANSPORT, AND SIMPLE MODELS

Indicators of focused fluid flow along fault zones include veins and thermal and geochemical anomalies centered at faults sampled by drilling (e.g., Vrolijk & Sheppard 1987; Blanc *et al.* 1991; Kastner *et al.* 1991), observations of seeps and mud mounds at the seafloor where faults outcrop (e.g., Carson *et al.* 1994; Henry *et al.* 2002; Hensen *et al.* 2004), and shoaling of the gas hydrate bottom-simulating reflector where it intersects faults in the subsurface (Zwart *et al.* 1996). Patterns of consolidation and inferred pore pressure within the underthrusting sediment section suggest upward drainage and thus also provide clear evidence for a permeable décollement or fractured hanging wall overlying the subducted sediment package (Fig. 18.4) (e.g., Moore 1989; Saffer 2003, 2007; Screaton & Saffer 2005; Skarbek & Saffer 2009). For the most part, these observations do not yield detailed quantitative constraints on fault zone permeability or flow rate, but do provide first-order estimates of hydraulic properties and key evidence for focused and transient flow.

The common observation of seafloor seeps and active mud mounds localized near outcrops of major fault zones illustrates that the master décollement, out-of-sequence thrusts, and splay faults are important pathways for the focused transport of fluids, as well as the advection of dissolved solutes that support chemosynthetic communities and form authigenic carbonate deposits (cf Fig. 18.2) (Moore *et al.* 1990; Carson *et al.* 1994; Henry *et al.* 2002; Hensen *et al.* 2004; Ranero *et al.* 2008; Sahling *et al.* 2008). Flow rates associated with fault zones are consistently 3–6 orders of magnitude higher than measured or modeled background rates (e.g., Carson & Screaton 1998; Sahling *et al.* 2008). Sites of seafloor seepage at or near fault outcrops, and geochemical and thermal indicators of active focused flow at faults sampled by drilling have been correlated with negative seismic polarity fault zone reflections at the Nankai (Henry *et al.* 2002), Cascadia (Moore *et al.* 1990, 1995), Costa Rica (e.g., Chan & Kastner 2000; Spinelli *et al.* 2006; Ranero *et al.* 2008), and Barbados (Gieskes *et al.* 1990; Shipley *et al.* 1994) subduction zones, consistent with hypothesized links between subsurface overpressure, fault zone dilation, and enhanced permeability (Fig. 18.5) (e.g., Brown *et al.* 1994; Saffer & Bekins 1999; Bourlange & Henry 2007).

Pore-water geochemical anomalies observed at shallow fault zones (less than approximately 1 km below seafloor) commonly include increased concentrations of B and Li, increased

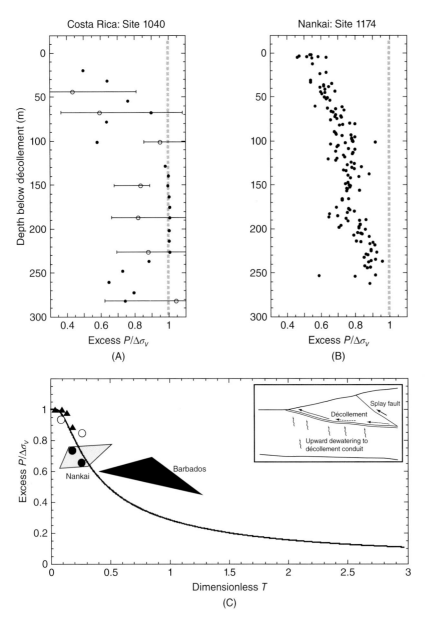

Fig. 18.4. Excess pore pressures within the underthrust sediment sections at (A) Costa Rica (ODP Site 1040) and (B) Nankai (Site 1174), normalized to the increase in overburden from burial by subduction (after Saffer (2007)). Small circles are from measured shipboard porosity; open circles and error bars are from reconsolidation tests. Values of unity indicate undrained behavior, whereas values of zero indicate fully drained (hydrostatic) conditions. At both margins, the pressure profiles suggest upward drainage. (C) Synthesis of normalized excess pore pressures at several drill sites from the Costa Rica (open circles), Barbados (triangles), and Nankai (filled circles) margins, and pressures estimated from seismic reflection interval velocities at the Nankai and Barbados margins (polygons, as labeled) (from Saffer 2007). Line shows predicted pore pressure in a top-drained layer (i.e., inset, top right) (Wissa *et al.* 1971; von Huene & Lee 1982). The agreement between model and observations at multiple margins suggests that underthrust sediments commonly dewater upward to a permeable décollement.

δ^{18} O, decreased CI and K concentrations attributed to clay transformation and other dehydration reactions at greater depth (i.e., pore-water freshening), and increased concentrations of thermogenic hydrocarbons (e.g., Kastner *et al.* 1991; Vrolijk *et al.* 1991; Chan & Kastner 2000; Morris *et al.* 2003b; Hensen *et al.* 2004; Ranero *et al.* 2008; Solomon *et al.* 2009). Collectively, these geochemical signatures require migration of fluids along permeable and hydraulically connected faults over lateral distances of 20 km or more, from source regions where temperatures are 80–150°C (Figs 18.2 and 18.5) (e.g., Moore *et al.* 1987; Vrolijk *et al.* 1991; Spinelli *et al.* 2006; Saffer & McKiernan 2009; Solomon *et al.* 2009). The finite width

of the anomalies provides evidence for transient flow focused along fault zones, combined with outward diffusion (e.g., Blanc *et al.* 1991). Thermal anomalies centered at fault zones penetrated by drilling provide further evidence for time-varying and localized flow from depths of at least a few kilometers, although the depth of fluid origin is not as well constrained as for geochemical anomalies (e.g., Vrolijk *et al.* 1991; Davis *et al.* 1995). The drilling and seep site observations also suggest that splay faults, in addition to the master décollement, are important pathways for solute and fluid transport (Cloos 1984; Vrolijk *et al.* 1991; Hensen *et al.* 2004; Teichert *et al.* 2005; Lauer & Saffer 2012).

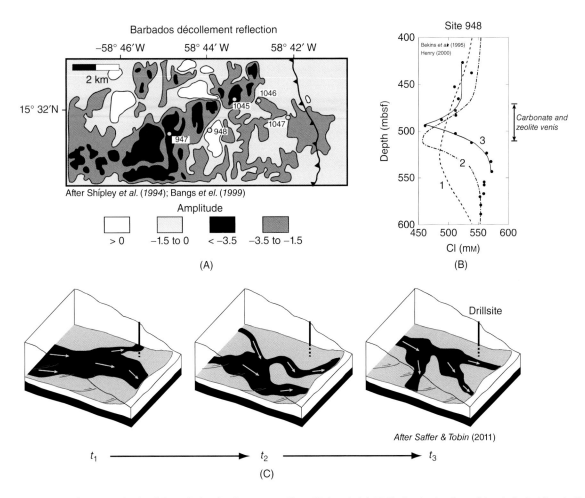

Fig. 18.5. (A) Seismic reflection amplitude of the Barbados décollement zone (from Shipley *et al.* (1994)), showing locations of trench (barbed line; teeth facing subduction direction) and ODP drill sites (circles). (B) Pore-water chloride profile at Site 948, with freshening centered at the décollement. Curves show model results from Bekins *et al.* (1995) for transiently increased décollement permeability of 10^{-13} m^2 (1; duration 30 kyr) and 10^{-11} m^2 (2; duration 300 year) and transient diffusion model from Henry (2000) (3; duration approximately 11 kyr). Distribution of veining within the décollement is shown at right. (C) Schematic showing conceptual model in which permeable channels along the décollement shift over time, leading to transient behavior at a given location, and spatially distributed zones of high permeability and flow (after Saffer & Tobin (2011)). (*See color plate section for the color representation of this figure.*)

Simple one-dimensional (1D) analyses of observed flow rates at seep sites and of thermal anomalies encountered during drilling have provided first-order estimates of transmissivity and permeability along subduction thrusts at several margins (Table 18.1; Fig. 18.6). For example, at the Costa Rican margin, Ranero *et al.* (2008) estimate the permeabilities required to accommodate observed flow rates at seafloor mud mounds and seeps correlated with splay faults cutting the upper plate. Assuming steady-state and 1D flow along fault or fracture conduits that feed the vents, a minimum estimated permeability of 1.3×10^{-14} to 2×10^{-12} m^2 is obtained by assuming a lithostatic (maximum) pressure gradient driving flow (Table 18.1; Fig. 18.6). At the Barbados margin, Fisher & Hounslow (1990) used a simple model to demonstrate that transient fluid flow along the décollement is required to

explain thermal observations at a suite of ODP boreholes. The estimated flow rates are consistent with a fault-parallel décollement permeability of approximately 10^{-12} m^2. Davis *et al.* (1995) conducted a similar analysis of a down-hole temperature anomaly centered at the frontal thrust in the Oregon accretionary prism (ODP Site 892) and showed that it requires transient fluid flow that initiated in approximately the last 400 years at rates of approximately 6×10^{-5} m s^{-1}, corresponding to permeabilities of approximately 1.9×10^{-12} m^2. Davis *et al.* (1995) also estimated a regional background flow rate of approximately 1×10^{-6} m s^{-1} along the frontal thrust required to perturb subsurface temperatures and deflect the gas hydrate bottom-simulating reflector upward (e.g., Zwart *et al.* 1996); this corresponds to a permeability of approximately 3×10^{-14} m^2. Henry *et al.* (1992) report comparable flow rates

Table 18.1 Reported permeabilities of active subduction faults

Location	Depth (mbsf)	Permeability (m²)	Type of flow/ in situ conditions	Technique	Source(s)
Barbados décollement	398–463 m	1 to 1.2×10^{-14}	In situ; σ_v' of approximately 1.7 MPa	Two-well test	Screaton et al. (2000)
Barbados décollement	398–538 m	Approximately 8×10^{-16} to 6×10^{-13}	In situ; at σ_v' from 0.1 to 2 MPa	Single-well injection tests	Fisher & Zwart (1996, 1997)
Barbados décollement	398–463 m	5×10^{-18} to 2×10^{-17}	In situ; at σ_v' from 2.25 to 2.7 MPa	Single-well, low-volume injection tests	Screaton et al. (1997)
Barbados décollement	398–538 m	$\log(k) = -14.8 - \log(\sigma_v')$	In situ; as $f(\sigma_v')$	Synthesis of well tests	Bekins et al. (2011)
Barbados décollement	250–1200 m	10^{-13} to 10^{-15}	Steady state	Numerical modeling	Screaton et al. (1990)
	250–6800 m	Approximately 10^{-14}	Steady state		Bekins et al. (1995)
	Approximately 0.2–20 km	Approximately 10^{-14}	Steady state		Cutillo et al. (2003)
	Approximately 0.2–5000 m	Approximately 10^{-15} to 10^{-14}	Steady state*		Henry & Le Pichon (1991)
	250–6800 m	10^{-13} to 10^{-11}	Transient		Bekins et al. (1995)
	Approximately 0.2–20 km	Approximately 10^{-12}	Transient		Cutillo et al. (2003)
	250–2000 m	Approximately 3×10^{-13} to 10^{-14}	Transient†		Henry (2000)
Barbados décollement	190–280 m	Approximately 10^{-12}	Transient	Analysis of thermal anomaly	Fisher & Hounslow (1990)
Barbados décollement	399–429 m	1.1×10^{-18} to 1.1×10^{-17}	At effective stresses approximately 0.1–0.8 MPa	Laboratory measurements on cores	Zwart et al. (1997)
Costa Rica décollement	133–371 m	2.5×10^{-16} to 7×10^{-15}	Background/steady state	Geochemical mixing constraint	Saffer & Screaton (2003)
Costa Rica décollement	0–19 km	$>4 \times 10^{-15}$ 10^{-13} to 10^{-11}	Steady-state transient	Numerical modeling	Spinelli et al. (2006)
	0–300 m	$>10^{-17}$	Steady state		Screaton & Saffer (2005)
Costa Rica shallow splay fault	281 m	6×10^{-19} to 2×10^{-15}	At effective stresses 0.05–0.9 MPa	Laboratory measurements on core	Bolton et al. (1999)
Costa Rica splay faults	0–10 km	1.3×10^{-14} to 2×10^{-12}	Focused discharge; unclear if steady or transient	Analysis of seep flow rates	Ranero et al. (2008)
Costa Rica splay faults	0–13 km	Approximately 10^{-12} to 10^{-14}	Steady state	Numerical modeling	Lauer & Saffer (2012)
Nankai décollement	800–4600 m	10^{-15} to 10^{-17}	Steady state	Numerical modeling	Saffer & Bekins (1998a,b)
		5×10^{-14} to 10^{-12}	Transient		
	940–7400 m	10^{-19} to 3×10^{-14}‡	Steady state		Saffer (2010)
Nankai décollement	700–5000 m	Approximately 7×10^{-16} to 8×10^{-14}§	Steady state	Underthrust dewatering; numerical modeling	Skarbek & Saffer (2009)
Nankai megasplay	389–407 m	$>2 \times 10^{-17}$	In situ; background	Response to tidal loading	Hammerschmidt et al. (2013)
Oregon décollement		10^{-16} to 10^{-14}	Steady state	Numerical modeling	Saffer & Bekins (1998b)
Oregon frontal thrust	92–116 m	1.9×10^{-12}	Transient	Analysis of thermal anomaly	Davis et al. (1995)
		3×10^{-14}	Background	Bottom-simulating reflector shoaling	
Oregon frontal thrust	92–116 m	6×10^{-13} to 6×10^{-12}	Transient	Thermal disequilibrium from vein fill	Sample (1996)
Oregon frontal thrust	92–116 m	6.3×10^{-14} to 5.7×10^{-13}	In situ; as $f(\sigma_v')$; approximately 0.30–0.35 MPa	Single-well injection tests	Screaton et al. (1995)
Oregon frontal thrust	92–116 m	2.5×10^{-16} to 1.8×10^{-15}	In situ; as $f(\sigma_v')$; approximately 0.59 MPa	Single-well, low-volume injection tests	Screaton et al. (1995)
Oregon frontal thrust	105–165 m	Approximately 6×10^{-17} to 3.5×10^{-15}	At estimated in situ σ_v'	Laboratory measurements on cores; fault-normal orientation	Brown (1995)

*Varies systematically from approximately 10^{-14} m² at the trench to 10^{-15} m² by 100 km landward.
†Model of solitary wave propagation based on permeability–effective stress relationship from packer tests reported by Fisher & Zwart (1997).
‡Assigned to vary linearly with depth from 3×10^{-14} m² at the trench to 10^{-19} m² at 60 km landward.
§Decreases systematically from approximately 6 to 8×10^{-14} m² at the trench to 7×10^{-16} m² at 38 km landward.

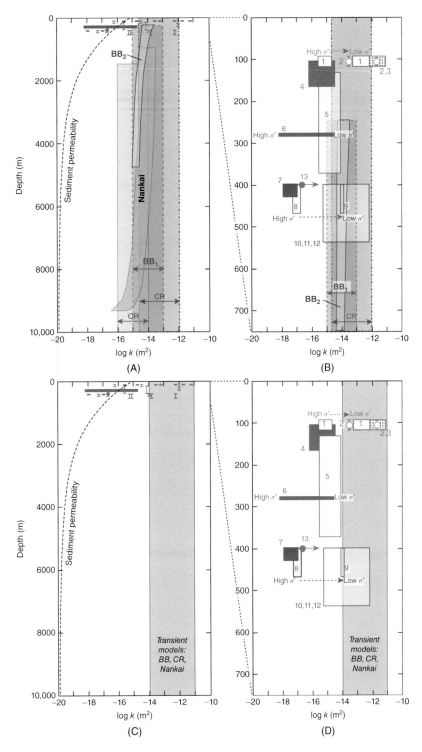

Fig. 18.6. Synthesis of published constraints on fault permeability reported in Table 18.1, shown as functions of depth below seafloor (mbsf). (A) and (B) show *in situ* permeability measurements or estimates described in the text and in Table 18.1, with permeability ranges derived from steady-state numerical modeling studies for the Barbados (BB_1: Bekins *et al.* (1995) and Screaton *et al.* (1990); BB_2: Henry & Le Pichon (1991)), Nankai (green; Saffer (2010); Skarbek & Saffer (2009)), Oregon (light brown; Saffer & Bekins (1998b)), and Costa Rican (blue; Spinelli *et al.* (2006)) margins. Mudstone sediment matrix permeability from Skarbek & Saffer (2009) (dashed line) is shown for comparison. (C) and (D) show the *in situ* permeabilities with transiently increased permeabilities from numerical models for the Nankai, Barbados, and Costa Rican margins (gray-shaded region). The *in situ* and laboratory measurements are blown up in panels (B) and (D). These include laboratory measurements (filled squares), *in situ* determinations from well testing (open squares), analyses of thermal anomalies and tidal responses in borehole observatories (open and closed circles, respectively), and analysis of vein chemistry (stippled square). These follow the same color scheme as for model results: brown = Oregon; maroon = Barbados; blue = Costa Rica; green = Nankai. Arrows show the range of permeabilities obtained over ranges of effective stress in well tests at the Oregon frontal thrust (*1*), Barbados décollement (*8–9*), and in laboratory measurements of fault zone core from Costa Rica (*6*). *1* = Screaton *et al.* (1995); *2* = Davis *et al.* (1995); *3* = Sample (1996); *4* = Brown (1995); *5* = Bolton *et al.* (1999); *6* = Saffer & Screaton (2003); *7* = Zwart *et al.* (1997); *8* = Screaton *et al.* (1997); *9* = Screaton *et al.* (2000); *10* = Bekins *et al.* (2011); *11* = Fisher & Zwart (1997); *12* = Fisher & Hounslow (1990); *13* = Hammerschmidt *et al.* (2013). (*See color plate section for the color representation of this figure.*)

(approximately 100 m year^{-1}, or 3.2×10^{-6} m s^{-1}) based on the analysis of shallow temperature probe measurements at the eastern Nankai accretionary prism.

Additional constraints on fault zone permeability have been extracted from detailed analyses of veins. Carbonate veins found within scaly shear zones in the Barbados décollement zone are characterized by O, Sr, and C isotope ratios indicative of a deep source, where temperatures are higher than those *in situ* (Vrolijk & Sheppard 1987; Labaume *et al.* 1997b). Similar observations from carbonate veins along the frontal thrust at the Oregon margin (Sample 1996) also suggest chemical and thermal disequilibrium between vein-forming fluids and the current thermal structure of the upper plate, with excess temperatures of up to approximately 100°C. To advect fluids along the fault zone and preserve this thermal signal, the flow rate must have been sufficiently high to limit diffusive heat loss to the wall rock during flow; Sample (1996) estimates that a linear water particle velocity (*v*) of approximately 1–10 km year^{-1} is required. For an assumed head gradient of unity driving flow (i.e., significant overpressure as suggested by Screaton *et al.* (1995) and Brown (1995)) and a porosity of 20%, these flow rates correspond to a permeability of approximately 6×10^{-13} to 6×10^{-12} m^2 (Table 18.1). These inferred permeabilities and flow rates are similar to those estimated by Davis *et al.* (1995) for the same fault.

At the Costa Rican margin, two boreholes that sample décollement fluids at different locations document a systematic increase in the magnitude of pore-water freshening with a distance from the trench (Morris et al. 2003b). Saffer & Screaton (2003) combined this observation with the known dewatering flux from underthrusting sediment to estimate the fluid flux in the décollement using a simple mixing model. Because the fluids entering the décollement via dewatering of the consolidating underthrust section below (i.e., Fig. 18.4C, inset) have a seawater composition, the change in décollement fluid composition (freshening) between the two drill sites defines the mixing between contributions from the décollement and underthrust fluid reservoirs and, thus, constrains the flux along the fault. Their analysis indicates flow rates of 4×10^{-3} to 1.1×10^{-1} m year^{-1}. For a 10-m-thick fault zone conduit and a head gradient driving flow defined by pore-fluid pressure measurements in a long-term borehole observatory (Solomon *et al.* 2009), these rates correspond to a permeability of approximately 2.5×10^{-16} to 7×10^{-15} m^2. A thinner flow conduit would result in proportionally higher permeability values.

The suite of observations from drilling, seafloor campaigns, and geophysical surveys underpin a conceptual model of fault zone hydrogeology in which permeability enhancement and pore pressure fluctuations are transient and potentially linked to fault slip events (e.g., Bekins *et al.* 1995; Solomon *et al.* 2009; Saffer & Tobin 2011). As discussed in more detail later (constraints from regional-scale numerical models), this is consistent with the magnitude and width of geochemical and thermal profiles centered at fault zones that suggest transient

focused flow (Fig. 18.5) (e.g., Fisher & Hounslow 1990; Gieskes *et al.* 1990; Blanc *et al.* 1991; Davis *et al.* 1995), and with analyses of fluid budgets showing that the high flow rates observed at seepage sites or required to explain geochemical and thermal anomalies cannot be sustained (e.g., Le Pichon *et al.* 1990; Carson & Screaton 1998; Saffer & Bekins 1999; Bekins & Screaton 2007).

QUANTITATIVE CONSTRAINTS ON FAULT ZONE PERMEABILITY FROM MEASUREMENTS AND FLOW MODELS

Direct quantitative constraints on fault zone permeability have come from active perturbation testing and passive monitoring at sealed ODP and IODP boreholes, and from numerical modeling studies that define plausible fault hydraulic properties through comprehensive sensitivity analyses. In general, these quantifications of fault zone permeability treat the fault itself as an effective porous medium over the scales of interest, typically a few to tens of meters in thickness and several kilometers along structure (e.g., Screaton *et al.* 1990; Bekins *et al.* 1995; Bense *et al.* 2013). Results from constant injection rate and slug testing at wells are well matched by solutions for Darcian flow in radial coordinates, supporting this general assumption (e.g., Fisher & Zwart 1997; Screaton *et al.* 2000). The geometry of perturbation tests and passive monitoring (vertical boreholes intersecting shallowly dipping faults) yield effective fault-parallel hydraulic properties and are insensitive to fault-normal permeability. In general, constraints derived from numerical modeling studies are also primarily sensitive to the permeability along structure.

Insights from well testing and borehole observatories

The most direct constraints on fault zone hydraulic properties have come from well testing in sealed boreholes that are either screened across fault zones or access them via an open hole below casing (e.g., Davis *et al.* 1995). Such direct constraints are generally scarce, but ODP and IODP drilling efforts have enabled targeted hydrologic experiments that have accessed fault zones in the Cascadia, Barbados, and Nankai subduction zones, and chemical/hydraulic monitoring at the décollement zone offshore Costa Rica (Table 18.1; Fig. 18.6) (e.g., Screaton *et al.* 1995, 1997, 2000; Fisher & Zwart 1997; Solomon *et al.* 2009; Bekins *et al.* 2011). Although these tests are limited to shallow depths (<1 km), where faults have formed within relatively high-porosity and poorly consolidated sediments and where total and effective stresses are low, they represent the best and only direct measurements of *in situ* fault zone permeability. These measurements have provided essential ground truth for numerical models that extend constraints to greater depths and larger scales, as described later (constraints from regional scale numerical models).

At the frontal thrust of the Oregon accretionary prism at ODP Site 892, a suite of shipboard packer injection tests was conducted immediately after drilling, and the site was visited approximately 1 year later by submersible to conduct low-volume pumping tests (Screaton *et al.* 1995). These tests were all conducted in an interval of open hole from 93.6 to 178.5 mbsf. On the basis of structural features described in cores and down-hole thermal data that identify an active flow zone, the thickness of the fault conduit sampled by the down-hole hydraulic tests was assumed to extend from 92 to 116 mbsf (24 m thickness). The tests provided an opportunity to measure fault zone hydraulic properties at high *in situ* pore pressure (pore pressures of approximately 65% of lithostatic; shipboard packer tests) and at low *in situ* pressures (near hydrostatic) during the follow-up submersible tests. The packer injection tests included both slug testing and constant rate injection tests and yielded permeabilities of 6.3×10^{-14} to 5.7×10^{-13} m^2. In contrast, the low-volume pump and recovery tests yielded permeabilities of 1.2 to 1.8×10^{-15} m^2. An additional permeability estimate of 2.5×10^{-16} m^2 was obtained from the formation pressure recovery following shut-in of the well. The large discrepancy between the two sets of tests indicates a strong sensitivity of permeability to effective normal stress (e.g., Rice 1992); the fault zone permeability is approximately 2–3 orders of magnitude higher at elevated *in situ* fluid pressures than at hydrostatic pore pressure. Observed seafloor seepage fluxes and flow rates inferred from shoaling of the gas hydrate bottom-simulating reflector are most consistent with the *in situ* permeabilities measured under increased pore pressure conditions, and suggest that the undisturbed fault zone in the subsurface is both overpressured and permeable (e.g., Tobin *et al.* 1994; Brown 1995; Davis *et al.* 1995; Screaton *et al.* 1995).

A similar set of tests was conducted in cased boreholes at the Barbados subduction zone that provided access to the décollement at ODP Sites 948 and 949, and monitored the formation over a depth range from 481 to 538 and from 398 to 463 mbsf, respectively (Fisher & Zwart 1996; Fisher & Zwart 1997; Screaton *et al.* 1997). Shortly after drilling and casing the borehole across the fault zone at each site, a series of packer tests – including pulse (slug) tests and constant flow rate injection tests – were carried out over a several-hour period. Over the course of each test, the background pore pressure in the tested depth interval rose such that the suite of packer tests provided permeability measurements over a range of effective vertical normal stress (σ_v') from approximately 0.1 to 1.6 MPa at Site 948 (Fisher & Zwart 1996) and approximately 0.3–2.0 MPa at Site 949 (Fisher & Zwart 1997). These tests yielded fault zone permeabilities that increased systematically from approximately 8×10^{-16} to 6×10^{-13} m^2 as pore pressure increased and σ_v' decreased. An additional suite of low-volume slug and recovery tests were conducted approximately 1.5 years later by submersible (Screaton *et al.* 1997). During these low-volume tests, pore pressures were lower than in the previous injection tests, and the effective normal stress was concomitantly higher, ranging from 2.25 to 2.7 MPa. These tests yielded permeabilities of 5×10^{-18} to 2×10^{-17} m^2, consistent with laboratory measurements on core samples (Zwart *et al.* 1997).

Monitoring of the pressure response at Site 949 to drilling of a nearby borehole approximately 2.5 years later provided additional constraints on fault zone permeability via an inadvertent two-well test (Screaton *et al.* 2000). Forward modeling of the pressure response yielded a best-fit hydraulic diffusivity of approximately 0.008 m^2 s^{-1} for the décollement zone, with the primary uncertainty associated with the exact distance between the two holes. For storativity values estimated from formation pressure response to seafloor tidal loading (e.g., Wang & Davis 1996), this diffusivity corresponds to a décollement permeability of 1 to 1.2×10^{-14} m^2, approximately 1 order of magnitude higher than expected from the relationship between permeability and effective stress defined by the suite of single-well tests (Screaton *et al.* 2000; Bekins *et al.* 2011). Bekins *et al.* (2011) revisited the analysis of the single-well tests to account for erroneously high reported permeabilities by invoking hydraulic fracturing induced by near-wellbore stress perturbations in the injection tests at low effective stresses, and by attributing erroneously low permeabilities reported for the low-volume slug tests to partial infilling of the borehole and screens. After accounting for these effects, the range of permeabilities is narrowed significantly from approximately 10^{-15} to 10^{-14} m^2; a best fit to the corrected well tests is defined by $\log(k) = -14.8 - \log(\sigma_v')$.

At the Nankai margin, Hammerschmidt *et al.* (2013) estimated formation hydraulic properties from the pressure response to ocean tidal loading in a sealed borehole (IODP Site C0010) spanning the megasplay, a major out-of-sequence thrust that reaches the surface approximately 30 km from the trench. For their analysis, pressures were monitored continuously within an isolated interval spanning the fault from 389 to 407 mbsf. Based on the minimal phase lag of measured pressures relative to the ocean tidal loading, combined with estimates of formation and wellbore storage, Hammerschmidt *et al.* (2013) show that the fault zone permeability is $>2 \times 10^{-17}$ m^2. This calculation is based on a wellbore storage effect (Sawyer *et al.* 2008), whereby if formation hydraulic diffusivity is sufficiently low, a phase lag will be introduced because a finite mass of fluid must move from the formation to the instrument to sense pressure changes. The analysis provides only a lower bound, however, because (i) larger permeability (or hydraulic diffusivity) would also result in a zero phase lag, and (ii) the estimate of instrument compliance includes only the borehole volume and the compressibility of pure water; if the instrument compliance is higher than reported, for example, due to damaged or remolded sediment in the annulus or within the rock volume in the near field of the hole, higher hydraulic diffusivity would be required.

Constraints from regional-scale numerical models

Steady-state models: temporally and spatially averaged permeability

Numerical modeling studies at individual margins have provided constraints on fault zone hydrogeologic properties through detailed analyses relating pore pressure to sediment and fault permeabilities (e.g., Screaton *et al.* 1990; Bekins *et al.* 1995; Saffer & Bekins 1998a; Spinelli *et al.* 2006). These models account for the distribution of fluid sources driven by sediment compaction and mineral dehydration, and incorporate constraints on sediment matrix permeability defined by laboratory experiments (e.g., Skarbek & Saffer 2009; Daigle & Screaton, 2015a). Model simulations define bounds on time-averaged regional-scale fault permeability that are consistent with a range of observations including (i) measured flow rates (e.g., Lauer & Saffer 2012), (ii) down-hole temperatures (e.g., Henry 2000), and (iii) pore-fluid pressures measured in boreholes, predicted from seismic interval velocities, or required at a regional scale to explain a narrowly tapered wedge

geometry that reflects low absolute shear strength along the base of the accretionary wedge (e.g., Davis *et al.* 1983; Matmon & Bekins 2006; Saffer & Bekins 2006).

Several 2D numerical hydrologic models of the Barbados accretionary complex have defined bounds on the décollement permeability across the forearc. Screaton *et al.* (1990) and Bekins *et al.* (1995) used a steady-state model driven by fluid sources from a prescribed sediment compaction field, computed from an estimated porosity distribution (Bray & Karig 1985; Bekins & Dreiss 1992), and illustrated that the time-averaged décollement permeability should lie between 10^{-15} and $10^{-13} \, m^2$ and must be $>10^3$–10^5 times higher than the surrounding sediment matrix to channelize flow (Table 18.1; Fig. 18.6). Although overall simulated pore pressures in the accretionary complex are not as sensitive to the décollement permeability as to that of the matrix (pore pressures are sensitive to décollement permeability only in the immediately adjacent wall rock; Fig. 18.7), flow *directions* are strongly sensitive to décollement permeability. Higher fault permeability would lead to substantial downward drainage from

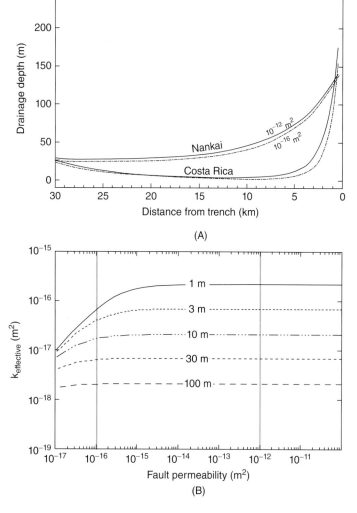

Fig. 18.7. (A) Drainage penetration depth (defined in text) as a function of distance from the trench for the Nankai and Costa Rican margins and for décollement permeabilities of $10^{-12} \, m^2$ (solid curves) and $10^{-16} \, m^2$ (dashed curves). (B) Effective permeability for a flow path from the matrix sediments 30 km landward of the trench to the trench at Costa Rica (cf. Fig. 18.4C, inset) as a function of fault conduit permeability, shown for sediment located at different distances away from the fault (as labeled). Vertical gray lines mark values of fault permeability shown by curves in panel (A). Both the drainage of adjacent sediments and the effective conductivity for flow paths to the seafloor are independent of fault permeability except in cases of low fault permeability or for sediments within several meters of the fault.

the accretionary prism inconsistent with Cl^- and CH_4 profiles observed at drill sites, whereas lower permeabilities would lead to unrealistically high overpressures in the décollement zone and adjacent sediments (Screaton *et al.* 1990).

Henry & Le Pichon (1991) simulated flow in the Barbados prism using a 2D model explicitly coupling loading, sediment consolidation, and fluid flow and obtained a similar result: the décollement permeability required to accommodate the dewatering flux from adjacent sediments should decrease from $10^{-14}\,m^2$ at the toe of the margin to $10^{-15}\,m^2$ by 100 km landward, where the décollement is ~5 km deep (Fig. 18.6). They also showed that to explain channelized flow along the décollement without significant vertical leakage, the décollement permeability must be approximately 10^4 times larger than that of the adjacent sediment.

Similar modeling studies have also been conducted for the Nankai (Saffer & Bekins 1998a; Bourlange & Henry 2007; Skarbek & Saffer 2009; Saffer 2010), Costa Rican (Spinelli *et al.* 2006), and Cascadia (Wang 1994; Saffer & Bekins 1998b) subduction zones (Fig. 18.6; Table 18.1). At the Nankai margin, a detailed study designed to explore the pore pressure distribution and patterns of pore-water freshening showed that steady-state décollement permeability should be at least 10^{-17} to $10^{-15}\,m^2$ in the outermost 50 km of the accretionary complex (Saffer & Bekins 1998a). More recent work by Saffer (2010) explored the effects of sandy layers in the accreted and underthrust sediment sections and showed that décollement permeabilities in a similar range, decreasing from $3 \times 10^{-14}\,m^2$ near trench to approximately $10^{-17}\,m^2$ 60 km landward, yield pore pressures consistent with the observed wedge taper angle. Models that couple loading, sediment consolidation, and fluid flow (following a similar approach to that of Henry & Le Pichon (1991)) illustrate that to accommodate the dewatering flux from underthrusting sediments while also sustaining pore pressures required to match both large-scale mechanical constraints from critical taper theory and those inferred along the wedge base from porosity and seismic wave speeds, the décollement permeability must decrease systematically from approximately 6 to $8.2 \times 10^{-14}\,m^2$ near trench to $7 \times 10^{-16}\,m^2$ by 40 km landward (Table 18.1; Fig. 18.6) (Skarbek & Saffer 2009).

A similar, although less extensive, suite of models have defined bounds on décollement zone permeability at the Costa Rican and Oregon margins, including coupled 2D simulations of consolidation and dewatering focused on the seaward-most 2.5 km of the Costa Rican margin (Screaton & Saffer 2005), and regional-scale models of both the Costa Rican (Spinelli *et al.* 2006) and Cascadia margins (Saffer & Bekins 1998b) (Table 18.1; Fig. 18.6). Coupled models of loading and sediment consolidation demonstrate that décollement permeabilities $>10^{-17}\,m^2$ can match observed porosity and inferred pore pressure profiles in the subducting section. The regional studies yield ranges of décollement permeability of 10^{-14} to $10^{-12}\,m^2$ and 10^{-16} to $10^{-14}\,m^2$ for the Costa Rican and Oregon margins, respectively, required to accommodate

dewatering fluid sources driven by porosity reduction (sediment transport through a prescribed porosity field; e.g., Bekins & Dreiss (1992)) while maintaining fluid pressures consistent with mechanical constraints from critical taper theory.

Modeling efforts at different locations yield steady-state décollement permeabilities in a strikingly similar range (10^{-15} to $10^{-13}\,m^2$) (Fig. 18.6), despite the fact that the geometries, rates of accretion and dewatering, and sediment matrix permeabilities differ substantially between margins. These permeability values are consistent with many of the *in situ* measurements at shallow depths (Fig. 18.6B). Values of décollement permeability at the Costa Rican margin extend above this range (to $10^{-12}\,m^2$), likely because (i) the steeply tapered upper plate increases the flow path length for fluids to escape diffusely through the prism and therefore confines larger fluxes in décollement than at other margins, and (ii) the combined effects of complete sediment subduction and a higher convergence rate (approximately 9 cm year^{-1} vs approximately 4 cm year^{-1} at Nankai and Cascadia) lead to larger dewatering fluxes from the subducting sediments that must be accommodated (e.g., Henry & Le Pichon 1991; Saffer & Bekins 2006). The lower permeability range for the Cascadia system can be explained by a similar logic. The Cascadia margin is characterized by a large overall sediment thickness (approximately 3–4 km), of which only a few hundred meters are subducted. As a result, the subducted section is mostly dewatered before reaching the trench, and a smaller dewatering flux is available to drive flow and elevated pore-fluid pressure along the décollement.

Modeling studies also illustrate the importance of major splay faults at some margins as agents of dewatering and solute or heat transport, as indicated qualitatively by geochemical and thermal signals of focused flow (e.g., Henry *et al.* 2002; Hensen *et al.* 2004; Teichert *et al.* 2005; Ranero *et al.* 2008). At the Costa Rican margin, where seafloor seepage associated with faults cutting the upper plate is well characterized, Lauer & Saffer (2012) used a 2D model to show that for realistic compaction- and dehydration-driven fluid source distributions at depth, reported flow rates in areas of focused discharge require splay fault permeabilities of approximately 10^{-14} to $10^{-12}\,m^2$ (Table 18.1). These permeabilities are generally consistent with those estimated by first-order analyses of thermal and geochemical anomalies in splay faults at both the Costa Rican and Oregon margins (e.g., Sample 1996; Ranero *et al.* 2008).

Time dependence and heterogeneous distribution of permeability

Models of coupled heat or solute transport and fluid flow illustrate that transiently increased permeability is required to explain observed geochemical and thermal profiles (e.g., Bekins *et al.* 1995; Henry 2000; Spinelli *et al.* 2006), down-hole temperature data (e.g., Fisher & Hounslow 1990), and seafloor seepage measurements (e.g., Saffer & Bekins 1998a) (Fig. 18.6C,D, Table 18.1). For example, to explain pore-water freshening anomalies centered at the décollement and other major fault zones at the Costa Rican, Nankai, and Barbados subduction zones, fluids must flow from zones of mineral dehydration at depth to the near-trench region where boreholes have been drilled, and flow must be rapid enough to preserve a chemical signature of the deeply seated fluids (cf., Figs 18.2 and 18.5) (e.g., Bekins et al. 1995; Spinelli et al. 2006). Similarly, rapid transient flow is needed to explain localized thermal anomalies spanning fault zones in the Cascadia and Barbados accretionary prisms (e.g., Fisher & Hounslow 1990; Davis et al. 1995; Sample 1996; Cutillo et al. 2003).

In 2D numerical models that simulate clay dehydration and coupled fluid flow and solute transport, décollement permeabilities of approximately 10^{-11} to 10^{-13} m^2 are required to reproduce observed profiles of pore-water freshening at drill sites near the trench at the Barbados, Nankai, and Costa Rican margins (e.g., Bekins et al. 1995; Saffer & Bekins 1998a; Henry 2000; Spinelli et al. 2006). The 2D and 3D models of coupled fluid flow and heat transport at the Barbados margin illustrate that fault-centered thermal anomalies observed in shallow boreholes are best reproduced by transient flow with fault permeabilities of approximately 10^{-12} m^2 (Cutillo et al. 2003). These permeabilities are approximately 1–3 orders of magnitude higher than those estimated from steady-state models (e.g., Screaton et al. 1990; Henry & Le Pichon 1991; Bekins et al. 1995; Saffer & Bekins 1998a,b; Spinelli et al. 2006) and are most consistent with in situ measurements conducted at increased pore pressure and low effective stress conditions (e.g., Davis et al. 1995; Screaton et al. 1995; Bekins et al. 2011) (Table 18.1; Fig. 18.6).

To explain the width of the anomalies, the modeling results also suggest that flow along the fault conduit must have been continuous (or quasi-continuous) over timescales of several hundred to thousands of years (Bekins & Screaton 2007) to as long as approximately 100 kyr (Saffer & Bekins 1998a, 1999; Henry 2000). Detailed recent studies of pore-water freshening at the Nankai (Brown et al. 2001; Saffer & McKiernan 2009) and Barbados (Fitts & Brown 1999; Bekins & Screaton 2007) subduction zones have accounted for the potential effects of in situ clay transformation or artifacts caused by pore-water extraction on reported pore-water geochemical profiles. The resulting corrected anomalies are shifted and more clearly centered on the décollement zone at both margins, but their widths are comparable to those originally reported.

As suggested on the basis of structural observations, analyses of vein textures and compositions, and theoretical arguments (e.g., Rice 1992), strong effective stress dependence of permeability is one likely mechanism to explain transient permeability enhancement, potentially associated with solitary pressure waves or seismic activity that generates an initial perturbation to pressure or permeability at depth (e.g., Rice 1992; Henry 2000; Bourlange & Henry 2007). This interpretation is consistent with recent work by Thorwart et al. (2013), who estimated the permeability of the Costa Rican subduction plate interface at approximately 15 km depth from time-dependent

migration of seismicity swarms, based on the assumption that their migration is linked to diffusion of fluid pressure along the fault (e.g., Shapiro et al. 2003). Their analysis yields fault zone permeabilities of 3.2×10^{-14} to 1.06×10^{-12} m^2, comparable to those required in transient models to explain pore-water freshening profiles at boreholes near the trench (Spinelli et al. 2006). Transient pulses of pore-fluid pressure at shallower depths are also required to satisfy mechanical constraints at the Nankai and Barbados subduction zones (e.g., Henry 2000; Skarbek & Saffer 2009), where measured or inferred present-day pressures near the trench are approximately 0.5–3 MPa lower than is needed for sliding along the décollement. This concept has been explored by modeling of pressure and permeability "fronts" migrating up-dip as predicted by models of solitary wave propagation in fault zones with strongly effective stress-dependent permeability (Henry 2000; Bourlange et al. 2003) and is consistent with the idea that regions of high seismic reflectivity and interpreted dilation along the décollement represent the leading edge of a migrating pulse of near-lithostatic pore pressure (Shipley et al. 1994).

Modeling of coupled consolidation and fluid flow at Barbados has further shown that the volume fluxes of fluid needed to maintain dilated faults greatly exceed those available within the incoming sedimentary section, implying that if migration of pressurized fluids is responsible for increased fault permeability, the permeable zones must be transient, heterogeneously distributed, or both (cf. Fig. 18.5) (Stauffer & Bekins 2001; Bekins & Screaton 2007). A similar argument arises at other locations, including Nankai and Costa Rica, from consideration of the fluid budget in combination with observed focused flow rates or those required in numerical models to explain chemical and thermal anomalies. These fluxes are approximately 10–500 times greater than the inventory of fluids, implying that increased permeability is present only approximately 0.2–10% of the time or that permeable channels occupy only this fraction of the fault surface at any given time (Fig. 18.5) (e.g., Le Pichon et al. 1990; Brown et al. 1994; Saffer & Bekins 1999; Spinelli et al. 2006).

Laboratory measurements

Laboratory measurements of permeability on intact fault zone core samples are scarce, owing to the difficulty of coring, recovering, and preserving intensely fractured and brecciated materials (Brown 1995; Zwart et al. 1997; Bolton et al. 1999). Additional insights have come from permeability measurements on experimentally sheared synthetic analogs or remolded sediment (e.g., Arch & Maltman 1990; Dewhurst et al. 1996; Ikari & Saffer 2012). In general, the laboratory measurements yield permeabilities at the lower bound or below those obtained from in situ testing or derived from numerical modeling studies.

For core samples obtained from the active frontal thrust in the Oregon accretionary prism, Brown (1995) measured permeabilities ranging from approximately 6×10^{-17}

to 3.5×10^{-15} m^2 under effective stresses similar to those estimated *in situ* (Table 18.1; Fig. 18.6). These values are below or at the lower end of the range obtained by well testing at the same fault (Screaton *et al.* 1995), and as much as 3 orders of magnitude lower than values of enhanced – and presumably transiently increased – fault permeability estimated on the basis of observed down-hole temperatures and vein chemistry (Davis *et al.* 1995; Sample 1996). Permeability measurements on core samples from the décollement zone at Barbados yield values of approximately 10^{-18} to 10^{-17} m^2, over a range of effective stresses of 96–768 kPa (Brückmann *et al.* 1997; Zwart *et al.* 1997). As is the case for the Oregon frontal thrust, these values fall below (Bekins *et al.* 2011) or overlap only the lower bound (Screaton *et al.* 1997) of permeabilities obtained by well testing.

Bolton *et al.* (1999) report on a series of triaxial tests conducted on cores from the toe of the Costa Rican margin, including one friable claystone from an active splay fault in the upper plate. Their tests illustrate that under normal consolidation conditions up to 900 kPa mean effective stress, the fault zone permeability is less than 10^{-18} m^2, whereas in an overconsolidated state achieved by reduction in effective mean stress to <100 kPa following initial consolidation to 900 kPa, its permeability is increased by more than 3 orders of magnitude to 2×10^{-15} m^2. As described earlier (fault zone architecture: inferences about hydraulic properties and behavior), geologic evidence suggests that similar processes may operate *in situ*; fracture networks within the décollement are interpreted to dilate due to shearing of overconsolidated mudstone under conditions of increased pore pressure and reduced effective stress (e.g., Karig 1990; Ujiie *et al.* 2003; Uehara & Shimamoto 2004).

The differences between laboratory measurements of permeability and those obtained from *in situ* well testing and regional-scale modeling are most likely explained by two effects. First, in most cases, the laboratory samples do not sample the most disturbed and intensely fractured or brecciated portions of the fault zone (cf. Fig. 18.3). This leads to a sampling bias due to preferential core recovery and a potential scale dependence because the core samples do not include the largest and most conductive fractures. As a result, the laboratory measurements are most compatible with *in situ* tests conducted at high effective normal stresses, during which connected fracture networks in the fault zone are partly closed (e.g., Screaton *et al.* 1995) (Fig. 18.6). Second, the core samples used in laboratory tests are oriented vertically (i.e., at a high angle to the fault); if the fault permeability is strongly anisotropic (e.g., Brown *et al.* 1994; Faulkner & Rutter 2001), then the *in situ* well tests, which sample permeability parallel to the fault, should yield higher values.

Permeability measurements on synthetic analogs, and on remolded or sheared sedimentary protolith and wall rock, also provide additional basic insights into the hydraulic behavior of subduction faults (e.g., Arch & Maltman 1990; Brown *et al.* 1994; Dewhurst *et al.* 1996; Ikari & Saffer 2012). These tests generally demonstrate an overall decrease in permeability and increase in permeability anisotropy (higher permeability parallel to the shearing direction) as a function of shearing, due to clay particle alignment, development of microstructures, and consolidation. The shear-parallel permeability is similar to that measured in unsheared sediments (e.g., Dewhurst *et al.* 1996) or in some cases slightly enhanced by alignment of fabric elements (e.g., Arch & Maltman 1990), whereas fault-normal permeability is reduced by one-half to as much as 3 orders of magnitude, with fault-normal permeability values as low as 10^{-21} to 10^{-20} m^2 (e.g., Dewhurst *et al.* 1996; Ikari & Saffer 2012).

This result may be relevant to the behavior of clay-rich fault cores that accommodate large localized shear strains (e.g., Vannucchi & Leoni 2007), but is probably not representative of regional-scale fault permeability that is dominated by connected fracture networks in the damage zone, as indicated by structural observations and logging data. If low-permeability sheared clays form a spatially continuous fault core, décollements could act as combined conduit – barriers with high permeability parallel to structure and low permeability across it (e.g., Tobin *et al.* 2001), leading to isolation of flow systems above and below the fault (e.g., Silver *et al.* 2000). However, there is little direct evidence for such continuity of sheared clays, in part, because direct sampling is restricted to a small number of boreholes. Moreover, both the drainage pattern in underthrusting sediments (Fig. 18.4) and geochemical observations (Saffer & Screaton 2003) indicate fluid communication between the subjacent sediment section and the fault zone, and therefore suggest that low-permeability zones of localized shear are not ubiquitous or continuous.

IMPLICATIONS AND KEY OUTSTANDING QUESTIONS

The nature of fault zone permeability and flow paths

Temporally and spatially averaged fault zone permeabilities (i.e., those represented by steady-state models and a subset of *in situ* measurements) are consistent across several margins worldwide and range from approximately 10^{-13} to 10^{-15} m^2. The similarity between margins is striking, considering that they are characterized by different convergence rates, wedge geometries, sediment thicknesses, thermal structures, and partitioning of sediment between accretion and underthrusting at the trench. Permeabilities above this range would lead to well-drained conditions within the fault and immediately adjacent wall rocks, inconsistent with a wide range of observations that indicate increased pore-fluid pressure and mechanical weakness along the décollement and in the uppermost subducted sediments (e.g., Davis *et al.* 1983). Lower permeabilities lead to décollement zone pore pressures in excess of lithostatic or which are inconsistent with predictions from critical taper theory (e.g., Brown *et al.* 1994; Bekins *et al.* 1995; Saffer & Bekins 1998a).

The fault zone permeabilities required to explain geochemical and thermal anomalies observed in drill holes and to match measured flow rates at focused seepage sites are higher than the range defined by steady-state models. They are also consistent between margins and range from approximately 10^{-11} to 10^{-13} m² (e.g., Fisher & Hounslow 1990; Bekins *et al.* 1995; Davis *et al.* 1995; Saffer & Bekins 1998a; Spinelli *et al.* 2006) (Fig. 18.6; Table 18.1). The shape of advective signals centered at fault zones is most consistent with transient permeability enhancement and flow, likely driven by pulses of increased pore pressure that reduce effective normal stress and drive dilation of fracture networks (e.g., Blanc *et al.* 1991; Henry 2000). However, these increased pressures and permeabilities cannot be continuous in time, nor can they be present across the entire fault surface at a given time; the former would violate the overall fluid budget (Le Pichon *et al.* 1990; Saffer & Bekins 1999; Bekins & Screaton 2007), whereas the latter would be inconsistent with the maintenance of a tapered wedge geometry (e.g., Brown *et al.* 1994). As an additional consideration, if permeable regions on the fault are heterogeneously distributed, to access fluids and allow dewatering of the subducted section, they cannot be stationary (e.g., Bekins & Screaton 2007). The emerging picture is that increased fault permeability is both transient and heterogeneous, with overpressured and permeable patches or "channels" restricted to only a portion of the fault surface and shifting over time (cf. Fig. 18.5C) (e.g., Carson & Screaton 1998; Saffer & Bekins 1999; Bekins & Screaton 2007).

The transiently increased permeabilities defined by numerical modeling studies overlap the upper range of measured values obtained at low effective normal stresses (Fig. 18.6C,D), consistent with the idea that transient flow is linked to the pressure dependence of permeability and dilation of fracture networks. Although lower, permeabilities estimated from steady-state models are most consistent with the middle to upper range of *in situ* borehole measurements, and are generally higher than the lowest point measurements (Fig. 18.6A,B). This may suggest that the long-term average fault permeability reflects modestly increased *in situ* pore pressure.

Recent work has also shown that in addition to the master décollement, splay faults cutting the upper plate are important egress pathways for fluids derived from sediment consolidation and dehydration reactions (e.g., Ranero *et al.* 2008; Sahling *et al.* 2008). Although focused studies have been limited to a few margins, mounting evidence from seafloor sampling, drilling, and numerical modeling demonstrates that these structures efficiently tap deep fluids by intercepting them from the décollement or deep interior of the wedge (Sample 1996; Teichert *et al.* 2005; Lauer & Saffer 2012). One implication is that fluids sampled at seafloor seep sites or by drilling into upper plate faults should exhibit a systematic increase in the deeply sourced contribution with distance landward (Fig. 18.2) (Mottl *et al.* 2004; Teichert *et al.* 2005). Translation of pore pressure from the plate interface to the base of layered slope sediments capping the upper plate wedge may also provide a mechanism to explain the locations of active mud volcanoes and mud mounds (Hensen *et al.* 2004).

Faults and the plumbing of subduction complexes

Dewatering and drainage state: the role of faults

Because faults are permeable, they carry focused fluxes of fluid derived from dewatering and drainage of underthrust and accreted sediments (e.g., Screaton *et al.* 1990; Saffer & Bekins 1998a). Yet despite their key role in carrying dewatering fluxes, the drainage state of the accretionary wedge and underthrusting section remains primarily controlled by the sediment matrix permeability (Screaton *et al.* 1990; Saffer & Bekins 2006). To simply and directly quantify the effect of fault permeability on drainage of the adjacent wall rock, I define a dewatering "penetration depth" (Z_p) as the distance into the wall rock (away from a fault conduit) at which the effective hydraulic diffusivity along a flow path to the trench allows pressure dissipation at a rate comparable to the tectonic loading rate.

To define the penetration depth, I adopt the criterion for maintenance of abnormal pressure introduced by Neuzil (1995), relating the strength of geologic forcing (Γ) to the characteristic path length for fluid escape (L) and the effective hydraulic conductivity (K_{eff}) along the flow path:

$$\frac{\Gamma L}{K_{eff}} > 1. \tag{18.1}$$

The geologic forcing (Γ; units of fluid volume per rock volume per time) results from processes that act as sources or sinks of fluid; these may represent actual fluid volume production (e.g., from hydrocarbon generation or dehydration reactions) or virtual sources that act to increase pressure (e.g., disequilibrium compaction or thermal pressurization). Although not a strict threshold for the generation or maintenance of increased pore pressure, the criterion given by Eq. 18.1 provides a framework to evaluate the potential for overpressures in a wide range of geologic settings, including subduction zones (e.g., Neuzil 1995; Saffer & Bekins 2006).

Assuming a drainage path through the sediment matrix and to the fault (cf. Fig. 18.4C, inset), L is given by ($x + Z_p$), where x is the distance along the fault to the seafloor. For the case of the décollement, x is the distance from the trench. The effective hydraulic conductivity (K_{eff}) from the penetration depth to the seafloor is given by:

$$K_{eff} = \frac{L}{\left(\frac{Z_p}{K_{sed}} + \frac{x}{K_{fault}} \right)}, \tag{18.2}$$

where K_{sed} and K_{fault} are the hydraulic conductivity of the sediment matrix and the fault zone, respectively. Rearranging Eq. 18.2 and combining with Eq. 18.1 defines an expression for the penetration depth:

$$Z_p = \frac{\left(\frac{\Gamma L x}{K_{fault}} - x \right)}{\left(1 - \frac{\Gamma L}{K_{sed}} \right)} \tag{18.3}$$

I solve Eq. 18.3 for Z_p as a function of distance from the trench (x) for a suite of fault permeabilities, using the Nankai and Costa Rican subduction zones as example cases (Fig. 18.7). The geologic forcing is defined on the basis of porosity loss with increasing depth and distance into the subduction zone (e.g., Spinelli *et al.* 2006). Sediment matrix permeability is defined by permeability–porosity relationships reported for sediments at each margin (Skarbek & Saffer 2009; Gamage *et al.* 2011; Daigle & Screaton 2015a). In computing both fault and sediment hydraulic conductivity from permeability, I account for variation in fluid viscosity with temperature (where dynamic viscosity $\mu = 2.4 \times 10^{-5}[10^{248.37/(T+133.15)}]$ and T is temperature in degrees Celsius) (Smith & Chapman 1983).

Penetration depth decreases systematically with burial and consolidation (from approximately 130 to 175 m near the trench to only tens of meters by 10 km landward), as sediment matrix permeability decreases (e.g., Daigle & Screaton 2015a) (Fig. 18.7A). A highly permeable fault ($k = 10^{-12}$ m^2; consistent with reported transient permeabilities) drains the adjacent sediments only very slightly more efficiently than a fault 4 orders of magnitude less permeable ($k = 10^{-16}$ m^2; consistent with the lower bound on reported steady-state permeabilities). At distances beyond approximately 25 km landward of the trench, the penetration depth increases slightly because with progressive burial and porosity loss the effects of decreasing geologic forcing and fluid viscosity outpace the decrease in sediment permeability.

Although simplified, this approach illustrates that permeable faults (or other conduits) will be efficient in dewatering sediments only locally over most of the outer forearc. This is because the matrix permeability at depths greater than a few kilometers is several orders of magnitude lower than fault permeability (cf. Fig. 18.6). As a result, fluid access to the permeable conduit is controlled by the matrix permeability. This is also illustrated by considering the sensitivity of effective permeability for drainage to the seafloor to conduit permeability (Fig. 18.7B; example for a location 30 km from the trench). For fault permeabilities $>10^{-15}$ m^2, further increases in fault permeability have no effect on the effective permeability. For lower fault permeabilities, the effective permeability is sensitive to variations in fault permeability only within less than 30 m of the fault. This behavior is also noted in 1D and 2D numerical modeling studies for individual margins (Barbados, Costa Rica, and Nankai) (Screaton *et al.* 1990; Saffer & Bekins 1998a; Spinelli *et al.* 2006; Skarbek & Saffer 2009) and those designed to explore the hydrologic behavior of subduction complexes more generally (Saffer & Bekins 2006), both of which show that permeable faults are in hydraulic communication with only the upper tens of meters of the underthrusting section, and that simulated overall pore-fluid pressures are far less sensitive to fault permeability than to the bulk sediment permeability.

Advection and preservation of localized anomalies along faults

In contrast to the drainage of pore-fluid pressures, faults, including both the master décollement and splay faults, are clearly key pathways for focused advective transport of heat and solutes. Commonly, first-order evaluation of transport behavior is framed in terms of the Peclet number, which relates rates of advection and diffusion in one dimension in the transport direction. However, for localized thermal or geochemical signals transported along conduits sufficiently rapidly that they cannot diffuse laterally (fundamentally a 2D problem), a slightly more restrictive formulation is appropriate (e.g., Fisher & Hounslow 1990; Sample 1996). In this case, the dissipation of an initial pulse localized in a flow conduit can be described by

$$C_{peak} = C_{max}[1 + 4tD/h^2]^{-1/2}, \tag{18.4}$$

where C_{peak} is the maximum concentration at the center of the conduit at any time t; C_{max} is the initial maximum concentration at the center of the conduit; D is the diffusivity; and h is the conduit width (Carslaw & Jaeger 1959). Adopting the simple criterion that to preserve observed anomalies centered at fault zones, flow must transport a pulse faster than it can be dissipated by diffusion, the problem can then be defined by the ratio of timescales required for fluid flow and chemical or thermal diffusion (Fig. 18.8) (e.g., Sample 1996). This approach may slightly overestimate the timescale of dissipation where matrix permeability is high enough to allow significant advection of solute or heat away from the potentially overpressured fault conduit; thus, the flow rates shown in Fig. 18.8 should be considered minima. However, the effect of fault-normal solute or heat advection is likely to be small (e.g., Henry 2000; Bekins & Screaton 2007) and restricted to shallow depths where sediment matrix permeability is greater than approximately 10^{-16} m^2.

This simple formulation yields basic insights into the conditions that favor rapid advection and preservation of focused anomalies along faults. In general, for realistic hydraulic gradients driving flow, fault zone permeability must be greater than approximately 10^{-14} m^2 to maintain focused pore-water geochemical signals as they are transported up-dip (Fig. 18.8). The required permeability scales inversely with flow distance: for longer migration paths, higher permeability is needed to deliver a pulse before it is dissipated by diffusion (Fig. 18.8, compare curves).

Geochemical anomalies observed in shallow fault zones generally indicate source regions located several tens of kilometers down-dip, where temperatures are >80–$150°$C (e.g., Kastner *et al.* 1991; You *et al.* 1996). To preserve focused anomalies over these migration distances requires fault zone permeabilities of approximately 10^{-12} to 10^{-13} m^2. This is in excellent agreement with the results of detailed 2D modeling studies at numerous margins, which suggest that (i) transiently increased permeabilities $>10^{-13}$ m^2 are needed to match the magnitude and shape of geochemical and thermal profiles, and (ii) steady-state

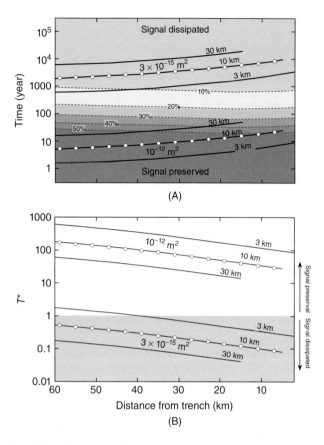

Fig. 18.8. (A) Time required for up-dip fluid migration over distances of 3, 10, or 30 km (black curves, as labeled) as a function of distance landward of the trench, for fault permeabilities of 10^{-12} m^2 (bottom, as labeled) and 3×10^{-15} m^2 (top, as labeled). Flow rates are computed assuming a fault zone porosity of 0.05, a hydraulic gradient equivalent to a pore pressure ratio ($\lambda = P_{fluid}/P_{lithostatic}$) of 0.80, and accounting for varying fluid viscosity as a function of temperature. Times required for the dissipation of a 1-m-wide initial solute concentration pulse are shown by color shading, computed assuming a molecular diffusivity (D) of 10^{-9} m^2 s^{-1}, and with effective diffusivity defined as a function of sediment matrix porosity by $D^* = D\eta/2$ and accounting for the effects of temperature and fluid viscosity by the Stokes–Einstein equation (Helfferich 1966). Dashed contours show the times at which the peak concentration of the pulse has decreased to a particular percentage of the initial peak value. (B) The ratio (T^*) of the times shown in panel (A) ($T_{diffusion}/T_{flow}$) as a function of distance from the trench, where $T_{diffusion}$ is defined as the time by which the pulse is reduced to 10% of its initial amplitude. For a fault zone permeability of 10^{-12} m^2, localized concentration pulses would dissipate by <50% over transport distances of <10 km, regardless of location. For fault permeability of 3×10^{-15} m^2, signals would dissipate by >90% in most scenarios. Thermal structure, geometry, and sediment matrix porosity distribution are defined using the Nankai accretionary complex as an example (Skarbek & Saffer 2009). (*See color plate section for the color representation of this figure.*)

permeabilities of approximately 10^{-15} to 10^{-14} m^2 lead to simulated profiles that are both too small in magnitude and too broad to match observations (e.g., Bekins *et al.* 1995; Saffer & Bekins 1998a; Spinelli *et al.* 2006). The simple approach illustrated in Figure 18.8 provides a useful and simple framework to derive constraints from field observations, without requiring construction and parameterization of a full 2D or 3D model

of flow and transport. Because thermal diffusivity is approximately 3 orders of magnitude higher than chemical diffusivity, preservation of temperature anomalies would require permeabilities > 10^{-12} m^2. This may also explain why most thermal signals observed in shallow boreholes are considered to reflect flow from only a few kilometers depth (i.e., short migration distances) (Fisher & Hounslow 1990; Sample 1996).

From Figure 18.8, it is also evident that the rates of both flow and diffusion vary with depth. As a consequence, advection of focused anomalies should be increasingly efficient with increased depth. This results from the combined effects of reduced effective diffusivity with depth, and decreased fluid viscosity that leads to faster flow rates for a given fault permeability. Fluid viscosity decreases by approximately 1 order of magnitude over the temperature range from 0 to 100°C. In contrast, the effective diffusivity decreases only modestly with depth because there is a trade-off between reduction in diffusivity due to porosity decrease and an increase in diffusivity as fluid viscosity decreases (i.e., as described by the Stokes–Einstein equation).

Key outstanding questions

Despite wide-ranging observations that constrain fault zone architecture, physical properties, flow rates, distances of flow and transport, and a basic model of temporally and spatially variable flow (e.g., Carson & Screaton 1998), the timescale and causes of permeability variations remain enigmatic (Saffer & Tobin 2011). Direct observations of geochemical and flow rate changes associated with earthquakes and slow slip events, both in borehole observatories (e.g., Solomon *et al.* 2009; Davis *et al.* 2011) and at the seafloor (e.g., Brown *et al.* 2005; Tsuji *et al.* 2013), document unequivocally that seismic activity drives pulses of fluid flow. Yet, the timescale of transient flow interpreted from the width of geochemical and thermal anomalies is as long as tens to hundreds of kyr – far longer than earthquake recurrence intervals.

This apparent discrepancy could be explained if permeability variations associated with fault slip and healing are superimposed on a longer-term cycle of permeability enhancement driven by pore-fluid pressure buildup and release or by chemical precipitation and sealing. Alternatively, the width of the down-hole geochemical profiles may simply document the initiation of continuous focused flow. This initiation time could reflect the point at which the fault is buried deeply enough and has low enough porosity to host brittle deformation that would lead to the generation of discrete fractures and conduit behavior (cf., Fig. 18.3A) (Caine *et al.* 1996; Saffer & Bekins 1999). In both of these scenarios, seismicity-driven flow pulses would result in quasi-continuous flow since the time of fault localization or superimposed on longer-timescale flow cycles, but would not be resolvable from down-hole profiles that record only the integrated signal from many events.

The details of feedbacks and causal relationships between permeability enhancement, pore pressure translation, and fault slip in subduction zones are also incompletely understood. Numerical and analytical models of perturbations in fault zones with strongly effective stress-dependent permeability demonstrate the possibility of solitary waves, in which pressure pulses migrate up-dip and modify fault permeability as they pass (e.g., Rice 1992; Bourlange & Henry 2007), raising the prospect that this process could also trigger the migration of seismicity or propagation of slow slip (e.g., Thorwart *et al.* 2013). Models specific to the Nankai subduction zone show that migration of pressure pulses by this mechanism is more than 500 times too slow to directly explain slip propagation inferred from migration of small earthquake swarms at rates of kilometers per day (Bourlange & Henry 2007). This suggests that pressure pulse migration may not explain triggering of fault slip. However, the propagation velocity of the pressure pulse depends strongly on the maximum value and form of the stress–permeability relation; thus, the parameter space explored by modeling may simply warrant further investigation. Sealing via cementation or vein filling and subsequent rupture of seals (i.e., by hydraulic fracturing or seismic slip) may also be important controls on permeability cycling. However, at shallow depths where faults have been cored most extensively, veining is noted but generally sparse (e.g., Labaume *et al.* 1997a,b; Vrolijk & Sheppard 1987); at greater depths where pressure and temperature are higher, chemical sealing may become increasingly important (e.g., Sibson 1981b).

Ultimately, better defining the sensitivity of permeability to effective stress (e.g., Bekins *et al.* 2011) and its variation as functions of depth and fault zone architecture is essential to better quantify the nature and magnitude of permeability changes in response to hypothesized perturbations in pressure or permeability. These include continuous processes of metamorphic fluid release, tectonic loading, and chemically driven porosity reduction that could drive a cyclic buildup and release of pressure (e.g., Sibson 1981b), as well as seismicity or slow slip events that perturb pressure or permeability at depth (e.g., Bourlange & Henry 2007). Direct measurements are limited to shallow depth (<1 km); down-hole experiments in deeper (5–6 km) boreholes will extend the range of effective stresses and P–T conditions where fault properties are sampled (e.g., Tobin & Kinoshita 2006). At depths beyond the reach of drilling, modeling studies constrained by shallow observations, or by the (presumed fluid-driven) migration of seismicity (Shapiro *et al.* 2003; Thorwart *et al.* 2013), provide a promising approach to quantify fault permeability behavior.

Finally, although advective signals measured at the seafloor and in the shallow subsurface have yielded clear evidence for long-distance migration of fluids, significant uncertainty remains in precisely defining fluid source regions, documenting the mixing of fluids derived from different dehydration reactions or diagenetic processes, and quantifying the behavior of nonconservative solute tracers. Considerably tighter constraints on the depth of origin for migrating fluids, and on the relative contributions from distinct source regions, may be possible with the use of multiple tracers that originate at different depths and temperatures (e.g., You *et al.* 1996; Mottl *et al.* 2004). Ultimately, multitracer numerical simulations of coupled flow and transport will lead to a refined and more quantitative understanding of volatile budgets, fluid migration pathways, flow rates, and the associated fault permeabilities.

ACKNOWLEDGMENTS

This research used data obtained as part of the ODP and IODP. This work was supported by postexpedition funding from the Consortium for Ocean Leadership (US-SSP) and NSF awards 0451602, 0623633, 1049591, and 0752114. Detailed discussions with Elizabeth Screaton, Barbara Bekins, and J. Casey Moore were instrumental in conceiving and developing this review. I also thank Katelyn Huffman (Fig. 18.3C), Tom Shipley (Fig. 18.5A), Peter Vrolijk (micrograph; Fig. 18.2), and Miriam Kastner (data for Fig. 18.5B) for providing graphics and data used in figures.

CHAPTER 19

Changes in hot spring temperature and hydrogeology of the Alpine Fault hanging wall, New Zealand, induced by distal South Island earthquakes

SIMON C. COX[1], CATRIONA D. MENZIES[2], RUPERT SUTHERLAND[3], PAUL H. DENYS[4], CALUM CHAMBERLAIN[5] AND DAMON A. H. TEAGLE[2]

[1] GNS Science, Dunedin, New Zealand; [2] Ocean and Earth Science, National Oceanography Centre Southampton, University of Southampton, Southampton, UK; [3] GNS Science, Lower Hutt, New Zealand; [4] School of Surveying, University of Otago, Dunedin, New Zealand; [5] School of Geography, Environment and Earth Sciences, Victoria University of Wellington, Wellington, New Zealand

ABSTRACT

Thermal springs in the Southern Alps, New Zealand, originate through penetration of fluids into a thermal anomaly generated by rapid uplift and exhumation on the Alpine Fault. Copland hot spring (43.629S, 169.946E) is one of the most vigorously flowing and hottest springs, discharging strongly effervescent CO_2-rich 56–58°C water at 6 ± 1 l s^{-1}. Shaking from the Mw7.8 Dusky Sound (Fiordland) 2009 and Mw7.1 Darfield (Canterbury) 2010 earthquakes, 350 and 180 km from the spring, respectively, resulted in a characteristic approximately 1°C delayed cooling over 5 days. A decrease in conductivity and increase in pH were measured following the Mw7.1 Darfield earthquake. Earthquake-induced decreases in Cl, Li, B, Na, K, Sr and Ba concentrations and an increase in SO_4 concentration reflect higher proportions of shallow-circulating meteoric fluid mixing in the subsurface. Shaking at amplitudes of approximately 0.5% g peak ground acceleration and/or 0.05–0.10 MPa dynamic stress influences Copland hot spring temperature, which did not respond during the Mw6.3 Christchurch 2011 aftershock or other minor earthquakes. Such thresholds should be exceeded every 1–10 years in the central Southern Alps. The characteristic cooling response at low shaking intensities (MM III–IV) and seismic energy densities (approximately 10^{-1} J m^{-3}) from intermediate-field distances was independent of variations in spectral frequency, without the need for post-seismic recovery. Observed temperature and fluid chemistry responses are inferred to reflect subtle changes in the fracture permeability of schist mountains adjacent to the spring. Permanent 10^{-7}–10^{-6} strains recorded by cGPS reflect opening or generation of fractures, allowing greater quantities of relatively cool near-surface groundwater to mix with upwelling hot water. Active deformation and tectonic and topographic stresses in the Alpine Fault hanging wall, where orographic rainfall, uplift and erosion are extreme, make the Southern Alps hydrothermal system particularly susceptible to earthquake-induced transient permeability.

Key words: Copland hot spring, earthquake, fluid flow, orogenic geothermal systems, permeability change, Welcome Flat

INTRODUCTION

Earthquake-induced fluid redistribution is of relevance in earth science as it provides an important mechanism for mineral deposit formation (Sibson 2001), and fluids have some control on the strength of faults and the crust (Townend & Zoback 2000; Faulkner et al. 2010; Saffer 2015). Fluids affect ground strength and shaking behaviour at near-field distances from earthquake epicenters, and transient stresses of seismic waves have potential to interact with fluids and dynamically trigger seismicity at far-field distances (Hill et al. 1993; Brodsky et al. 2000; Husen et al. 2004b; Taira et al. 2009). Post-seismic fluid flow can have implications for groundwater supply and quality, contaminant transport, underground repository safety, and hydrocarbon production (Wang et al. 2013; Wells et al. 2013). Earthquake-induced hydrological responses have the potential to provide information on crustal permeability and tectonic processes at spatial and temporal scales that are otherwise difficult to study, helping to highlight that crustal permeability is not a fixed quantity but an ever evolving, dynamic parameter (Wang & Manga 2010a; Manga et al. 2012).

Given that earthquakes can change groundwater flow, changes in temperature and composition of groundwater are

Crustal Permeability, First Edition. Edited by Tom Gleeson and Steven E. Ingebritsen.
© 2017 John Wiley & Sons, Ltd. Published 2017 by John Wiley & Sons, Ltd.
Companion Websites: www.wiley.com/go/gleeson/crustalpermeability/
http://crustalpermeability.weebly.com/

to be expected, but to date there are few comprehensive observations at appropriate epicentral distances and length scales to constrain the various operational mechanisms (e.g. Wang *et al.* 2004a). At far-field (many fault lengths) distances from an epicenter, static stresses due to the earthquake are small, so sustained changes in ground water are thought to be the result of processes that can convert (the larger) transient, dynamic strains into medium- to long-term changes in fluid connectivity and flow. Both laboratory and field observations indicate that stresses too small to produce new cracks or pathways are able to change permeability, but laboratory experiments appear to require larger strain amplitudes than observed in natural systems (Manga *et al.* 2012).

In solid rocks, sustained far-field changes have been mostly attributed to transient changes in permeability caused by local changes in the aperture of fractures and/or mobilisation of small particles into the fluid phase (e.g., Brodsky *et al.* 2003; Montgomery & Manga 2003; Shi *et al.* 2015). Many questions still remain as to the exact pore- and fracture-scale mechanisms that create permeability changes, which can be difficult to decipher in natural systems because of their complexity and the inaccessibility of the subsurface (Manga *et al.* 2012). However, depending on geological context, different mechanisms of hydrological response to earthquakes should have different characteristic timescales of response and recovery or occur over different length scales (Montgomery & Manga 2003) or shaking thresholds (Wang & Manga 2010a,b). By developing empirical catalogues of earthquake-induced changes, it may be possible to elucidate the processes and driving mechanisms by testing hypotheses against scale-appropriate observations (e.g. Rojstaczer *et al.* 1995; Roeloffs, 1996, 1998; Wang *et al.* 2004b, 2013; Elkhoury *et al.* 2006, 2011; Liu & Manga 2009; Manga & Rowland 2009). Here, we contribute observations from the Copland hot spring in the Southern Alps of New Zealand. This system is a prime candidate for investigations of geothermal responses to seismic activity because this tectonically active region on a plate boundary is the host to a dense network of seismometers and continuous and semi-continuous Global Positioning System (cGPS) stations (see Supporting information for details of the GeoNet seismometer network and cGPS stations), and the South Island has recently experienced a series of moderate magnitude earthquakes.

Rapid uplift and strong erosion in the Southern Alps of New Zealand has perturbed the thermal structure of the upper crust, leading to geothermal activity and a series of thermal springs in the Pacific plate hanging wall of the Alpine Fault (e.g. Allis *et al.* 1979; Koons 1987; Allis & Shi 1995; Beavan *et al.* 2010a; Sutherland *et al.* 2012). Elevated topography locally drives circulation and cooling from the surface, whereas rapid uplift and conduction generate buoyancy-driven circulation from depth (Koons & Craw 1991; Koons *et al.* 1998; Menzies *et al.* 2014). In this contribution, we document transient changes in the hydrothermal circulation in the shallow crust, which are observed through characteristic changes

in temperature and chemistry of the spring fluids that were induced by seismic shaking from large distal earthquakes. We present observations from continuous monitoring of Copland hot spring temperatures and local rainfall from 2009 to 2011, with less frequent fluid conductivity measurements and chemical analyses spanning periods before and after earthquakes, together with observations from seismometers and cGPS stations. Drawing on surface site observations as a guide, we discuss constraints on the geometry of subsurface flow, nature of observed responses and potential mechanisms for change.

Although we have yet to obtain absolute values of permeability change, our contribution quantifies the low thresholds of shaking needed to produce observable changes in the Southern Alps hydrothermal flow regime. A corollary is that transience in permeability and changes to the flow regime must be relatively frequent, with occurrence and recovery over timescales of years to decades. We also document earthquake-related motion of nearby GPS stations, showing that observed dynamic earthquake-induced perturbations of the flow regime are associated with small permanent strains at amplitudes of 10^{-7} to 10^{-6} that are near the lower limit of strains previously known to cause hydrological effects (Manga *et al.* 2012). We propose that the setting of Copland spring in the mountainous central Southern Alps, on the actively deforming hanging wall of the Alpine Fault plate boundary, is likely to be particularly sensitive to earthquake-induced permeability change and thermal spring response.

SETTING AND CONTEXT

Tectonics

The Southern Alps are a topographic expression of oblique $39.7 \pm 0.7 \, \mathrm{mm \, a^{-1}}$ collision between the Pacific and Australian plates (DeMets *et al.* 2010) (Fig. 19.1). The largest late Quaternary fault displacement rates in the region occur on the central Alpine Fault ($27 \pm 5 \, \mathrm{mm \, a^{-1}}$ strike-slip and approximately $10 \, \mathrm{mm \, a^{-1}}$ dip-slip rates – Norris & Cooper 2001; Sutherland *et al.* 2006) along the western side of the alps. Although it has not produced any major earthquakes or measureable creep during New Zealand's < 200 year written history, palaeoseismologic evidence suggests this mature transpressive fault ruptures regularly in great (magnitude M~8) earthquakes at recurrence intervals of 329 ± 68 years (Sutherland *et al.* 2007; Berryman *et al.* 2012). The most recent rupture was approximately 1717 AD, so the fault appears to be late in the cycle of stress accumulation that will lead to a future large earthquake (Wells *et al.* 1999; Howarth *et al.* 2012). Regional GPS campaigns have revealed that the highest geodetic shear strain rates in New Zealand ($>0.3 \, \mathrm{ppm \, a^{-1}}$) are presently occurring within the Southern Alps, above the down-dip extension of the locked uppermost portion of the Alpine Fault (Beavan *et al.* 2007; Wallace *et al.* 2007). An array of continuous and semi-continuous GPS stations across the central Southern Alps

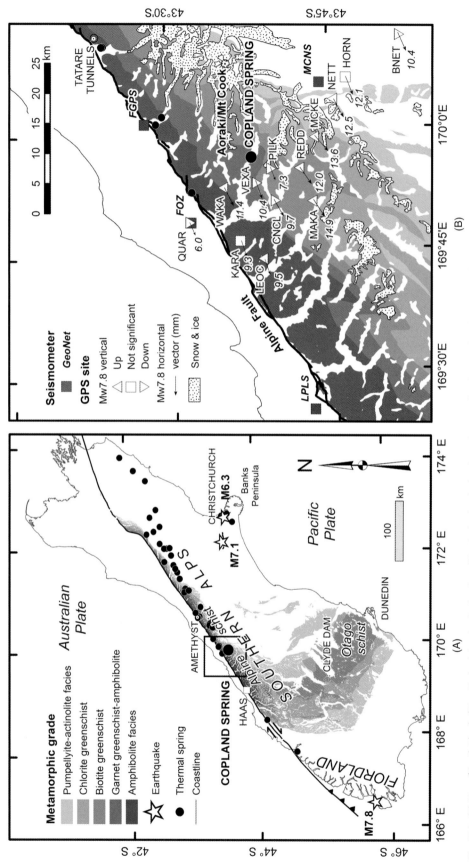

Fig. 19.1. (A) Map of South Island of New Zealand showing the location of the Copland and other thermal springs, the Alpine Fault and other active faults, and the epicenters of major South Island earthquakes during 2009–2011. The area of the Haast Schist Group, including some sub-greenschist facies semi-schist, is shaded by metamorphic grade. (B) Detailed map of the central Southern Alps, showing the location of Copland hot spring relative to nearby seismometers (filled squares) and cGPS stations (open triangles and squares). Scaled vectors show horizontal displacement of cGPS stations during the Mw7.8 earthquake, and the triangles are oriented according to the direction of vertical motion.

records approximately $4\,mm\,a^{-1}$ vertical (uplift) rates 10–20 km south–east of the fault (Beavan *et al.* 2010a). The extent to which this strain is elastic or permanent, and where or when this stored energy may be released, is poorly known. Present-day microseismicity occurs at depths $< 15\,km$ within a seemingly uniform crustal stress field in the Alpine Fault hanging wall (Wallace *et al.* 2007; Boese *et al.* 2012, 2013; Cox *et al.* 2012a; Townend *et al.* 2012).

Pacific plate rocks are uplifted in the Alpine Fault hanging wall into the path of the prevailing westerly winds from the Tasman Sea and Southern Ocean to produce a strongly asymmetric pattern of orographic weather and erosion (Hicks *et al.* 1996), with heavy ($>10\,m\,a^{-1}$) rainfall on the windward western side of the mountains and near arid conditions ($<1\,m\,a^{-1}$) in the east (Tait *et al.* 2006; Cox & Sutherland 2007). A near-continuous mid-upper crustal geological section has been exposed by differential uplift and erosion resulting in amphibolite facies Alpine schist adjacent to the Alpine Fault, grading through greenschist facies schist and pumpellyite–actinolite facies semi-schist at the main drainage divide (Main Divide), to prehnite–pumpellyite facies greywacke sandstone in the south–east (Fig. 19.1; Cox *et al.* 1997; Cox & Barrell 2007). The asymmetric rainfall, rapid uplift, erosion and exhumation result in high heat flow, thermal weakening and strong focusing of deformation at or near the Alpine Fault (Koons 1989; Koons *et al.* 2003; Herman *et al.* 2009; Sutherland *et al.* 2012). Topographic-driven infiltration of meteoric waters through fractures intersects the shallow thermal anomaly, producing hydrothermal circulation in the shallow crust that emerges as thermal springs at the surface (Allis *et al.* 1979; Koons & Craw 1991; Koons *et al.* 1998; Reyes *et al.* 2010; Menzies *et al.* 2014).

Copland hot spring

Copland hot spring (43.629S, 169.946E, 440 m asl) is located 12 km east of the Alpine Fault and 12 km west of the Main Divide, where Copland valley is deeply incised between $>2000\,m$ peaks and glaciated ridges of the central Southern Alps (Fig. 19.1A,B). The spring is of interest as it has the strongest discharge of both water and gas and is the second hottest and the most chemically evolved of about 40 thermal springs along the rapidly uplifting part of the central Alpine Fault (Fig. 19.1A). Copland hot spring is unusual in the region in that it effervesces abundant CO_2 and has precipitated a large travertine deposit. The spring is within the Westland National Park, and bathing pools downstream from the main upwelling vent are a popular destination for hikers.

The spring emanates through a thin cover of debris fan deposits and alluvial gravels (Fig. 19.2), estimated to be $<50\,m$

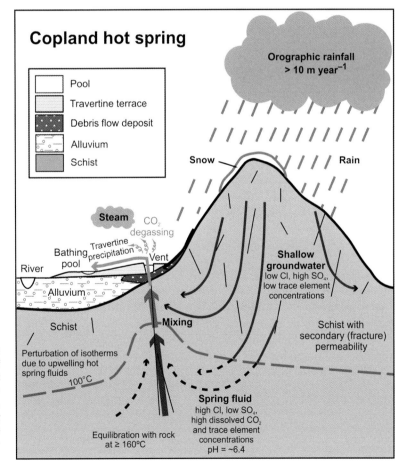

Fig. 19.2. Cartoon of the hydrothermal circulation that results in Copland hot spring. Rapid uplift and strong erosion of the Southern Alps in the hanging wall of the Alpine Fault have perturbed the thermal structure of the upper crust, leading to geothermal activity and a number of thermal springs. Elevated topography locally drives circulation and cooling from the surface, whereas rapid uplift and conduction drive buoyancy-driven circulation from depth. Horizontal and vertical dimensions are not drawn to scale.

thick at the site, overlying fractured garnet-zone schist (Alpine schist) that forms the dominant rock type along the western Southern Alps (Cox & Barrell 2007). Strongly effervescent CO_2-rich water emerges at approximately 56–58°C into a1.5×1×1 m vent pool, then discharges at $6 \pm 11 s^{-1}$ through a small restricted channel (cross-sectional area of 300 cm² with an average flow velocity of 0.2 m s⁻¹) and out across a terrace of travertine. We estimate over 90% of the fluid emerges through the source pool, but seeps of CO_2 gas and hot water emanate from other small vents nearby. The temperature of the ground approximately 0.7 m below surface, mostly within travertine or underlying debris–flow deposits or alluvium, was mapped using a calibrated thermistor (±0.1°C). There is a strong thermal anomaly in the ground with an area >50 m² associated with the upwelling vent pool, and a secondary, less intense anomaly near the bathing pools associated with downwards seepage of hot water into the travertine terrace (Fig. 19.3). Continuous monitoring of fluid discharge at Copland spring has yet to be attempted, but is complicated by the two-phase flow, our inability to constrain flow at subsidiary vents, heat of the fluid and the limited site engineering that is permitted due to site conservation values.

Hydrogeology

There is a paucity of information available on the *in situ* permeability of rocks in the Southern Alps, and none of it was formally published. Although work has been completed recently in the Alpine Fault and its damage zone (Sutherland *et al.* 2012), to our knowledge, only one investigation has quantified hydraulic properties of fractured Alpine schist in the hanging wall further from the fault. Hydraulic testing was recently completed for a hydroelectric power project near Amethyst Stream, Wanganui River, 1.5 km from the Alpine Fault (see Supporting information, Table SI, Fig. 19.1A) where garnet-zone schists contain spaced fractures and shears and are likely to be directly comparable with Alpine schist at Copland valley. These rocks were regionally metamorphosed to either greenschist or amphibolite facies during the Jurassic-Cretaceous and uplifted during the Neogene (Little *et al.* 2005). They form part of the Haast Schist Group, which is also exposed in the Otago region (Fig. 19.1A), but where quartzofeldspathic-dominated schist was regionally metamorphosed in the Jurassic and uplifted in Early Cretaceous (Mortimer 1993).

On the basis of regional observations in both Otago and the Southern Alps (Supporting information, Table SI), Alpine schist forming the mountains around Copland valley is expected to have >10 m scale (intrinsic) fracture permeability of between 10^{-15} and 10^{-12} m². Corresponding groundwater hydraulic conductivities are estimated to be approximately 10^{-8} to 10^{-5} m s⁻¹, assuming saturated conditions and standard values of viscosity and specific gravity (based on water at 10°C and 1 atm). An effective fracture porosity of between 10^{-3} and 10^{-1} (i.e. 0.1–10%) would cover the range expected from fracture

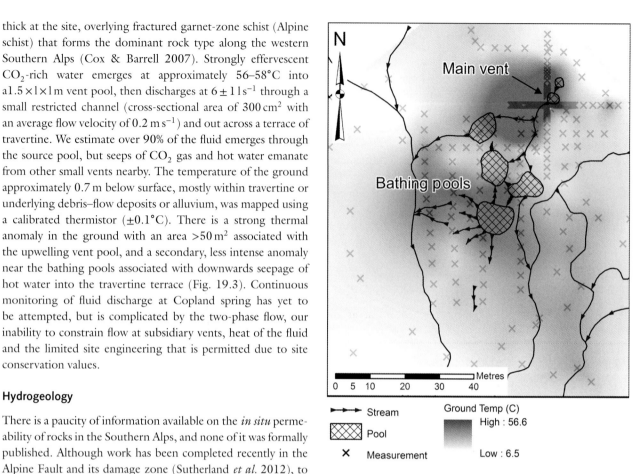

Fig. 19.3. Thermal map of ground temperature at depths of approximately 0.7 m, as measured on 2 September 2009. X symbols show the location of 298 data-points, collected by hammering a steel rod into the ground, quickly removing it and inserting a fibre glass probe with a thermistor on its tip, then reading the temperature once the probe had thermally equilibrated. The principal anomaly is adjacent to the main upwelling vent pool, where temperatures reach 56.6°C. Hot water flows out from the vent across the ground into bathing pools, where we interpret a secondary thermal anomaly from seepage down into the travertine. Over 8000 m² is anomalous above 10°C mean annual air temperature, but much of which seems to reflect surface flow or shallow subsurface flow immediately beneath the travertine.

densities encountered in Amethyst drill cores and textbook values (e.g. Freeze & Cherry 1979; Domenico & Swartz 1990; Supporting information). Fractured Alpine schist permeability is therefore similar to damage-zone mylonite (approximately 10^{-14} m²), but higher than cataclasite (approximately 10^{-16} to 10^{-17} m²) and the principal slip zone (approximately 10^{-19} m²) of the Alpine Fault (Sutherland *et al.* 2012).

Recent earthquakes

Three major earthquakes occurred in the southern South Island during 2009–2013 (Fig. 19.1A). A moment magnitude (Mw) 7.8 earthquake occurred in Dusky Sound Fiordland at 21:22 (NZST) on 15 July 2009, centred 350 km southwest of the Copland hot spring. Low-angle thrusting on the Australian–Pacific plate interface produced shaking intensities not felt in South Island for at least 80 years (Fry *et al.* 2010). Reverse slip of 5–6 m at the hypocentre produced shifts at cGPS monitoring stations, with over 300 mm of coseismic horizontal motion and approximately 50 mm of post-seismic motion at the station nearest the hypocentre, and discernable coseismic offsets throughout the lower South Island (Beavan *et al.* 2010a,b; see later). The Mw7.8 Dusky Sound earthquake was notable for the relatively small amount of high-frequency (>5 Hz) shaking given the size of the event (Fry *et al.* 2010).

On 4 September 2010 at 04:36 (NZST), an Mw7.1 earthquake occurred near Darfield, Canterbury, centred 180 km east of the Copland hot spring. The event ruptured the 30-km-long Greendale Fault (Quigley *et al.* 2010), previously concealed beneath the Canterbury alluvial outwash plains, within a network of faults that accommodates distributed deformation east of the Alpine Fault (Cox *et al.* 2012a; Litchfield *et al.* 2013). Right-lateral strike-slip displacement reached 5.3 m at the surface, averaging 2.5 ± 0.1 m (Quigley *et al.* 2012), initiating a prolonged aftershock sequence that lasted throughout 2011 and 2012. The largest and most damaging aftershock was the Mw6.3 Christchurch earthquake at 12:51 (NZST) on 22 February 2011, which occurred on an oblique thrust fault beside the volcanic rocks of Banks Peninsula (Beavan *et al.* 2011; Kaiser *et al.* 2012) at a distance of 220 km from Copland hot spring.

The Canterbury earthquakes radiated anomalously high levels of seismic energy relative to their magnitudes. The Mw7.1 Darfield earthquake had an energy magnitude (Me) of 7.4, with peak vertical ground accelerations reaching 1.3 g near the epicenter, whereas the Mw6.3 Christchurch aftershock had an Me of 6.7 and produced vertical accelerations of up to 2.2 g (Cousins & McVerry 2010; Gledhill *et al.* 2011; Reyners 2011). Attenuation with distance was stronger for the Mw6.3 than in the Mw7.1, as expected for the approximately one magnitude step difference between the two events, with correspondingly greater attenuation of accelerations than shaking velocities or displacements (Kaiser *et al.* 2012). However, the close proximity of the Mw6.3 earthquake to the city of Christchurch, the shallow source and exceptionally strong vertical ground motions resulted in high levels of damage and human casualties (e.g. Bradley & Cubrinovski 2011; Fry *et al.* 2010; Orense *et al.* 2011). In stark contrast, the Mw7.8 Dusky Sound earthquake of 2009 occurred in a sparsely populated region in the remote area of Fiordland, radiated a similar amount of seismic energy (Me7.3) in a predominantly offshore (south-westward) direction and produced relatively little damage and no casualties (Reyners 2009). Hydrological responses to these South Island earthquakes were observed in monitoring wells at near-, intermediate- and far-field distances (Cox *et al.* 2010, 2012b; Gulley *et al.* 2013).

The seismicity rate in the central Southern Alps increased significantly after the Mw7.8 Dusky Sound and Mw7.1 Darfield earthquakes, but not after the energetic Mw6.3 Christchurch aftershock (Boese *et al.* 2014). Delayed-triggered swarms of low-magnitude earthquakes occurred beneath the Main Divide 20–25 km north–east of Copland valley, but were not detected in the immediate vicinity of the Copland hot spring. The largest number of events occur in swarms delayed several hours to tens of hours after arrival of the remotely sourced seismic waves. Seismicity triggered by the Mw7.8 Dusky Sound earthquake continued for approximately 5 days in the central Southern Alps and at least 2 days after the Mw7.1 Darfield earthquake.

COPLAND HOT SPRING TEMPERATURE OBSERVATIONS

Rainfall, air and pool temperatures at Copland hot spring were continually monitored between March 2009 and July 2011 using a Hobo U12 temperature logger (relative precision $\pm 0.04°C$, accuracy $\pm 0.4°C$) recording at 15-min intervals, and a Hobo RG3-M tipping bucket rain gauge (0.2 mm per tip; air temperature precision $\pm 0.1°C$, accuracy $\pm 0.5°C$) situated 80 m south of the spring. The U12 logger was submerged at a depth of 1 m in the vent pool formed where fluid emanates at the surface. This pool is quite distinct from pools where the downstream flow is dammed for bathing.

Water temperatures fluctuated about background values of 57–58°C, with up to $\pm 0.1°C$ diurnal variation, but fell dramatically during episodes of heavy rainfall. Abrupt temperature drops occurred as soon as it started raining with more gradual recovery to background values 3–5 h after the rain stopped (Fig. 19.4). Spring water cooled to as low as 36.6°C when 535 mm of rain fell in a single storm between 25 and 28 April 2009. There is little surface run-off directly into the vent pool,

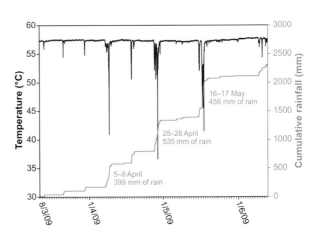

Fig. 19.4. Graph of cumulative rainfall (mm, thin grey line, right axis) and pool water temperature (°C, black line, left axis) during the period 8 March–13 June 2009. A strong correlation exists between the main rainfall events, marked by large steps, and major water temperature depression. A subtle diurnal $\pm 0.1°C$ fluctuation occurs when temperatures returned to background values of 57–58°C between the rainfall events.

so cooling is primarily a result of subsurface flow, as reflected by the quick recovery.

A distinct and very different cooling occurred following the Mw7.8 Dusky Sound earthquake on 15 July 2009. After a delay of 180 ± 15 min, there was first a slight $0.20 \pm 0.05°C$ rise in spring temperature to 58°C, followed by a slow $0.90 \pm 0.05°C$ cooling over a 5-day period to a new lower background value approximately 57°C (Fig. 19.5A). Fifteen months later on 4 September 2010, the Mw7.1 Darfield earthquake centred 180 km east of the spring produced a very similar response. After a delay of 140 ± 15 min, there was a $1.1 \pm 0.2°C$ cooling (Fig. 19.5B) that accentuated the long-term temperature depression, but was then overprinted by rainfall-related cooling between 5 and 7 September.

The background pool temperature remained depressed at approximately 56°C from September 2010 until February 2011 when the Mw6.3 Christchurch aftershock occurred 220 km to the east. The pool temperature was already lower at the time of the earthquake due to a rainstorm that had started the previous day. Consequently, any shaking-induced signal is masked by the cooling caused by 104 mm of rain (Fig. 19.5C). The pool had been at a background temperature at 56.74°C on 21 February prior to the rain, but post-rainfall recovery was to a new background at 56.38°C by 16:45 on 22 February, only 4 h after the earthquake and 2 h after it stopped raining. The earthquake may have induced a small $0.36 \pm 0.05°C$ temperature depression, accomplished over a period <4 h, but masked by the effects of rainfall. If real, this was comparatively short and steplike compared with the cooling induced by the Mw7.8 Dusky Sound and Mw7.1 Darfield earthquakes. Due to concerns over battery life, a new temperature logger was deployed on 24 February 2010 that performed with lower precision and measured slightly cooler temperatures that reflect instrument performance/calibration issues (Fig. 19.5C – see annotation). Consequently, when combined with rainfall masking, we are not confident that there was any lasting temperature departure caused by the Mw6.3 Christchurch earthquake, but certainly the response was different if any occurred at all. By 24 March 2011, the spring temperature had recovered to background temperatures of approximately 57°C, equivalent to those prior to the Mw7.1 Darfield earthquake. In October 2013, the background spring temperature was still at approximately 57.1°C, yet to recover to the state it was prior to the Mw7.8 Dusky Sound earthquake in July 2009.

Earthquake-related cooling responses have been time-shifted with regard to earthquake origin time, so they can be compared (Fig. 19.6). These responses are quite distinct from the sharp drops and recoveries resulting from rainfall (Fig. 19.4). The Mw7.1 Darfield earthquake cooling in September 2010 started from approximately 1°C lower background temperature (at approximately 57°C), but followed a trajectory very similar to the cooling caused by the Mw7.8 Dusky Sound earthquake in July 2009. The main differences were as follows: (i) the Mw7.8 Dusky Sound earthquake showed a slight temperature

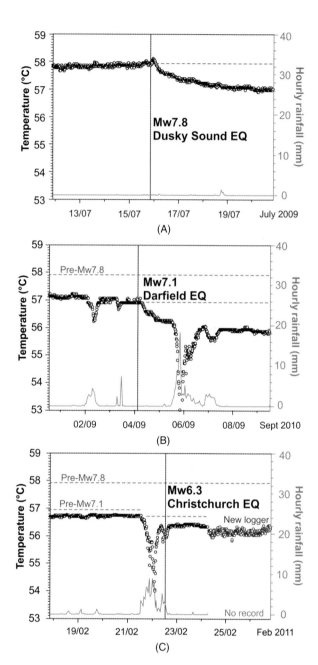

Fig. 19.5. Graphs of Copland spring temperatures (open circles) and hourly rainfall (grey line) at the time of recent earthquakes: (A) Mw7.8 Dusky Sound (15 July 2009), (B) Mw7.1 Darfield (4 September 2010) and (C) Mw6.3 Christchurch (22 February 2011). The temperature logger is submerged 1 m below the pool surface and is not directly affected by surface waters. A new U12 logger was deployed in the late February 2011, when equipment problems also resulted in missing rainfall data.

increase before cooling, which was not observed following the Mw7.1 Darfield earthquake; (ii) the response to the Mw7.1 Darfield earthquake was overprinted with rainfall-related cooling to as low as 45.6°C. We consider these to represent minor departures from what was otherwise a characteristic temperature response that predominantly reflects cooling

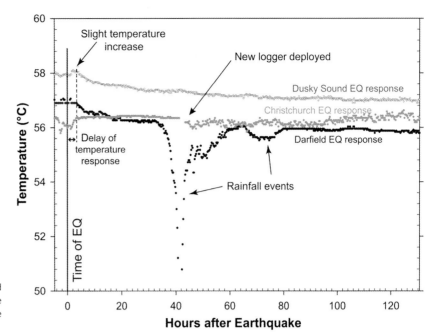

Fig. 19.6. Copland spring temperatures, time-shifted relative to the source time of each earthquake, so the nature of earthquake-related responses can be more closely compared.

caused by earthquake-induced changes in the flow regime, with temperature decay that is related to thermal capacitance of the hydrothermal system (see Discussion).

FLUID CHEMISTRY OF COPLAND HOT SPRING

Copland hot spring waters have the highest concentration of dissolved solids of thermal springs along the central Alpine Fault and the second highest temperature (Reyes *et al.* 2010; Menzies 2012). The waters are slightly acidic (pH = 6.1–6.6), have high alkalinity (approximately 17,000 μeq$\,l^{-1}$) and degas CO_2 as they upwell. They have elevated concentrations of Li, B, Na, Mg, Si, Cl, K, Mn, Rb, Sr, Ca and Ba and low concentrations of SO_4 (Table 19.1) relative to other springs in the central Southern Alps. Stable oxygen and hydrogen isotopic signatures plot on or near the global meteoric water line (Barnes *et al.* 1978; Reyes *et al.* 2010; Menzies 2012), indicating fluids are principally derived from deep infiltration of precipitation (rainfall or snow melt) into the schist mountains. Application of the silica (no steam loss) geothermometer of Truesdell (1976) to fluid chemical analyses returned equilibration temperatures of 153–164°C (Table 19.1). This suggests circulation to depths of 2–3 km if the local geothermal gradient is similar to the 62.6°C km^{-1} gradient measured at the Alpine Fault (Sutherland *et al.* 2012), although thermal modeling suggests that geothermal gradients should be higher in the hanging wall away from the Alpine Fault due to the cooling influence of relatively static footwall rocks (Koons 1987). Clearly, the deep fluids have cooled during upwelling, but low sulphate concentrations indicate that this is principally by conduction rather than dilution by cool shallow meteoric fluids.

Conductive cooling may result in silica precipitation that would result in an underestimation of the maximum equilibration temperature using the silica geothermometer. Unfortunately, major cation concentrations (Na, K, Mg, Ca) indicate that the Copland hot spring fluids are immature waters far from equilibrium with the host rock Alpine schists (following Giggenbach 1988) precluding the use of a range of geothermometers (e.g. Giggenbach *et al.* 1993; Menzies 2012). However, this lack of equilibration probably excludes deep fluid–rock interaction at temperatures much >200°C (see also Reyes *et al.* 2010).

Having observed cooling of Copland hot spring in response to rainfall and the Mw7.8 Dusky Sound (July 2009) earthquake, a series of temperature, conductivity and pH readings were taken at the vent pool throughout 2010 and 2011. The spring fluids showed a subtle shift to lower conductivity and higher pH after the Mw7.1 Darfield earthquake (Fig. 19.7), changing from a mean (± standard error) conductivity of 2311 ± 56 to 2074 ± 27 μS cm^{-1} and pH 6.43 ± 0.01 to 6.53 ± 0.02. Together with the records of cooling, these chemical observations are interpreted to reflect a change in the dilution of deep upwelling thermal waters (more acidic, more conductive) by near-surface (more basic, less conductive) groundwater.

Samples of Copland spring fluid (Table 19.1) were collected for geochemical analysis on 8 March 2009, 3 October 2010, 8 November 2010 and 23 February 2011, spanning the period of the recent earthquakes, and can be compared with published analyses of fluids collected about 1975 and 27 November 2005 (Barnes *et al.* 1978; Reyes *et al.* 2010 respectively). Concentrations of some cations and anions have changed systematically over time (Fig. 19.8, Table 19.1). Cl concentrations were highest (164, 167 and 154 μg g^{-1}) in samples collected before the recent earthquakes, compared with 147 μg g^{-1} in a sample from

Table 19.1 Chemical analyses of Copland spring, rainwater, nearby side streams in Copland valley and groundwater in >300-m-long tunnels cut through schist at Tatare stream, Franz Josef (Fig. 19.1B).

Date	Units	Copland warm spring						Rain	Rivers		Groundwater Tatare tunnels	
		Barnes 1978	Reyes 27/11/05	HS6 8/03/09	HS29 3/10/10	HS28 8/11/10	WF2 23/02/11	Karangarua 18/02/10	Shiels Ck 8/03/09	Rough Ck 8/03/09	Average 2009/2010	Range 2009/2011
Temp. surface	°C	56.0	56.0	56.9	55.5	56.0	56.1	nd	7.3	11.4	10.3	10.0–10.7
Temp. U12 logger	°C	nd	nd	nd	56.3	57.3	56.4	na	na	na	na	Na
T of Si eq.*	°C	162	157	164	159	160	153	na	na	na	na	Na
pH		6.1	6.4	6.3	6.6	6.5	6.5	6.3	8.1	7.5	7.3	6.8–7.9
Conductivity	$\mu S\ cm^{-1}$	nd	nd	2,100	2,360	2,100	2,040	4	56	nd	0.08	0.03–0.1
Total alkalinity	$\mu eq\ l^{-1}$	18,770	nd	nd	nd	nd	16,850	19	nd	233	485	123–666
F	$\mu g\ g^{-1}$	1.0	0.74	1.28	nd	1.23	1.29	bd	0.01	0.08	0.04	0.02–0.1
Cl	$\mu g\ g^{-1}$	164	167	154	nd	147	137	bd	1.03	1.68	2.24	1.88–2.41
Br	$\mu g\ g^{-1}$	nd	0.24	0.37	nd	0.35	0.33	bd	bd	0.004	0.01	0.008–0.012
SO$_4$	$\mu g\ g^{-1}$	bd	0.63	0.91	nd	0.81	1.07	bd	4.88	3.48	6.00	3.11–8.50
Li	$ng\ g^{-1}$	2,300	2,200	2,110	2,040	2,030	1,940	0.01	1.32	0.93	2.37	0.57–4.00
B	$ng\ g^{-1}$	5,950	6,400	6,250	6,050	5,980	5,590	0.01	2.29	1.12	2.29	2.01–2.86
Na	$\mu g\ g^{-1}$	392	381	396	365	369	349	2.04	0.78	1.53	2.43	1.67–2.83
Mg	$\mu g\ g^{-1}$	4.50	4.70	4.72	4.52	4.64	4.29	0.004	0.26	0.44	1.28	0.29–1.91
Al	$ng\ g^{-1}$	nd	nd	0.62	8.23	7.04	2.75	0.005	6.88	13.6	4.33	1.02–13.1
Si	$\mu g\ g^{-1}$	71.4	65.8	59.5	67.6	68.3	66.1	0.17	1.33	2.46	3.63	2.04–4.82
K	$\mu g\ g^{-1}$	28.0	29.0	27.8	25.5	25.9	24.6	0.06	1.63	1.52	2.18	0.80–3.10
Ca	$\mu g\ g^{-1}$	87.0	85.0	66.2	87.3	89.5	82.6	0.97	4.73	4.09	14.7	2.14–23.0
Mn	$ng\ g^{-1}$	nd	nd	245	252	219	241	0.02	0.24	0.25	0.16	0.06–0.33
Fe	$ng\ g^{-1}$	nd	nd	1.62	15.1	25.2	94.9	0.27	2.86	3.74	1.46	0.52–3.46
Rb	$ng\ g^{-1}$	170	160	228	221	170	214	bd	3.45	3.75	0.58	0.58–0.34
Sr	$ng\ g^{-1}$	1,900	1,900	1,840	1,780	1,810	1,680	0.02	13.7	12.2	35.3	8.95–52.5
Cs	$ng\ g^{-1}$	167	bd	175	189	157	183	bd	0.19	0.06	bd	bd
Ba	$ng\ g^{-1}$	nd	190	175	171	176	161	0.02	8.80	9.49	9.16	3.79–12.6

nd, not determined; na, not applicable; bd, below detection.

*T of Si eq. = equilibration temp using the Truesdell (1976) silica geothermometer.

Source: Data from Barnes et al. (1978), Reyes et al. (2010) and Menzies (2012). Details of field sampling methods and analytical procedures are outlined in the Supporting information.

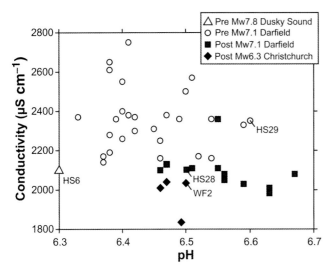

Fig. 19.7. Graph of specific conductivity versus pH at Copland hot spring, showing earthquake-induced changes. Measurements are classified according to time relative to earthquakes, showing a clear lowering of specific conductivity (i.e. total dissolved solids) and pH increase from pre- (open circles; 31 December 2009–22 August 2010) to post- (filled squares; September–October 2010) Mw7.1 Darfield earthquake.

2 months after the Mw7.1 Darfield earthquake and $137\,\mu g\,g^{-1}$ in a sample collected 1 day after the Mw6.3 Christchurch earthquake. Concentrations of Li, B, Na, K, Sr and Ba were also higher in samples collected before the earthquakes than those during and after the earthquakes. SO_4 concentrations are variable and lower in 1978 and 2005 analyses (Barnes *et al.* 1978; Reyes *et al.* 2010) than those of this study. Such systematic differences between samples corroborate the notion of earthquake-induced changes in fluid chemistry observed in pH and conductivity over time. Other elements (F, Si, Mg, Ca, Rb, Fe, Mn, Br and Cs) show no regular changes with time and earthquake events. This may be because the temporal variability in concentration of these elements (e.g. due to seasonal variation, subsurface precipitation of calcite or rainfall) is higher than any earthquake-induced changes or because any changes were insufficient to exceed analytical uncertainties for these elements. Concentrations of most elements except SO_4 are considerably lower in surface waters such as rain, river waters and shallow groundwater in schist (Table 19.1), commensurating with conductivities that are 2 orders of magnitude lower than spring fluid. Surface waters are more oxygenated so carry sulphur as SO_4, whereas deeper hydrothermal fluids tend to be sulfide-bearing (Barnes *et al.* 1978). Mixing of surficial water and shallow groundwater into deep hydrothermal fluid should therefore decrease the concentration of anions and cations, as well as temperature, increasing only SO_4 and pH. Assuming the concentration of SO_4 is near-zero in the deep hydrothermal fluid and shallow fluid has a $6.0\,\mu g\,g^{-1}$ SO_4 equivalent to the average groundwater in Tatare tunnels (Table 19.1), mass balance between the pre-earthquake sample

HS6 ($SO_4 = 0.91\,\mu g\,g^{-1}$) and average Tatare groundwater suggests about 15% of Copland hot spring prior to the earthquakes was derived from shallow fluid. Assuming the spring did not experience major flow rate changes as a result of seismic events, corroborated by the lack of any significant change reported by Department of Conservation staff and visitors present at the time, mass balance between post-earthquake sample WF2 ($SO_4 = 1.07\,\mu g\,g^{-1}$) and Tatare tunnel groundwater using SO_4 as a tracer suggests the earthquake increased the proportion of shallow end-member fluid to 18%. Alternate mixing models are also feasible, but yield mixing ratios around 5:1 and similarly small changes in response to earthquakes. In the absence of information on absolute changes in discharge, none are able to distinguish whether there was an increase in volume of shallow fluid mixing into the hot spring fluid or decrease in deep hydrothermal fluid.

DYNAMIC SHAKING

National seismometers closest to, and surrounding, Copland hot spring are FOZ (22.2 km north–west), MCNS (16.6 km southeast), FGPS (19.4 km north–east) and LPLS (43.3 km south–west) (GeoNet 2014; see also Supporting information). Peak ground accelerations (PGA) and peak ground velocities (PGV) are listed in Table 19.2, reported also in the text as absolute acceleration (%g) values. Shaking from the Mw7.1 Darfield earthquake (peak ground acceleration = 0.5–2.8% g) was more intense than the Mw7.8 Dusky Sound event (peak ground acceleration = 0.4–2.3% g), which in turn was more intense than the Mw6.3 Christchurch aftershock (peak ground acceleration =0.1–0.5% g). Peak ground accelerations mostly occurred at periods of 0.4–1.0 s. Simplified interpolation between seismometers, using the nearest-neighbours function (Sibson 1981a) in ArcGIS, suggests that peak ground accelerations in the vicinity of the Copland hot spring are very unlikely to have exceeded approximately 2% g during these earthquakes, corresponding to New Zealand Modified Mercalli (MM) intensities (Dowrick 1996) of III–IV (weak to light). The seismic energy densities at Copland hot spring, estimated from the empirical relation of Wang & Manga (2010b), are from 10^{-2} to 10^{-1} J m^{-3} (Mw7.8 = 0.24 J m^{-3}, Mw7.1 = 0.18 J m^{-3}, Mw6.3 = 0.01 J m^{-3}; Table 19.2). The static stress changes at Copland hot spring (Table 19.2) are estimated to have been approximately 0.48 kPa for both the Mw7.8 and Mw7.1 earthquakes, but only 0.03 kPa for the Mw6.3, based on estimations from the published seismic moment (Mo) values of 2.72×10^{20} N m (Fry *et al.* 2010), 3.46×10^{19} N m (Gledhill *et al.* 2011) and 3.72×10^{18} N m (Kaiser *et al.* 2012), using the equation

$$\text{Stress} = \frac{\text{Mo}}{(4\pi D^3)},$$

where D is the epicentral distance (Brodsky *et al.* 2003).

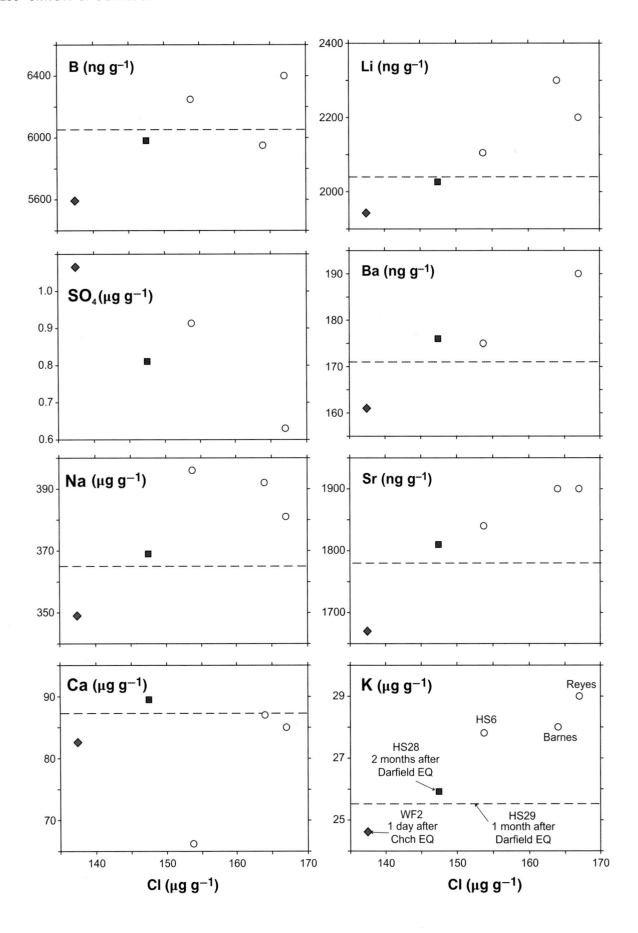

Absolute acceleration response spectra differ markedly between the Mw7.8 Dusky Sound and the Mw7.1 Darfield earthquakes (Fig. 19.9A,B). There was high power at 1- to 2.5-s periods during the Mw7.1, which appears to have been a characteristic of spectra recorded elsewhere in South Island (Cousins & McVerry 2010), and many seismometers recorded relatively small amounts of low period (<0.2 s) shaking during the Mw7.8 (Fry *et al.* 2010) including FOZ, FGPS and MCNS. Peak ground accelerations were higher during the Mw7.1 Darfield earthquake (see also Table 19.2), particularly well exemplified by the FOZ N horizontal (north–south) channel (Fig. 19.9C). However, none of the accelerations were particularly strong in the seismometers surrounding Copland hot spring, which is not surprising given the epicentral distances involved. In these recent earthquakes, FOZ recorded lower spectral accelerations than MCNS, FGPS and LPLS, which we interpret to be largely a site response effect (Fig. 19.9A,B). FOZ is situated on a solid bedrock site (subsoil category Type B – Rock, as defined in NZS 1170.5 by Standards New Zealand 2004), whereas the other seismometers are situated on thick sequences of compact glacial gravels (Type D – deep or soft soils). It is therefore likely that any of the GeoNet seismometers, the national network station FOZ, situated 15 km from Copland hot spring, best represents the shaking of bedrock and mountains surrounding Copland valley, through which the thermal spring waters flow (see also Supporting information).

The 2007–2012 record of shaking at FOZ, plotted as the maximum peak ground acceleration measured from earthquakes > Mw4 (Fig. 19.9D), shows that the Mw7.8 Dusky Sound, Mw7.1 Darfield and Mw6.3 Christchurch earthquakes produced the highest levels of shaking observed over the period when the spring temperatures were being monitored (March 2009–July 2011). However, only the Mw7.8 and Mw7.1 earthquakes were able to produce a clear response in Copland hot spring that followed characteristic cooling and did so despite very clear differences in source location, directivity, energy magnitude and spectral behaviour. The ground motion of the Mw7.8 Dusky Sound and Mw7.1 Darfield earthquakes exceeded peak ground acceleration values of 45 mm s^{-2} and PGV values of 10 mm s^{-1} in both horizontal and vertical directions for nearly all seismometers surrounding Copland hot spring, corresponding to dynamic stresses over 0.1 MPa (Table 19.2). In contrast, during the Mw6.3 Christchurch earthquake, shaking was nearly all measured at <35 mm s^{-2} and <5 mm s^{-1} (dynamic stresses < 0.05 MPa) (Table 19.2). The maximum static stress changes in the vicinity of Copland hot spring, derived from seismic moment of the earthquakes, were far smaller, being approximately 0.5 kPa for the Mw7.8 Dusky Sound and Mw7.1 Darfield earthquakes and 0.03 kPa for the Mw6.3 Christchurch aftershock. We conclude there is some form of shaking threshold for temperature changes at Copland hot spring, at peak ground acceleration approximately 0.5% g and dynamic stresses somewhere between 0.05 and 0.1 MPa. These are essentially equivalent to thresholds that triggered nearby seismicity beneath the Main Divide 20–25 km north–east of Copland valley (Boese *et al.* 2014). Over the longer period of 2007–2012, there were five earthquakes that produced peak ground acceleration at FOZ > 0.2% g (Fig. 19.9C) that may have been sufficient to produce a response at Copland hot spring. Records from the FOZ seismometer, as well as the more comprehensive National Seismic Hazard Model (Stirling *et al.* 2002, 2012), indicate that this level of shaking can be expected every 1–10 years (see also Boese *et al.* 2014).

PERMANENT DEFORMATION

A network of cGPS stations transects the Southern Alps, including sites on mountains immediately adjacent to Copland hot spring, and HAAS 100 km to the south–west (Fig. 19.1). Contemporary deformation is dominated by deep-seated slip on the Alpine Fault and locking from depths shallower than approximately 13–18 km (see also Beavan *et al.* 2007; Wallace *et al.* 2007). During the Mw7.8 Dusky Sound earthquake, there were clear offsets of GPS station horizontal positions (easting or northing) ranging from 6 to 15 mm (Table 19.3; see also Supporting information; Beavan *et al.* 2010a,b). These are similar to, but much smaller than, offsets for cGPS stations closer to the epicenter (Beavan *et al.* 2010b), yet are still significant above uncertainties at 95% (2σ) confidence levels. Smaller offsets occurred during the Mw7.1 Darfield earthquake in 2010, with 11 stations having horizontal offsets of 2–8 mm that appear significant (Table 19.3). There were no discernable changes in the Southern Alps cGPS station positions (horizontal or vertical) as a result of the Mw6.3 Christchurch aftershock in February 2011. During the Mw7.8 Dusky Sound earthquake, the coseismic horizontal motion was consistently towards the south–west (Fig. 19.1B), deviating from the long-term secular motion of South Island sites towards the north–west. South–west shifts are compatible with the direction of motions modelled for slip on the subduction interface beneath Fiordland during the Mw7.8 earthquake (Beavan *et al.* 2010b,c). In addition to horizontal shifts, there

Fig. 19.8. Evolution of Copland hot spring fluid chemistry. Concentration of trace elements in Copland spring waters plotted against chloride concentration for samples taken before the Mw7.8 Dusky Sound earthquake (open circles), after Mw7.1 Darfield earthquake (filled square and grey dashed line) and after Mw6.3 Christchurch earthquake (filled diamond). Pre-Mw7.8 Dusky Sound earthquake data taken from Barnes *et al.* (1978) and Reyes *et al.* (2010) and HS6 in March 2009 – see Table 19.1.

Table 19.2 Records of peak ground acceleration (PGA) and peak ground velocity (PGV; hor = horizontal, vert = vertical) for seismometers FOZ, MCNS, FGPS and LPLS during the Mw7.8 Dusky Sound, Mw7.1 Darfield and Mw6.3 Christchurch earthquake

Code Instrument	Unit	FOZ Episensor	MCNS Etna	FGPS Etna	LPLS Etna	COP_SPR Interpolation
Site latitude S	°	−43.5338	−43.7381	−43.4649	−43.7163	−43.6295
Site longitude E	°	169.8154	170.0971	170.0197	169.4230	169.9464
Site elevation	M	54	701	181	23	440
Distance to Copland	km	22.2	16.6	19.4	43.3	0.0
Site geology and strength	ASNZS 1170.5	B, Rock	D, Deep or soft soil	D, Deep or soft soil	D, Deep or soft soil	B, Rock
Site geology and strength	NZS4203	A, Rock or very stiff soil	C, Flexible or deep soil	C, Flexible or deep soil	C, Flexible or deep soil	A, Rock or very stiff soil
Mw7.8 Dusky Sound, 15/07/09 09:22:29						
Epicentral distance	km	358	359	375	321	356
PGA_vert	mm s⁻¹ s⁻¹	47.9	38.0	58.0	83.5	46.0
PGA_horMAX	mm s⁻¹ s⁻¹	52.7	100.2	118.8	229.5	91.0
PGV_vert	mm s⁻¹	11.8	8.8	10.8	15.5	10.4
PGV_horMAX	mm s⁻¹	15.7	17.0	25.7	36.5	18.3
Peak dynamic stress*	MPa	0.13	0.15	0.22	0.31	0.16
Static stress change	kPa					0.48
Energy density†	J m⁻³					0.24
Mw7.1 Darfield, 3/09/10 16:35:41						
Epicentral distance	km	191	169	175	223	180
PGA_vert	mm s⁻¹ s⁻¹	62.1	77.6	127.9	52.4	75.1
PGA_horMAX	mm s⁻¹ s⁻¹	121.0	204.5	278.6	180.7	178.8
PGV_vert	mm s⁻¹	18.2	13.9	22.5	7.8	16.0
PGV_horMAX	mm s⁻¹	27.3	27.4	43.5	26.6	28.8
Peak dynamic stress*	MPa	0.23	0.23	0.37	0.23	0.25
Static stress change	kPa					0.48
Energy density†	J m⁻³					0.18
Mw6.3 Christchurch, 21/2/11 23:51:42						
Epicentral distance	km	233	209	217	265	221
PGA_vert	mm s⁻¹ s⁻¹	11.7	13.9	24.7	13.0	14.0
PGA_horMAX	mm s⁻¹ s⁻¹	17.0	47.0	48.9	32.9	35.2
PGV_vert	mm s⁻¹	2.3	2.6	2.8	1.6	2.5
PGV_horMAX	mm s⁻¹	3.9	3.8	9.5	4.2	4.4
Peak dynamic stress*	MPa	0.03	0.03	0.08	0.04	0.04
Static stress change	kPa					0.03
Energy density†	J m⁻³					0.01

The column COP_SPR contains a simplistic nearest-neighbours interpolation (italics) that accounts for distance to the surrounding seismometers, but not site conditions nor any differences in station instrument types.
*Estimated PGV max times the shear modulus (3×10^{10} Pa), divided by shear wave velocity (3.5 km s⁻¹).
†Seismic energy density after Wang & Manga (2010b).

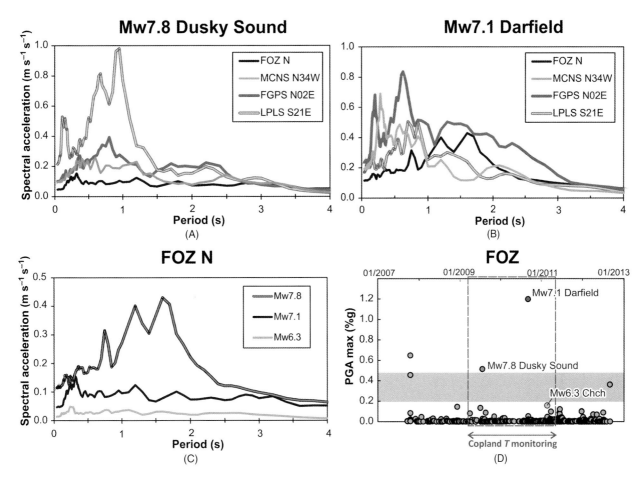

Fig. 19.9. Earthquake shaking recorded at seismometers FOZ, MCNS, FGPS and LPLS of the GeoNet national network. (A) Selected spectral accelerations measured during the Mw7.8 Dusky Sound earthquake for comparison with (B) records of the same channels during the Mw7.1 Darfield earthquake. (C) Horizontal north–south spectral accelerations at FOZ during all three recent earthquakes. (D) A longer-term 5-year record of peak ground acceleration max (% g) at FOZ, marked with dashed lines showing the period that Copland spring was monitored, and grey shading for possible threshold levels above which shaking produces a response at Copland hot spring.

are clear earthquake-related changes in vertical position at the time of the earthquakes, albeit within a higher degree of uncertainty. Stations near Copland hot spring (VEXA, PILK, CNCL) decreased in elevation during the Mw7.8 Dusky Sound earthquake, but were surrounded by stations that decreased in elevation (Fig. 19.1B, Table 19.3). Most stations were uplifted during the Mw7.1 Darfield earthquake.

The earthquake-induced motions recorded by the cGPS stations are small and of a scale that would be impossible to confirm by campaign-style geodetic studies. By plotting the horizontal offsets versus distance across the GPS transect from the Alpine Fault, we obtain a sense of the regional variation, which shows some consistency between the Mw7.8 Dusky Sound and the Mw7.1 Darfield earthquakes on the western side of the Southern Alps (Fig. 19.10). In both the earthquakes, there are differences of at least 6 mm in the offsets over the 30-km distance west of the Main Divide, with smallest offsets west of the Alpine Fault at QUAR. This indicates that

both earthquakes induced small, permanent, dextral shear strains at amplitudes of between 10^{-7} and 10^{-6}. There are also local differences in vertical shifts, a local aberration at PILK (Table 19.3), or differences in horizontal offsets in a NW–SE direction, which point at volumetric strains of a similar (10^{-7} to 10^{-6}) magnitude along the Southern Alps. But given the strong alignment of the stations in a transect, with motions near the lower limits of resolution, we remain cautious not to over interpret these data. Importantly, we can be confident that both the earthquakes imparted permanent dextral strains of 10^{-7} to 10^{-6} through rocks in the vicinity of Copland hot spring, which are at the lower limit of strains where permeability changes have been previously demonstrated to occur (Elkhoury *et al.* 2006; Manga *et al.* 2012). Although dynamic strains associated with seismic waves may have been higher, it means there is no requirement to invoke mechanisms that operate without permanent strains, such as pore unblocking, to explain changes observed at the spring.

Table 19.3 Offsets in cGPS station north, east and vertical positions caused by the Mw7.8 Dusky Sound and Mw7.1 Darfield earthquakes

SITE (Units)	E_NZTM (m)	N_NZTM (m)	EQdist (km)	East (mm)	1σ (mm)	North (mm)	1σ (mm)	Vert up (mm)	1σ (mm)	Horiz (mm)	Dir (°)	AFdist (km)
Position shift: Mw7.8 Dusky Sound earthquake, 15 July 2009												
QUAR	1,342,705	5,175,214	356	-5.49	0.31	-2.43	0.18	-5.68	0.61	6.00	246	-3.2
KARA	1,339,750	5,166,567	348	-8.78	0.26	-3.02	0.18	0.64	0.38	9.28	251	2.7
LEOC	1,336,803	5,162,603	343	-8.87	0.28	-3.31	0.30	2.23	0.40	9.47	250	4.0
WAKA	1,348,538	5,169,611	356	-10.13	0.25	-5.30	0.26	2.18	0.39	11.43	242	4.8
VEXA	1,349,399	5,163,674	353	-9.72	0.51	-3.69	0.38	-9.38	1.62	10.40	249	10.1
CNCL	1,346,505	5,160,392	348	-8.54	0.26	-4.64	0.17	-4.08	0.45	9.72	241	11.3
PILK	1,351,779	5,161,217	353	-6.83	0.27	-2.58	0.20	-8.71	0.42	7.30	249	13.4
MAKA	1,346,467	5,154,322	344	-12.40	0.56	-8.27	0.34	5.16	0.75	14.90	236	16.6
REDD	1,351,911	5,156,610	349	-9.64	0.50	-7.08	0.33	2.88	0.70	11.96	234	17.3
MCKE	1,357,123	5,154,075	351	-10.27	0.56	-8.96	0.41	2.48	0.82	13.63	229	22.4
NETT	1,363,418	5,151,015	354	-10.84	0.65	-6.12	0.34	3.00	0.72	12.45	241	28.6
HORN	1,367,066	5,148,785	355	-11.17	0.47	-4.56	0.25	0.02	0.95	12.06	248	32.6
BNET	1,374,198	5,139,562	355	-9.45	0.35	-4.32	0.21	-2.88	0.60	10.39	245	44.3
HAAS†	1,262,516	5,111,354	256	-11.69	0.25	-6.09	0.22	0.77	0.38	13.18	242	4.4
Position shift: Mw7.1 Darfield earthquake, 4 September 2010												
QUAR	1,342,705	5,175,214	191	-1.24	0.40	-1.86	0.24	0.66	0.79	2.24	214	-3.2
KARA	1,339,750	5,166,567	194	-1.96	0.33	-1.98	0.22	4.88	0.48	2.79	225	2.7
LEOC	1,336,803	5,162,603	197	-2.57	0.33	-4.63	0.36	2.98	0.48	5.30	209	4.0
WAKA	1,348,538	5,169,611	185	-3.40	0.32	-4.52	0.33	4.72	0.51	5.66	217	4.8
VEXA	1,349,399	5,163,674	184	-4.23	0.46	-3.84	0.34	3.34	1.49	5.71	228	10.1
CNCL	1,346,505	5,160,392	188	-3.45	0.33	-1.00	0.21	4.41	0.56	3.59	254	11.3
PILK	1,351,779	5,161,217	182	-6.54	0.37	-4.30	0.28	5.04	0.57	7.83	237	13.4
MAKA*	1,346,467	5,154,322	188									16.6
REDD*	1,351,911	5,156,610	183									17.3
MCKE	1,357,123	5,154,075	178	-5.09	0.52	-5.18	0.35	-2.54	0.78	7.26	224	22.4
NETT	1,363,418	5,151,015	172	-2.22	0.58	0.81	0.30	6.33	0.64	2.36	290	28.6
HORN	1,367,066	5,148,785	169	-4.11	0.66	-5.03	0.35	2.33	1.32	6.50	219	32.6
BNET	1,374,198	5,139,562	164	-3.83	0.38	-3.72	0.24	0.93	0.66	5.34	226	44.3

Note that due to the longer time series available for calculating secular trends in this study, values for HAAS differ slightly from Beavan *et al.* (2010b) and Mahesh *et al.* (2011). NZTM, New Zealand Transverse Mercator; EQdist, epicentral distance between station and the earthquake; 1σ, standard errors; Horiz, total horizontal displacement; AFdist, distance between the station and the Alpine Fault. Shifts in italics/shaded are not significant at 2σ confidence levels.

*Semi-continuous sGPS site. Cannot estimate offsets due to insufficient data.

†HAAS is 100 km south–west of the Copland spring (Fig. 19.1A).

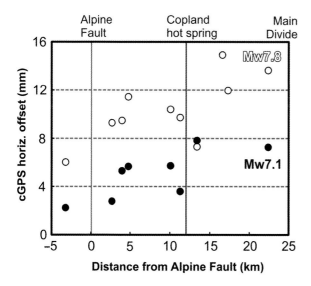

Fig. 19.10. Calculated cGPS station horizontal offsets (in mm) during the Mw7.8 Dusky Sound (open circles) and Mw7.1 Darfield (filled circles) earthquakes, plotted against distance from the Alpine Fault in a transect south-eastwards across the Southern Alps, a high angle to the displacement vectors (see Fig. 19.1B). The regional variation in horizontal offset reflects permanent dextral shear strains imparted by the earthquakes. Copland hot spring is situated in the centre of the transect, 12 km from the Alpine Fault.

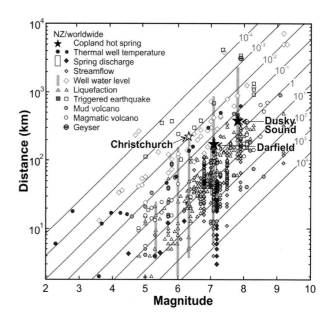

Fig. 19.11. The Mw7.8 Dusky Sound and Mw7.1 Darfield earthquake-induced changes at Copland hot spring (star symbols) on an earthquake magnitude versus distance plot showing worldwide earthquake-triggered hydrological changes collated by Wang & Manga (2010a,b). Hydrological responses to New Zealand (NZ) earthquakes have been collated from Hancox et al. (2003), Cox et al. (2012b), Gulley et al. (2013) and Boese et al. (2014). The temperature change in Copland spring following the Mw6.3 Christchurch earthquake is plotted for reference, although the response was not definitive. Contours of constant seismic energy density follow Wang & Manga (2010a,b), and thermal well temperature changes (Mogi et al. 1989) are distinguished from spring discharge changes (King et al. 1994a; Sato et al. 2000; Manga & Rowland 2009) that were not shown in the original figures of Wang & Manga (2010a,b).

DISCUSSION

Hydrological changes associated with moderate-to-large earthquakes around the world have been compiled as a function of earthquake magnitude and epicentral distance and related to seismic energy density as a measure of the maximum energy available to do work at a given location during an earthquake (Wang & Manga 2010a,b; Fig. 19.11). Relatively few studies have quantified earthquake-induced temperature changes in thermal springs and wells, although the phenomenon is long-recognised. Mogi et al. (1989) outlined temperature increases of approximately 1°C in a 600-m deep, cased, geothermal well on Izu Peninsula (Japan) following nearby (<600 km distant) large earthquakes. These temperature changes, generally sharp rises in a steplike pattern with gentle linear cooling between seismic events, were interpreted to reflect increased flow rates in the well, which occurred once seismic energy density reached approximately 10^{-2} J m^{-3} (Wang & Manga 2010b; see also Fig. 19.11). Other changes in thermal spring discharge, temperature or hydrochemistry have occurred at higher energy thresholds approximately 10^{-1} J m^{-3} (for streamflow) to approximately 1 J m^{-3} (for spring discharge) (e.g. King et al. 1994a; Sato et al. 2000; Manga & Rowland 2009; Liu et al. 2010; see Wang & Manga 2010a,b). Increased discharge, conductivity or ionic concentrations of springs and streams after earthquakes are usually attributed to changes in permeability and post-seismic release of deep fluids (e.g. Rojstaczer & Wolf 1992; Sato et al. 2000; Favara et al. 2001). Observed thermal well and spring responses to earthquakes are most commonly opposite to the cooling and chemical dilution we describe at Copland hot spring (although see Howald et al. 2015).

Copland hot spring is derived from mixing of deep, upwelling, thermally heated meteoric water and a minor component of shallow, cool, infiltrating meteoric water. Our interpretation of the observed earthquake-related temperature and fluid chemistry responses is that shaking has altered the permeability structure of adjacent mountains or shallow crust, somehow opening or closing the aperture of fractures in schist bedrock. The earthquake-induced change in permeability alters the proportions of deep thermal to shallow meteoric water emerging at the surface, thereby creating a change in the pH, conductivity and trace element chemistry of spring water and contributing to the cooling. From observations immediately at hand, however, we cannot be entirely certain whether there was an increase in permeability that allowed greater quantities of cool near-surface groundwater to be driven downwards by topographic head to mix with upwelling hot water, or a decrease in permeability at depth that restricted the amount of upwelling hot water. We infer the former to be more likely, supported by the following arguments.

Of importance is that the present flow regime in the Southern Alps can be affected at relatively low shaking intensity, hundreds of kilometres from the epicenter of an earthquake. The observed earthquake-related responses at Copland spring occurred in association with shaking at peak ground accelerations that are very unlikely to have exceeded approximately 2% g during these earthquakes, corresponding to Modified Mercalli (MM) intensities of III–IV (weak to light) and seismic energy densities of approximately 10^{-1} J m^{-3} (Fig. 19.11). Observed responses occurred when shaking exceeded peak ground acceleration approximately 0.5% g and dynamic stresses exceeded 0.05–0.1 MPa. There was an absence in a clear thermal response to the Mw6.3 Christchurch earthquake at seismic energy density approximately 10^{-2} J m^{-3}. The threshold for response at Copland may be close to the approximately 10^{-2} J m^{-3} required the artesian thermal well in Japan (Mogi *et al.* 1989) to respond (Fig. 19.11) and certainly lower than that recorded to produce changes in spring discharge elsewhere (King *et al.* 1994a; Sato *et al.* 2000; Manga & Rowland 2009; Fig. 19.11). Records from the nearby FOZ seismometer and the more comprehensive National Seismic Hazard Model (Stirling *et al.* 2002, 2012) suggest this level of shaking can be expected at every 1–10 years. Correspondingly, subtle changes in permeability can be expected on an annual to decadal timescale in the central Southern Alps.

Temperature and chemistry constraints on hydrological flow models

When guided by site surface observations, spring discharge measurement and the mapped thermal anomaly (Fig. 19.3), the spring water temperature and chemistry observations provide information on the subsurface geometry of the Copland spring hydrothermal system. The emergence of water with hot temperature at the surface requires flow to be confined to a relatively narrow pathway such that its flow rate is sufficient to retain heat advected from depth. The response time associated with rainfall-related cooling provides information about near-surface hydrology and heat exchange. The time lag of cooling onset and the subsequent decay of temperature in response to earthquakes provide information about deeper hydrological change and heat exchange. Water chemistry observations confirm that cooling after earthquakes is associated with addition of cold meteoric fluids to the hydrothermal conduit.

Advected heat in the spring at the surface has an equivalent power of 1.4 MW, based on the total estimated flow rate (7 kg s^{-1} – includes main source vent and other small peripheral discharge seeps), the difference between spring temperature and ambient surface conditions (50.4°C), and specific heat capacity of water. Spring water chemical analyses indicate that the deep fluid source equilibrated at temperatures of at least approximately 160°C, which is estimated to be found at a depth of $L = 2$–3 km beneath the ground at this location, on the basis of modelled and measured geothermal gradients in the region

(Allis *et al.* 1979; Koons 1987; Sutherland *et al.* 2012). The surface temperature is approximately 10°C, so ambient conditions are approximately 150°C hotter at this depth. Fluid emerges at 56–58°C, so about one-third of the heat is retained in the fluid and the rest must be conducted away laterally along its pathway to the surface. Flow from depth presumably occurs within a relatively porous, fractured zone in the schist.

Rain-related transient changes in temperature are rapid, responding and recovering in just a few hours. Finite difference models of a shallow reservoir that fills according to rain input and empties at a rate proportional to its fill height are able to explain the time lags and amplitudes of temperature change. The physical explanation is probably a combination of two factors: (i) dilution of spring water by near-surface groundwater from within the thin, underlying layer of alluvium and debris–flow sediment, including surface addition during times of particularly heavy rain (Fig. 19.2); and (ii) heat advection from the shallow thermal boundary layer around the hot spring during times of high near-surface cold-water flow through surface sediments. The rapid response reflects high near-surface permeability and very shallow depth of the processes (<10 m).

Water chemistry, conductivity and pH measurements indicate earthquakes are associated with connection of cold-water input to the hot spring conduit at depth. Thermal responses to the Mw7.8 Dusky Sound and Mw7.1 Darfield earthquakes were delayed after arrival of seismic waves by 180 and 140 ± 15 min, respectively. It is possible that there was a delay between earthquake shaking and connection of the new cold input, similar to the lag between triggered swarms of seismicity observed beneath the Main Divide (Boese *et al.* 2014). However, we assume that the primary reason for the delay of temperature change onset relates to the time it takes to drain existing fluids between the new input at depth and the surface. At discharge of 6 ± 1 l s^{-1}, between 38 and 82 m^3 of fluid was expelled during the 140- to 180-min delay. If the zone of fractured schist has porosity within the extremes of 10–25% and forms a conduit with the cross-sectional area of between 1 m^2 (the dimension of the vent pool) and 10 m^2, the depth of intersection for the cold-water input will be in the range of 15–410 m.

A small increase in temperature was observed following the Mw7.8 Dusky Sound earthquake. This might be explained by an increase in flow rate of the spring caused by an increase in permeability, allowing greater volumes of hot water to carry additional heat to the pool from within the fracture system at depth. Alternatively, it is possible that shaking increased CO_2 exsolution or shaking released CO_2 bubbles from fracture walls, resulting in an increased volume of gas and fluid rising to the surface, driving a greater flow rate, and hence an increased amount of heat supplied to the pool. The spring fluid and gas discharge rate were not continuously monitored.

We interpret the decay of temperature after an earthquake, which is approximately exponential, to reflect thermal capacitance of the rock within a thermal boundary layer in and around the hot spring conduit. The temperature decay has a time constant of about 2 days (see Supporting information Fig. S2), and the temperature change was entirely accomplished over 5–8 days. Assuming that there is an instantaneous change in source water temperature caused by dilution by a new cold-water pathway, as indicated from chemistry results, and integrating the Mw7.8 Dusky Sound earthquake-induced change in temperature through time, we find that the thermal capacitor of the rock absorbs 4400 MJ. We assume an instantaneous change in input temperature caused by a new mixing ratio at the start of the exponential decay; a constant input temperature at the base of the capacitor during the decay that is equal to the final equilibrated temperature; a flow rate of 7 kg s^{-1}; and assume that the anomalous temperature of the fluid above the input temperature is derived from heat extracted from the capacitor. If the entire volume of the rock thermal buffer changed by the full 1°C temperature change, and assuming schist has specific heat capacity between 850 and 1000 J kg^{-1} K^{-1} and 2.67 kg m^{-3} density (Hatherton & Leopard 1964), then a rock volume of 1500–2000 m^3 is implied, although it is likely that some of the boundary layer changes by less. If we assume that the cross-sectional area of the thermal boundary layer is similar to the surface elevated temperature anomaly area (Fig. 19.3), between 10 and 50 m^2, then the depth of connection implied is in the range 30–200 m. Whilst the values adopted are selected with ranges that are deliberately extreme, the result is consistent with the time-delay calculation result. Importantly, their corroboration implies that the input of shallower meteoric water that produces the characteristic cooling is a shallow process, occurring in the uppermost few hundred metres of schist bedrock.

Key observations and interpreted mechanism

Although observations at Copland spring show that distal earthquakes can change the flow regime, fully connecting the field observations to mechanisms will require continuous monitoring of spring discharge, fluid chemistry and CO_2 degassing. Some pertinent observations are as follows:

(1) Despite variable overprinting effects of rainfall, the Mw7.8 Dusky Sound and Mw7.1 Darfield earthquakes produced a characteristic style of temperature response.

(2) Cooling following the Mw7.1 Darfield earthquake occurred from a background temperature that was already (still) depressed, 14 months after the Mw7.8 Dusky Sound earthquake had affected the system. The causal mechanism was capable of stacking one cooling on top of the next – the system temperature, and by inference permeability, it did not need to fully recover before being influenced by another large earthquake.

(3) Earthquakes with quite different source characteristics (location, Mw, Me, direction) and resultant offsets recorded locally by cGPS, both produced small permanent dextral shear strains of 10^{-7} to 10^{-6} and a similar characteristic response.

(4) Nearest seismometers recorded quite different spectral patterns during the Mw7.8 Dusky Sound and Mw7.1 Darfield earthquakes, so causal mechanisms are independent of shaking frequency.

(5) There are a number of similarities between observed earthquake-induced responses in Copland spring and nearby delayed-triggered microseismicity (Boese et al. 2014). Although no specific triggered swarms were noted in the vicinity of Copland valley, swarms occurred several hours after the earthquakes, lasting for approximately the same period it takes for Copland spring to adjust to a new cooler background temperature.

(6) The smaller Mw6.3 Christchurch earthquake produced either no response or at most a short-lived, small step like temperature drop – despite having almost recovered to the mid-2010 (pre-Mw7.1) background temperatures. Given the Mw7.1 Darfield earthquake had previously influenced the system's temperature without the need for recovery, we suggest the 5-month period between the Mw6.3 Christchurch and the Mw7.1 Darfield earthquakes, and partial recovery would have been sufficient for a response to have occurred and been observed had the Mw6.3 shaking been sufficiently large. We therefore infer that seismic amplitude is important for disturbing the subsurface fluid flow in the Southern Alps. A shaking threshold at around peak ground acceleration approximately 0.5% g and/or with dynamic stresses of 0.05–0.1 MPa and seismic energy density approximately 10^{-1} J m^{-3} needs to be met to produce a characteristic response at Copland hot spring.

(7) Partial recovery had occurred around 1 year after the Mw7.1 Darfield earthquake, but total recovery had yet to occur by December 2013. Recovery does not seem to be as linear/characteristic as cooling decays, but appears likely to occur at annual to decadal timescales. Variable recovery times might not be unexpected given the Mw7.8 Dusky Sound earthquake was the largest seismic event experienced in the South Island for about 80 years.

(8) There is some degree of stability of the hydrothermal system, such that the spring has maintained a temperature at approximately 56°C for nearly four decades (since Barnes et al. 1978). The spring was visited by Māori in pre-European time and has been present at the site for sufficient time for a large deposit of travertine calcite to precipitate.

It has been shown in a recent contribution outlining meteoric infiltration into mountains of crystalline rock in Fiordland (Fig. 19.1A) that permeability cannot be a simple function of depth (Upton & Sutherland 2014). Instead, upper crustal permeability is modelled as a function of strength and the state of stress in the rock, which correlates with the proximity of the rock mass to brittle failure criteria (see also Townend & Zoback 2000; Zoback 2010). Topography creates spatial heterogeneities in the proximity to failure that extend to depths similar to the height of adjacent mountains. The study proposes three processes that load the rock mass to failure and hence create permeability: long-term regional tectonic loading; short-term transient loading during phenomena such as earthquakes; and topography. Following this conceptual framework, we suspect that active deformation and topography must have an important influence on the processes observed in Copland spring and propose the setting of the Southern Alps in the hanging wall of the Alpine Fault may be particularly sensitive to influences such as distal earthquakes. The notion that tectonic and hydrogeological settings could be an important factor in determining the scale of groundwater response to distal earthquakes, with greater changes occurring at stress-sensitive sites, has also been proposed on the basis of continental-scale observations in China by Shi *et al.* (2015).

The central Southern Alps region is a locus for contemporary deformation, with geodetic-determined shear strain rates > 0.3 ppm a^{-1} and maximum uplift rate 10–20 km south-east of the Alpine Fault (Wallace *et al.* 2007; Beavan *et al.* 2010a). The contemporary deformation, however, does not correspond with the long-term (geologic) pattern of rock exhumation and deformation, in which the most strongly deformed, highest metamorphic grade mylonite and schist have been most recently exhumed immediately beside Alpine Fault at rates around 10^{-2} m a^{-1} (Little *et al.* 2005; Herman *et al.* 2009). Observed contemporary deformation in the upper crust is therefore thought to be largely elastic and expected to be released when the Alpine Fault next ruptures (Ellis *et al.* 2006; Beavan *et al.* 2007). The rapidly uplifting mountains appear to be in a state of incipient gravitational (Allen *et al.* 2011) as well as tectonic failure such that over multiple-earthquake timescales, there is a balance between uplift and erosion. Landslides, rock-falls and erosion presently generate annual suspended sediment yields in the rivers that are among the highest in the world and broadly match long-term 10^{-2} m a^{-1} rates of exhumation (see Hicks *et al.* 1996; Cox & Sutherland 2007; and refs therein). The central Southern Alps should be considered to be in a state of tectonic and topographic criticality (Koons 1994), or near-criticality, such that shaking at relatively low thresholds is potentially capable of generating perturbations to shallow fluid circulation.

GPS monitoring showed that earthquakes induced permanent shear strains of 10^{-7} to 10^{-6} across the Southern Alps, and potentially local volumetric strains of a similar magnitude, that are towards the lower limits of resolution. Given the rate and state of elastic strain accumulation, it is conceivable that the passage of seismic waves could have triggered the release of stored elastic deformation or perturbed the topographic stress state, thereby contributing to the motions and strain recorded by the GPS network. It may imply that the local ambient stress state of the rock mass is important in the evolution of permeability and hydrological responses at intermediate- and far-field distances, so therefore warrants further research. Given also that the Alpine Fault is late in a regular cycle of stress accumulation

(Sutherland *et al.* 2007; Berryman *et al.* 2012), some form of rupture seems imminent and much may soon be learnt about the evolution of permeability and the flow regime in the Southern Alps.

Flow of fluid through tight schist rocks will be highly sensitive to small changes in the apertures of the fractures and connected networks. Based on the situation presented earlier, we suspect that the passage of seismic waves exploits either topographic and/or stored tectonic stress to enhance permeability through the physical opening of fractures, or minor generation of new fractures, in the upper most few kilometres. The strains are likely to be small at a regional scale, at least 10^{-7} or 10^{-6} as recorded by motion of the cGPS network, although locally could be potentially larger reflecting the presence of critically oriented fractures. Closure of the fracture network might then occur through the regional plate-tectonic-related strain, which accumulates at between 10^{-7} and 10^{-6} a^{-1} in the Southern Alps, so that it has the potential to facilitate physical recovery within about 1–10 years.

The explanation of fracture opening, above, draws on the observation of permanent strain induced by the earthquakes. However, there are alternative explanations, or other physical and chemical mechanisms at Copland hot spring that could also contribute to observed responses and changes in permeability. For example, PGV values (Table 19.2) suggest the passage of seismic waves imparted dynamic stresses exceeding 0.05–0.10 MPa, which might enable the fractures to become unclogged and/or held open by some form of chemical process, cleaned by an increase in turbidity, or poroelastic pressuring (Manga *et al.* 2012). Copland hot spring fluid, at least in the near surface, is a two-phase mixture in which CO_2 bubbles are dispersed in suspending water. The CO_2 bubbles potentially block channels through capillary forces, inhibiting the bulk flow and fluid mobility. So a change in the mobilisation or nucleation of gases by seismic waves could conceivably lead to an increase in bulk flow rate and an apparent increase in permeability. However, for such a mechanism to continue over periods of the observed responses at Copland spring, it would also need some form of sustained local disequilibrium through changes in temperature, fluid phase or invasion of new fluids, or allow flow that generates a poroelastic response that keeps fractures open for a sustained period. Based on evidence of prolonged, earthquake-induced, infiltration of cool, unexchanged, meteoric water into a fault-controlled geothermal system, Howald *et al.* (2015) suggested that crustal-permeability changes can last much longer lived than previously thought ($>10^3$ years).

The rate of recovery of the flow regime at Copland, extending over years or even tens of years, is also consistent with the notion of effects limited by kinetics or diffusive mass transport. The spring actively exsolves CO_2 and has deposited a travertine terrace at the surface. During upwelling, the drop in pressure causes CO_2 to degas from the fluid, raising its pH and causing precipitation of calcite. Calcite is a common mineral phase within joint coatings, fault gouge and veins in the schist, with calcite precipitation inferred to have occurred at a range of depths in the subsurface (Koons *et al.* 1998; Boulton *et al.* 2014; Menzies *et al.* 2014). Recovery of the hydrothermal system following an earthquake could therefore involve a slow clogging of fractures and aperture closure by precipitation of calcite, rather than physical closure by accumulation of tectonic-related stress.

Without monitoring the exact discharge and state of two-phase flow, we cannot fully decipher the interplay of chemical and physical processes to conclusively elucidate the causal mechanism(s). However, the hydrothermal flow regime was clearly perturbed, and observations at Copland spring provide some fresh insight. Observations of Copland spring responses also correspond with observations of nearby-triggered seismicity (Boese *et al.* 2014), so we propose that the transient permeability is a function of stress and strength of the rock (Upton & Sutherland 2014) and proportional to the proximity of failure. Copland spring provides empirical evidence that mountains at active plate boundaries are susceptible to infiltration, advection of heat and transient changes in permeability.

SUMMARY AND CONCLUSIONS

Thermal springs in the Southern Alps, New Zealand, originate through deep penetration of meteoric fluids into a thermal anomaly generated by rapid uplift and exhumation in the hanging wall of the Alpine Fault. The Copland hot spring (−43.629S, 169.946E) is one of the most vigorously flowing of the springs, discharging strongly effervescent CO_2-rich water at $6 \pm 1 \text{ls}^{-1}$. Rainfall, air and vent water temperatures were continuously monitored at 15-min intervals from March 2009 to July 2011. The spring fluctuated at background values between 56 and 58°C, with minor diurnal variation, but fell dramatically to approximately 40°C during heavy rainfall when deep upwelling fluids are diluted near surface by cool meteoric water. Recovery to background temperatures typically took 3–5 h after rain stopped falling, depending on the rainstorm magnitude. Smaller, but more sustained temperature decreases occurred in response to shaking from distal earthquakes. The Mw7.8 Dusky Sound (Fiordland) earthquake on 15 July 2009, centred 350 km south–west of the Copland spring, produced weak to light shaking intensities in the area (approximately MM III–IV) and low peak ground accelerations (0.4–2.3% g) at nearby seismometers and horizontal south–west displacements of 6–12 mm at cGPS stations nearby. After 180 ± 15 min delay, the spring showed a warming of 0.20 ± 0.05°C, followed by a 0.9 ± 0.2°C cooling over a 5-day period. Temperature remained depressed until the Mw7.1 Darfield (Canterbury) earthquake on 4 September 2010, centred 180 km east of Copland spring, when shaking at approximately MMIV (light) intensities and peak ground acceleration of 0.5–2.8% g resulted in a further 1.1 ± 0.2°C cooling delayed 140 ± 15 min after

the earthquake. Comparison of the temperature responses suggests they followed a characteristic cooling that did not require recovery between earthquakes and perhaps that similar features were activated by the ground motion. Shaking during the Mw6.3 Christchurch aftershock on 22 February 2011 was mostly <0.5% g at nearby seismometers and was unable to produce the same response, but possibly induced a 0.36 ± 0.05°C cooling masked behind effects of rainfall.

The Copland hot spring appears to respond characteristically to distal (intermediate-field) large earthquakes once shaking at peak ground acceleration approximately 0.5% g and/or a dynamic stress threshold of approximately 0.05–0.10 MPa has been overcome, following an exponential temperature decay that is quite distinct from more rapid rainfall-related cooling. The passage of seismic waves, even at low MMIII–IV shaking intensities, peak ground acceleration <2% g and seismic energy density around approximately 10^{-1} J m^{-3}, can affect spring temperature, producing responses independent of variations in spectral frequency. Cooling at Copland spring was accompanied by a fluid pH increase and electrical conductivity decrease, reflecting a change in the mixture of deep-circulated hydrothermal meteoric fluid and shallow-circulating meteoric fluid. cGPS stations indicate Mw7.8 Dusky Sound and Mw7.1 Darfield earthquakes caused permanent shifts with motion near the lower limits of detection, which were sufficient to induce static strains of 10^{-7} to 10^{-6} across the central Southern Alps. Such shaking thresholds will be exceeded and potentially affect the hydrothermal flow regime at annual to decadal timescales.

Deciphering the exact mechanisms responsible for Copland spring response to earthquakes has been hampered by not knowing the exact discharge of the hot spring through time. We infer the temperature and hydrochemical responses reflect subtle earthquake-induced changes in the 10^{-15} to 10^{-12} m^2 fracture permeability of schist mountains adjacent to the spring, allowing greater quantities of cool near-surface groundwater to mix with upwelling hot water. Our favoured mechanism is a physical opening or generation of new fractures in the uppermost few hundred metres of the schist bedrock, perhaps later closing under gradual readjustment to regional stress, but there is potential for combined physico-chemical mechanisms to play a role in the process, particularly the deposition of calcite during post-earthquake recovery. We propose that active deformation and tectonic and topographic stresses make the hanging wall of the Alpine Fault particularly susceptible to earthquake-induced transient permeability changes.

ACKNOWLEDGEMENTS

Samples from the Westland National Park were collected under Department of Conservation permit WC-22994-GEO. This project was funded under GNS Science's 'Impacts of

Global Plate Tectonics in and around New Zealand Programme' (PGST Contract C05X0203). Geochemical analyses were funded by a Natural Environmental Research Council-CASE PhD studentship award NE/ G524160/1 to CDM (GNS Science CASE Partner) and NERC grants NE/ H012842/1, NE/J024449/1 and IP-1187-0510 to DAHT. We acknowledge the New Zealand GeoNet project and its sponsors EQC, GNS Science and LINZ, as well as the late John Beavan, for providing data used in this study. We wish to thank our colleagues John Townend, Jim Cousins, John Haines, Carolin Boese, Rachael James, Dave Craw, Delia Strong and Agnes Reyes for discussions and helpful comments, although not necessarily implying they agree with all of our interpretations and conclusions. We also thank our reviewers and editors Tom Gleeson and Steve Ingebritsen.

SUPPORTING INFORMATION

Additional supporting information may be found in the online version of this chapter, *Geofluids* (2015) 15, p. 216–239:

Data SI. Schist permeability, chemical methods, seismic and geodetic data, thermal decay observations and model.

Figure S1. Decomposed time series of easting, northing and height measurements at continuous GPS stations CNCL and KARA, plotted relative to their secular modes.

Figure S2. Temperature data from Copland hot spring following the Dusky Sound earthquake on Wednesday 15 July 2009 at 9:22 pm NZST (red line).

Table S1. Permeability and hydraulic conductivity data from Lugeon packer tests (PT), head response tests (HT), or other methods found in unpublished engineering and/ or consultancy reports for the Haast Schist Group, which encompasses both Alpine and Otago schists.

CHAPTER 20

Transient permeability in fault stepovers and rapid rates of orogenic gold deposit formation

STEVEN MICKLETHWAITE[1], ARIANNE FORD[1], WALTER WITT[1] AND HEATHER A. SHELDON[2]

[1] *Centre for Exploration Targeting, The University of Western Australia, Crawley, WA, Australia;* [2] *Australian Resources Research Centre, Commonwealth Scientific and Industrial Research Organisation (CSIRO), Kensington, WA, Australia*

ABSTRACT

Fault stepovers are features where the main trace of a fault steps from one segment to the next in either an underlapping or overlapping manner. Stepovers exert a critical influence on crustal permeability and are known to control phenomena such as migration of hydrocarbons and location of geothermal fields. In the Kalgoorlie–Ora Banda greenstone district, Western Australia, we demonstrate a spatial association between stepovers and gold deposits. It is shown that although underlapping stepover geometries are typically rare in fault systems, they are anomalously associated with gold deposits. Furthermore, the along-strike and across-strike dimensions of both underlapping and overlapping fault stepovers fit, to a first-order approximation, the same self-similar trend. Boundary element modeling of Coulomb failure stress changes is used to explain these observations in terms of damage generated by rupture events on the bounding fault segments and associated aftershock sequences. Our models indicate that a larger region of damage and permeability enhancement is created around underlapping stepovers than around overlapping stepovers. By taking into account both the enhancement and decay of permeability during the seismic cycle, it is estimated that a 5 Moz goldfield could feasibly form in 1–16 earthquake–aftershock sequences, potentially representing durations of just 10–8000 years. The existence of supergiant gold deposits is evidence that crustal permeability attains transiently high values in the order of 10^{-12} m^2. It should be expected that transient and time-integrated permeability values have a distinct three-dimensional structure in continental crust due to stepover-related channels.

Key words: fluid flow, gold, permeability, scaling, static stress change, stepovers, time-integrated fluid flux

INTRODUCTION

In earth materials, permeability is a dynamic and changing parameter that may potentially vary by 5–10 orders of magnitude over short timescales (Ingebritsen & Manning 2010; Micklethwaite *et al.* 2010; Rowland & Simmons 2012; Miller 2013). Active deformation of faults and shear zones is widely acknowledged as a critical agent for transient enhancement of permeability in the Earth's crust (e.g. Rojstaczer *et al.* 1995; Claesson *et al.* 2007). One important outcome is the migration of large volumes of fluid through narrow domains, resulting in the formation of hydrothermal ore deposits (Cox *et al.* 2001). In particular, fault stepovers have a documented spatial association with the distribution of ore deposits (e.g. Sibson 1987; Connolly & Cosgrove 1999; Ford *et al.* 2009; Micklethwaite *et al.* 2010; Witt *et al.* 2013), which act as a proxy for high time-integrated fluid flux (Sheldon & Micklethwaite 2007). Indeed, fault stepovers are recognised as playing a role

in a wide range of fluid-related geological phenomenon, such as post-seismic fluid redistribution (Peltzer *et al.* 1996), the migration of hydrocarbons (Fossen *et al.* 2010) and the location of geothermal fields (e.g. Faulds *et al.* 2011). In this contribution, we provide a systematic examination of the relationship between fault stepovers and ore deposits, to provide insight into a primary control on permeability within continental crust at length scales of tens to thousands of kilometres, integrated over timescales of ore deposit formation (previously estimated at 10^3 to 10^5 years; Sanematsu *et al.* 2006; Rowland & Simmons 2012).

This study combines field observations, analysis and modeling to identify first-order principles controlling gold mineralisation and to obtain estimates for absolute permeability values and their temporal variation. As such, we take the unusual step of presenting the study in three parts, each with a 'methods' and 'results', with each part building on the last. In the first part, we establish a spatial relationship between fault stepovers and gold mineralisation, using data from the Eastern Goldfields Superterrane (EGS), Western Australia. In the second part of this study, we use accurate deposit-scale maps of stepovers associated with gold deposits from around the world to examine the geometry, aspect ratios and scaling properties of those stepovers. The data are compared against published fault stepover data, documented from non-mineralised, active and inactive fault systems (Aydin & Schultz 1990; Kim *et al.* 2004). In the third part, the results from the geometric analysis are used to inform 3D boundary element numerical models. Here, we assess the influence of stepover geometries on the shear and normal stress changes that occur during earthquake or fault creep events and the implications these have for fracturing. Finally, we review and discuss our results in terms of transient enhancement of permeability through the Earth's crust and the profound implications they have for time-integrated fluid fluxes.

ASSOCIATION BETWEEN STEPOVERS AND MINERALISATION

Methods

Ad hoc observations of ore deposits from many different mineral provinces suggest that there is a spatial association between deposits and stepovers on adjacent shear zones or faults (e.g. Hagemann *et al.* 1992; Garza *et al.* 2001; Richards 2003; Weinberg *et al.* 2004; Bateman *et al.* 2008), although this relationship has never been quantitatively measured. We first establish whether this association has a statistical significance by an assessment of the proximity of ore deposits to stepovers.

It should be noted that the term stepover is not common in economic geology literature but subtypes of stepovers, such as jogs, pull-aparts or relay zones, are routinely referred to and all these features are classified as stepovers in this study. In two dimensions, stepover zones may or may not be linked (hard and soft linkages, respectively) but adjacent segments kinematically interact with one another (Soliva & Benedicto 2004) and sometimes merge into a single fault surface in three dimensions (Walsh *et al.* 2003).

Fault map data from the Kalgoorlie–Ora Banda district of the EGS, Yilgarn Craton, Australia (Fig. 20.1), were analysed using unpublished digital files from the Geological Survey of Western Australia (representing faults at the 1:100,000 scale), combined with 1:10,000 scale structural maps from a database originally derived by the exploration company AurionGold Ltd (Witt *et al.* 2013). The data set comprises thousands of individual fault segments, and stepovers were automatically identified with the geographic information system package MapInfo™. Deposit locations were derived from MINE-DEX (the web URL for this data source is provided in the references).

Fault stepovers were identified using the geometries summarised in Figure 20.2. Specifically, the parameters used to define the stepovers included the following: the maximum separation or width ($2s$) between adjacent fault segments was

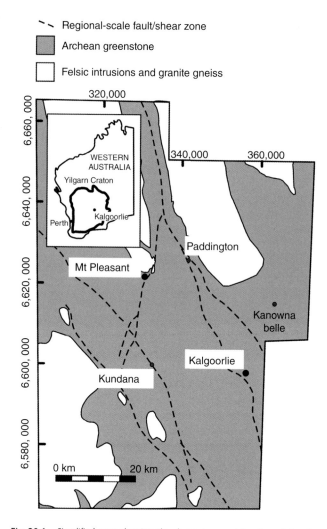

Fig. 20.1. Simplified map showing the dimensions and location of the Ora Banda–Kalgoorlie district, Western Australia. Inset shows the location of the district within the Yilgarn Craton. Included are the locations of selected major gold deposits present in the data set. Deposits labelled in grey and black are part of the data set used for the spatial analysis in the first part of this study. A map of the full data set is available as Supporting Information (Fig. S1). Deposits labelled in black have map data of high enough quality to be incorporated in the scaling and geometry analysis (second part of this study). Australian grid GDA94.

2500 m; the maximum underlap/overlap distance (2λ) was 2500 to −2500 m; the allowable deviation in angle (α) linking two segments relative to the normal to the strike of the regional structure was 5°–45°. The process automatically identified 2656 stepovers within the district (see Fig. S1 in Supporting Information). Following the identification of stepovers, the number of deposits within a given buffer distance (radius) of the stepovers was measured in successive increments. Buffer distances were concentric around the centroid point of each stepover (e.g. Fig. 20.2C). Buffer distances began at 100 and 250 m and then increased in 250 m increments up to a buffer distance of 1 km. Beyond 1 km, buffer distances were incremented at every kilometre. This process of defining successive buffer distances was based on the examination of statistical

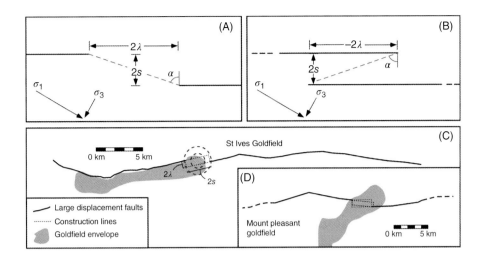

Fig. 20.2. Stepover geometries used for both the GIS-based spatial analysis and the geometric scaling analysis. Parameters explained in text. (A) Underlapping fault stepover, with positive along-strike distance (2λ). (B) Overlapping fault stepover, with negative along-strike distance (-2λ). (C,D) Examples of stepovers associated with orogenic-type gold mineralisation from the Kalgoorlie greenstone terrain. Two sets of construction lines for the geometric scaling analysis are given for the Mount Pleasant example, due to different strikes of the fault segments either side of the stepover. Concentric circles in the St Ives example are a schematic representation of the buffer distances used in the spatial analysis at distances of 1000 and 2000 m.

measures derived during analysis of data in the Yilgarn (Witt *et al.* 2013). In this manner, the spatial relationship between deposits and stepovers was quantified per square kilometre.

There are two main limitations to the analysis presented here. First, results are dependent on the quality of the original structural data. Although outcrop is poor, the structural data sets are considered reliable because they are derived from multiple approaches (field mapping, interpretation of potential field data and direct drill core sampling) and multiple sources (industry and Geological Survey of Western Australia). Second, the analysis does not take into account the size of deposit (endowment) but is a statistical measure of the proximity of deposits to a stepover. For example, a result of 2.5 deposits km^{-2} with a 100 m buffer is a measure of the average density of deposits within that buffer distance. Additional statistics that take into account endowment are presented in Witt *et al.* (2013). A map of the district documenting faults, deposits, endowments, stepovers and buffers is available in Figure S1, in the Supporting Information to this study.

Results

Results (Fig. 20.3) are presented in terms of (i) the number of deposits per square kilometre relative to stepovers and (ii) a normalised ratio of observed (O) to expected (E) deposits relative to stepovers (e.g. $O-E/E$, where E is the number of deposits expected within a buffer region if deposits are randomly distributed). Values of $O-E/E > 1$ indicate a positive relationship between the location of stepovers and deposits.

Figure 20.3A demonstrates an increase in the density of gold deposits at distances <3 km from stepovers and that this relationship is statistically significant compared with a random distribution at distances less than ~1.5 km (Fig. 20.3B). These

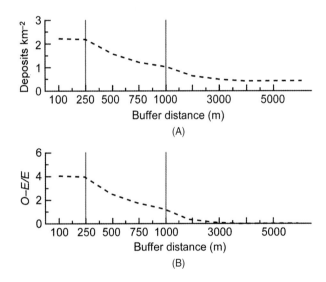

Fig. 20.3. Evidence for a strong spatial relationship between gold deposits and fault stepovers, Kalgoorlie–Ora Banda district. (A) Stepover curves for deposit density relative to distance from the stepover. (B) Ratio of observed (O) to expected (E) deposits relative to stepovers. Values of $O-E/E > 1$ indicate a positive correlation between the location of stepovers and deposits. Grey vertical lines mark changes in the scale of the horizontal axis, for ease of visualising the results.

results indicate a strong positive relationship between the presence of stepovers and the location of gold deposits within the Ora Banda–Kalgoorlie district. The relationship equates to a maximum endowment of 83.8% of the total gold in 30.8% of the area at distances <1 km from stepovers.

Although the absence of gold deposits does not imply the absence of hydrothermal fluid flow, gold deposits are themselves good proxies for hydrothermal fluid flow. Given the spatial association between stepovers and gold deposits, the analysis indicates that, on the scale of a mineralised district (tens of thousands square kilometres), stepovers are important controls on fluid flux through the Earth's continental crust.

GEOMETRY AND SCALING PROPERTIES OF STEPOVERS

Methods

The geometry and scaling properties of stepovers were investigated following a methodology modified from Aydin & Schultz (1990) and data from fault systems mostly dominated by strike-slip movement. In the first instance, data were compiled from stepover studies of active fault systems (Aydin & Schultz 1990) and inactive but well-exposed fault systems (e.g. Kim *et al.* 2004). These data are used as a control population for comparison against data compiled from well-constrained structural maps of ore camps (e.g. Fig. 20.2C,D; Table 20.1). The sample size is necessarily small as high-quality structural maps of gold camps associated with stepovers are rare in the published literature. The data used in this geometric analysis of mineralised stepovers comprise goldfields from the Kalgoorlie–Ora Banda domain (orogenic-type deposits) and also the North Carlin Trend, Nevada (Carlin-type deposits), and a porphyry deposit, Chile, which is associated with a strike-slip fault (Garza *et al.* 2001).

The domain between interacting fault segments was quantified by the modified aspect ratio (R) of the along-strike (2λ) and across-strike separation ($2s$) between the tips of the segments. The aspect ratio here is modified from previous conventions (e.g. Aydin & Schultz 1990; Acocella *et al.* 2000; Gürbüz 2010; Long & Imber 2011) in that underlap configurations have positive values and overlap configurations have negative values, as indicated in Figure 20.2.

Three problems affect the quality of the results when quantifying the geometric properties of stepovers:

(1) Ore fields are typically poorly exposed such that interpretations are affected by an incomplete understanding of structural relationships and are dependent on drill core and geophysical imaging. As a result, a survey of stepovers associated with ore fields is subject to small population statistics due to the limited number of published studies with sufficient quality.

(2) Maps of non-mineralised fault systems, used as the control population, vary in accuracy and scale. The data sourced in this study were from high-resolution maps (e.g. Kim *et al.* 2004) and subject to strict quality control (e.g. Aydin & Schultz 1990).

(3) A degree of ambiguity is introduced during analysis due to the non-parallelism of adjacent fault segments. For example, the strike of fault segments on either side of a stepover may show small differences in orientation. In these cases, more than one length (2λ) and width ($2s$) measures were taken from the stepovers, determined by the orientation of each fault tip (e.g. Fig. 20.2D). Measurements are straightforward for fault traces that are parallel (e.g. Fig. 20.2C).

Results

Several key results emerge from the geometric analysis of the field data. (i) For any given fault system, overlapping stepover geometries dominate, with releasing stepovers being the most common. (ii) However, the mineralised stepovers are anomalously associated with underlap stepover configurations (Fig. 20.4A). (iii) The overlapping stepover configurations follow a linear trend over several orders of magnitude. When directly compared with these overlapping geometries, mineralised underlapping stepovers fit the same trend (Fig. 20.4B). (iv) The aspect ratio of the mineralised underlapping stepovers is approximately 3:1 (a modified aspect ratio value of $R = -3$), which matches that for overlapping stepovers. Whether a stepover is restraining or releasing does not appear to affect the aspect ratio.

These observations raise the issue of whether there is something special about underlapping fault stepover geometries in particular, which are conducive for the enhancement of crustal permeability in relatively localised zones.

Table 20.1 Dimensions of well-constrained fault stepovers associated with gold camps

Deposit	Country	Along-strike separation, 2λ (m)	Across-strike separation, $2s$ (m)	Stepover type	Reference
Mt Pleasant 1	Australia	2369	822	u, rel.	Micklethwaite *et al.* (2010)
Mt Pleasant 2	Australia	3071	823	u, rel.	Micklethwaite *et al.* (2010)
New Celebration	Australia	6310	380	u	Based on propriety industry data from SM's files
Golden Mile	Australia	4400	1550	u	Based on propriety industry data from SM's files
St Ives	Australia	2460	700	u, restr.	Micklethwaite *et al.* (2010)
North Carlin 1	USA	2311	536	u	Thompson *et al.* (2002)
North Carlin 2	USA	2487	986	u	Thompson *et al.* (2002)
Escondida	Chile	9521	2419	u, rel.	Garza *et al.* (2001)

All the stepovers examined in this study are underlapping (u) but a mixture of both releasing (rel.) or restraining (restr.) modes.

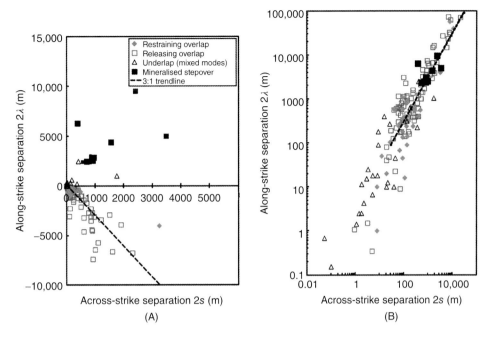

Fig. 20.4. Comparison of along-strike (2λ) versus across-strike ($2s$) separation between mineralised stepovers and examples from well-constrained strike-slip fault data (Aydin & Schultz 1990; Kim *et al.* 2004). (A) 2λ versus $2s$ demonstrating that mineralised stepovers investigated were all underlapping (positive R) in contrast to the majority of data (negative R). The mineralised stepovers for the most part had larger dimensions than other underlapping stepovers previously measured. (B) log 2λ versus log $2s$. When compared directly with one another, underlapping mineralised stepovers maintain an aspect ratio ~3:1, similar to their overlapping counterparts (for along-strike dimensions >100 m). Restraining or releasing overstep data show no appreciable difference in geometry and scaling properties. For the sake of brevity, the releasing or restraining modes of underlapping stepovers are not distinguished. (*See color plate section for the color representation of this figure.*)

NUMERICAL ANALYSIS OF THE RELATIONSHIP BETWEEN STEPOVER GEOMETRY AND FAULT DAMAGE

Methods

The field data show a multi-scale, linear trend for stepover aspect ratios, which is one that is well documented for non-mineralised fault systems (e.g. Aydin & Schultz 1990; Soliva & Benedicto 2004; Long & Imber 2011). Fault systems are also dominated by overlapping stepover configurations (Aydin & Schultz 1990; Acocella *et al.* 2000). For pull-apart basins between strike-slip fault segments, typical geometries are overlapping with an aspect ratio of 3:1 (Gürbüz 2010), for example $R \approx -3$ in this study. For relay zones in normal fault systems, geometries are also overlapping with $R \approx -4.2$ (Long & Imber 2011). Linear elastic fracture mechanics provides a mechanical rationale, suggesting that linked overlap fault configurations are a natural outcome arising from interaction between two propagating fault segments, over geological time (e.g. Aydin & Schultz 1990; Burgmann & Pollard 1994; Willemse 1997), and that stepovers should evolve from underlapping geometries to overlapping ones.

We recognise that stepovers from gold mineralised districts have, in our examples, the unusual characteristic of being underlapping. In this section, we use these results to inform boundary element modeling of fault slip events, in order to better

understand how fault systems transiently generate permeability and explain the observed relationship between underlapping geometries and mineralisation.

We use the boundary element code COULOMB 3 to simulate three-dimensional deformation around two fault segments, according to the configurations in Figure 20.5. In particular, we test for Coulomb failure stress changes ($\Delta\sigma_f$) in and around fault stepovers arising from fault rupture events or creep. The methodology is fully described in King *et al.* (1994a) and only briefly outlined here. The faults are planar and discretised using rectangular elements (e.g. Fig. 20.5A). The mean slip (0.4 m) is appropriate to fault rupture events on 10-km-long segments (Wells & Coppersmith 1994) and corresponds roughly to an M6 earthquake. We prescribe tapered slip distributions that match compilations of earthquake data (Manighetti *et al.* 2005), which are triangular and asymmetric, with a maximum slip on each segment at ~30% of the distance from the stepovers. Micklethwaite & Cox (2006) previously calculated $\Delta\sigma_f$ for two orogenic gold camps associated with stepovers and showed that $\Delta\sigma_f$ was an excellent proxy for mineralisation as opposed to mean stress (e.g. Connolly & Cosgrove 1999), because it maps regions where damage was triggered by static stress changes following ancient fault rupture or creep events.

We test the impact of changes in stepover geometry on $\Delta\sigma_f$ and achieve this by varying a number of parameters (Fig. 20.5B). Both releasing and restraining stepovers are

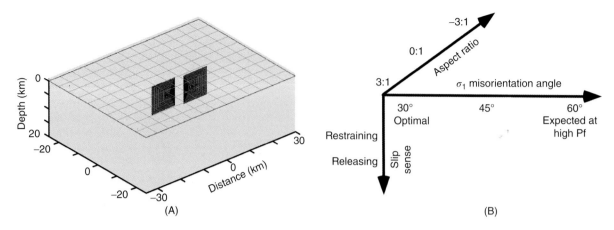

Fig. 20.5. (A) Example model of the three-dimensional boundary element modeling configuration. In this case, two underlapping fault segments are shown with aspect ratio $R = 3$ and a tapered slip distribution with maximum slip ~30% of the strike length from the stepover. (B) A 3-3-2 matrix of parameters was investigated to understand the impact of stepover geometry on $\Delta\sigma_f$, including the orientation of maximum principal stress (σ_1) relative to the fault strike. In some cases, we also ran additional models with aspect ratio $R = 1$. (*See color plate section for the color representation of this figure.*)

examined and geometries are systematically altered from underlapping, to neutral, to overlapping. The data from stepovers associated with gold mineralisation are presently too small to quantify their aspect ratio with any statistical significance. However, the data are consistent with the 3:1 trend for strike-slip fault systems (e.g. Fig. 20.4B) defined by Gürbüz (2010), indicating that underlapping stepovers may be self-similar to a first-order approximation. On this basis, we choose to test stepover geometries that have aspect ratio values of $R = -3$, 0, 3. In some cases, we tested additional stepover

geometries with aspect ratio values of $R = 1$. Boundary stress conditions are such that the angle between the maximum principal stress and the faults vary in different models between (i) 30° for conditions where faults are optimally oriented for failure, (ii) 45° for conditions at the upper limits of where poroelastic changes may drive fault failure (Micklethwaite 2008) and (iii) 60° for conditions at high fluid pressures (Sibson *et al.* 1988; Cox *et al.* 2001; Sibson 2007; Micklethwaite 2008).

In addition, one important feature of A calculations is that they can be performed for faults of a chosen orientation and

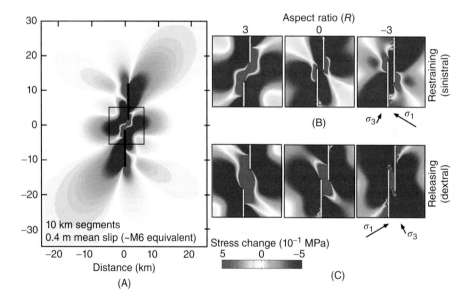

Fig. 20.6. Horizontal distributions of change in Coulomb failure stress ($\Delta\sigma_f$) for two models, with stepover geometries varying from underlapping, to neutral, to overlapping (modified aspect ratios of $R = 3$, 0, −3). The remote maximum and minimum principal stresses are σ_1 and σ_3, respectively, with $\sigma_1 = 60°$ relative to fault strike, representative of mineralised systems affected by high fluid pressures (e.g. Sibson *et al.* 1988; Cox *et al.* 2001; Micklethwaite 2008). In both cases, $\Delta\sigma_f$ is calculated for triggering of strike-slip faults. (A) Results for an underlapping, restraining stepover. Subsequent stresses are sampled from the more restricted domain shown around the stepover. (B) Model 1 – restraining stepovers (sinistral fault slip). (C) Model 2 – releasing stepovers (dextral fault slip). In Models 1 and 2, underlapping geometries have significantly larger areas of positive $\Delta\sigma_f$, relative to neutral and overlapping geometries. (*See color plate section for the color representation of this figure.*)

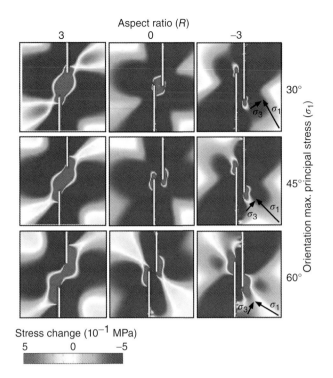

Aspect ratio (R)

3   0   −3

30°

45°

60°

Orientation max. principal stress (σ_1)

Stress change (10^{-1} MPa)

5   0   −5

Fig. 20.7. The impact of stepover geometry (modified aspect ratios of $R = 3$, 0, −3) on $\Delta\sigma_f$, under conditions of varying remote stress orientation. The remote maximum and minimum principal stresses are σ_1 and σ_3, respectively. In the examples shown, all models are for restraining stepovers. (*See color plate section for the color representation of this figure.*)

Results

We investigated a broad range of models, and selected results are presented in Figures 20.6–20.8. Figure 20.6 highlights the influence of stepover mode (restraining or releasing) on damage zone distribution, relative to stepover geometry. Figure 20.7 reproduces the results of Figure 20.6B to demonstrate the influence of variations in the orientation of the maximum principal stress.

The most critical result to emerge is that the area of positive stress change differs dramatically between underlap and overlap configurations. Underlapping stepovers are able to generate positive stress changes over areas that are one order of magnitude greater in size relative to overlapping stepovers, irrespective of slip sense or orientation of the remote principal stress. Therefore, consistent with our field observations, underlapping stepovers may favour mineralisation because they have the ability to generate static stress–triggered damage and permeability enhancement over a larger area than other stepover geometries. The results also show that releasing stepovers tend to generate $\Delta\sigma_f$ over greater areas than restraining stepovers (Figs 20.6B,C, 20.8A). Similarly, larger areas of positive $\Delta\sigma_f$ are associated with large angles of misorientation of the maximum principal stress (σ_1), relative to the strike of the fault segments (Figs 20.7, 20.8B). This result is important, because fault systems associated with orogenic gold mineralisation are typically misoriented due to high fluid pressures and are therefore likely to develop larger regions of damage around stepovers. Figure 20.8A also indicates that damage zone faults of normal dip-slip characteristics can be triggered over even larger surface areas around underlapping releasing stepovers, although strike-slip damage zone faults are expected to dominate in strike-slip systems. In all the underlapping models, positive stress changes are not confined to the domain between

rake. This functionality allowed stress changes to be calculated to test for activation of adjacent damage zone structures with reverse, normal and strike-slip fault characteristics, depending on whether the model was a releasing or restraining stepover.

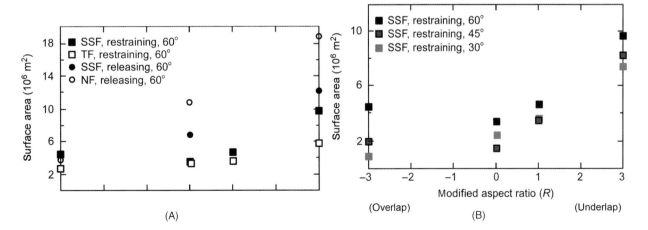

(A)

(B)

(Overlap)   (Underlap)

Fig. 20.8. Areas of positive $\Delta\sigma_f$ relative to stepover geometry. In each case, underlapping geometries of $R = 3$ potentially trigger damage over an area that is an order of magnitude greater than neutral or overlapping geometries. Results were derived from models where $\Delta\sigma_f$ was calculated for the triggering of strike-slip (SSF), normal (NF) and thrust faults (TF). Areas were measured for stepover stresses directly connected to each stepover at $\Delta\sigma_f > 4 \times 10^{-1}$ MPa. (A) Comparison between restraining and releasing stepovers, respectively, under uniform remote principal stress (60°). (B) Influence of changes in the orientation of the remote principal stress for a restraining stepover at different aspect ratios.

fault segments but have a footprint that includes the rocks surrounding the fault stepover.

DISCUSSION

In this study, we have reviewed the relationship between fault stepovers and gold mineralisation as a proxy for fluid flux in the Earth's crust. It has been demonstrated that stepovers have a positive relationship with the distribution of orogenic gold deposits over tens of thousands of square kilometres. Careful studies of fault stepovers, from well-constrained goldfields, indicate that such stepovers have the unusual characteristic of being underlapping. We have shown, using boundary element modeling of static stress changes, that the underlapping nature of fault stepovers associated with gold mineralisation can be understood when damage development around stepovers (and therefore permeability enhancement) is linked to earthquake or creep events and the seismic cycle. Indeed, linkages between gold mineralisation and seismogenic processes have been demonstrated in many independent studies (Sibson 1987; Sibson *et al.* 1988; Robert *et al.* 1995; Cox *et al.* 2001). We conclude that stepovers are critical 'channels' for the communication of fluid between crustal reservoirs at the plate scale, acting as zones of high time-integrated fluid flux, and that underlapping stepovers may be particularly important because of their ability to enhance permeability over larger volumes during fault slip events.

The connection between mineralisation and the seismic cycle is a crucial one because it links two competing processes that relate to the concept of 'dynamic permeability' (Cathles & Adams 2005). Earthquakes or creep events trigger aftershocks or related seismicity that clearly enhance permeability (Miller 2013), but it is also well understood that hydrothermal fluids have a significant impact in healing a system, leading to permeability decay. Permeability decay can be very rapid, as evidenced by hydrothermal deformation experiments (Kay *et al.* 2006; Tenthorey & Fitz Gerald 2006; Giger *et al.* 2007), post-seismic monitoring of groundwater geochemistry (Rojstaczer *et al.* 1995; Claesson *et al.* 2007) and observations of changing seismic wave speeds around recently ruptured faults (Tadokoro & Ando 2002; Hiramatsu *et al.* 2005). Individual faults may heal in as little as 2–3 years but aftershock sequences around large earthquakes continue for at least a decade (Rolandone *et al.* 2004), potentially maintaining permeability for that period. Kitagawa *et al.* (2007) found that permeability decay lasted for a minimum of a 6-year period following the 1995 Hyogoken-Nanbu M7.3 earthquake on the Nojima fault zone, while Vidale & Li (2003) used changes in seismic wave speed around the Landers fault to infer healing over ∼10 years.

Given that fault systems heal rapidly, we estimate here the time-integrated fluid fluxes possible through damage zones developed in and around underlapping stepovers during a seismic cycle (i.e. involving transient enhancement then decay of

permeability). The results are compared with the fluid volumes required to form the range of gold deposit sizes observed in regions such as the Kalgoorlie Terrane from Western Australia. On this basis, conclusions are drawn on the transient permeabilities that are necessary to form such gold deposits.

Fluid flux in stepovers

We estimate the time-integrated fluid flux through a fault stepover by assuming that fluid flow is governed by Darcy's law. In the case of underlapping stepovers in regional-scale strike-slip fault systems, the fluid pressure gradient is vertical and close to lithostatic, such that fluid flow is also vertical. The vertical fluid flux is given by

$$q_z = -\frac{k}{\mu}\left(\frac{\partial P}{\partial z} - \rho_f g\right), \tag{20.1}$$

where k is permeability (m^2), μ is fluid viscosity (Pa s), P is fluid pressure (Pa), z is the vertical coordinate (m; positive upwards), ρ_f is fluid density (kg m^{-3}) and g (m s^{-2}) is acceleration due to gravity (Table 20.2). The vertical fluid pressure gradient is equivalent to $\rho_s g$, (where ρ_s is average density of overlying rock, kg m^{-3}), provided it remains lithostatic throughout the period of flow, which requires a large, well-connected, overpressured fluid reservoir. This assumption is justified because large volumes of potentially aqueous fluid have been geophysically imaged in the mid-crust (e.g. Li *et al.* 2003), and overpressured fluid reservoirs at depth are known to contribute to earthquake–aftershock seismic sequences (e.g. Miller *et al.* 2004; Antonioli *et al.* 2005).

Following an earthquake, permeability in the stepover damage zone is assumed to decay exponentially from a seismically enhanced value k_0 to a background value k_1 over a time period N years, according to the following equation:

$$k = k_0 e^{-rt}, \tag{20.2}$$

where k (m^2) is the permeability at time t (s) after the earthquake and r (1/s) is the decay rate, given by

$$r = -\frac{1}{31,536,000 N} \ln\left(\frac{k_1}{k_0}\right). \tag{20.3}$$

Experimental measurements of the decay rate for pervasive flow through quartz fault gouges under hydrothermal conditions follow the form given in Eqs 20.2 and 20.3 (Giger *et al.*

Table 20.2 Parameter values for fluid flux calculations

Parameter	Value
Rock density, ρ_s	2800 kg m^3
Fluid density, ρ_f	1000 kg m^3
Fluid viscosity, μ	0.0005 Pa s
Acceleration due to gravity, g	−9.81 m s^2

2007). Substituting Eq. 20.2 into Eq. 20.1 and integrating with respect to time yields

$$Q_z = \frac{k_0}{\mu r}\left(\frac{\partial P}{\partial z} - \rho_f g\right)(e^{-rt_1} - 1), \qquad (20.4)$$

where Q_z (m^3/m^2) is the time-integrated vertical fluid flux through the stepover damage zone between $t=0$ and $t=t_1$. For this analysis, Q_z is calculated for the period N years (i.e. with $t_1 = 31,536,000\ N$) after an earthquake. We investigated a range of values of N ($N=2, 5, 10, 20, 30$ years).

Typical values of 'background' permeability in various crustal lithologies, at pertinent depths of 2–15 km, are in the range 10^{-22} to $10^{-18}\ \text{m}^2$ (e.g. Brace 1980; Rowland & Simmons 2012). However, where the crust is critically stressed and close to failure, deep drilling and induced seismicity experiments indicate values of 10^{-17} to $10^{-16}\ \text{m}^2$ (Townend & Zoback 2000) and permeability follows a profile linked to crustal depth (Manning & Ingebritsen 1999). In contrast, very large hydraulic changes occur during an earthquake sequence. Coseismic permeability increases have been measured by linking the migration of hypocentres over time to diffusion of a fluid pressure front. In situations where mainshocks have ruptured overpressured fluid reservoirs, permeability routinely attains 10^{-13} to $10^{-10}\ \text{m}^2$ (Miller *et al.* 2004; Waldhauser *et al.* 2012) and in extreme cases may reach values as high as $10^{-8}\ \text{m}^2$ (Noir *et al.* 1997). On the basis of these field constraints, time-integrated fluid fluxes were calculated using permeability values that decay from coseismic values of 10^{-13} to $10^{-9}\ \text{m}^2$ back to intrinsic background value of $10^{-18}\ \text{m}^2$.

The calculations show that an enormous range of time-integrated fluid fluxes are possible (Fig. 20.9, Table 20.3), which are highly sensitive to the initial coseismic permeability value and to a lesser extent the period of time over which healing occurs (i.e. the value of N). There is an order-of-magnitude difference in flux depending on the order of magnitude of permeability enhancement (Fig. 20.9). Furthermore, even in scenarios where healing occurs over a 30-year period, 90% of the fluid flux has accessed the system within ~5 years or less (~15% of the healing period).

Duration of goldfield formation and transient permeability in the Earth's crust

Table 20.3 shows the time-integrated fluid fluxes that could arise due to transient coseismic changes in permeability from single earthquake–aftershock sequences. These constraints allow us to calculate the number of seismic cycles and overall time duration that may be necessary to form a goldfield associated with a stepover, and draw conclusions regarding which values of coseismic permeability are sensible. In what follows, we base our estimates on a 10-year permeability decay period, because this matches the healing periods observed for large earthquake sequences.

Gold solubility in hydrothermal fluids typical of orogenic gold deposits is expected to exist across the range 1–1000 ppb

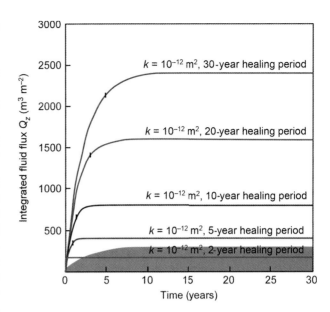

Fig. 20.9. Time-integrated fluid flux derived from initial coseismic permeabilities of $10^{-12}\ \text{m}^2$ decaying to $10^{-18}\ \text{m}^2$ over time periods of 2, 5, 10, 20 and 30 years (N). For comparison, the field in grey is the range of flux values if coseismic permeabilities were $10^{-13}\ \text{m}^2$, illustrating the sensitivity of time-integrated fluid flux to transient coseismic permeability. The marked inflection points are 90% of the total time-integrated fluid flux for each curve – which occurs at ~15% of the healing period (N). (*See color plate section for the color representation of this figure.*)

(Micucki 1998), whereas Simmons & Brown (2007) measured slightly undersaturated solubilities directly from geothermal fluids in the order of 10 ppb. Under these conditions, aqueous fluid volumes of $1.6 \times 10^{10}\ \text{m}^3$ are required to form a moderately large goldfield with 5 Moz of gold. Given that the area of damage triggered by static stress changes around an underlapping stepover is in the order of $10–12 \times 10\ \text{m}^2$ (Fig. 20.8), ranging up to $\sim16 \times 10^6\ \text{m}^2$ for results from the Mount Pleasant and St Ives goldfields (Micklethwaite & Cox 2006), then it is possible to use the results of Table 20.3 to estimate the number of seismic sequences necessary to generate a 5 Moz goldfield. We divide the fluid volume required by the product of the time-integrated flux and the surface area (i.e. calculating for permeability integrated uniformly over the area of the damage zone).

For transient permeabilities at even moderate values of the geophysically determined spectrum ($k_0 = 10^{-12}\ \text{m}^2$), 5 Moz of gold can precipitate extremely rapidly, within one or two earthquake sequences (using the directly measured but conservative gold solubility value of 10 ppb). This is assuming that >99% of dissolved gold precipitates, which has been observed in geothermal fields (Brown 1986). Even for coseismic permeabilities at the lower end of the spectrum ($k_0 = 10^{-13}\ \text{m}^2$), the 5 Moz goldfields still form rapidly, within 10–16 earthquake sequences. Therefore, it is feasible that many orogenic goldfields form in as little as 10–8000 years.

Table 20.3 Time-integrated fluid flux due to different values of coseismic permeability enhancement (k_0) that decay to a background value (k_1), over different periods of time (N)

k_0 (m²)	k_1 (m²)	N (years)	Q_z (m³ m⁻²)
10^{-9}	10^{-18}	30	1,612,283
10^{-9}	10^{-18}	20	1,074,855
10^{-9}	10^{-18}	10	537,428
10^{-9}	10^{-18}	5	268,714
10^{-9}	10^{-18}	2	107,486
10^{-10}	10^{-18}	30	181,382
10^{-10}	10^{-18}	20	120,921
10^{-10}	10^{-18}	10	60,461
10^{-10}	10^{-18}	5	30,230
10^{-10}	10^{-18}	2	12,092
10^{-11}	10^{-18}	30	20,729
10^{-11}	10^{-18}	20	13,820
10^{-11}	10^{-18}	10	6,910
10^{-11}	10^{-18}	5	3,455
10^{-11}	10^{-18}	2	1,382
10^{-12}	10^{-18}	30	2,418
10^{-12}	10^{-18}	20	1,612
10^{-12}	10^{-18}	10	806
10^{-12}	10^{-18}	5	403
10^{-12}	10^{-18}	2	161
10^{-13}	10^{-18}	30	290
10^{-13}	10^{-18}	20	194
10^{-13}	10^{-18}	10	97
10^{-13}	10^{-18}	5	48
10^{-13}	10^{-18}	2	19

In order to assess the validity of these estimates, it is important to understand the concept of stepover lifespan. Fault stepovers are ephemeral geometric features within a fault system and are not expected to exist for the duration of a fault system. On regional-scale faults, stepovers may exist for 10^5 years (Parsons *et al.* 2003; Walker *et al.* 2005) although constraints are poor. Recurrence intervals of individual fault segments on regional-scale strike-slip fault systems, such as the San Andreas fault system, are in the order of 100–500 years (Smith-Konter & Sandwell 2009), which suggests a stepover may be shortcut within hundreds of regional-scale fault rupture events. Therefore, even the lower end of estimated coseismic permeability ($k_0 = 10^{-13}$ m²) is enough to allow a 5 Moz goldfield to form around stepovers before the stepover ceases to exist.

A further consideration in this argument is the existence of truly supergiant deposits, such as Kalgoorlie with >68 Moz gold. Based on the same aforementioned logic, supergiant deposits provide a further constraint on coseismic damage zone permeabilities. At coseismic permeability of 10^{-13} m², they require several hundred earthquake sequences and are unlikely to form within the lifespan of a stepover, unless derived from high gold solubility fluids. In contrast, they are comfortably generated if their hosting damage zone structures attain transient permeability values of 10^{-12} m², when triggered by >30 mainshock events on the regional-scale structure.

CONCLUSIONS

This study has demonstrated that fault stepovers have a distinct and quantifiable spatial relationship with goldfields, over distances approaching the plate scale. Underlapping stepovers, although rarer components of fault systems, appear to have a particular association with mineralisation. The examples studied here hold to an approximate aspect ratio of 3:1, which fits a first-order self-similar global trend for both overlapping and underlapping stepovers. Boundary element modeling indicates that underlapping stepovers may favour mineralisation because rupture events around such stepovers have the ability to generate static stress–triggered damage and permeability enhancement over a larger area than overlapping or neutral stepover geometries.

In fault systems undergoing mineralisation, slip events on regional-scale fault segments trigger damage around stepovers and rupture overpressured fluid reservoirs at depth. A pulse of gold-bearing fluid, under high pressure, then migrates through the system at remarkably high but transient permeabilities, during multiple fracture and fault slip events in the damage zone.

Time-integrated fluid flux calculations that take into account both the transient enhancement and decay of permeability indicate that 90% of the fluid flux is attained in ~15% of the healing period, when fluid flow is driven by a lithostatic fluid pressure gradient. The fluid fluxes involved suggest that moderately large goldfields of 5 Moz could form in as little as 1–16 seismic sequences, representing a potential duration of 10–8000 years. Similar rapid durations for goldfield formation have been inferred from different environments (e.g. volcano sector collapse, Lihir, Papua New Guinea, Heinrich 2006; convective circulation, Taupo Volcanic Zone, Rowland & Simmons 2012; earthquake events, North Carlin goldfield, Nevada, Hickey *et al.* 2014). We have shown that supergiant goldfields can be explained if crustal permeability transiently reaches values of 10^{-12} m² or less. Extreme permeability values in the upper limits of those estimated from seismological studies of earthquake sequences (e.g. 10^{-10} to 10^{-8} m²) are not required to explain the formation of orogenic goldfields.

The permeability values discussed in this study are likely most relevant for the mid- to upper-crust, at depths of 15 to ~2 km. In stepover regions, permeability enhanced by fault rupture events is expected to be channel-like and independent of depth, until the fracturing related to seismicity has begun to heal and decrease in intensity. Instead, the permeability enhancement will depend on the degree of coseismic damage. As demonstrated by Figures 20.6 and 20.7, these permeability channels will not necessarily be confined to the volume of rock between overlapping or underlapping fault stepovers but also extend into the wall rock adjacent to the stepover.

ACKNOWLEDGEMENTS

We are extremely grateful for the diligence of the editors Steven Ingebritsen and Mark Person. Philipp Weis substantially improved the manuscript and detected an error in one of the initial calculations. Both Weis and Tom Gleeson (guest editor) provided excellent and thorough reviews. S.M. was supported by the Hammond-Nisbet Endowment at the Centre for Exploration Targeting during completion of this study.

SUPPORTING INFORMATION

Additional supporting information may be found in the online version of this chapter, *Geofluids* (2015) 15, p. 240–251.

Figure S1. Map of the Kalgoorlie–Ora Banda district, showing the distribution of faults and deposits used in this analysis, as well as different buffer distances. Greenstone rocks are in black, granite-gneiss in pink.

CHAPTER 21

Evidence for long-timescale (>10³ years) changes in hydrothermal activity induced by seismic events

TREVOR HOWALD[1], MARK PERSON[1], ANDREW CAMPBELL[1], VIRGIL LUETH[2], ALBERT HOFSTRA[3], DONALD SWEETKIND[3], CARL W. GABLE[4], AMLAN BANERJEE[5], ELCO LUIJENDIJK[6], LAURA CROSSEY[7], KARL KARLSTROM[7], SHARI KELLEY[2] AND FRED M. PHILLIPS[1]

[1] *Department of Earth & Environmental Science, New Mexico Institute of Mining and Technology, Socorro, NM, USA;* [2] *New Mexico Bureau of Geology and Mineral Resources, Socorro, NM, USA;* [3] *U.S. Geological Survey, Denver, CO, USA;* [4] *Los Alamos National Laboratory, Los Alamos, NM, USA;* [5] *Indian Statistical Institute, Kolkata, India;* [6] *Department of Structural Geology and Geodynamics, Georg-August-Universität Göttingen, Göttingen, Germany;* [7] *Department of Earth and Planetary Sciences, University of New Mexico, Albuquerque, NM, USA*

ABSTRACT

The pollen ^{14}C age and oxygen isotopic composition of siliceous sinter deposits from the former Beowawe geyser field reveal evidence of two hydrothermal discharge events that followed relatively low-magnitude (<M5) earthquakes of Holocene and Late Pleistocene age along the Malpais fault zone in Whirlwind Valley, Nevada, USA. The observed 20‰ trend of decreasing δ^{18}O over about a 5000–7000-year period following each earthquake is consistent with a fault-controlled groundwater flow system that, following initial discharge of deep and hot groundwater, contains increasing amounts of cool meteoric water through time. Model simulations of this hydrothermal system can only match trends in the isotope data if we include a 1000-fold increase in fault permeability (from <10^{-14} m² to >10^{-11} m²) following each earthquake. However, the timescale for the onset of thermal convection implied by an overturned temperature profile in a geothermal well 300 m from the Malpais fault is much shorter: 200–1000 years. We speculate that individual segments of the Malpais fault become clogged on shorter timescales and that upward flow of groundwater subsequently follows new routes to the surface.

Key words: fault, hydrothermal, oxygen isotope, permeability, sinter

INTRODUCTION

Numerous studies have reported centimeter- to meter-scale water-level fluctuations in wells (King *et al.* 1999; Wang *et al.* 2004a; Chia *et al.* 2001; Manga *et al.* 2012), temperature changes (Dziak *et al.* 2003), increases in stream flow (Muir-Wood & King 1993), and changes in geyser periodicity (Husen *et al.* 2004a) in response to seismic events. The hydrologic response to seismic waves is thought to be due to either deep-seated fracture closure/dilation (Muir-Wood & King 1993; Sibson 1996) or shallower crustal permeability changes (Rojstaczer & Wolf 1992; Brodsky *et al.* 2003; Elkhoury *et al.* 2006; Manga *et al.* 2012). Both mechanisms call upon significant permeability changes. The hydrogeologic response to seismicity can be complex (Wang *et al.* 2004a) with water levels rising or falling in different locations of the stress field proximal to the earthquake foci. Hydrologic changes have been observed to occur at great lateral distances from the earthquake epicenter (10^3 km; Husen *et al.* 2004a). In the vast majority of the aforementioned studies, the hydrologic system responded to seismic events are on timescales no greater than several years (Manga *et al.* 2012). Curewitz & Karson (1994) proposed that stress changes at fault tips and fault kinematics could induce transient hydrothermal behavior on much longer timescales. Quantitative paleohydrologic reconstructions of ore-deposit formation and diagenesis have also supported hydrologic transients associated with seismicity and fault-permeability changes on timescales of 10^3 years (e.g., Garven *et al.* 2001; Appold *et al.* 2007). The objective of this study is to constrain seismicity-induced fault-permeability changes and their effects on hot-spring/geyser isotopic composition over timescales of 10^3 years within the Beowawe geothermal system, Nevada. The unique contribution of this study is a new data set of sinter-deposit ages and isotopic compositions over timescales of thousands of years.

Within Whirlwind Valley, Nevada, a large sinter terrace about 65 m high and 1600 m long (Rimstidt & Cole 1983) crops

Crustal Permeability, First Edition. Edited by Tom Gleeson and Steven E. Ingebritsen.
© 2017 John Wiley & Sons, Ltd. Published 2017 by John Wiley & Sons, Ltd.
Companion Websites: www.wiley.com/go/gleeson/crustalpermeability/
http://crustalpermeability.weebly.com/

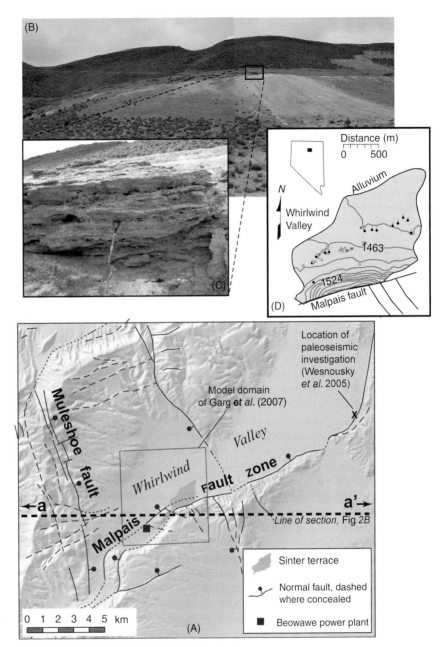

Fig. 21.1. (A) Structural map showing location of Beowawe sinter deposits, paleoseismic investigation of Wesnousky *et al.* (2005) along the Malpais fault zone, the Beowawe power plant (black square). (B) Photo mosaic of Beowawe sinter deposits, Whirlwind Valley, Nevada. The photo was taken looking southeast. (C) Outcrop scale photo of sinter deposits. (D) Sinter topographic map *with locations of historic and paleo hot springs (black circles, as well as locations of sinter samples (red and yellow circles). Multiple samples were collected at locations with yellow dots. (See color plate section for the color representation of this figure.)*

out along the Malpais fault zone, which is a high-angle normal fault with an east–northeast strike (Fig. 21.1A, B, D). The sinter deposits document geothermal discharge of the former Beowawe geyser field. The Malpais fault zone forms the southeastern edge of a graben that underlies Whirlwind Valley; the Muleshoe fault forms the northwestern edge (Fig. 21.1A). The sinter deposit formed within an area of exceptionally high heat flow at shallow depth (dashed orange lines, Fig. 21.2A). Zoback (1979) estimated that the sinter deposits contain 1.76×10^{10} kg of silica and that the time required to form the Beowawe sinter deposits was 210,000 years assuming a constant flow rate (Renner *et al.* 1975) and concentration of dissolved silica.

As part of this study, we analyzed 63 sinter samples from the Beowawe geyser field for $\delta^{18}O$ (red and yellow dots in Fig. 21.1D). Pollen extracted from a subset of the sinter-deposit samples were dated using ^{14}C (Fig. 21.3). The dated samples revealed isotopic depletion in fluid $\delta^{18}O$ of about 20‰ over 5000–7000 years that appear to be correlated with Holocene and Late Pleistocene seismic activity along the Malpais fault (Fig. 21.3). Isotopic fractionation associated with boiling can only produce approximately a 2‰ shift in composition (Harris 1989) and cannot be called upon to explain the 20‰ depletion. Similarly, the effects of Quaternary climate change on the isotopic composition of precipitation in the southwestern United

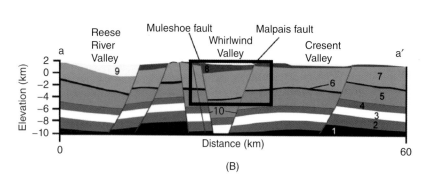

Heatflow (mW m⁻²) — – – –
Water table elevation (m) ———
Geothermal power plant ■
Sinter deposits 🌿

(A)

Fig. 21.2. (A) Heat flow (orange dashed contour lines; in mW m⁻²) and water-table elevation (solid light blue lines; m) contour map across Whirlwind Valley, Nevada. Location of regional hydrothermal model cross section is depicted by a–a'. Dark blue box indicates the location of the Beowawe sinter terrace shown in Figure 21.1B. Heat-flow and water-table contour maps from Olmsted & Rush (1987) and Faulder *et al.* (1997). (B) Hydrogeologic cross section across Reese, Whirlwind, and Crescent Valleys. The inset (solid black rectangle) denotes the portion of the hydrothermal model domain presented in Figure 21.3A–F. Permeability data for the different numbered units in Figure 21.1C are presented in Table 21.1 and described in the supplemental materials section. (*See color plate section for the color representation of this figure.*)

(B)

Table 21.1 Radiocarbon dates of pollen taken from sinter deposits

Sample number	Age (yrs)	Con. int. (yr)	$\delta^{13}C$ (‰)	Primary minerals	Latitude	Longitude	$\delta^{18}O_{sinter}$	$\delta^{18}O_{fluid}$ @ 90°C
B-1c	10,454	±40	−25.3	Opal, Qtz, kaolinite	N40°33''44.5''	W116°35'25.9''	20.10	−2.20
B-8	4,706	±25	−20	Opal	N40°33'41.4''	W116°35'29.4''	24.93	2.63
B-14a	893	±25	−23.6	Opal, plagioclase	N40°33'36.4''	W116°35'40.8''	15.1	−7.2
B-15	2,684	±25	−24.5	Opal	N40°33'38.0''	W116°35'34.2''	21.4	−0.9
B-19c	1,137	±20	−22.3	Opal, Qtz	N40°33'43.1''	W116°35'10.7''	12.5	−9.8
B-3	8,797	±60	−29.2	Opal, Qtz, kaolinite	N40°33'42.8''	W116°35'26.9''	6.93	−15.37
B-5	14,375	±70	−29.6	Opal	N40°33'38.7''	W116°35'29.1''	12.56	−9.74
B-6 btm	10,774	±50	−29.1	Opal	N40°33'40.50''	W116°35'33''	8.77	−13.54
B-7	10,580	±50	−29.3	Opal	N40°33'41.5''	W116°35'29.3''	8.90	−13.41
B-11 wht	3,693	±30	−26.3	Calcite, cristobalite, sanidine	N40°33'30.29''	W116°35'21''	16.90	−5.40
B-10b	1,322	±40	−26.1	Opal	N40°33'38.7''	W116°35'29.1''	13.4	−8.9

States are only on the order of 4‰ (Asmerom *et al.* 2010). We believe the simplest explanation for these isotopic trends is that seismicity along the Malpais fault increased fracture permeability, initiating rapid discharge of geothermal water within a single-pass hydrothermal system that, as time passed, received an increasing proportion of isotopically light meteoric water. The system gradually shut down due to mineralization and incursion of cool, dense meteoric water. One of the main goals of this study is to use thermal and isotopic modeling to test this hypothesis.

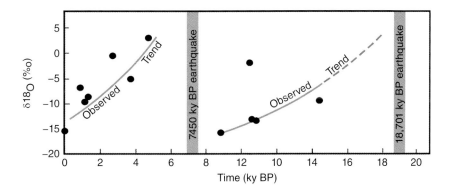

Fig. 21.3. Reconstructed temporal evolution of fluid compositions based on calculated $\delta^{18}O_{H_2O}$ (at 93°C) in equilibrium with sinter and radiocarbon dates on pollen in sinter (circles). The best-fit dashed gray lines drawn through the sample points show that $\delta^{18}O_{H_2O}$ progressively decreased during two distinct episodes of fluid flow through the Beowawe geothermal system. Timing of earthquakes reported by Wesnousky *et al.* (2005) suggests that each episode of fluid flow was triggered by seismicity. Initially, hot fluids are in isotopic equilibrium with the carbonate reservoir rocks at 5 km depth discharge. Introduction of isotopically depleted meteoric water through time results in the progressive decrease in isotopic composition of discharging fluids along the Malpais fault. Sample point on the origin of the Y-axis (−15‰) is the present-day average $\delta^{18}O_{H_2O}$ value at Beowawe for precipitation (John *et al.* 2003).

Linkages between seismicity and geyser/hot-spring activity have been documented in modern geothermal systems (Wang & Manga 2010b). Husen *et al.* (2004a) documented temporal changes in Yellowstone geyser eruption periodicity and linked them to distant earthquakes. However, no linkages between seismicity and surface water geochemistry have been found within the Yellowstone National Park watershed (Hurwitz *et al.* 2007). Mogi *et al.* (1989) were among the first to report geothermal temperature increases in response to seismic events: a 1–2°C increase following seismic activity at Usami Hot Springs, Izu Peninsula in Japan. More recently, King *et al.* (1994a) and Manga & Rowland (2009) proposed that permeability increases following an M 5.5 earthquake led to a 1–2°C temperature decrease in Alum Rock spring, California. Diagenetic studies have also suggested episodic fluid flow events associated with seismicity along shallow fault systems. Boles & Grivetti (2000) and Boles *et al.* (2004) reported relatively high fluid-inclusion homogenization temperatures (80–125°C) at shallow depths (<300 m) along faults that crop out on the onshore region of the Santa Barbara Basin. These authors hypothesize that Pleistocene paleoseismicity increased the permeability associated with a series of faults, resulting in an episodic hydrothermal system. Paleohydrologic modeling by Appold *et al.* (2007) indicated that the combination of pressure declines with a deep overpressured reservoir and permeability changes associated with calcite cementation led to the eventual decline of hydrothermal activity over a period of about 100–1000 years. However, we are unaware of any seismicity study that has provided a high-resolution temporal record of changes in hydrothermal activity over geologic timescales.

Our study is one of the first to assess the isotopic evolution of hydrothermal fluids within a geothermal system using pollen-dated sinter deposits. We are only aware of one other study that has applied ^{14}C dating in the study of sinter deposits within the Great Basin (Lynn *et al.* 2008). They document changes in silica mineralogy (Opal-A to quartz) as a function of sinter age from the Steamboat Springs sinter deposits, NV. They analyzed terrace and distal apron samples as well as a 13.1 m-deep drill core. AMS ^{14}C pollen ages from the drill core revealed discordant ages with depth. The authors suggested that this was the result of physical mixing of older and younger sinter fragments.

STUDY AREA

Background

The Beowawe geothermal system is one of the highest temperature geothermal reservoirs in the Basin and Range physiographic province (Fig. 21.2A), with a temperature of about 220°C at 3 km depth (Olmsted & Rush 1987). The system is liquid dominated; boiling is localized at very shallow depths. The most recent volcanism in the area occurred during the middle-Miocene and is not considered a feasible heat source. The low helium R/R_A value of 0.46 reported by Welhan *et al.* (1988) is also indicative of an amagmatic system. The Beowawe geothermal system is located within the Battle Mountain heat-flow high, which has a background heat flow of about 110 mW m⁻² (Smith 1983). This is believed to be partially due to a thinned crust (19–23 km; Heimgartner et al 2006). Shallow temperature-gradient measurements indicate heat flow greater than 2000 mW m⁻² (Olmsted & Rush 1987). The large shallow heat flow is thought to be associated with deep-seated and permeable normal faults that allow hot fluids to be brought up to the surface quickly. Geothermal waters discharge at a high rate (18 kg s⁻¹) along the Malpais fault zone, where volumetrically significant sinter deposits have accumulated (Zoback 1979).

The permeability of the Malpais fault zone has been estimated by several prior studies. Faulder *et al.* (1997) reported well

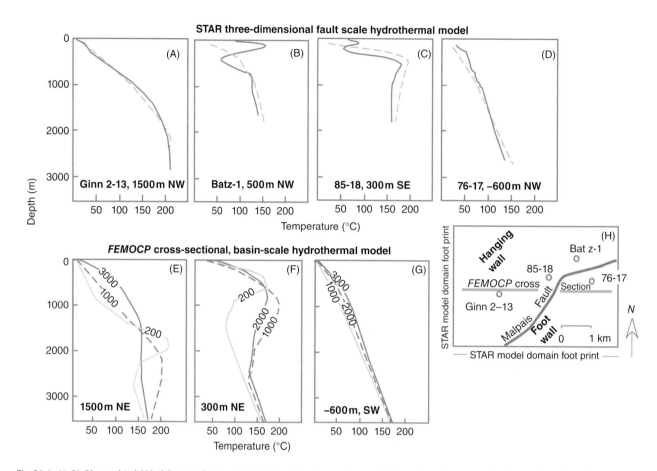

Fig. 21.4. (A–D) Observed (solid black lines) and computed (red dashed lines) temperature profiles from three-dimensional, fault-scale, hydrothermal STAR model of Malpais fault zone. Well distances from Malpais fault and well names are listed in the lower left-hand corner of each plot. All computed temperatures are after 25,000 years (Adapted from Garg *et al.* (2007)). (E–G) Time-dependent evolution of computed temperature profiles extracted from FEMOCP cross-sectional, basin-scale hydrothermal model at different distances from the Malpais fault zone using "best-fit" Malpais zone fault permeability of 10^{-12} m^2. Location of wells and the STAR model footprint are shown in Figure 21.2H along with the location of the FEMOCP cross-sectional transect.

pressure tests that were used to estimate the Malpais fault-zone permeability. They reported the fault-zone conductance to be between about 6×10^{-11} and 1.2×10^{-10} m^3. Person *et al.* (2008) developed a paleohydrothermal model of gold mineralization to assess the plumbing of the Miocene-age Mule Canyon gold deposits in Whirlwind Valley. They assigned a vertical permeability of 10^{-11} m^2 to the Muleshoe and Malpais faults in their basin-scale, cross-sectional hydrothermal model of Whirlwind Valley. Garg *et al.* (2007) developed a three-dimensional, single-phase, fault-scale (about 3 km × 3 km × 3.5 km; blue box in Fig. 21.1A) hydrothermal and solute-transport model of the Malpais fault zone to reproduce direct-current resistivity, magnetotelluric, and well temperature surveys. Two of the six wells studied by these authors, those closest to the Malpais fault zone (Balz-1, 85-15; see Fig. 21.4B,C) displayed pronounced temperature overturns with depth consistent with a transient geothermal flow system (Ziagos & Blackwell 1986). In order to match observed temperature data (Fig. 21.4A–D), Garg *et al.* (2007) assigned a permeability of 2×10^{-13} m^2 to a 1.2-km-long and 3-km-deep section of the Malpais fault zone,

while the remaining segments of the Malpais fault were assigned 10^{-15} m^2. The model included a specified fluid flux of 20 kg s^{-1} along the base of the Malpais fault (at about 3 km depth) to match observed geothermal discharge and was run for 25,000 years, sufficiently long to approach quasi-steady-state thermal conditions. While the Garg *et al.* (2007) model was able to simulate temperatures that were a good match to four of the six wells, the observed thermal overturns (Fig. 21.4B,C) in the wells closest to the Malpais fault zone could not be reproduced. This may be because the model represents late time (25,000 years), quasi-steady-state conditions, and the overturned temperature profiles are likely the result of a short-term transient fluid pulse (Ziagos & Blackwell 1986).

History of geothermal development at Beowawe

The Beowawe geothermal system has been under scientific investigation since 1934. Detailed reports on the geology, structure, and hypothesized hydrothermal fluid flow pathways can be found in Zoback (1979) and Struhsacker (1980). Both these papers provide detailed geologic maps of the area, lithologic descriptions of the rock units, cross sections derived from geothermal wells, and summaries of the geologic history.

The Beowawe geothermal system was once the second-largest site in the United States for active geyser, fumarole, and hot-spring activity, second only to Yellowstone National Park. Today, these geothermal discharges have subsided and Beowawe is host to only a large sinter terrace and a few fumaroles (Fig. 21.1B,C). This change is likely due to geothermal fluid production, which began in 1985 with a 16.7 MW dual-flash power plant (square, Fig. 21.1A).

White (1998) published a detailed report on the Beowawe geothermal system before, during, and after geothermal drilling and development. The report includes pictures of the formerly active geysers and springs, geologic and hydrologic maps of the geysers and Whirlwind Valley area, and a table summarizing water geochemistry from previous studies of the Beowawe geothermal system. This report, along with the article published by Nolan & Anderson (1934), provides a detailed characterization of the Beowawe geothermal system before geothermal development.

Conceptual model of transient hydrothermal flow within the Beowawe geothermal system

We propose that the isotopic trends presented in Figure 21.3 are the result of a transient, single-pass hydrothermal system of the type described by Lowell (1991) (Fig. 21.5). We hypothesize that within the discharge area (Malpais fault, Figs 21.1A and 21.2B), initially hot, isotopically enriched fluids discharged at the surface following a seismic event. Discharge temperature and isotopic composition peaked shortly after the onset of flow. The brief lag time is required for deep reservoir fluids to make their way up to the surface. In the recharge area, in response to temporarily enhanced fault permeability, we hypothesize that isotopically depleted meteoric water flowed downward along fault conduits (here the Muleshoe fault zone,

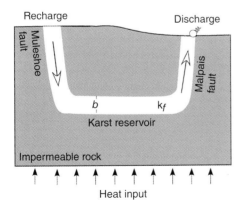

Fig. 21.5. Conceptual diagram depicting single-pass model of a hydrothermal system having a conduit width of "b" and a fault-zone permeability of k_f. In this diagram, circulation is driven by a hydraulic head gradient between the uplands and lowlands. (Adapted from Lowell (1991)).

Figs 21.1A and 21.2B), moved laterally through the karst reservoir (thin dark blue unit in Fig. 21.2B) picking up heat, and then discharged at the surface along the Malpais fault. The $\delta^{18}O$ of the meteoric water increased along the flow path due to mixing with geothermal reservoir fluids (karst aquifer, see unit 6, Fig. 21.2B) and fluid–rock isotope exchange. However, as the flow system evolved over time, the discharging fluids trended toward a meteoric composition. Lowell *et al.* (1993) showed that discharge temperature is controlled by the product of fault-zone width and permeability or conductance (kb), and showed that hot-spring temperature peaks at intermediate values of fault conductance. This is because at the highest flow rates (or widest fault zones), the recharge increases to the point that it has a net cooling effect on the hydrothermal system. For lower permeability conditions, heat conduction dominates. In the conceptual model presented in Figure 21.5, the groundwater flow is driven by the water-table topographic gradient. It is likely that density effects also contribute significantly to driving fluid flow up the Malpais fault zone at Beowawe (Person *et al.* 2008).

Another independent line of evidence that argues for transient hydrothermal fluid flow within the Beowawe geothermal system comes from the analysis of borehole temperature profiles. Present-day borehole temperature profiles proximal to the Malpais fault zone show temperature overturns (solid black lines in Fig. 21.4B,C). Ziagos & Blackwell (1986) showed that such temperature overturns result from transient lateral flows of geothermal fluids (Fig. 21.6). These authors developed analytical and numerical solutions for transient temperature profiles associated with lateral fluid flow into a shallow aquifer. They considered the consequences of hydrothermal fluids moving up a fault zone and then laterally into a shallow aquifer (Fig. 21.6A), and showed that after the onset of lateral flow temperature profiles in wells displayed a distinctive thermal overturn that decreased with increasing distance from the fault zone (Fig. 21.6B). At late time, however, the overturns dissipate due to downward heat conduction (Fig. 21.6C). Ziagos & Blackwell (1986) also presented numerical results using the fast Fourier transform method to investigate time-dependent changes in the enthalpy of the discharging fluids. In this scenario, high enthalpy fluids were input for 1000 years, followed by lower enthalpy fluid input. Fluid temperatures within the water-table aquifer rose quickly at first, followed by a more gradual temperature increase. Once cool fluids were introduced after 1000 years, the opposite pattern developed: rapid cooling followed by a more gradual temperature decline. We argue later that the post-1000-year temperature declines shown in Figure 21.6D are analogous to time-dependent changes in isotopic composition of geothermal fluids within the Beowawe geothermal system (Fig. 21.3). One issue we must address in this study is that the onset of convection we estimate from the temperature profile data in Figure 21.4B,C is considerably shorter than the duration of flow revealed by the isotopic data in Figure 21.3.

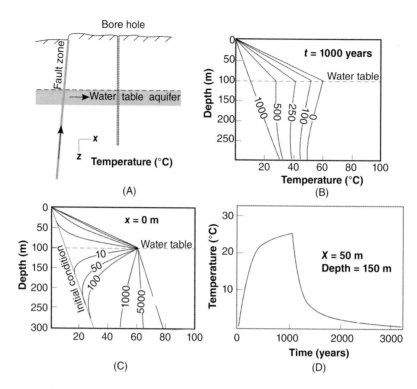

Fig. 21.6. (A) Schematic diagram depicting conceptual model for time-dependent analytical solution of thermal transients presented by Ziagos & Blackwell (1986). The Laplace-transform analytical solution was developed to represent temperature changes in wells resulting from lateral outflow of hot geothermal fluids that move up a fault zone and out into a shallow water-table aquifer. (B) Changes in temperature along a well bore after 1000 years of lateral flow rate at 1 m year^{-1}. The initial geothermal gradient was 100°C km^{-1}. Profile labels are the distance of the well bores from the fault zone in meters. (C) Temperature changes through time at a well bore located at the fault zone ($x = 0$ m). Profile labels denote time since the onset of flow. Note that at early times, temperature overturns. At late time, the temperature inversions are erased by downward heat conduction. (D) Fast Fourier transform solution for time-dependent changes in temperature 50 m from the fault zone at a depth of 150 m below the water table. Changes in temperature are due to time-dependent changes in the temperature of the fluid entering the shallow water-table aquifer. A step function is used to represent the inflow of initially hot water for 1000 years followed by cooler water inflow. (Adapted from Ziagos & Blackwell (1986)).

The onset of convection appears to be correlated with Holocene and Late Pleistocene seismic events along the Malpais fault zone (see X, Fig. 21.1A), which hosts the sinter deposits (Fig. 21.3; Wesnousky *et al.* 2005). The age of each paleoseismic event along the Malpais fault was estimated using ^{14}C dates on carbonaceous material in soil layers that are truncated by the fault (Wesnousky *et al.* 2005). The small size (<0.5 m) of the offsets between colluvium layers along the fault (Wesnousky *et al.* 2005) suggests that these were low-energy earthquakes (<M5) (Friedrich *et al.* 2004). We hypothesize that flow commences following permeability increases along subvertical faults following seismicity (Wang & Manga 2010b; Fig. 21.5). Hot, silica-rich fluids within a deep (~5 km depth) carbonate geothermal reservoir move up the Malpais fault and discharge at the surface (Fig. 21.2B). Progressive silica mineralization and influx of cool meteoric fluids along the fault zone eventually shut the convective flow system down or at least reduces rates of sinter formation (Lowell *et al.* 1993). To test the aforementioned hypotheses, we constructed a series of geologically referenced hydrothermal models that include fluid–rock oxygen-isotope transport and exchange (Bowman *et al.* 1994) within the Beowawe geothermal system.

We systematically vary the permeability of the fault zones in order to constrain the magnitude of permeability changes following a seismic event in a manner that is consistent with the sinter-deposit data. We address the following questions: (i) What fault permeability is consistent with present-day conditions within the Beowawe geothermal system? (ii) How large of a permeability increase is required to enhance hydrothermal activity within Beowawe geothermal system? (iii) Does a single-pass geothermal system with progressive influx of meteoric water produce a trend of decrease isotopic composition consistent with the sinter data we collected? (iv) Over what timescales does this occur? (v) Assuming that the flow system shuts down at late time, how long does it take for the isotopic fluid composition in the deep carbonate reservoir to recover?

METHODS

Sample preparation

We collected 63 sinter samples between 2009 and 2012 along the Beowawe siliceous sinter terrace within Whirlwind Valley

(Fig. 21.1D) to establish the periodicity of hydrothermal activity in the Beowawe geothermal field. Almost all the samples were collected at the surface and along a single road cut. All the sinter samples were analyzed for mineralogy, crystallinity, and oxygen-isotope composition. Pollen and charcoal extracted from 11 sinter samples were dated using radiocarbon methods (Table 21.1).

In order to separate organic matter from mineral grains, the sinter was treated with a 50% hydrogen peroxide solution to remove any organic material trapped in pores and outer surfaces. The samples were rinsed with water, oven-dried, and crushed. The crushed material was treated in 10% HCl to remove carbonates and then in 50% HF to remove silicates. The sinter material was then washed three times in 10% HCl and then in water. The remaining material was separated using a 150 μm sieve. The organic matter larger than 150 μm was collected in a centrifuge tube. The material <150 μm was sieved at 6 μm and particle sizes between 6 μm < X < 150 μm were collected. Multiple gravity separations were also performed using sodium polytungstate (density $2.0\,\mathrm{g\,ml^{-1}}$) to remove remaining insoluble minerals. Plant fragments that floated at $2.0\,\mathrm{g\,ml^{-1}}$ were then sieved using a 6 μm filter and rinsed with water. The float material was loaded into a combustion tube. The radiocarbon activity of the pollen/charcoal samples were then analyzed at Rafter Radiocarbon Laboratory in New Zealand using an accelerator mass spectrometry.

Sinter mineralogy analysis

The mineralogy and crystallinity of the sinter samples were determined using X-ray diffraction. Samples were dried in an oven at 60°C to remove any water present and then hand ground to pass through a 150 μm mesh sieve. The composition of the sinter deposits was determined using a PanAnalytical XPert Pro diffractometer and methods outlined in Herdianita *et al.* (2000). The sinter samples were analyzed using CuKα radiation for 15 minutes each over a range of 10°–40° 2Θ and a step size of 0.0170°. X-ray diffraction patterns were interpreted using PANalytical X'Pert HighScore Plus software to determine the crystallinity (order/disorder) of the opaline phase of the sinters. Crystallinity can be measured as the full-width half-maximum (FWHM) by manually fitting a profile and baseline to the diffraction pattern and then measuring the width of the band at half intensity (Herdianita *et al.* 2000). The sinter terrace at the Beowawe geothermal system consists primarily of Opal-A, Opal-A/CT, Opal-CT/A, or Opal-CT and indicates that geothermal fluids were saturated with respect to amorphous silica. Rimstidt & Cole (1983) argue that the amorphous silica comprising the sinter deposits at Beowawe precipitated at temperatures of 88–98°C. Measured temperatures of geothermal fluids in the "Frying Pan" geyser and another small geyser were about 95 and 98°C, respectively, prior to exploitation (Mariner *et al.* 1983; Nolan & Anderson 1934).

Sinter oxygen-18 analysis

Stable-isotope analysis of oxygen from the opal samples was performed using New Mexico Tech's fluorination line following Borthwick & Harmon (1982) and Baertschi & Silverman (1951). Powdered samples were reacted with dilute HCl to remove any carbonate, then rinsed with distilled water, and dried in an oven at 60°C for 12 hours. Samples were prefluorinated with ClF_3 in nickel reaction vessels for 10 minutes at room temperature to ensure that all remaining water was removed from the samples and reaction chambers, and then reacted with ClF_3 at 500°C for 8 hours. The liberated oxygen gas was converted into CO_2 with a carbon electrode. The CO_2 gas was analyzed via dual inlet on the Delta XP IRMS spectrometer for $\delta^{18}O$ using an OZ-Tech CO_2 gas standard. Samples that produced yields below 60% were not used for interpretation. Samples that produced yield values between 60% and 80% were analyzed two or three times and then averaged with a reproducibility of ±0.8. The isotopic composition of water ($\delta^{18}O_{H_2O}$) was calculated from that of sinter ($\delta^{18}O_{sinter}$) at 93°C using the equilibrium fractionation factor of Kita *et al.* (1985) ($1000\ln\alpha_{(amorphous\,SiO_2\to H_2O)} = 21.9$). The temperature selected is within the limits of the fractionation equation and the reported range of fluid temperatures in the geyser field where sinter was deposited.

Single-phase, single-pass hydrothermal model description

In order to interpret the stable-isotope history of the sinter deposit and the present-day temperature trends, we constructed a relatively simple, cross-sectional, single-phase hydrothermal model of the Beowawe geothermal system. In this model, permeability is instantaneously increased along the Muleshoe and Malpais fault zones (unit 10, Fig. 21.2B) to create a single-pass hydrothermal system. Manga *et al.* (2012) argued that this type of permeability change can be caused by the removal of small particles that block fault apertures. The faults are connected to permeable geothermal reservoir at depth, a karst subunit within the Great Basin Paleozoic aquifer system (unit 6, Fig. 21.2B). Based on geophysical, geologic, and drilling information, the source reservoir for the geothermal fluids is interpreted to be Paleozoic carbonate 3–5 km below the surface (Watt *et al.* 2007). This interpretation is confirmed by the $\delta^{13}C$ compositions of dissolved inorganic carbon in Beowawe geothermal fluids (−2 to 1‰), which is typical of Paleozoic marine limestones (Day 1987). The isotopic compositions of water and dissolved inorganic carbon at Beowawe reported by Day (1987) were further discussed by John and coworkers (2003, p. 460–461). The δD of the water (−115‰) is about 10‰ less than modern meteoric water in the area. Such values are common in active hot springs across the Basin and Range Province, USA, and have been interpreted to reflect the involvement of Pleistocene meteoric water in these systems (Smith *et al.* 2002). Day (1987) attributed the high $\delta^{13}C$ values (−2 to 1‰) to dissolution of Paleozoic limestone (−2.5 to

3‰) situated at deep levels (~3000 m) below siliciclastic rocks of the Roberts Mountains thrust. Given the high reservoir temperature (~220°C), the small $\delta^{18}O$ shift from the meteoric water line (<3‰) requires large water–rock ratios (~10; John *et al.* 2003) and a permeable fault and/or karst-controlled flow system. Early Paleozoic marine carbonate rocks in Nevada have $\delta^{18}O$ values that range from ~20‰ to 30‰ (Hofstra & Cline 2000). Over time, fluids in a high-temperature reservoir will equilibrate with the surrounding rocks and shift to higher $\delta^{18}O_{H_2O}$ values. The aforementioned range of carbonate rock compositions and the fractionation equation of Kim & O'Neil (1997) ($1000 \ln \alpha_{calcite-H_2O} = 4.14@220°C$) show that an equilibrated reservoir fluid will have $\delta^{18}O_{H_2O}$ values between 16‰ and 26‰, comparable to the higher $\delta^{18}O_{H_2O}$ values from the sinter deposits (Fig. 21.3). The lowest $\delta^{18}O_{H_2O}$ values in Figure 21.3 appear to require a component of meteoric water.

Our paleohydrologic model includes calculations of groundwater flow rates, temperatures, and fluid–rock isotopic interactions/transport. The Beowawe hydrogeologic cross section was constructed along an east–west transect across the southern part of Whirlwind Valley (a–a', Figs 21.1A and 21.2A,B). It extends to the adjacent Reese River and Crescent Valleys, which enables us to consider the possibility of interbasin transfers of groundwater. These were found to be negligible (Person *et al.* 2008). Model results presented here are focused on Whirlwind Valley, which hosts the Beowawe geyser field.

The governing transport equations are presented in Table 21.2. Similar to that of Garg *et al.* (2007), our model is not capable of representing boiling conditions. In addition, mineral dissolution–precipitation reactions and their effects on fault permeability are not considered (e.g., Lowell *et al.* 1993). A detailed stratigraphy along this section was generalized into a limited number of hydrogeologic units comprised of stratigraphic units inferred to have similar hydrologic properties. Geologic structure was similarly generalized such that the only structures portrayed on the sections are either those that have sufficient offset to juxtapose different hydrogeologic units or structures that were explicitly included as part of the flow scenario being tested. In certain cases, structures that were subparallel to the section were omitted or generalized as a fault more nearly perpendicular to the section trace. Although based on surface and subsurface geologic and geophysical data, the hydrogeologic cross sections are intended to represent generalized scenarios that contain the salient geologic features needed to test a flow hypothesis; they are not intended to be literal representations of all that is known concerning subsurface geology.

The hydrothermal model used in this study considered the effects of both topography- and density-driven single-phase groundwater flow (Raffensperger & Garven 1995). The fluid flow (Eq. 21.1, Table 21.2) and heat-transport equations (Eq. 21.2, Table 21.2) were coupled using temperature- and pressure-dependent density and viscosity relations from Batzle & Wang (1992). In this study, the transients are driven mainly

Table 21.2 Governing transport equations

Stream functions

$$\nabla_x \cdot \left[\frac{k}{|k|} \nabla_x \psi \right] = -\frac{\partial \rho_r}{\partial x} \tag{21.1}$$

Cauchy–Riemann equations

$$q_x = \frac{\partial \psi}{\partial z} \qquad q_z = \frac{\partial \psi}{\partial x} \tag{21.2}$$

Heat transport

$$\nabla_x \cdot [\lambda \nabla_x T] - \vec{q} \rho_f c_f \nabla_x T = [\phi \rho_f c_f + (1 - \phi) \rho_s c_s] \frac{\partial T}{\partial t} \tag{21.3}$$

Fluid–rock isotope transport

$$\nabla_x \cdot [D \nabla_x R_f] - \frac{\vec{q}}{\phi} \nabla_x R_f = \frac{\partial R_f}{\partial t} + \frac{\partial R_{rk}}{\partial t} \frac{X_{rk}}{X_f} \tag{21.4}$$

Kinetic fluid–rock isotope exchange

$$\frac{\partial R_{rk}}{\partial t} = \bar{A} r_{rk} [\alpha_{rk} R_f - R_{rk}] \tag{21.5a}$$

$$\alpha_{rk} = \sum_{m-1}^{M} f_m \alpha_m \tag{21.5b}$$

$$r_{rk} = \sum_{m=1}^{M} A_o^m \exp \left[-\frac{E_a^m}{Rt} \right] \tag{21.5c}$$

Thermal conduction–dispersion tensor

$$\lambda_{xx} = \rho_f c_f \alpha_L \frac{q_x^2}{|q|} + \rho_f c_f \alpha_T \frac{q_z^2}{|q|} + \lambda_f \phi + \lambda_s (1 - \phi)$$

$$\lambda_{zz} = \rho_f c_f \alpha_T \frac{q_x^2}{|q|} + \rho_f c_f \alpha_L \frac{q_z^2}{|q|} + \lambda_f \phi + \lambda_s (1 - \phi)$$

$$\lambda_{xz} = \lambda_{zx} = \rho_f c_f (\alpha_L - \alpha_T) \frac{q_x q_z}{|q|} \tag{21.6}$$

Solute diffusion–dispersion tensor

$$D_{xx} = \alpha_L \frac{v_x^2}{|v|} + \alpha_T \frac{v_z^2}{|v|} + D_d$$

$$D_{zz} = \alpha_T \frac{v_x^2}{|v|} + \alpha_L \frac{v_z^2}{|v|} + D_d$$

$$D_{zx} = D_{xz} = (\alpha_L - \alpha_T) \frac{v_x v_z}{|v|} \tag{21.7}$$

k is permeability tensor; ∇_x is the gradient operator; ρ_r is the relative density ($\rho_r = (\rho_f - \rho_0)/\rho_0$); ρ_0 is the reference fluid density; ρ_f is fluid density at elevated temperature and pressure; x, z are the spatial coordinates; Ψ is the stream function; $|k|$ is the determinant of the permeability tensor; T is temperature; q_x, q_z are the components of Darcy flux in x- and z-directions; c_f is the specific heat capacity of the fluid; c_s is the specific heat capacity of the solid; φ is porosity; λ_{xx}, λ_{zx}, λ_{xz}, λ_{zz} are the components of the thermal conductivity–dispersion tensor of the porous medium; D_{xx}, D_{zx}, D_{xz}, D_{zz} are the components of dispersion–diffusion; $|v|$ is the magnitude of velocity, v_x and v_z are the components of specific discharge in the x- and z-directions; D_d is the molecular diffusion coefficient for porous media; q_x and q_z are the components of Darcy flux in the x- and z-directions, respectively; α_L and α_T are the longitudinal and transverse dispersivities, respectively; R_f is the fluid isotopic ratio; R_{rk} is the bulk rock isotopic ratio (averaged over all mineral phases present); t is time; X_{rk} is the fractional abundance of oxygen in the bulk rock phase; X_f is the fractional abundance of oxygen in water; $\bar{A}$ is the bulk rock reactive surface area; r_{rk} is the bulk rock reaction rate for fluid–rock isotope exchange; α_{rk} is the bulk fluid–rock equilibrium isotope-exchange factor averaged over all oxygen-bearing mineral phases; M is the total number of oxygen-bearing mineral phases for a given rock; A_o^m is the preexponential factor of the mth mineral phase; E_o^m is the activation energy for the exchange reaction; and R is the ideal law constant. Note that temperature is in Kelvin for isotope-exchange reactions.

Table 21.3 Mineral isotopic parameters

X_m	c_m	d_m	e_m	f_m	ρ_m	δ_0	A_o	E_a	Initial $\delta^{18}O$ composition	Mineral
0.5325	0.0	3.306	0.0	−2.71	2650	18	3.46E-05	11	10	Quartz[#]
0.13	−0.388	5.538	−11.35	3.132	2850	−2	4.50E-08	6.5	26	Dolomite[$]
0.48	−0.891	8.557	−18.11	8.27	2710	1	4.50E-08	6.5	26	Calcite[$]
0.1343	0.0	2.76	0.0	−6.75	2600	16	0.0000047	9.5	16	Kaolinite[*]
0.2747	0.0	3.904	−5.47	1.86	4500	20	0.0475	13.4	20	Barite
0.4180	0.0	4.12	−7.5	2.24	2700	20	1.39E-07	26.2	20	Anorthite

A_o is the preexponential factor in moles $m^{-2} s^{-1}$; E_a is the activation energy in $kcal\,mol^{-1}$; X_m is the fraction of oxygen in a given mineral phase; c_m–f_m are empirical constants used to estimate the equilibrium temperature-dependent fractionation factor for the mth oxygen-bearing mineral phase using the following relationship:

$$10^3 \ln(\alpha_m) = \frac{10^9 c_m}{(T + 273.15)^3} + \frac{10^6 d_m}{(T + 273.15)^2} + \frac{10^3 e_m}{(T + 273.15)} + f_m.$$ The bulk equilibrium fractionation factor for a given lithologic unit (α_{rk}) in our model is given by

$$\alpha_{rk} = \sum_{m=1}^{M} fr_m \alpha_m,$$ where fr_m is the fraction of the mth oxygen-bearing mineral phase that occurs within a given lithologic unit.

[#]Zhang et al. (1994)et al.
[$]Zheng (1999).
[*]Sheppard & Gilg (1996).

Table 21.4 Initial percentages of oxygen-bearing minerals in each stratigraphic layer

Unit numbers	Unit name	Qtz	Dol	Cal	Kaol	Bar	Anor
10	Fault zone	75	5	–	10	10	–
9	Quaternary basin fill	36	5	50	–	–	9
8	Volcanoclastics (tuffs, alluvium)	36	5	50	–	–	9
7	Middle Miocene volcanics	70	–	–	20	–	10
6	High-permeability karst unit	–	10	80	10	–	–
5	Low-permeability carbonates	–	10	80	10	–	–
4	Low-permeability Proterozoic siliciclastics	75	5	–	19	1	–
3	High-permeability Proterozoic siliciclastics	75	5	–	19	1	–
2	Low-permeability Proterozoic siliciclastics	75	5	–	19	1	–
1	Proterozoic metamorphic rocks	80	–	–	5	–	15

Abbreviations: Qtz – quartz, Dol – dolomite, Cal – calcite, Kaol – kaolinite, Bar – barite, Anor – anorthite.

by permeability changes. We represented advective–dispersive isotope transport and fluid–rock isotope exchange (Eqs 21.4 and 21.5, Table 21.2) using a kinetic-based approach described by Bowman et al. (1994). We assumed that each lithologic unit was composed of seven different minerals, each having its own isotopic fluid–rock exchange parameters and initial isotopic composition (Tables 21.3 and 21.4). Temperature-dependent equilibrium isotope fractionation factors, preexponential factor (A_o), and the activation energy (E_a) for each of the mth mineral phases were taken from experimental data reported by Zhang et al. (1994), Zheng (1999), and Sheppard & Gilg (1996).

For the groundwater flow equation, we specified a constant value ($\Psi = 0$) stream-function boundary along the sides and base of the domain consistent with a no-flux boundary condition. We specified a flux ($d\Psi/dz$) boundary condition across the top of the model domain with the magnitude of the flux being proportional to the lateral water-table gradient (Fogg & Senger 1985). For heat transfer, a specified temperature of 10°C was imposed along the top boundary except at the Malpais fault zone, where groundwater discharges. Here, a no-flux boundary condition was imposed ($dT/dz = 0$). This boundary condition implies that vertical flow rates are high enough that heat is not lost by conduction at the water table. This boundary condition was used by Appold et al. (2007) to emulate spring discharge along Refugio-Carneros and Coast faults in the Santa Barbara Basin. A basal heat flux was also imposed at the base of the solution domains ($80\,mW\,m^{-2}$), based on heat flux estimates from the northern Nevada rift (Blackwell 1983). We fixed the $\delta^{18}O$ composition of meteoric recharge at −15‰ at the water table. A no-flux boundary condition was imposed ($dR_f/dz = 0$) at the Malpais fault zone for these two tracers. No-flux boundary conditions were imposed along the sides and base of the solution domain for isotope transport.

Thermal conductivity, solute diffusivity, and longitudinal and transverse dispersivities were fixed (Table 21.5) for all lithological units. We assumed a conductive geothermal gradient as an initial condition (about 40°C km^{-1}). Initial fluid isotope composition (R_f) values were assigned using these initial conductive temperatures. Each stratigraphic unit was assigned an initial rock isotope composition depending on its mineral assemblages (Table 21.3).

Hydrostratigraphic framework model

The permeabilities of the hydrostratigraphic units shown in Figure 21.2B are listed in Table 21.6. The permeability values

Table 21.5 Rock thermal and transport properties

Variable	Symbol	Value/units
Thermal conductivity of fluid	λ_s	$0.58\,W\,m\,°C^{-1}$
Thermal conductivity of solids	λ_s	$2.5\,W\,m\,°C^{-1}$
Heat capacity of the fluid phase	c_f	$4180\,J\,kg^{-1}$
Heat capacity of the solid phases	c_s	$800\,J\,kg^{-1}$
Longitudinal dispersivity	α_L	$1.0\,m$
Transverse dispersivity	α_T	$0.1\,m$
Solute diffusivity	D_d	$10^{-10}\,m^2\,s^{-1}$
Reactive surface areas	A	$10^{-4}\,m^2\,mol^{-1}$

Table 21.6 Hydrologic rock properties

Unit	$\log_{10}(k)$	Description
10	−11 to −13	Faults (black lies)
9	−16	Basin fill (yellow)
8	−17	Tuffs (brown)
7	−17	Volcanics (orange)
6	−10	Karst zone (blue)
5	−17	Carbonates (purple)
4	−17	Proterozoic siliciclastics (green)
3	−14	Proterozoic shales (white)
2	−17	Proterozoic sandstones (green)
1	−17	Proterozoic bed rocks (dark blue)

k_x (horizontal permeability, m^2), unit numbers are listed in Figure 21.1C.

used in our model are consistent with values reported in the literature for Basin and Range sedimentary rocks (Blankennagel & Weir 1973; Winograd & Thordarson 1975; Plume & Carlton 1988; Maurer *et al.* 1996; Bredehoeft 1997; Belcher 2004; Welch *et al.* 2007). We used published permeability data sets for crustal rocks as a guide in selecting these values (Brace 1980, 1984; Clauser 1992, Manning & Ingebritsen 1999). Because these data sets present a range of permeability conditions for a given lithologic unit, we relied on model calibration to match Beowawe heat-flow and temperature profiles (Person *et al.* 2008). The hydrogeologic section includes the following hydrogeologic units: a deep, low-permeability unit that includes Proterozoic crystalline basement rocks (Unit 1; dark blue strata in Fig. 21.2B); two low-permeability intervals of Upper Proterozoic to Early Cambrian siliciclastic rocks (Units 2, 4; green strata in Fig. 21.2B) that bracket a middle 1.5-km-thick interval of higher permeability, highly fractured quartz sandstone (Unit 3; white unit in Fig. 21.2B); Cambrian–Devonian dominantly carbonate rocks (Unit 5, purple strata in Fig. 21.2B); low-permeability strata of the Roberts Mountain allochthon (Unit 7, orange strata in Fig. 21.2B); Miocene volcanic rocks, including lava flows and ash-flow tuff (Unit 8, brown strata in Fig. 21.2B); alluvial-basin fill (Unit 9, yellow strata in Fig. 21.2B); and fault zones (Unit 10, red lines in Fig. 21.2B). A 100-m-thick interval of potentially enhanced horizontal permeability was added at the top of the Paleozoic carbonate unit (Unit 6, blue interval in Fig. 21.2B). This interval represents a sequence boundary that commonly contains karst features and is highly permeable. Such sequence boundaries are present at various stratigraphic levels within the Paleozoic section (Cook & Corboy 2004). The inferred subsurface structural geometry portrayed on section a–a′ is generalized from published sections presented in John & Wrucke (2003) and Watt *et al.* (2007). Major structural elements include west-dipping range-bounding faults on the west side of the Shoshone Range, and steep faults in the vicinity of the Mule Canyon deposit that proxy for the east-dipping Muleshoe fault (John *et al.* 2003) and for highly anisotropic permeability within dike swarms of the northern Nevada rift (John & Wrucke 2003). A single steep west-dipping fault is portrayed in the vicinity of Beowawe and generalizes the Malpais fault and other ENE-trending and NNW-trending faults that localize geothermal fluids (Zoback 1979; Struhsacker 1980). All of these faults are portrayed as cutting both low-permeability strata of the Roberts Mountain allochthon and underlying Cambrian–Devonian dominantly carbonate rocks. Gentle dips result in the most permeable part of the Paleozoic carbonate rock section, only attaining maximum depths of about 6 km below land surface.

The subvertical faults are relatively thin (∼40 m) and are assigned isotropic fault properties. Subvertical faults are connected hydrologically to a thin aquifer unit at the top of the Paleozoic carbonates and, potentially, by a deep, thick, more porous aquifer in the middle of the siliciclastic sequence ($10^{-13}\,m^2$). Practical considerations regarding simulation time restricted our ability to represent the fault zones using grid refinement on the submeter scale, which would have been desirable.

Sensitivity study

We conducted a sensitivity study to assess what values of fault permeability could explain both the modern temperature overturns and the isotopic evolution of geothermal discharge. We ran the simulations using fault permeabilities ranging from 10^{-14} to $10^{-11}\,m^2$. Computed temperatures simulated using fault permeability less than $10^{-13}\,m^2$ were clearly too low and are not presented. Computational limitations prevented us from running simulations with fault permeabilities greater than $10^{-11}\,m^2$. After 5000 years of hydrothermal fluid flow, we lowered the subvertical fault and karst reservoir permeability to $10^{-19}\,m^2$ in order to allow the system to return to conductive equilibrium. This was done to represent the effects of fault clogging by silica mineralization. We then monitored the thermal and isotopic recovery for an additional 5000 years. The rationale for including a thermal-recovery phase was to see whether the fluids could return to isotopic equilibrium after 5000 years. We monitored simulated temperatures at the base of the Muleshoe fault in the recharge area and at the water table along the Malpais fault within the discharge area.

RESULTS

Isotopic composition and age of the sinter deposits

Calculated $\delta^{18}O_{H_2O}$ values and ^{14}C dates on pollen recovered from sinter samples provide evidence for two periods of geothermal fluid discharge that each exhibit a progressive decrease in $\delta^{18}O_{H_2O}$ through time (Fig. 21.3). Some of our data are from a single flowstone layer extending down slope from a vent lacks the progressive changes that might be expected if Rayleigh fractionation during progressive precipitation of silica was important. The highest $\delta^{18}O_{H_2O}$ values reflect isotopic exchange between fluid and sedimentary rocks in a geothermal reservoir at low water/rock ratios (e.g., 0.1) and temperatures in excess of 200°C (Sheppard 1986). The trends of decreasing $\delta^{18}O_{H_2O}$ through time suggest increasing contributions of unexchanged meteoric water. Modern geothermal fluids at Beowawe have an isotopic composition ($\delta^{18}O_{H_2O} = -16.5‰$, $\delta D = -120‰$) (John *et al.* 2003) that is representative of unexchanged meteoric water and that plots near (within 4‰ $\delta^{18}O$) the meteoric water line.

Model results

For a fault permeability of 10^{-12} m^2, computed temperatures after 200 years (Fig. 21.7A) show strong convective cooling within the recharge area along the Muleshoe fault. By 1000 years, temperature declines extend into the karst aquifer and pronounced thermal overturns developed (Fig. 21.7B). The rate of cooling within the Muleshoe fault is strongly controlled by fault permeability (Fig. 21.8A). Within the discharge area along the Malpais fault, temperatures increased to a peak value about 180°C before declining (Fig. 21.8C). The timing of peak discharge temperature along the Malpais fault zone is quite sensitive to fault permeability (Fig. 21.8C). For a fault permeability of 10^{-13} m^2, the temperatures peaked after about 5000 years, whereas for a fault permeability of 10^{-11} m^2, they peaked in less than 100 years. A fault permeability of $10^{-12.3}$ m^2 (5.0×10^{-13} m^2) produced peak temperatures of about 180°C after 1000 years. The highest permeability fault zone did not correspond to the highest discharge temperature because of the cooling effect of low-temperature meteoric water (Forster & Smith 1989). Vertical flow rates along the Malpais fault varied between about 60 and 3 m year^{-1} for the range of permeabilities considered (10^{-11} to 10^{-13} m^2). We estimate that for this range of fault permeabilities, the time required for groundwater to pass through the single-pass system by pure advection is between 300 and 4500 years.

When the flow system is shut off at 5000 years, a relatively cool region with a vertical width of 1000 m is present above and below the karst zone (Fig. 21.7C). Conductive thermal recovery occurs at a much slower rate than the advective cooling (Figs 21.7D–F and 21.8A). This prevented the recovery back to initial thermal conditions over a 5000-year period and slows temperature-dependent isotopic exchange rates

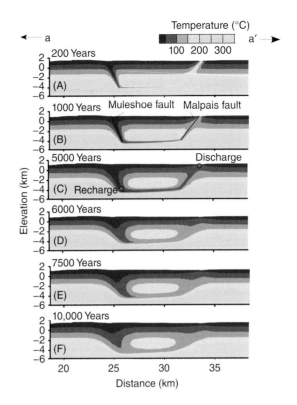

Fig. 21.7. Computed temperature contour maps illustrating thermal evolution along the Malpais fault, Muleshoe fault, and karst aquifer. Fault zones were assigned a uniform permeability of 10^{-12} m^2. Fluid flow is active for 5000 years (A-C) followed by 5000 years (6000–10,000 years; D-F) of conductive cooling. Shallow meteoric fluid infiltration occurs down the Muleshoe fault zone and then along a thin, permeable karst zone. Upflow occurs along the Malpais fault zone. Points labeled "recharge" and "discharge" are the locations within the Beowawe geothermal system where temperature and $\delta^{18}O_{fluid}$ were plotted through time in Figure 21.8A,C. (*See color plate section for the color representation of this figure.*)

(Fig. 21.8B,D). Based on the vertical thickness of the thermally disturbed region above and below the karst zone (1000 m), we estimate that it would require at least 10^4 additional years to approach initial conductive conditions, assuming a thermal diffusivity of 1.25×10^{-6} m^2 s^{-1}. If the flow system were reinitiated following the 5000-year period of conductive recovery, the discharging fluid along the Malpais fault zone would be only moderately enriched.

Figure 21.5E–G presents computed, time-dependent evolution of temperature profiles extracted from our cross-sectional model at different distances from the Malpais fault. Early-time (200–1000 years since the onset of flow) temperature profiles within the hanging wall close to the Malpais fault resulted in temperature overturns that are similar in form to the observed conditions within the Batz-1 and 85-15 wells (Fig. 21.5E–G). The timescale is much shorter than the geochemical transients inferred from the sinter deposits (Fig. 21.3).

By 5000 years, fluid temperatures were below 100°C at a depth of 5 km for the most permeable fault scenarios

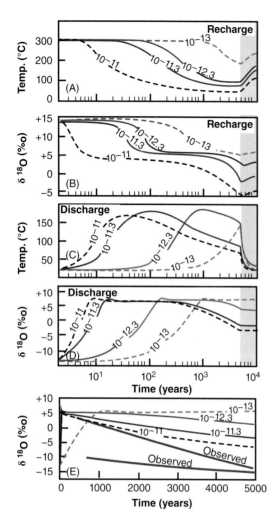

Fig. 21.8. Effect of fault permeability variations on simulated temporal temperature changes (A) and (B) oxygen isotopic composition along the Muleshoe fault zone at a depth of about 5 km. The numbers on the lines denote the permeability assigned to the Muleshoe and Malpais faults (in m²). Effect of fault permeability variations on simulated temporal changes in temperature (C) and (D) oxygen isotopic composition along the Malpais fault zone at the water table. The shaded area of the graphs represents conductive recovery following 5000 years of convection and cooling of fluids moving through the system. Locations within the recharge and discharge area are shown in Figure 21.2c (red circles). (E) Comparison of observed trends (gray dashed lines) and computed changes in $\delta^{18}O_{H_2O}$ versus time for the Beowawe sinter terrace. (*See color plate section for the color representation of this figure.*)

(Fig. 21.8A). These lower temperatures restrict isotopic fluid–rock interactions. Once the flow system ceased at 5000 years, computed temperatures within the recharge area along the Muleshoe fault rebounded toward their initial conditions (Figs 21.7 D–F and 21.8A,C). Computed $\delta^{18}O_{H_2O}$ composition within the recharge area along the Muleshoe fault underwent rapid initial depletion as meteoric water began to descend (Fig. 21.8B). As progressively lighter fluids invaded the fault zone, reaction rates increased under isothermal conditions, causing stabilization in $\delta^{18}O_{H_2O}$ levels. The amount

of isotopic depletion depends strongly on permeability. The isotopic composition of fluids entering the karst aquifer was lightest for the most permeable conditions. Isotopic enrichment between the bottom of the Muleshoe fault zone and the discharge area at the top of the Malpais fault was only slight (~2‰ Fig. 21.8D). Enrichment in the isotopic composition in the discharge area (Fig. 21.8D) along the Malpais fault peaked earlier than computed temperatures (Fig. 21.8C). Computed $\delta^{18}O_{H_2O}$ composition dropped from about +6 to −3‰ but never reached the meteoric-water end member (−15 $\delta^{18}O$). The slight bump in the initial peak of $\delta^{18}O_{H_2O}$ is due to a downward tilt in the karst reservoirs to the west; the initial isotopic composition along the west edge of the karst zone was slightly enriched relative to the east side.

The time required for the simulated $\delta^{18}O_{H_2O}$ composition within the Malpais fault zone to peak (+7‰) ranged between 10 and 1000 years after the onset of flow, for the permeability conditions considered (Fig. 21.8D). These values are more depleted than the initial equilibrium conditions with the karst reservoir at 5 km depth (+13‰), likely because of mixing of relatively unexchanged meteoric water and carbonate reservoir fluids along the flow path. Simulated oxygen-isotope composition of the discharging fluids decreased from about +7 to −3‰ for fault permeabilities between 10^{-12} and 10^{-11} m² (Fig. 21.8D). This is about 50% of the decline observed (Figs 21.3 and 21.8E). Once flow terminated, recovery of computed temperatures (Fig. 21.8A) and $\delta^{18}O_{H_2O}$ began (Fig. 21.8B). However, the fluid isotopic composition did not return to equilibrium levels over the simulated time frame due to the slow rates of thermal recovery.

DISCUSSION AND CONCLUSIONS

Our cross-sectional, single-pass model is clearly too idealized to account for all features of the observed thermal and isotopic data. In order to match the observed thermal overturns in temperature profiles near the Malpais fault (Fig. 21.5B,C), the model requires that hydrothermal fluid flow persist for only 1000 years or less. This is much briefer than the duration of declining isotopic composition within the sinter deposits we sampled (5000–7000 years). How can these two observations be reconciled? We propose that the Malpais fault zone is composed of several pathways to the surface, each pathway having different fault-zone apertures. As the most permeable conduit is sealed off by mineralization, alternative conduits are accessed along the Malpais fault. This is slightly analogous to channel avulsion in braided river systems. Published timescale estimates for fault clogging by Lowell *et al.* (1993) support this conceptual model. These authors developed simple analytical models for fault sealing by silica precipitation using a wide range of fault permeabilities (10^{-10} to 10^{-16} m²) and geothermal reservoir temperatures (50°C and 300°C). For these two temperatures, the time required for fault sealing ranged from less than 1 year (300°C) to more than 300 years (50°C) for a fault permeability of $10^{-12}\,\mathrm{m^2}$. The notion of changes in the locus of hot-spring discharge is supported, in part, by the observation that peak shallow heat flow today is to the south of the Beowawe sinter deposits, which was clearly the locus of discharge in the past (Fig. 21.2A). This conceptual model of fault clogging and rerouting of flow through fault zones is not a new idea. Saffer (this issue) presents thermal/geochemical data from the Barbados and Nankai accretionary prism to argue that relatively high rates of fluid fluxes through permeable fault systems could only be active between 0.2 and 10% of the time due to clogging by fluid–rock geochemical reactions (Saffer, Fig. 21.4C, this issue).

Another complexity that is not accounted for in our single-pass model is the observed abrupt shift from nearly meteoric fluid composition ($-15‰\ \delta^{18}O$) at about 7.5 ka to isotopically enriched discharge at 4.5 ka. We speculate that different reservoirs may be tapped during different seismic events. Alternatively, the conflict between modeled and measured values might be resolved if the discharging geothermal fluids are actually part of a two-component mixture of a deeper, slowly circulating, hot fluid with enriched $\delta^{18}O$ and a shallow-circulating, warm meteoric fluid with lighter $\delta^{18}O$. The fault rupture might preferentially alter the permeability of the deeper part of the fault, and thus increase the amount of deeper fluid in discharging water. After fault slip, the deeper portion of the fault might "heal up" and gradually decrease the deeper fluid contribution to very low levels, while the shallow fluid contribution might remain more constant. Thus, the deeper fluid temperature and $\delta^{18}O$ oscillates only moderately with time (and could perhaps fully recover between earthquakes).

Our interpretation of the thermal and isotopic evolution of the Beowawe geothermal system must be tempered by parameter uncertainty. There is a significant uncertainty in the reactive surface area and kinetic coefficients used in our fluid–rock isotope-exchange models. Surface area for fluid–rock isotope-exchange models can be computed using mineral grain size (Brantley & Mellott 2000) or fracture spacing (Rimstidt & Barnes 1980). Surface areas based on mineral grains can be as high as $1\,\mathrm{m^2\,mol^{-1}}$. The value we used ($10^{-4}\,\mathrm{m^2\,mol^{-1}}$) was based on fracture spacing and would be representative of a fracture spacing of about 70 cm (assuming the rock is comprised entirely of feldspar). Our model of fault permeability was static. Other studies (Roberts *et al.* 1996; Garven *et al.* 2001; Appold *et al.* 2007) have allowed fault permeability to decay through time due to changes in fault stress, mineralization, and fluid pressure decline.

For a fault permeability of $10^{-12}\,\mathrm{m^2}$, our model results indicate that discharge peaked at temperatures of 180°C along the Malpais fault about 30 years after the onset of convection. Then, discharge temperatures began to cool. The most isotopically enriched discharge fluids break through in about 10 years. The finding of transient heat pulses at hot springs is not new. Numerous researchers have observed similar phenomena in their models of fault-controlled hydrothermal systems

(e.g., Forster & Smith 1989; Lopez & Smith 1995). This is due to the high rate of enthalpy extraction allowed by the high fault permeability and consequent high flow rates. It is worth noting that the magnitude of the temperature change presented in this study is far larger than that observed in studies of modern earthquake-influenced hot springs (e.g., Mogi *et al.* 1989; Manga & Rowland 2009). This may be due to the extraordinary high extensional deformation rates along the Basin and Range faults (Kennedy & Van Soest 2007).

Our oxygen isotope analysis of ^{14}C-dated sinter deposits (Fig. 21.3) show that large (up to 20‰) time-dependent decreases in the isotopic composition of the geothermal fluids occurred along the Malpais fault zone over geologic timescales. We interpret this decline as the result of progressive influx of meteoric fluids into a single-pass, liquid-dominated, geothermal system. Other potential mechanisms that could account for isotopic changes (e.g., boiling, climate-driven changes in precipitation composition and isotopic composition during the Late Pleistocene, or Rayleigh distillation of the geothermal fluids during sinter-deposit formation) cannot account for such a large shift in $\delta^{18}O$. The single-pass hydrothermal/isotope transport model of the Beowawe geothermal system seems to be the most plausible mechanism for the observed data, although it is clearly too idealized. If our conceptual model of the Beowawe geothermal system is correct, our analysis provides insights into how transient permeability changes occur along fault zones over geological timescales. Estimated rates of silica clogging of faults (as fast as 10 years; Lowell *et al.* 1993) are much shorter than those indicated by the sinter-deposit isotope data.

Groundwater flow must be three-dimensional in nature. Person *et al.* (2012) showed how hydrothermal flow frequently focuses at relatively high-permeability intersections of fault zones, as observed across the Basin and Range Province by Coolbaugh *et al.* (2005). Furthermore, Garg *et al.* (2007) found that only narrow regions of the Malpais fault zone were permeable. Nevertheless, any three-dimensional model must still incorporate downward fluid movement on the Muleshoe or some other fault system, lateral flow through a reservoir, and upward flow along the Malpais fault.

Our studies have described transient hydrothermal systems within the Basin and Range Province. McKenna *et al.* (2005) concluded that the high temperatures (280°C at 3.8 km depth) along the base of the Dixie Valley fault within the Dixie Valley geothermal system must be due to hydrothermal flow transients in the order of 30,000–50,000 years. These authors also speculated that the time-dependent fault permeability was related to seismicity. Temperature profiles from wells at Dixie Valley, however, show little evidence of thermal overturns.

Low-magnitude earthquakes result in slip increments of millimeters to centimeters along faults. Hill (1977) and Sibson (1996) proposed that low-magnitude (<4M) earthquake swarms can induce significant amounts of fluid movement capable of generating economic gold deposits along fracture networks. Varying the permeability of the fault zones in our model provided some insight into the magnitude of permeability change following each seismic event. While none of the models produced the full 20‰ range observed in Figure 21.3, the fault permeability had to have been on the order of $10^{-11}\,\mathrm{m}^2$ or greater. The highest permeability scenario ($5 \times 10^{-11}\,\mathrm{m}^2$) produced about a 10‰ decrease in $\delta^{18}O_{H_2O}$ over a few thousand years, while a permeability of $10^{-13}\,\mathrm{m}^2$ produced almost no decrease (Fig.21.8D). Numerical considerations prevented us from assigning higher fault permeability to the Muleshoe and Malpais faults. It is possible that representing isotope transport through distinct fractures, or greater mesh refinement, would have improved the model fit to observed data. Extrapolating the trends in the rate of decline in computed $\delta^{18}O_{H_2O}$ in Figure 21.8E, we suspect that a fault permeability of $10^{-9}\,\mathrm{m}^2$ would likely produce a close match with observed data. Prior to the onset of hydrothermal circulation, the permeability of the Malpais and/or Muleshoe fault must have been at or below $10^{-14}\,\mathrm{m}^2$, producing low-temperature discharge that could not have carried much dissolved silica to form sinter deposits. Fault conductance controls the magnitude of convective heat flow, so our reported fault permeabilities are to some degree nonunique. A wider fault zone would require a lower fault permeability in order to produce the same convective thermal anomaly.

Isotope data collected as part of this study suggest that Holocene geothermal systems in the Basin and Range Province are episodic in nature on timescales of about 5000–7000 years. Zoback (1979) estimated that about 210,000 years were required to account for all of the silica (1.28×10^{11} kg) along the Beowawe geyser field terrace. She assumed continuous flow at modern Beowawe geyser field discharge rates ($611\,\mathrm{s}^{-1}$) and modern silica concentrations. In our model, each hydrothermal flow event discharged about 9.1×10^9 kg of silica (Person *et al.* 2008). Assuming a fault-zone width of 40 m, a length of 2 km, and that 10% of the dissolved silica is available for sinter formation, more than 100 episodic flow events would be required to form the 1.28×10^{11} kg Beowawe geyser field sinter terrace.

ACKNOWLEDGMENTS

We thank Andy Manning of the USGS and two anonymous reviewers for their constructive criticism of an earlier draft of this manuscript. This work was supported by an NSF grant to Mark Person and Albert Hofstra (NSF-EAR 0809644). We also acknowledge the support of the National Science Foundation (EPSCoR) under Grant No. IIA-1301346 to Mark Person and Laura Crossey.

CHAPTER 22

The permeability of crustal rocks through the metamorphic cycle: an overview

BRUCE YARDLEY

School of Earth and Environment, University of Leeds, Leeds LS2 9JT, UK

ABSTRACT

The controls on the permeability and fluid regime of the crust more than a few kilometres below the surface are fundamentally different according to whether it is composed of sediments that are actively undergoing prograde metamorphism, with or without the intrusion of magma bodies, or is composed of crystalline rocks that have cooled below their original temperature of formation and are susceptible to retrograde metamorphic reactions when fluid is introduced. For the prograde case, it is likely that fluid pressures are high and therefore permeability is low during metamorphism. Permeability is not an independent variable but is coupled to the rate of fluid production by reaction and the tendency of the rock fabric to recrystallise, reducing permeability. The ubiquitous high fluid pressures inferred for prograde metamorphism from field studies imply very low permeability controlled by the rate of endothermic reactions. Under retrograde conditions, the equilibrium fluid fugacities are very low, and therefore the fluid that is introduced tends to be consumed by reaction with the relic high-T mineral assemblages. Transient retrograde permeability is associated with structural features such as faults, veins and shear zones. While fracturing may lead to significant permeability initially, the influx of water leads to mineral reactions and, at appropriate temperatures, to water weakening and ductile deformation. These processes lead to further changes in permeability until the rocks fracture again; how permeable the rocks are overall will depend on the rate of deformation that they experience.

Key words: metamorphic permeability, fluid production, prograde metamorphism, retrograde metamorphism, transient permeability

INTRODUCTION

Metamorphism involves the recrystallisation of existing mineral grains in a rock to form more stable mineral assemblages and textures. The most widespread form, regional metamorphism, involves the burial and heating of an original sequence of supracrustal rocks and is followed by their return to the surface, but metamorphism can also occur in very different settings such as geothermal systems. Specifically, for the purpose of this chapter, prograde metamorphism is defined as metamorphism proceeding as a result of changes in temperature and/or pressure that result in devolatilization reactions. In the course of a simple metamorphic cycle, the initial sedimentary rocks containing abundant fluid-filled pores and volatile-bearing minerals (e.g. clays, carbonates) are transformed by prograde metamorphism into very low-porosity crystalline rocks with much lower contents of volatiles; these are almost entirely bound in mineral lattices. As we discuss later, release of metamorphic fluid mainly takes place very slowly, under strongly overpressured conditions, and through rocks that have very low porosity. However, since metamorphic fluid is generated throughout reactive layers, a pervasive fluid phase must be present through much of the pore network of a rock undergoing prograde devolatilization.

Fluid behaviour in retrograde metamorphism is very different. Once rocks begin to cool, pore water is consumed by retrograde reactions, and from being pervasively wet, rocks become very dry. Not only are the metamorphosed rocks now a sink, rather than a source, for fluid, they also no longer contain a pervasive fluid phase. When fluid is able to infiltrate them it does so along discrete fractures and reacts with the immediate fracture walls, rather than pervasively, altering the fracture walls much more than the intervening rock (e.g. Holness 2003). Where older, high-grade rocks are caught up in a later orogeny, then they will similarly behave as a fluid sink initially, while adjacent sediments undergoing prograde metamorphism act as sources (McCaig 1997).

In this chapter, I seek to explore how this evolution in hydrogeologic characteristics takes place and what it tells us about the permeability of the deep crust in different tectonic settings.

PERMEABILITY AND FLUID FLOW IN METAMORPHIC ROCKS

Permeability is a measure of how easily fluid moves through rock in the presence of a hydraulic head to drive it. The relationship between fluid flow through a porous medium and the forces that drive it is provided by Darcy's law:

$$q = -K \left(\frac{dh}{dL} \right),$$

where q is the volumetric flow rate through a unit cross section of a flow path, (dh/dL) or the hydraulic gradient is the difference in hydraulic head between two points on the flow path divided by the distance that separates them and K is hydraulic conductivity. Hydraulic conductivity contains terms that relate to the properties of both the fluid and the rock through which it is moving, and so for deep geological applications where the fluid properties may vary, it is normally broken down into its component terms, with the rock properties described by the permeability. Darcy's law is believed to be valid for all but the highest flow rates, although it is difficult to demonstrate that it remains valid in very low permeability materials (Ingebritsen *et al.* 2006). For the purposes of this chapter, I assume that fluid flux continues to reflect hydraulic gradient, rock permeability and fluid properties, but with the proviso that under metamorphic conditions rock and fluid properties will not be constant. Rock permeability is linked to porosity, texture, mineralogy and the way in which the porosity is distributed and interconnected, for example, as discrete pores, on grain boundaries or as open fractures. With increasing reactivity between minerals and rocks, permeability will vary as the volume and distribution of porosity changes and at low porosity it may also vary with the nature of the mineral surfaces in contact with fluid.

With increasing depth, most sedimentary sequences become overpressured as low permeability layers provide seals above underlying formations, and from this point on into metamorphism, pore water behaviour changes. Overpressured rocks lose fluid irreversibly as a result of compaction and/or mineral devolatilization, but are not recharged except from more highly pressured rocks.

There are three possible approaches to investigating overall fluid fluxes during orogeny. Petrology-based models do not consider permeability explicitly, and instead estimate fluid flux from the inferred reaction history via a series of assumptions (e.g. Ferry 1987, 1994). These models typically invoke large fluid fluxes without addressing the possible sources of the fluid or the permeability required, and focus on the specific stage in the metamorphic history during which the mineral assemblages preserved for study, were formed.

A second general approach is to assign inferred permeability values to metamorphic rocks. If assumptions are then made about the variation in fluid pressure with depth, fluid flux can be estimated but these are similarly not constrained to reflect the amount of fluid that might actually be available. Manning & Ingebritsen (1999) and Ingebritsen & Manning (2010) drew on a wide range of lines of evidence to develop a relationship between depth and permeability that is generally applicable to the lower crust, but recognised that rocks undergoing prograde metamorphism might not lie on this trend. Nevertheless, the approach is applicable in conditions where high fluid pressure is not required by considerations of mineral equilibria. More recently, Lyubetskaya & Ague (2009) took a low baseline permeability for prograde metamorphism, consistent with the results of Manning & Ingebritsen (1999), but then modified it locally in the model to reflect reactions that were taking place.

An alternative approach that is specifically valid for rocks undergoing devolatilization reactions is to obtain the minimum permeability that a rock would need to have to allow the fluid that is generated within it to escape, based on the supply of heat to drive endothermic reactions (Yardley 1986; Hanson 1992). Support for this end-member model comes from the widespread evidence for high fluid pressures during prograde metamorphism, which implies minimal permeability. This approach is described in more detail later.

Fluid flow under retrograde metamorphic conditions has generally received little attention from petrologists, although there is an extensive literature on the hydration and carbonation of mantle ultramafic rocks in the upper mantle or emplaced in the crust, a closely analogous case (e.g. Ranero *et al.* 2003; Power *et al.* 2013). There is, however, a growing body of information on fluids in upper crustal crystalline rocks, which provides important insights into the behaviour of deeper fluids (Stober & Bucher 2015).

Approaches to understanding permeability in rocks undergoing metamorphism may appear to be contradictory, but differences often simply reflect differences in perspective. For example, petrographic and geochemical studies of specific rocks that have clearly experienced an episode of focussed fluid flow provide information about the behaviour of a specific suite of rocks during a specific period of time, whereas calculations about the overall thermal and fluid budget of an orogenic belt give an overview of a much wider range of lithologies through a much longer period of time, without capturing the diversity of behaviour in different rock types and at different times. The results are complementary, and together provide a much better understanding of the role of fluids in orogenesis than either approach alone.

PERMEABILITY DURING DEVOLATILIZATION

It is extremely difficult to conceive of a circumstance in which fluid being released by devolatilization reactions in a rock or sediment would not migrate away, because fluids are less dense than rocks and considerably less viscous, while under most P–T conditions, devolatilization reactions lead to an overall volume increase. It follows that rocks that experience devolatilization must be permeable to some degree in order for the fluid to escape at comparable rate overall to that at which it is generated. This section explores the meaning of permeability in a rock that is undergoing prograde metamorphism.

Fluid pressure and permeability

The porosity and permeability of a porous material depend on the effective pressure, that is, the difference between the confining pressure holding grains together (normally equivalent to the overburden pressure or lithostatic pressure in geology) and the fluid pressure in the intervening pores or cracks, which tends to push them apart. In the crust, there are three possible end-members that might represent equilibrium between these balancing forces. If the fluid phase is effectively continuous to the surface, confining pressure is equal to lithostatic pressure while fluid pressure is equal to hydrostatic pressure. This is a common equilibrium end-member in shallow settings, but requires the difference between fluid pressure and confining pressure to be too small to cause the solid grains to deform and reduce porosity. Situations in which fluid pressure exceeds hydrostatic pressure are referred to as overpressured, and this may arise because of the properties of the rock itself, or because it is overlain by an impermeable cap rock. For overpressured rocks, compaction and reaction result in a rise in fluid pressure, and pressure may evolve to approach an alternative equilibrium, in which confining pressure is equal to fluid pressure and the effective pressure is zero. Any further increase in fluid pressure, or a drop in lithostatic confining pressure due to erosion, is likely to lead to hydraulic fracturing of the overpressured rock. The third possible equilibrium is one in which hydrous and anhydrous phases coexist in the rock, providing a thermodynamic buffer for water fugacity. In this case, the water fugacity can be equated to a real or notional water pressure; how this then relates to fluid pressure depends on the composition of the fluid phase. In prograde metamorphism, it is commonly assumed that rocks are close to fulfilling both this thermodynamic equilibrium condition and the physical equilibrium condition for confining pressure to be equal to fluid pressure. In contrast, in retrograde metamorphism the thermodynamically buffered fugacities of fluid species are so low that a free fluid phase cannot be present (Yardley 1981; Frost & Bucher 1994; Yardley & Valley 1997). Figure 22.1 is a plot from Yardley & Bodnar (2014), showing the different values of water fugacity that might be present as a function of temperature and depth according to which of these three equilibrium end-member conditions prevailed.

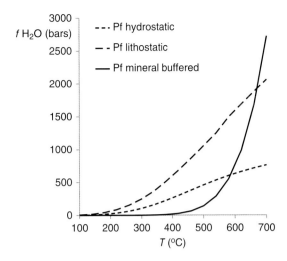

Fig. 22.1. Possible variations in water fugacity with temperature in the crust. The broken lines represent values for water-saturated crust with fluid pressure equal to either lithostatic or hydrostatic pressure, while the solid line represents values buffered by the coexistence of muscovite, K-feldspar, sillimanite and quartz and does not assume that a free fluid phase is present. All curves are calculated for a geothermal gradient of $30°$ km^{-1}. (Yardley & Bodnar (2014)).

How do these relationships come into play in the course of the metamorphic cycle? Below a few kilometres depth, sedimentary rocks are already likely to be strongly overpressured prior to the onset of metamorphism, judging by oilfield analogues. Metamorphic petrologists commonly assume that the effective pressure is already zero at the onset of metamorphism, and although in detail this may be an oversimplification, many authors have documented evidence in support of this assumption, ranging from the occurrence of veins (evidence of hydraulic fracturing) to the coincidence of P–T estimates from fluid-dependent and fluid-independent thermobarometers. If fluid pressure is linked to lithostatic pressure and is not an independent variable, then permeability is also not an independent variable or an intrinsic property of a particular rock since it must allow the relationship between Pf and Pl to be maintained as reactions proceed. The permeability must be such that the rate of fluid escape matches the rate of fluid generation, and if we neglect any effects from expulsion of original pore fluid, this means that the permeability is a function of the rate of dehydration or, more generally, the rate of devolatilization.

In detail, the system is dynamic and fluctuations must occur. If fluid pressure drops because of an increase in permeability, then the rate of reactions that release fluid is likely to increase, resulting in a gradual restoration of the fluid pressure. Similarly, if permeability is too low for fluid to escape at the rate it is generated, then fluid pressure will rise so that effective pressure decreases and permeability increases until a balance is reached between generation of fluid and its loss by flow.

Figure 22.2 is a Mohr diagram to illustrate how for rocks subject to only a small deviatoric stress, the effective pressure may become very small without rupture, while for rocks subject to

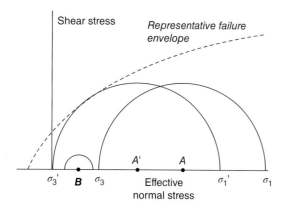

Fig. 22.2. Mohr diagram to illustrate the effects of deviatoric stress and effective pressure on whether rocks hydrofracture during prograde metamorphism. The broken line is a representative failure envelope and the semicircles represent stress in the plane normal to σ_2. Semicircles with centres (mean stress) at A and A′ have the same deviatoric stress ($\sigma_1-\sigma_3$) but in the case of A′ fluid pressure is higher, resulting in a reduction in the effective normal stresses so that it now intersects the failure envelope. Compared with A, an increase in fluid pressure has led to brittle failure. In contrast, B represents a rock subject to a much smaller deviatoric stress. The effective mean stress is much smaller than for A (likely leading to enhanced permeability), but the rock is not close to the failure envelope. For this rock, fluid may escape as it is generated without the formation of hydrofractures.

a larger deviatoric stress, increase in fluid pressure is likely to lead to brittle failure. As a result, in the course of a dehydration reaction, permeability may increase by hydraulic fracturing if the rock is subject to a significant deviatoric stress, but if the deviatoric stress is small, progressive inflation of the pore network as the effective pressure decreases may create a sufficient increase in permeability to permit fluid loss without failure. It is notable that metamorphic veins are typically sparse or absent in contact metamorphism (low deviatoric stress) but are common in deformed metamorphic rocks, implying that, despite the lower heating rate in regional metamorphism, the rising fluid pressure often leads to fracturing before the matrix permeability can rise to allow escape of fluid as it is generated.

The oxygen isotopic composition of metamorphic quartz veins typically reflects that of the immediate host rock, and for this reason, and because metasomatic vein margins are generally absent, it is likely that such veins are predominantly the result of segregation of quartz from wall rocks by repeated inflation and deflation of fractures rather than simple precipitation from migrating fluid that was released elsewhere. Veins may mark fractures that allowed fluid to be redistributed within rock units in response to increments of deformation (Yardley 2009).

The role of reaction rate in controlling permeability

Reactions that release volatiles are strongly endothermic, and so overall the rate of dehydration is driven by the rate of heat input (Yardley 1986; Hanson 1992). If a reaction must be overstepped before products nucleate, then it can proceed more quickly than heat is supplied, by drawing down temperature

towards the equilibrium value, while if a reaction fails to proceed as rapidly as latent heat is supplied, the temperature of the rock mass will continue to rise, increasing the degree of overstepping and thereby, for near-equilibrium conditions, increasing the reaction rate. Thus, heat supply must determine the overall rate of fluid production in a suite of rocks experiencing prograde metamorphism although locally changes in reaction rate are possible to reflect local conditions. Since heat supply dictates the rate at which dehydration reactions proceed, it also determines permeability. Several studies have attempted to estimate the permeability of pelitic rocks undergoing metamorphism, based on the assumption that it is just sufficient to allow fluid to escape at the rate it is generated for a specific heat flow (Hanson 1992; Yardley 1986). The results obtained using this methodology are extremely low, even for contact metamorphism, around $10^{-18}\,\mathrm{m}^2$ or less (Connolly 2010). These values are within the ranges of shales and mudrocks, and of laboratory measurements of crystalline rocks (Brace 1980). In most cases, loss of fluid pressure also increases the degree of overstepping of devolatilization reactions and so their rates can transiently increase as a result of uplift and erosion, lowering the local temperature and restoring high fluid pressure.

Of course, this approach is based on a large-scale approximation, but it has some interesting implications. First, most metamorphosed sedimentary rocks have similar heat capacities, but they may vary significantly in their contents of reactive hydrous minerals. Dehydration reactions being endothermic, reactive rocks require additional heat to change their temperature by a given amount than an unreactive rock would require to be heated by the same amount. As a result, the heat required to raise the temperature of a siliciclastic sediment undergoing metamorphism depends on the amount of clay minerals and carbonates initially present, that is, the occurrence of devolatilization reactions effectively increases the heat capacity of the more reactive layers. It is possible that local temperature differences will develop between reactive and unreactive layers, leading to lateral conduction of heat into the reactive units. How significant this effect might be will depend on the thickness and relative abundance of the reactive and unreactive layers and their orientation relative to the overall direction of heat transport.

Second, although the heating rate of reactive layers such as pelite will tend to be slower than that of less reactive interbeds, the rate of fluid release from pelite is faster. It follows that, during prograde metamorphism, pelitic schists must have a higher transient permeability than interbedded rocks that were originally sandstones. The reactions in pelites are typically continuous, that is, they take place over a range of P–T conditions with reactants and products coexisting but changing composition as the reaction proceeds, and so progressive fluid release takes place over an extended temperature range. As far as siliciclastic rocks are concerned, however, the effect of composition on permeability is probably not very large. The difference in original clay content between mudrock and very pure sandstone is less than two orders of magnitude. This would result in a maximum of two log units difference in permeability as a result of metamorphic dehydration reactions, which is quite modest compared with the likely original difference in permeability of the sediments before metamorphism and suggests that most siliciclastic rocks undergoing regional metamorphism probably have a permeability close to 10^{-20} m^2, with slightly higher values possible for reactive rocks heated rapidly, but lower value for less reactive lithologies and slow heating rates.

The previous discussion has not considered the effect of different rock layers in directing flow, but the importance of lithology in controlling prograde metamorphic fluid flow has been appreciated for many years. Dynamic contrasts in permeability leading to layer-parallel flow have been demonstrated in regional metamorphism by Oliver (1995, 1996), Skelton *et al.* (1995, 2000) and Yardley et al. (1991a), among others. The role of decarbonation reactions in establishing layer-parallel flow was also modelled by Balashov & Yardley (1998). At shallower levels, in contact aureoles, there is similar considerable evidence for layer-parallel flow associated with the presence of carbonate beds (e.g. Baumgartner *et al.* 1997; Baumgartner & Valley 2001; Heinrich *et al.* 1995; Marchildon & Dipple 1998; Nabelek 2002; Roselle *et al.* 1999), although permeability-creating reactions may require the introduction of fluid from nearby sources (Bucher-Nurminen 1981). In an overpressured system, fluid can be expelled along convoluted paths as it moves towards a region of lower hydraulic head, exploiting more permeable pathways, as these studies have documented. Layer-parallel flow is likely to be ubiquitous during prograde metamorphism, with breakouts between layers being localised by structural and/or petrological features.

Although fluid flow considered on a large scale can be treated as steady state, it is perhaps likely that permeability and fluid flow may be episodic for any specific rock volume. The concept of porosity waves, developed for understanding large-scale metamorphic fluid migration by Connolly (1997) (see also Connolly 2010; Connolly & Podladchikov, 2015), predicts that buoyant fluid will be released in waves rising through the overlying rock mass, rather than simply moving through a uniform porous medium. This approach also emphasises the importance of flow controlled by lithology.

Extreme volatile loss during prograde metamorphism

While the overall behaviour of fluid in rocks undergoing metamorphism may be constrained by mineral–fluid equilibria, the input of heat and deformation, there are circumstances where particular combinations of rock types and fluid composition can result in distinctive local behaviour that may have a major impact on how fluid flows through the rock column. In particular, reactions that lead to volume change can result in the development of localised zones of concentrated fluid flow, especially since the overpressuring of a succession undergoing progressive metamorphism means that it is always susceptible to leakage if a pathway can be found.

A few rocks encountered in prograde metamorphic terranes show evidence for local, large-scale fluid movement under peak metamorphic conditions. There are some situations in which endothermic devolatilization reactions may take place much more rapidly than heat flow constraints would suggest, because they occur far from equilibrium and can proceed by drawing down temperature. Where this is the case, permeability must also be enhanced.

The clearest example of permeability enhancement during prograde metamorphism is skarn formation. Breakdown of carbonate minerals gives rise to significant transient porosity (Yardley & Lloyd 1989; Milsch *et al.* 2003) and this in turn leads to infiltration of water, giving a positive feedback to mixed volatile reactions by driving them further from equilibrium (Yardley et al., 1991a, 1991b; Yardley 2009). The development of monomineralic assemblages is indicative of extensive metasomatism by infiltrating fluids (Korzhinskii, 1970), while skarn textures themselves appear to document the transient enhancement of porosity, with oscillatory zoning patterns documenting euhedral and subhedral growth into vugs in rocks, which now have very little porosity remaining (e.g. Cann *et al.* 2015). Skarn beds that require infiltration of fluid to form must be part of a larger overall flow system. For a bed-like morphology, anomalously large layer-parallel flows may be fed by very modest flows perpendicular to layering, but the focussed flow along the bed must eventually break out and cross other rock types before dissipating. Indeed, it is likely that it is the opening of fractures through adjacent rock that triggers the episode of skarn formation, with enhanced permeability migrating upstream along the reactive lithology until starved of fluid. Yardley et al. (1991a, 1991b) presented evidence for such a flow system giving rise to coupled diopside-, olivine- or tremolite-bearing skarn beds in original dolomitic marbles and andradite–epidote skarn pipes in adjacent siliciclastic units.

The formation of skarns requires large fluxes of overpressured fluids to flow through a rock with which they are sufficiently out of equilibrium to drive metasomatic reactions. While it is not possible to calculate a permeability or a duration for the flow event, it is clear that the permeability of the reacting rock must be very much higher than that of the adjacent units, and the duration of skarn formation is sufficiently short that it occurs as a discrete event within the history of regional or contact metamorphism and deformation.

Another example of metamorphic reactions taking place more rapidly than would be the case if their rate was limited by heat input occurs where rapid uplift drives low-grade dehydration reactions. This may lead to the development of a permeable fracture network through the combined effects of reduction in lithostatic pressure by erosion and increase in fluid pressure by reaction. I argued earlier that the effect of reducing effective stress is to increase permeability and allow fluid to escape more rapidly. In the context of uplift, this means that if uplift and erosion are slow, fluid can leak out of an overpressured rock without any change in the permeability network. Indeed, once

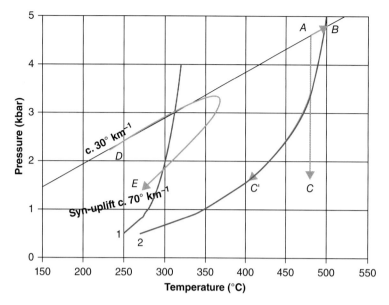

Fig. 22.3. *P–T* diagram contrasting the effects of rapid uplift with those of heating for a period of 4 Ma. Curves 1 and 2 represent generic dehydration equilibria. Path *A–B* shows the change in conditions along an average crustal gradient with a heating rate of 10° Ma⁻¹. Path *A–C* shows the effect of uplift at 2.5 mm a⁻¹ over the same time period, without heat loss or reaction, while *A–C'* is the path for the same uplift with temperature buffered by the progress of a dehydration reaction. These paths cross the dehydration reaction in a prograde sense. Path *D–E* is a reference path for slow uplift and cooling to show that under most conditions, dehydration reactions are crossed in a retrograde sense. (Yardley & Cleverley (2013).)

the rock begins to cool, fluid overpressure will rapidly reduce through retrograde reactions as described earlier. With rapid uplift, however, not only is there little cooling but confining pressure reduces through erosion at the same time that fluid pressure increases though dehydration reactions driven by the drop in confining pressure.

The significance of uplift for metamorphism is that it is a much more rapid process than conductive heating and so can result in faster changes in ΔG. Figure 22.3 illustrates this by contrasting the change in *P–T* conditions that arises due to uplift and erosion at a rate of 2.5 mm a⁻¹ with that which results from heating at 10°C Ma⁻¹ over the same time period. Unlike skarn formation, there is no positive feedback between the driving process and the creation of permeability, and the amount of reaction that can take place is limited by the extent to which reactions can drawdown the temperature of the rock mass before this brings the conditions back to equilibrium. Nevertheless, for appropriate mixtures of rock types, rapid uplift may drive metamorphism and uplift-driven metamorphism may be an important process in the formation of orogenic gold deposits (Yardley & Cleverley 2013).

Permeability accompanying hydration

For most of the time, the rocks of the middle to lower crust are experiencing retrograde conditions relative to those at which they formed. Retrograde metamorphism, including geothermal field metamorphism, presents different challenges for understanding permeability, as noted in the introduction to this chapter. Retrograde reactions are normally associated with structural features such as joints, faults or shear zones and so these must be the main pathways that determine permeability and allow ingress of fluid. In detail, the evolution of metamorphic fluids and their relation to metamorphism and deformation under retrograde conditions are both complex

and of considerable importance for understanding how the crust works.

Once rocks have cooled below their temperature of formation (assuming that there is not also a sufficient drop in pressure to restore equilibrium), pore water is consumed by retrograde reactions and, from being pervasively wet, rocks become very dry. Not only are the metamorphosed rocks now a sink, rather than a source, for fluid, they also no longer contain a pervasive fluid phase. When fluid is able to infiltrate them it does so along discrete fractures and reacts with the immediate fracture walls rather than the bulk of the rock. Only if there is such extensive infiltration that the high-grade assemblages become completely hydrated can a fluid phase remain stable, because hydration reactions are rather fast under most crustal conditions (Yardley *et al.* 2010, 2014). The fluid moving through rocks undergoing retrograde metamorphism may be surface-derived rather than a product of deep processes but this proof of this, for alteration under mid to lower crustal conditions, remains elusive.

In the retrograde case, water–rock reactions will also have an impact on permeability, but the driver for the initial creation of permeability is likely to be deformation rather than reaction. This has been argued for the serpentinization of mantle rocks by Ranero *et al.* (2003). Since fluid accesses the rock along fractures and shear zones, the permeability reflects the rate at which these form and the wider context of regional deformation. A further factor under mid to lower crustal conditions is that the change in fluid regime as rocks begin to cool also has enormous consequences for rock strength (Figure 22.4). Water weakening is pervasive during prograde metamorphism from low greenschist facies conditions but must be much more limited during cooling since free water is no longer present (Yardley & Baumgartner 2007). In summary, after peak metamorphism, rocks evolve from being pervasively wet and weak to being dry and strong.

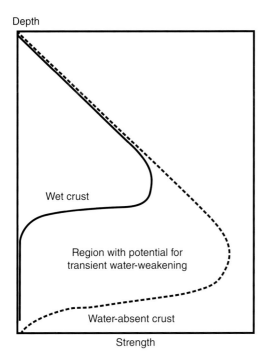

Fig. 22.4. Schematic representation of the variation in the strength of the crust with depth (Yardley & Baumgartner (2007)). The solid line represents water-saturated rocks, the dashed line dry rocks. In much of the middle crust, dry rocks are brittle and can sustain large deviatoric stresses, but will yield in a ductile manner if water is introduced.

Simple hydration (or carbonation), replacing or partially replacing a higher grade assemblage by a lower grade one, generally results in an increase in solid volume, and this may result in a decrease in porosity and permeability. Since retrograde products often form low-density mats of platy or acicular crystals the effect on permeability may be marked after just a very small amount of reaction, and will create a negative feedback to the hydration reaction. However, precisely because retrograde reactions commonly take place under far-from-equilibrium conditions, in response to infiltration of fluid from an external source, the large ΔG of the hydration reactions means that mineral products may be able to do work on the surrounding grains and exert a 'force of crystallisation' on the grains around them. This effect has been documented in the work of Renard et al. (2000b), Jamtveit *et al.* (2009) and Ulven *et al.* (2014) for a range of examples of hydration in the upper crust.

At greater depths and higher temperatures, there is less evidence of permeability enhancement by crystal growth, but rocks become weaker in response to fluid infiltration as deformation mechanisms that require water (e.g. pressure solution) become effective. It was noted earlier that under mid-crustal conditions cooled high-grade rocks have very low water fugacities, are therefore very strong and may support large deviatoric stresses. They are very much weaker if water is introduced, even at constant temperature. Fusseis *et al.* (2006) and Fusseis & Handy (2008) have documented the development of a

greenschist facies shear zone and demonstrated that the onset of deformation was marked by the formation of microfractures, but these enhanced permeability and permitted introduction of water, as a result of which the cracked rocks became weaker and underwent ductile deformation with dynamic recrystallisation.

From these observations, we can conclude that throughout much of the crystalline crust, where peak metamorphic conditions have been passed, permeability is closely linked to deformation and at any particular site can vary greatly through time, although at a larger scale a continuum approach is probably valid. Deformation zones acquire permeability, as a result of which they become weaker and continue to concentrate strain. As the ductile stage of deformation continues, grain boundary processes may create a pumping effect, enhancing the permeability of the zone. However, for most high-grade host rocks, water will be consumed by hydration reactions (Yardley *et al.* 2014), the rocks will become stronger, and deformation may be intermittent.

Metasomatism and permeability

Metasomatism – the chemical alteration of rock by fluids – is a widespread marker of rocks that have experienced enhanced permeability, and many examples are from sedimentary basins (e.g. dolomitisation of limestone, albitisation of detrital feldspar) where the rocks are clearly permeable at the time of the alteration. Similarly, metasomatism is a ubiquitous feature of geothermal field alteration. On the other hand, most metasediments have bulk compositions remarkably similar to their likely precursors; after allowing for loss of water, many analyses of high-grade pelites reveal compositions close to those of illite clays. The localised nature of any metasomatism, together with the evidence for fluid overpressuring, suggests that metamorphism is mainly accompanied by irreversible loss of fluid, which probably becomes focussed into structurally or lithologically controlled pathways, rather than by convective recirculation of fluid.

In addition to these near-surface examples, many metamorphic rocks with long histories of lower crustal residence have experienced metasomatism at specific stages in their evolution. This is true of many high-temperature quartzofeldspathic crystalline rocks as well as the skarns discussed earlier. Often, metasomatism has been achieved predominantly by exchange of cations between an infiltrating fluid or melt and pre-existing feldspars, but sometimes more extensive alteration and recrystallisation have taken place. Modern studies increasingly show that metasomatism has affected such rocks as one or more discrete events, but the exact nature and causes of those events, and in particular the sources of fluids, are often unclear. Engvik *et al.* (2014) have reviewed a wide range of examples of albitisation in southern Norway, while Oliver *et al.* (2008) and Rubenach (2013) discuss metasomatic events in the basement of the Mount Isa Inlier, also widely associated with albitisation.

Despite the uncertainties, it seems likely that metasomatism of basement rocks is the result of one or more discrete events, rather than background processes affecting the crust through an extended period of time. For example, the studies in Norway and Queensland document strong structural controls on the development of metasomatic albite, so fluid flow was dominated by fractures and shear zones reflecting specific tectonic regimes. It is often assumed that metasomatic events in basement rocks reflect deep crustal processes but this is unlikely to always be the case. For example, in some cases, there are albitised breccias, suggestive of a relatively shallow environment. Some metasomatic albitisation appears to be associated with contemporaneous igneous activity, and association with minerals such as clinopyroxene supports a high-T origin. However, other examples of albitisation appear to be retrograde (e.g. Munz et al. 1995) and to be associated with brines penetrating a fracturing basement from an overlying sedimentary basin (Munz et al. 2002; Gleeson et al. 2003). Infiltrating shallow, retrograde fluids are also implicated in the formation of retrograde K-feldspar (e.g. Holness 2003). There is a vast range in the K/Na ratios of modern oilfield formation brines, and while some have very low values, reflecting dissolution of halite, and will drive albitisation of any feldspars that they encounter (in many oilfield reservoir sandstones, albite is the only feldspar present), others are K-rich bittern brines and would drive alteration of other feldspars to K-feldspar.

Based on the diversity of ways in which high-grade igneous and metamorphic rocks can undergo retrograde hydrothermal alteration, it seems that there are several mechanisms by which fluid can access them. Relatively high-grade hydration may be associated with expulsion of fluids from crystallising magmas, whether locally or externally derived, while under moderate crustal temperatures hydration occurs during orogenesis, when basement slices are interleaved with younger sediments undergoing prograde metamorphism, or at a later stage as fluids penetrate fracturing, cooling crust. At temperatures around 300°C and lower, veining and hydration of fracture walls can in some cases be linked directly to the introduction of fluids derived from overlying sedimentary basins (Gleeson et al. 2003), and further introduction of surface-derived waters continues through exhumation (e.g. Milodowski et al., 1998) to weathering. It appears that for much of this extended history, rocks may be sufficiently impermeable that the amount of fluid infiltration is negligible, but during specific tectonic events brittle deformation creates transient enhanced permeability. While at a large scale, the development of fractures may be the key step in initiating hydration, work on the specific processes involved makes it clear that the reactions by which mineral replacement takes place may themselves play a central role in enhancing permeability. Putnis & Johns (2010) recently reviewed the development of porosity during mineral replacement, while the nature of permeability enhancement during deep granite alteration has been explored by Sardini et al. (1997).

THE CONTRIBUTION OF METAMORPHISM TO THE PERMEABILITY STRUCTURE OF THE CRUST

The permeability of the deep crust is clearly a very difficult topic to investigate because not only is direct access not possible but also theoretical approaches to investigating permeability often require assumptions that will in turn dictate the result.

There is a fundamental difference between regions of the deep crust, which are actively undergoing metamorphic reactions that lead to release of fluid, and regions of crust that have cooled below their peak temperatures and are a sink for fluid. While water is being released, it forms a pervasive pore fluid and pore-fluid pressure approaches lithostatic pressure; as a result, permeability must be very low (generally $\leq 10^{-18}\,\mathrm{m}^2$). In most rock types, for most of prograde metamorphism, permeability is only sufficient to allow fluid to escape as it is generated, but there are rock types and specific settings where rapid devolatilization is possible and rocks can acquire enhanced permeability through reaction or fracturing. Nevertheless, even when permeability is enhanced it allows overpressured fluids to escape but offers limited opportunity for recirculation. Despite being weak, metamorphosing rocks are also potentially brittle if subjected to a deviatoric stress, and so fluid loss may be by pervasive flow through a pore network or may exploit fractures.

In contrast, cooled rocks that make up typical crystalline basement are intrinsically dry and therefore strong. They can develop transient permeability in response to deformation, and under most crustal conditions the initial deformation is brittle. However, once fractures develop, they permit ingress of water-rich fluids and this causes the fractured rocks to become weaker and to deform in a ductile manner. At a local scale, the development of permeability in basement rocks is transient, but at a larger scale it may be a continuous process throughout the life of a deformation zone. There is, however, no evidence to suggest that there is pervasive background infiltration of fluid through the deep crust. While some examples of structurally controlled permeability enhancement develop at high temperatures and pressures this is not always the case, and some of the metasomatic features that mark permeability enhancement almost certainly develop late in the exhumation history of basement rocks, especially when they are overlain by overpressured sedimentary basins.

ACKNOWLEDGEMENTS

I am indebted to the editors for the opportunity to contribute to this volume and also to Kurt Bucher and Philipp Weiss for their patient and insightful comments on an earlier version.

CHAPTER 23

An analytical solution for solitary porosity waves: dynamic permeability and fluidization of nonlinear viscous and viscoplastic rock

JAMES A. D. CONNOLLY[1] AND YURY Y. PODLADCHIKOV[2]

[1] *Department of Earth Sciences, Swiss Federal Institute of Technology, Zurich, Switzerland;* [2] *Earth Sciences Department, University of Lausanne, Lausanne, Switzerland*

ABSTRACT

Porosity waves are a mechanism by which fluid generated by devolatilization and melting, or trapped during sedimentation, may be expelled from ductile rocks. The waves correspond to a steady-state solution to the coupled hydraulic and rheologic equations that govern the flow of the fluid through the matrix and matrix deformation. This chapter presents an intuitive analytical formulation of this solution in one dimension that is general with respect to the constitutive relations used to define the viscous matrix rheology and permeability. This generality allows for the effects of nonlinear viscous matrix rheology and disaggregation. The solution combines the porosity dependence of the rheology and permeability in a single hydromechanical potential as a function of material properties and wave velocity. With the ansatz that there is a local balance between fluid production and transport, the solution permits the prediction of dynamic variations in permeability and pressure necessary to accommodate fluid production. The solution is used to construct a phase diagram that defines the conditions for smooth pervasive flow, wave-propagated flow, and matrix fluidization (disaggregation). The viscous porosity wave mechanism requires negative effective pressure to open the porosity in the leading half of a wave. In nature, negative effective pressure may induce hydrofracture, resulting in a viscoplastic compaction rheology. The tube-like porosity waves that form in a matrix with this rheology channelize fluid expulsion and are predicted by geometric argumentation from the one-dimensional viscous solitary wave solution.

Key words: analytic solution, dynamic permeability, fluidization, lower crust, nonlinear viscous, porosity waves

INTRODUCTION

Many geological processes involve the expulsion of pervasively distributed fluids, as in the case of fluids trapped during sedimentation or fluids generated by partial melting and metamorphic devolatilization. Given the high elastic strength characteristic of rocks, efficient fluid expulsion requires irreversible deformation by time-dependent (viscous) or time-independent (plastic) mechanisms (Neuzil 2003; Gueguen *et al.* 2004). Models for ductile plastic compaction (cataclasis) reproduce the near-surface porosity profiles of sedimentary basins (Shi & Wang 1986; Audet & Fowler 1992), but creep is regarded as essential to deeper compaction processes (McKenzie 1987; Birchwood & Turcotte 1994; Fowler & Yang 1999). For such processes, permeability is a dynamic property that is determined by the interaction between rheology and the inherent gravitational instability of the intermingling of rock and fluids with different densities. This interaction may give rise to a hydrologic regime in which flow is accomplished by self-propagating domains of fluid-filled porosity (Fowler 1984; Richter & McKenzie 1984; Scott & Stevenson 1984). These domains, or porosity waves, correspond to steady-state solutions, that is, solutions in which the waves propagate with unchanging form, of the equations governing fluid flow through a viscous matrix. The intent of this study is to develop and explore an intuitive analytical solution for these waves that is general with respect to the constitutive laws chosen to characterize the viscous rheology and permeability of the rock matrix.

There are numerous formulations of the equations governing compaction of a two-phase viscous system consisting of a porous rock matrix saturated with a less viscous interstitial fluid in the geological literature (McKenzie 1984; Scott & Stevenson 1984; Bercovici *et al.* 2001). These formulations differ primarily in the choices of constitutive relations and independent variables. Analytical solutions for solitary waves that develop according

Crustal Permeability, First Edition. Edited by Tom Gleeson and Steven E. Ingebritsen.
© 2017 John Wiley & Sons, Ltd. Published 2017 by John Wiley & Sons, Ltd.
Companion Websites: www.wiley.com/go/gleeson/crustalpermeability/
     http://crustalpermeability.weebly.com/

to these formulations have been published (Barcilon & Richter 1986; Rabinowicz *et al.* 2002; Richard *et al.* 2012), but assume linear viscous behavior. This limitation is potentially important in the context of fluid flow in the lower crust because the viscous response of crustal rocks is expected to be nonlinear (Kohlstedt *et al.* 1995; Ranalli 1995). A nonlinear viscous formulation is of broad interest because it defines the length scale on which lower crustal fluid flow patterns may deviate from gravitationally controlled flow. The compaction length is also important in that it defines the spatial scale for porosity. Thus, although porosity conjures up an image of grain-scale structures, it may apply to substantially larger features, such as fractures, provided these features are hydraulically connected and small in comparison with the compaction length.

Shortly after the recognition of porosity waves as a potential fluid transport mechanism in geologic systems, it was shown that the planar, sill-like, waves predicted from one-dimensional (1D) formulations of the governing equations were unstable with respect to spherical waves in two-dimensional (2D) and three-dimensional (3D) space (Scott & Stevenson 1986; Barcilon & Lovera 1989; Wiggins & Spiegelman 1995). Despite this result, the present analysis is restricted to the 1D case because it provides a lower limit on the efficacy of compaction-driven fluid expulsion and because the characteristics of the 1D waves converge rapidly with those of 3D waves for the geologically interesting case of large amplitude waves (Connolly & Podladchikov 2007). With this restriction, there are three compaction-driven flow regimes that may arise as a consequence of a perturbation to the flux of an initial regime of steady flow through a matrix with uniform porosity (Fig. 23.1). If the perturbation is small, the steady-state solution is a periodic wave that degrades to a uniform increased porosity once the perturbation is eliminated. For larger perturbations, the steady state is a solitary wave. Once nucleated, the solitary wave is nondissipative and independent of its source. Thus, the solitary wave solution defines a steady-state regime in which flow is accomplished by self-propagating waves. Because the solitary wave amplitude is proportional to the intensity of perturbation, large perturbations may cause the matrix to disaggregate to a fluidized suspension. The analytic solution outlined here is used to define the conditions for these regimes.

The solitary wave regime is one where the fluid flows through a coherent solid matrix, whereas in the fluidized regime the solid is suspended within, and carried by, the fluid, as in a granitic magma so that the suspension behaves as a single phase that is transported through dikes or as diapirs (Vigneresse *et al.* 1996). Although peripheral to the scope of this study, the generation of granitic melts by partial melting of the lower crust is a prominent example of fluidization in ductile rock. Unless melting is so extensive that it forms a magmatic suspension directly, a segregation mechanism is required to amplify melt fractions to the level required for transport by dikes or diapirs. Given the common occurrence of oriented vein-like segregations (e.g., Brown 2010), attention in the compaction literature has focused on the role of shear-enhanced melt segregation (Stevenson 1989; Holtzman *et al.* 2003; Rabinowicz & Vigneresse 2004). While melt segregation is not explored in this study, the mechanism of fluidization discussed here is distinct in that low melt fractions can be amplified to a suspension even under isostatic conditions. Such a mechanism may be relevant in the formation of large-scale diatexite migmatites, which record a wholesale evolution from unmelted source rock to granitic magma (Sawyer 1998; Milord *et al.* 2001).

The mechanism responsible for porosity waves in a viscous matrix is implicit in the conventional view of the mechanics responsible for compaction profiles in active sedimentary basins (Hunt 1990; Japsen *et al.* 2011) and the partially molten region beneath mid-ocean ridges (Forsyth *et al.* 1998). In both of these settings, which span the physical conditions of the lower crust, compaction is thought to maintain near-eustatic porosity–depth profiles in rocks that are moving relative to the Earth's surface. In the case of sedimentary basins, the rock matrix moves downward relative to surface due to burial, whereas at mid-ocean ridges, the rock matrix rises toward the surface as a consequence of mantle upwelling. The distinction between these scenarios and that of a porosity wave is no more than a matter of reference frame, in that in the former porosity is eustatic and the rock matrix moves, while in the latter, the rock matrix is largely stationary and the porosity moves. Indeed, mechanical models that explain eustatic porosity in sedimentary basins (Fowler & Yang 1999; Connolly & Podladchikov 2000) and at mid-ocean ridges (Katz 2008) differ trivially from the simple viscous formulation employed here. Specifically, for sedimentary basins a viscoplastic rheology is introduced to account for near-surface compaction, and for mid-ocean ridges, the model is modified to account for fluid production. The existence of porosity waves in a viscous matrix has also been demonstrated experimentally by mechanical analog (Olson & Christensen 1986; Scott *et al.* 1986; Helfrich & Whitehead 1990). Thus, suggestions that porosity waves act as agents for compartmentalization and fluid migration in sedimentary basins (McKenzie 1987; Connolly & Podladchikov 2000; Appold & Nunn 2002) and the lower crust (Suetnova *et al.* 1994; Connolly 1997; Gliko *et al.* 1999; Tian & Ague 2014) are less exotic than they might seem at first sight. We make no attempt to make the case for the role of porosity waves in the lower crust in this study; rather, the reader is referred to the aforementioned works and recent reviews (Connolly & Podladchikov 2013; Ague 2014) that consider the relevance of the model and the hydraulic properties of the crust in greater detail.

In addition to neglecting 3D effects, the viscous solitary wave solution neglects the potential roles of plasticity (e.g., brittle failure), elasticity (e.g., fluid compressibility and poroelasticity), and thermal activation. To a first approximation, many of these effects can be inferred from the 1D viscous model as addressed in the "Discussion" Section of this chapter. Elastic effects are the exception. The poroelastic limit, relevant to fluid flow in the upper crust, admits a solitary porosity wave solution that is manifest by fluid pressure surges (Rice 1992). This solitary

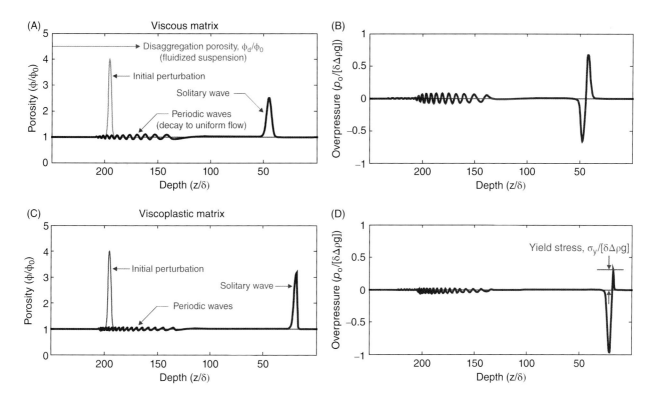

Fig. 23.1. Numerically simulated porosity wave evolution from a region of increased porosity within an otherwise uniform flow regime through viscous and viscoplastic porous media. (A) Porosity versus depth for the viscous case. The first wave to initiate from the flow perturbation (i.e., the region of elevated porosity) is a solitary wave that propagates above the background porosity in the same direction as fluid flow through the unperturbed matrix. The solitary wave is nondissipative in the steady-state limit; thus, it propagates infinite distance with essentially unchanging form. Due to transient effects, a periodic wave train initiates behind the solitary wave; the periodic wave train propagates both in and against the direction of fluid flow through the unperturbed matrix. The wave train corresponds to a periodic steady state in which the porosity oscillates about the background level. Porosity wave velocities are proportional to amplitude; thus, with time the solitary wave becomes isolated from the periodic wave train, which has no significant excess volume and degrades to the original uniform flow regime as it spreads. A narrow region of increased porosity was chosen for the initial conditions to emphasize the periodic solution. This choice leads to solitary waves that have lower porosity than the source region. Wider source regions tend to generate waves with porosities that are higher than in the source region; if these porosities exceed the disaggregation porosity (indicated schematically), the matrix disaggregates to a fluidized suspension. This range of porosity wave behavior can also be induced by perturbing the background fluid flux, as might occur as a consequence of devolatilization (melting). (B) Fluid overpressure (negative effective pressure) versus depth for the viscous case, the porosity dependence of the matrix permeability causes fluid pressure anomalies that are responsible for dilating and compacting the porosity during the passage of a wave. Although fluid flow from low to high overpressure may seem to contradict Darcy's law, if the overpressures are converted to hydraulic head, it is apparent that the direction of fluid flow in porosity waves is consistent with Darcy's law. (C) Porosity and (D) overpressure profiles for a viscoplastic scenario in which plasticity is manifest by hydrofracture when overpressure exceeds the brittle yield stress. This rheology affects wave symmetry, but does not fundamentally change porosity wave behavior because the rate of fluid expulsion remains limited by viscous compaction. In the numerical simulation, the effect of the viscoplastic rheology on the periodic waves with overpressures below the yield stress is an artifact of the reduced effective shear viscosity used to simulate brittle failure. Porosity (ϕ), depth (z), overpressure (p_0), and time are scaled relative to the background porosity (ϕ_0), viscous compaction length (δ), characteristic pressure ($\delta|\Delta\rho g|$), and the speed of fluid flow through the unperturbed matrix ($|v_0|$) discussed later in the text. The 1D volume of the initial perturbation is the same (8.862 $\delta\phi_0$) in both simulations, and the transient profiles (blue curves) are for the same model time (50 $\delta/|v_0|$). The numerical simulations were obtained by finite difference methods for the small porosity formulation (Appendix) with $m_\sigma = 1$, $n_\phi = 3$, and $n_\sigma = 1$. (See color plate section for the color representation of this figure.)

solution, which is consistent with a variety of upper crustal phenomena (Revil & Cathles 2002; Miller *et al.* 2004; Joshi *et al.* 2012), contrasts with the viscous case in that it is dissipative and does not require supra-lithostatic fluid pressures. Viscoelastic compaction formulations show that there is a continuum of periodic wave solutions between the viscous and elastic solitary wave limits (Connolly & Podladchikov 1998; Chauveau & Kaminski 2008). However, in numerical simulations, thermal activation of the viscous rheology leads to a rapid variation between a lower crustal regime in which viscoelastic porosity

wave solutions show no appreciable elastic character (Connolly 1997) and an upper crustal regime lacking any significant viscous character (Connolly & Podladchikov 1998). These results are taken as justification for the neglect of elastic phenomena in the present treatment of lower crustal fluid expulsion. In this regard, it is important to distinguish fluid expulsion from fluid flow as, particularly in the noncompacting limit, thermoelastic expansivity of the fluid may create pressure gradients responsible for fluid circulation (Hanson 1997; Nabelek 2009; Staude *et al.* 2009).

MATHEMATICAL FORMULATION

Darcian flow of an incompressible fluid through a viscous matrix composed of incompressible solid grains is considered here largely following the formulation of Scott & Stevenson (1984). Although the solid and fluid components are incompressible, the matrix is compressible because fluid may be expelled from the pore volume. Conservation of solid and fluid mass requires

$$\frac{\partial(1-\phi)}{\partial t} + \nabla \cdot ((1-\phi)\mathbf{v}_s) = 0 \qquad (23.1)$$

and

$$\frac{\partial \phi}{\partial t} + \nabla \cdot (\phi \mathbf{v}_f) = 0, \qquad (23.2)$$

where ϕ is porosity and subscripts f and s distinguish the velocities, $\mathbf{v}$, of the fluid and solid (see Table 23.1 for notation). From Darcy's law, the force balance between the solid matrix and fluid is

$$\phi(\mathbf{v}_f - \mathbf{v}_s) = -\frac{k}{\eta_f}(\nabla p_f - \rho_f \mathbf{g}\mathbf{u}_z), \qquad (23.3)$$

where k is the hydraulic permeability of the solid matrix, an unspecified function of porosity; ρ_f and η_f are the density and shear viscosity of the fluid, respectively; and $\mathbf{u}_z$ is a downward-directed unit vector. Identifying the mean stress $\bar{\sigma}$ as the vertical load

$$\bar{\sigma} = \int_0^z [(1-\phi)\rho_s + \phi \rho_f] g \mathbf{u}_z dz. \qquad (23.4)$$

Thus, in terms of the fluid overpressure

$$p_o = p_f - \bar{\sigma}, \qquad (23.5)$$

(i.e., negative effective pressure) Eq. 23.3 is

$$\phi(\mathbf{v}_f - \mathbf{v}_s) = -\frac{k}{\eta_f}(\nabla p_o + (1-\phi)\Delta\rho g \mathbf{u}_z), \qquad (23.6)$$

where $\Delta\rho = \rho_s - \rho_f$. The divergence of the total volumetric flux of matter is the sum of Eqs 23.1 and 23.2:

$$\nabla \cdot (\mathbf{v}_s + \phi(\mathbf{v}_f - \mathbf{v}_s)) = 0, \qquad (23.7)$$

and substituting Eq. 23.6 into Eq. 23.7 gives

$$\nabla \cdot \left(\mathbf{v}_s - \frac{k}{\eta_f}(\nabla p_o + (1-\phi)\Delta\rho g \mathbf{u}_z)\right) = 0. \qquad (23.8)$$

It is assumed that the bulk viscosity of a pure phase is infinite, an assumption necessary to assure that the individual phases do not have time-dependent compressibility (Nye

Table 23.1 Frequently used symbols

Symbol	Meaning
A; A_ϕ	Viscous flow coefficient, Eq. 23.9; wave amplitude, ϕ_{max}/ϕ_0
a_ϕ	Permeability function geometric factor, Eq. 23.17
a_σ	Compaction rate function geometric factor, Eq. 23.19
b_ϕ	Permeability function solidity exponent, Eq. 23.17
f_ϕ	Compaction rate function, Eq. 23.19
f_1; f_2	Hydraulic function, Eq. 23.36; rheological function porosity dependence, Eq. 23.37
H; ΔH	Hydraulic potential, Eqs 23.39, 23.54, 23.62, and 23.76; $H(\phi) - H(\phi_0)$
k; k_0	Permeability, Eq. 23.17; background value
m_σ	Compaction rate function porosity exponent, Eq. 23.19
n_ϕ	Permeability function porosity exponent, Eq. 23.17
n_σ	Viscous flow law stress exponent, Eq. 23.9
p_o; p_f; p	Fluid overpressure, $p_f - p$; fluid pressure; total pressure ($\bar{\sigma}$)
q; q_0	Fluid flux; background value
$\bar{q}_e$; q_s	Time-averaged excess flux, Eq. 23.61; 1D fluid production rate
Q; Q_p	Fluid transport rate for a spherical viscous wave, Eq. 23.63; 3D fluid production rate
Q_p	Fluid transport rate for a 3D viscoplastic wave, Eq. 23.65
V_e	1D wave excess volume, Eq. 23.60
v_0; v_ϕ	1D Darcy fluid velocity at $\phi = \phi_0$, $p_o = 0$, Eq. 23.20; 1D wave velocity, Eqs 23.55 and 23.57,
v_ϕ^{crit}	1D solitary wave critical velocity, Eq. 23.43
$\mathbf{v}$	3D velocity
z	Depth coordinate, positive downward
δ	Viscous compaction length scale, Eq. 23.51
$\Delta\rho$; $\Delta\sigma$	$\rho_s - \rho_f$; differential stress
η_s; η_f	Solid shear viscosity; fluid shear viscosity
λ, λ_p	Viscous wavelength; viscoplastic wavelength
ϕ; ϕ_0; ϕ_d	Hydraulically connected porosity; background value; value at disaggregation
ϕ_1; ϕ_{max}	Focal point porosity, a real root of $f_1 = 0$; maximum wave porosity
ρ_s; ρ_f	Solid density; fluid density
$\bar{\sigma}$; σ_y	Mean stress (p); failure stress
τ	Compaction timescale, $\delta/|v_0|$
∇, $\nabla \cdot$	Gradient, divergence
$f\|_{x=x_0}$	Value of a function f at $x = x_0$

1953); therefore, viscous compaction must be accomplished by grain-scale shear deformation that eliminates porosity. For solid grains that deform according to a power-law constitutive relation,

$$\dot{\epsilon} = A|\Delta\sigma|^{n_\sigma - 1}\Delta\sigma, \tag{23.9}$$

where $\dot{\epsilon}$ is the uniaxial strain rate in response to differential stress $\Delta\sigma$, n_σ is the stress exponent, and A is the coefficient of viscous flow; matrix rheology is then introduced through Terzaghi's effective stress principle as

$$\nabla \cdot v_s = f_\phi A |p_o^{n_\sigma - 1}| p_o. \tag{23.10}$$

where f_ϕ includes an unspecified porosity dependence and a geometric factor that relates the uniaxial strain rate of the solid to the bulk strain rate of the matrix. In the limit $n_\sigma \to 1$, the shear viscosity of the solid is $\eta_s = 1/(3A)$, and the bulk viscosity of the solid matrix is η_s/f_ϕ The divergence of the solid velocity is identical to the bulk strain rate of the matrix and related to the compaction rate by

$$\dot{\phi} \equiv -\frac{1}{\phi}\frac{d\phi}{dt} = \nabla \cdot v_s \frac{1 - \phi}{\phi}. \tag{23.11}$$

The power-law form of Eq. 23.9 precludes certain less common viscous constitutive relations; for example, the exponential form appropriate for low temperature plasticity (Kameyama *et al.* 1999), a completely general derivation, follows if the term $A|p_o^{n_\sigma - 1}|$ in Eq. 23.10 is replaced by a generic function of the magnitude of the overpressure.

Equations 23.1, 23.8, and 23.10 form a closed system of equations in the unknown quantities (ϕ), p_o, and v_s. To avoid the unnecessary complication associated with the use of vector notation for a 1D problem, in the remainder of this analysis vector quantities are represented by signed scalars and the gradient and divergence operators are replaced by $\partial/\partial z$.

The 1D steady state

For analytical purposes, the existence of a 1D solitary porosity wave solution is assumed in which the wave propagates with unchanging form and velocity through a matrix with an initial fluid-filled porosity ϕ_0 at zero overpressure. In a reference frame that travels with the wave, integration of Eq. 23.1 gives the solid velocity as

$$v_s = v_\infty \frac{1 - \phi_0}{1 - \phi}, \tag{23.12}$$

where $v_\infty(1 - \phi_0)$ is the solid flux at infinite distance from the wave. After substitution of Eq. 23.12, the integrated form of Eq. 23.8 can be rearranged to

$$\frac{\partial p_o}{\partial z} = \left(q_t - v_\infty \frac{1 - \phi_0}{1 - \phi}\right)\frac{\eta_f}{k} - (1 - \phi)\Delta\rho g, \tag{23.13}$$

where $q_t = \phi v_f + (1-\phi)v_s$ is the constant, total, volumetric flux of matter through the column, which evaluates in the limit $\phi \to \phi_0$ and $p_o \to 0$ as

$$q_t = v_\infty - (1 - \phi_0)\frac{k_0}{\eta_f}\Delta\rho g, \tag{23.14}$$

where k_0 is the permeability at ϕ_0. Using Eq. 23.14, Eq. 23.13 is rewritten

$$\frac{\partial p_o}{\partial z} = v_\infty \frac{\eta_f}{k}\frac{\phi - \phi_0}{1 - \phi} - \Delta\rho g\left(1 - \phi - (1 - \phi_0)\frac{k_0}{k}\right). \tag{23.15}$$

Likewise, after substitution of Eq. 23.12, Eq. 23.10 can be rearranged to

$$\frac{\partial \phi}{\partial z} = \frac{(1 - \phi)^2}{1 - \phi_0}\frac{f_\phi}{v_\infty}A|p_o|^{n_\sigma - 1}p_o. \tag{23.16}$$

For a given wave velocity, Eqs 23.15 and 23.16 form a closed system of two partial differential equations in two unknown functions (ϕ and p_o) of depth.

Constitutive relations and scales

Although general forms are retained where possible, to place the analysis in context, it is useful to specify possible constitutive relations for permeability and the porosity dependence, f_ϕ of the rheological constitutive relation (Eq. 23.10). To describe the variation in permeability due to compaction, the theoretical Carman–Kozeny porosity–permeability relationship (Carman 1939) is generalized as

$$k = a_\phi \frac{\phi^{n_\phi}}{(1 - \phi)^{b_\phi}}, \tag{23.17}$$

where a_ϕ is a grain-size-dependent material constant and the formal values of b_ϕ and n_ϕ, 2 and 3, respectively, imply that the first-order control on the porosity dependence of the permeability at small porosity is determined by n_ϕ. From analysis of *in situ* rock permeability, Neuzil (1994) shows that pore geometry and grain size give rise to variations in permeability that span eight orders of magnitude, but that porosity dependence is approximately cubic. A cubic dependence is predicted from theory irrespective of whether flow is intergranular or fracture controlled (Norton & Knapp 1977; Gavrilenko & Gueguen 1993). Accordingly, a cubic dependence, that is, $n_\phi = 3$, is considered to be most relevant. Higher exponents are observed in rocks where the degree of hydraulic connectivity varies strongly with porosity (Zhu *et al.* 1995, 1999). The solidity (i.e., $1 - \phi$) exponent b_ϕ is constrained by considering the settling of a single grain through a static fluid. In this case, both the effective pressure and its gradient vanish, and substitution of Eq. 23.17 into Eq. 23.6 yields the settling velocity

$$v_s = \frac{a_\phi}{\eta_f}\frac{\phi^{n_\phi - 1}}{(1 - \phi)^{b_\phi - 1}}\Delta\rho g, \tag{23.18}$$

which is an increasing function for all allowed values of porosity only if $1 \leq b_\phi < n_\phi$ and finite at $\phi \to 1$, as required by Stoke's law, only if $b_\phi = 1$.

The porosity dependence f_ϕ of the rheological equation (Eq. 23.10) must satisfy two physical constraints. In the limit $\phi \to 0$, f_ϕ must similarly vanish so that effective bulk viscosity becomes infinite to assure that the pure solid does not compact; and in the limit $\phi \to \phi_d$, f_ϕ must be infinite to assure that the solid and fluid pressure fields converge when the matrix has no cohesive strength, that is, when the matrix is fluidized. In detail, this transition is likely to be complex and material dependent, but theoretical and experimental considerations suggest that the transition occurs at $\phi_d \sim 20\%$ (Arzi 1978; Auer *et al.* 1981; Ashby 1988; Vigneresse *et al.* 1996). To account for these limits, the expressions of Wilkinson & Ashby (1975), derived explicitly for compaction by dislocation creep, are generalized here as

$$f_\phi = a_\sigma \phi^{m_\sigma} (1-\phi)/(1-\phi^{1/n_\sigma})^{n_\sigma} (\phi_d/|\phi_d - \phi|)^{n_\sigma - 1/2}, \quad (23.19)$$

where formally $m_\sigma = 1$ and, for spherical pores, $a_\sigma = n_\sigma^{-n_\sigma} (3/2)^{n_\sigma+1}$. For diffusion-controlled compaction, a_σ is strongly dependent on grain size and the exponent m_a varies between $1/2$ and $5/6$ (Ashby 1988). In practice, the numerous compaction formulations in the geological literature differ primarily in the choice of m_σ. For simplicity, early studies neglected the porosity dependence ($m_\sigma = 0$; McKenzie 1984; Barcilon & Richter 1986), whereas most recent formulations (Sumita *et al.* 1996; Connolly 1997; Bercovici *et al.* 2001) take $m_\sigma = 1$.

Many of the material properties relevant to geological compaction vary over orders of magnitude and/or are extraordinarily uncertain (Neuzil 2003); for this reason, no attempt is made to parameterize the relations used here. Rather, results are given relative to the background porosity, Darcian fluid velocity through the unperturbed matrix

$$v_0 = -\frac{k_0}{\eta_f \phi_0}(1-\phi_0)\Delta\rho g \quad (23.20)$$

or its speed $c_0 = |v_0|$, and the compaction length scale

$$\delta = \sqrt[n_\sigma+1]{\left(\frac{2}{3}\right)^{n_\sigma+1} \frac{k_0 n_\sigma^{n_\sigma}}{A\eta_f \phi_0^{m_\sigma}} |\Delta\rho g|^{1-n_\sigma}} \quad (23.21)$$

suggested by nondimensionalization of Eqs 23.15, 23.16 and 23.19 in the small porosity limit (Appendix), a limit that has the consequence that the solutions are independent of the absolute porosity. The scales for pressure, time, and fluid flux are then $p_0 = \delta|\Delta\rho g|$, $\tau = \delta/|v_0|$, and $q_0 = \phi_0 v_0$, respectively. Parameter ranges relevant to lower crustal fluid flow are reviewed elsewhere (Connolly & Podladchikov 2013; Ague 2014).

For a linear viscous matrix with shear viscosity $\eta_s = 1/(3A)$, Eq. 23.21 simplifies to

$$\delta = \sqrt{\frac{4}{3} \frac{\eta_s}{\phi_0^{m_\sigma}} \frac{k_0}{\eta_f}}, \quad (23.22)$$

which, accounting for differences in the formulation of the bulk viscosity of the matrix, is identical to the viscous compaction length of McKenzie (1984). In the linear viscous case with $m_\sigma = 1$, δ is the length scale over which the bulk strain rate would change by a factor of e, the base of the natural logarithm, for the characteristic overpressure gradient $\Delta\rho g$. It is, therefore, the length scale over which compaction processes would generate an e-fold variation in porosity. For the present formulation, the analytical significance of δ is less clear, but it emerges that δ remains a reasonable estimate of the length scale for an e-fold variation in porosity.

ANALYTICAL SOLUTION FOR THE 1D STEADY STATE

The steady-state wave solutions to the compaction Eqs 23.13 and 23.16 are best understood by analogy with the solutions to the equations of motion of an initially stationary ball on a frictionless 1D curved surface in response to gravitational acceleration. To exploit this analogy, the solution for an oscillating ball (Fig. 23.2) is recapitulated here.

The oscillating ball

The equations of motion for the ball are its acceleration due to gravity

$$\frac{\partial v}{\partial t} = -\frac{\partial h}{\partial x}g, \quad (23.23)$$

and the definition of velocity

$$\frac{\partial x}{\partial t} = v, \quad (23.24)$$

where x is the horizontal position of the ball, v is its velocity, and h is a shape function that describes the height of the surface as a function of x. Combining Eqs 23.23 and 23.24 to eliminate time

$$\frac{\partial v}{\partial x} = -\frac{g}{v}\frac{\partial h}{\partial x} \quad (23.25)$$

and rearranging Eq. 23.25 yields

$$0 = v dv + g\frac{\partial h}{\partial x}dx. \quad (23.26)$$

The indefinite integral of Eq. 23.26 defines a property

$$u \equiv \frac{v^2}{2} + gh, \quad (23.27)$$

the energy per unit mass, which is conserved by the ball. The solutions to Eqs 23.23 and 23.24 correspond to contours of u as a function of v and x, where, through Eq. 23.27 at $v = 0$, the initial height of the ball (Fig. 23.2A) defines the contour of interest for a particular problem (Fig. 23.2B). Closed contours (e.g., the red contour in Fig. 23.2B) correspond to a wave solution in which the ball oscillates between its initial position, x_i and

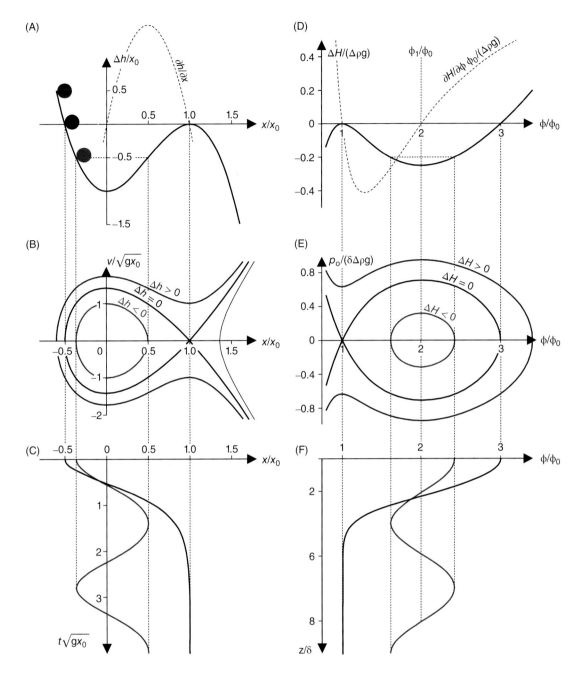

Fig. 23.2. Analogy of porosity waves with the oscillatory movement of a ball on a frictionless surface of variable height with a local maximum (i.e., a saddle point). In the analogy, the porosity wave properties H, ϕ, p_o, and z map to h, x, v, and t of the oscillating ball. If the ball is placed on the surface (e.g., the red ball in A) at a height below and to the left of the local maximum, the ball oscillates between points of equal height. In this case, the velocity–position trajectory of the ball defines a closed path around the focal point at $x = v = 0$ (e.g., the red curve in B) and its velocity–time (or position–time) trajectory defines a periodic wave train (e.g., the red curve in C). The location of the focal point ($x/x_0 = 0$) for the waves is dependent only on the shape of the surface, whereas the wave frequency is controlled by the gravitational acceleration. The solitary wave solution for the oscillating ball corresponds to the case that a ball (i.e., the black ball in A) is placed on the surface at the height of the saddle point. In this case, the ball would have no kinetic energy if it reached the saddle point; however, because the ball decelerates as it approaches the saddle point, the wave solution has an infinite period. If the ball is released from a height above the saddle point (i.e., the blue ball in A), it rolls continuously away from its initial position and there is no wave solution to the governing equations. The analogy of the oscillating ball to porosity wave solution is imperfect only in that the shape of function, or hydraulic potential, H of the porosity wave is dependent on wave velocity (Fig. 23.3), which is related to the intensity of the perturbation responsible for generating waves. For the velocity specified for H as illustrated in (D), the only wave solution capable of stably connecting the background porosity to a region of increased porosity is the solitary wave (i.e., the black trajectories in E and F); the periodic solutions (e.g., the red trajectories E and F) correspond to waves in which the porosity would oscillate about the focal point ($\phi/\phi_0 = 2$) between a porosity that is less than the maximum porosity of the solitary wave and greater than the background porosity. The solitary wave shape function illustrated in (D) is from the small porosity formulation of the compaction equations (Appendix) with $m_\sigma = 0$, $n_\phi = 3$, and $v_\phi/v_0 = 7$. (See color plate section for the color representation of this figure.)

a position of equal height at which its kinetic energy vanishes, and open contours (e.g., the blue contour in Fig. 23.2B) correspond to an aperiodic (i.e., non-wave) solution in which the ball rolls indefinitely away from its initial position. Because u is a monotonic function of v and is directly proportional to h, the focus of any closed contour must lie along the $v = 0$ axis and correspond to an extremum in h, that is, a real root *of $\partial h/\partial x = 0$*. From physical considerations, it is evident that the solution is only stable if this root is a minimum, that is, $\partial^2 h/\partial x^2 > 0$. A well-known implication of this solution is that shape function h entirely determines both the amplitude and stability of oscillation, while the gravitational constant controls the velocity of the ball and, therefore, the period of oscillation.

For a ball placed at position x_i with $v_i = 0$, its velocity as it accelerates from its initial position can be computed by rearranging the definite integral of Eq. 23.26 as

$$v = \sqrt{-2g\Delta h},\tag{23.28}$$

where $\Delta h = h - h_i$ and h_i is height of the ball at x_i. The time dependence of the solution is then recovered by substitution of Eq. 23.28 into Eq. 23.24 and inverting the result to obtain

$$t = \frac{1}{\sqrt{2g}}\int_{x_i}^{x}\frac{dx}{\sqrt{\Delta h}}.\tag{23.29}$$

For oscillatory solutions, Eq. 23.29 gives the time dependence of the solution for half the period of the oscillation. Thus, the oscillatory solution corresponds to a periodic wave (e.g., the red curve in Fig. 23.2C) in position (or velocity) as a function of time.

To quantify the preceding discussion, consider an arbitrarily chosen shape function such that

$$\frac{\partial h}{\partial x} = \frac{6}{x_0^2}x(x_0 - x),\tag{23.30}$$

which integrates to $h = x^2(3x_0 - 2x)/x_0^2$. The roots of Eq. 23.30 define the local extrema of h; thus, the surface has extrema at the structural root $x = 0$, at which $h = 0$, and the general root $x = x_0$, at which $h = h_0 = x_0$. Restricting attention to the case $x_0 < 0$, for which the general root is a maximum, the structural root $x = 0$ is the focus of all possible wave solutions and the general root x_0 defines the maximum value of x_i for which these solutions are possible (Fig. 23.2A). The minimum value of x_i, that is, $-x_0/2$, at which a wave solution is possible is obtained by solving $h = h_0$ for $x_i < x_0$. The closed contour of u as a function of position and velocity that demarcates the boundary between periodic and aperiodic solutions corresponds to a solitary solution. The existence of such a solution requires that h has at least two extrema. In the present example, the solution corresponds to the portion of the contour of u that emanates from the saddle point located by the structural root x_0 at $v = 0$ (the black contour in Fig. 23.2B at $x/x_0 < 1$). In distinction to the periodic solutions, for the solitary wave

solution, after the ball is released it does not return to its initial position, but rather comes to rest at the saddle point (at $x/x_0 = 1$), where its height is identical to its initial height. The physical reason for this behavior is that both the acceleration of the ball and its kinetic energy vanish at the saddle point. This trajectory (the black curve in Fig. 23.2C) is half the solitary wave solution; the complete solution would be obtained if the motion of the ball initiated from the saddle point. The absence of both kinetic energy and acceleration at the saddle point strictly precludes the occurrence of the complete solitary solution; however, the negative curvature of the surface at the saddle point has the consequence that a ball placed at the saddle point would be unstable with respect to infinitesimally small perturbations.

There are two types of solitary solution distinguished on the basis of whether the period of the solution is finite or infinite. The origin of these solutions can be understood by a thought experiment in which the initial conditions are chosen to coincide exactly with the conditions at which the trajectory of the solitary solution crosses the $v = 0$ axis, that is, in the present example at $x/x_0 = -1/2$ (Fig. 23.2A). As the ball has no kinetic energy and is at height $h = h_0$, the ball has exactly the energy necessary to reach the saddle point x_0, but because the acceleration of the ball becomes vanishingly small as the saddle point is approached, the time required for the ball to reach the saddle point may be infinite. In the present example, it can be verified by the analytic integration of Eq. 23.29 that the time required for the ball to reach the saddle point is infinite; this result can be deduced more generally by observing that it is only necessary to consider the motion of the ball in the immediate vicinity of the saddle point. Accordingly, taking the first non-zero term of a Taylor series expansion of Eq. 23.28 about x_0

$$v \approx \pm(x - x_0)\sqrt{-g\left.\frac{\partial^2 h}{\partial x^2}\right|_{x=x_0}}.\tag{23.31}$$

Equation 23.29 then evaluates as

$$t \approx \frac{\ln(X)}{\sqrt{-g\left.\dfrac{\partial^2 h}{\partial x^2}\right|_{x=x_0}}},\tag{23.32}$$

where $X = x - x_0$ is the distance from the saddle point and noting that $\partial^2 h/\partial x^2|_{x=x_0} < 0$ is a necessary condition for x_0 to be a saddle point, it follows that $t \to \infty$ as $X \to 0$. Thus, solitary solutions to Eqs 23.23 and 23.24 are of infinite period regardless of the details of the shape function. Rearranging Eq. 23.32 to express the distance of the ball from the saddle point as a function of time

$$X \approx e^{t/\tau},\tag{23.33}$$

where $\tau = 1\left/\sqrt{-g\partial^2 h/\partial x^2|_{x=x_0}}\right.$ provides a natural timescale for the motion of the ball.

Wave solutions to the compaction equations

To make the analogy between the wave solutions of the compaction equations and the oscillating ball apparent, Eqs 23.15 and 23.16 are abbreviated as

$$\frac{\partial p_o}{\partial z} = f_1 \tag{23.34}$$

and

$$\frac{\partial \phi}{\partial z} = f_2 \frac{A}{v_\phi} |p_o|^{n_\sigma - 1} p_o, \tag{23.35}$$

where $v_\phi = -v_\infty$ is the velocity of a wave relative to a fixed point in the unperturbed matrix and f_1 and f_2 are

$$f_1 = -v_\phi \frac{\eta_f}{k} \frac{\phi - \phi_0}{1 - \phi} - \Delta\rho g \left[1 - \phi - (1 - \phi_0) \frac{k_0}{k} \right] \tag{23.36}$$

and

$$f_2 = \frac{(1 - \phi)^2}{1 - \phi_0} f_\phi. \tag{23.37}$$

Combining Eqs 23.34 and 23.35 to eliminate z, and rearranging, yields

$$0 = |p_o|^{n_\sigma - 1} p_o \, dp_o - \frac{v_\phi}{A} \frac{f_1}{f_2} \, d\phi, \tag{23.38}$$

which must be satisfied by the ϕ–p_c trajectory of any steady-state solution to Eqs 23.15 and 23.16. Defining a function H such that

$$H \equiv -\int \frac{f_1}{f_2} \, d\phi \tag{23.39}$$

or $\partial H / \partial \phi = -f_1 / f_2$, Eq. 23.38 is rewritten as

$$0 = |p_o|^{n_\sigma - 1} p_o \, dp_o + \frac{v_\phi}{A} \frac{\partial H}{\partial \phi} \, d\phi. \tag{23.40}$$

Comparison of Eqs 23.40 and 23.26 reveals that wave solutions to the compaction equations, at constant phase velocity, are a mathematical analog to the equations of motion for the oscillating ball, wherein the compaction variables $[\phi, p_o, z]$ map to $[x, v, t]$; the shape function H can be thought of as a hydromechanical potential that corresponds to the height h of the ball, and the factor v_ϕ / A has the same role as the gravitational constant g. Integration of Eq. 23.40 defines a property

$$U \equiv \frac{|p_o|^{n_\sigma - 1} p_o^2}{n_\sigma + 1} + \frac{v_\phi}{A} H \tag{23.41}$$

akin to the mass normalized energy u for the oscillating ball, which is conserved by porosity waves. Closed contours of U define the wave solutions to the compaction equations as a function of p_o and ϕ for a given velocity. As in the case of the oscillating ball, closed contours define periodic solutions where the porosity oscillates between two values,

characterized by equal H, at which p_o vanishes (e.g., the red contour in Fig. 23.2E). The periodic solutions are bounded by the contour that defines the solitary solution (black contour, Fig. 23.2E). Similarly, analogous to the oscillating ball solution, because U is directly proportional to H and increases with both negative and positive overpressure, the focus of any closed contour must lie along the $p_o = 0$ axis and correspond to an extremum in H, for example, the real root, ϕ_1, of $\partial H / \partial \phi = 0$. The porosity dependence (f_2) of the rheologic equation (Eq. 23.35) must be finite and positive if the matrix is coherent; consequently, the roots of $\partial H / \partial \phi = 0$ are independent of the rheologic constitutive relationship and identical to the roots of the hydraulic equation (Eq. 23.34), that is, the porosities that satisfy $f_1 = 0$. Although the number of roots cannot be determined without specifying the porosity–permeability relationship, the formulation of Eq. 23.15 is such that ϕ_0 is always a root, analogous to x_0 in Eq. 23.30. Consequently, if

$$\left. \frac{\partial^2 H}{\partial \phi^2} \right|_{\phi = \phi_0} = \frac{\Delta\rho g}{f_2 |_{\phi = \phi_0}} \left(\frac{1 - \phi_0}{k_0} \left. \frac{\partial k}{\partial \phi} \right|_{\phi = \phi_0} - 1 + \frac{v_\phi}{\phi_0 v_0} \right) \tag{23.42}$$

is greater than zero, then ϕ_0 is a stable level of porosity and small flow perturbations to a uniform flow regime will lead to periodic oscillations in the porosity about ϕ_0 (Fig. 23.3A). In contrast, if H is a maximum at ϕ_0, then ϕ_0 is a saddle point and solitary wave solutions are possible (Fig. 23.3B). Equating Eq. 23.42 to zero and solving for v_ϕ yields the critical velocity at which the background porosity ϕ_0 switches from focal to saddle point

$$v_\phi^{crit} = v_0 \phi_0 \left(1 - \frac{(1 - \phi_0)}{k_0} \left. \frac{\partial k}{\partial \phi} \right|_{\phi = \phi_0} \right) \tag{23.43}$$

such that ϕ_0 is a saddle point for waves with $v_\phi / v_0 > v_\phi^{crit} / v_0$, which is thus a necessary condition for the existence of the solitary wave solution. Although Eq. 23.43 appears to admit the possibility of solitary waves that propagate against the direction of buoyancy-driven fluid flow through the unperturbed matrix, substituting the explicit function for permeability given by Eq. 23.17 in Eq. 23.43 yields

$$v_\phi^{crit} / v_0 = [n_\phi (1 - \phi_0) + (b_\phi - 1)\phi_0], \tag{23.44}$$

which is positive for any plausible choice of n_ϕ and b_ϕ, as discussed earlier.

Provided a solitary wave solution exists, that is, $v_\phi / v_0 > v_\phi^{crit} / v_0$ and $H(\phi_d) > H(\phi_0)$ (as in Fig. 23.3B), then solving $\Delta H = H(\phi) - H(\phi_0) = 0$ yields the maximum porosity of the wave ϕ_{max}. As the overpressure vanishes at ϕ_{max}, the dependences of p_o and z on ϕ are obtained in exactly the same manner as the dependence of v and t on x for the oscillating ball (Eqs 23.28 and 23.29). Thus, from the definite integral of Eq. 23.40

$$p_o = \pm \sqrt[n_\sigma + 1]{(n_\sigma - 1) \frac{v_\phi}{A} \Delta H} \tag{23.45}$$

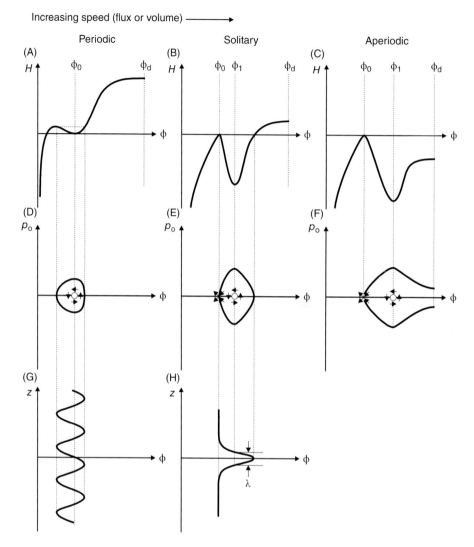

Fig. 23.3. Shape functions, phase portraits, and porosity wave patterns as a function of phase velocity illustrating the periodic, solitary, and aperiodic (fluidized) compaction-driven flow regimes. In general, phase velocity can be taken as a measure of the intensity of a flow perturbation. At low velocities (A, D, G), the background porosity ϕ_0 is the focal point of a periodic solution. At intermediate velocities (B, E, H), the focal point, that is, the minimum in H, shifts to $\phi_1 > \phi_0$; the potential H recovers to $H(\phi_0)$ at a porosity intermediate between ϕ_1 and the disaggregation porosity ϕ_d; and the relevant solution is a solitary wave. At still higher velocities (C, F), $H(\phi_d) < H(\phi_0)$ so there is no steady-state wave solution as there is no closed path in $p_o - \phi$ space connecting ϕ_0 to an elevated level of porosity; and the perturbation causes the matrix to disaggregate.

and inverting the result of substitution of Eq. 23.45 into Eq. 23.35,

$$z = \pm \sqrt[n_\sigma+1]{\frac{v_\phi}{A}} \int_{\phi_{max}}^{\phi} \frac{d\phi}{p_o^{n_\sigma} f_2}, \qquad (23.46)$$

where z is the depth relative to the wave center at which $\phi = \phi_{max}$ and $p_o = 0$. Any porosity of the solitary solution is associated with both positive and negative values of p_o and z, corresponding to the upper and lower halves of the wave. For this reason, the sign of factors in expressions for p_o and z has no significance; rather than explicitly indicating this with magnitude notation in Eqs 23.45 and 23.46, and subsequent equations, it is to be

understood that any negative term is to be replaced by its absolute value.

Nonlinear rheology creates a distinction between the solitary solution of the compaction equations and that of the oscillating ball in that it is possible to obtain a solitary porosity wave of finite wavelength. This behavior is demonstrated by linearizing Eq. 23.46 about ϕ_0 to obtain

$$z \approx \pm \left(\frac{v_\phi}{A f_2|_{\phi=\phi_0}} \left(\frac{n_\sigma + 1}{2} \frac{\partial f_1}{\partial \phi}\Big|_{\phi=\phi_0} \right)^{-n_\sigma} \right)^{\frac{1}{n_\sigma+1}} \int_{\Phi}^{0} \Phi^{-\frac{2n_\sigma}{n_\sigma+1}} d\Phi, \qquad (23.47)$$

where $\Phi = \phi - \phi_0$. The bifurcation between finite- and infinite-wavelength solutions is determined entirely by the stress exponent in Eq. 23.47, such that finite solutions can exist only for $n_\sigma < 1$, and is independent of the details of the hydraulic potential. As $n_\sigma \geq 1$ is characteristic of viscous behavior in rocks, it is expected that solitary porosity waves only develop in viscous rocks with finite initial porosity. Finite-wavelength solitary waves propagate, by definition, through a matrix with no initial porosity. It has been shown elsewhere that finite-wavelength solutions exist for viscoelastic compaction rheology (Connolly & Podladchikov 1998); the present analysis raises the possibility that shear-thickening viscous mechanisms ($n_\sigma < 1$) could also operate at the zero porosity limit in natural systems.

Just as the linearized equation for time in the oscillating ball problem (Eq. 23.33) provides a natural timescale for the movement of the ball near x_0, the linearized equation for depth in the solution of the compaction equations provides a characteristic length scale for variations in porosity near ϕ_0. By rewriting the integral in Eq. 23.47 in terms of $d\ln\Phi$, and differentiating, this length scale is

$$\delta' \sim \frac{\partial z}{\partial \ln \Phi} \approx \left(\Phi^{1-n_\sigma} \frac{v_\phi}{Af_2|_{\phi=\phi_0}} \left(\frac{n_\sigma+1}{2} \frac{\partial f_1}{\partial \phi}\bigg|_{\phi=\phi_0} \right)^{-n_\sigma} \right)^{\frac{1}{n_\sigma+1}}, \tag{23.48}$$

the depth interval over which porosity decays from $e\phi_0$ to ϕ_0 within a solitary wave. The derivative on the right-hand side of Eq. 23.48

$$\frac{\partial f_1}{\partial \phi}\bigg|_{\phi=\phi_0} = \frac{\Delta\rho g}{\phi_0} \left(\frac{v_\phi^{\text{crit}}}{v_0} - \frac{v_\phi}{v_0} \right) \approx -\frac{\Delta\rho g}{\phi_0} \frac{v_\phi}{v_0} \tag{23.49}$$

is zero at $v_\phi = v_\phi^{\text{crit}}$, but decreases monotonically in v_ϕ; thus, the approximate form is valid for large speeds. Adopting this approximation, substituting $\Phi = (e-1)\,\phi_0$ in Eq. 23.48, and expanding f_2 at ϕ_0 as $a_\sigma f_\phi|_{\phi=\phi_0}(1-\phi_0)$ yields

$$\delta' = \frac{\partial z}{\partial \ln \Phi} \approx \left(\frac{v_\phi[\phi_0(e-1)]^{1-n_\sigma}}{Aa_\sigma f_\phi|_{\phi=\phi_0}(1-\phi_0)} \left[\frac{n_\sigma+1}{2} \frac{\Delta\rho g}{\phi_0} \frac{v_\phi}{v_0} \right]^{-n_\sigma} \right)^{\frac{1}{n_\sigma+1}}, \tag{23.50}$$

effectively a lower bound on the wavelength of the solitary solution. Using the constitutive relations and scales given by Eqs 23.17, 23.19 and 23.20, and estimating wave speed as the magnitude of v_ϕ^{crit} ($\sim n_\phi|v_0|$, Eq. 23.44), then in the small porosity limit

$$\delta' = \sqrt[n_\sigma+1]{n_\sigma^{n_\sigma} \left(\frac{2}{3} \right)^{n_\sigma+1} \frac{a_\phi \phi_0^{n_\sigma-m_\sigma}}{A\eta_f} |\Delta\rho g|^{1-n_\sigma}}$$
$$\sqrt[n_\sigma+1]{\left(\frac{2}{n_\sigma+1} \right)^{n_\sigma} (n_\phi[e-1])^{1-n_\sigma}}, \tag{23.51}$$

where the first factor is the scale δ obtained by dimensional analysis (Eq. 23.21) and the second factor is unity for the linear viscous case and close to, but less than, one for the nonlinear case with typical values of n_ϕ and n_σ. This result confirms that δ is a reasonable estimate of the compaction length scale and suggests, unsurprisingly, that increasing the nonlinear character of the viscous rheology generally leads to stronger spatial variations in porosity. In view of the minor difference between δ and δ', δ is preferred here because of its simplicity.

Solitary wave properties in the small porosity limit

To illustrate the features of solitary waves explicitly, the solution is considered in conjunction with the constitutive relations given by Eqs 23.17 and 23.19 in the small porosity limit ($1 - \phi \rightarrow 1$, $\phi_d - \phi \rightarrow \phi_d$). Equations 23.34 and 23.35 are then

$$\frac{\partial p_o}{\partial z} = \Delta\rho g \left(\frac{v_\phi}{v_0} \frac{[(\phi/\phi_0)-1]+1}{[\phi/\phi_0]^{n_\phi}} - 1 \right) \tag{23.52}$$

and

$$\frac{\partial \phi}{\partial z} = Aa_\sigma \frac{\phi^{m_\sigma}}{v_\phi} |p_o|^{n_\sigma-1} p_o, \tag{23.53}$$

respectively. Using these forms, the wave hydraulic potential H can be arranged as the sum of two integrals

$$H = \frac{\Delta\rho g}{\phi_0^{m_\sigma} a_\sigma} \int \left(\frac{\phi_0}{\phi} \right)^{m_\sigma} - \left(\frac{\phi_0}{\phi} \right)^{n_\phi+m_\sigma} d\phi$$
$$+ \frac{\Delta\rho g}{\phi_0^{m_\sigma} a_\sigma} \frac{v_\phi}{v_0} \int \left(\frac{\phi_0}{\phi} \right)^{n_\phi+m_\sigma} - \left(\frac{\phi_0}{\phi} \right)^{n_\phi+m_\sigma-1} d\phi, \tag{23.54}$$

where both integrands are zero at $\phi = \phi_0$, and for $\phi > \phi_0$, $n_\phi > 1$, and $m_\sigma \geq 0$, the first integrand is positive and the second integrand negative. Furthermore, for conditions at which the solitary solution is possible, that is, $v_\phi/v_0 > v_\phi^{\text{crit}}/v_0$, H is a maximum at $\phi = \phi_0$, and H must have a minimum at the ϕ_1, the focal point where $\partial p_o/\partial z$ (Eq. 23.52) vanishes and the magnitude of the overpressure is a maximum (Fig. 23.3B). At $\phi > \phi_1$, H recovers to the background value $H(\phi_0)$ at the maximum porosity of the wave ϕ_{max}. It follows from the form of Eq. 23.54 that for a fixed choice of exponents, the rate at which H recovers to the background value $H(\phi_0)$ at $\phi > \phi_1$ decreases with wave speed. Thus, wave amplitude must increase with wave speed (Fig. 23.4A). For specified v_ϕ the leading term of the first integrand will dominate the rate at which H recovers to $H(\phi_0)$ at $\phi > \phi_1$. Consequently, increasing m_a increases amplitude (cf. solid black and cyan curves, Fig. 23.4A); this result is intuitive because increasing the nonlinearity of the effective bulk viscosity leads to a weakening of the matrix with increasing porosity. A less intuitive consequence of Eq. 23.54 is that increasing the nonlinearity of the porosity–permeability relationship, that is, increasing n_ϕ decreases wave amplitude (cf. orange and black curves, Fig. 23.4A). This occurs because at $\phi > \phi_1$ the rate at which the sum of the integrands decays with porosity increases with n_ϕ.

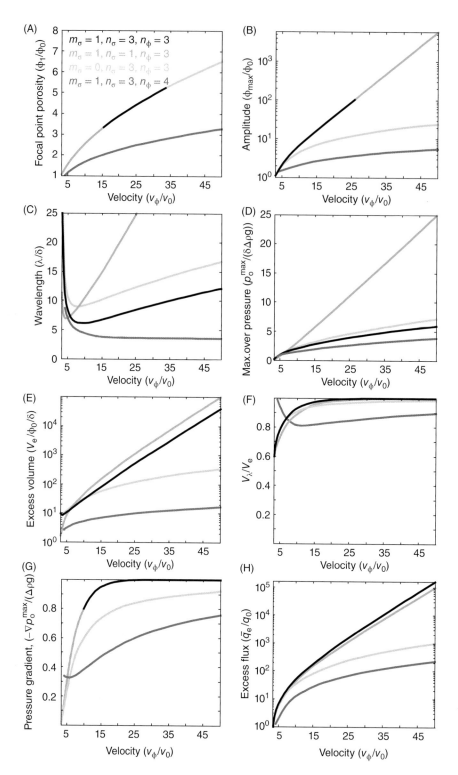

Fig. 23.4. Porosity wave properties in the small porosity limit ($\phi \ll \phi_d$) as a function of relative wave velocity for choices of the porosity exponents in the constitutive relations as indicated by color coding. Typical values for the exponents characterizing the porosity dependences of the permeability and effective bulk viscosity are $n_\phi = 3$ and $m_\sigma \leq 1$; a stress exponent $n_\sigma = 3$ is characteristic of dislocation creep, the viscous deformation mechanism commonly assumed for the lower crust (Ranalli 1995). The minimum relative velocity considered for each choice of exponents is $v_\phi/v_0 = n_\phi + 0.2$, slightly above the critical value $v_\phi/v_0 = n_\phi$, for the existence of the solitary solution (Eq. 23.44). Continuous curves drawn in different colors indicate that the properties are identical for the corresponding exponent choices. Focal point porosity (A) and amplitude (B). Focal point porosity, ϕ_1, is dependent only on n_ϕ, and the divergence of amplitude from ϕ_1 shows that the effect of increasing the nonlinearity (i.e., m_σ) of the effective bulk viscosity is to increase amplitude. In contrast, increasing nonlinearity of the porosity–permeability relationship decreases amplitude, which is independent of n_σ. (C) Wavelength, λ, is the distance separating the overpressure extrema within a wave (Fig. 23.3H). (D) Maximum fluid overpressure, the maximum underpressure (i.e., effective stress) is $-p_o^{max}$, both extrema occur at ϕ_1 (Fig. 23.3E). (E) Excess volume (Eq. 23.60) correlates with amplitude except at low wave speed, whereupon it decreases with speed in solutions for high n_σ/n_ϕ. That the excess volume for the nonlinear viscous cases ($n_\sigma = 3$) becomes larger than that for the linear viscous case ($n_\sigma = 1$) at low speeds ($v_\phi/v_0 \sim 6$) indicates a shifting of the porosity toward the tails of the nonlinear viscous solution. (F) Fraction of the volume that occurs within $\pm\lambda/2$ of ϕ_{max}. (G) Maximum $-p_0$ gradient, which occurs at ϕ_{max}, unity corresponds to a hydrostatic fluid pressure gradient. (H) Average excess fluid flux associated with wave passage. (*See color plate section for the color representation of this figure.*)

The relationship between wave velocity and amplitude $(A_\phi = \phi_{max}/\phi_0)$ for $m_\sigma \neq 1$ obtained by solving $\Delta H = 0$ is

$$v_\phi = v_0 \frac{n_\phi + m_\sigma - 2}{1 - m_\sigma} \frac{([m_\sigma - 1][A_\phi^{-n_\phi} - 1] - n_\phi)A_\phi^{1-m_\sigma} + n_\phi}{A_\phi^{1-n_\phi-m_\sigma}([n_\phi + m_\sigma - 1][A_\phi - 1] + 1) - 1},$$

(23.55)

which for the specific case $n_\phi = 3$ and $m_a = 0$ reduces to the linear relationship $v_\phi = v_0(2A_\phi + 1)$ obtained by Scott & Stevenson (1984, Barcilon & Richter 1986). The integrated form of Eq. 23.54 for general values of m_a is singular at $m_a = 1$, the value typically assumed in compaction literature. For this less general, but more widely used, case

$$H = \frac{\Delta \rho g}{a_\sigma} \left(\frac{1 - v_\phi/v_0}{n_\phi(\phi/\phi_0)^{n_\phi}} + \frac{v_\phi/v_0(\phi/\phi_0)^{1-n_\phi}}{n_\phi - 1} + \ln(\phi/\phi_0) \right),$$

(23.56)

$$v_\phi = v_0(n_\phi - 1)\frac{A_\phi^{n_\phi}(\ln A_\phi^{n_\phi} + 1) - 1}{1 + (A_\phi - 1)n_\phi - A_\phi^{n_\phi}},$$

(23.57)

which similarly reduces to Scott & Stevenson's (1984) result for $n_\phi = 3$. Although Eq. 23.57 cannot be solved analytically for amplitude, it is apparent that in the small porosity limit, amplitude is not a function of n_σ. Evaluation of the integral in Eq. 23.46 gives the two values of pressure at any porosity within a solitary wave (Fig. 23.3E) as

$$p_o = \pm \delta \Delta \rho g (n_\sigma + 1) \left(\frac{1}{2^{n_\sigma} n_\phi} \frac{v_\phi}{v_0} \right)^{\frac{1}{n_\sigma + 1}}$$

$$\left(\frac{v_\phi/v_0 - 1}{(\phi/\phi_0)^{n_\phi}} - \ln(\phi/\phi_0)^{n_\phi} + 1 - \frac{v_\phi}{v_0} \frac{\left(1 - \frac{n_\phi}{(\phi/\phi_0)^{n_\phi} - 1} \right)}{n_\phi - 1} \right)^{\frac{1}{n_\sigma + 1}}.$$

(23.58)

The corresponding integral for depth

$$z = \pm \delta \left(\frac{n_\phi}{2} \right)^{\frac{n_\sigma}{n_\sigma + 1}} \left(\frac{v_\phi}{v_0} \right)^{\frac{1}{n_\sigma + 1}} \int_{\phi_{max}}^{\phi} \frac{1}{\phi}$$

$$\left(\left[1 - \frac{v_\phi}{v_0} \right] \left[\frac{\phi_0}{\phi} \right]^{n_\phi} + \ln \frac{\phi}{\phi_0} - 1 - \frac{v_\phi}{v_0} \frac{\left[1 - \left(\frac{\phi_0}{\phi} \right) \right]^{n_\phi - 1}}{n_\phi - 1} \right)^{-\frac{n_\sigma}{n_\sigma + 1}} d\phi$$

(23.59)

must, in general, be evaluated numerically (a Fortran computer program for this purpose is available from the author).

Because the matrix recovers to the background porosity asymptotically in a steady-state solitary wave (for $n_\sigma \geq 1$), the true wavelength is infinite (cf. Eq. 23.47). For practical purposes, it is desirable to define an effective wavelength, which defines the extent of the wave that includes the bulk of the anomalous porosity. To this end, the wavelength λ is taken to be the interval between the points of minimum and maximum overpressure (Figs 23.3H and 23.4C). The ratio of the excess volume, that is, the total volume of fluid associated with the passage of a wave (Fig. 23.4E),

$$V_c = \int_{-\infty}^{\infty} (\phi - \phi_0) dz$$

(23.60)

to that obtained by integrating over $\pm \lambda/2$ shows that even at low speeds, >80% of the porosity of a wave occurs within the interval $\pm \lambda/2$ about the center of wave (Fig. 23.4F).

The effect of nonlinear viscous rheology is best understood in terms of the overpressure at the focal point porosity ϕ_1 (Fig. 23.4D). The magnitude of the overpressure gradient is limited by the hydrostatic pressure gradient for the fluid phase, a limit that is approached rapidly with increasing velocity at the center of a porosity wave (black–green curve, Fig. 23.4G); thus, at the velocity at which the maximum pressures of the linear and nonlinear viscous solutions are equal ($\sim 5.9\ v_0$, black and green curves, Fig. 23.4D), the dilational strain rate must fall more rapidly in the nonlinear case between ϕ_1 and ϕ_{max}, and as both ϕ_1 and ϕ_{max} are independent of n_σ, this must lead to a relatively flat-topped porosity distribution in which a greater proportion of the porosity lies within the interval $\pm \lambda/2$ about ϕ_{max}. Conversely, as speeds fall below that at which the overpressures at the focal point porosities of the solutions are equal, a greater proportion of the porosity shifts to the tails of the porosity distribution for the nonlinear case, leading to broad, poorly defined waves. This behavior is confirmed by linearization of the integral for the second moment of the solitary wave porosity distribution, which shows that in the limit $A_\phi \to 1$ or, equivalently, $v_\phi \to v_\phi^{crit}$, the moment becomes infinite if $n_\sigma \geq 3$ and explains the minima in V_c as a function of velocity for nonlinear viscous matrix rheology (black and cyan curves, Fig. 23.4E). The existence of the minima is of little practical consequence, because it occurs at velocities at which wavelengths are so long that the solitary waves would be indistinguishable from uniform fluid flow. Increasing n_ϕ counters this effect so that for $n_\sigma = 3$ and $n_\phi = 4$, waves are well formed at all velocities (orange curve, Fig. 23.4E). The instantaneous excess fluid flux, that is, the flux in excess of the background value $q_0 = v_0 \phi_0$ within a wave is $q_c = v_\phi(\phi - \phi_0)$, and time-averaged fluid flux associated with wave passage (Fig. 23.4H) is estimated as

$$\overline{q}_c = \frac{v_\phi}{\lambda} V_c.$$

(23.61)

In the limit $v_\phi \to v_\phi^{crit}$, $\lambda \to \infty$; therefore, $\overline{q}_c/q_0$ must fall monotonically to zero with velocity, implying that there is a solitary wave solution for any value of $\overline{q}_c/q_0 > 0$.

Dynamic permeability in response to external forcing

There is no fundamental principle that dictates a balance between fluid production and transport in geological environments, but for the range of conditions investigated by numerical simulations of metamorphic compaction-driven fluid flow, this balance does develop locally (Connolly 1997, 2010). Assuming such a balance in conjunction with the solitary porosity wave solution provides a means of predicting the dynamic variations in permeability that develop from an initially steady hydrologic regime in response to metamorphic fluid production (Connolly & Podladchikov 2013). This model amounts to no more than assuming that the time-averaged permeability of a compacting system is that necessary to accommodate fluid flux associated with an external forcing (Ingebritsen & Manning 1999). The information gained by implementing the solitary wave solution in this context is insight into the instantaneous variations in porosity and pressure that develop in response to the forcing.

In a 1D compacting system, a requirement for a balance between wave-propagated fluid transport and fluid production is that the magnitude of the time-averaged flux associated with the passage of a wave (Eq. 23.61) must be greater than or equal to the vertically integrated production q_s, because a wave with $|\bar{q}_c| < q_s$ would be unable to separate from its source. If $|\bar{q}_c| > q_s$, then the waves must be separated by a depth interval of $\Delta z = \lambda(|\bar{q}_c/q_s| - 1)$. In numerical simulations, the transient

dynamics of wave separation are such that $|\bar{q}_c/q_s|$ is typically <1 (Connolly 1997). This result suggests that the properties of waves expected in metamorphic environments can be predicted by equating $\bar{q}_c$ to q_s and exploiting the monotonic relationship between $\bar{q}_c$ and v_ϕ (Figs 23.4H and 23.5). In earlier works (Connolly & Podladchikov 2013), it was asserted incorrectly that solitary wave solutions do not exist for $\bar{q}_c/q_0 > 2$; in fact, solitary solutions exist for all $\bar{q}_c/q_0 > 0$, but, as remarked previously, waves that develop at small excess flux magnitudes have such long wavelengths that it is unlikely they would be distinguishable from uniform fluid flow in natural environments.

While the scenario outlined here seems the most relevant to fluid flow in ductile portions of the Earth's crust, it is conceivable that fluid production may occur so rapidly, that is, on a timescale $\ll \delta/|v_0|$, that compaction mechanisms cannot accommodate fluid production. The effect of such an imbalance may be to produce a region of increased porosity bounded by unreacted and, presumably, compacted rocks. In this scenario, the response of the system is dependent on the vertical extent, Δz, of the region of increased porosity. If the extent is small ($\Delta z \sim \delta$), then a single solitary wave will evolve from the source region in such a way to carry the excess volume of the source region (as in Fig. 23.1A). The minimum in excess volume as

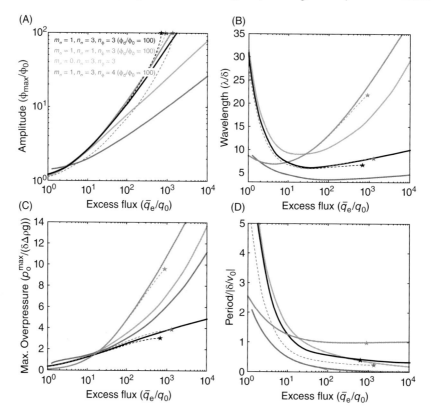

Fig. 23.5. Porosity wave properties as a function of the average excess fluid flux associated with wave passage for the exponent choices indicated by the legend and color coding. Solid curves are for the small porosity limit ($\phi \ll \phi_d$) as in Figure 23.4; dashed curves show the influence of a disaggregation in the intermediate porosity limit ($\phi_d \gg 1 - \phi$) for the specific case that $\phi_d/\phi_0 = 100$, discussed in the text (Fig. 23.36). Star symbol at the end of dashed curves indicates the disaggregation condition. Properties of waves likely to be generated in natural settings in response to fluid production can be estimated by equating vertically integrated fluid production to the average excess flux. (*See color plate section for the color representation of this figure.*)

a function of velocity for strongly nonlinear viscous rheology gives rise to potential ambiguity for such initial conditions, because the same excess volume may be accommodated in a wave with either low or high velocity. It is speculated here that the high-velocity solution dominates. If the extent of the reacted porosity is large ($\Delta z \gg \delta$), then the multiple waves that evolve from the region can be expected to carry the excess flux $q_c \approx q_2 - q_0$ where q_2 is the flux through the increased porosity ϕ_2 at $p_o = 0$. Spiegelman (1993) demonstrated that waves that nucleate at the boundary between an infinite source and unreacted rocks are periodic waves in which porosity oscillates about the focal point ϕ_1 of the solitary solution (e.g., the red curve in Fig. 23.2E,F). However, for large porosity contrasts, that is, $\phi_1 \gg \phi_0$, the distinction is unimportant.

Example #1: 1D VISCOUS WAVE

To illustrate the application of the 1D viscous solitary wave solution, consider an initial state characterized by $\delta \sim 100$ m, $\phi_0 \sim 10^{-4}$, $\Delta\rho g \sim 10^3$ kg m^{-3}, $v_0 \sim -10^{-9}$ m s^{-1} ($q_0 = v_0\phi_0 = -10^{-13}$ m s^{-1}), as might be appropriate for dehydration in the lower crust at amphibolite facies conditions (i.e., temperatures of 773–923 K, Connolly & Podladchikov 2013). Taking $n_\sigma = n_\phi = 3$ and $m_\sigma = 1$ as the most probable values for the constitutive exponents, a fluid production rate of $q_s = 10^{-11}$ m s^{-1} will generate solitary waves with (black curves, Fig. 23.5) $\phi_{max} = 10.0$ $\phi_0 = 1.0 \times 10^{-3}$, $\lambda = 6.3$ $\delta = 630$ m, $p_o^{max} = 2.4$ $\delta\Delta\rho g = 0.24$ MPa, and a period of 1.0 $\delta/|v_0| = 1.6 \times 10^3$ years. From the period (0.50 $\lambda/|v_0|$), or the relationship between flux and velocity (Fig. 23.4H), $v_\phi = -\lambda/\text{period} = -0.39$ m year^{-1} (12.2 v_0) and the maximum overpressure gradient is $\nabla p_o = -0.84\,\Delta\rho g$ (Fig. 23.4E), that is, the fluid pressure gradient within the wave is nearly hydrostatic (cf. Eq. 23.5). Holding all other parameters constant, the effect of changing from power-law viscous ($n_\sigma = 3$) to linear viscous ($n_\sigma = 1$, green curves in Figs 23.4 and 23.5) matrix rheology is to double the speed, amplitude, and maximum overpressure of the waves. This effect reflects that at $v_\phi/v_0 > 5.9$ (the crossing of the green and black curves in Fig. 23.4E), the nonlinear wave has a greater excess volume; thus, slow waves in the nonlinear viscous case are capable of accommodating the same flux as faster waves in the linear viscous case.

Disaggregation and the compaction-driven flow regimes

Wave amplitude grows monotonically with speed in the small porosity approximation because the $1 - \phi$ and $\phi_d - \phi$ terms in Eqs 23.15–23.17 and 23.19 that limit the possible values of the porosity are neglected; thus, the formulation has no upper bound on porosity. Given that a granular matrix is expected to disaggregate at $\phi_d \sim 20\%$ (Arzi 1978; Auer *et al.* 1981; Ashby 1988; Vigneresse *et al.* 1996), the $\phi_d - \phi$ term is likely to dominate wave behavior before the dampening effects of the $1 - \phi$ terms become significant. Elsewhere, it has been shown that for constitutive relations that do not account for disaggregation, the $1 - \phi$ terms are unimportant at absolute porosities of $\sim 25\%$ for typical choices of the exponents n_σ, n_ϕ, b_ϕ, and m_σ (Connolly & Podladchikov 2000; a Fortran computer program that solves the large porosity formulation is available upon request). Accordingly, the effect of disaggregation is assessed here by an intermediate porosity approximation in which the $\phi_d - \phi$ term of Eq. 23.19 is retained, but porosity terms of order 1 (i.e., $1 - \phi$ and $1 - \phi^{1/n_\sigma}$) are dropped to obtain

$$H = \frac{\Delta\rho g}{\phi_0^{m_\sigma} c_\sigma}$$

$$\int \frac{\left[\left(\dfrac{\phi_0}{\phi}\right)^{m_\sigma} - \left(\dfrac{\phi_0}{\phi}\right)^{n_\phi+m_\sigma} + \dfrac{v_\phi}{v_0}\left(\left[\dfrac{\phi_0}{\phi}\right]^{n_\phi+m_\sigma} - \left[\dfrac{\phi_0}{\phi}\right]^{n_\phi+m_\sigma-1}\right)\right]}{\left[\dfrac{\phi_d}{|\phi_d - \phi|}\right]^{n_\sigma-1/2}} d\phi$$

$$(23.62)$$

for the properties of waves with porosities approaching ϕ_d. As the denominator of the integrand in Eq. 23.62 becomes infinite at $\phi = \phi_d$, H must have a maximum at ϕ_d and if $H(\phi_d) < H(\phi_0)$ (Fig. 23.3C), then there is no closed contour of the function U that connects the background porosity to an increased level of porosity (Eq. 23.41, Fig. 23.3F) and no solitary solution is possible. Because the integrand of Eq. 23.62 is simply the combined integrands of Eq. 23.54, scaled by the disaggregation term, the effects of varying the exponents n_ϕ, n_σ, and m_σ are readily separated. Specifically, lowering n_σ or raising n_ϕ or m_σ increases $H(\phi_d)$ relative to $H(\phi_0)$, extending the range of solitary wave velocities that the matrix can sustain without disaggregating (Fig. 23.6A). In contrast to the small porosity limit where H, and therefore wave amplitude, is independent of the stress exponent n_σ, in the intermediate porosity limit, although the focal point porosity ϕ_1, and therefore p_o^{max} (Fig. 23.6C), remains independent of n_σ, the relation between amplitude and velocity is dependent on n_σ. For $n_\sigma = n_\phi = 3$, this dependence is prominent for $\phi/\phi_d > 0.1$ (black curves, Fig. 23.6A) and is even more pronounced with increasing nonlinearity in the porosity–permeability relationship (orange curves, Fig. 23.6A). Because the effect of the disaggregation term is to weaken the matrix with increasing porosity, its effect is to sharpen the porosity distribution within solitary waves, akin to the result of increasing m_σ, leading to an increase in excess volume compared with models that do not account for disaggregation.

By solving for the solitary wave velocity at which $H(\phi_d) = H(\phi_0)$ and computing the corresponding value of $\bar{q}_e$ (Eq. 23.61), it is possible to estimate the range of fluid production rates that can be sustained without causing the solid matrix to become fluidized. For example, taking $n_\phi = 3$ and $\phi_d/\phi_0 = 100$, fluid production rates of 700–900 $|q_0|$ are adequate to induce fluidization (Fig. 23.7); for comparison, to cause fluidization by, albeit unstable, uniform flow, the required fluid production rates are $|q_0|(\phi_d/\phi_0)^{n_\phi}$, that is, 10^6 $|q_0|$. Thus, porosity waves have the potential to strongly enhance weak flow perturbations. In terms of fluxes, the lower limit of the solitary wave regime corresponds to $\bar{q}_e = 0$; thus, the periodic regime can only be induced by a negative vertically integrated fluid production rate such as would result from the consumption of fluids by retrograde hydration reactions. Alternatively, for waves induced

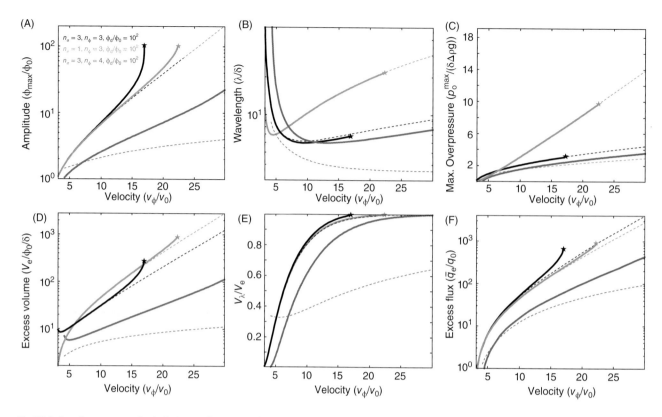

Fig. 23.6. Porosity wave properties in the intermediate porosity limit ($\phi_d \ll 1 - \phi$) as a function of velocity for a disaggregation porosity ϕ_d a hundred times greater than ϕ_0. All examples are for $m_\sigma = 1$ with other exponents of the formulation as indicated by the legend and color coding. The star symbol indicates the disaggregation condition. Dashed curves show the corresponding properties for the small porosity limit as in Figure 23.4. The solutions show that the effect of disaggregation is strongly dependent on both n_ϕ and n_σ; for $n_\phi = 3$, the disaggregation effect becomes significant if ϕ_{max} is within an order of magnitude of ϕ_d and is amplified by increasing n_σ. Properties for these solutions are shown as a function of $\bar{q}_e$ in Figure 23.5. (*See color plate section for the color representation of this figure.*)

by a perturbation defined in terms of an excess volume (e.g., Fig. 23.1), the periodic solution requires negative excess volume, that is, an obstruction (Spiegelman 1993) to a region of uniform flow. The appearance of periodic waves in numerical simulations (e.g., Fig. 23.1A or Connolly 1997) reflects the dynamics of solitary wave separation, in which over-compaction of the matrix obstructs the background flow.

DISCUSSION

This study has explored the behavior of the solitary porosity wave solution to the compaction equations in 1D viscous media. The solution provides a simple means of estimating the scales of pressure and porosity (or permeability) variations as a function of fluid production rates and constitutive relations. Although porosity waves have been posited as a mechanism for fluid flow in the lower crust (Suetnova *et al.* 1994; Connolly 1997; Gliko *et al.* 1999; Ague 2014; Tian & Ague 2014), their expression in nature would be complicated by a number of factors. These factors, which include geometry, lithological heterogeneity, tectonic stress, and rheological variations, have been reviewed elsewhere (Connolly & Podladchikov 2004, 2013). Here, some aspects of this earlier review, which

are particularly relevant to potential applications of the 1D solitary wave solution, are recapitulated in the context of a conceptual model for compaction-driven fluid flow in the lower crust (Fig. 23.8).

Linear or nonlinear viscous rheology?

Even if the viscous deformation mechanism is nonlinear, in rocks undergoing simultaneous compaction and shear deformation, it does not necessarily follow that effective viscous rheology for the compaction process is nonlinear. Both compaction and macroscopic shear deformation are accomplished by microscopic shear. Thus, if a rock is simultaneously subject to both modes of deformation, then they must be accommodated by the same microscopic mechanism. This mechanism is determined by the largest of the stresses responsible for deformation, $|\Delta\sigma|$ or $|p_o|$, with the result that if the stresses are of different magnitude, the viscous response to the inferior stress is approximately linear and determined by effective viscosity resulting from the deformation induced by the superior stress. Regardless of magnitude, far-field tectonic stress facilitates compaction by lowering the effective viscosity of the solid matrix (Tumarkina *et al.* 2011).

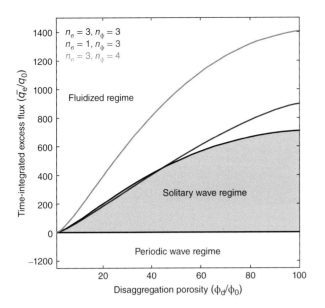

Fig. 23.7. Phase diagram depicting the hydrologic regimes predicted by the intermediate porosity limit solitary wave solution as a function of the disaggregation porosity and excess flux. The regime that develops in response to fluid production can be predicted by equating the magnitude of the excess flux carried by the waves to the vertically integrated fluid production $\bar{q}_s$. The boundary between the solitary wave and fluidized regime is dependent on the exponents m_σ, n_σ, and n_ϕ; it is shown for $m_\phi = 1$, with other exponents as indicated by the inset and color coding. Solitary waves become progressively more diffuse and indistinguishable from uniform flow as $\bar{q}_e/q_0 \to 0$. Periodic wave trains develop in response to negative fluid production, for example, the consumption of fluids by hydration reactions, or an obstruction to the background porosity (Spiegelman 1993). In nature, such periodic wave trains would decay to uniform flow. (*See color plate section for the color representation of this figure.*)

3D geometry and nonviscous rheology

As remarked earlier, the 1D solitary wave solution is unstable with respect to spherical solitary waves in three dimensions (Wiggins & Spiegelman 1995). However, as wave speeds increase, the overpressure gradient in solitary waves rapidly approaches the limit (i.e., $-[1-\phi]\Delta\rho g$) imposed by the fluid hydrostat (Fig. 23.4G). At this condition, the velocity and porosity distribution along the vertical axis of the 1D and 3D waves are essentially identical, and the excess volume of the 3D wave can be estimated by applying spherical symmetry to porosity distribution of the 1D wave (Connolly & Podladchikov 2007). Transient models of multidimensional waves suggest that they collect fluid from a source region of area $\sim\pi\lambda^2$ (Wiggins & Spiegelman 1995). Thus, the properties of the 3D waves that would initiate in response to fluid production can be predicted by equating the product of vertically integrated fluid production rate and the source area, $Q_s = \pi\lambda^2 q_s$ with the volumetric transport rate

$$Q = \frac{v_\phi}{\lambda} \int_0^\infty 4\pi r^2 (\phi - \phi_0)\mathrm{d}r, \qquad (23.63)$$

where the integral is the 3D excess fluid volume associated with the wave and is approximated by using the 1D solitary wave solution for the radial porosity distribution. In practice, because λ varies as a function of q_s, solving $Q_s = Q$, is an iterative problem.

Example #2: 3D VISCOUS WAVE

To illustrate the consequences of 3D geometry, consider the same parameters as in Example #1. Taking the 1D wavelength, $\lambda = 6.3 \, \delta$, as an initial estimate for the 3D solution, the required fluid transport rate is $Q/|q_0\delta^2| = q_s\pi\lambda^2/|q_0\delta^2| = 100 \, \pi \, 6.3^2 = 1.25 \times 10^4$. For this value of Q, $v_\phi/v_0 = 29$ (black curve, Fig. 23.10) and $\lambda/\delta = 9.1$ (black curve, Fig. 23.4C). Using this revised estimate of wavelength, $Q/|q_0\delta^2| = 2.6 \times 10^4$, which in turn yields new velocity and wavelength estimates of $v_\phi/v_0 = 37.6$ and $\lambda/\delta = 11.0$. After three iterations, successive refinement of the estimates for fluid transport rate, velocity, and wavelength by this method yields $Q = 37.9 \times 10^{-5} \, \mathrm{m^3\,s^{-1}}$, $\lambda = 11.0 \, \delta = 1100\,\mathrm{m}$, $v_\phi = 41.4$ $v_0 = -1.31 \, \mathrm{m\,year^{-1}}$, $p_o^{max} = 5.35$ $\delta\Delta\rho g = 5.35\,\mathrm{MPa}$, and $\phi_{max} = 1390$ $\phi_0 = 0.139$ for the 3D wave. This result demonstrates that increased spatial focusing of fluid flow caused by 3D effects has the capacity to generate both the large porosities necessary to cause disaggregation and/or the overpressures necessary to induce brittle (plastic) failure.

Thermal activation

Thermal activation will, generally, lead to an upward increase in the effective shear viscosity of the lower crust on a length scale l_A that is dependent on the activation energy of the viscous mechanism and the geothermal gradient, but typically $\sim 1\,\mathrm{km}$ (Connolly & Podladchikov 2013). Consequently, all other factors being equal, the compaction length scale δ will increase upward through the crust as $^{n_\sigma+1}\sqrt{\exp(z/l_A)}$, that is, by a factor of ~ 10 over a vertical interval distance of 6–$8 \, l_A$. This variation has consequences for the relevance of the steady-state solution, which assumes a constant effective shear viscosity. Provided $\delta < l_A$, the variation in shear viscosity due to thermal activation is weak on the porosity wave length-scale. In this case, quasi-steady-state waves that closely approximate the steady-state solution can be expected to develop. The evolution of such quasi-steady-state waves can be anticipated from the steady-state solution given that the waves are likely to conserve excess volume (Fig. 23.4E, Connolly & Podladchikov 2013). As δ becomes comparable to l_A, multidimensional waves flatten to sill-like structures. Although these structures superficially resemble the 1D steady-state solitary wave solution, their vertical dimension is dictated by l_A and they slow exponentially as they propagate upward (Connolly 1997; Connolly & Podladchikov 1998). This behavior suggests that if porosity waves develop on a geologically relevant length scale at depth within the crust, then, in the absence of other deformation mechanisms, they will tend to stagnate below the brittle–ductile transition (Fig. 23.8).

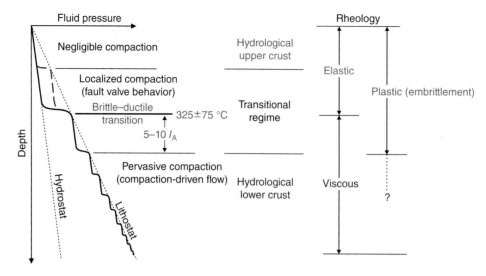

Fig. 23.8. Conceptual model of the hydrologic regimes that would result from superimposing thermally activated compaction on crustal column with heterogeneous permeability (Connolly & Podladchikov 2013). In the upper crustal regime, faulting maintains such high permeabilities that negligible deviation from hydrostatic fluid pressure is adequate to drive fluid circulation (Zoback & Townend 2001). This regime is limited at depth by the conditions at which localized compaction becomes an effective mechanism for sealing fault-generated permeability (Gratier *et al.* 2003; Tenthorey & Cox 2006). At greater depth, pervasive compaction and/or metamorphic fluid production may generate transient fluid overpressure that is periodically relieved by faulting (Sibson 1992). At the brittle–ductile transition (i.e., the base of the seismogenic zone), it is improbable that pervasive compaction can keep pace with metamorphic fluid production; thus, the transitional hydrologic regime is likely to persist over an interval that extends $\sim$10 l_A below the brittle–ductile transition, where l_A is the characteristic length scale for variation in the ductile rheology (typically $\sim$1 km, Connolly & Podladchikov 2013). Beneath the transitional regime, pervasive compaction is capable of generating hydraulic seals, and fluid, if present, is at near-lithostatic pressure. Within this lowermost regime, fluid flow is truly compaction driven. In the absence of fluid production, the tendency of both time and depth is to decrease the wavelength of the fluid pressure compartments, resulting in a near-steady state. Barring the possibility of a subcrustal fluid source, the flux in this near-steady regime must decrease with depth. Thus, the magnitude of the perturbation caused by fluid production to the lower crustal regime is dependent on its depth. In the deepest portion of the crust, the rheology is viscous as assumed in the formulation presented here. Upward strengthening of the viscous rheology would cause porosity waves to provoke elastic and plastic deformation mechanisms at shallower levels. Because viscous porosity waves are associated with negative effective pressure anomalies, $\sim\lambda\Delta\rho g/2$, the first deviation from viscous behavior is likely to be viscoplastic. Viscoplastic rheology causes fluid flow to be focused into tube-like channels (Connolly & Podladchikov 2007; Connolly 2010). At still shallower depths, viscous compaction becomes entirely ineffective, leading to a viscoelastic transition. In numerical models, such a viscoelastic transition causes lower crustal solitary porosity waves to dissipate as porosity–pressure surges in the upper crust (Connolly & Podladchikov 1998). (*See color plate section for the color representation of this figure.*)

Example #3: THERMAL ACTIVATION

Consider a 1D solitary porosity wave with initial properties $\delta_i \sim 100$ m, $\lambda = 630$ m, $p_o^{max} = 0.24$ MPa, and $v_\phi = 12.2\, v_0$, as in Example #1, which propagates upward through a cooling, but otherwise uniform crust, characterized by $l_A \sim 1$ km. The initial dimensionless excess volume ($V_e/\delta_i/\phi_0$) of the wave is 41.4 (black curve, Fig. 23.4H). After the wave rises 5 km, the local compaction length increases to $\delta = \delta_i\,^{n_\sigma+1}\sqrt{\exp(\Delta z/l_A)} = 350$ m. If the wave conserves its dimensional excess volume (v_e), then the dimensionless excess volume $V_e/\delta/\phi_0$ must decrease to 11.9. For this new dimensionless excess volume, the wave velocity is $v_\phi/v_0 = 5.6$ (Fig. 23.4H), and its wavelength and maximum overpressure increase to 2200 m (Fig. 23.4C) and 0.40 MPa (Fig. 23.4D), respectively. As this wavelength is greater than l_A, the steady-state solution most likely overestimates both velocity and wavelength (Connolly & Podladchikov 1998).

Viscoplastic rheology

In the viscous limit, a solitary wave is associated with a maximum fluid overpressure of $\sim\lambda\Delta\rho g/2$ that grows as the wave propagates upward into cooler rocks. As rocks have little tensile strength (e.g., Gueguen *et al.* 2004), it is probable that such fluid overpressures would induce hydrofracture and/or other plastic dilational mechanisms. Brittle deformation associated with active metamorphism (Etheridge *et al.* 1984; Simpson 1998) is broadly consistent with the notions that embrittlement occurs at high fluid pressure and on spatial scales $\ll\delta$. In this scenario, the effect of plastic weakening can be simulated by reducing the coefficient of viscous flow by a factor of $R^{n_\sigma+1}$ for $p_o > 0$. The *ad hoc* factor R can be adjusted to match the presumed yield stress, σ_y, of the plastic mechanism. In 1D numerical models that use this approximation, asymmetrical, steady-state solitary waves develop in which a small overpressured region is fed by a much larger underpressured region (Fig. 23.1C–D). In the small porosity limit, such solutions are permitted because the hydraulic potential H (Eq. 23.54), which determines the shape of the viscous solitary wave, is independent of A and n_σ. Thus, both the upper and lower portions of the viscoplastic solitary wave are given by the viscous solitary wave solution with the sole modification that the compaction length scale in the overpressured region is

$$\delta_p = \delta R. \tag{23.64}$$

(A)

(B)

Overpressure > 0
(tensile stress)

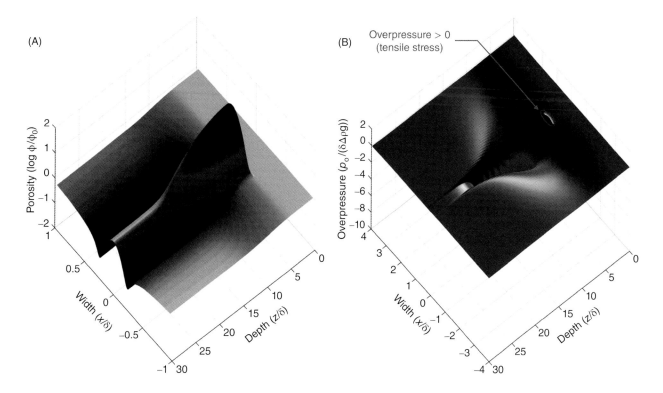

Fig. 23.9. Two-dimensional numerical simulation of a solitary porosity wave in a viscoplastic matrix. (A) Porosity; (B) fluid overpressure. The axial porosity and overpressure profiles of the wave are identical to the 1D case (Fig. 23.1C,D), but in the 2D case, the asymmetric overpressure distribution causes compaction of the background porosity on either side of the wave, leaving a tube-like channel that localizes subsequent fluid flow; the logarithmic scale for porosity emphasizes this effect. Numerical simulations (B. J. P. Kaus, personal communication 2005) have confirmed that 3D solitary waves in a viscoplastic matrix have radial symmetry orthogonal to the direction of propagation as in the viscous limit (Wiggins & Spiegelman 1995); thus, the 2D wave shown here corresponds to the axial section of a 3D wave. (*See color plate section for the color representation of this figure.*)

As the effect of weakening is to reduce the timescale during decompaction to $\tau_p = \delta R/|v_0|$, it is unsurprising that in three dimensions the overpressured region develops the spherical porosity distribution of the viscous solution on the length scale δ_p, which then recovers in the underpressured region to ϕ_0 on the length scale δ, giving rise to a wave shape similar to that of a cigar aligned in the direction of flow with the lit end upward (Connolly & Podladchikov 2007). In contrast to the 1D case, the asymmetry of the pressure distribution for such a wave obviates a true steady state. Specifically, numerical simulations (Fig. 23.9) show that the underpressured lower portion of the wave drains more fluid from surrounding matrix than is expelled into the matrix by the overpressured upper portion with the result that viscoplastic solitary waves grow with time. The imbalance in fluxes has the consequence that waves leave a tube-like channel, with porosities slightly ϕ_0, in their wake. This channel localizes subsequent fluid flow because it is surrounded by an interval of compacted matrix radius δ.

Although the 3D viscoplastic solitary wave solution is not steady state, at any point in time its properties are well represented by a geometric transformation of the viscous steady-state solution. Neglecting the small fraction of the excess volume associated with the overpressured portion of the wave, the

porosity distribution of the viscoplastic wave approximates the lower half of a prolate ellipsoid with semi-major axis $\lambda_p = \lambda/2$ and semi-minor axis of $R\lambda/2$, where λ is the wavelength of a viscous wave with the same velocity as the viscoplastic wave (Fig. 23.4C). As the velocity–amplitude relation (Fig. 23.4B) is, for $v_\phi/v_0 > {\sim}2n_\phi$, essentially independent of the dimension of the solution, the fluid transport rate for the viscoplastic case is

$$Q_p = \frac{Q}{2}R^2, \qquad (23.65)$$

where Q is the transport rate for the spherical viscous solitary wave (Eq. 23.63). In contrast to the 3D viscous case, where fluid is collected from an area proportional to λ, in the viscoplastic case, the horizontal radius of the wave is small in comparison with δ (Connolly & Podladchikov 2007; Connolly 2010). Thus, 3D viscoplastic waves collect fluid from a source area of ${\sim}\pi(\delta/2)^2$ regardless of the vertically integrated fluid production rate q_s. Consequently, for a given q_s, the initial velocity of a viscoplastic wave can be estimated by equating the fluid production likely to be collected by the wave, $q_s\pi(\delta/2)^2$, with Q_p. Using Eq. 23.65, the fluid transport rate of the viscous solution with the same velocity–amplitude relation as

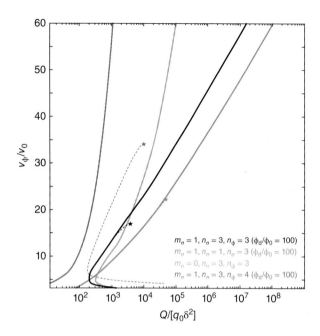

Fig. 23.10. Wave velocity as a function of volumetric fluid transport rate (Q) by spherically symmetrical 3D porosity waves (Wiggins & Spiegelman 1995) for exponent choices as indicated by the legend. For velocities at which the fluid pressure gradient (Fig. 23.4G) is nearly hydrostatic, the properties of the 1D and 3D solution are essentially identical; thus, at these conditions, the velocity can be used to predict 3D wave properties from the 1D solution (Fig. 23.4, and Examples #2 and #3). Where velocity is not a monotonic function of Q, it is probable that the high-velocity (short-wavelength) solution dominates. (*See color plate section for the color representation of this figure.*)

the viscoplastic wave is then

$$Q = \frac{\pi}{2} q_s \left(\frac{\delta}{R}\right)^2. \tag{23.66}$$

From this value of Q the relation between Q and v_ϕ for the 3D viscous solution (Fig. 23.10) yields the velocity of the viscoplastic wave. The remaining properties of the wave are then recovered from the 1D viscous solution as a function of this velocity with the modifications: $\lambda_p = \lambda/2$, $p_o^{min} = -p_o^{max}$, and the maximum overpressure, ostensibly σ_y, is $-Rp_o^{min}$.

It is possible to derive rigorous expressions for the effective viscosity resulting from various types of plastic yielding and the corresponding 1D solitary wave solutions (Yarushina 2009), but the viscoplastic solution for brittle yielding is well represented by the simple model presented here (Eq. 23.65) if the parameter R is adjusted to match σ_y. In particular, that viscoplastic matrix rheology causes solitary waves to grow with time and channelize fluid flow is likely to be a robust prediction. Unfortunately, the nonsteady character of the 3D solution simulated by the simple model creates an unrealistic situation in which the implied brittle yield strength grows with time. Although a multidimensional transient model with true brittle yielding remains to be investigated, in such a model p_o^{max} is constrained by σ_y, while the volume of the overpressured portion

at σ_y can be expected to increase with time. Such effects could substantially alter the results of the geometric model (i.e., Eq. 23.65) used to estimate the fluid transport rate here.

Example #4: 3D VISCOPLASTIC WAVE

To quantitatively illustrate the consequences of the foregoing model for viscoplastic waves, consider parameters as in Example #1, but with $R = 0.1$. The fluid transport rate of the corresponding 3D viscous wave (Eq. 23.66) is $Q/|q_0\delta^2| = (q_s/|q_0|)\pi/(2R^2) = 1.57 \times 10^4$. From the relationship between Q and v_ϕ (Fig. 23.10), $v_\phi = 25.3$ $v_0 = -0.79$ m year^{-1}, and for this velocity (Fig. 23.4), $\phi_{max} = 94.5$ $\phi_0 = 9.45 \times 10^{-3}$, $\lambda_p = \lambda/2 = 4.24$ $\delta = 424$ vm, and $p_o^{min} = -p_o^{max} = -3.99$ $\delta\Delta\rho g = -3.99$ MPa. From the model geometry, the actual fluid transport rate (Eq. 23.65) $Q_p = 1.57 \times 10^4$ $(R^2/2)$ $|q_0\delta^2| = 2.48$ m^3 year^{-1}; the maximum fluid overpressure $\sigma_y = -Rp_o^{min} = 0.399$ MPa; the radius of the wave and the channel left in its wake is $\lambda_p = \lambda/2 = 4.24$ $\delta = 424$ m; and the channels would have a spacing $\sim\delta = 100$ m.

Viscoelastic rheology

The omnipresent elastic response of the rock matrix (or pore fluid) becomes significant as the effective bulk viscosity of the matrix increases. Such an increase is to be expected as the crust strengthens upward toward the brittle–ductile transition and also locally in response to decreases in porosity. In general, steady-state solutions for Maxwell viscoelastic porous media take the form of heteroclinic shock waves that connect two distinct levels (ϕ_0 and ϕ_1 in the present formulation) of porosity (Rice 1992; Connolly & Podladchikov 1998). These can be understood in terms of the oscillating ball analogy to viscous solution (Fig. 23.2) in that elasticity acts similarly to friction on the motion of the ball, which dampens the oscillations of the ball so that it comes to rest at the focal point. Fluid compressibility and poroelasticity have opposite effects (Connolly & Podladchikov 1998): in a system composed of a viscous, inelastic, matrix and a compressible fluid the focal point porosity ϕ_1 is at the leading edge of the shock and the background porosity ϕ_0 is in its wake, whereas in a system composed of a viscoelastic matrix and a incompressible fluid the background porosity ϕ_0 is at the leading edge of the shock and the focal point porosity ϕ_1 is in its wake. Thus, the relative magnitude of the fluid and matrix elastic compressibilities controls whether the elevated porosity ϕ_1 is at the leading edge or in the wake of the viscoelastic. Most applications of elastic and viscoelastic porosity wave solutions in the geological literature (Rice 1992; Revil & Cathles 2002; Miller *et al.* 2004; Chaveau & Kaminski 2008; Joshi *et al.* 2012) assume negligible fluid compressibility. The heteroclinic character of viscoelastic solutions has the peculiar implication that the nondissipative elastic rheology leads to dissipative porosity shock waves. In transient models, a viscoelastic transition caused by upward strengthening provokes a rapid transition from lower crustal solitary waves, which are well approximated by the viscous limit, to porosity fluid pressure surges in the upper crust (Connolly & Podladchikov

1998). Even if both solid and fluid constituents are considered to be incompressible, surface tension as incorporated in the formulation of Bercovici *et al.* (2001) has the effect of generating a Kelvin viscoelastic compaction rheology. The Kelvin limit may be of relevance at the small porosities thought to be characteristic of the lower crust (Connolly & Podladchikov 2013; Ague 2014) at which surface tension may inhibit compaction.

CONCLUDING REMARKS

There is no smoking gun as evidence for the existence of porosity waves as a mechanism for fluid flow in the lower crust. The porosity wave model is the mathematical consequence of a set of physical assumptions that are generally thought to apply to lower crustal processes. Most prominent among these assumptions are that lower crustal rocks compact by viscous creep and that fluid flow is described by Darcy's law. The virtue of the porosity wave model is that it represents a physically consistent steady state and provides a simple means of anticipating the hydrodynamic response of the lower crust to perturbations such as fluid production. The formulation developed here has small ($\phi \ll \phi_d$) and intermediate ($\phi_d \ll 1 - \phi$) porosity approximations that are dependent only on relative porosity (ϕ/ϕ_0); material properties or, alternatively, scales (ϕ_0, $|v_0|$, and δ); two exponents (m_σ and n_ϕ) that characterize the porosity dependence of the effective bulk viscosity and permeability of the rock matrix; and an exponent (n_σ) that characterizes the stress dependence of effect shear viscosity of the rock matrix. In particular, the role of the stress exponent n_σ has not been considered in previous studies. The most surprising feature resulting from this nonlinearity is that it appears to admit a finite-wavelength solitary solution for shear-thickening ($n_\sigma < 1$) viscous mechanisms. Finite-wavelength solitary porosity waves are of interest because they permit deformation-propagated fluid flow through an initially impermeable matrix (Connolly & Podladchikov 1998). For the shear-thinning ($n_\sigma > 1$) viscous mechanisms thought to be characteristic of the lower crust (Kohlstedt *et al.* 1995; Ranalli 1995), the stress exponent does not fundamentally change the behavior described for the linear viscous case (Fowler 1984; Richter & McKenzie 1984; Scott & Stevenson 1984). However, somewhat counterintuitively, at low velocities ($v_\phi/v_0 < \sim 6$) nonlinearity results in poorly defined waves in which a greater proportion of the porosity lies in the tails of the waves compared to the porosity distribution of the linear viscous case. At higher velocities, this trend reverses so that a greater proportion of the fluid occurs near the center of mass of a wave in the nonlinear case. Disaggregation effects and increasing the nonlinearity of the effective bulk viscosity also lead to more sharply defined porosity distributions.

The ansatz that porosity waves evolve to accommodate the vertically integrated fluid production rate q_s in natural systems has the trivial consequence that in the 1D limit the effective permeability resulting from the porosity wave mechanism, $k_{effective} \approx k_0 \bar{q}_e/q_0$, is $\sim k_0 q_s/|q_0|$, where k_0 and q_0 are the background permeability and fluid flux, respectively, and $\bar{q}_e$ is time-averaged flux carried by a wave (Fig. 23.4H). Local variations in permeability are significantly larger, for example, in the 1D quantitative example considered here (Example #1), the maximum local permeability $k_0[\phi_{max}/\phi_0]^{n_\phi}$ is an order of magnitude greater than the effective permeability and three orders of magnitude greater than k_0. Spatial effects associated with 3D porosity waves lead to substantially higher effective permeability. In the quantitative example of the 3D viscous case (Example #2), $k_{effective} \approx k_0 Q/|q_0 \pi(\lambda/2)^2| = 3990 \, k_0$, and for the viscoplastic case (Example #4), $k_{effective} \approx k_0 Q_p/|q_0 \pi(R\lambda_p)^2| = 139 k_0$. These results are dependent on highly uncertain, but plausible, values for q_s and the scales ϕ_0, $|v_0|$, and δ (Connolly & Podladchikov 2013). In general, q_s can be estimated from the knowledge of the lithology of interest and the geodynamic scenario responsible for fluid production. The background porosity ϕ_0 and fluid velocity v_0 are roughly constrained from relatively well-known physical properties and theoretical considerations, leaving the compaction length scale δ as the greatest source of uncertainty in that it combines the hydraulic and rheological properties of the combined fluid–rock system. At present, it seems that the spatial scales of compaction-driven flow phenomena offer the most accurate means of estimating the compaction length in natural environments.

ACKNOWLEDGMENTS

This chapter was improved by reviews from Jay Ague and Martin Appold and by the editorial direction of Tom Gleeson, Steve Ingebritsen, and Craig Manning. The original version of this chapter was written while the author was a guest of the Centre of Advanced Studies at the Norwegian Academy of Science and Letters for the "Dynamics of Fluid Rock Systems" project led by Bjorn Jamtveit between 2000 and 2001.

APPENDIX: NONDIMENSIONALIZATION

For typical constitutive relations, the compaction equations admit a dimensionless form in the small porosity limit ($1 - \phi \to 1$, $\phi_d - \phi \to \phi_d$) that is independent of the absolute porosity (Scott & Stevenson 1984). In this limit, the constitutive relations given by Eqs 23.17 and 23.19 are

$$k = a_\phi \phi^{n_\phi} \tag{23.67}$$

$$f_\phi = n_\sigma^{-n_\sigma}(3/2)^{n_\sigma+1}\phi^{m_\sigma}. \tag{23.68}$$

Using these relations, and substituting $v_\phi = -v_\infty$, the dimensional forms of Eqs 23.15 and 23.16 simplify to

$$\frac{\partial p_o}{\partial z} = -v_\phi \frac{\eta_f}{k}(\phi - \phi_0) - \Delta\rho g\left(1 - \left[\frac{\phi_0}{\phi}\right]^{n_\phi}\right) \tag{23.69}$$

and

$$\frac{\partial \phi}{\partial z} = -\left(\frac{3}{2}\right)^{n_\sigma+1} \frac{\phi^{m_\sigma}}{n_\sigma^{n_\sigma} v_\phi} A |p_o|^{n_\sigma-1} p_o. \tag{23.70}$$

Taking the small porosity limit for the Darcy velocity through the unperturbed matrix

$$v_0 = -\frac{a_\phi \phi_0^{n_\phi-1}}{\eta_f} \Delta \rho g, \tag{23.71}$$

ϕ_0, $|\Delta \rho g|$, and δ as characteristic scales for velocity, porosity, pressure gradient, and length, respectively, the nondimensional wave velocity, porosity, overpressure, hydraulic potential, and depth are $v_\phi' = v_\phi/v_0$, $\phi' = \phi/\phi_0$, $p_o' = p_o/(\delta|\Delta \rho g|)$, $H' = H\phi_0^{m_\sigma-1}/|\Delta \rho g|$, and $z' = z/\delta$. Inverting these relations to express the dimensional variables in terms of the scales and nondimensional variables, the nondimensional forms of Eqs 23.69 and 23.70 are

$$\frac{\partial p_o'}{\partial z'} = [1 + v_\phi'(\phi' - 1)]/\phi'^{n_\phi} - 1 \tag{23.72}$$

and

$$\frac{\partial \phi'}{\partial z'} = \left(\frac{3}{2}\delta\right)^{n_\sigma+1} \frac{a_\phi \phi_0^{n_\phi-m_\sigma}}{n_\sigma^{n_\sigma} |\Delta \rho g|^{n_\sigma-1}} A \frac{\phi'^{m_\sigma} |p_o'|^{n_\sigma-1} p_o'}{v_\phi'}. \tag{23.73}$$

Defining the compaction length scale as

$$\delta \equiv \sqrt[n_\sigma+1]{\frac{n_\sigma^{n_\sigma} a_\phi \phi_0^{n_\phi-m_\sigma}}{A\eta_f |\Delta \rho g|^{n_\sigma-1}} \left(\frac{2}{3}\right)^{n_\sigma+1}}, \tag{23.74}$$

Eq 23.73 reduces to

$$\frac{\partial \phi'}{\partial z'} = \frac{\phi'^{m_\sigma} |p_o'|^{n_\sigma-1} p_o'}{v_\phi'}. \tag{23.75}$$

The dimensionless hydraulic potential is then

$$H' = \int \frac{1 - [1 + v_\phi'(\phi' - 1)]/\phi'^{n_\phi}}{\phi'^{m_\sigma}} d\phi'. \tag{23.76}$$

The hydraulic potential and solitary wave solution in Figure 23.2D–F are computed from Eqs 23.72, 23.75 and 23.76 with $m_\sigma = 0$, $n_\phi = 3$, and $v_\phi' = 7$.

CHAPTER 24

Hypocenter migration and crustal seismic velocity distribution observed for the inland earthquake swarms induced by the 2011 Tohoku-Oki earthquake in NE Japan: implications for crustal fluid distribution and crustal permeability

TOMOMI OKADA[1], TORU MATSUZAWA[1], NORIHITO UMINO[1], KEISUKE YOSHIDA[1], AKIRA HASEGAWA[1], HIROAKI TAKAHASHI[2], TAKUJI YAMADA[2], MASAHIRO KOSUGA[3], TETSUYA TAKEDA[4], AITARO KATO[5], TOSHIHIRO IGARASHI[5], KAZUSHIGE OBARA[5], SHINICHI SAKAI[5], ATSUSHI SAIGA[6], TAKASHI IIDAKA[5], TAKAYA IWASAKI[5], NAOSHI HIRATA[5], NORIKO TSUMURA[7], YOSHIKO YAMANAKA[8], TOSHIKO TERAKAWA[8], HARUHISA NAKAMICHI[9], TAKASHI OKUDA[8], SHINICHIRO HORIKAWA[8], HIROSHI KATAO[9], TSUTOMU MIURA[9], ATSUKI KUBO[10], TAKESHI MATSUSHIMA[11], KAZUHIKO GOTO[12] AND HIROKI MIYAMACHI[12]

[1] Research Center for Prediction of Earthquakes and Volcanic Eruptions, Graduate School of Science, Tohoku University, Sendai, Japan; [2] Institute of Seismology and Volcanology, Graduate School of Science, Hokkaido University, Sapporo, Japan; [3] Graduate School of Science and Technology, Hirosaki University, Hirosaki, Japan; [4] National Research Institute for Earth Science and Disaster Prevention, Tsukuba, Japan; [5] Earthquake Research Institute, University of Tokyo, Tokyo, Japan; [6] Tono Research Institute of Earthquake Science, Association for the Development of Earthquake Prediction, Mizunami, Japan; [7] Faculty of Science, Chiba University, Chiba, Japan; [8] Earthquake and Volcano Research Center, Graduate School of Environmental Studies, Nagoya University, Nagoya, Japan; [9] Disaster Prevention Research Institute, Kyoto University, Kagoshima, Japan; [10] Kochi Earthquake Observatory, Faculty of Science, Kochi University, Kochi, Japan; [11] Institute of Seismology and Volcanology, Faculty of Sciences, Kyushu University, Shimabara, Japan; [12] Graduate School of Science and Engineering, Kagoshima University, Kagoshima, Japan

ABSTRACT

After the occurrence of the 2011 magnitude 9 Tohoku earthquake, seismicity in the overriding plate changed. Seismicity appears to form distinct belts. From the spatiotemporal distribution of hypocenters, we can quantify the evolution of seismicity after the 2011 Tohoku earthquake. In some earthquake swarms near Sendai (Nagamachi-Rifu fault), Moriyoshi-zan volcano, Senya fault, and the Yamagata–Fukushima border (Aizu-Kitakata area, west of Azuma volcano), we can observe temporal expansion of the focal area. This temporal expansion is attributed to fluid diffusion. Observed diffusivity would correspond to the permeability of about 10^{-15} (m^2). We can detect the area from where fluid migrates as a seismic low-velocity area. In the lower crust, we found seismic low-velocity areas, which appear to be elongated along N–S or NE–SW, the strike of the island arc. These seismic low-velocity areas are located not only beneath the volcanic front but also beneath the fore-arc region. Seismic activity in the upper crust tends to be high above these low-velocity areas in the lower crust. Most of the shallow earthquakes after the 2011 Tohoku earthquake are located above the seismic low-velocity areas. We thus suggest fluid pressure changes are responsible for the belts of seismicity.

Key words: crustal fluid, crustal permeability, hypocenter migration, induced earthquake, seismic low-velocity area, the 2011 Tohoku-Oki earthquake

INTRODUCTION

The northeastern Japan arc is a typical subduction zone. Seismicity is high not only along the plate boundary but also in the overriding plate. Shallow seismic activity and crustal deformation are strongly affected by the water dehydrated and upwelling from the subducting Pacific plate (Hasegawa *et al.* 2009).

On March 11, 2011, the M9 megathrust earthquake (the 2011 off the Pacific coast of Tohoku earthquake or the 2011 M9.0 Tohoku-Oki earthquake) ruptured the plate boundary east of northern Japan beneath the Pacific Ocean. Shallow seismic activity in the crust of the overriding plate west of the source area changed significantly after the 2011 M9.0

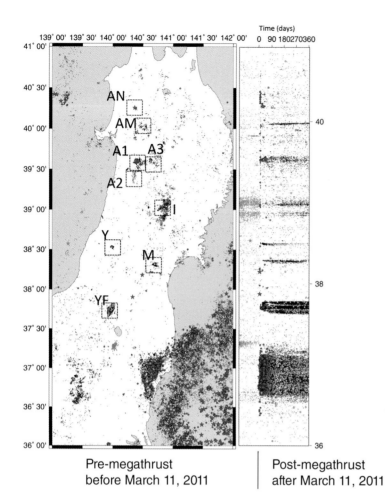

Fig. 24.1. Epicenter distribution of shallow microearthquakes (depth < 40 km) in the Tohoku region. Gray and black dots denote the earthquakes before and after the 2011 Tohoku-Oki earthquake. Circles and stars denote earthquakes with a magnitude of >3 and >5, respectively. Right side shows the time–latitude plot for these earthquakes. Horizontal axis is days from the 2011 Tohoku-Oki earthquake. Gray bold lines are the surface trace of active faults.

Pre-megathrust before March 11, 2011

Post-megathrust after March 11, 2011

Tohoku-Oki earthquake (e.g., Hirose *et al.* 2011; Okada *et al.* 2011; Toda *et al.* 2011; Kosuga *et al.* 2012).

Figure 24.1 shows the time–latitude distribution. The change of seismicity after the 2011 Tohoku earthquake can be seen. Most of the earthquake clusters were activated just after the 2011 Tohoku earthquake and subsequently decreased in activity, although some of them activated gradually. Yoshida *et al.* (2012) show that the seismogenic stress field changed because of the large slip caused by the 2011 Tohoku earthquake (Hasegawa *et al.* 2011, 2012; Sibson 2013). However, some of the earthquakes in the central part of Tohoku have focal mechanisms that do not correspond to the stress change by the 2011 Tohoku-Oki earthquake. Another factor that controls the occurrence of earthquakes is pore-fluid pressure (Okada *et al.* 2011; Tong *et al.* 2012; Kato *et al.* 2013; Terakawa *et al.* 2013). In particular, hypocenter migration can be explained as the fluid diffusion process (Shapiro *et al.* 1997; Zoback & Harjes 1997; Hill & Prejean 2005; Yukutake *et al.* 2011; Chen *et al.* 2012; Hardebeck 2012). Hypocenter migration has been observed for induced earthquakes (Shapiro *et al.* 1997; Zoback & Harjes 1997), tectonic earthquakes (Miller *et al.* 2004; Chen *et al.* 2012; Hardebeck 2012), and volcanic earthquakes (Hill & Prejean 2005; Yukutake *et al.* 2011).

In this study, we consider some possible evidence of the influence of crustal fluid/water on the occurrence of the triggered seismicity after the 2011 Tohoku-Oki earthquake. We also estimate the permeability from the observed hypocenter diffusion.

DATA AND METHOD

We used data from the temporary seismic network deployed by the Group for the aftershock observations of the 2011 off the Pacific coast of Tohoku earthquake. The Group consists of the members of Hokkaido University, Hirosaki University, NIED, University of Tokyo, Chiba University, Nagoya University, Kyoto University, Kochi University, Kyushu University, Kagoshima University, and Tohoku University. Seismometers (short-period: 1 Hz, 2 Hz or broad band: 20, 80, and 120 s) and off-line data loggers (sampling frequency: 100, 200, and 250 Hz) are deployed at most of the stations. Satellite or cell phone telemetry systems are deployed at some stations. This network was established after the 2011 Tohoku-Oki earthquake, and most of stations continue as of January 2014. We obtained precise hypocenter distribution not only for the earthquakes after the seismic network was activated but also for the earthquakes before the network was installed (e.g., Okada *et al.* 2003). We also used other temporary stations of Tohoku University and routinely operated stations of the Japan Meteorological Agency (JMA), Hi-net/National Research Institute for Earth Science and Disaster Prevention (NIED), University of Tokyo, Hirosaki University, the Association for the Development of Earthquake Prediction (ADEP), and Tohoku University.

For relocating the hypocenters, we adopted the double-difference location method (Waldhauser & Ellsworth 2000). We used manually picked arrival time data and differential arrival time calculated by cross-correlation. Assumed velocity structure is from Hasegawa *et al.* (1978). Because of limitations on the memory of calculation, we divided the study area into two subareas: northern (38.0N–0.5N) and southern (36.5N–38.0N).

We also estimate three-dimensional seismic velocity structure using the double-difference tomography method (Zhang & Thurber 2003). Because of the limited memory of calculation, we divided the study area into three subareas: northern (39.3N–0.8N), central (38.0N–39.7N), and southern (36.8N–38.5N). The central area corresponds to a previous study (Okada *et al.* 2011). Grid interval is 6 km. Initial velocity structure was derived from the study by Hasegawa *et al.* (1978).

Maps of stations, earthquakes, and grid distributions are shown in Appendix A. Results of a checkerboard resolution test are shown in Appendix B.

RESULTS

From the obtained hypocenter distribution, we identify temporal expansion of focal areas for some swarms. Figures 24.2–24.7 show the hypocenter distribution of each earthquake swarm. Color of the hypocenter indicates the elapsed time since the 2011 Tohoku-Oki earthquake.

In Appendix C, we show the hypocenter distributions of swarms for which temporal expansion of focal area is not clear, possibly because of the preexisting aftershock distribution for the 1914 Akita-Senboku earthquake (Sato *et al.* 2004), the 2008 Iwate–Miyagi Nairiku earthquake (Okada *et al.* 2012), and a Mjma5.1 earthquake on April 1, 2011 in northern Akita (Okada *et al.* 2011), or because of the small and complex

hypocenter distribution beneath Mt. Gassan volcano (Nakazato *et al.* 1996), in Appendix C.

Figure 24.2 shows the case of a seismic swarm in the central part of Miyagi Prefecture (Region M in Fig. 24.1). We show the focal mechanism of the Mjma 3.2 event that occurred on April 30. This mechanism is a reverse fault type, with a small amount of left lateral strike-slip. We determined the hypocenter distribution of this cluster. Events of the cluster appear to align with a strike of NNE–SSW and an eastward dip, which corresponds with one nodal plane of the focal mechanism. We assume that this plane with a strike of NNE–SSW is the fault plane of the earthquake swarm. A few kilometers east of this swarm is an earthquake cluster that aligns with a dip to the west. This cluster is the aftershock area of the 1998 M5.0 earthquake (Umino *et al.* 2002). A shallower extension of this aftershock alignment seems to reach the surface trace of the Nagamachi-Rifu fault. The swarm is not located on the shallower/deeper extension of this 1998 cluster and possible Nagamachi-Rifu fault. Figure 24.3 shows the seismicity around this earthquake swarm. In this earthquake swarm, we can observe temporal expansion of the focal area, which is clearer for first 80 days (see also Fig. 24.8).

Figure 24.4 shows the hypocenter distribution of the swarm near Mt. Moriyoshi-zan (Nakagawa 1983) in northern Akita (Region AM in Fig. 24.1). Mt. Moriyoshi-zan is a Quaternary volcano. In 1982, a distinct seismic swarm occurred and a reflected phase from the possible magma body was found (Hori & Hasegawa 1991). There are many small clusters, and the largest one shows clear temporal expansion of the focal area in their first 2 months. Figure 24.5 shows the seismicity around the major earthquake swarm. The focal mechanism of Mjma 2.4 event that occurred on May 11, 2011, is left lateral strike-slip with E–W oriented *P*-axis.

Figure 24.6 shows the hypocenter distribution of the swarm in southern Akita (Region A3 in Fig. 24.1). This swarm is located just beneath the surface trace of the Senya fault (Sato *et al.* 2002). The Senya fault is the seismogenic fault of the 1896 Riku-u earthquake. This cluster shows clear temporal expansion of the focal area. It expands to the northern part of the cluster.

Figure 24.7 shows the hypocenter distribution of the swarm in the Aizu-Kitakata area near the border between Yamagata and Fukushima prefectures (Region YF in Fig. 24.1). This swarm is located near the Aizu-Bonchi fault zones, along which there have been M7-class earthquake such as the 1611 Aizu earthquake (M7.3) (Sangawa 1987; The Headquarters for Earthquake Research Promotion 2008), and also is located west of Azuma volcano. This swarm consists of several subswarms, which apparently aligned in an approximately north–south direction. In this earthquake swarm, we can observe temporal expansion, which is clearer for about first 70 days, of the focal area.

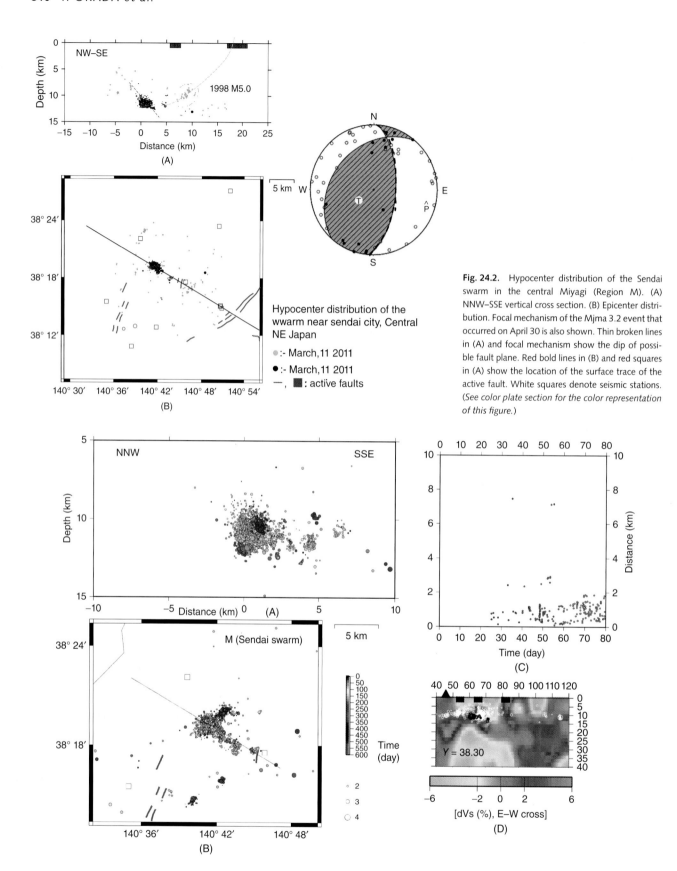

Fig. 24.2. Hypocenter distribution of the Sendai swarm in the central Miyagi (Region M). (A) NNW–SSE vertical cross section. (B) Epicenter distribution. Focal mechanism of the Mjma 3.2 event that occurred on April 30 is also shown. Thin broken lines in (A) and focal mechanism show the dip of possible fault plane. Red bold lines in (B) and red squares in (A) show the location of the surface trace of the active fault. White squares denote seismic stations. (*See color plate section for the color representation of this figure.*)

Hypocenter distribution of the wwarm near sendai city, Central NE Japan

● :- March,11 2011
● :- March,11 2011
—, ■ : active faults

Fig. 24.3. Hypocenter distribution of the Sendai swarm in the central Miyagi (Region M). (A) NNW–SSE vertical cross section. Colors indicate lapse times after the 2011 Tohoku-Oki earthquake. (B) Epicenter distribution. (C) Time–distance plots. Distance from the first event of this cluster is shown. (D) E–W vertical cross section of Vs perturbation. White and colored circles show the hypocenters before and after the 2011 Tohoku-Oki earthquake. Red triangles show the volcano. Red bold lines in (B) and red squares in (D) show the location of the surface trace of the active fault. White squares in (B) denote locations of seismic stations. (*See color plate section for the color representation of this figure.*)

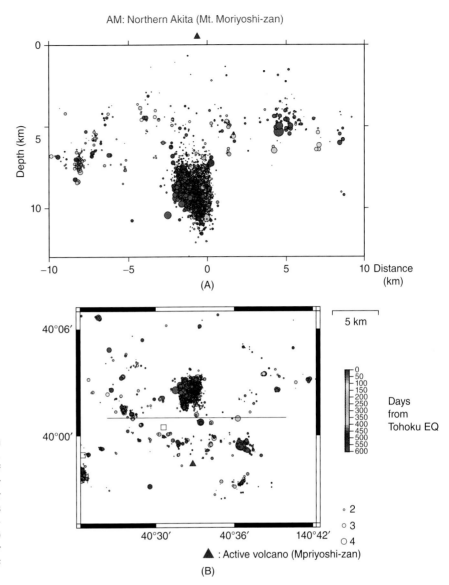

Fig. 24.4. Hypocenter distribution of the swarm near Mt. Moriyoshi in Northern Akita (Region AM). (A) E–W vertical cross section with a width of 10 km. Color of circle means the lapse time after the 2011 Tohoku-Oki earthquake. (B) Epicenter distribution. Line denotes the location of the cross section (A). Red triangle shows the volcano (Geological Survey of Japan 2013). White square in (B) denotes the location of a seismic station. (*See color plate section for the color representation of this figure.*)

DISCUSSION

Spatiotemporal expansion of hypocenter areas of some earthquake swarms can be attributed to fluid diffusion (e.g., Shapiro *et al.* 1997; Zoback & Harjes 1997). Figure 24.8 shows the r (distance)–t (time) plot of the hypocenter of the swarm in central Miyagi, northern Akita, and southern Akira, and at the Fukushima–Yamagata border. In Figure 24.8, r means the distance to each event from the location of the first event. We plot the expansion of the diffusion front as

$$r = \sqrt{4 \cdot \pi \cdot D \cdot t}$$

for D, which is the hydraulic diffusivity ($m^2\,s^{-1}$) (Shapiro *et al.* 1997). The front of temporal expansion of the swarm could

be well explained by this synthetic diffusion curve with D of 0.05–0.1 $m^2\,s^{-1}$. This diffusivity is in the range of previous studies (Ingebritsen & Manning 2010). Note that hypocentral error is smaller than 100 m and has little effect on the estimate of D. This diffusivity corresponds to a permeability of about 10^{-15} (m^2) assuming fluid compressibility of $4.8 \times 10^{-10}\,m^2\,N^{-1}$ (e.g., Shapiro *et al.* 1997; Ingebritsen & Manning 2010). This permeability could be a fault permeability because the earthquakes occur along fault zones. The value is in the range summarized by previous studies (Sibson & Rowland 2003; Ingebritsen & Manning 2010). The earthquake swarms occurred at similar depths of 4–10 km and yield similar permeability. This similarity may owe to the depth dependency of permeability (e.g., Ingebritsen & Manning 2010).

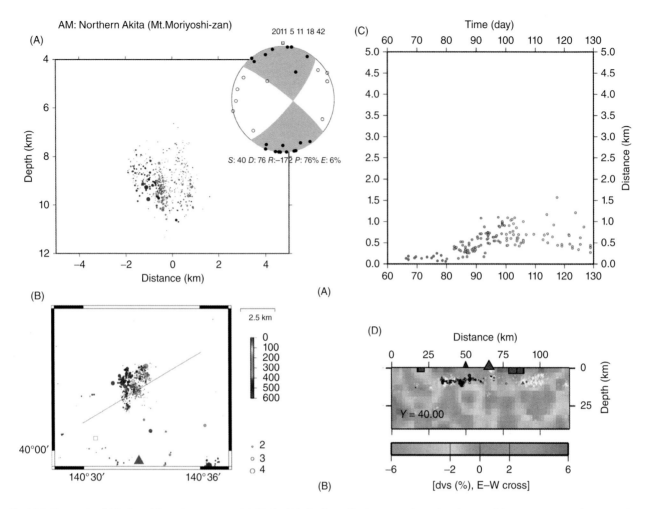

Fig. 24.5. Hypocenter distribution of the major swarm near Mt. Moriyoshi in Northern Akita (Region AM). Focal mechanism of the Mjma 2.4 event that occurred on May 11 2011 is also shown. (A) NE–SW vertical cross section. Color of circle means the lapse time after the 2011 Tohoku-Oki earthquake. (B) Epicenter distribution. Arrow shows the approximated front of hypocenters. (C) Time–distance plots. Distance from the first event of this cluster is shown. (D) E–W vertical cross section of Vp perturbation. Gray and colored dots show the hypocenters before and after the 2011 Tohoku-Oki earthquake. Red triangle shows the volcano. Red square in (D) shows the location of the surface trace of the active fault. White square in (B) denotes the location of a seismic station. (*See color plate section for the color representation of this figure.*)

We used a simplified method assuming linear diffusion (Shapiro *et al.* 1997). Some previous studies (e.g., Rice, 1992; Miller *et al.* 2004) suggested that permeability depends on effective normal stress; earthquakes can change permeability. Further studies can consider these nonlinear problems.

We calculated Coulomb stress change (delta-CFF) for each plausible fault plane (Okada *et al.* 2011). Hypocenter distributions were used to differentiate the fault plane from the auxiliary plane of the focal mechanisms for those earthquake sequences. Some of the plausible fault planes are not correlated with nearby previously identified active faults. Figure 24.9 shows the focal mechanisms with plausible fault plane and the calculated Coulomb stress change for the earthquakes. Some of the data are from Okada *et al.* (2011). For most of the sequences, the calculated Coulomb stress change is

positive, assuming an effective frictional coefficient of 0.65. This positive Coulomb stress change is mainly derived from the decrease of normal stress. This stress change could trigger pore pressure redistribution, although stress change on the order of 0.1 MPa might be too small to cause pore pressure redistribution. Strong seismic waves are another possible cause of pore pressure redistribution (e.g., Miyazawa 2011). We also estimated the minimum value of the effective frictional coefficient that would result in positive Coulomb stress change for each sequence (Cattin *et al.* 2009). Some of the resulting frictional coefficients are slightly >0.6, although most are <0.6.

We can detect areas with crustal fluid as seismic low-velocity areas. We also estimated a detailed seismic velocity structure in the central part of NE Japan. Figure 24.10 shows the obtained

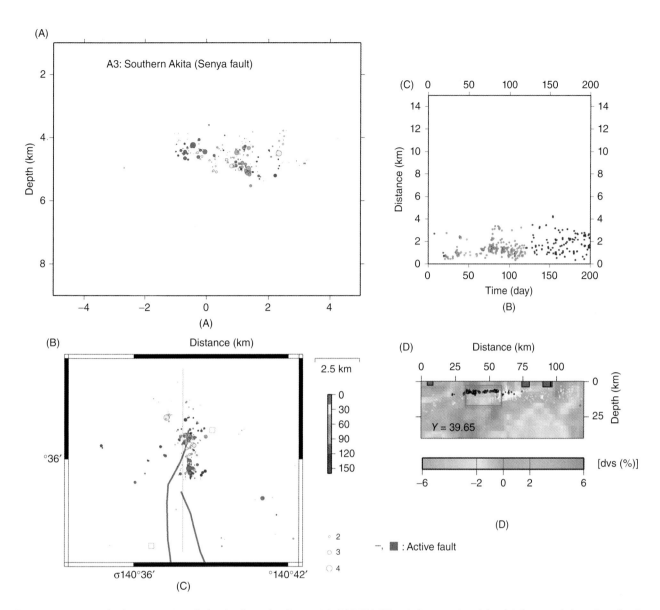

Fig. 24.6. Hypocenter distribution near Senya fault in Southern Akita (Region A3). (A) NNW–SSE vertical cross section. Color of circle means the lapse time after the 2011 Tohoku-Oki earthquake. (B) Epicenter distribution. (C) Time–distance plots. Distance from the first event of this cluster is shown. (D) E–W vertical cross section of Vs perturbation. Gray and colored dots show the hypocenters before and after the 2011 Tohoku-Oki earthquake. Red bold lines in (B) and red squares in (D) show the location of the surface trace of the active fault. White squares in (B) denote the locations of seismic stations. (*See color plate section for the color representation of this figure.*)

seismic velocity structure. The Vp/Vs distribution is shown in Figure 24.11.

In the upper crust (Fig. 24.10A), there are many distinct seismic low-velocity areas. These low-velocity areas are distributed in and around the active faults and the volcanoes. Some of these low-velocity areas seem to be spatially correlated with large extensional strain areas (e.g., Latitude 38.5, Longitude 140.3) produced by the 2011 Tohoku-Oki earthquake (Ohzono *et al.* 2013). In the upper crust, the earthquakes seem to be distributed in seismic high-velocity areas rather than in seismic low-velocity areas.

In the lower crust (Fig. 24.10B), we found seismic low-velocity areas that appear to be elongated along N–S or NE–SW, the strike of the island arc. These seismic low-velocity areas are located not only beneath the volcanic front but also beneath the fore-arc region. Seismic activity in the upper crust tends to be high above these low-velocity areas in the lower crust.

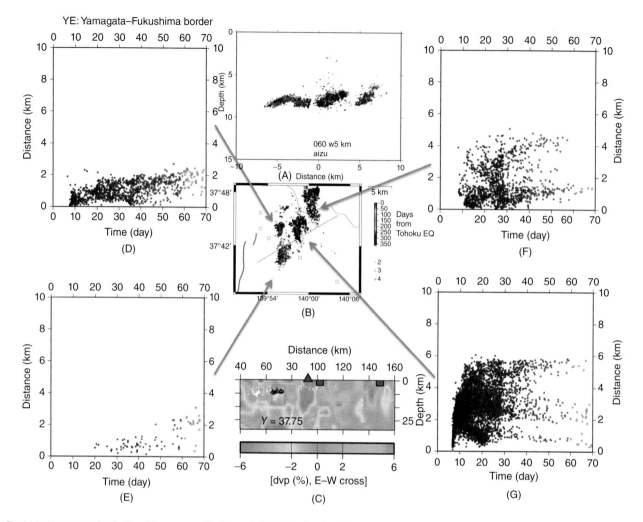

Fig. 24.7. Hypocenter distribution of the swarm on the Yamagata–Fukushima border. (A) NE–SW vertical cross section. Color of circle means the lapse time after the 2011 Tohoku-Oki earthquake. (B) Epicenter distribution. (C) E–W vertical cross section of Vs perturbation. Black and white dots show hypocenters before and after the 2011 Tohoku-Oki earthquake. Red triangle shows the volcano. Red bold lines in (B) and red squares in (C) show the location of the surface trace of the active fault. White square in (B) denotes the location of a seismic station. (D–G) Time–distance plots. Distance from the first event of each cluster is shown. (*See color plate section for the color representation of this figure.*)

Vp/Vs is a useful information for understanding the water/fluid distribution (e.g., Okada *et al.* 2012). Figure 24.11 shows the Vp/Vs distribution. In the upper crust (A), low Vp/Vs areas are distributed in and around the earthquakes. Low Vp/Vs areas can be interpreted as high-aspect-ratio pores with water or gas. In the lower crust (B), high Vp/Vs areas are distributed beneath the volcanoes and earthquakes. These high Vp/Vs areas can be interpreted as melt. Water or other crustal fluids are released upward from the melt, which is cooled and solidified. Note that the low-velocity zone in the fore-arc lower crust is more obvious for Vp (Fig. 24.12) with low Vp/Vs. This might mean that the low-velocity zone in the fore-arc

lower crust can be interpreted as high-aspect-ratio pores with water or gas rather than melt.

We found a distinct seismic low-velocity area below the seismically active areas (the seismic belt) along the volcanic front and the back-arc and fore-arc region. This seismic low-velocity area could correspond to an area with overpressurized fluids, and migration of overpressurized fluids from there would promote the occurrence of earthquakes. Most of the shallow earthquakes, including the normal fault earthquakes in northern Ibaraki and southeastern Fukushima (Tong *et al.* 2012; Kato *et al.* 2013) after the occurrence of the 2011 Tohoku earthquake, are also located above the

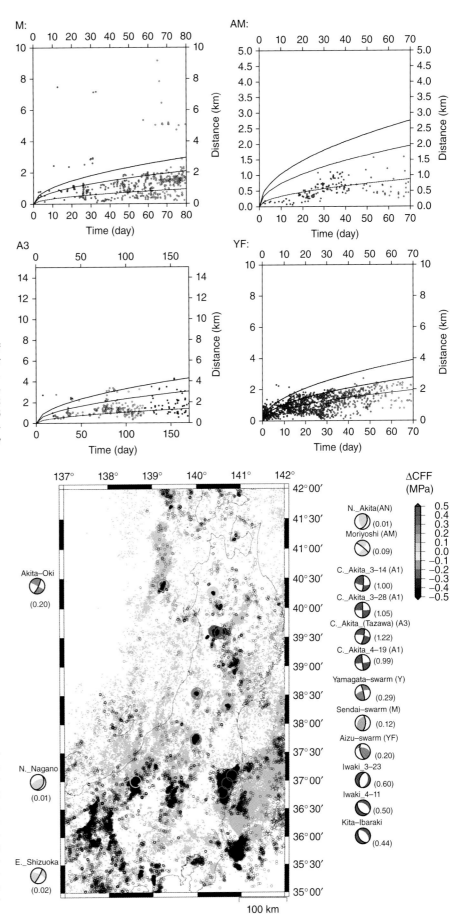

Fig. 24.8. Examples of temporal expansion of earthquake swarms. Lines are the solution for the front of fluid diffusion by Shapiro *et al.* (1997) with D of 0.1 (bold line), 0.05 (thin line), 0.01 (thin broken line) (m² s¹) for M, AM, A3; 0.2 (bold broken line), 0.1 (bold line), and 0.05 (thin line) for YF. The front of temporal expansion of the swarm would be well explained by hydraulic diffusivity of 0.05–0.1 m² s⁻¹.

Fig. 24.9. Focal mechanisms with the plausible fault plane defined by bold great circles. Color of circle indicates the amount of Coulomb stress change (numerals in parenthesis) caused by the 2011 Tohok-Oki earthquake for each fault assumed in this study. Hypocenters of inland earthquakes that occurred in NE Japan based on the Japan Meteorological Agency (JMA) catalog are also shown. Inland earthquakes before and after the 2011 Tohoku earthquake are distinguished by gray and black circles, respectively. Stars denote earthquakes with magnitude >5. (*See color plate section for the color representation of this figure.*)

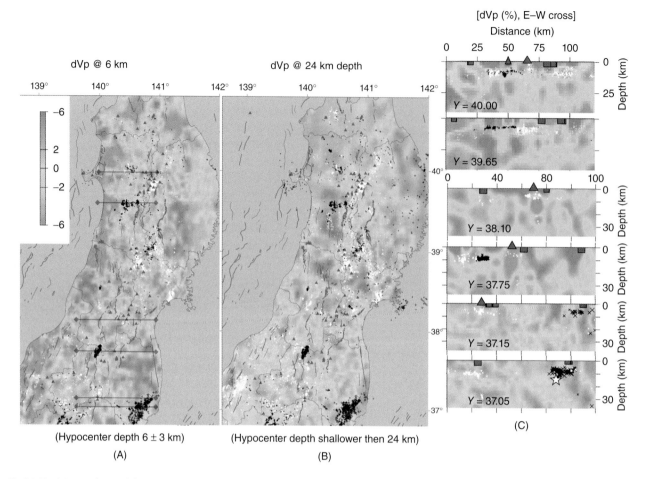

Fig. 24.10. Seismic velocity of the crust in NE Japan. (A) Vp perturbation at a depth of 6 km. Hypocenters of events at depths of 3–9 km are also shown. (B) Vs perturbation at a depth of 24 km. Hypocenters of events at depths <24 km are also shown. (C) E–W vertical cross section of Vp perturbation at various latitudes. Locations of cross sections are shown by gray lines in (A). White and black dots show hypocenters before and after the 2011 Tohoku-Oki earthquake. Large and small stars denote the hypocenters of a Mjma 7.0 earthquake on April 11, 2011 and a Mjma 6.4 earthquake on April 12, 2011 respectively. Red triangles show volcanoes (Geological Survey of Japan 2013). Red bold lines in (A) and (B), and red squares in (c) show the location of surface traces of active faults. (*See color plate section for the color representation of this figure.*)

seismic low-velocity area (Figs 24.3D, 24.5D, 24.6D, 24.7C, and IOC).

CONCLUSIONS

We obtained the detailed hypocenter distribution of inland earthquakes in NE Japan after the 2011 Tohoku-Old earthquake as well as the seismic velocity structure. From the results, certain earthquake swarms are thought to be affected by both the stress change and a possible fluid pressure change. From the temporal expansion of some earthquakes swarms, the permeability would be about 10^{-15} in the seismogenic crust. Further monitoring can test this conclusion.

ACKNOWLEDGMENTS

This study was supported by the Ministry of Education, Culture, Sports, Science and Technology (MEXT) of Japan, under its Observation and Research Program for Prediction of Earthquakes and Volcanic Eruptions. This work was conducted with the support of a Grant-in-Aid for Special Purposes (23900002), MEXT, Japan. This work was partly conducted with the support of the Scientific Research Program on Innovative Areas, "Geofluids: Nature and Dynamics of Fluids in Subduction Zones," at the Tokyo Institute of Technology (21109002). This study is a part of the "multidisciplinary research project for high strain rate zone" promoted by MEXT. We used data from the Japan Meteorological Agency (JMA), Hi-net/NIED,

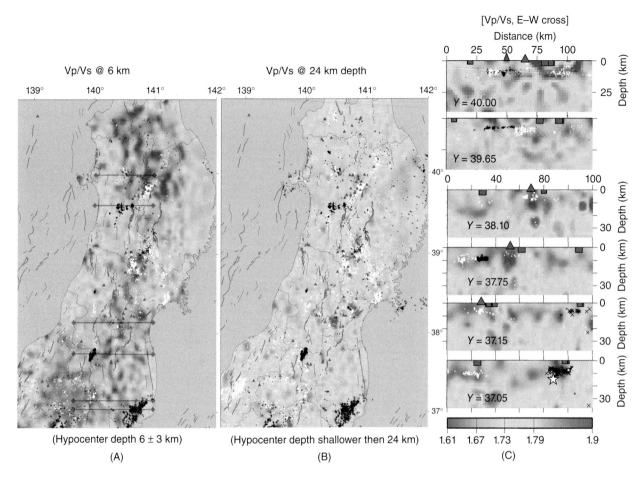

Fig. 24.11. Vp/Vs distribution of the crust in NE Japan. (A) At a depth of 6 km. Hypocenters of events at depths from 3 to 9 km are also shown. (B) At a depth of 24 km. Hypocenters of the events at depths <24 km are also shown. (C) E–W vertical cross section of Vp perturbation at various latitudes. Locations of cross sections are shown by gray lines in (A). White and black dots show hypocenters before and after the 2011 Tohoku-Oki earthquake. Large and small stars denote the hypocenters of a Mjma 7.0 earthquake on April 11, 2011 and a Mjma 6.4 earthquake on April 12, 2011, respectively. Red triangles show volcanoes. Red bold lines in (A) and (B), and red squares in (C) show the locations of surface traces of active faults. (*See color plate section for the color representation of this figure.*)

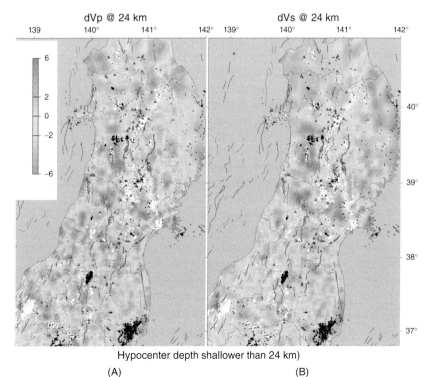

Fig. 24.12. Seismic velocity distribution of the crust in NE Japan. (A) dVp at a depth of 24 km. (B) dVs at a depth of 24 km. Hypocenters of events at depths <24 km are also shown. (*See color plate section for the color representation of this figure.*)

University of Tokyo, Hirosaki University, and the Association for the Development of Earthquake Prediction (ADEP). Some of the temporary stations are operated cooperatively with the Japan Nuclear Energy Safety Organization (JNES). The authors have declared no conflicts of interest. We thank Sadato Ueki, Satoshi Miura, Shinji Toda, Dapeng Zhao, Takeyoshi Yoshida, Yasuo Yabe, Junichi Nakajima, Naoki Uchida, Yusaku Ohta, Mako Ohzono, Masa'aki Ichiki, Ryota Takagi, Shin'ya Sakanaka, Jun Muto, Bun'ichiro Shibazaki, and Richard Sibson for their valuable comments. We acknowledge the efforts toward seismic observation by all the members of the Group for the aftershock observations of the 2011 off the Pacific coast of Tohoku earthquake, and Shuichiro Hori, Toshio Kono, Satoshi Hirahara, Takashi Nakayama, Toshiya Sato, Kenji Tachibana, Syuichi Suzuki, Tomotsugu Demachi, and Toshiki Kaida. The comments by Prof. C. Manning, Prof. T. Gleeson, S. Ingebritsen and two anonymous reviewers improved the manuscript well.

APPENDIX A – MAPS FOR STATION AND HYPOCENTER, AND GRID FOR TOMOGRAPHIC INVERSION

See Figure 24.A1.

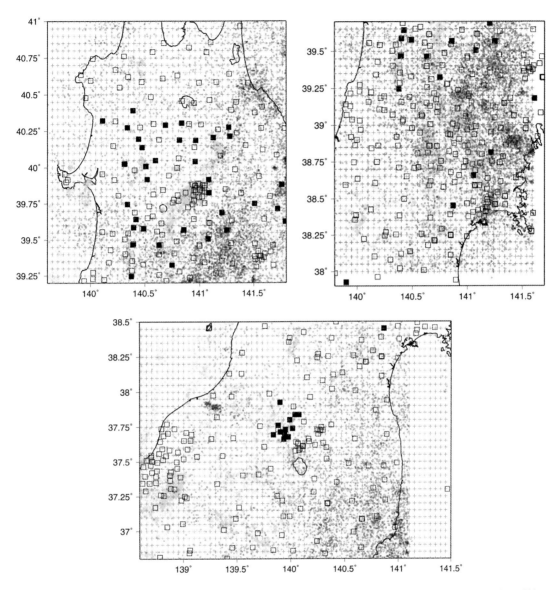

Fig. 24.A1. Distribution of stations and earthquakes used in this study. We show three subareas for seismic tomography. Square, green circle, and blue circle denote station, shallow earthquake, and deep earthquake, respectively. Black square denotes the station deployed by the Group for the aftershock observations of the 2011 off the Pacific coast of Tohoku earthquake. Grid used in this study is also shown by crosses. (*See color plate section for the color representation of this figure.*)

APPENDIX B – CHECKERBOARD RESOLUTION TEST

In this study, we estimate three-dimensional seismic velocity structure by the double-difference tomography method (Zhang & Thurber 2003). To confirm the resolution of an obtained image, we have performed the checkerboard resolution test. In the checkerboard resolution test, we make a synthetic structure with a 20-km checkerboard pattern, calculate the synthetic travel time data for the synthetic structure, and invert them as the real data to determine how the synthetic structure can be recovered. Figure 24.B1A–C shows the recovered image of dVp at a depth of 6 km, which corresponds to Figure 24.10A. The patterns are well recovered. Figure 24.B1D–F shows the recovered image of dVs at a depth of 24 km, which corresponds to Figure 24.10B. The patterns are recovered enough to be discussed.

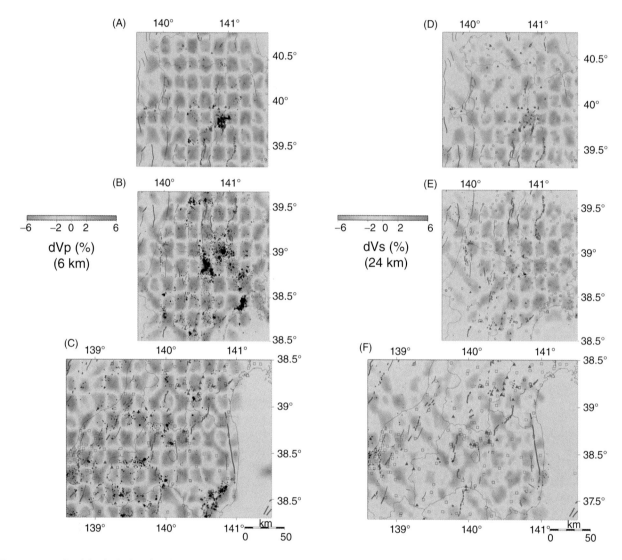

Fig. 24.B1. Results of the checkerboard resolution test. (A) dVp at a depth of 6 km for the northern part, (B) the central part, and (C) the southern part. (D) dVs at a depth of 24 km for the northern part, (E) the central part, and (F) the southern part. Black dots show the hypocenters before the 2011 Tohoku-Oki earthquake. Red triangles show the volcano. Red open square shows the location of the station used in this study. Red bold line denotes the surface trace of active fault. (*See color plate section for the color representation of this figure.*)

APPENDIX C – HYPOCENTER DISTRIBUTION OF THE SWARM FOR WHICH TEMPORAL EXPANSION OF FOCAL AREA IS NOT CLEAR

Figure 24.C1A,B shows the hypocenter distribution of the swarm in southern Akita (Region Al). This swarm is a few kilometers north of the hypocenter of the 1914 Akita-Senboku earthquake (shown by star). Many subswarms align in the ENE–WSW directions. Temporal expansion of the hypocenter after the 2011 Tohoku-Oki earthquake is difficult to see. Figure 24.C1C,D shows the hypocenter distribution of the swarm in southern Akita (Region A2). This swarm is a few kilometers south of the hypocenter of the 1914 Akita-Senboku earthquake (shown by star). Pre-megathrust seismic activity (shown in gray circles) would correspond with the aftershocks

of the 1914 Akita-Senboku earthquake (Sato *et al.* 2004). Most of them are the aftershocks of the 1914 Akita-Senboku earthquake, and temporal expansion of the hypocenter after the 2011 Tohoku-Oki earthquake cannot be seen.

Figure 24.C2 shows the hypocenter distribution in the aftershock area of the 2008 Iwate–Miyagi earthquake (Region I). The earthquakes after the 2011 Tohoku-Oki earthquake forms the westward-dipping and crossing eastward-dipping alignments. They are the aftershock distribution of the 2008 Iwate–Miyagi Nairiku earthquake (Okada *et al.* 2012), as well as those before the 2011 Tohoku-Oki earthquake. Most of them are the aftershocks of the 2008 Iwate–Miyagi earthquake, and temporal expansion of the hypocenter after the 2011 Tohoku-Oki earthquake cannot be seen.

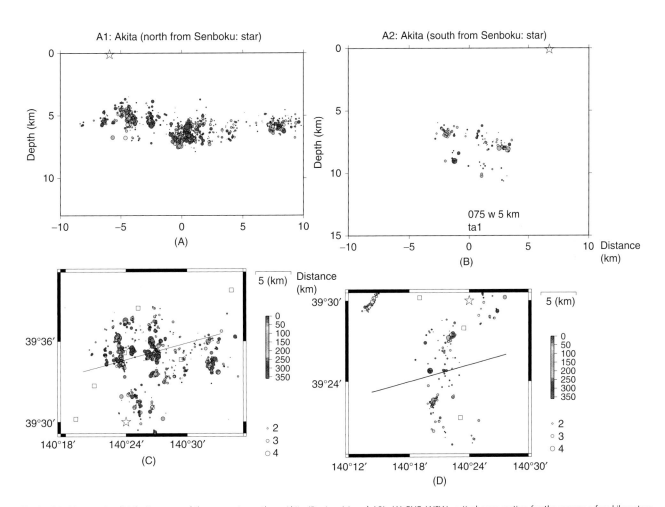

Fig. 24.C1. Hypocenter distribution near of the swarm in southern Akita (Region A1 and A2). (A) ENE–WSW vertical cross section for the swarm a few kilometers north of the hypocenter of the 1914 Akita-Senboku earthquake. Color of circle means the lapse time after the 2011 Tohoku-Oki earthquake. (B) Epicenter distribution for (A). (C) ENE–WSW vertical cross section for the swarm a few kilometers south of the hypocenter of the 1914 Akita-Senboku earthquake. Color of circle means the lapse time after the 2011 Tohoku-Oki earthquake. Gray circles are the earthquake within 1 year before the 2011 Tohoku-Oki earthquake. (D) Epicenter distribution for (C). White square denote the location of seismic station. (*See color plate section for the color representation of this figure.*)

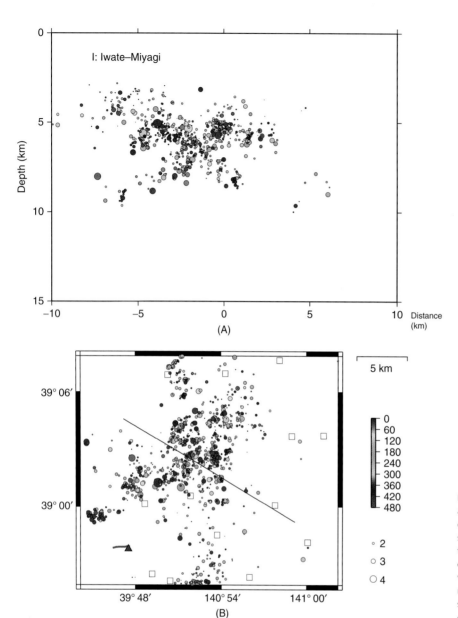

Fig. 24.C2. Hypocenter distribution in the focal area of the 2008 Iwate–Miyagi Nairiku earthquake (Region I). (A) NW–SE vertical cross section. Color of circle means the lapse time after the 2011 Tohoku-Oki earthquake. (B) Epicenter distribution. Red triangle and red bold line denote volcano and the surface trace of active fault, respectively. White square denote the location of seismic station. (*See color plate section for the color representation of this figure.*)

Figure 24.C3 shows the hypocenter distribution of the swarm near Mt. Gassan volcano, southern Yamagata prefecture (Region Y). The swarm is a few kilometers southwest of the summit of Gassan volcano shown by red triangle. Hypocenters are distributed at a depth of 5–10 km around the seismic low-velocity area (B). It seems that temporal expansion is to shallower depths, although it is not clear because of small and complex hypocenter distribution.

Figure 24.C4 shows the hypocenter distribution of the earthquake in northern Akita prefecture (Region AN). The earthquakes are the aftershocks of an Mjma 5.1 earthquake on April 1, 2011. They align on a southeastward dipping plane which corresponds to the fault plane of the Mjma 5.1 earthquake (Okada *et al.* 2011). The aftershock area seems to be slightly extended, but it is not distinct.

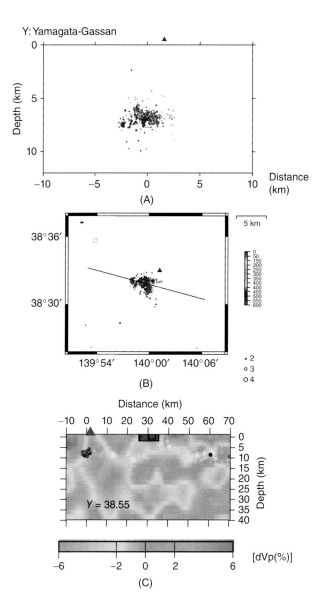

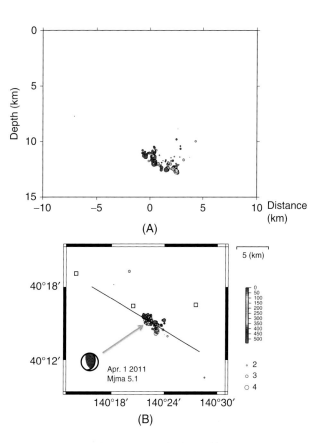

Fig. 24.C4. Hypocenter distribution in the focal area of the Mjma 5.1 earthquake in northern Akita (Region AN). (A) NW–SE vertical cross section. Color of circle means the lapse time after the 2011 Tohoku-Oki earthquake. (B) Epicenter distribution. AQUA-CMT (2011-04-01 19:49:47/NORTHERN AKITA PREF on http://www.hinet.bosai.go.jp/AQUA/aqua_catalogue.php?y=2011™m=04™ LANG=en) is also shown. Whites square denote the locations of seismic stations. (*See color plate section for the color representation of this figure.*)

Fig. 24.C3. Hypocenter distribution near Mt. Gassan volcano in Yamagata (Region Y). (A) NNW–SSE vertical cross section. Color of circle means the lapse time after the 2011 Tohoku-Oki earthquake. (B) Epicenter distribution. (C) E–W vertical cross section of Vs perturbation. Black and white dots show the hypocenters before and after the 2011 Tohoku-Oki earthquake. Red triangles show the volcano (Gassan). Red square shows the location of the surface trace of the active fault. White square denote the location of seismic station. (*See color plate section for the color representation of this figure.*)

CHAPTER 25

Continental-scale water-level response to a large earthquake

ZHEMING SHI[1,2], GUANG-CAI WANG[1], MICHAEL MANGA[2] AND CHI-YUEN WANG[2]

[1] *School of Water Resources and Environment, China University of Geosciences, Beijing, China;* [2] *Department of Earth and Planetary Science, University of California, Berkeley, CA, USA*

ABSTRACT

Coseismic groundwater-level changes induced by earthquakes have been reported for thousands of years. The M8.0 Wenchuan earthquake caused coseismic groundwater-level responses across the Chinese mainland. Three types of changes were recorded in 197 monitoring wells: coseismic oscillations ranging in amplitude from 0.004 to 1.1 m, immediate coseismic step changes ranging from 0.0039 to 9.188 m, and more gradual postseismic changes ranging from 0.014 to 1.087 m. We find that the coseismic groundwater-level response is complex. There is neither a clear relationship between the response amplitude and the distance from the epicenter nor a clear relationship between the groundwater response and lithology at the continental scale. Both the sign and amplitude of water-level changes are random at the continental scale, and a poroelastic response to the coseismic static strain cannot explain most of the coseismic changes. However, wells located near the boundaries of tectonically active blocks have larger response amplitudes than those in the middle of these active blocks. Considered together, these observations indicate that permeability enhancement caused by the earthquake is a significant or dominant mechanism causing water-level changes. These data indicate that large earthquakes can cause widespread permeability changes in the shallow crust although the magnitude of permeability change is uncertain.

Key words: coseismic strain, groundwater level, permeability, Wenchuan earthquake

INTRODUCTION

Coseismic hydrological responses following earthquakes have been documented for thousands of years (Institute of Geophysics-CAS 1976). These include changes in the water level, temperature, and chemical composition in wells, disappearance or formation of springs, increasing streamflow, and changes in the activity of mud volcanoes and geysers (Roeloffs 1996; Montgomery & Manga 2003; Manga & Wang 2007; Wang & Manga 2010a). Understanding the origin of these hydrological phenomena may have significant impacts on understanding the occurrence of liquefaction (Cox *et al.* 2012a), water supply and quality (Gorokhovich & Fleeger 2007), and underground waste storage (Carrigan *et al.* 1991; Wang *et al.* 2013). Furthermore, the hydrological responses may provide unique insight into the interaction between hydrogeologic and tectonic processes at scales in space and time that help us to understand the long-term evolution of groundwater flow. Understanding hydrological responses to stresses may also provide a framework to assess proposed hydrological precursors to earthquakes (Wang & Manga 2010b).

In the past decades, a large number of hydrological responses have been documented. Among these, changes in water level have attracted the most attention, and many mechanisms have been put forward to explain these phenomena. These include the following: (i) the coseismic groundwater-level response can be explained by the coseismic static strain and pore-pressure change predicted by poroelastic theory (Wakita 1975; Roeloffs 1996; Ge & Stover 2000; Jonsson *et al.* 2003; Shi *et al.* 2013b); (ii) undrained dilatation and consolidation of saturated sediment lead to the step-like changes in water level (Wang 2001; Manga *et al.* 2003; Wang & Chia 2008); (iii) permeability enhancement of the crust in the intermediate and far fields (Rojstaczer *et al.* 1995; Sato *et al.* 2000; Brodsky *et al.* 2003; Wang *et al.* 2009; Shi *et al.* 2013c; Xue *et al.* 2013); and (iv) oscillations caused by the seismic waves in the far field produced by resonant coupling of flow between the well and the aquifer (Cooper *et al.* 1965; Liu *et al.* 1989; Kano & Yanagidani 2006; Wang *et al.* 2009).

The majority of these investigations focused on the coseismic response in a small region or the response of a small set of wells to several earthquakes. Response over large alluvial fans was

Crustal Permeability, First Edition. Edited by Tom Gleeson and Steven E. Ingebritsen.
© 2017 John Wiley & Sons, Ltd. Published 2017 by John Wiley & Sons, Ltd.
Companion Websites: www.wiley.com/go/gleeson/crustalpermeability/
http://crustalpermeability.weebly.com/

documented following the 1999 M7.1 Chi-Chi earthquake in Taiwan (Chia *et al.* 2001, 2008; Wang 2001; Wang & Chia 2008) and the 2011 M7.1 Christchurch earthquake, New Zealand (Cox *et al.* 2012a; Gulley *et al.* 2013). Here, we consider coseismic response in consolidated rocks to the great Wenchuan M8.0 earthquake. Many studies have investigated the relationship between water-level changes and the well–epicenter distance, and several have found that the coseismic response amplitude is correlated with the earthquake magnitude and well–epicenter distance (Roeloffs 1998; Sil & Freymueller 2006; Chia *et al.* 2008). The present data set allows us to reexamine the relationship between the well–epicenter distance and the water-level response amplitude at the continental scale.

THE WENCHUAN EARTHQUAKE AND THE GROUNDWATER-LEVEL MONITORING NETWORK

The Wenchuan earthquake

On May 12, 2008, M8.0 Wenchuan earthquake occurred in the Longmenshan fault zone on the eastern margin of the Tibetan plateau. Three major subparallel faults constitute the northeast trending Longmenshan fault zone: the Pengguan fault to the east along the mountain front fault; the Beichuan fault about 10–15 km to its west; and the Wenchuan–Maowen fault about another 30 km west of the Beichuan fault (Shen *et al.* 2009). The Wenchuan earthquake was the largest earthquake in the mainland of China in the past 60 years. It killed 69,227 people and destroyed many towns and villages (Burchfiel *et al.* 2008). The epicenter was 31.02°N, 103.37°E, with a focal depth of 19 km (USGS, http://comcat.cr.usgs.gov/earthquakes/event-page/pde20080512062801570_19#summary). Coseismic fault scarps reveal a complicated pattern of slip, with the Wenchuan earthquake rupturing 240 km of the Beichuan fault and 72 km of the parallel Pengguan fault (Xu *et al.* 2009). InSAR analysis shows that the coseismic slip on the fault can be divided into two parts with the boundary in Beichuan: thrust fault-slip dominated along the southwest part of the fault while dextral fault-slip dominated along the northeast part of the fault (Hao *et al.* 2009).

Groundwater monitoring network in China

Groundwater monitoring is an important component of the earthquake prediction research program in China. Many monitoring wells were constructed, and some abandoned oil exploration wells have also been utilized (Wang 1985; Roeloffs 1988). Since the 1960s, the groundwater monitoring network has experienced four development stages: (i) creation stage (1966–1978), (ii) development stage (1979–1989), (iii) enhancement stage (1991–2000), and (iv) overall modernization stage (2000–). Today, there are 670 monitoring wells (including springs), of which more than 400 wells are

specifically for groundwater-level monitoring (Shi *et al.* 2013a). These wells are maintained at different administrative levels: (i) the basic wells and the regional wells are supported by the State Seismological Bureau and managed by the State Seismological Bureau and provincial seismological bureaus, respectively; (ii) the local wells are supported by the local government and managed by the local government seismological bureaus; (iii) the enterprise wells (springs) are supported and managed by enterprises such as petroleum companies (Huang *et al.* 2004). After two digital upgrades of the monitoring wells, most of the basic and regional monitoring wells sample with periods of 1 h or 1 min, and the accuracy of water-level measurement can reach 1 mm. The data from the basic wells, regional wells, and some local wells are sent to the China Earthquake Network Center (CENC) each day for analysis and prediction (all the data were raw records, without corrections for barometric pressure or tides). The rest are sent to the local earthquake administration or saved by the specialists who work at the monitoring stations. All the wells are constructed along faults or points of special tectonic interest in consolidated rock and tap confined aquifers (Huang *et al.* 2004).

COSEISMIC GROUNDWATER-LEVEL CHANGES INDUCED BY THE WENCHUAN EARTHQUAKE

The Wenchuan earthquake caused large-scale groundwater-level changes across the Chinese mainland. We collected groundwater-level data from the CENC for a total of 336 monitoring wells. Among these, 76 wells were out of order at the time of the earthquake and 63 wells exhibited no coseismic response to the earthquake. The remaining 197 monitoring wells did record a coseismic response: groundwater level went up in 88 wells, went down in 67 wells, and showed only oscillations in 42 wells (Fig. 25.1).

Figure 25.1 shows that almost all the monitoring wells that showed coseismic responses are located along or near the boundary of active blocks. Active tectonic blocks are defined as geological units separated by active tectonic boundaries across which there are high gradient of differential movement. These are zones of late Cenozoic to the present tectonic deformation, which likely have a series of fault zones. Almost all earthquakes of magnitude >8 and 80–90% of earthquakes of $M > 7$ occur along these boundaries (Zhang *et al.* 2003).

Groundwater levels rose in the Chuandian block, north and east of the South China block, and on the junction boundary of the Ordos block and North China block, the Yanshan block and North China block. Water levels decreased near the epicenter, along the boundary of South China block and South China Sea block, and near the southwest boundary of Shandong Yellow Sea block. We find that most wells exhibiting only groundwater oscillations are located far from the epicenter, with epicenter distance larger than 1000 km, especially in the northeast of the North China block, Northeast block, and the southwest of the East Shandong Yellow Sea block. Several stations in Chuandian

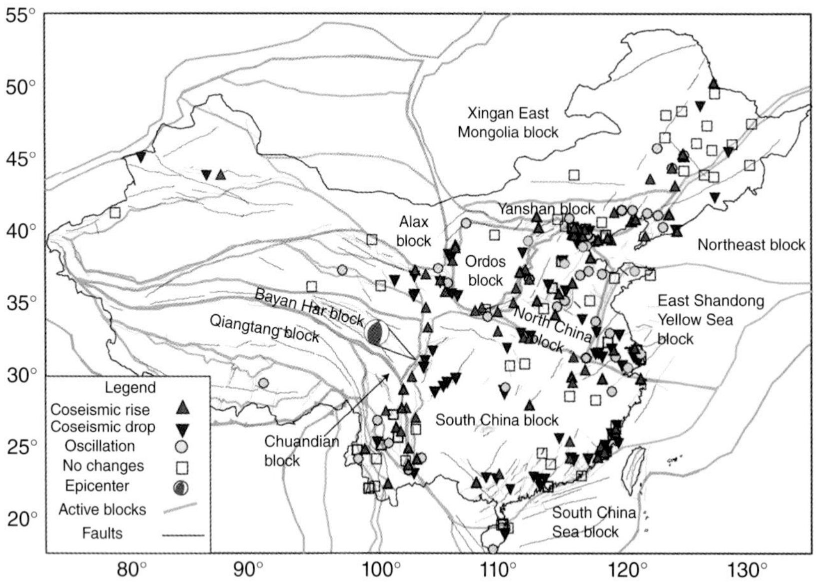

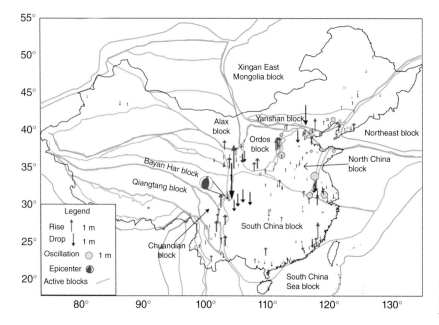

Fig. 25.1. Coseismic water-level changes in response to the Wenchuan earthquake. (*See color plate section for the color representation of this figure.*)

Fig. 25.2. Distribution of coseismic response amplitudes induced by the Wenchuan earthquake. (*See color plate section for the color representation of this figure.*)

block also show oscillations in groundwater level. The wells which have no response are mostly located in the interior of the blocks and at large distances from the epicenter.

Spatially, the response amplitudes are variable (Fig. 25.2). The maximum water-level rise occurred near the epicenter (JY well with an epicenter distance of 161 km), but large increases also occurred in the southeast of China. Monitoring wells located within 80 km of the junction of active tectonic blocks (junction of Ordors and North China block, junction of Yanshan and Northeast block) also showed large coseismic responses. For coseismic water-level decline, the maximum excursion also occurred near the epicenter; however, near the junction of some of the active blocks there were also large

decreases. Wells in Northeast China, at the junction between the North China block and the South China block, showed the largest amplitude of water-level oscillations (as large as 1.1 m at an epicenter distance of 1400 km) despite the large well–epicenter distance.

We classified the coseismic water-level responses into three major types (gradual, step-like, and oscillation) and nine subtypes (Fig. 25.3): (i) gradual drop, (ii) gradual rise, (iii) gradual-drop oscillation, (iv) gradual-rise oscillation, (v) step drop, (vi) step rise, (vii) step-drop oscillation, (viii) step-rise oscillation, and (ix) oscillation. For the gradual change, we take the maximum amplitude of changes in 24 h as the response amplitude. This classification is similar to that adopted by

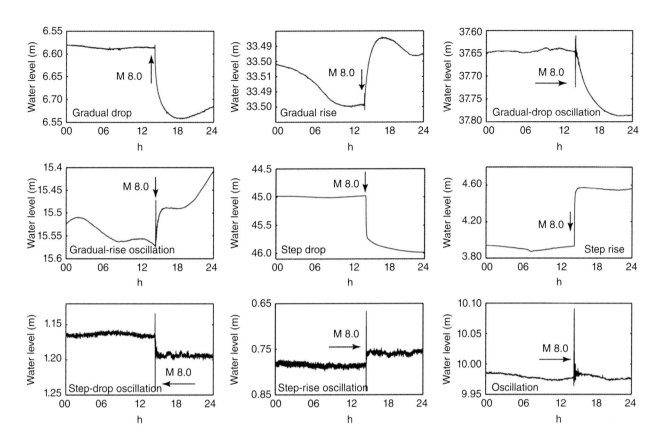

Fig. 25.3. Nine distinct response patterns of water-level changes induced by the Wenchuan earthquake. All the examples shown have sampling intervals of 1 min. (*See color plate section for the color representation of this figure.*)

Cox *et al.* (2012a). We use the amplitude of steps as the coseismic response amplitude for the step-like changes, and we take the maximum oscillation amplitude as the coseismic change for oscillations. Because the sampling interval for most wells was 1 min (see table in the supplemental material), coseismic oscillations were recorded. We further divide the locations of wells into two categories: the near field, within approximately 1.5 rupture fault lengths, and intermediate field, 1.5–10 rupture fault lengths (Shi *et al.* 2014). Figure 25.4 summarizes the relationship between coseismic groundwater-level response, so-defined, and the distance from the epicenter.

The relationship between the coseismic groundwater-level response and distance to the Wenchuan epicenter is not as strong as that reported for the Chi-Chi earthquake (Chia *et al.* 2008). In fact, the amplitude and sign of coseismic water-level changes show only a weak relationship with distance from the epicenter. The best-fit regression curve to the coseismic rise h as a function of well–epicenter distance x is $h = -0.27\log(x) + 2.25$ with squared correlation coefficient (R^2) of 0.14, indicating a weak correlation between coseismic water-level rise and the well–hypocenter distance. For the coseismic water-level drop, the R^2 is only 0.074, an even weaker correlation. This can be largely attributed to the relative heterogeneity of the geological and tectonic settings of the two earthquakes. For the Chi-Chi

earthquake, almost all the wells were installed in an alluvial fan consisting of unconsolidated sand and gravel (Wang 2001); hence, the documented water-level changes reflect the response of these materials to dynamic shaking and the attenuation of seismic stress with distance. The monitoring wells in this study, however, are located in many different geological and tectonic settings, and most of them are located in consolidated sedimentary and crystalline rocks. We compared the coseismic responses to the Wenchuan and Chi-Chi earthquakes as shown in Figure 25.5. The coseismic responses following the Chi-Chi earthquake display a more systematic variation with distance than the responses to the Wenchuan earthquake.

Comparison of the water-level responses during the two events may thus highlight the effect of geological settings on the coseismic response. Various coseismic response types occurred in the near and far fields and differ from those reported in previous studies. As summarized by many previous studies, step-like changes are usually observed in the near field (Chia *et al.* 2001; Wang 2001; Jonsson *et al.* 2003), gradual changes in intermediate field (Roeloffs 1998; Brodsky *et al.* 2003), and oscillations in the far field (Cooper *et al.* 1965; Liu *et al.* 1989; Kano & Yanagidani 2006). In the near field of the Wenchuan earthquake, most of the wells show step-like changes, but four of them show gradual changes. In the intermediate field,

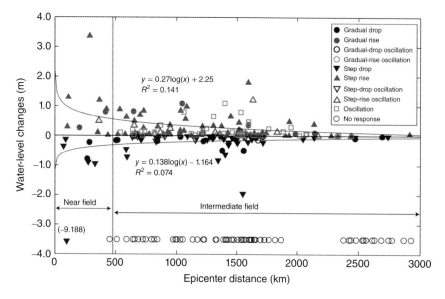

Fig. 25.4. Coseismic groundwater-level responses as a function of distance from the epicenter. The black circles indicate the locations of wells that recorded no coseismic response. (*See color plate section for the color representation of this figure.*)

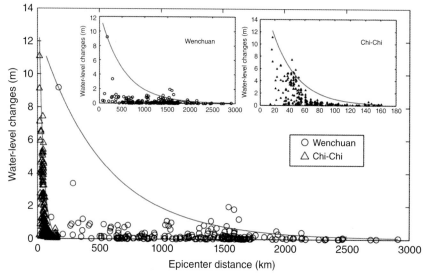

Fig. 25.5. Envelopes of the responses to the Wenchuan and 1999 Chi-Chi earthquakes. Triangles and circles show the coseismic responses following the 1999 Chi-Chi and the Wenchuan earthquakes, respectively. The red and blue curves are the envelopes for the Chi-Chi and the Wenchuan earthquakes, respectively. (*See color plate section for the color representation of this figure.*)

step-like changes and oscillations are observed, in addition to gradual changes (Fig. 25.4). The step-like changes have larger response amplitudes than the gradual changes and oscillations.

MECHANISMS OF THE COSEISMIC WATER-LEVEL CHANGE

Relationship between volumetric strain and water-level change

Several studies have suggested that coseismic water-level changes are due to the coseismic volumetric strain caused by slip on the ruptured fault (Roeloffs 1996; Ge & Stover 2000; Jonsson *et al.* 2003). For wells installed in unconsolidated sedimentary aquifers, however, the changes in water level have often shown the opposite sign to those predicted by the coseismic static strain change (Wang 2001; Koizumi

et al. 2004). As noted earlier, water-level changes after the Wenchuan earthquake were complicated by the geological and hydrological settings. Nevertheless, it may still be informative to assess the relationship between the coseismic water-level changes and the volumetric strains. Here, we calculate the static strain change induced by an earthquake assuming a fault dislocation in a uniform half space with the analytical expression of Okada (Okada 1992; Lin & Stein 2004; Toda *et al.* 2005). We use 31.100°N, 103.300°E as the location of the Wenchuan epicenter and the finite fault model obtained from the USGS, which consists of 168 subfaults, each with a length of 15 km and a width of 5 km. Details of the fault model are available at http://www.geol.ucsb.edu/faculty/ji/big_earthquakes/home.html. The computed coseismic static volumetric strain is shown in Figure 25.6.

The sign of the groundwater-level response is generally consistent with the static strain changes when the strain is

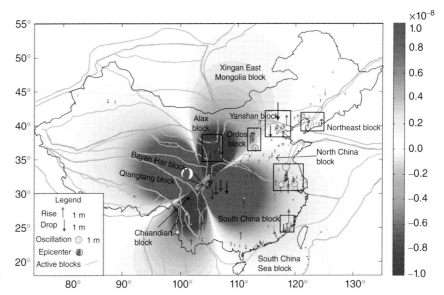

Fig. 25.6. Coseismic groundwater-level changes as a function of volumetric strain calculated from the dislocation model. Red colors indicate dilatation, and blue colors indicate contraction. The black rectangles highlight sensitive monitoring sites that have large coseismic responses to the earthquake. (*See color plate section for the color representation of this figure.*)

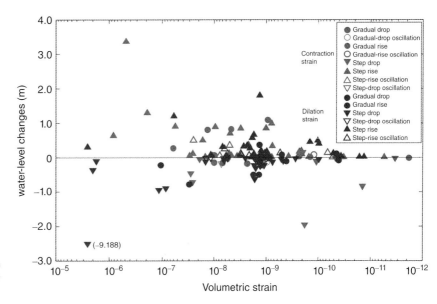

Fig. 25.7. Relationship between volumetric strain and the water-level response. (*See color plate section for the color representation of this figure.*)

larger than 5×10^{-8} (Fig. 25.7). The well–hypocenter distance that results in strains larger than 10^{-8} is approximately 1000 km for the Wenchuan earthquake, consistent with the result derived from Dobrovolsky *et al.* (1979). The volumetric strain ε_{kk} can be related to pore-pressure change p in a well by $p = -(2GB/3)[(1+\nu_u)/(1-2\nu_u)]\,\varepsilon_{kk}$ where G is the shear modulus of the material, B is Skempton's coefficient, and ν_u is the "undrained" Poisson's ratio. Typical values of the coefficient $((2GB/3)[(1+\nu_u)/(1-2\nu_u)])$ on the right side are 5–50 GPa (assuming G of 5–30 GPa, B of 0.5–0.8, $\nu_u = 0.3$ (Detournay & Cheng 1993)). If the coefficient equals 10 GPa, volumetric strain of 10^{-6} can produce a water-level change of 1 m, whereas a strain of 10^{-8} would produce a 1 cm change (Roeloffs 1996; Montgomery & Manga 2003). Thus, for the coseismic water-level changes for the Wenchuan

earthquake, most amplitudes are larger than predicted from this scaling. There are 71 wells that show response amplitudes larger than 0.1 m in areas with $<10^{-8}$ static strain, and many wells show coseismic responses despite static strain $<10^{-9}$ (Fig. 25.7). Thus, the coseismic static strain may only explain a small fraction of the water-level changes. For most of the water-level changes, the amplitude dose not simply scale with the static strain, and the relationship between response amplitude and static strain is random for many wells. Especially when the static strain is $<10^{-8}$, the direction of many water-level changes is opposite to that predicted from the static strain field, implying that static strain is not responsible for most coseismic responses. Other factors such as dynamic stress, geological, and hydrogeological conditions must play important or dominant roles.

Relationship between ground motion and coseismic water-level change

Dynamic strain, induced by the passage of seismic waves, can also cause groundwater-level changes (Montgomery & Manga 2003; Itaba *et al.* 2008). Several studies have investigated the relationship between coseismic water-level changes and ground shaking in Taiwan and Japan (Wang *et al.* 2003; Lai *et al.* 2004; Wong & Wang 2007; Itaba *et al.* 2008). As the peak ground velocity is proportional to the dynamic strain (Brodsky *et al.* 2003; Jiang *et al.* 2010a; Wang & Manga 2010a), we plot peak ground velocity versus groundwater-level change at 21 locations where the seismic stations are <25 km from our groundwater monitoring wells (although only 21 data points were selected, they are distributed throughout the Chinese mainland) (Fig. 25.8). The plot shows no clear relationship between water-level changes and peak ground velocity, similar to the result reported by Lai *et al.* (2004), indicating that the local geological or tectonic factors may have a more significant effect on the magnitude of coseismic water-level changes.

The effect of lithology to the coseismic water-level response

Both Figures 25.4 and 25.7 show that some groundwater-level changes have large amplitude in the intermediate field despite small static strains ($<10^{-8}$). We selected 20 wells with response amplitude larger than 0.3 m, epicenter distances larger than 1000 km, and static strains $<10^{-8}$ (Table 25.1). Of these wells, 15 are located in sedimentary rocks (i.e., sandstone, limestone), and the others are in igneous rocks. Thus, lithology also seems to have little impact on the coseismic groundwater changes. The static stresses caused by the Wenchuan earthquake are too small to cause these large changes. Thus, the dynamic stresses from the passage of seismic waves must be responsible for the changes or there must be some additional mechanism that allows small static stress changes to cause large responses. One common explanation is that the dynamic or static stresses cause an increase in permeability that leads to changes in water level as hydraulic heads adjust to the permeability changes (Brodsky *et al.* 2003; Elkhoury *et al.* 2006; Wang & Chia 2008; Manga *et al.* 2012).

In order to assess the overall influence of lithology on the coseismic response, we divide well–aquifer systems into three types: sandstone aquifers (105 wells), limestone aquifers (42 wells), and igneous rock (including granite) aquifers (47 wells), and plot the response amplitude and epicenter distance for each lithology (Fig. 25.9). Step-like changes are the most common in all the three kinds of lithology. Only three wells in igneous rock show oscillations, compared with 18 in limestone and

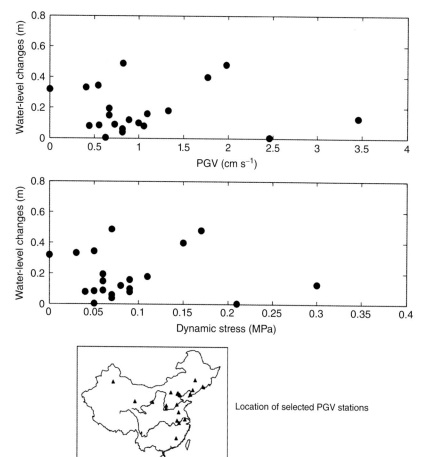

Location of selected PGV stations

Fig. 25.8. Relationship between coseismic water-level changes and peak ground velocity (PGV) and dynamic stress; 21 PGV stations were selected from 271 stations of Chinese Digital Seismic Network (CDSN). The selected seismic stations are <25 km from groundwater monitoring wells. (PGV data from Jiang *et al.* 2010a.)

Table 25.1 Monitoring wells with larger response amplitude

Well name	Amplitude (m)	Epicenter distance (m)	Strain	Lithology
XY	0.845	1034	−1.01E-09	Sandstone
JX	1.087	1045	−9.34E-10	Sandstone
GP	0.57	1073	1.84E-09	Sandstone
QX	0.984	1082	−7.90E-10	Pyroclastic
HUS	0.32	1213	2.66E-09	Sandstone
AQ	0.345	1305	2.29E-09	Sandstone
SZ	−0.829	1344	−1.43E-11	Limestone
ZC	0.333	1373	−7.63E-10	Basalt
CH	−0.502	1379	1.75E-09	Sandstone
HS	−0.63	1404	1.65E-09	Sandstone
WH	−0.307	1431	1.58E-09	Limestone
WX	0.66	1449	1.72E-09	Gravel
HY	0.453	1517	1.44E-10	Limestone
ZJZ	−1.952	1548	−1.80E-10	Limestone
ZZNJ	−0.5	1555	1.37E-09	Granite
YC	1.8	1593	1.33E-09	Granite
AX	0.363	1593	1.34E-09	Rhyolite
TSCS	1.2125	1631	8.29E-12	Limestone
LF	0.489	1936	1.02E-10	Sandstone
DDB	0.4	2140	9.69E-11	Pegmatite

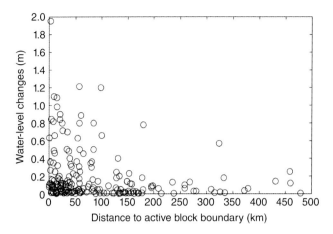

Fig. 25.10. Relationship between the coseismic water-level response and tectonic blocks. The distance to block boundaries is the minimum distance from the well to the nearest active block boundary.

21 in sandstone. The relationship between coseismic response amplitude and epicenter distance are all weak for all the three types of rocks: for igneous rocks, the square correlation coefficient (R^2) between water level and epicenter distance is 0.11 for coseismic rise and 0.10 for coseismic drop; for the limestones, it is 0.35 for the coseismic rise and 0.10 for coseismic drop; for the sandstones, it is 0.15 for coseismic rise and 0.07 for coseismic drop. The poor correlation indicates that the coseismic groundwater-level responses are not dominated by lithology.

Effect of tectonic setting

Finally, we test whether the tectonic setting affects the coseismic response. To reduce the effect of static strain caused by the slip in the near field, we choose the wells beyond the near field to plot the relationship between coseismic water-level response (absolute value of water-level change) versus distance to active block boundary (Fig. 25.10). We find that most of the wells that had large coseismic response (larger than 0.5 m) are located <80 km from a boundary. We thus infer that tectonic setting may have a significant effect on coseismic water-level responses.

Furthermore, in Figure 25.6, we identify six areas that show large amplitudes of coseismic response at large epicenter distances. One common feature is that they are all located near

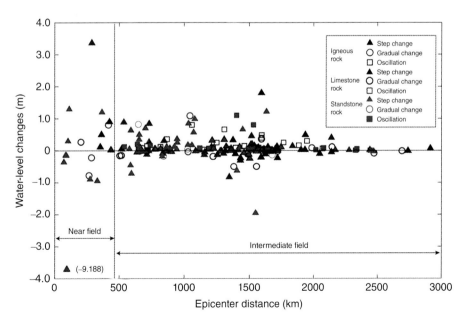

Fig. 25.9. Relationship between coseismic groundwater-level changes and lithology. (See color plate section for the color representation of this figure.)

the junction of two or three active blocks. Discontinuous deformation due to repeated accumulation and release of tectonic stress near the junctions of these blocks may have damaged and weakened the rocks that make up these aquifers (Zhang *et al.* 2003). Those wells that showed large response at great distances may be located at sites that are sensitive to seismic shaking (King *et al.* 1999) or at sites where seismic waves or shaking focuses.

Previous studies indicate that the fault permeability structure is complex and variable by increasing and/or decreasing permeability (Chester *et al.* 1993; Evans *et al.* 1997). Faults can be hydraulically permeable (conduit faults) or hydraulically permeable parallel to the fault and less permeable normal to the fault (conduit-barrier fault), which Bense *et al.* (2013) suggest are the most common. As active faults form the boundaries of active blocks, the junctions of active blocks are expected to have higher permeability, and the larger amplitude of water-level changes may reflect this higher permeability (Manga & Wang 2007; Shi *et al.* 2013c). Other possibilities are that the wave-guide effect of active fault zones (Li *et al.* 1997) may focus seismic energy, enhancing shaking near the junction of active blocks, thus enhancing response amplitudes. Alternatively, fault zone material along active faults zones may be more elastic than the surrounding material, and wells located in such zones would exhibit larger pressure changes in response to stress changes. More observational data are needed to test these hypotheses.

DISCUSSION

After collecting data recorded at 260 groundwater monitoring wells, which were operational on May 12, 2008, we identified 197 wells with coseismic responses. The tectonic setting of the wells seems to control the coseismic response. The signs of groundwater-level changes are roughly consistent with the coseismic static strain field when the wells are located within 1000 km of the epicenter and static strains are $>10^{-8}$. However, the relationship between the response amplitude and epicenter distance or static strain is not consistent with a poroelastic response in most wells (Shi *et al.* 2014). In the near field, hydrogeological responses are generally caused by a combination of static and dynamic stresses (Lai *et al.* 2014). Because both the static and dynamic stresses (strains) caused by the earthquake are significant (Manga & Brodsky 2006), both may have effects on the water level that are not easily distinguished. Indeed, some wells that located within an epicentral distance <300 km showed static strain effects after the Wenchuan earthquake (Shi *et al.* 2013b). Overall, however, even in the near field (about 500 km for the Wenchuan earthquake), the coseismic water-level response is not consistent with the static strain, especially the predicted magnitude of water-level changes (Shi *et al.* 2014). It should be noted, however, there are few wells located very near to the epicenter, and the tidal signals in these

wells are not clear, which makes it difficult to distinguish the effect of static and dynamic strains.

Our conclusion that static strains have only a small effect on water-level change is quite different from those of some previous studies that found a good relationship between well–hypocenter distance and coseismic response amplitude (Roeloffs 1998; Chia *et al.* 2008). In these studies, the wells were mostly located in unconsolidated sediments and similar tectonic settings, so the coseismic responses are largely controlled by sediment consolidation (Wang 2001) and the attenuation of seismic energy. However, because of the heterogeneity of geological and tectonic settings at the continental scale, coseismic response amplitudes are highly variable. Many of the responses that we document are likely responses to changes in permeability, which can lead to either positive or negative changes in the water level because enhanced permeability can occur either upgradient or downgradient of a well. If the enhanced permeability is located upgradient of the well, then the water level in the well will exhibit a coseismic rise. If the enhanced permeability occurs downgradient of the well, then a coseismic fall in water level would be expected. At the continental scale, permeability enhancement may occur either in upgradient or downgradient of wells. If a sufficiently large number of observations are available, the model of enhanced permeability would predict a statistically random occurrence in the sign of the water-level change as a function of distance from the epicenter (Wang & Chia 2008), and this is the case in the intermediate field after the Wenchuan earthquake.

CONCLUSIONS

In this chapter, we report the coseismic continental-scale groundwater-level changes following the Wenchuan earthquake. Three types of changes were recorded in 197 monitoring wells, with coseismic oscillations ranging in amplitude from 0.004 to 1.1 m, immediate coseismic step changes ranging from 0.0039 to 9.188 m, and more gradual postseismic changes ranging from 0.014 to 1.087 m. The coseismic water-level changes on the continental scale are rather complex, and there is great variability in the relationship between water-level changes, epicentral distance, and static strain. Hydrogeological and tectonic settings are dominant factors in determining the coseismic response. Wells located near the boundaries of active blocks can have large coseismic responses even when the epicenter distance is large. Both the sign and amplitude of water-level changes are random at the continental scale. Poroelastic response to the coseismic static strain cannot explain most of the coseismic changes. Permeability changes caused by stress changes (either static or dynamic) may explain the large variability of the coseismic response amplitude. These results also suggest that earthquakes can cause the widespread permeability changes in the shallow crust.

ACKNOWLEDGMENTS

This work is supported by the National Natural Science Foundation of China (40930637, 41272269), The Specialized Research Fund for the Doctoral Program of Higher Education of China (20100022110001), Fundamental Research Funds for the Central Universities (2652013088), with financial support from the China Scholarship Council, and the US National Science Foundation.

SUPPORTING INFORMATION

Additional supporting information may be found in the online version of this article, *Geofluids* (2015) 15, p. 310–320.

Table S1. Well information for 197 groundwater wells that showed coseismic response to the 2008 M8.0 Wenchuan earthquake.

PART III(D): Effects of fluid injection at the scale of a reservoir or ore-deposit

Development of connected permeability in massive crystalline rocks through hydraulic fracture propagation and shearing accompanying fluid injection

GIONA PREISIG[1], ERIK EBERHARDT[2], VALENTIN GISCHIG[3], VINCENT ROCHE[4], MIRKO VAN DER BAAN[5], BENOÎT VALLEY[6], PETER K. KAISER[7], DAMIEN DUFF[7] AND ROBERT LOWTHER[8]

[1] *Swiss Geological Survey, Bundesamt für Landestopografie Swisstopo, Wabern bei Bern, Switzerland;* [2] *Geological Engineering, EOAS, The University of British Columbia, Vancouver, BC, Canada;* [3] *Swiss Competence Centre on Energy Research (SCCER-SoE), ETH Zurich, Zurich, Switzerland;* [4] *School of Earth Sciences, University College Dublin, Dublin, Ireland;* [5] *Department of Physics, University of Alberta, Edmonton, AB, Canada;* [6] *Center for Hydrogeology and Geothermics (CHYN), University of Neuchâtel, Neuchâtel, Switzerland;* [7] *CEMI - Centre for Excellence in Mining Innovation, Sudbury, ON, Canada;* [8] *Newcrest Mining Limited, Cadia Valley Operations, South Orange, NSW, Australia*

ABSTRACT

The ability to generate deep flow in massive crystalline rocks is governed by the interconnectivity of the fracture network and its permeability, which in turn is largely dependent on the *in situ* stress field. The increase in stress with depth reduces fracture aperture, leading to a decrease in rock mass permeability. The frequency of natural fractures also decreases with depth, resulting in less connectivity. The permeability of crystalline rocks is typically reduced to about 10^{-17} to $10^{-15}\,\text{m}^2$ at targeted depths for enhanced geothermal systems, that is, >3 km. Therefore, fluid injection methods are required to hydraulically fracture the rock and increase its permeability. In the mining sector, fluid injection methods are being investigated to increase rock fragmentation and mitigate high-stress hazards due to operations moving to unprecedented depths. Here as well, detailed understanding of permeability and its enhancement is required. This chapter reports findings from a series of hydromechanically coupled distinct-element models developed in support of a hydraulic fracture experiment testing hypotheses related to enhanced permeability, increased fragmentation, and modified stress fields. Two principal injection designs are tested as follows: injection of a high flow rate through a narrow-packed interval and injection of a low flow rate across a wider packed interval. Results show that the development of connected permeability is almost exclusively orthogonal to the minimum principal stress, leading to strongly anisotropic flow. This is because of the stress transfer associated with the opening of tensile fractures, which increases the confining stress acting across neighboring natural fractures. This limits the hydraulic response of fractures and the capacity to create symmetric isotropic permeability relative to the injection wellbore. These findings suggest that the development of permeability at depth can be improved by targeting a set of fluid injections through smaller packed intervals instead of a single longer injection in open boreholes.

Key words: fracture network, hard rocks, hydraulic fracturing, numerical modeling, permeability, reservoir enhancement, shearing, stress transfer

INTRODUCTION

Rock mass permeability is the foremost hydromechanical parameter for industries concerned with geofluids extraction, including groundwater, geothermal water, oil, and gas. In massive crystalline rocks, often favored for enhanced geothermal system projects, permeability is governed by fracture connectivity and aperture. However, the dependency of fracture permeability on mechanical stresses limits the accessibility of geofluids located in reservoirs at substantial depths. The increase in stress with depth leads to the closure of fracture aperture, which results in the following: (i) reduced permeability, (ii) decreased fracture network connectivity, and

Crustal Permeability, First Edition. Edited by Tom Gleeson and Steven E. Ingebritsen.
© 2017 John Wiley & Sons, Ltd. Published 2017 by John Wiley & Sons, Ltd.
Companion Websites: www.wiley.com/go/gleeson/crustalpermeability/
http://crustalpermeability.weebly.com/

(iii) increased rock stiffness (Louis 1969; Tsang & Witherspoon 1981; Durham 1997; Rutqvist & Stephansson 1996; Ingebritsen & Manning 2010; Preisig *et al.* 2012). There are exceptions related to the presence of highly conductive fractures; however, such structures are sparse.

In this context, injection of pressurized fluid for hydraulic treatments is critical for enhancing the interconnectivity of fracture permeability in tight rock masses. This form of preconditioning, that is, altering the rock mass properties for engineering purposes, is widely used in the development of deep geothermal power production and shale gas extraction, where focus is placed on enhancing the rock mass permeability. Similarly, hydraulic fracturing is being utilized in the mineral industry to ensure suitable fragmentation in block caving operations (Fairhurst 2013; Jung 2013; Kaiser *et al.* 2013), as well as being investigated as a means to mitigate high-stress hazards, for example, rock bursting, in deep mining operations.

Two injection procedures may be employed in this context: hydraulic fracturing and hydraulic shearing (HS). The main difference is that hydraulic fracturing aims to initiate and propagate new tensile fractures through injection, whereas HS tries to shear preexisting natural discontinuities. It should be noted that hydraulic fracturing and HS are conceptual end members and will often act to varying degrees in combination. To initiate a new hydraulic fracture, the injection pressure must exceed the so-called breakdown pressure, which is driven by the stress concentration around the borehole wall and the tensile strength of the rock. The magnitude of the breakdown pressure will depend on the stress ratio and will typically be larger than minimum principal stress σ_3. The injection pressure to propagate a hydraulic fracture or to propagate in tension a preexisting natural fracture in which borehole pressure may have infiltrated is typically less than the breakdown pressure but still has to exceed σ_3. Consequently, it is conservative to state that the minimum fluid injection pressure p_f required to perform hydraulic fracturing is

$$p_f > \sigma_3. \tag{26.1}$$

It is also expected that the hydraulic fracture will close when the pressure in the opened fracture dissipates, resulting in a small net permeability increase. In the case of HS, the objective is to induce slip, which assuming zero cohesion along a rough tensile fracture surface can be expressed using the Mohr–Coulomb shear failure criterion

$$|\tau| \geq \mu(\sigma_n - p_f), \tag{26.2}$$

where τ is the shear stress, μ is the coefficient of friction of the fracture, that is, $\mu = \tan(\phi)$, and σ_n is the stress acting normal to the fracture plane. The fluid injection pressure p_f required to mobilize shear slip along the fracture is generally less than the *in situ* σ_3 and consequently less than the pressure needed for hydraulic fracturing if the fracture is favorably oriented for shearing, that is, if the fracture makes an angle of about 30° with

the maximum principal stress σ_1 (Pine & Batchelor 1984). It is also assumed that dilation associated with shear failure, owing to the roughness and irregularity of the fracture surface, leads to a permanent gain of aperture and fracture permeability, a mechanism referred to as self-propping (Hsiung *et al* 2005).

Taking into account the aforementioned theoretical aspects, the injection of pressurized fluid for hydraulic fracturing and/or HS will lead to different geometries depending on the tectonic regime. According to Anderson's (1951) classification of tectonic regimes, a thrust-fault (TF) regime is characterized by a vertical σ_3, and horizontal σ_1 and σ_2 referring to the minor, major, and intermediate principal stresses, respectively. In such an environment, hydraulic treatments will promote the creation and enhancement of structures with horizontal and subhorizontal geometries (Jeffrey *et al.* 2009; Bendall *et al.* 2014). Normal-fault (NF) regimes involve a vertical σ_1 and horizontal σ_2 and σ_3, and strike-slip (SS) regimes are characterized by a vertical σ_2 and horizontal σ_1 and σ_3. In both regimes, hydraulic treatments will therefore promote the creation/reactivation of vertical and subvertical discontinuities (Evans *et al.* 2005; Häring *et al.* 2008). Such considerations imply that, theoretically, horizontal boreholes can affect a larger rock volume in NF and SS regimes, whereas in TF regimes vertical boreholes could affect a larger volume.

The optimal deployment of hydraulic fracturing and stimulation is impaired because our understanding of the key processes involved, including hydromechanical coupling in fractured rock and the associated generation of seismicity, is still poorly understood (Jung 2013; Kaiser *et al.* 2013). This limits our ability to design and optimize reservoir enhancement operations and to mitigate any environmental impact on groundwater quality and induced seismicity associated with rock mass response, that is, slip and tensile opening of fractures (Dusseault & McLennan 2011; Fairhurst 2013; Vincent 2013). Despite different geological settings, rock properties, local site conditions, and operational objectives, the ability to develop connected rock mass permeability by means of hydraulic treatments is a shared challenge faced by enhanced geothermal systems, shale gas, and deep mining projects. In enhanced geothermal systems, the enhancement of permeability at depth is necessary for initiating long-term circulation of water between an injection and a pumping well at volumetric flow rates and temperatures of commercial interest, that is, >80–100 l s^{-1} at 200 °C (Evans *et al.* 2005; Polski *et al.* 2008). This enhancement should preferably occur across a large volume and multiple fractures distributed throughout the reservoir to ensure an optimal exchange of heat between the rock and the fluid, as well as to avoid a rapid deterioration of reservoir permeability if major flowing fractures are clogged by mineral precipitation. In shale gas, the development of connected permeability is necessary for enhancing well productivity and maximizing resource recovery in tight reservoir rocks. Similar issues arise regarding closure or collapse of induced fractures leading to rapid deterioration of reservoir permeability and declining well production. In the mining industry, increased fracture connectivity is also of interest, especially for increased fragmentation with use of the block caving mining method (Araneda *et al.* 2007). Another issue facing deep mines with the targeting of deeper ore bodies is the management of high stresses and associated hazards, such as rock bursting. One of the primary current research objectives is to verify the capacity to modify the stress field prior to mining by means of fluid injection and induced HS. Reactivation of natural fractures via hydraulic shear/slip has the potential to relax local concentrations of stress and mitigate related ground-control hazards (Kaiser *et al.* 2013). However, the effectiveness of an HS injection relies on the presence of sufficient connected permeability to allow the diffusion of fluid pressure. These permeable paths are fewer and poorly connected in massive crystalline rocks where many deep mines are located, and must first be generated.

In this context, a series of hydraulic fracturing and HS injection experiments are planned to be carried out in a deep mine in New South Wales, Australia. Extensive monitoring of the rock mass response will be carried out, including microseismicity, stress change, and tilt deformations. The injections will be designed to test two central hypotheses: (i) hydraulic fracturing and/or HS can be promoted by adjusting fluid injection parameters and (ii) hydraulic fracturing and/ or HS can permanently modify the rock mass properties. Indeed, hydraulic fracturing alone does not generate significant permanent changes in permeability or stress because of the narrow zone of influence and closure of aperture and asperity locking after injection ceases. Permanent changes in rock mass permeability/stress can be achieved through HS by causing slip and dilation along natural fractures, possibly aided by injecting a strength-reducing agent (low friction grout). Installation of the monitoring network has been completed with the injection sequences scheduled to begin in late 2014. A detailed site and experiment description can be found in Kaiser *et al.* (2013). This experiment produces data under field-scale conditions on the following: (i) stress field modification/relaxation, (ii) rock mass deformation, (iii) induced seismicity, and (iv) increasing rock mass fragmentation and permeability. The experiment consists of multiple injections with varying flow rates, injection interval lengths (promoting hydraulic fracturing and HS), and absence or presence of stress shadows from earlier adjacent injections.

This chapter reports the findings from a detailed set of numerical models performed as part of the experiment design. These analyses have been used to help define the fluid injection magnitudes and rates, optimal locations of monitoring sensors, and preliminary estimates of expected response (magnitude and sensitivity analyses). Specifically, these models aim to investigate and quantify the capacity to develop interconnected permeability via different designs of fluid injection in deep, massive, crystalline rocks populated by a poorly connected network of natural fractures. Focus is also placed on investigating the dominant hydromechanical processes promoting or inhibiting the development of permeability, by comparing the numerical outcomes

with past field experiments focusing on the development of permeability. This quantification issue has not been addressed by previous numerical modeling studies.

The chapter is organized in three parts as follows: the first introduces the numerical approach; the second focuses on field properties, modeling strategy, and design; and the third presents the results and discussion of their interpretation.

NUMERICAL APPROACH

Currently, no modeling approach is readily available that fully captures all aspects of the hydromechanically coupled processes involved in hydraulic fracture initiation, propagation, and interaction with preexisting natural fractures. As our focus is the development of interconnected permeability in a fractured rock mass in response to hydraulic fracturing and HS injections, the fully coupled hydromechanical distinct-element code UDEC (Itasca 2013) was selected because of its ability to capture in detail the governing mechanisms: (i) the tensile and shear response of a natural discrete fracture network to fluid pressure changes and (ii) the relevant physical processes related to the hydromechanical response of flow in fractures (Miller 2015). Within this context, thermal and chemical couplings are neglected. A key advantage of using UDEC is that it allows for the explicit modeling of an invaded zone (Dusseault & McLennan 2011) ahead and around a hydraulic fracture, together with tensile opening of preexisting natural fractures favorably oriented for hydraulic fracturing (i.e., orthogonal to σ_3) and tensile breakage of intact rock bridges represented by preferential paths of weakness (referred to here as incipient fractures; Fig. 26.1).

The main limitation of the chosen numerical technique is that the blocks comprising the problem domain are indivisible once time stepping begins; accordingly, hydraulic fracture propagation is limited to the predefined discrete fracture network. To mitigate this, strength properties are assigned to segments of the fracture network to represent either preexisting natural fractures or intact rock bridges (incipient fractures), thus providing the necessary degrees of freedom for the propagation of a hydraulic fracture (Zangeneh *et al.* 2012).

The network is defined through the vertices of randomly sized polygonal blocks generated via a Voronoi tessellation discretization scheme. This algorithm arbitrarily distributes a set of points within the domain of discretization that are then moved iteratively until reaching a uniform spacing, to which Voronoi polygons are fitted (Itasca 2013). It should be noted that the Voronoi approach increases the computational time of a coupled hydromechanical analysis. Moreover, Voronoi blocks include a large number of segments that will be perpendicular to the major principal stress, effectively stopping the hydraulic fracture from propagating further by forcing it to open against the major principal stress.

To overcome this limitation, an alternative approach was developed for this study consisting of "directional polygons."

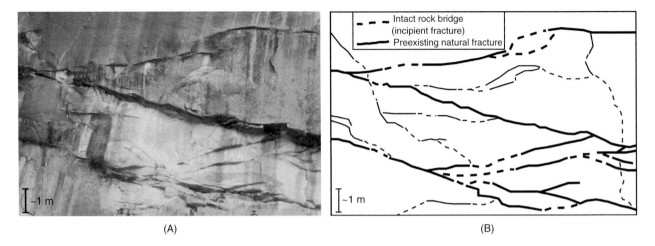

Fig. 26.1. (A) Picture showing a fractured crystalline rock mass (British Columbia, Canada) and (B) its illustrative conceptualization as a fracture network composed of cohesionless natural preexisting fractures interconnected with intact rock bridges behaving as preferential paths of weakness (incipient fractures).

These control the direction of incipient fractures so that fracture propagation directions, that is, intact rock bridges between adjacent nonpersistent (stopping against a rock block), nonconnected, preexisting natural fractures, align at a favorable orientation for fracture initiation ($\pm15°$–$30°$ relative to σ_3), in contrast to the random directions resulting from the Voronoi tessellation. This work simulates a fully coupled hydromechanical analysis at field scale, that is, greater than 100 m and incorporating a detailed fracture network geometry. To date, these types of distinct-element models are intractable in 3D and can only be practically achieved through 2D analyses.

EXPERIMENT AND NUMERICAL ANALYSIS DESCRIPTION

Geological setting

The experiment design was performed for the case of injection at depths between 1400 and 1430 m within a sparsely fractured (massive) monzonite. Mapping observations made in an access tunnel, and in other parts of the mine, indicate that the natural fracture network is weakly interconnected and consists of three main fracture families plus some random orientations as shown in Table 26.1. The stress state at the site was determined by multiple overcoring stress measurements and back analyses of excavation performance. These indicate a thrust regime where the major and intermediate principal stresses, σ_1 and σ_2 are horizontal and the minor principal stress, σ_3 is vertical. The horizontal to vertical stress ratio, K, is approximately 1.7. The rock mass is assumed to be under zero initial pore pressures, in accordance with field observations. No information is available on initial fracture apertures. However, based on the *in situ* stress state, fracture families 1 and 2 should initially be tightly closed and fracture family 3 more open because it is more orthogonal to σ_1. A series of development tunnels and niches provide access to install the monitoring network, together with

a vertical 96-mm-diameter borehole that will be used to inject fluids following a schedule alternating between hydraulic fracturing and HS treatments. Several observation boreholes will be used to complete the monitoring network and to directly observe hydraulic fracturing and HS fracture responses intersecting the boreholes.

Injection intervals and volumetric flow rates proposed here are our starting strategy for inducing dominant hydraulic fracturing or dominant HS within the rock. One of the objectives of the modeling exercise is to assess whether they will likely lead to the desired rock mass response. The injection design adopts current hydraulic fracturing practices at the site, that is, injection within a small packed interval of 2 m into which a volumetric flow rate of $400\,l\,min^{-1}$ will be pumped to exceed the breakdown pressure and initiate a new tensile hydraulic fracture. The use of HS has not yet been explored at the site; thus, the optimal conditions in this case are uncertain. Injection metrics for HS include a larger packed interval (15–30 m) and a lower injection rate, $<250\,l\,min^{-1}$. These are based on the assumption that to achieve HS without hydraulic fracturing, the injection interval must be long enough to straddle several multiple natural fractures and the injected flow rate must be controlled so that the pressure in the borehole is kept below σ_3. The injection design also includes consideration of logistical and equipment constraints.

Modeling strategy

As previously stated, modeling of hydraulic fracturing was carried out to explicitly represent a discrete fracture network composed of the following: (i) natural preexisting cohesionless fractures superimposed on top of (ii) incipient fractures behaving as intact rock bridges and having intact rock properties. In UDEC, the transient flow equation for a compressible fluid is fully coupled with kinematic equations for a discontinuum medium. In such a case, the breakage of an incipient fracture or

Table 26.1 Field data and parametric inputs used in numerical models.

	Dip direction (°)	Dip angle (°)
Discrete fracture network (DFN)		
Family 1	035	85
Family 2	340	80
Family 3	260	15

	Dip direction (°)	Dip angle (°)	Spacing (m)
Intersection between DFN and the 2D vertical model oriented E–W			
Family 1	090	81	2.4
Family 2	270	63	3.6
Family 3	270	15	4.2

Rock properties		Fracture properties		
Persistence of fractures: fully or variable (see Fig. 26.2)				
			Incipient	Natural
Young modulus E (Pa)	60×10^9			
Poisson ratio v (-)	0.25	Normal stiffness k_n (Pa m^{-1})	1.3×10^{11}	1.3×10^{11}
Density ρ (kg m^{-3})	2700	Shear stiffness k_s (Pa m^{-1})	1.3×10^{10}	1.3×10^{10}
Bulk modulus K (Pa)	$K(E, v)$	Tensile strength T (Pa)	0.5×10^6	0.0
Shear modulus G (Pa)	$G(E, v)$	Cohesion C (Pa)	1.0×10^6	0.0
Fluid properties and constants		Friction angle Φ (°)	30	45
Viscosity μ (Pa s)	0.001	Aperture at zero effective	2.0×10^{-5}	2.0×10^{-5}
Density ρ_w (kg m^{-3})	1000	Normal stress a_0 (m)		
Bulk modulus K_w (Pa)	0.1×10^9	Residual aperture a_{res} (m)	4.0×10^{-6}	4.0×10^{-6}
Gravity g (m s^{-2})	9.81	Dilation angle ψ (°)	0	0

Stress σ (MPa), depth Z (m)	Orientation
In situ stress state	
$\sigma_1 = 5 + 0.0479\ Z$	Horizontal E–W
$\sigma_2 = 0 + 0.0344\ Z$	Horizontal N–S
$\sigma_3 = 0 + 0.0297\ Z$	Vertical

the slipping of a natural fracture under increasing pore pressure depends not only on Eqs 26.1 and 26.2, but also on the entire deformation response of the fractured rock mass, including rotation, wedging, and elastic strain of the intact rock blocks as well as the opening/closure and slip along the segments of the fracture network. A linear elastic constitutive model and an elastoplastic Coulomb slip model are applied to the rock blocks and fractures, respectively.

Pore pressure propagation through the fracture network is modeled using nonlinear stress-dependent fracture aperture. Once a natural or an incipient fracture slips or opens in response to the disturbed stress field, fluid flow, and pressure diffusion

take place conforming to the cubic law (Witherspoon *et al.* 1980)

$$Q = \frac{a^3}{12}\ \frac{\rho_f g}{\mu_f} \nabla H, \tag{26.3}$$

where Q is the flow rate parallel to the fracture, a is the fracture hydraulic aperture, ρ_f is fluid density, g is gravitational acceleration, μ_f is the fluid viscosity, and ∇H is the hydraulic head gradient. In Eq. 26.3, the fracture is conceptualized as a pair of parallel surfaces whose orthogonal distance corresponds to the hydraulic aperture a. In this model, the hydraulic aperture matches the mechanical aperture and results in a parallel fracture permeability and transmissivity of $k = a^2/12$ and $t_f = a^3/12$, respectively. The aperture and parallel fracture permeability are controlled by the following hydromechanical processes: (i) pore pressure effects, where changes in pore pressures (effective stresses) result in a mechanical deformation affecting fracture aperture and permeability; (ii) stress transfer, where a change in applied stresses results in a change in fluid pressure and stiffness. Depending on the magnitude of these two processes, fracture aperture will vary linearly between a residual hydraulic aperture a_{res}, a hydraulic aperture at zero normal effective stress a_0, and optionally, a maximum hydraulic aperture a_{max}; finally, (iii) hydraulic shearing, accompanied by the permanent opening of the fracture controlled by the dilation angle specified (Itasca 2013). Note that in this work, rock blocks are considered to be impervious and flow only occurs in the fractures.

Due to the accuracy of the governing algorithms, the analysis of hydromechanical processes is time consuming and becomes intractable at large scales, in 3D, and for long fluid flow times; one minute of injection time for a field-scale model can take up to one day of computation time on an Intel i7 3.2 GHz machine with 64 GB RAM. Given that hydraulic fractures propagate orthogonal to σ_3 the use of the directional polygon discretization technique developed here and shown in Figure 26.2 lends itself to more efficient solution times compared with the random orientations derived from using UDEC Voronoi. For large-scale models, that is hundreds of meters or greater, another means to reduce computation time is to separate the mesh into refined and nonrefined zones. The refined zone is designated around the injection well and along the expected path of the hydraulic fracture and invaded zone, incorporating the network of preexisting natural fractures and intact rock bridges. The nonrefined zone helps to extend the model boundaries away from the zone of interest and allows for the investigation of the large-scale mechanical response (strain field) of the fractured rock mass.

Model geometry and mechanical conditions

Figure 26.2 and Table 26.1 summarize information described in this section. The 2D model corresponds to a vertical slice 200 m wide and 170 m high in the σ_1–σ_3 plane oriented east–west with

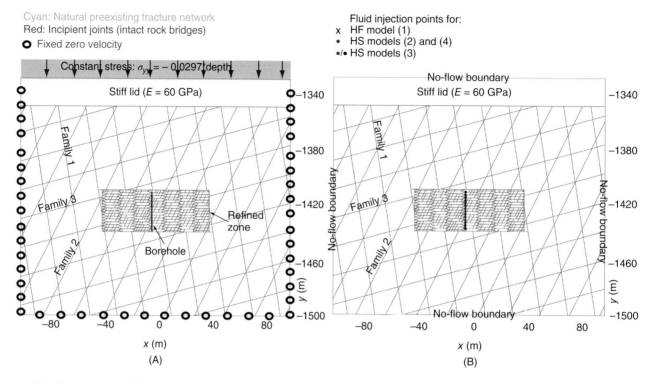

Fig. 26.2. (A) Mechanical and (B) hydraulic boundary conditions with the model geometry and fracture network implementation. Refer to Figures 26.3 and 26.5 for an enlargement of the refined zone. (*See color plate section for the color representation of this figure.*)

a refined zone 80 × 30 m in the middle, and a stiff cover/beam along the top of the problem domain. This stiff cover is used in conjunction with a constant stress condition assigned along the top of the model to add a bending stiffness to this boundary, simulating the influence of more than 1300 m of overburden above the modeled domain. Rollers (zero normal displacement) are specified for the remaining boundaries. The lateral boundaries are restricted in the *x*-direction and the bottom boundary restricted in the *y*-direction. The *in situ* stress state is compressive with a major principal stress σ_1 (horizontal) of 73 MPa and a minor principal stress σ_3 (vertical) of 42 MPa at the level of the injection (1412 m depth). The *in situ* stress state is imposed to increase linearly with depth according to the equations presented in Table 26.1.

The introduction of fractures in the model requires a compromise between the desire to capture the disconnected nature and approximate geometry of the natural fracture network at the site and the need to minimize complexity and associated computation times. Initially, the three fracture families mapped in the mine tunnels are considered for implementation in the model by computing their intersection with the vertical model plane (see Table 26.1). In the refined zone, the disconnected nature of the fracture network is represented by inserting nonpersistent horizontal fractures (approximating fracture family 3) connected via intact rock bridges (incipient fractures) dipping at 15°. These two elements form the main fabric of the model on which a west dipping set (representing family 2) is added. Family 1, which

is subvertical and perpendicular to σ_1, is omitted because it is unfavorably oriented, highly compressed, and will not respond to the fluid injection (i.e., open against σ_1). This simplification helps to reduce computational time and avoids numerical problems related to excessive fluid stiffness when subject to substantial compression. The limitations of the modeled geometry are that (i) the horizontal fractures forming the main fabric of the model are exactly aligned with the principal stress axis, reducing the ability for these fractures to shear (hydroshear) and (ii) the rock bridges are geometrically aligned, potentially forming a barrier to fracture propagation. The implications of these limitations are discussed in "Results" Section of this chapter. In the nonrefined zone, the discrete fracture network is introduced by considering persistent cohesionless fractures having a spacing of 20 m. Across the entire model domain, the rock blocks are modeled as being elastic.

Fluid injection and hydraulic conditions

Fluid injection is simulated by specifying a constant volumetric flow rate entering the model at points where the vertical borehole intercepts the fracture network. The borehole itself and the associated stress perturbation are not included in this model. Four injection designs are tested as follows: hydraulic fracturing model (1) includes simulation of 400 l min^{-1} injected over a 2-m packed interval for 60 min; hydraulic shearing model (2) includes simulation of 50 l min^{-1} injected over a 15-m packed interval for 90 min; hydraulic shearing model (3) includes simulation of $50 \, l \, min^{-1}$ injected over a 30-m packed interval for 90 min; and hydraulic shearing model (4) includes simulation of $250 \, l \, min^{-1}$ injected over a 15-m packed interval for 90 min. Hydraulic fracturing model (1) mainly focuses on the capacity to develop interconnected permeability by means of tensile hydraulic fractures. The hydraulic shearing models mainly focus on the capacity to enhance interconnected permeability by means of hydraulic shearing. The fluid injections are simulated with full consideration given to the logistical and equipment constraints at the mine. As previously noted, computational constraints limit the length of the injection times modeled to those <120 min. Nevertheless, these still allow the governing hydromechanical processes to be captured and are considered to be representative of longer injections (i.e., several hours to days).

Applied injection rates need to be scaled from 3D to 2D according to

$$Q_{2D} = Q_{3D}\alpha, \qquad (26.4)$$

where Q_{2D} stands for the scaled injection rate in $m^3 \, s^{-1} \, m^{-1}$, Q_{3D} is the volumetric injection rate, and α is a scaling reduction factor. The value of α depends almost exclusively on two points as follows: (i) the scaling from 3D to 2D and (ii) the anisotropy of fluid flow in fractured rocks due to *in situ* stress and the intrinsic properties of the natural fractures network. In 3D, volumetric fluid injection is commonly considered as a radial process. A 3D radial process cannot be scaled to a 2D vertical configuration. There is thus no clear solution for deriving α. A parametric analysis was carried out retuning a value of $1/70$, which was subsequently assigned to all injection metrics. This value yields hydraulic fracture lengths that are in agreement with those observed during preconditioning treatments at the mine. Note that $\alpha = 1/70$ is specific to the stress environment of field study site and to the model size. Note also that despite the scaling injection rates and 2D nature of the model, for practical purposes, the model input and output will be expressed in volumetric terms, that is, $l \, min^{-1}$ and m^3, throughout the chapter.

No flow boundaries surround the model, and the rock mass is under zero initial pore pressures throughout the model. This agrees with field observations. The fluid properties correspond to water, except for the bulk modulus which is one order of magnitude lower than standard values for water. This lower bulk modulus allows for the consideration of small changes in fluid volume if subjected to high stress, thus avoiding numerical instabilities related to an excessive fluid stiffness. The lower bulk modulus also works to slightly decrease computational times. Note that the contribution of the bulk modulus of the fluid is almost irrelevant with regard to the pressure diffusion and fluid penetration distances. These depend mainly on the failure of fractures and rock mass deformation. Again, rock blocks are considered impervious, and unsaturated flow processes are neglected. This is justified by the fact that water can only flow after rupture events, implying that generated pore space is instantaneously saturated. After the injection phase, backflow to the wellbore is simulated by specifying a constant fluid pressure in the borehole.

Parametric inputs

Parametric inputs are based on field data, literature values, and personal communications with mine staff (see Table 26.1). Initial fracture apertures are in the order of tens of micrometers, which is reasonable for such depths and stresses (e.g., see Snow 1970; Luthi & Souhaité 1990). UDEC considers a fracture as open when its strength is exceeded or the fracture has slipped (Itasca 2013). Model results are considered here to be acceptable as long as the hydraulic fractures generated and corresponding fluid flow remains limited to the refined zone.

RESULTS

Growth, persistence, and aperture of hydraulic fractures

Length and shape

As expected, for hydraulic fracturing model (1), hydraulic fractures grow orthogonal to σ_3 along a path linking horizontal preexisting cohesionless fractures and failed subhorizontal rock bridges. After an injected volume of $16 \, m^3$ (40 min of injection at $400 \, l \, min^{-1}$), hydraulic fractures reach a total lateral extent of 65 m (Fig. 26.3A). This agrees with previous observations of hydraulic fractures generated at the study site (Bunger *et al.* 2011), providing a measure of model validation and confirming that the input parameters are reasonable.

Incipient fractures (intact rock bridges) first slip and then are broken in tension. Horizontal natural fractures are opened normal to the horizontal plane. Only one fracture belonging to family 2 (steeply dipping to the west) is activated, close to the injection point. The largest hydraulic apertures, in the range of millimeters, occur close to the injection well and progressively decrease toward the tip of the hydraulic fracture. Fracture growth occurs both toward the east (model right) and west (model left). The most important observation emerging from Figure 26.3A is that the developing hydraulic fracture remains constrained within a quasi-planar geometry and does not develop additional branches. A similar behavior holds for the hydraulic shearing models (Fig. 26.3B), where out of the seven branches activated along the wider injection interval, only three branches (two to the west and one to the east) continue at some distance from the well with others converging and merging. This observation is in agreement with laboratory results from Bunger *et al.* (2011). A plausible explanation for this behavior is that tensile opening of a hydraulic fracture increases the confining stresses seen by the adjacent branches (stress transfer/shadowing), limiting/arresting their development. This mechanism could also explain observations from hydraulic stimulation tests related to enhanced geothermal

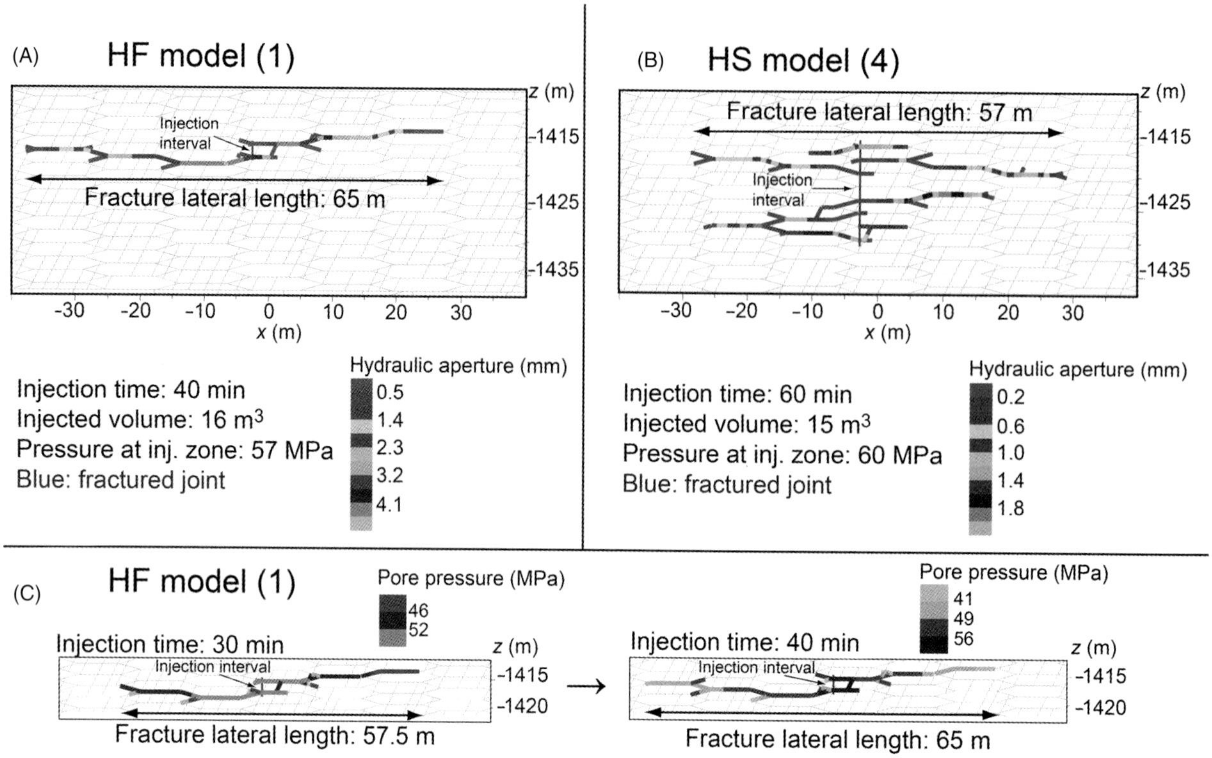

Fig. 26.3. Model enlargements showing total lateral extent and opening of hydraulic fractures for (A) hydraulic fracturing model (1) after 40 min of fluid injection at 400 l min⁻¹ into a 2-m packed interval, and (B) hydraulic shearing model (4) after 60 min of fluid injection at 250 l min⁻¹ into a 15-m packed interval. (C) Cyclical increase and decrease of pore pressures accompanying hydraulic fractures growth. (*See color plate section for the color representation of this figure.*)

systems, where different lengths of packed intervals have led to the propagation of only a few fractures instead of a pervasive stimulation of a rock mass volume (Jung 2013). However, an alternate explanation involving the influence of the preexisting permeability field could also be invoked (Evans *et al.* 2005).

Pressure–time response and cyclic growth

Figure 26.3C shows the pore pressure behavior during hydraulic fracture growth. Before a failure event, pore pressure increases, leading to decreasing pressure gradients and flow velocities between the well and hydraulic fracture fronts. Beyond the front, the pressure gradient is high, but flow is null due to the very low permeability of the incipient fractures before failure. Failure happens when pore pressures at the hydraulic fracture front exceed the incipient/natural discontinuity strength leading to increasing fracture volume and permeability. This allows fluid flow and the release of accumulated pore pressures, leading to increasing pressure gradients and flow velocities between the injection well and hydraulic fracture fronts. This sequence repeats itself in a cyclical manner as indicated by the repeating peaks in the blue pore pressure curve of Figure 26.4A. The process is much more pronounced at early stages of injection because initiation (first breakages) coincides with a shorter hydraulic fracture length and therefore limited system compliance. This cyclic growth of hydraulic

fractures is supported by microseismic signals recorded during hydraulic fracturing (Eaton *et al.* 2014). Unfortunately, in Figure 26.4A, the resolution of the model output tracking hydraulic fracture growth with injection time is too low compared with that for the pore pressure response, and the growth cannot be directly related to each pore pressure peak. Early stages of injection are also characterized by rapid growth of the fracture. Propagation velocity decreases as the fracture enlarges. This is primarily because the pore pressure gradient between the wellbore and hydraulic fracture front decreases with increasing fracture length. Thus, it becomes increasingly more difficult to increase pore pressure at the fracture front and exceed the fracture tip rock strength. In contrast, the hydraulic fracture aperture profile indicates that as the hydraulic fracture develops laterally, it is harder to open. Thus, the hydraulic fracture begins to open considerably only when the growth decelerates and pressures increase. It is important to note that these normal dislocations are fully reversible due to the elasticity of fractures if slip and dilation do not occur (Tsang & Witherspoon 1981; Cappa 2006; Preisig *et al.* 2012). Thus, if pore pressure is significantly decreased after injection, the aperture of the hydraulic fracture is much reduced. A phase of proppant injection, comprised of fluid and sand, is commonly employed to avoid elastic closure of hydraulic fractures and ensure permanent apertures.

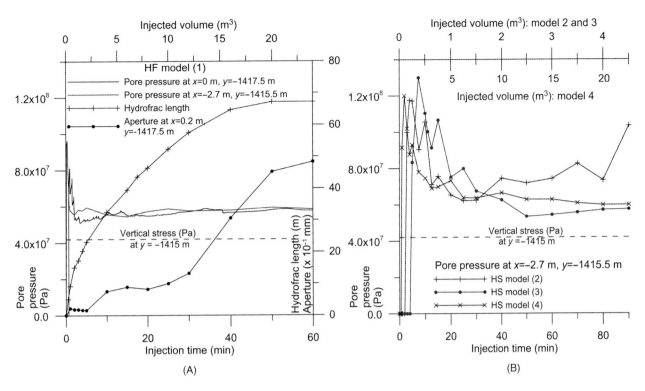

Fig. 26.4. (A) Pore pressure at and close to injection, hydraulic fracture length, and aperture as a function of injection time and volume (in m³) for hydraulic fracturing (HF) model (1): 400 l min⁻¹ into a 2-m interval. (B) Pressure–time response for hydraulic shearing (HS) models at injection point $x = -2.7$ m, $y = -1415.5$ m; HS model (2): 50 l min⁻¹ into a 15-m packed interval; HS model (3): 50 l min⁻¹ into a 30-m packed interval; HS model (4): 250 l min⁻¹ into a 15-m interval. (*See color plate section for the color representation of this figure.*)

Figure 26.4 shows pore pressure as a function of injection time and volume. In all cases, there is a substantial build up of pore pressure related to the initial impervious character of the massive rock, regardless of injection design. For hydraulic fracturing model (1), compared with the hydraulic shearing models, the pressure build up is more rapid because of the higher injection rate. After this peak, the pore pressure stabilizes around 55 MPa. The cyclical increase and decrease of pore pressure with hydraulic fracture growth is then responsible for localized pore pressure peaks. In Figure 26.4, it is also interesting to note that the increase of packed injection interval length leads to increasing pressure build up; see the hydraulic shearing models compared with the hydraulic fracturing model and/or the hydraulic shearing model (3) compared with the hydraulic shearing models (2)/(4). In fact, even for the hydraulic shearing models, tensile opening dominates, regardless of the injection design, principally because of the fracture network geometry and parametric inputs (Fig. 26.5D). Thus, the increase of packed interval length results in an increasing number of natural and incipient fractures opening and breaking, leading to interaction between fractures in the form of stress transfer/shadows. This increases the normal stresses acting on the adjacent fracture planes, which results in local increases in fracture strength. As soon as the hydraulic fractures initiate and begin to develop, pore pressures

are consequently decreased. This pressure behavior is not realistic compared with that observed in major hydraulic stimulation tests for enhanced geothermal systems (Evans *et al.* 2005; Häring *et al.* 2008), because the modeled injection is performed via different points. However, this behavior clearly illustrates the influence of stress transfer during the injection.

Shear displacements on fractures

As previously noted, tensile breakage of the incipient fractures may be preceded by the shear slip. However, for all models, this is largely the major manifestation of shearing directly related to the pressure perturbation induced by the injection (Fig. 26.5A,B). In fact, shear displacements on natural preexisting fractures of family 2 mainly occur as a consequence of movement and rotation of rock blocks (Fig. 26.5C,D), related to sinistral or dextral shear movement between the tensile hydraulic fractures. The weak presence of hydroshearing is explained by the absence of favorably oriented long-persistent fractures, the input parameters selected, that is, zero dilation angle, and the stress transfer accompanying tensile opening, which increases the confining stress acting across the neighboring fractures, reducing their ability to slip. Despite the effort to promote more hydroshear events by changing the injection design in the hydraulic shearing models, long fractures favorably oriented for hydroshearing are sparse in the refined zone.

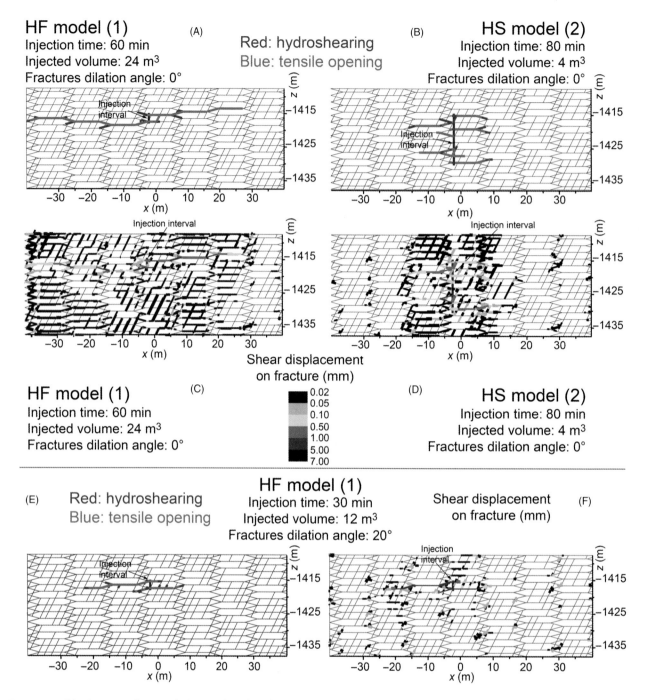

Fig. 26.5. Model enlargements illustrating hydroshear/slip and tensile opening and shear displacement on fractures for (A)–(C) hydraulic fracturing (HF) model (1) after 60 min of fluid injection at 400 l min⁻¹ into a 2-m packed interval, (B)–(D) hydraulic shearing model (2) after 80 min of fluid injection at 50 l min⁻¹ into a 15-m packed interval, and (E)–(F) HF model (1) after 30 min of fluid injection with fractures having nonzero dilation angle. Note that for shear displacement, fracture line thickness increases when its value is close to the upper boundary of its color grade. (*See color plate section for the color representation of this figure.*)

Moreover, the natural rock mass conditions are considered to be impermeable, thus preventing pressure diffusion, and implying that the rock must first be fractured. Nevertheless, the models show that the first rupture events are associated with pressure build up developing along the incipient and pre-existing subhorizontal fractures nearest the injection. Then, as

the initiated hydraulic fractures advance and permeable paths develop, pore pressures also begin to diffuse into the subvertical fractures of family 2.

Figure 26.5E,F illustrates the response when a high dilation angle is applied $(\psi = 20°)$. Subvertical fractures of family 2 fail in shear when pressurized (Fig. 26.5E). However, hydroshear is rapidly inhibited along these fractures because they are not sufficiently persistent, and because of the increasing shear strength associated with fracture dilation and stress transfer. No remarkable gain in fracture connectivity occurs for the high dilation angle case, and tensile opening remains the dominant process. These results confirm previous observations that shear displacement and associated gain in permeability through hydroshearing require the presence of natural fractures sufficiently long and favorably oriented within the stress field (Rutqvist 2015). Figure 26.5E,F also shows that the addition of a dilation angle has resulted in hydraulic fracture paths that differ slightly compared with the case with a zero dilation angle (Fig. 26.5A). Regardless of the dilation angle assumed, the geometry of the fractured zone still reflects the preexisting stress field and fracture network.

Impact on the rock mass

Rock mass deformation
Tensile opening of horizontal hydraulic fractures results in vertical compressive strain of the adjacent fractured rock, with the highest displacement vectors being close to the injection zone (Fig. 26.6A). These deformations are partly attenuated by elastic deformation of the rock blocks, leading to lower displacements away from the hydraulic fracture. Despite this, the upper boundary is still subject to millimeter-scale uplift and bending. The displacement field presented in Figure 26.6A for hydraulic fracturing model (1) is vertically asymmetric, with a highly attenuated zone below the hydraulic fracture due to the fixed bottom boundary condition. Note that a slight vertical asymmetry is also expected due to the increase of *in situ* stresses with depth. Deformation of the fractured rock mass under fluid

injection also results in strains and block rotation, which induces tilt relative to a horizontal plane. Not surprisingly, the highest tilt magnitudes occur close to the injection zone and decrease along the vertical axis (Fig. 26.6B). Again, there is a strong vertical asymmetry between tilts located above and below the hydraulic fracture. The weak horizontal asymmetry results from the difference in shape between the hydraulic fracture propagating eastward and westward. In Figure 26.6B, another important observation is that the location of maximum tilt is laterally offset from the center of the hydraulic fracture, and this offset increases with the vertical distance away from the fracture.

Stress change
Tensile opening of horizontal hydraulic fractures under fluid injection leads to vertical strains, and accordingly, increasing vertical stresses above and below the hydraulic fracture. In contrast, at the tips of the hydraulic fractures, vertical opening results in decreasing stresses (Fig. 26.7C–F). This overall behavior is partly due to the horizontal orientation of the major principal stress, as well as a Poisson's ratio effect where the vertical shortening strains adjacent to the hydraulic fracture produce expanding strains in the horizontal direction and therefore increased horizontal stresses (Fig. 26.7A–D). The change in shear stress illustrated in Figure 26.7B–E reflects the general right-lateral shear displacement affecting the system. Note that for the hydraulic shearing models, the shape of change in stresses is similar to that presented in Figure 26.7 for the hydraulic fracturing model.

On the one hand, the increase in stress above and below the primary hydraulic fractures leads to a stress shadow/transfer around the adjacent preexisting and incipient fractures, limiting

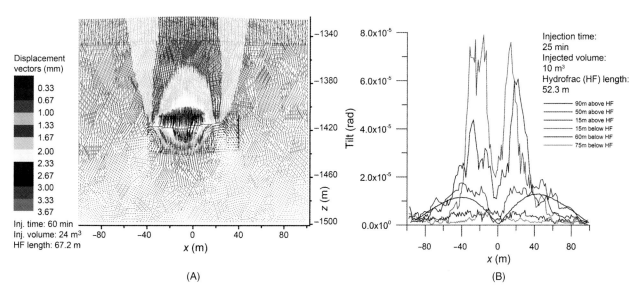

Fig. 26.6. Hydraulic fracturing model (1): induced (A) displacement field after 60 min of fluid injection at 400 l min^{-1} into a 2-m packed interval and (B) tilt after 25 min of fluid injection for six observation planes located above and below hydraulic fractures. (*See color plate section for the color representation of this figure.*)

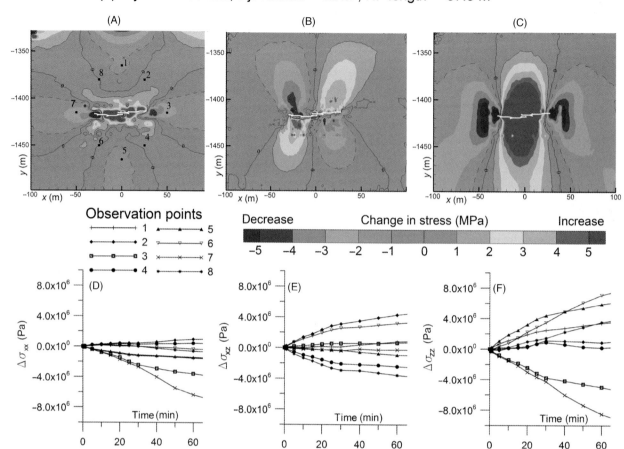

Fig. 26.7. Hydraulic fracturing model (1): change in (A) horizontal, (B) shear, and (C) vertical stress after 30 min of fluid injection at 400 l min⁻¹ into a 2-m packed interval. Note that change in stress is computed via a kriging of data obtained by subtracting stress at a given injection time from the initial stress. (D)–(F) illustrate stress responses during fluid injection for eight observation points whose location is shown in (A). (*See color plate section for the color representation of this figure.*)

their possibility to slip or open. On the other hand, the decrease in stress at the hydraulic fracture tip fronts creates a preferential path, allowing the hydraulic fracture to continue propagating far away from the injection zone.

DISCUSSION ON THE DEVELOPMENT OF CONNECTED PERMEABILITY

The hydraulic fracturing and hydraulic shearing models presented here capture the hydromechanical processes accompanying injection of pressurized fluid. Of course, quantitative output derived from these analyses should be treated with caution due to the underlying model simplifications (e.g., 2D vs. 3D), and variability and uncertainty associated with the parametric inputs. However, these findings provide considerable insights into the mechanisms, responses, and interactions involved, which can have important implications, in particular for permeability development.

The numerical analyses suggest that the response of a fractured rock mass to fluid injection includes both tensile opening and hydroshear/slip, which are mainly governed by the orientations of the preexisting natural fracture network and principal stresses, rather than the injection design. However, this finding requires further verification by exploring fluid injection response in other stress-discontinuity configurations. Under the injection metrics and stress-discontinuity discretization used here, tensile opening has been more pronounced than hydroshear. Other configurations, such as the discretization of fracture family 3 through horizontal intact rock bridges and ±15° dipping preexisting natural fractures, might have resulted in more hydraulic shear. The analysis also illustrated that the stress transfer/shadows accompanying fluid injection are a limiting factor for the development of connected permeability and reservoir enhancement through multiple, adjacent hydraulic fracturing treatments or across an injection interval, because stress transfer serves to confine nearby natural and incipient fractures, limiting their response. This mechanism focuses the development of permeability into a relatively thin layer of rock, instead of across a large volume.

The enhancement of rock permeability due to fluid injection can be illustrated by means of equivalent permeability tensors. The equivalent permeability tensor for the discrete fracture network in the refined area of the proposed model is computed as follows: first, average apertures for each fracture family are transformed into parallel permeabilities through the cubic law; next, geometrical properties of the fracture families are combined in space, resulting in a tensor describing the equivalent permeability of the rock mass

$$\mathbf{k} = \sum_{i=1}^{m} \frac{f_i a_i^3}{12} (\mathbf{I} - \mathbf{n}_i \otimes \mathbf{n}_i), \qquad (26.5)$$

where for each fracture family i, until the total number of fractures families m, a is the fracture aperture, f is the frequency of the fracture family i, $\mathbf{I}$ is the identity matrix, $\mathbf{n}$ is the unit vector normal to the fracture family i, and $\otimes$ denotes a tensor product. In matrix form, Eq. 26.5 leads to

$$\mathbf{k}_{3D} = \begin{bmatrix} k_{xx} & k_{xy} & k_{xz} \\ k_{yx} & k_{yy} & k_{yz} \\ k_{zx} & k_{zy} & k_{zz} \end{bmatrix}; \quad \mathbf{k}_{2D} = \begin{bmatrix} k_{xx} & k_{xy} \\ k_{yx} & k_{yy} \end{bmatrix}. \qquad (26.6)$$

One major advantage of permeability tensors is that their eigenvalues match the magnitude and direction of maximum k_{max} and minimum k_{min} permeability (Király 1969; Berkowitz 2002). Note that permeability is expressed in this work in m^2 and it can be linked to hydraulic/fluid conductivity in $m\,s^{-1}$ via the fluid properties as follows: $K = k(\rho_f g / \mu_f)$.

Before fluid injection, the equivalent permeability tensor for the discrete fracture network in the refined area of the given model is

$$\mathbf{k} = \begin{bmatrix} 7.5 & 2.0 \\ 2.0 & 1.8 \end{bmatrix} \cdot 10^{-18} m^2,$$

$$k_{max} = 8.1 \cdot 10^{-18}; k_{min} = 1.2 \cdot 10^{-18}; \theta = 17, \quad (26.7)$$

where θ is the counterclockwise angle between the horizontal plane and the direction of k_{max}, in degrees. The initial shape of the permeability tensor reflects that of the initial stress tensor, with k_{max} and k_{min} almost orthogonal to σ_3 and σ_1, respectively.

After injection of $20\,m^3$ of fluid, the equivalent enhanced permeability tensors for hydraulic fracturing model 1 ($t = 50\,min$) and hydraulic shearing model 4 ($t = 80\,min$) are, respectively,

HF model 1 :   HS model 4 :

$$\mathbf{k} = \begin{bmatrix} 181.9 & 9.2 \\ 9.2 & 2.5 \end{bmatrix} \cdot 10^{-15} m^2 \quad \mathbf{k} = \begin{bmatrix} 27.0 & 8.3 \\ 8.3 & 8.5 \end{bmatrix} \cdot 10^{-15} m^2$$

$$k_{max} = 182.4 \cdot 10^{-15}; \qquad k_{max} = 30.2 \cdot 10^{-15};$$

$$k_{min} = 2.0 \cdot 10^{-15}; \theta = 1 \quad k_{min} = 5.3 \cdot 10^{-15}; \theta = 15, \quad (26.8)$$

and gains in permeability in the pressurized rock mass are

HF model 1 :   HS model 4 :

$$\mathbf{k}/\mathbf{k}_{init} = \begin{bmatrix} 24.3 & 4.6 \\ 4.6 & 1.4 \end{bmatrix} \cdot 10^3 \quad \mathbf{k}/\mathbf{k}_{init} = \begin{bmatrix} 3.6 & 4.2 \\ 4.2 & 4.7 \end{bmatrix} \cdot 10^3,$$

$$k_{max}/k_{max}^{init} = 22.5 \cdot 10^3 \qquad k_{max}/k_{max}^{init} = 3.7 \cdot 10^3$$

$$k_{min}/k_{min}^{init} = 1.7 \cdot 10^3 \qquad k_{min}/k_{min}^{init} = 4.4 \cdot 10^3 \quad (26.9)$$

$$\Delta\theta = -16 \qquad \Delta\theta = -2$$

where $\Delta\theta$ expresses tensor rotation. These data indicate that for both cases, fluid injection is only able to enhance preexisting magnitudes of the equivalent permeability tensor with limited impact on its direction. This leads to strongly anisotropic preferential flow instead of simple isotropic pressure diffusion, limiting the possibility of volume stimulation. Note that for the hydraulic fracturing case, there is a tensor rotation associated with the increase of anisotropy because of the substantial increase of k_{xx}.

If after fluid injection, the rock mass is suddenly and completely depressurized, gains in permeability reduce to

HF model 1 :   HS model 4 :

$$\mathbf{k}/\mathbf{k}_{init} = \begin{bmatrix} 3.9 & 1.0 \\ 1.0 & 1.0 \end{bmatrix} \quad \mathbf{k}/\mathbf{k}_{init} = \begin{bmatrix} 1.2 & 1.0 \\ 1.0 & 1.0 \end{bmatrix},$$

$$k_{max}/k_{max}^{init} = 3.7 \qquad k_{max}/k_{max}^{init} = 1.1$$

$$k_{min}/k_{min}^{init} = 1.4 \qquad k_{min}/k_{min}^{init} = 1.0 \quad (26.10)$$

$$\Delta\theta = -16 \qquad \Delta\theta = -8$$

These gains in residual rock permeability result from rotation and wedging of neighboring, irregularly shaped rock blocks during fluid injection, which are more pronounced in hydraulic fracturing model 1. In these examples, the dilation angle was set to zero; thus, no permanent gains in permeability occur due to fracture dilation during shear, especially for the hydraulic shearing models. Permeability tensors in Eq. 26.10 clearly illustrate that the gain in permeability is a reversible process without dilation due to the elastic stiffness of the hydraulic fractures. This is in good agreement with some enhanced geothermal systems where fluid injection led to very small enhancements of permeability, such as in Ogachi, Japan: $k_{stimulated}/k_{initial} \approx 20$ (Kaieda *et al.* 2005). In other enhanced geothermal systems, for example Basel, Switzerland, and Soultz-sous-Forêt, France, the permanent enhancement of permeability reached factors ranging between 200 and 400 (Evans 2005; Häring *et al.* 2008). These enhancement factors reflect permanent gains in permeability associated with fracture dilation during shear along preexisting natural fractures. These permeable paths are often comprised of a few major long fractures where the enhancement of permeability is focused (Evans 2005). This leads also to increased anisotropic preferential flow, instead of developing permeability uniformly across a large rock volume. Similar effects can be inferred from the seismic cloud obtained for the Cooper Basin project, Australia (Bendall *et al.* 2014).

The enhanced geothermal systems cited here also suggest that the reactivation of long-persistent fractures via hydraulic shearing is likely to produce significant induced seismicity. Permeability tensors in Eq. 26.10 also suggest that proppant injection is critical for achieving permanent apertures.

The numerical analyses performed, together with the points discussed earlier, suggest that in tight rock masses, it will be difficult to develop connected permeability across a large volume through a single hydraulic fracturing (HF) or hydraulic shearing (HS) injection. This may be solved by first inducing a stack of hydraulic fractures, as is commonly done during multistage hydraulic fracturing (Dusseault & McLennan 2011), and then performing a hydraulic shearing injection. Doing so will permit pressure diffusion across a larger volume of rock during the second-stage injection, with possible reactivation in shear of favorably oriented connected natural fractures. Aside from logistical constraints, such a strategy merits verification through the *in situ* experiment planned. Moreover, in the first stage of injection, the development of a stack of tensile hydraulic fractures can be optimized by utilizing small packed intervals where the borehole intersects natural fractures orthogonal to σ_3 (identified in borehole televiewer logs). Again, this type of control is not possible via a single fluid injection in an open borehole, where the rock mass response to injection will mostly depend on the geometrical and hydraulic properties of the natural pre-existing fracture network. Although these findings are specific

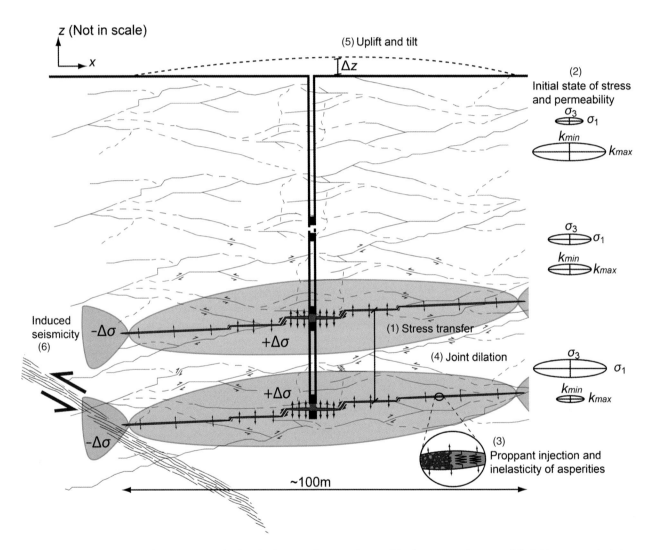

Fig. 26.8. Schematic cross section illustrating issues related to development of connected permeability by injection of pressurized fluid: (1) stress transfer between principal hydraulic fractures and its attenuation by intact rock blocks; (2) *in situ* stress state and initial permeability; (3) permanent aperture of tensile hydraulic fractures due to inelasticity of asperities or because of proppant injection; (4) permeability enhancement associated with fracture dilation under shearing; (5) generation of uplift and tilt, as well as (6) induced seismicity. (*See color plate section for the color representation of this figure.*)

to massive crystalline rocks, which are primarily encountered in deep mining and enhanced geothermal systems, they also merit consideration in future research related to the improvement of reservoir permeability in tight sedimentary rocks, especially where fluid flow in fractures significantly exceeds flow into the rock matrix. Despite simulated injection times on the order of 60 min, the highlighted processes are also likely to develop during longer fluid injections of hours, days, or weeks, as is common practice in enhanced geothermal systems. For example, the orientation and direction of development of anisotropic permeability highlighted in the first stages of a hydraulic treatment is expected to persist during a longer injection. A change in the direction of enhancement may occur if the propagating front encounters and pressurizes a zone of high preexisting natural permeability. This mechanism explains the deviation of a seismic cloud associated with an injection of pressurized fluid, as experienced at Basel (Häring *et al.* 2008). Based on the results presented in this chapter, a conceptual model is presented in Figure 26.8 illustrating important interactions and responses related to the enhancement of reservoir permeability, especially for a multistage framework. Key considerations to be further investigated through the *in situ* experiment and continued numerical modeling are as follows:

(1) Investigate the capacity of a rock mass to attenuate stress transfer in order to propose a critical distance between the stack of hydraulic fractures.
(2) Consider different initial permeability and stress states, that is, extensional and strike-slip regimes, as well as highly connected fracture networks with different geometries.
(3) Consider the ability to generate permanent apertures with and without proppant injection. This includes the role of asperities and dilation in the elastic or inelastic behavior of fractures.
(4) Assess the possibility of hydroshearing and associated fracture dilation between the stack of hydraulic fractures, as well as their ability to increase connectivity and allow fluid flow at full-size reservoir scale.
(5) Confirm model results showing that rock mass deformations in the form of uplift and tilt are low for a single hydraulic fracturing or hydraulic shearing treatment and that displacement is attenuated by the deformation of the adjacent rock blocks. Massive enhancement can generate non-negligible uplifts and tilts and therefore needs to be further investigated.
(6) Assess the georisk related to induced seismicity, such as the pressurization of critically stressed faults during and after preconditioning, leading to fault slip and seismicity.

Finally, regarding the long-term exploitation of an enhanced geothermal system, it is also critical to consider the time-dependent deterioration of reservoir permeability due to fracture closure associated with fluid pressure leak off and

dissipation, clogging of fracture apertures due to mineral precipitations, and other impacts on permeability associated with thermal depletion and compaction of the reservoir. These issues need to be approached through implementation of more advanced thermal–hydraulic–mechanical–chemical (THMC) modeling, which can account for long-term performance over year to decade timescales. Simulation of all these processes during long periods of fluid flow may be achieved by selecting only the dominant, governing mechanisms (Miller 2015; Rutqvist 2015; Weis 2015). Based on these models, a long-term exploitation design should minimize these problems and enable maximum resource extraction.

CONCLUSIONS

Stress transfer associated with fluid injection is a key limiting factor for developing interconnected rock mass permeability and reservoir enhancement at depth, in particular for a network of natural, nonpersistent fractures. Tensile opening of a hydraulic fracture will generate an increase in stress, which limits the response of neighboring fractures in both tensile opening and hydraulic shearing. The result is that hydraulic stimulation across a wide interval will be prone to produce a thin layer of enhanced permeability instead of a large volume. This will lead to strongly anisotropic flow. Moreover, there are limited options via the injection design to influence the rock mass response, for instance by promoting hydraulic shearing over tensile opening, especially when long, optimally oriented natural fractures are sparse. Instead, the system response will mainly depend on the geometrical characteristics of the pre-existing natural fracture network and orientation of the *in situ* stress field. These findings suggest that deep reservoir enhancement for geofluids extraction and circulation can be better approached by targeting fluid injections in small packed intervals.

However, additional work is required to assess the effectiveness of permeability enhancement in deep, fractured rock masses where the *in situ* stress state, fracture network geometry, and initial connectivity differ. Further testing is also required to investigate stress transfer between hydraulic fractures in the case of a multistage design, as well as the potential for shearing between the stack of tensile hydraulic fractures. Verification and validation of these results will be explored through the *in situ* experiments for which this modeling was performed. Other related issues that will be explored include stress field modification for managing high stresses during deep mining and the minimization of induced seismicity accompanying fluid injection for geothermal and shale gas production. These issues underscore the challenges faced in the design of deep reservoir enhancement and its exploitation, and the need for continued research.

ACKNOWLEDGMENTS

This work was carried out within the framework of the CEMI Newcrest mine-back experiment and was sponsored in part by the Swiss National Science Foundation (Project No. 146075) and Natural Sciences and Engineering Research Council of Canada. The authors wish to thank Newcrest for provision of the rock mass property data and permission to publish, in particular Dr Geoff Capes. Constructive comments and time allocated to this work of two anonymous reviewers were also much appreciated by the authors.

CHAPTER 27

Modeling enhanced geothermal systems and the essential nature of large-scale changes in permeability at the onset of slip

STEPHEN A. MILLER

Center for Hydrogeology and Geothermics (CHYN), University of Neuchâtel, Neuchâtel, Switzerland

ABSTRACT

The permeability structure resulting from high fluid pressure stimulation of a geothermal resource is the most important parameter controlling the feasibility and the viability of enhanced geothermal systems, yet is the most elusive to constrain. Linear diffusion models do a reasonably good job of constraining the front of the stimulated region because of the $t^{1/2}$ dependence of the perturbation length, but triggering pressures resulting from such models and the permeability inferred using the diffusivity parameter drastically underestimate both permeability and pressure changes. This leads to incorrect interpretations about the nature of the system, including the degree of fluid pressures needed to induce seismicity required to enhance the system. Here, I use a minimalist approach to modeling and show that all of the observations from Basel (Switzerland) fluid injection experiment are well matched by a simple model where the dominant control on the system is a large-scale change in permeability at the onset of slip. The excellent agreement between observations and these simplest of models indicates that these systems may be less complicated than envisaged, thus offering strategies for more sophisticated future modeling to help constrain and exploit these systems.

Key words: crustal permeability, permeability of enhanced geothermal systems, transient permeability

INTRODUCTION

Injecting fluids into the earth is both ubiquitous and a necessary component of future energy extraction and sequestration. This includes enhanced geothermal systems (Evans *et al.* 2005), geologic carbon sequestration, and hydraulic fracturing of tight shales. In the cases of enhanced geothermal systems and hydraulic fracturing, high-pressure fluids injected into low-permeability horizons enhance the permeability to either extract hydrocarbons or open permeable pathways for advection and extraction of heat. In the case of geologic carbon sequestration, injecting CO_2 into existing crack and pore networks is designed for long-term geological storage. The benefits of CO_2 sequestration in reducing atmospheric greenhouse gases require that CO_2 remains stored over very long time periods and can remain trapped only when cracks or the development of permeable pathways to the surface are prevented. In all cases, understanding the permeability structure and its evolution is essential to characterize the system, yet little is known about the permeability structure at depth (except around the borehole), and

thus the permeability structure must be inferred from matching a range of geophysical measurements to numerical models.

The future of geothermal energy exploitation, geologic carbon sequestration, and hydraulic fracturing may be better assessed with models sufficiently sophisticated to lend confidence to the success of a particular site, especially when large initial capital investments are required for their development. Hydraulic fracturing, despite its negative publicity to date, will likely continue unabated because of the enormous energy potential and economic benefits it provides. Therefore, understanding the complex processes of fracture, flow, and stress changes is necessary to both optimize extraction and minimize or mitigate the associated seismic and environmental hazards.

Enhanced geothermal systems and geologic carbon sequestration fall into the problem category of thermal–hydrological–mechanical–chemical systems. To date, the most sophisticated models focus on the thermal–hydrological–chemical aspects, and either ignore the mechanical or alternatively employ either some stochastic (and static) fracture pattern (Bruel 2007) or some elementary elasticity models. Many models exist (Kolditz

1995; Kohl & Megel 2005), and a nice summary of the latest numerical simulators is given elsewhere (Rutqvist 2011).

In general, coupling thermal, hydrological, and chemical systems to the underlying evolving mechanical system is numerically cumbersome. To date, only low-resolution 3D models are available using commercial codes such as FLAC3D coupled to flow and chemical codes such as TOUGH and TOUGH-REACT (Taron & Elsworth 2009, 2010a). One drawback of existing simulators, in addition to inadequate resolution, is that they can quickly became overly sophisticated without an appreciation of the dominant aspects of the underlying physics that actually control the system. Consequently, these models can become enveloped in an intractable parameter space that quickly confounds, thus losing their utility. There are other approaches to studying enhanced geothermal systems, including Coulomb stress transfer studies (Catalli *et al.* 2013), where each event perturbs the stress field. However, these studies can become intractable because of the many thousands of events, each of which perturbs the stress field, and are complicated by the inherent uncertainty of the orientation of the slip plane and the requirement of defining receiver fault planes.

At the other end of the spectrum, very simple models can be useful for developing insight, but care should taken because results from simple models may sometimes be inadequate, or even misleading. For example, linear diffusion models have long been used in fluid injection studies (Shapiro *et al.* 2002; Shapiro & Dinske 2009b) because of the apparent match between the seismicity front and the perturbation length in the diffusion formulation. Permeability is then inferred from the diffusivity (Delepine *et al.* 2004), resulting in values significantly lower than those to be expected in an evolving fracturing environment. Although the diffusivity can reasonably well constrained, the fluid pressure, and thus the inferred earthquake triggering pressures, may be orders of magnitude lower than fluid pressures likely triggering the earthquakes. This would significantly underestimate poroelastic stresses developed during stimulation, and affect ideas of the critical state of stress in the Earth's crust.

In this study, I attempt to reduce enhanced geothermal systems to the dominant controlling processes. As is always the case in numerical modeling, the goal is not necessarily to find the answer, but rather use the models to develop insight and determine which aspects of the myriad of processes operating actually control the system. Using this minimalist approach, I apply the model to the Basel fluid injection experiment and compare numerical results with observations.

THE BASEL FLUID INJECTION EXPERIMENT

The goal of the Basel fluid injection experiment was to open up a permeable pathway from the injection well to a production well for eventual mining of geothermal energy. This experiment has been described in detail elsewhere (Haring *et al.* 2008;

Deichmann & Giardini 2009), so only a short description is given here for completeness.

In December 2006, about $11,500 \, m^3$ of water was injected at a depth of around 4700 m near the city of Basel, Switzerland. The injection produced thousands of minor earthquakes in the $M = 0$–2 range, as designed to stimulate permeability in the reservoir. After 3 days of injection, the maximum wellhead pressure was measured at about 30 MPa, with maximum flow rates of close to $4000 \, l \, min^{-1}$. At maximum pressure, an earthquake ($M = 2.8$) occurred and was felt by the local population. Although some reports suggested widespread damage, most damage was limited to a few cracks and some broken dinnerware. In any case, the earthquake was larger than a predefined limit, and consequently, the maximum fluid pressure was reduced. Some hours later, an even larger earthquake occurred ($M = 3.4$), resulting in the experiment being aborted and the bleeding off of excess fluid pressure from the system. Smaller earthquakes in the range of $M = 0$–2 continued for months after the experiment was stopped, and occasional minor events are still sporadically observed.

Initial attempts to model this system relied on linear diffusion models aimed at matching the space–time evolution of hypocenters (Goertz-Allmann *et al.* 2011). Linear diffusion models, which implicitly assume homogeneity and isotropy, are invalid on first order because they ignore the highly nonlinear switch in permeability associated with an earthquake event. That is, the very nature of an earthquake is to create a new slip surface, or frictionally slide on an existing fault, which then must locally increase permeability in the neighborhood of the event (Miller *et al.* 1996, 2004). The strong nonlinearity and coupling between slip and permeability increases (Min *et al.* 2004; Hsiung *et al.* 2005 Cappa & Rutqvist 2011) have long been recognized (Shapiro & Dinske 2009a), but have only recently been incorporated into numerical models (Ortiz *et al.* 2011; Gischig & Wiemer 2013; Preisig *et al.* 2015; Weis 2015). This path should be followed in future studies.

THE MODEL

I use the simplest of models for a 2D cross section and include the essential ingredients controlling enhanced geothermal systems – namely, (i) a Mohr–Coulomb frictional failure criterion, (ii) an effective stress-dependent permeability, (iii) a stepwise increase in permeability when the failure criterion is satisfied, and (iv) a range of existing fault orientations relative to the far-field stress state.

I assume a simple Mohr–Coulomb failure condition for a cohesionless material with a friction coefficient of 0.8 (Byerlee 1978). Note that the friction coefficient in this model plays only a minor role in the system behavior. A grid point "fails" when $\tau - \mu\bar{\sigma}_n > 0$, and the time and location of each model "earthquake" is cataloged.

The dependence of permeability on the effective normal stress is needed to approximate the change in crack aperture that accompanies increasing (or decreasing) fluid pressure. Permeability takes the form (Rice 1992; Miller *et al.* 2004):

$$k = k_0 e^{-(\bar{\sigma}_n/\sigma^*)}, \qquad (27.1)$$

where k is permeability [m^2], k_0 is a baseline permeability, σ^* is a normalizing constant, and $\bar{\sigma}_n$ [Pa] is the effective normal stress, defined as

$$\bar{\sigma}_n = \sigma_{\text{eff}} = \sigma_n - P_f, \qquad (27.2)$$

where σ_n is the normal stress and P_f is the fluid pressure. The effective normal stress $\bar{\sigma}_n$ and shear stress τ acting on a plane through a point are

$$\bar{\sigma}_n = \frac{\sigma_1 + \sigma_3 - 2P_f}{2} \cos 2\theta, \qquad (27.3)$$

$$\tau = \frac{\sigma_1 - \sigma_3}{2} \sin 2\theta, \qquad (27.4)$$

where σ_1 and σ_3 are the maximum and minimum principal stresses acting in the far field, and θ is the angle measured from the normal to the σ_1 plane.

The stress state in the Basel region is predominantly strike slip, so the maximum σ_1 and minimum σ_3 principal stresses are horizontal and orthogonal (Fig. 27.1), while σ_2 is vertical and assumed to be the lithostatic stress. The stress ratio in the Basel region (Valley & Evans 2009) is well constrained, with $R = \frac{\sigma_1 - \sigma_2}{\sigma_1 - \sigma_3} = 0.36$, and I adopt these values in the simulations. To compare with the model with the locations of well-constrained hypocenters (Deichmann & Giardini 2009), I choose to view this problem in the σ_1, σ_2 plane, with σ_3 directed outward from the plane. In this configuration, the optimum failure plane is oriented about 30° (out of the plane) from σ_1. A range of orientations is chosen to reflect heterogeneity in the *in situ* fracture orientations, and I assume a normal distribution of angles centered about the angle of optimal orientation (Fig. 27.1 inset). The approximately linear trend of epicenters in map view (see Haring *et al.* 2008) justifies the 2D approximation assumed in the model.

The equation to be solved is a nonlinear form of the diffusion equation

$$\frac{\partial P_f}{\partial t} = \frac{1}{\phi\beta} \left(\nabla \cdot \left(\frac{k_0}{\eta} \exp^{-\frac{\sigma_n}{\sigma^*}} \nabla P_f \right) + \Gamma \right), \qquad (27.5)$$

where P_f *is* the fluid pressure above hydrostatic, ϕ is porosity, η is the viscosity [Pa s], β [Pa^{-1}] is the lumped compressibility of the fluid and pore space (Segall & Rice 1995), and Γ [s^{-1}] is a source term. Pressure-dependent storativity, which may also have influence on the behavior, is currently not considered.

INITIAL CONDITIONS

The distribution of failure planes results in a range of initial shear and effective normal stresses acting on incipient slip planes

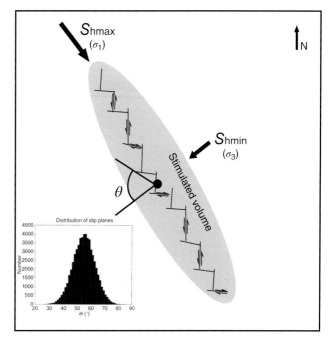

Fig. 27.1. Map view of the configuration of the Basel fluid injection experiment. (Adapted from Haring *et al.* 2008.) In this strike-slip environment, σ_1 and σ_3 are horizontal and σ_2 is vertical. The model is in the σ_1–σ_2 plane, so a distribution of virtual planes at angle θ (inset histogram) from the normal to σ_1 is input into the model. These planes fail in the model when the Mohr–Coulomb failure condition is reached. The two fractures sets shown in the conceptual stimulated volume are optimally oriented for hydroshear/slip.

(Eqs 27.2 and 27.3). This results in a heterogeneous initial permeability field because of effective stress (and depth) dependence of the permeability. Figure 27.2 shows the initial permeability field, where the average permeability at the start of the simulation is consistent with the measured *in situ* permeability of about 10^{-17} m^2 (Haring *et al.* 2008). I assume that the permeability out of plane along virtual fractures is the same as in the σ_1, σ_2 plane.

The simulation proceeds by applying the recorded pressure–time history (Fig. 27.3) to the internal boundary nodes (approximated by the white in Fig. 27.2), and solving Eq. 27.5 using an explicit finite difference scheme on a regular grid with 300 nodal points in the x direction, and 200 nodal points in the y direction.

The initial conditions result in a complex stress state, shown in Mohr–Coulomb space (Fig. 27.4A), where each dot represents the state of stress acting on the plane at angle θ from the normal to σ_1 and passing through the point represented by a computational node. When fluid pressures rise due to injection, the effective stress state moves (at different rates) toward the failure condition (Fig. 27.4B–D). The rate of stress state migration depends on the rate of fluid pressure increase, which itself depends on the permeability structure within the numerical domain. In this simple model, I ignore stress changes associated with failure, and the manner in which stress changes

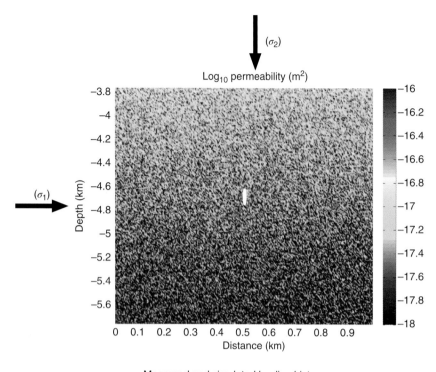

Fig. 27.2. Initial permeability field of the numerical model of a 2D cross section of the Basel experiment. The pressure–time history (Fig. 27.3) is applied as a boundary condition in the region approximated by the white ellipse in the center of the model. The average permeability of this configuration is about 10^{-17} m^2. No flow boundary conditions are applied on all external boundaries, and shear and normal effective normal stresses acting on virtual slip planes extending out of the page are calculated using Eqs 27.3 and 27.4. (*See color plate section for the color representation of this figure.*)

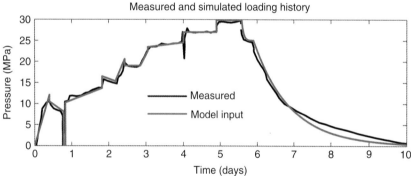

Fig. 27.3. Comparison of the recorded pressure–time history and the pressure–time history applied in the model. (*See color plate section for the color representation of this figure.*)

influence the results is not known until more sophisticated models are developed to monitor the evolving stress state within the model. The most important aspect of the current model is a stepwise change in permeability when a failure condition is reached (Miller & Nur 2000). That is, when the failure condition is reached, then k_0 in Eq. 27.1 adopts $k = xk_0$, where x is a multiplication factor. For the simulations presented here, k_0 is taken as 4×10^{-16} m^2, $\sigma^* = 10$ [MPa], and $x = 1000$. The multiplication factor x controls the evolution of the permeability, and many simulations were performed with differing values. The best qualitative fit to the data was achieved using $x = 1000$. The geological basis for stepwise change in permeability (Miller & Nur 2000) rests with the strong aperture dependence of permeability (e.g., Nemcok *et al.* 2002), where small changes in aperture result in very large changes in permeability. Examples where earthquake-induced increases in rock mass permeability (with increases up to 10^{-13} m^2) have been inferred include the Colfiorito earthquake sequence (Miller *et al.* 2004), seismicity along a splay fault in Sumatra (Waldhauser *et al.* 2012), and

postseismic fluid flow within a subduction zone (Koerner *et al.* 2004). Stress changes associated with slip on the failure plane are ignored in this simple model, so stress triggering and stress shadows are not considered and can only be investigated using more sophisticated models.

MODELING RESULTS

Many observations were recorded for the Basel fluid injection experiment, among which were (i) very well-constrained hypocenters (Deichmann & Giardini 2009), (ii) earthquake rates, (iii) fluid injection rates, and some estimate of the average permeability prior to, and after, injection. What is not known from data, but can be inferred from the modeling results, are the average permeability and fluid pressures within the stimulated region, the fluid pressure of triggered "model events," and the angle of failure relative to the applied boundary condition of the far-field stress.

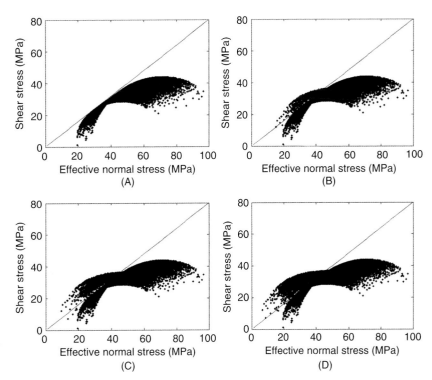

Fig. 27.4. Evolution of the stress state. (A) A dot represents the state of stress on each numerical grid point, with the entire model beneath the Mohr–Coulomb failure condition. As fluid pressure increases in the model (B–D), numerical grid points "fail" when they reach the failure condition and a model "earthquake" is recorded. Times are (A) 0 days, (B) 0.9 days, (C) 1.8 days, and (D) 2.7 days.

The pressure-induced effective stress shift (Fig. 27.4), and the consequent failure at many numerical grid points, gives rise to complex permeable pathways (Fig. 27.5) from the combined effects of nonlinear diffusion and the stepwise change in permeability (Miller & Nur 2000). The model identifies the stimulated region as that part of the numerical domain that has reached the failure criterion. The mean permeability is simply the average permeability at the numerical grid points that have reached the failure condition. The average permeability of the stimulated region (Fig. 27.6) shows an evolution from about 10^{-17} m^2 to a maximum of about $10^{-14.5}$ m^2. This compares very well with permeability measurements determined by *in situ* tests at the beginning of stimulation, and at the end of the stimulation phase where a 400-fold increase in transmissibility was determined (Haring *et al.* 2008). Permeability reduces after the peak because the reduction in fluid pressure is associated with a loss in permeability from increased effective normal stress (Eq. 27.1). Figure 27.6 is an important result because such data are needed to constrain more sophisticated models that solve the governing equations for fluid flow, heat flow, and chemical reactions to assess the viability, efficiency, and longevity of the geothermal resource.

Figure 27.7 compares the evolution of fluid pressure within the model with the hypocenter locations, showing a very good correlation between the enhanced fluid pressure front and the outward migration of hypocenters. The calculated fluid pressure field shows that the large increase in permeability accommodates high fluid pressure propagation to far distances from the source, with elevated fluid pressures on the order of MPa and in sharp contrast to kPa range predicted by linear

diffusion models. The average excess fluid pressure of the stimulated region (Fig. 27.8) peaks around 13 MPa, indicating substantial overpressures and efficient fluid pressure propagation from the source. Fluid pressures of this magnitude would impart significant poroelastic stresses in addition to dramatically lower effective normal stresses acting on incipient slip planes, thus substantially increasing seismic hazard. The relatively rapid fluid pressure reduction after the peak reflects the decreasing pressure in the reservoir as the system drained. This is addressed later in relation to the observed earthquake rates.

The hypocenter evolution between the model and observations is further compared (Fig. 27.9) in the x and y directions through time. Both the model and the observations show a triggering front similar to the r–t plots often used (Shapiro *et al.* 2002), but show a more linear trend that reflects the shock-like solutions associated with a permeability formulation of Eq. 27.1. Persistent seismicity behind the front is observed in both the model and the observations. In the model, this occurs because, as fluid pressures continue to rise, faults that were not optimally oriented relative to the principal stresses are triggered. These planes fail in the model at later times (behind the diffusion front) due to the monotonically increasing fluid pressure. In addition, both the model and observations show persistent triggering (but at a lower rate) after pressure reduction in the well, because the high-pressure field remaining in the reservoir continues to propagate outward as long as a pressure gradient persists.

Figure 27.10 compares the observed and modeled earthquake rates and shows an excellent match both during loading and unloading. To arrive at the good agreement after the peak

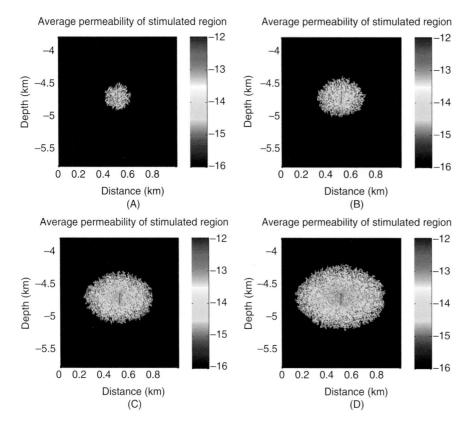

Fig. 27.5. Evolution of the permeability field in the model showing the stimulated region defined as those parts of the numerical domain that have reached the failure condition. Permeability field at (A) 1.4 day, (B) 2.8 days, (C) 4.2 days, and (D) 5.5 days of model stimulation. The scale is log(k), shown in the color bar. (*See color plate section for the color representation of this figure.*)

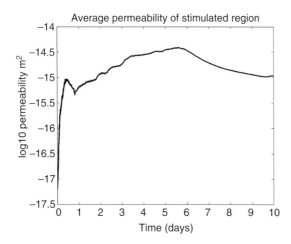

Fig. 27.6. Time history of the average permeability of the stimulated region, reaching a peak of about $10^{-14.5}$ m². The initial and peak permeabilities in the model compare very well with that observed at Basel (Haring *et al.* 2008). The stimulated region is defined as those parts of the numerical domain that have reached the failure condition.

introduced, the earthquake rates after the peak were persistently too high.

Probably, the best constraint on the model is a comparison between observed and modeled flow rates. This is difficult to achieve because flow rates are very sensitive to small changes in the model, especially when permeability can vary by orders of magnitude. The flow rate into the reservoir was calculated using

$$Q_{in} = vA = \frac{2\pi r h(t) k_{max}}{\eta} \frac{\partial P}{\partial x}, \qquad (27.6)$$

where Q_{in} is volume flow rate, defined as the Darcy velocity v across the cross-sectional area A of the open-hole section $(2\pi r h)$. In the model, I assume that $h = h(t)$ because the initial hypocenters showed limited entry points along the borehole, while subsequent seismicity indicated a larger area of the open-hole section accessing the reservoir. For a good match to the data, I chose $h(t) = 20\,\text{m} + 30\,\text{m day}^{-1}$. The maximum permeability k_{max} is taken from the model and represents the most permeable pathway for injecting fluids. The pressure gradient $\partial P/\partial x$ is approximated by the pressure measured in the borehole acting over the length scale to the time-dependent diffusion front.

Flow-back to the borehole is computed through a similar formulation and is calculated using

$$Q_{out} = vA = \frac{-2\pi h k_{mean}}{\eta}\left[\frac{\sigma_{mean} - P_{reservoir}}{dx}\right], \qquad (27.7)$$

rate, I allowed the fluid pressure in the stimulated reservoir to decline by $20\,\text{Pa s}^{-1}$ to account for fluid loss associated with opening the system. That is, a negative source Γ is put into Eq. 27.5, where $\frac{\Gamma}{\varphi\beta} = -20$ Pa s^{-1}, resulting in the relatively rapid fluid pressure reduction seen in Figure 27.6. In simulations where no fluid pressure reduction (negative source term) was

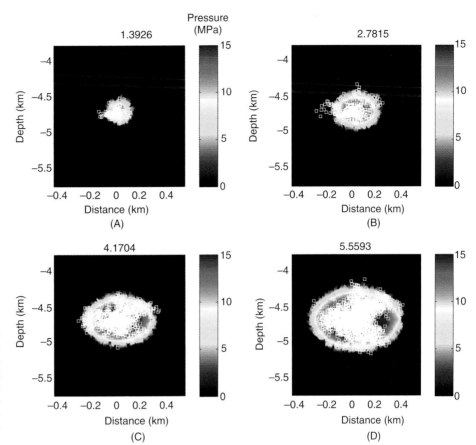

Fig. 27.7. Evolution of the fluid pressure field (above hydrostatic) and comparison with measured earthquake hypocenters, showing that the diffusing front compares well with the hypocenter migration. Note that the color bar is in MPa, showing high pressures propagating through the system. *(See color plate section for the color representation of this figure.)*

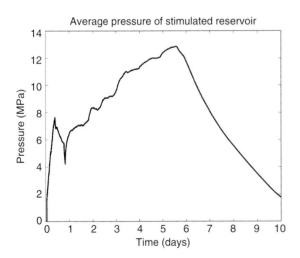

Fig. 27.8. Time history of the average fluid pressure (above hydrostatic) of the stimulated region, reaching a peak of about 13 MPa.

where Q_{out} is the volume flow rate into the borehole (of constant b). Here, I use k_{mean} instead of k_{max}, and the pressure gradient is determined differently – namely, when the system is open, the regional mean stress is the driving force that squeezes fluids out of the permeable reservoir with permeability k_{mean} through

the open-hole section. Therefore, the effective mean stress is the driving pressure, and the length scale defining the gradient is the thickness of the stimulated reservoir volume.

Figure 27.11 compares the observed flow rates and the calculated flow rates using Eqs 27.6 and 27.7 and shows satisfactory agreement between the model and observations. Importantly, this means that both the maximum permeability k_{max} and the average permeability k_{mean} evolved in the model are close to those developed in the stimulated reservoir. Any reasonable match of the flow rates at all, considering the simplicity of the model and its highly nonlinear behavior, suggests that the dominant processes driving this system have been captured.

Validation of the model against all observables thus supports investigating other model output. Figure 27.12 shows histograms of the fluid pressure (above hydrostatic) at the time of failure within the model (Fig. 27.12A), and the angle of the slip plane relative to the normal of σ_1 at failure (Fig. 27.12B). The earthquake triggering pressures in the model are predominantly in the 1–2 MPa range, but a significant number of events are triggered at fluid pressures as high as 10–12 MPa. This has important implications for seismic hazard assessment and indicates that a focus of future modeling should include direct coupling between the evolving stress state and the evolving fluid pressure state. This is also consistent with the pressure

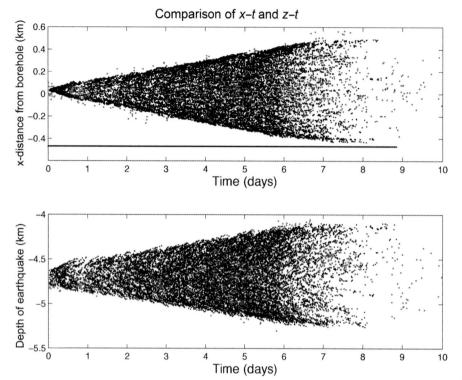

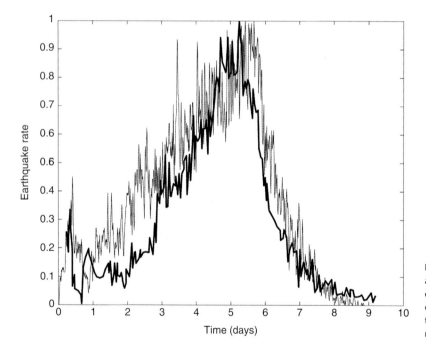

Fig. 27.9. Comparison between model (black) and observations (red) of hypocenter migration away from the borehole (top) and with depth (bottom). (*See color plate section for the color representation of this figure.*)

Fig. 27.10. Comparison between model (lighter curve) and observations (darker curve) of earthquake rates. Earthquake rates in the model were determined by the number of earthquakes occurring during a constant time interval (in this case about 15 min). Earthquake rates have been normalized by the maximum rate.

levels needed to reconcile earthquake focal mechanisms that indicate significant misorientation of slip planes relative to the regional stress state (Terakawa *et al.* 2012).

The orientation of slip planes relative to remote stress applied to the model also indicates a range of failure planes (Fig. 27.12), which slip in the model when sufficient fluid pressures are achieved. The variation in angle reflects to some degree the input normal distribution of failure planes, but may provide a basis for understanding variations in earthquake focal mechanisms observed at Basel (Deichmann & Ernst 2009). In principle, failure angles from the model might reflect variations of earthquake focal mechanisms, but it is not clear whether the distribution of failure planes is within the resolution of the focal mechanism catalog.

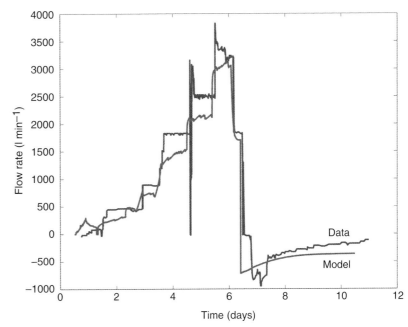

Fig. 27.11. Comparison between model (lighter curve) and observations (darker curve) of flow rates, showing satisfactory agreement during injection and withdrawal.

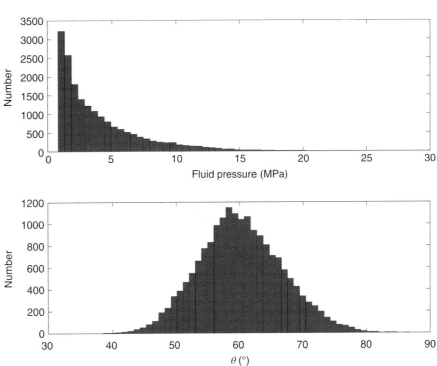

Fig. 27.12. Model output histograms (number of occurrences) of the fluid pressure (above hydrostatic) at the time of failure (top), and the angle of failure relative to the normal to σ_1.

DISCUSSION

I set out in this study to strip the problem to its basics and determine to what degree a simple model could explain the observations. The surprisingly good match to all of the observations suggests that the dominant processes are captured with this model, setting a baseline for more sophisticated modeling in the future. The mechanics are rudimentary, and each component of this model is oversimplified and therefore can be improved upon, but the model clearly demonstrates that the most important ingredient in simulating enhanced geothermal systems is the large-scale change in permeability coincident with slip. This is intuitive and satisfying and suggests that future efforts should focus on the complexities associated with this strong and dominant coupling. There are certainly a myriad of other processes and interactions in play in enhanced geothermal systems, but the advantage of this minimalist approach to modeling is that there are relatively few parameters to adjust, making the problem tractable on an intuitive level and avoiding the large parameter spaces of more sophisticated mathematical formulations. This study suggests that the mechanical and hydrological effects dominate early-time behavior of the system, and therefore subsequently control the thermal–chemical components of thermal–hydrological–mechanical–chemical systems.

The success of exploiting future enhanced geothermal systems will rely heavily on numerical models because of their complexity. The model I used was simple, and although it matched well all of the observations, ultimately its utility is limited because the actual state of stress within the system was neither known nor modeled. One of the first challenges proposed for successful modeling of enhanced geothermal systems (Fairley *et al.* 2010) is the development of fully coupled thermal–hydrological–mechanical models that can reasonably reproduce borehole stimulation. I demonstrated that even simple models can reasonably reproduce borehole stimulation. But the quote continues "and makes reliable predictions of short- and medium-term reservoir performance." This is where an explicit treatment of crack nucleation, growth, and coalescence of evolving fracture networks, coupled to fluid mechanics and heat flow, is required. The problem with developing such models, however, is that the computational power needed increases with each additional level of physics brought into study. An approach that is showing great promise is the use of the inherent massively parallel architecture of the graphics processing unit (Galvan & Miller 2013). Programming the graphics processing unit card may allow substantially more physics to be simulated at the high resolutions needed to adequately model localization (e.g., fractures), and then be coupled to the hydrological, thermal, and chemical processes associated with enhanced geothermal systems, geologic carbon sequestration, and hydraulic fracturing.

CONCLUSIONS

A simple model simulates the Basel enhanced geothermal systems and captures the dominant processes operating at depth.

The model shows that the essential ingredient to simulate these systems is the large increase in the local permeability at the onset of seismic slip. This effect, in conjunction with an effective stress-dependent permeability, results in a highly nonlinear diffusion process. The very good comparison between model and observations shows that this simple model captures this essential behavior and also provides inferences about the general properties needed to exploit the system, specifically the average permeability of the stimulated region, and the average fluid pressure within the reservoir. Knowing the average permeability of the stimulated region allows the use of more sophisticated models to determine the efficacy of the resource. Model prediction of the average fluid pressure of the stimulated region provides constraints needed to determine the poroelastic stresses and fluid pressure-induced earthquake hazard, a very significant impediment to large-scale exploitation of enhanced geothermal systems.

This model shows that future developments in numerical modeling of enhanced geothermal systems should concentrate on the explicit treatment of the evolving stress state within the system, allowing fractures to nucleate, grow, and coalesce in response to far-field stresses and stress perturbations arising from the evolving fracture network. Although models currently exist, such as UDEC and 3DEC (Itsaca), and ELFEN, these models are computationally cumbersome and of limited use for in-depth studies of evolving fracture networks. Thus, new algorithms and numerical approaches need to be developed. Once developed to simulate evolving fracture networks with appropriate resolution and computational speed, these models can be coupled to the hydrological, thermal, and chemical processes in play.

ACKNOWLEDGMENTS

I thank S. Ingebritsen and two anonymous reviewers for suggestions that substantially improved this manuscript. I thank Geo-Energie Suisse for providing the data.

CHAPTER 28

Dynamics of permeability evolution in stimulated geothermal reservoirs

JOSHUA TARON, STEVEN E. INGEBRITSEN, STEPHEN HICKMAN AND COLIN F. WILLIAMS

U.S. Geological Survey, Menlo Park, CA, USA

ABSTRACT

Spatially and temporally evolving permeability fields are fundamentally associated with the operation of enhanced geothermal systems. Indeed, permeability will evolve dynamically in any rock mass subjected to strong alteration of stress and temperature. During nonisothermal fluid injection, thermoelastic stress and fluid pressure changes act upon partially open or hydrothermally altered fracture sets to enhance or degrade formation permeability. The physical processes that drive this behavior are highly nonlinear and interdependent. In this work, we explore the resulting magnitude and patterns of permeability alteration. We utilize a thermal–hydrological–mechanical simulator capable of coupling the dominant physics of shear stimulation. Permeability is allowed to evolve under constitutive models tailored to fractured rock, considering the influence of thermal–hydrological–mechanical stress and elastoplastic shear and dilation in a ubiquitously fractured medium. On this basis, we explore the coupled physical processes that control the evolution of permeability during shear stimulation and long-term evolution of a geothermal reservoir.

Key words: permeability, geothermal, numerical simulation, thermohydromechanical coupling, reservoir stimulations

INTRODUCTION

Exchanges between mass, momentum, and energy in hydrothermal systems generate complex physical couplings (Fig. 28.1). In many cases, permeability is *the* communicative mechanism that determines how these systems evolve and to what extent they are coupled. Thus, while there are a multitude of considerations in the design of enhanced geothermal systems, most relate to the behavior of critically stressed fractures and how their characteristics control the evolution of formation permeability. Minimizing short-circuiting potential (thermal breakthrough), increasing heat-transfer area and fluid residence time, controlling working fluid losses, and mitigating the harmful effects of induced seismicity are primary operational considerations in enhanced geothermal systems, while the mechanisms underlying permeability change are of general scientific interest. Laboratory studies indicate that chemical–mechanical compaction mechanisms (such as pressure solution) may be important at representative enhanced geothermal systems conditions on the order of months (e.g., Polak *et al.* 2003), while bulk chemical reactions can inhibit or enhance permeability over several years (e.g., Taron & Elsworth 2009). Clearly, shear stimulation and elastic compaction driven

by pressure and thermal perturbations are critical concerns at any timescale (e.g., Dempsey *et al.* 2013).

To address these issues, we perform simulations using a modified version of OpenGeoSys (opengeosys.org). The new simulator is well suited for enhanced geothermal system scenarios. Realistic physics is incorporated with flexible degrees of coupling between the dominant energy conservation, momentum, and mass balance relationships. New fluid equations of state bring the simulation range well above the typical pressure and temperature conditions for enhanced geothermal systems and allow us to consider the influence of the larger scale hydrothermal system, with smooth consideration of liquid/vapor equilibrium conditions at temperatures up to 700 °C and pressures to 100 MPa. Energy, fluid, and solid momentum balance are flexibly combined in any number of "continua" or between materials in local thermal nonequilibrium (i.e., fracture and matrix and/or fluid and solid).

We explore mechanisms involved in the enhancement of fracture permeability, targeting injection pressures that encourage self-propping shear stimulation alongside elastic pressure and thermal dilatation of fracture sets. The elastoplastic response is governed by a Mohr–Coulomb failure criterion and flow rule utilizing a full Newton stress integration scheme. Fracture

Crustal Permeability, First Edition. Edited by Tom Gleeson and Steven E. Ingebritsen.
© 2017 John Wiley & Sons, Ltd. Published 2017 by John Wiley & Sons, Ltd.
Companion Websites: www.wiley.com/go/gleeson/crustalpermeability/
                    http://crustalpermeability.weebly.com/

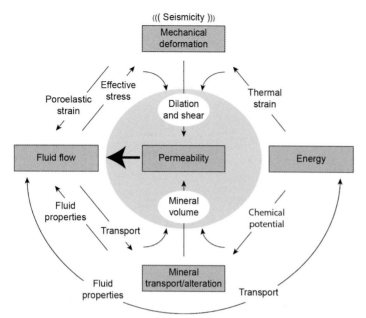

Fig. 28.1. Physical exchanges between mass, momentum, and energy, and the central communicative importance of permeability in these exchanges.

shear and dilation are linked directly to the failure model with a nonlinear model for dilation angle. Mechanical stiffness of the fractures is modified in response to dilation and relative to a nonlinear fracture compaction model, subsequently impacting bulk stiffness of the system.

Because fractures have a dominant influence on bulk rock properties, several decades of research have focused on their hydrological–mechanical properties. Topics include compressive strength of fractured rock (Goodman 1976; Brown & Scholz 1986; Barton *et al.* 1985), the effect of normal displacements and fracture roughness on permeability (Goodman 1976; Elsworth & Goodman 1986; Gangi 1978), the effect of shear displacement on permeability (Olsson & Barton 2001; Yeo *et al.* 1998), and relationships between mechanical and hydraulic aperture (Zimmerman & Bodvarsson 1996; Renshaw 1995; Piggott & Elsworth 1993). Mineral precipitation and dissolution have been studied as well. Two notable works relate to localized dissolution (Detwiler 2008) and heterogeneity of mineral assemblages (Peters 2008a).

Fractures under elevated stress

In enhanced geothermal system and other environments, permeability change is dominated by thermoelastic compression, creep (chemical and/or mechanical), shearing, and mineral infilling or dissolution along fractures. Fracture elastic compaction under stress is generally addressed with (i) a laboratory-derived stress-closure model, or (ii) an elastic contact model based on estimates of real contact area between surfaces of prescribed roughness. Both methods are empirical; the former relies on laboratory information regarding stress and closure and the latter on laboratory information regarding contact area and surface roughness (fracture profiles) and elastic rock properties.

As fractures dilate or compact (by any mechanism), the contact area between opposing fracture surfaces is altered, leading to changes in fracture apparent stiffness. This is accommodated directly in contact models, and indirectly in stress-closure models. Here, a fracture set is considered to respond nonlinearly to an evolving stress field through the often-used relationship for *mechanical* fracture aperture,

$$b = b_r + (b_m - b_r)\exp(-\alpha\sigma'), \qquad (28.1)$$

dating at least to Goodman (1976), in a slightly different form, where b_r is residual aperture, b_m is maximum aperture, σ' is the effective normal stress, and α is the stiffness coefficient. Similar forms have been introduced to include thermal–chemical behaviors (*Min et al.*, 2009). Moreover, Liu et al. (2009, 2013) have derived such an exponential relationship theoretically and shown strong agreement across a wide range of laboratory data.

In addition to Eq. 28.1, another important product of Goodman's (1976) experiments is the observation that "mated" (zero shear offset) and "unmated" (finite shear offset) fractures exhibit the same closure behavior (i.e., can be represented by Eq. 28.1), only differing proportionally with the constants, b_r and b_m (see also Brown & Scholz 1985). This suggests that the contact area relationship for any fracture (plot of contact area versus mechanical closure) after some amount of shear will be an extension of the mated solution. Taron & Elsworth (2010a) performed numerical experiments (utilizing profiles of several laboratory-scale fractures, roughly $20\,\mathrm{cm} \times 15\,\mathrm{cm}$) to quantify this behavior. Fracture surface profiles were separated to zero overlap, numerically sheared at small increments, and closed again to calculate a new relationship for closure at each degree of offset (Fig. 28.2). The closure curve at each offset (as a function of shear distance, x) can be fit with the exponential function, which is identical in form to

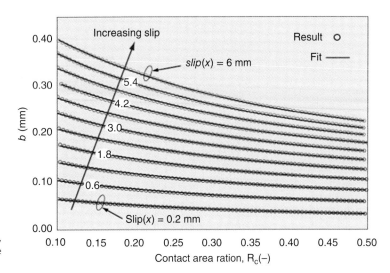

Fig. 28.2. Fracture surface profiles separated to zero overlap, numerically sheared, and closed again under stress. The fit lines are produced by Eq. 28.2 (modified from Taron & Elsworth 2010a).

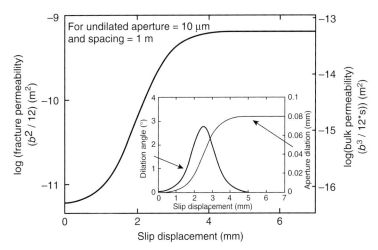

Fig. 28.3. General relationship between fracture dilation and shear displacement used in our model. The dilatation parameters illustrated here are consistent with (but slightly lower than) those obtained in the laboratory experiments of Lee and Cho (2002), and were allowed to vary in our model.

the relationship presented by Rutqvist et al. (2002) and shown in Eq. 28.1, with effective stress substituted for R_c (ratio of contact area to total square area of the fracture),

$$b = a_1(x) + a_2(x) \exp(-R_c/a_3) \tag{28.2}$$

Eq. 28.2 is akin to introducing an angle of dilation for fracture shear, but the correction is nonlinear and bimodal. Laboratory experiments show similar behavior; intuitively, fractures dilate (elastically) under a nonlinear and reproducible rule (e.g., Lee & Cho 2002), while another work (Rutqvist 2015) has successfully matched a similar relationship to *in situ* data. In this work, we use the direct aperture form of this relationship, which is quite similar to the relationship of Rutqvist (2015),

$$b = b_r(x) + (b_m(x) - b_r(x)) \exp(-\alpha\sigma'). \tag{28.3}$$

Several relationships are available to produce such a response. In this chapter, we utilize a hyperbolic tangent Heaviside type function, with results illustrated in Figure 28.3 for one potential parameter set,

$$b(x) = b_0 + \mathrm{d}b_{max} \tanh(\mathrm{del} \times |x|), \ |x| = 2\frac{x - x_{min}}{x_{max} - x_{min}} - 1,$$
$$\tag{28.4}$$

where x_{min} is the shear distance to begin dilation (break in), x_{max} is the shear distance after which dilation no longer occurs, del is a factor controlling the nonlinear angle of dilation during this period, b_0 is the aperture with zero shear, and $\mathrm{d}b_{max}$ is the amount of dilation at x_{max}. An alternative, but similarly behaving, relationship was expressed in Lee & Cho (2002), and, as a simplification, the same dilation is applied to b_r and b_m. Shear slip is computed from inelastic shear strain in the mechanical model. For a discussion of the scale dependence of such behaviors, see Taron & Elsworth (2010a), who utilized fractal upscaling, and Rutqvist (2015) for a more detailed discussion.

Consider one further numerical experiment. As a fracture is offset, its contact area is decreased, leading to a decrease in apparent stiffness and, in the presence of chemical and/or mechanical creep (i.e., Polak *et al.* 2003; Tada & Siever 1989; Beeler & Hickman 2004; Taron & Elsworth 2010a,b), an increased tendency for irreversible closure due to increased

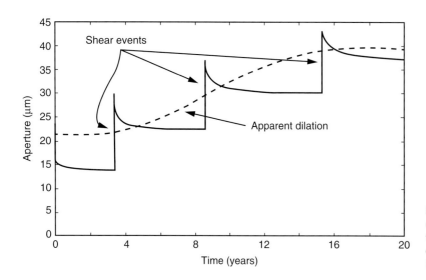

Fig. 28.4. Representative fracture profile offset in increments and allowed to creep under new conditions of contact area. Results are directly analogous to elastic closure for each exponential closure segment (modified from Taron & Elsworth 2010a).

contact stress. In Figure 28.4, a single representative fracture is offset discretely and allowed to creep under its new conditions of contact area (using the model from Taron & Elsworth (2010b). Because the contact area decreases with each offset, it becomes more susceptible to compaction and ultimately achieves a new equilibrium closure. While creep is not considered here, the results are directly analogous to elastic closure, where each exponential closure segment represents a range of stress, rather than equilibration time. In other words, as the reservoir undergoes shear stimulation, compliance and general characteristics of the responsible fracture sets are both reversibly and irreversibly altered.

Under an anisotropic stress state, the resulting aperture may have directional components of permeability. Under such anisotropy, the cubic law may be expressed as (Bear 1972),

$$B_i = b_i^3/s_i, \quad k_x = (B_y + B_z)/24, \quad k_y = (B_x + B_z)/24, \quad k_z = (B_x + B_y)/24 \tag{28.5}$$

for fracture spacing s and the diagonal components of the permeability tensor $\mathbf{k}$. The derivative of Eq. 28.1 with respect to stress gives fracture stiffness. Here, b should be the hydraulic, rather than mechanical, aperture (e.g., Zimmerman & Bodvarsson 1996). The subscript of b and s refers to the direction orthogonal to the respective quantity, and the subscript on k refers to the direction of flux. Thus, permeability in x depends on the apertures on the xy (b_z) and xz (b_y) planes. Note that for the isotropic case this reduces to the cubic law, $k_x = k_y = k_z = b^3/12s$. Later, we assume isotropic spacing in an orthogonal network, with apertures derived from the stress acting orthogonal to their respective plane.

METHODOLOGY

Conservation of mass

Here, we present, for brevity, only the basic forms for the case of single porosity. The general mass balance for a fluid of multiple phases is, for phase ψ,

$$\frac{\partial}{\partial t}(\phi \rho^f) + \nabla \cdot \left(\phi \sum \rho^\psi \mathbf{u}^\psi \right) = q_p \tag{28.6}$$

for bulk fluid density $\rho^f = S^l \rho^l + S^v \rho^v$, and density, ρ, saturation, S, porosity, ϕ, and fluid velocity $\mathbf{u}$, and where superscripts l and v refer to the liquid and vapor phases, respectively, and q_p is a pressure source term. Considering Darcy's law for flow and substituting the solid mass balance into the aforementioned relationship, we can write

$$\mathbf{u}_r^\psi = \phi \mathbf{u}^\psi - \mathbf{u}^s = \frac{\mathbf{k} k_r^\psi}{\mu^\eta}(\nabla P - \rho^\psi \mathbf{g}), \tag{28.7}$$

$$\phi \frac{\partial \rho^f}{\partial t} + \rho^f M_P \frac{\partial P}{\partial t} + \rho^f M_T \frac{\partial T}{\partial t} + \phi \rho^f \frac{\partial \varepsilon_v}{\partial t}$$
$$+ \nabla \cdot \left(\sum \rho^\psi \mathbf{u}_r^\psi \right) = q_p, \tag{28.8}$$

where $M_P = (\alpha - \phi)/K_g$ and $M_T = (1 - \alpha)\beta_T - (1 - \phi)\beta_{T_g}$, for the Biot–Willis coefficient α, grain bulk modulus K_g, bulk linear coefficient of thermal expansion β_T, and granular thermal expansion β_{T_g}, where $\mathbf{u}_r^\psi$ is relative fluid velocity with respect to the solid for solid velocity $\mathbf{u}^s$. The form of the density relationship is important, especially when working with enthalpy, h, whose derivative is not merely temperature dependent,

$$\frac{\partial \rho^f}{\partial t} = \frac{\partial \rho^f}{\partial P}\frac{\partial P}{\partial t} + \frac{\partial \rho^f}{\partial T}\frac{\partial T}{\partial t} = \frac{\partial \rho^f}{\partial P}\frac{\partial P}{\partial t} + \frac{\partial \rho^f}{\partial h}\left(\frac{\partial h}{\partial t} - \frac{\partial h}{\partial P}\frac{\partial P}{\partial t} \right). \tag{28.9}$$

Note that in multiphase conditions (i.e., if the current state of the system is that of liquid–vapor equilibrium) the density derivative must include the derivative of vapor saturation with respect to pressure and enthalpy. All fluid properties and their derivatives in water systems are calculated directly from the IAPWS industrial formulation (IAPWS 1997).

Conservation of energy

Both enthalpy (h)- and temperature (T)-based derivations begin with the general balance of specific internal energy (e), where superscript s refers to the solid, considering both viscous dissipation and pressure–volume work terms (e.g., Garg & Pritchett 1977; Faust & Mercer 1979),

$$\frac{\partial}{\partial t}(\phi \rho^f e^f) + \frac{\partial}{\partial t}((1-\phi)\rho^s e^s) + \nabla \cdot (\phi \rho^f e^f \mathbf{u}^f) + \nabla \cdot (\mathbf{u}^f P)$$
$$+ \nabla \cdot (\boldsymbol{\lambda} \nabla T) = q_h. \qquad (28.10)$$

Now taking $\rho e = \rho h - P$ and canceling terms,

$$\frac{\partial}{\partial t}(\phi \rho^f h^f) - \frac{\partial}{\partial t}(\phi P) + \frac{\partial}{\partial t}((1-\phi)\rho^s h^s) + \nabla \cdot (\phi \rho^f h^f \mathbf{u}^f)$$
$$+ \nabla \cdot (\boldsymbol{\lambda} \nabla T) = q_h. \qquad (28.11)$$

Lastly, by writing $\rho^f h^f = S^l \rho^l h^l + S^v \rho^v h^v$, and substituting the fluid mass balance Eq. 28.8 into Eq. 28.8 to remove many expansions,

$$\frac{\partial h^f}{\partial t}\left(\phi \rho^f + (1-\phi)\rho^s \frac{C_p^s}{C_p^f}\right) - \frac{\partial P}{\partial t}\left(\phi + (1-\phi)\rho^s \frac{C_p^s}{C_p^f}\frac{\partial h^f}{\partial P}\right)$$
$$+ h^f \mathbf{u}_r^f \nabla h^f + \nabla \cdot (\boldsymbol{\lambda} \nabla T) + \eta(T - T_\eta) + h^f q_p = q_h, \qquad (28.12)$$

which maintains thermal equilibrium between fluid and solid. Alternatively, the solid temperature balance may be extracted and treated separately. Here, C_p^s is solid specific heat capacity at constant pressure, and similarly C_p^f for the fluid.

The temperature formulation is obtained from Eq. 28.11 by recognizing $\mathrm{d}h = C_p^f \mathrm{d}T + \partial h / \partial p \mathrm{d}P$. This leads also to the fact that, in enthalpy form, $\nabla \cdot (\boldsymbol{\lambda} \nabla T)$ must be numerically discretized with respect to both enthalpy and pressure. Note the left-hand side term that arises in Eq. 28.12 accounting for any fluid mass sources, $h^f q_p$. This term is highly nonlinear and must be linearized when pressure and energy are solved in the same global equation. To consider multiple continua or local nonequilibrium, transfer functions must be included in the fluid and energy balance relationships. For energy, the exchange is governed by the interphase or intercontinuum heat-transfer coefficient, η, where T_η in Eq. 28.12 is the result of an additional energy balance partial differential equation for that continuum or phase. With enthalpy as a primary variable, since the interphase transfer is temperature based, a linearization is required for this term, unless it is to be placed on the right-hand side. Both of these terms are strongly coupled between respective balance equations, and so left-hand side coupling (monolithic) is important in most hydrothermal cases. See, for example, Galet (2011), for a thorough discussion of interphase transfer.

Conservation of momentum

In the usual manner (i.e., Zienkiewicz & Taylor 2005; Jaeger *et al.* 2007), we can write solid momentum equilibrium as,

$$\nabla \cdot \boldsymbol{\sigma} + \rho \mathbf{g} = 0, \; \boldsymbol{\sigma} = \boldsymbol{\sigma}' + \mathbf{M}P, \; \mathrm{d}\boldsymbol{\varepsilon} = 0.5(\nabla \mathrm{d}\mathbf{u} + (\nabla \mathrm{d}\mathbf{u})^T)$$
$$= \mathbf{B}^T \mathrm{d}u, \; \boldsymbol{\sigma} = \boldsymbol{D}_e(\boldsymbol{\varepsilon} - \boldsymbol{\varepsilon}_T - \boldsymbol{\varepsilon}_P) + \mathbf{M}P, \; \boldsymbol{\varepsilon}_T$$
$$= = \mathbf{I}\beta_T(T - T_0), \qquad (28.13)$$

where $\boldsymbol{\varepsilon}_T = \mathbf{I}\beta_T(T - T_0)$ and $\mathbf{M} = \mathbf{I} - (\boldsymbol{D}_e \mathbf{I})/(3K_g)$ determines the effective stress relationship, where $\mathbf{M} = \mathbf{I}\alpha \approx \mathbf{I}(1 - K/K_g)$ for isotropic elasticity using the Biot–Willis coefficient α along with moduli for the bulk solid, K, and grains, K_g. Also introduced is the stress–strain constitutive relationship $\boldsymbol{D}_e$ and the strain–displacement matrix $\mathbf{B}$, for the total, $\boldsymbol{\sigma}$, and effective, $\boldsymbol{\sigma}'$, stress, and the total, ε, thermal, ε_T, and plastic, ε_p, strain. Note that we solve this system as incremental in strain and displacement (in keeping with small strain approximations), but that thermal strains are relative to a fixed thermodynamic initial state, plastic strain to an undamaged state, and elastic strain relative to a state of zero stress. In other words, the stress–strain relationship is *not* incremental, which is important when moduli evolve with strain and which we find introduces less numerical error. For a Mohr–Coulomb plasticity model, the basic relationships are nonassociated in shear and given by

$$N_f = (1 + \sin(\theta))/(1 - \sin(\theta)), \; N_d = (1 + \sin(\varphi))/(1 - \sin(\varphi)),$$
$$f = \sigma_3 N_f - \sigma_1 - 2C\sqrt{N_f}, \; \frac{\partial f}{\partial \boldsymbol{\sigma}} = [-1, 0, N_f], \; \frac{\partial g}{\partial \boldsymbol{\sigma}} = [-1, 0, N_d]. \qquad (28.14)$$

Tensile flow follows the usual associated form (i.e., Jaeger *et al.*, 2007), while for the case when two flow functions are simultaneously active (i.e., two shear yield criteria are met, which is common in two dimensions), we adopt the Koiter (Koiter 1953; De Borst 1987) composite flow rule:

$$\Delta \lambda_\alpha = \frac{k_{\beta\beta}f_\alpha - k_{\alpha\beta}f_\beta}{k_{\alpha\alpha}k_{\beta\beta} - k_{\alpha\beta}k_{\beta\alpha}}, \; k_{\alpha\beta} = \left(\boldsymbol{D}_e \frac{\partial f_\alpha}{\partial \boldsymbol{\sigma}} - \frac{\partial f_\alpha}{\partial \boldsymbol{\varepsilon}_P}\right)^T \frac{\partial g_\beta}{\partial \boldsymbol{\sigma}}, \; \Delta \boldsymbol{\sigma}_p$$
$$= -\Delta \lambda_1 \boldsymbol{D}_e \frac{\partial g_1}{\partial \boldsymbol{\sigma}} + -\Delta \lambda_2 \boldsymbol{D}_e \frac{\partial g_2}{\partial \boldsymbol{\sigma}}. \qquad (28.15)$$

In these relationships, f is the failure criterion (for a friction angle θ) and g the flow rule (nonassociated reproduction of f, with N_d in place of N_f, and with dilation angle φ), while $\Delta \boldsymbol{\sigma}_p$ is the "plastic" stress change required for return to the yield surface, $\boldsymbol{\varepsilon}_P$ is plastic strain, k are cross-coupling coefficients for return to the α and β surfaces, $\boldsymbol{D}_e$ is the matrix of elastic coefficients, and λ is the flow multiplier resulting from the plasticity integration. The Drucker–Prager form exhibits only a single smooth yield surface, amenable to an elegant solution in deviatoric stress invariant space (Westergaard (1952) coordinates), while we thus far find the multiple surface Mohr–Coulomb (MC) model to be preferable to a smooth surface or corner smoothing MC model (see Zienkiewicz & Taylor 2005; Sloan

& Booker 1986) for smoothing model suggestions). All results here use the MC form shown earlier.

COUPLING STRATEGY

A fully coupled thermal–hydrological–mechanical–chemical simulator is often discussed as a critical step toward reliable prediction both of the results of enhanced geothermal system stimulation and of long-term enhanced geothermal system reservoir performance. However, "fully coupled" should refer to the physics of the problem (i.e., strong adherence to coupled terms between partial differential equations, as presented earlier) rather than to the numerical scheme used to combine necessary equations or the number of simulators that must be combined to achieve this. Although there are certain advantages to a single, self-contained simulator, in most scenarios we have explored, a full monolithic solution (all equations solved by a single call to a linear solver) is required only between those equations that are most strongly coupled in a given scenario (typically fluid mass and energy balance). Often a sequential approach is valid or even preferable for *some* of the equations (i.e., mixed sequential approach). To allow for these various approaches, we allow flexibility in how and where coupling occurs. In a general residual form of finite element method temporal discretization,

$$(\mathbf{M}^{\theta}/\Delta t + \theta(\mathbf{K}^{\theta} + \mathbf{J}))(\mathbf{x}^k - \mathbf{x}^1) = -\mathbf{K}^{\theta}(\mathbf{x}^1\theta + \mathbf{x}^0(1 - \theta))$$
$$- (\mathbf{M}^{\theta}/\Delta t)(x^1 - x^0) + \mathbf{R}, \qquad (28.16)$$

or

$$\mathbf{Ax} = \mathbf{b}, \qquad (28.17)$$

where $\mathbf{M}$ is the mass accumulation matrix, $\mathbf{J}$ the Jacobian (not present if using a Picard iterative solution), $\mathbf{K}$ the stiffness matrix, and $\mathbf{R}$ any right-hand-side terms (not residual) of all active equations in a local element. Note that $\mathbf{M}$ and $\mathbf{K}$ must be evaluated at time θ. In this formulation, a particular partial differential equation coefficient can be applied either on the left- or right-hand side of the balance equations presented earlier. For instance, if fluid and heat flow are to be solved in the same global equation, the term $M_T \partial T/\partial t$ from Eq. 28.8 can either be placed in $\mathbf{M}$ (left-hand side) or in $\mathbf{R}$ (right-hand side). There is no additional computational overhead in choosing one or the other, other than the potential need for a greater number of iterations due to the nonlinearity that is now weakly coupled if placed in $\mathbf{R}$. Furthermore, any number of $\mathbf{Ax} = \mathbf{b}$ equations can be solved in an iterative loop. Thus, any coupling strength can be dealt with in the appropriate manner. A full Newton–Raphson finite element method solution is rarely necessary in our experience, provided the left-hand side expansion includes at least all first derivatives. Instead, a Picard iteration works quite well provided that the time step control considers the error in an appropriate manner and the full

forms of the equations in "Methodology" Section are utilized. Equation 28.16 can obviously be found in many sources, but Bergamaschi and Putti (1999) provide a good discussion.

RESULTS

Utilizing this system of finite element method discretized equations and the permeability constitutive models discussed previously, we now explore the permeability response of a critically stressed, fractured geothermal reservoir. We take a single geometric conceptualization for all scenarios (Fig. 28.5). A radial symmetry model is adopted, with a 300-m-tall open-hole injection interval represented as a line source. Owing to vertical variations in porosity, with constant solid density and state-dependent fluid density, the initial stress magnitudes and fluid pressures are nonlinear functions of depth and must be equilibrated numerically.

A lateral constraint boundary is used to first determine a vertical stress profile and, simultaneously, a pore-pressure distribution with depth assuming rock matrix density of $2600\,\mathrm{kg\,m^{-3}}$, depth-varying porosity, an initial temperature gradient of $100\,°\mathrm{C\,km^{-1}}$, and properties of pure water calculated at Gaussian quadrature points from the IAPWS 1997 Industrial Formulation equation of state (IAPWS 1997). Then, a friction coefficient of 0.6 is used to apply horizontal stress to the outer boundary consistent with zero cohesion and stress limited by frictional failure on optimally oriented normal faults (e.g., Hickman & Davatzes 2010), wherein both horizontal principle stresses are assumed equal in magnitude. The lateral constraint is then removed, while the equilibrated fluid pressure is also applied to the outer boundary and the surface pressure removed to allow open flow at the surface. Because we are using an enthalpy formulation for the energy balance, after setting up an initial temperature profile the geothermal gradient is recalculated at each equilibration step to ensure an accurate inversion from temperature to enthalpy. This geothermal gradient is maintained at the outer boundary as well. The rather high initial geotherm of $100\,°\mathrm{C\,km^{-1}}$ is used to generate temperatures similar to enhanced geothermal systems in the reservoir zone ($250\,°\mathrm{C}$ at the center of injection), resulting in temperatures and pressures in the lower basement that are supercritical (Fig. 28.5).

Fracture parameters are highly variable and should be chosen with reference to field data if a particular site is to be examined. Our baseline fracture parameters were selected to be representative of very weathered surfaces – not highly compliant initially, but susceptible to large (relative) aperture increases when subjected to shear dilation. Unstimulated reservoir characteristics include low compliance fractures ($\sim$15–20 μm max aperture and potential closure of $\sim$10 μm) on a 1 m orthogonal spacing to produce a desired *in situ* permeability of $\sim$5 × $10^{-17}\,\mathrm{m}^2$ at the center of the injection interval. Laboratory evidence suggests that natural features of this scale

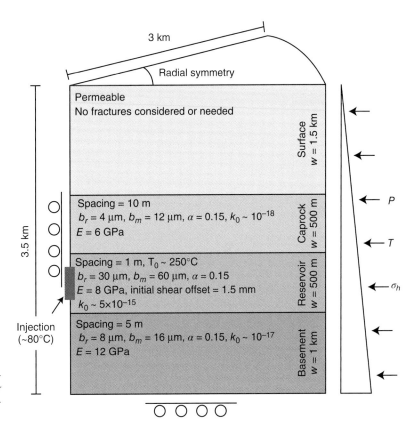

Fig. 28.5. Simulation domain: Reservoir is more highly fractured than caprock or basement. Parameters are shown for the previously stimulated reservoir ("Long-term observations" Section).

remain permeable and compliant to stresses well above those examined here (Pyrak-Nolte *et al.*, 1987). Alternative combinations are possible, such as larger fractures occluded by mineral deposition or simply with larger spacing, but these alternate parameterizations are not explored here.

Short-term stimulation behavior

Here, we explore the near-well region during a high-injection-rate shear stimulation exercise. The fluid injection temperature for these tests is 70 °C and fractures are cohesionless with a friction angle of 32° (coefficient of friction of 0.6). A constant pressure/temperature boundary is used at injection. This allows tight control on the fluid pressure, while staying below the least principal stress, and mimics what is sometimes attempted during a controlled shear stimulation enhanced geothermal system exercise (Chabora *et al.* 2012). However, this approach introduces a discontinuity in the pressure field, which can cause stability problems with plasticity. The injection pressure is applied on a mild Gaussian taper over the injection zone to reduce the discontinuity at each end of the open-hole interval. Injection pressure is set at 6 MPa above the *in situ* pore-fluid pressure at the center of the injection zone, or roughly 2 MPa below the least principal stress. The initial, unstimulated target formation permeability is roughly $5 \times 10^{-17}\,\mathrm{m}^2$, about 5 times larger than the prestimulation values at Soultz and Basel (Evans *et al.* 2005; Haring *et al.* 2008).

Dual temperature control is used, consisting of a fracture and intact rock domain. Because the timescale is short, heat transfer between fracture and rock becomes important. For pressure, we assume zero storage in the matrix: either because the pressure diffusion into the matrix is rapid relative to other processes, and thus equilibrates quickly with fracture pressure, or because the matrix is of such low porosity and/or permeability that it does not contribute strongly as a storage mechanism on short timescales. To do this, we introduce an additional term in the pressure equation that considers the different temperatures, and thus different fluid properties, in the fractures relative to the matrix; this is a mass accumulation term. This is a hybrid approach falling between a local thermal nonequilibrium simulation and a full dual-porosity simulation. The thermal–mechanical response to cooling in such a simulation is controlled by the temperature of the matrix, not the temperature of fluid in the fractures.

A first analysis shows the relative changes of parameters at two locations on a centerline outward from the injection well (Fig. 28.6). As expected, the fluid pressure increase and resulting dilation is rapid and significant while the thermal response is delayed and clearly distinguishable following passage of the pressure front. Importantly, a second, smaller pressure increase slightly precedes the second dilation phase (at roughly 1.2 days 2 m from injection and 1.5–2 days at 4 m). This indicates that permeability is increasing nearer the well, allowing the advance of a higher-pressure front, and corresponds closely

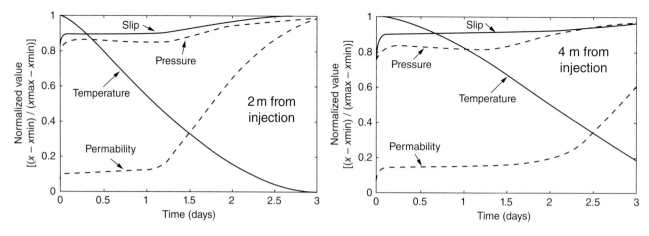

Fig. 28.6. Parameter evolution over time for the shear stimulation (short-term) case. At 2 m: temperature (250–70 °C), pressure (21.4–27.3 MPa), slip (0–3.0 mm), and permeability ($5.3 \times 10^{-17} - 7.0 \times 10^{-15}$). At 4 m: temperature (250–98 °C), pressure (21.4–27.2 MPa), slip (0–2.6 mm), and permeability ($5.3 \times 10^{-17} - 3.1 \times 10^{-15}$).

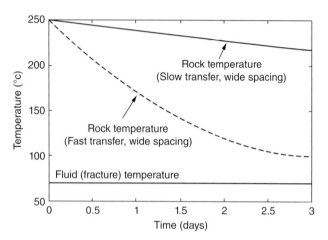

Fig. 28.7. Efficiency of cooling, expressed in terms of fluid and rock temperature very near the injection point.

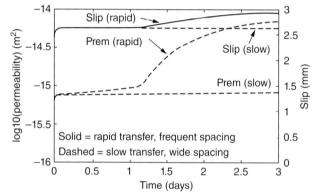

Fig. 28.8. Shear stimulation at two heat-transfer efficiencies, 2 m from injection. Rapid transfer (frequent spacing or higher efficiency) is capable of creating secondary slip through rapid cooling. Slow transfer (wide spacing or lower efficiency) cannot create the necessary thermal gradient for secondary slip.

with the arrival of colder injection fluids. Permeability increase is allowing greater rates of heat and pressure diffusion, which feeds back to higher rates of permeability increase due to both thermal and hydrological–mechanical effects.

Next, in Figures 28.7 and 28.8, we compare two cases with differing thermal transfer efficiencies between fracture and rock. Both have *identical bulk permeability*, but in one case heat transfer from fracture to matrix is more efficient and in the other the heat-transfer coefficient is reduced by a factor of 10. This latter case is similar to what would occur with larger fracture spacing, creating more inefficient or slower heat transfer.

Figure 28.7 shows the difference in temperature between fracture and matrix at the injection point. Note that the fluid temperature in the fractures is immediately constant at the injection value, while slow diffusion into the matrix introduces a delayed thermal response. Predictably, the more efficient simulation allows for much more rapid cooling of the matrix than the inefficient simulation.

Figure 28.8 explores the inelastic slip and permeability response in these two cases. Both cases display immediate shear failure and resulting permeability gain from the hydraulic reduction in effective normal stress, amounting to about 60% of the total permeability change realized during the simulation with better heat-transfer efficiency. The thermally inefficient case, however, experiences no additional dilation after this initial shear failure. The rate of heat transfer to the matrix is not sufficiently rapid to induce an inelastic response. The efficient heat-transfer case (corresponding to tighter fracture spacing) experiences an additional 40% gain in permeability due to the combined effects of shear failure and thermal contraction. It has a secondary slip mechanism as a result of the sharp thermal gradient in the rock, despite the *identical* initial permeability field.

Long-term observations

Now, the baseline values are made more compliant by introducing a small amount of initial shear, effectively assuming

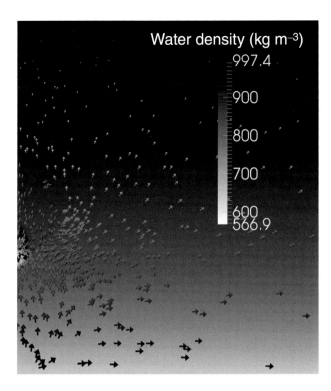

Fig. 28.9. Simulation domain: Injection process showing fluid density within the domain of Figure 28.5 after 20 years of injection; note the transition to lower supercritical densities beneath the reservoir. Arrows represent displacement of the *solid* (elastic and inelastic); rock is contracting toward the cool injection area creating areas of shear.

that the entire reservoir has been previously stimulated. Initial bulk permeability at the center of the injection interval is roughly $5 \times 10^{-15}\,\mathrm{m^2}$. This value is near the lower end of commercial geothermal reservoirs, which range from roughly $1 \times 10^{-15}\,\mathrm{m^2}$ to $1 \times 10^{-13}\,\mathrm{m^2}$, with permeability generally decreasing with temperature over an observed temperature range of 200–350 °C (Bjornsson & Bodvarsson 1990). This value is also similar to the poststimulation permeability at 2.85–3.4 km depth at Soultz ($3 \times 10^{-15}\,\mathrm{m^2}$: Evans *et al.* 2005) and at 4.6–5.0 km depth at Basel ($4 \times 10^{-15}\,\mathrm{m^2}$: Haring *et al.* 2008). Fractures are considered cohesionless with a friction angle of 34°, corresponding to a coefficient of friction of 0.7.

Fluid pressures are kept below the minimum principle stress and not greatly in excess of values required for Coulomb (frictional) failure, at an injection rate of $10\,\mathrm{kg\,s^{-1}}$. This injection rate is maintained at a temperature of 80 °C (not as a thermal Dirichlet boundary, but as a corresponding energy flux). Local thermal equilibrium is assumed for the simulations in this section, in that the same temperature is applied to fracture and adjacent matrix and fluid and rock. The injection is a line source that is slightly smoothed to zero flux at the ends (on a mild Gaussian taper), which removes some of the tendency for shear slip to concentrate at the lower and upper ends of the well; some failure does initiate here nonetheless. The overall behavior is visible in Figure 28.9, where arrows show the behavior

of solid displacement throughout the domain as the injection region cools and contracts.

The first areas to fail in shear are the ends of the injection interval, albeit over only a small area, followed by failure along the full length of the injection interval, and then mild shear advancing radially away from the well. Over time, the cold zone becomes large enough to disturb the stress field in a more significant way and plasticity begins to concentrate at the outer edge of the thermally perturbed zone (Fig. 28.10B). There is no gap in plasticity from the well to the outer edge of the advance (Fig. 28.10), only a larger accumulation at greater distance. Slip occurs in areas exhibiting the highest thermal gradient: the advancing cold boundary. The shape of the thermal plume influences the spatial concentration of frictional failure; a point injection with radial advance will look different from the line injection with linear advance shown here.

Figure 28.11 shows reservoir response in the center of the reservoir at two different distances from the injection well. The pressure pulse has a minimal impact on total permeability change at any considerable distance from injection. The thermal/plastic response is delayed, but leads to increases in permeability starting at about 1–2 years. In the most active areas, permeability increases by a factor of ~3. Adjustment of fracture stiffness and the fracture dilation curve would, of course, modify this value. In this stimulation, the fractures are initially sheared to represent an already stimulated system; the plastic dilation response would be different if this were not the case.

We highlight a few primary observations regarding the permeability response. As seen in Figures 28.10 and 28.11, shear failure concentrates at the boundary of the advancing cool region. Thus, the onset of shear failure and dilation occurs before a particular point cools significantly. The ensuing permeability change is associated with the magnitude of temperature change, not with the onset of fracture dilation. While dilation opens the fracture, it also increases its compliance, resulting in minimal net observable increase in permeability. However, once the cold front arrives fully, the fracture is more complaint due to its previous slip, and thus dilates more easily as the thermal stress declines. It is safe to assume that larger scale fractures will behave differently due to their tendency to dilate at a greater rate.

CONCLUSION

Simulation of a short-term, high-injection-rate enhanced geothermal system stimulation explored the importance of hydraulic *versus* thermal effects and the impact of a temperature differential between fluid in the fractures and the adjoining rock matrix. Under less efficient thermal transfer (corresponding to larger fracture spacing), thermal effects are relatively unimportant during short stimulation operations, inducing no significant additional shear-enhanced permeability gain. Alternatively, if the matrix cools more rapidly (due to

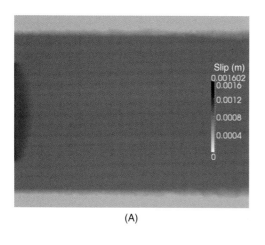
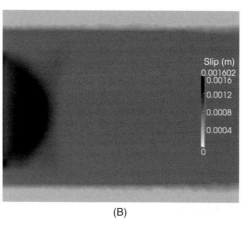

Fig. 28.10. Total inelastic slip distance (m) on the fractures within the domain of Figure 28.5 after 6 months (A) and 20 years (B). Area shown is roughly 500 m in height and 650 m in width.

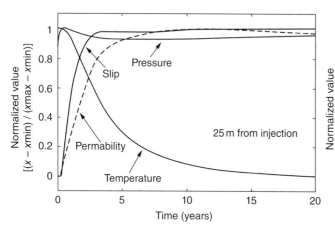
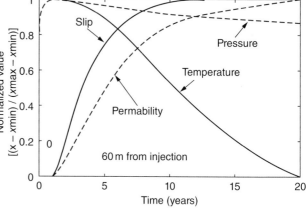

Fig. 28.11. Evolution of parameters with time at two radial distances from the center of the injection interval. At 25 m: temperature (250–85 °C), pressure (21.4–21.7 MPa), slip (0–0.6 mm), and permeability $(4.1 - 7.5 \times 10^{-15})$. At 60 m: temperature (250–136 °C), pressure (21.4–21.6 MPa), slip (0–0.8 mm), and permeability $(4.1 - 8.4 \times 10^{-15})$.

tighter fracture spacing), additional shear stimulation and stress relaxation from thermal effects add another order of magnitude to the total permeability change. This is a two-stage slip mechanism. Importantly, these two very different results occurred within a reservoir represented by an identical initial permeability field. Thus, the effects of variations in fracture spacing on the evolution of permeability is at least as important as the magnitude of initial permeability, and quantifying the driving mechanisms is critical to understanding reservoir response. Thermal effects are probably always important. Even without significant mechanical influence, they drive large changes in fluid properties; their impact on fluid density can introduce buoyancy effects, promoting downward migration of colder injection fluids, while viscosity differences influence the inelastic response via pressure diffusion.

Simulation of long-term enhanced geothermal system operation demonstrates the ability of sustained pressure/thermal perturbation to drive fracture slip and corresponding permeability gain. The most active inelastic areas exist on the boundary of the advancing cold injection fluid, at the location of the highest thermal gradient. In the case of small-scale fractures, inelastic slip does not significantly impact permeability, at least directly, at any significant distance from the injection point. However, such slip may weaken fractures in shear by altering their contact area, allowing them to respond significantly to sustained cooling by relaxing the thermal stress field.

CHAPTER 29

The dynamic interplay between saline fluid flow and rock permeability in magmatic–hydrothermal systems

PHILIPP WEIS[1,2]

[1] *Institute of Geochemistry and Petrology, ETH Zürich, Zürich, Switzerland;* [2] *GFZ German Research Centre for Geosciences, Potsdam, Germany*

ABSTRACT

Magmatic–hydrothermal ore deposits document the interplay between saline fluid flow and rock permeability. Numerical simulations of multiphase flow of variably miscible, compressible H_2O–NaCl fluids in concert with a dynamic permeability model can reproduce characteristics of porphyry copper and epithermal gold systems. This dynamic permeability model uses values between 10^{-22} and 10^{-13} m^2, incorporating depth-dependent permeability profiles characteristic for tectonically active crust as well as pressure- and temperature-dependent relationships describing hydraulic fracturing and the transition from brittle to ductile rock behavior. In response to focused expulsion of magmatic fluids from a crystallizing upper crustal magma chamber, the hydrothermal system self-organizes into a hydrological divide, separating an inner part dominated by ascending magmatic fluids under near-lithostatic pressures from a surrounding outer part dominated by convection of colder meteoric fluids under near-hydrostatic pressures. This hydrological divide also provides a mechanism to transport magmatic salt through the crust. With a volcano at the surface above the hydrothermal system, topography-driven flow reverses the direction of the meteoric convection as compared with a flat surface, leading to discharge at distances of up to 7 km from the volcanic center. The same physical processes at similar permeability ranges, crustal depths, and flow rates are relevant for a number of active systems, including geothermal resources and excess degassing at volcanoes. The simulations further suggest that the described mechanism can separate the base of free convection in high-enthalpy geothermal systems from the magma chamber as a driving heat source by several kilometers in the vertical direction in tectonic settings with hydrous magmatism. These root zones of high-enthalpy systems may serve as so-called supercritical geothermal resources. This hydrology would be in contrast to settings with anhydrous magmatism, where the base of the geothermal systems may be closer to the magma chamber.

Key words: brittle–ductile transition, goethermal energy, hydraulic fracturing, magmatic–hydrothermal ore deposits, numerical modeling, permeability

INTRODUCTION

Fluids in the Earth's crust have a critical control on a large number of geological processes on widely varying temporal and spatial scales (e.g. Ingebritsen & Gleeson 2015). The presence of fluids may decide whether and how a physical or chemical process takes place, ranging from slow metamorphism in orogenic roots to catastrophic volcanic eruptions from hydrous magma chambers. Wherever fluids play a major role, observed or inferred processes exhibit a strong interplay between fluid and rock, often leading to episodic behavior or a tendency toward self-organization, such as fault-valve behavior in fractures, compaction-driven porosity waves, or hydraulic fracturing (e.g., Sibson *et al.* 1988; Connolly & Podladchikov 2004; Cox 2005; Preisig *et al.* 2015; Saffer 2015). In most cases, however,

a quantitative and comprehensive understanding of how these processes act in natural systems is still in its infancy.

Magmatic–hydrothermal ore deposits are fossil witnesses of dynamic permeability creation and reduction in response to forced fluid flow (Hedenquist & Lowenstern 1994; Cox 2005; Sillitoe 2010). Both open pit and underground mining of mineral deposits provide valuable three-dimensional insights into several kilometers of the crust with outcrops, drill cores, and geochemical mapping. Cross-cutting relationships, alteration patterns, and fluid inclusions can document the temporal and spatial evolution of the hydrothermal systems. The variety of different hydrothermal alteration overprints and vein sequences with different mineral assemblages may result from a unique, complex history of ore formation for any individual deposit.

Crustal Permeability, First Edition. Edited by Tom Gleeson and Steven E. Ingebritsen.
© 2017 John Wiley & Sons, Ltd. Published 2017 by John Wiley & Sons, Ltd.
Companion Websites: www.wiley.com/go/gleeson/crustalpermeability/
http://crustalpermeability.weebly.com/

However, the classification of distinct deposit types with characteristic patterns and their global abundance in typical geological settings (Kesler 1994; Heinrich & Candela 2014) suggest that their formation results from recurrent combinations of hydrological and geological processes, resulting from the interplay between fluid flow and rock permeability.

Due to its extent over several kilometers of the crust and its lifetime over multimillennia, the spatial and temporal evolution of magmatic–hydrothermal systems cannot be assessed with observation techniques alone. Therefore, mass- and energy-conserving numerical methods describing the interactions between fluid and rock provide a powerful tool for quantifying the key processes controlling these dynamic hydrothermal systems (e.g., Ingebritsen et al. 2010). In numerical simulations, permeability has traditionally been considered as a static material property determining how fast groundwater can flow through a porous rock formation according to Darcy's law (Ingebritsen et al. 2006). Extending the concept of this continuum, porous medium approach to the characterization of fluid flow through a fractured rock mass, however, requires a revised concept of permeability as a dynamic parameter.

This study presents results from numerical simulations of the hydrology of porphyry copper systems. The model includes multiphase flow of compressible, variably miscible H_2O–NaCl fluids with an accurate thermodynamic description of phase relations and fluid properties as well as a dynamic permeability model describing the transition from brittle to ductile rock behavior and hydraulic fracturing. Building on previous simulation results (Weis et al. 2012), this study investigates the influence of permeability range, salinity, and volcano topography on the hydrology of porphyry systems and discusses its relation to active systems such as geothermal fields and volcanoes.

PORPHYRY COPPER AND EPITHERMAL GOLD DEPOSITS

Porphyry-type deposits form in continental collisional arcs within the upper crust and are characterized by typical sequences of intense veining and alteration due to ascending fluids (Gustafson & Hunt 1975; Sillitoe 2010; Richards 2013; Wilkinson 2013). Geologic observations show that the deposits are closely associated with volcanic activity and that the veins form from magmatic fluids expelled from a large upper crustal magma chamber (Hedenquist & Lowenstern 1994; Wilkinson 2013). Intense stockwork veining indicates that the host rock has been fractured and sealed in multiple cycles, which episodically created permeability to release the fluids expelled from the magma chamber. Cathodoluminescence images of copper-bearing quartz veins indicate that the early stages of vein formation are barren and that mineralization is associated with a generation of later quartz (Landtwing et al. 2005).

Some porphyry copper systems are inferred to be linked to epithermal gold deposits at shallower crustal levels (Henley & Ellis 1983; Hedenquist et al. 1998; Sillitoe 2010; Richards 2013). At low-sulfidation epithermal deposits, layers of different mineral assemblages reflect the temporal evolution of the system and may document several cycles. Within the vein deposit, mineralization with metal concentrations of economic grade is usually restricted to thin layers, which are often associated with boiling of the ore-forming hydrothermal fluid (Izawa et al. 1990).

Fluid inclusions trapped in vein minerals can document the state of the hydrothermal system before, during, and after ore formation. Magmatic fluids expelled from the underlying cooling magma are injected into the host rock as high-temperature, high-pressure single-phase fluids with an intermediate density (around 500–600 kg m^{-3}) and a salinity of about 10 wt% NaCl (Redmond et al. 2004; Landtwing et al. 2005). Upon ascent toward the surface, this pristine fluid phase-separates into a low-salinity, low-density vapor phase and a high-salinity, high-density brine phase. These fluid phases can physically separate from each other and flow with different phase velocities. Numerous fluid inclusion studies document brines and vapors with variable salinities and homogenization temperatures, often occurring as the so-called boiling assemblages where a two-phase fluid mixture has been trapped within a single crack, which later becomes preserved as a trail of fluid inclusions (e.g., Redmond et al. 2004; Landtwing et al. 2005; Bodnar et al. 2014).

Copper mineralization in porphyry deposits is associated with a decrease in temperature of the fluids from initially 600–700 °C to values between 350 and 450 °C (Hezarkhani & Williams-Jones 1998; Hezarkhani et al. 1999; Landtwing et al. 2005). Epithermal deposits are generally characterized by temperatures below 300 °C (Bodnar et al. 2014). To ensure enrichment to economic grades over time, the physical hydrology has to provide a robust cooling and ore precipitation mechanism.

Numerical simulations of the hydrology of porphyry systems suggest that ore formation occurs at a self-sustaining hydrological divide related to the interplay of fluid flow and rock mechanics (Weis et al. 2012). The inner part of the magmatic fluid plume is dominated by hot magmatic fluids ascending under near-lithostatic fluid pressures. The ascending fluids heat the host rock above the inferred transition temperature from brittle to ductile (plastic, viscous) behavior of dry rock (Fournier 1999), leading to a permeability reduction. This effect is counteracted by the incoming fluids, because fluid pressure can cause transient embrittlement by bringing rock to failure and creating permeable pathways. As a result, magmatic fluids ascend in overpressure–permeability waves through this inner part (Weis et al. 2012). From the outside, this magmatic fluid plume is cooled by convection of meteoric fluids under near-hydrostatic fluid pressures. This hydrology stabilizes a sharp temperature–pressure front that matches with conditions for porphyry copper mineralization, while conditions favorable for epithermal deposits develop in shallower depths where the convecting meteoric water admixes in variable proportions with surges of magmatic fluids, inducing episodic boiling events.

The concept of a hydrological divide will be an essential part of the work discussed here. The term "divide" is used to indicate a boundary between two domains in the system. The first is dominated by magmatic fluids, near-lithostatic pressures, high temperatures, ductile rock behavior, and low permeabilities. The second is dominated by meteoric fluids, near-hydrostatic pressures, lower temperatures, brittle rock behavior, and higher permeabilities. The divide is marked by a stable pressure–temperature front, which is controlled by the heat balance between the different flow regimes in the two domains. However, the dynamic flow behavior of ascending magmatic fluids causes the nominally ductile rock to temporarily become more permeable due to episodic embrittlement by hydraulic fracturing. These surges also result in magmatic fluids flowing across the hydrological divide, which leads to temporarily dynamic behavior at the otherwise stable pressure–temperature front. For better readability, the term ductile will occasionally be used to identify the transiently overpressured regime, even though it is characterized both by ductile and episodic brittle behavior.

METHODS

Fluid flow in magmatic–hydrothermal systems can be described with a continuum, porous medium approach according to Darcy's law. To quantify the interplay of saline fluid flow and rock permeability, the numerical model requires (i) the implementation of a thermodynamic model that can accurately describe the phase relations and fluid properties over the relevant ranges of temperature, pressure, and composition; (ii) the implementation of a dynamic permeability model that maps existing permeability-related models to the continuum approach; and (iii) a numerical method that can handle multiphase flow of compressible, variably miscible H_2O–NaCl fluids in the presence of large source terms that arise from fluid expulsion and phase separation.

Multiphase flow of H_2O–NaCl fluids

Subsurface fluid flow in porous and fractured media can be described with an extended form of Darcy's law with

$$v_i = -k \frac{k_{ri}}{\mu_i}(\nabla p - \rho_i g) \qquad i = \{v, l\}, \tag{29.1}$$

where v_i is the Darcy velocity of the mobile phases i, which can be vapor v and liquid l, the permeability k, the relative permeability k_r of the phase indicated, the fluid viscosity μ, the total fluid pressure p, the fluid density ρ, and the acceleration due to gravity g (Ingebritsen *et al.* 2006). Saline fluids can also saturate in a solid halite phase h, which is assumed to reduce the available

permeability for the mobile phases by applying a linear relative permeability model, which ensures that

$$k_{rv} + k_{rl} = 1 - S_h, \tag{29.2}$$

where S is the volumetric saturation of the phase indicated, and a residual saturation of $R_v = 0.0$ for the vapor and $R_l = 0.3(1 - S_h)$ for the liquid phase (Weis *et al.* 2014).

Conservation of fluid mass is calculated as

$$\frac{\partial(\phi(S_l\rho_l + S_v\rho_v + S_h\rho_h))}{\partial t} = -\nabla \cdot (v_l\rho_l)$$
$$-\nabla \cdot (v_v\rho_v) + Q_{H_2O+NaCl}, \tag{29.3}$$

with the porosity ϕ, the time t, and a source term $Q_{H_2O+NaCl}$. Conservation of salt mass is described by

$$\frac{\partial(\phi(S_l\rho_lX_l + S_v\rho_vX_v + S_h\rho_hX_h))}{\partial t} = -\nabla \cdot (v_l\rho_lX_l)$$
$$-\nabla \cdot (v_v\rho_vX_v) + Q_{NaCl}, \tag{29.4}$$

with X being the mass fraction of NaCl and a source term Q_{NaCl}. Conservation of energy is obtained by

$$\frac{\partial[(1-\phi)\rho_r h_r + \phi(S_l\rho_l h_l + S_v\rho_v h_v + S_h\rho_h h_h)]}{\partial t} = \nabla \cdot (K\nabla T)$$
$$-\nabla \cdot (v_l\rho_l h_l) - \nabla \cdot (v_v\rho_v h_v) + Q_e, \tag{29.5}$$

with the subscript r indicating the rock, the specific enthalpy h of the phase indicated, the thermal conductivity K, the temperature T, and a source term Q_e.

Phase relations and fluid properties are calculated by a thermodynamic model according to the IAPS-84 equation of state (Haar *et al.* 1984) for pure H_2O, and the effect of adding NaCl is described by the model of Driesner & Heinrich (2007) and Driesner (2007). Fluid and rock are assumed to be in local thermal equilibrium, which requires that the total enthalpy is distributed over the fluid and rock contained in a control volume and fluid properties are updated by the numerical method at every modeling time step (Weis *et al.* 2014).

Dynamic permeability

The calculation of fluid flow in a continuum approach with Darcy's law (Eq. 29.1) requires that permeability can be described with values for the bulk property k that are representative for the elements of the computational mesh of the modeling domain (Weis *et al.* 2012). For modeling the hydrology of porphyry copper deposits, simulations with static, homogeneous permeabilities were unable to reproduce a hydrothermal system that could explain all geological observations. Only after assembling a dynamic permeability model by mapping existing models for permeability, stress state, and failure conditions of the continental upper crust, a hydrothermal system with a steady, self-sustaining enrichment mechanism in agreement with observations developed (Weis *et al.* 2012). This permeability model uses five general assumptions:

(1) Permeability generally follows a depth-dependent profile characteristic for a tectonically active crust (Manning & Ingebritsen 1999).

(2) For the brittle crust, this profile is related to a nearly critically stressed crust (Zoback *et al.* 2002), resulting in a failure criterion for fractures at near-hydrostatic fluid pressure conditions (Cox 2010).

(3) With increasingly ductile behavior of the rock due to heating, permeability decreases and differential stress relaxes at elevated temperatures (Fournier 1999), leading to failure criteria at near-lithostatic fluid pressures (Cox 2010). Depending on the rock type and strain rate, this brittle–ductile transition may start at a temperature of 360 °C (Hayba & Ingebritsen 1997).

(4) Elevated fluid pressures can counteract the effect of temperature-dependent permeability closure, ensuring that permeability has the value of its characteristic profile when fluid pressures reach the failure criterion (Manning & Ingebritsen 1999; Cox 2005, 2010; Ingebritsen & Manning 2010).

(5) Hydraulic fracturing temporarily and locally increases permeability by up to two orders of magnitude higher than the characteristic permeability profile (Ingebritsen & Manning 2010), whenever fluid pressures exceed the respective stress state–dependent failure criterion.

Permeability ranges are generally limited to a maximum of 10^{-14} m^2, which is the reference permeability at 1 km depth in the study of Manning & Ingebritsen (1999). Sensitivity studies use permeabilities of up to 10^{-13} m^2.

Background permeability

Compilations of permeability estimates representative for tectonically active continental crust follow a characteristic, depth-dependent profile, which can be described as

$$\log k = -14 - 3.2 \log z, \qquad (29.6)$$

with the permeability k in m^2 and the depth z in km (Manning & Ingebritsen 1999). This profile is interpreted to represent time averages, and the dynamic permeability model presented here uses this relation as background values for permeability evolution (Fig. 29.1A). (Note: color versions of Figures 29.1–29.12 can be found in the online version of this chapter in *Geofluids* 15, pp. 350–371.) At brittle conditions, these values are maintained as long as fluid pressures are below or equal to the failure criterion. With the beginning of ductile behavior, the values are used as reference permeability for the case when fluid pressure equals the failure criterion (Fig. 29.2A). In simulations with a volcano topography, the depth-dependent permeability is calculated in relation to the level of the flat surface at the foot of the volcano, while the volcano is assigned the maximum permeability of 10^{-14} m^2 (or 10^{-13} m^2 for sensitivity studies).

Critically stressed brittle crust

Measurements from the KTB scientific research borehole show that differential stress increases with depth and can have values close to hydrostatic fluid pressures for the uppermost 10 km at temperatures following a normal geothermal gradient (Zoback & Harjes 1997; Zoback *et al.* 2002). For brittle conditions, the dynamic permeability model assumes a constant, homogeneous near-critical stress state of the crust under normal-faulting conditions (Fig. 29.1B). The maximum principal stress σ_1 is equal to the vertical stress σ_v. As a simplification, the vertical stress is calculated as being equal to the rock overburden given by the lithostatic pressure gradient. Differential stress given by $\sigma_{\text{diff}} = \sigma_1 - \sigma_3$ (with the minimum principal stress σ_3) is assumed to be 5 MPa above the initial hydrostatic pressure. The failure criterion for brittle shear expressed with the pore fluid factor

$$\lambda_v = \frac{p_{\text{fluid}}}{\sigma_v} \qquad (29.7)$$

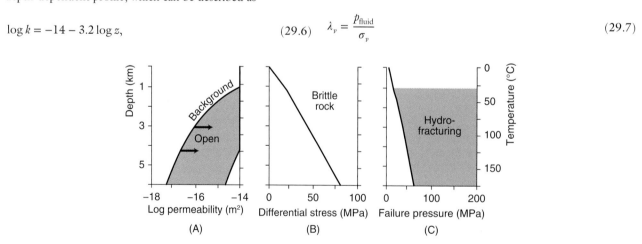

Fig. 29.1. Dynamic permeability model for the brittle crust at temperatures imposed by a normal geothermal gradient. (A) Depth-dependent permeability profile for average (background; lower limit of light gray area) and perturbed (light gray area) continental crust. (B) Stress state of the crust at near-critical conditions. (C) Hydrofracturing (light gray area) is described by incrementally increasing permeability for fluid pressures above a stress state–dependent failure criterion. The light gray area marks the field of fluid pressures that are not sustainable, because they exceed the pressure at failure. (Note: color versions of Figures 29.1–29.12 can be found in the online version of this chapter in *Geofluids* 15, pp. 350–371.)

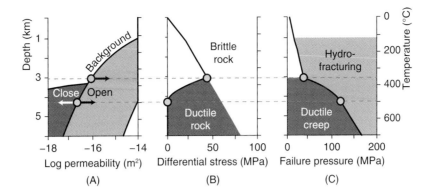

Fig. 29.2. Transition from brittle to ductile rock behavior within the dynamic permeability model at elevated temperatures imposed by a magmatic heat source. The gray dashes mark the inferred temperature interval for the brittle–ductile transition (360 to 500 °C). The linear temperature profile with depth schematically illustrates the constitutive relationships. (A) Permeability profile as in Figure 29.1A with superimposed permeability closure at high temperatures (dark gray area; Fig. 29.3A). (B) Differential stress (Fig. 29.1B) relaxes with elevated temperatures. (C) The failure criterion at nominally ductile conditions increases to values at near-lithostatic pressure conditions. Fluid pressure–dependent permeability counteracts the temperature-dependent permeability closure at conditions of ductile creep (dark gray area; Fig. 29.3B). (Adapted from Weis *et al.* (2012).)

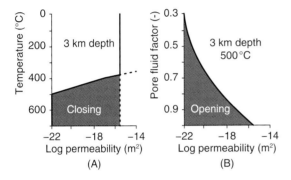

Fig. 29.3. Constitutive relationships for permeability at ductile conditions. (A) Permeability closes with elevated temperatures. (B) Permeability closure is counteracted by elevated fluid pressures; pore fluid factor is fluid pressure divided by failure pressure (see Fig. 29.2C). (Adapted from Weis *et al.* (2012).)

is given by the relation

$$\lambda_v = \frac{(4C - \sigma_1 + 4\sigma_3)}{3\sigma_v}, \tag{29.8}$$

where C is the cohesive strength of the rock (Eq. 8 in Cox 2010). Rearranging the equation gives the fluid pressure at failure ($p_{fluid} = p_{fail}$) for brittle shear at normal-faulting conditions with

$$p_{fail} = \sigma_v - \frac{4(\sigma_{diff} - C)}{3}. \tag{29.9}$$

Extension failure is given by the relation

$$\lambda_v = \frac{\sigma_3 + T}{\sigma_v}, \tag{29.10}$$

with the tensile strength of the rock T (Eq. 12 in Cox 2010). Rearranging this equation leads to a failure pressure of

$$p_{fail} = \sigma_v - \sigma_{diff} + T. \tag{29.11}$$

Assuming a cohesive strength of 10 MPa and a tensile strength of 5 MPa, the two curves intersect at $\sigma_{diff} = 5T = 2.5C$. Above this value, the failure criterion of brittle shear is reached first, while extension failure happens below this threshold (Fig. 29.1C).

The dynamic permeability model used for this study calculates this stress field once at the beginning of the simulation from the initial conditions. During the simulation, the stress field and its variations are not explicitly evaluated for the domain, except for a temperature-dependent relation describing the transition from brittle to ductile behavior of the rock.

Brittle–ductile transition

With a temperature profile of a regular geothermal gradient (as in Fig. 29.1), the brittle–ductile transition is located at about 10 km depth. At deeper crustal levels, the stress state gradually relaxes (Zoback *et al.* 2002). This ductile region is related to low permeabilities where conductive heat transfer dominates over advective heat transport (Hayba & Ingebritsen 1997; Ingebritsen *et al.* 2006). The presence of upper crustal magmatism can locally move this brittle–ductile transition to shallower depths (Fig. 29.2) around the magmatic intrusion and can be approximated depending on rock properties, temperature, and strain rate (Fournier 1999).

The dynamic permeability model used for this study adopts the temperature-dependent permeability approach of Hayba & Ingebritsen (1997), where k is assumed to decrease log-linearly with temperatures above 360 °C to essentially impermeable values ($k = 10^{-22}\,\mathrm{m^2}$) at 500 °C (Fig. 29.2A). In this study, temperature-dependent permeability is calculated in relation to a reference value of $10^{-14}\,\mathrm{m^2}$ (or $10^{-13}\,\mathrm{m^2}$ for sensitivity studies) and starts to dominate as soon as the temperature-dependent value is lower than the depth-dependent permeability (Fig. 29.3A).

The transition from brittle to ductile behavior of the rock also leads to a decrease in differential stress within the same temperature interval as the temperature-dependent permeability (Fig. 29.2B). The local stress state–dependent failure criterion is then recalculated according to Eqs 29.9 or 29.11 (Fig. 29.2C). At temperatures above 500 °C, this relationship leads to $\sigma_{diff} = 0$, which results in a failure pressure above lithostatic with

$$p_{fail} = \sigma_v + T. \tag{29.12}$$

With higher strain rates, the temperature interval of the brittle–ductile transition may move to higher values (Fournier 1999). For simulations of this study, however, we keep the strain rate constant. The assumption $\sigma_{diff} = 0$ implies that all stresses are uniform and equal to the rock overburden, which is a further simplification. Repeated brittle shear failure may also lead to a progressive reduction in differential stress (Cox 2010) but is neglected here because it would not be straightforward to quantify and parameterize.

Pressure-dependent permeability

The temperature-dependent decrease in permeability is assumed to be counteracted by an elastic opening of permeable structures where elevated fluid pressures work against the loss of interconnected pore space (Fig. 29.2C). If temperature-dependent permeability dominates over the depth-dependent value and tends to decrease permeability, permeability is increased with fluid pressures above hydrostatic depending on the modified pore fluid factor λ in relation to the failure pressure instead of the lithostatic pressure,

$$\lambda = \frac{p_{fluid}}{p_{fail}} \tag{29.13}$$

in a log-quadratic manner (Fig. 29.3B). This parameterization mimics results from permeability experiments (Cox 2005) and is similar to other constitutive relationships for fluid pressure–dependent permeabilities (Rice 1992).

Hydraulic fracturing

Compilations of transient permeability-creating events suggest that permeability can temporarily be increased by two to three orders of magnitude higher than the characteristic profile used as a background curve (Ingebritsen & Manning 2010). Whenever fluid pressure exceeds the stress state–dependent failure pressure determined by Eqs 29.9, 29.11, or 29.12 ($p_{fluid} > p_{fail}$) permeability is incrementally increased until fluid overpressure is released ($p_{fluid} \leq p_{fail}$). For the sake of consistency between the different components of the dynamic permeability model, the λ^2-dependence is inherited (see Eq. 29.13) to incrementally increase permeability with

$$\frac{k_{t+\Delta t}}{k_t} = \lambda^2 = \left(\frac{p_{fluid}}{p_{fail}}\right)^2, \tag{29.14}$$

where the subscript t indicates the current time level and $t + \Delta t$ denotes the following modeling time step. Permeabilities are limited to values exceeding the background profile by two orders of magnitude or the maximum permeability (Figs 29.1A and 29.2A). After release of the overpressure, permeability is immediately reduced to the background value. This parameterization of hydraulic fracturing accounts for field observations documenting that rocks are fractured and sealed by vein formation in multiple cycles. The relation further allows to use a uniform mechanism for permeability creation independent of whether it occurs at near-hydrostatic or near-lithostatic fluid pressures.

Other numerical simulations use different approaches for permeability creation and reduction. For (near-)lithostatic failure criteria, fluid pressure can be fixed at the values of failure and Darcy's law is solved for permeability instead of pressure (Hanson 1995). However, for (near-)hydrostatic failure criteria, this relationship can lead to unfavorable constellations where unreasonably drastic permeability changes are calculated because the driving force of Darcy's law, $\nabla p - \rho g$, is close or equal to zero. Other approaches scale permeability depending on effective normal stress, that is, on fluid pressure and stress state (Rice 1992; Miller et al. 2004), which is also the foundation of the calculation of failure pressure in this study. However, these formulations are not designed for fluid pressures exceeding lithostatic pressure and will not necessarily lead to a reduction in fluid pressure below failure pressure as long as the fluid supply rate can maintain a fluid pressure gradient that is larger than the lithostatic pressure gradient.

The reduction to fluid pressures below the failure criterion can be assured by defining a dependency of permeability change on overpressure as given by Eq. 29.14 rather than a permeability–pressure relation. This approach requires more *ad hoc* parameterizations and aims at mimicking the effect of hydraulic fracturing as a response to forced fluid flow rather than at resolving the actual process of rock mechanics (e.g. Rojstaczer et al. 2008).

Control volume finite element method

The numerical model used for this study is an implementation of the Complex Systems Modeling Platform (CSMP++, Matthai et al. 2007) for thermohaline convection, and the governing equations are solved with a control volume finite element method as described in detail by Weis et al. (2014). The method enables the use of unstructured meshes with varying resolution, which allows for mesh refinement in the most important parts of the domain. The approach further ensures that sharp solute and thermal fronts in advection-dominated flow systems as well as the physical separation of fluid phases can be resolved by using all fluid properties as calculated by the thermodynamic model and avoiding interpolations or averaging.

Fluid properties are defined as nodal values at the center of a respective control volume, while permeability is an elemental property. As finite elements include several nodes at their corners, the constitutive relationships of the dynamic permeability model are averaged, interpolated, or accumulated. Depth-dependent permeability is calculated in reference to the vertical coordinate of the element's barycenter. For the calculation of temperature-dependent permeability values, first the respective nodal permeability is evaluated and then averaged to the element variable. The same holds for the counteracting fluid pressure dependence for fluid pressures below the failure criterion. For the calculation of hydraulic fracturing, the effects of permeability change according to Eq. 29.14 are added up to ensure that elements where all corner nodes experience overpressure are more intensely fractured than elements where only one corner node is overpressured.

Model configuration

The modeling domain represents a two-dimensional section of the upper continental crust with a magma chamber of initially 900 °C emplaced at 5 km depth (Fig. 29.4). The host rock porosity is kept constant with 0.05, and the pore space is

initially saturated with pure water under hydrostatic pressure and a thermal gradient of 22.5 °C km^{-1}, which is maintained by a bottom heat flux of 45 mW m^{-2}, and 10 °C at the top boundary, representing the Earth's surface. During the simulation, the top boundary is open to any fluid outflow at atmospheric pressure and inflow of salt-free liquid water of 10 °C. The left, right, and bottom boundaries are no-flow boundaries. Rock properties are kept constant with a density of 2700 kg m^{-3}, a thermal conductivity of 2 W m^{-1} °C^{-1}, and a heat capacity of 880 J kg^{-1} °C^{-1}. The magma chamber uses the same values but starts with a doubled heat capacity, which is gradually reduced with the cooling of the magma chamber to account for the release of latent heat during crystallization (e.g. Hayba & Ingebritsen 1997).

Expulsion of magmatic fluids

The release rate of magmatic fluids to the host rock is calculated as being proportional to the crystallization rate of the magma chamber. At every modeling time step, the volume of the magma chamber that has cooled below an inferred solidus temperature of 700 °C is evaluated. All elements that have one node below this solidus and one above it add to the total

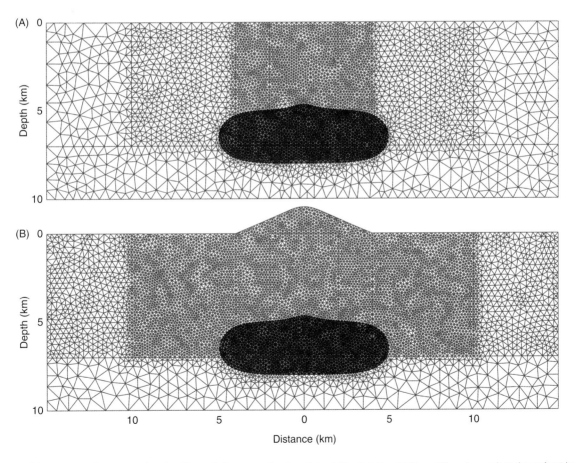

Fig. 29.4. Model geometry and unstructured mesh with an elliptic magma chamber of about 3 km height and 10 km width and a cupola in the roof at 5 km depth with flat topography (A) and a central volcano of about 1.5 km height (B). The dimensions are taken from field observations of the Yerington Batholith (Dilles 1987).

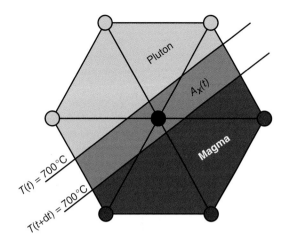

Fig. 29.5. Crystallization of the magma chamber. The dark gray points indicate nodes of the modeling mesh with temperatures above the solidus temperature of 700 °C and are considered to be part of the magma. The light gray nodes have already cooled below 700 °C and are considered to be part of the solidified pluton. The black dot shows a node that cooled below the solidus temperature within one modeling time step dt as indicated by the 700 °C-isotherms at time t and $t + dt$. The newly crystallized area $A_x(t)$ at the magma–pluton interface is calculated as the area of the respective element lying between these two isotherms.

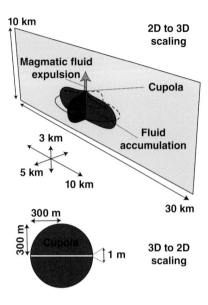

Fig. 29.6. Fluid accumulation from a three-dimensional magma chamber. Fluids accumulate at the highest point underneath the roof of the magma chamber, which is approximated as a horizontal, circular cupola. The total amount of fluids released from a three-dimensional magma chamber is scaled back to the one-dimensional outline of the cupola in the two-dimensional domain with a nominal thickness of 1 m.

fluid production and the newly crystallized area within one two-dimensional element during the last time step is calculated as $A_x(t)$ (Fig. 29.5). If all nodes eventually cooled below the solidus temperature, the contributions $A_x(t)$ distributed over the modeling time steps have added up to the total area of the respective element and are subsequently considered to be part of the host rock. To also account for fluid accumulation in the third dimension, the magma chamber is inferred to have an elongated shape with half the width in the dimension perpendicular to the modeling section (Fig. 29.6).

The total volume of newly crystallized magma is calculated as

$$V_{cryst}(t) = \sum_{\text{elements}} f_{3D} \cdot A_x(t), \qquad (29.15)$$

with a 2D-to-3D scaling factor

$$f_{3D} = \pi \cdot dx \cdot w_y, \qquad (29.16)$$

where dx is the horizontal distance from the vertical center axis of the magma chamber to the barycenter of the element and w_y is the aspect ratio of the horizontal dimensions of the magma chamber, that is, $w_y = 0.5$ for the elongated shape of the magma chamber (Fig. 29.6). For an aspect ratio of $w_y = 1.0$, the scaling factor gives the exact mathematical solution of the volume of an element rotating around an axis in a circle. As the calculation for the volume along an elliptical curve with a distorted element is not trivial, the simple scaling for $w_y \neq 1.0$ in Eq. 29.16 rather reflects a best estimate for the 3D volume.

Assuming a water content of 5% ($C_{water} = 0.05$) of the rhyolitic melt that cannot be incorporated in the solidified igneous rock and is therefore released as a fluid phase gives the total mass of magmatic fluid exsolved from this newly crystallized volume as

$$m_{fluid}(t) = C_{water} \cdot \rho_{rock} \cdot V_{cryst}(t), \qquad (29.17)$$

with the density of the rock ρ_{rock}. This fluid mass is assumed to be expelled to the host rock through a cupola in the roof of the magma chamber during the following modeling time step. Assuming a circular, horizontal cross-section of this cupola with a radius of $R_{cupola} = 300\,\text{m}$ and a uniform distribution of the expelled fluid mass across the cupola, the total amount of fluid scaled back to the two-dimensional modeling domain can be calculated as

$$m_{source}(t) = f_{2D} \cdot m_{fluid}(t), \qquad (29.18)$$

with a 3D-to-2D scaling factor

$$f_{2D} = \frac{2}{\pi \cdot R_{cupola}}. \qquad (29.19)$$

The source term $m_{source}(t)$ is then evenly distributed across the nodes that represent the current location of the cupola and enter the system of conservation equations with the source term $Q_{H_2O+NaCl}$ in Eq. 29.3. Assuming a salinity of 10 wt% NaCl for the magmatic fluids results in a source term of $Q_{NaCl} = 0.1 \cdot Q_{H_2O+NaCl}$ in Eq. 29.4. The thermodynamic model for H_2O–NaCl fluids then calculates the specific enthalpy $h_{H_2O+NaCl}$ of a fluid of 700 °C, 10 wt% NaCl and the current fluid pressure at the injection node, resulting in a source of $Q_e = Q_{H_2O+NaCl} \cdot h_{H_2O+NaCl}$ in Eq. 29.5. The location of the cupola can retreat to greater depth following the isotherms of the solidus temperature.

Proxies for vein formation and copper precipitation

The formation of vein quartz is estimated by translating the accumulated permeability increases Δk induced by hydraulic fracturing, and assuming a simple cubic law relationship (Walder & Nur 1984) to a virtual void space $\Delta\phi$, assuming

$$\frac{k_0 + \Delta k}{k_0} = \left(\frac{\phi_0 + \Delta\phi}{\phi_0} \right)^3, \qquad (29.20)$$

where k_0 is the background permeability (Fig. 29.1A) and ϕ_0 is the model's porosity ϕ, which is kept constant during the simulations. The vein volume potential calculates the volume of quartz that could have precipitated if all the virtual open space would have been filled during vein sealing as

$$\psi_{\text{veins}} = \frac{\Delta\phi}{1 + \Delta\phi} \cdot 100 \qquad (29.21)$$

and therefore represents an upper limit estimate (in %) for vein density. The spatial and temporal distribution pattern of this proxy for vein density can be compared with field observations from porphyry copper deposits.

Copper is introduced as a tracer to the system of equations as

$$\frac{\partial(\phi(S_l\rho_l + S_v\rho_v + S_h\rho_h)C_{\text{Cu}})}{\partial t} = -\nabla \cdot (v_l\rho_l C_{\text{Cu}})$$
$$-\nabla \cdot (v_v\rho_v C_{\text{Cu}}) + Q_{\text{Cu}}, \qquad (29.22)$$

with the copper concentration in the fluid C_{Cu} and a source term $Q_{\text{Cu}} = C_{\text{Cu}}^{\text{initial}} \cdot Q_{\text{H2O+NaCl}}$, reflecting a copper concentration of $C_{\text{Cu}}^{\text{initial}} = 500$ ppm in the pristine fluid exsolved from the magma chamber. Before precipitation, copper is advected as a passive tracer with the fluids and has no preferential fractionation behavior into the vapor, liquid, or halite phase. In reality, fractionation behavior may depend on whether Cu is preferentially forming complexes with Cl or S (Pokrovski *et al.* 2008).

Experimental and fluid inclusion studies suggest that copper in porphyry copper systems is precipitated upon cooling of the copper-bearing fluid within the temperature interval between 450 and 350 °C (Hezarkhani *et al.* 1999; Landtwing *et al.* 2005). Copper precipitation is therefore calculated as a function of temperature change. The copper enrichment potential within a control volume of the modeling mesh is calculated as

$$\psi_{\text{Cu}} = \frac{m_{\text{Cu}}}{C_{\text{Cu}}^{\text{initial}} \cdot C_{\text{water}} \cdot \rho_{\text{rock}} \cdot V}, \qquad (29.23)$$

with the mass of precipitated copper m_{Cu} and the control volume V, which can be used for a rough estimate of the potential ore grade (in wt%) as

$$G_{\text{Cu}} = \frac{m_{\text{Cu}}}{m_{\text{rock}} + m_{\text{Cu}}} \cdot 100 = \frac{\psi_{\text{Cu}} \cdot C_{\text{Cu}}^{\text{initial}} \cdot C_{\text{water}}}{1 + \psi_{\text{Cu}} \cdot C_{\text{Cu}}^{\text{initial}} \cdot C_{\text{water}}} \cdot 100, \qquad (29.24)$$

with the mass of the rock m_{rock}. The proxy for the formation of copper ore represents an upper limit estimate, as copper precipitation may be less efficient in nature. Furthermore, this estimate neglects sulfur availability, chemical equilibrium constants, and chemical potential gradients that will affect the precipitation efficiency.

RESULTS

The magma chamber conductively heats the surrounding host rock and induces convection of ambient fluids. As soon as heat conduction and advection have cooled the outer rim of the magma chamber below the solidus temperature, magmatic fluids are injected at the cupola and rapidly flow upward (Fig. 29.7A). The source rate of magmatic fluid injection in these early phases of the simulation is highest because the outermost layer of the three-dimensional magma body has the largest surface area (see Fig. 29.6). The initial expulsion event is accompanied by a pronounced pressure anomaly to accommodate the additional magmatic fluids, which leads to a quasi-radial outward propagation involving an area much wider than the width of the cupola and affects the entire domain between magma chamber and surface (Fig. 29.7A). The critically stressed brittle crust almost immediately reaches the failure criterion and hydraulic fracturing enhances permeability to a few kilometers distance around the cupola. In the beginning, the high source rate maintains a pressure anomaly that keeps permeability at the maximum value of 10^{-14} m^2 in the area surrounding the cupola.

The propagations of the thermal and solute front are decoupled, because the initially hot magmatic fluids lose heat to the surrounding colder host rock. While the thermal front has only progressed about 1 km from the cupola (Fig. 29.7E), magmatic salt has already reached the surface (Fig. 29.7I). With continuous expulsion, the area above the cupola is heated toward magmatic temperatures and the host rock starts to increasingly behave in a ductile manner resulting in decreasing permeability. As a consequence, the fluids exsolving from the magma are prevented from leaving the cupola, which leads to further increase in fluid pressure. With the transition from brittle to ductile behavior of the host rock, the stress state of the crust changes by relaxing differential stress, which leads to a transition of the failure criterion from near-hydrostatic to near-lithostatic conditions, meaning that the host rock is not fractured as easily as the near-critically stressed brittle crust.

The elevated fluid pressures lead to an increase in permeability, or rather work against the temperature-dependent permeability decrease, so that hydraulic fracturing starts from the same background permeability as in the brittle case, albeit now at near-lithostatic fluid pressures. As a result, the model develops a vertically extended zone dominated by high-pressure, high-temperature magmatic fluids and predominantly ductile rock behavior (Fig. 29.7A) – hereafter

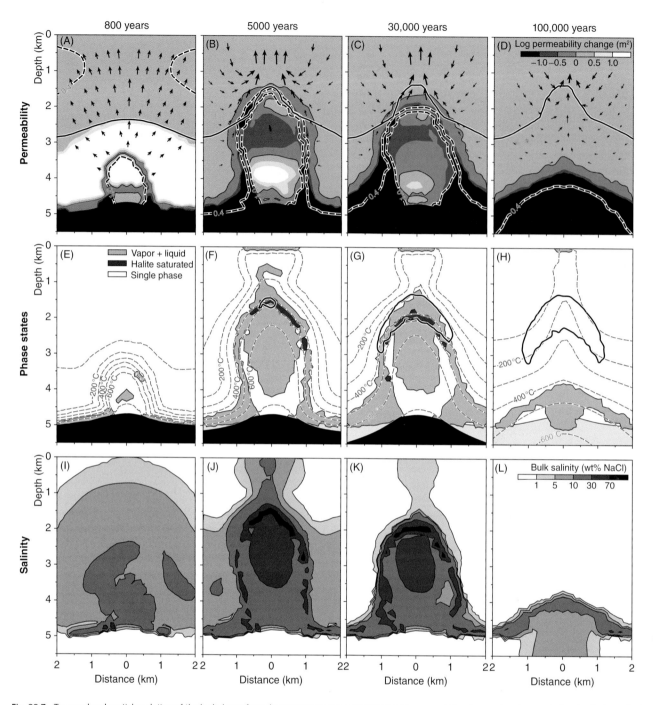

Fig. 29.7. Temporal and spatial evolution of the hydrology of porphyry copper systems. (A–D) Dynamic permeability changes in reference to the depth-dependent background profile (Fig. 29.1A) after 800, 5000, 30,000, and 100,000 years of simulation time. The black dashed lines show the pore fluid factor (fluid pressure divided by lithostatic pressure) with values below 0.4 indicating near-hydrostatic fluid pressures and values of 1.0 indicating near-lithostatic fluid pressures. Black arrows schematically show fluid flow in the dominantly brittle domain. The black solid lines show a vein volume potential of 10%. The black area indicates the magma chamber. (E–H) Temperature distribution and multiphase fluid states with light gray areas indicating liquid–vapor coexistence and dark gray areas halite-saturated fluids (vapor–halite, liquid–halite or vapor–liquid–halite coexistences). The black solid lines show a copper enrichment potential of 500. The black area indicates the magma chamber and the very light gray area the solidified pluton. (I–L) Variations in bulk salinity of the fluid mixture including the solid halite phase.

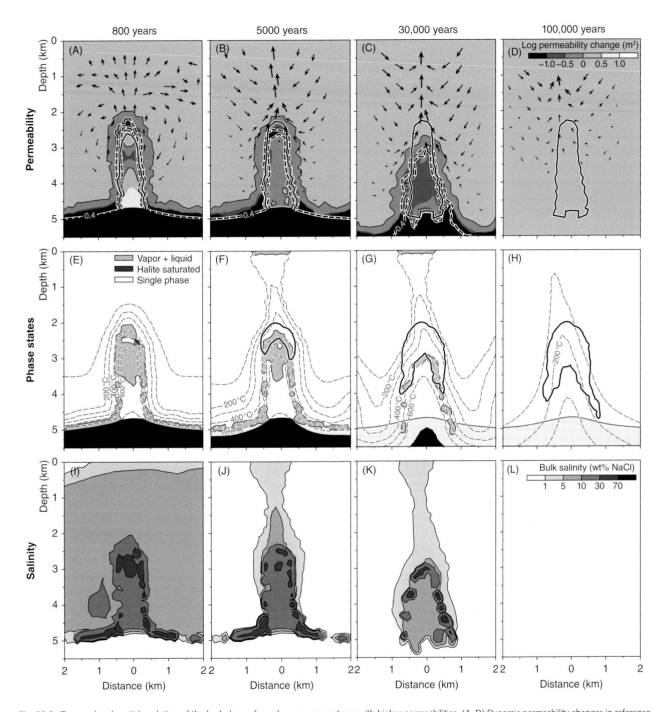

Fig. 29.8. Temporal and spatial evolution of the hydrology of porphyry copper systems with higher permeabilities. (A–D) Dynamic permeability changes in reference to the depth-dependent background profile of one order of magnitude higher than in Figure 29.1A after 800, 5000, 30,000, and 100,000 years of simulation time. The black dashed lines show the pore fluid factor. Black arrows schematically show fluid flow in the dominantly brittle domain. The black solid lines show a vein volume potential of 10%. The black area indicates the magma chamber. (E–H) Temperature distribution and multiphase fluid states with light gray areas indicating liquid–vapor coexistence and dark gray areas halite-saturated fluids. The black solid lines show a copper enrichment potential of 500. The black area indicates the magma chamber and the very light gray area the solidified pluton. (I–L) Variations in bulk salinity of the fluid mixture including the solid halite phase.

referred to as the "inner" part of the hydrothermal system with a hydrological divide. Fluid pressures episodically reach the failure criterion and release overpressure–permeability waves that ascend from the cupola to the current location of the hydrological divide. The snapshot after 800 years of simulation time (Fig. 29.7A) shows the establishment of this inner part of the magmatic fluid plume with one permeability wave developing at the cupola and an earlier one just about to release a surge of magmatic fluids to the brittle region surrounding it.

The wave-like behavior results from the parameterization of hydraulic fracturing by incremental permeability increase as a function of overpressure. Unless fluid supply can permanently maintain fluid pressures above lithostatic with the maximum permeability of $10^{-14}\,m^2$, fluid overpressure will be released periodically and permeability will be reduced to the background value or lower, followed by the next pressure build-up that will trigger the next wave. The frequency of the permeability waves depends on the fluid supply rate, which controls the rate of pressure increase at the cupola, and the exact parameterization of hydraulic fracturing, which describes the degree of permeability response to overpressure: changing the λ^2-dependence in Eq. 29.14 to a λ- or λ^3-dependence leads to lower or higher frequencies, respectively. However, the location of the brittle–ductile transition and hydrological divide within the host rock is governed by the larger-scale balances of heat and mass fluxes and is therefore less dependent on the exact implementation.

The initial expulsion phase leads to intense fracturing within a wide radius around the cupola and a predicted vein volume potential of 10% reaching out to a depth of about 2 km. With the continuous exsolution of magmatic fluids, the upper boundary of the ductile region extends to shallower depths, reaching a peak at about 5000 years of simulation time (Fig. 29.7B, F, and J). By that time, the pressure anomaly in the brittle part of the domain has relaxed and a convection of predominantly meteoric fluids cools the overpressured high-temperature part of the magmatic fluid plume from the sides. This brittle domain with meteoric convection is hereafter referred to as the "outer" part of the hydrothermal system. As the background permeability increases with shallower depth, this cooling effect increases toward the surface and the heat balance of ascending magmatic fluids and meteoric convection determines the round top of the predominantly ductile domain.

In the deeper parts of the system, the two hydrological systems are separated by a low-permeability zone that conductively transfers heat from the magmatic- to the meteoric-dominated part, where it is removed by convection of fluids at temperatures of 350–400 °C.

The snapshot after 5000 years captures an overpressure–permeability wave that ascends from the cupola toward the surface (Fig. 29.7B). At shallower levels, an earlier wave is about to reach the hydrological divide where the surge of magmatic fluids mixes with the convecting meteoric water. Here, the magmatic fluids experience a relatively sudden drop in pressure

(from near-lithostatic to near-hydrostatic) and temperature (from about 500 to 300 °C). While the ascent of the magmatic fluids is predominantly vertical, the surges at the hydrological divide can also have a significant horizontal component as the pressure gradient from lithostatic to hydrostatic is similarly sharp at the sides of the overpressured plume.

The repeated sealing–fracturing cycles lead to an increased vein density potential in the central high-temperature upflow zone (Fig. 29.7B). Copper enrichment potential starts to increase at the hydrological divide where conditions for porphyry mineralization at the stable temperature–pressure front are maintained with the sharp temperature drop from above 450 to below 350 °C (Fig. 29.7F).

Due to the high fluid pressures at the cupola, the magmatic fluids are injected in a single-phase state (Fig. 29.7F). Upon ascent, these primary fluids phase-separate into a low-salinity, low-density vapor and a high-salinity, high-density brine phase. As the two phases can physically separate from each other with the vapor phase rapidly ascending toward the surface, the bulk salinity of the fluid mixture increases to values between 30 and 70 wt% NaCl (Fig. 29.7J). However, the high-pressure conditions allow a further upward movement of the heavy brine phase. At the hydrological divide, the high-salinity fluids saturate in solid halite due to the pressure drop, which moves the fluids from vapor–liquid coexistence to vapor–halite coexistence.

Solid halite is assumed to be immobile and to reduce the permeability available for the mobile phases liquid and vapor (Eq. 29.2), which can lead to a blocking effect triggering pulsating fluid flow behavior and salt accumulation in the crust (Weis *et al.* 2014). With the hydrological divide in place, however, the precipitated salt can subsequently be dissolved by the convecting meteoric fluids and transported toward the surface as moderately saline mixtures of magmatic and meteoric fluids. The upflow zone above the divide is close to the boiling curve of these moderately saline fluids (Fig. 29.7J), fluctuating between a two-phase zone that extends from the divide to the surface and two distinct two-phase fields with one right above the divide and one just underneath the surface (Fig. 29.7F).

While the magma chamber is generally cooled from the outer rim of the original outline inwards, the hydrothermal system above the cupola of the magma chamber prevents this part of the magma to be cooled and the location of the cupola remains constant for almost the entire fluid expulsion phase. With further inward crystallization, the surface area of the magma chamber decreases, which leads to a smaller rate of fluid production. In response, the heat balance of magmatic fluid supply and convective cooling moves the hydrological divide and therefore also the temperature–pressure front to greater depths (Fig. 29.7C, G, and K). As a consequence, hydraulic fracturing becomes less intense and vein volume potential is only increased by minor amounts, while copper enrichment potential continues to accumulate, albeit at greater depths, extending the ore shell toward deeper levels. With this retreat, the model predicts that areas that have been overprinted by intense, barren stockwork veining at temperatures above 500 °C in the early phases are now overprinted by copper-bearing veins forming at lower temperatures, which explains why copper is introduced at a later stage than the first generation of barren quartz (Landtwing *et al.* 2005).

This process continues until the magma chamber is fully crystallized (Fig. 29.7D, H, and L). The remaining heat of the pluton drives thermal convection at moderately elevated temperatures and maintains an upflow zone passing through the deposit, which may lead to overprints of low-temperature alteration or redistribution of ore minerals.

The influence of permeability range

Permeability ranges at porphyry copper systems may be significantly higher than the ones obtained from tectonically active crust (Manning & Ingebritsen 1999) and the average permeability value may rather be characterized by a depth-dependent profile of the perturbed crust (Ingebritsen & Manning 2010). To investigate the influence of permeability range on the results, a simulation with the same general setup but permeability values of one order of magnitude higher have been conducted (Fig. 29.8). The background permeability profile is now calculated in reference to a log k of −13 at 1 km depth instead of −14 as in Eq. 29.6. Similarly, temperature-dependent permeability is calculated in reference to a permeability of $10^{-13}\,m^2$, which is now used also as a maximum permeability value.

After 800 years, the thermal front has already progressed further than in the previous simulation (Fig. 29.8A, E, and I). The snapshot captures the transition from the initial expulsion phase to the establishment of the hydrological divide separating the two regimes of fluid flow. In the uppermost part, fluids are still expelled radially outward, while some magmatic fluids traversing the hydrological divide at the top are cooled and sink back to greater depth where they become involved in the cooling mechanism that stabilizes the hydrological divide.

Fluid flow in the inner, high-pressure part is again dominated by overpressure–permeability waves, which now has a significantly smaller horizontal extent as in the simulations with lower permeabilities (Fig. 29.8A). The area of 10% vein volume potential is limited to this upflow zone, showing that hydraulic fracturing during the initial fluid expulsion phase is less drastic, because the overall higher permeability values are able to release the fluids more easily.

As in the simulation with lower permeabilities, the magmatic fluids are injected in the single-phase state and phase-separate upon ascent (Fig. 29.8E), leading to an increase in bulk salinity (Fig. 29.8I). The simulation also develops the same sharp temperature–pressure front at the hydrological divide. However, halite saturation is less pronounced at the transition, because convection in the outer, brittle part is more effective in removing the magmatic salt injected with the incoming fluid surges from the inner, high-pressure part due to the higher permeability of the host rock.

After 5000 years, the characteristic hydrology with the inner part being dominated by magmatic fluids and the outer part dominated by meteoric fluids is fully developed (Fig. 29.8B, F, and J). Copper starts to become enriched around the temperature–pressure front (Fig. 29.8F), which is already retreating to deeper levels, while vein density (Fig. 29.8B) only increased slightly as compared with Figure 29.8A. The expelled magmatic fluids still phase-separate upon ascent. However, the increase in bulk salinity has become less dramatic and fluids do not saturate in solid halite anymore.

After 30,000 years, phase separation is limited to the relatively narrow region of the hydrological divide (Fig. 29.8G). The gradual retreat of the associated precipitation front leads to an elongated, cylindrical shape of the copper enrichment zone (Fig. 29.8G). With decreasing source rate and due to the higher background permeability, the addition of magmatic fluids can be accommodated with the given background permeability profile and no hydraulic fracturing, and consequently, also no overpressure–permeability waves develop anymore. After 100,000 years of simulation time, the magma chamber has fully crystallized, all the magmatic salt has been flushed through the surface and the remaining heat only maintains a low-temperature hydrothermal system of a single-phase fluid (Fig. 29.8D, H, and L).

The influence of salinity

The hydrology of porphyry copper systems as described previously is controlled by nonlinear fluid properties and nonlinear feedbacks of rock permeability, and the complexity of the system makes it difficult to single out the influence of individual components. NaCl is known to be essential for ore formation and all fluid inclusion studies at porphyry deposits document low-salinity vapors and high-salinity brines (Bodnar *et al.* 2014). To single out the effect of salinity, the following simulation investigates to which extent the permeability model now governs the hydrology when fluid properties of pure water instead of salt water are used. The results also provide an end-member case for magmas with less saline fluid contents. The salinity of exsolved fluids may also vary in time as a function of composition and intrusion depth, with either the early or the late phase including more dilute fluids than the 10 wt% NaCl used in the previous simulations (Cline & Bodnar 1991).

The pure water simulation starts with a similar fluid expulsion phase as in the salt-water simulation (Fig. 29.7A) with quasi-radial outward propagation of the fluids and hydraulic fracturing of the brittle domain in a wide area around the cupola (Fig. 29.9A). The inner part with near-lithostatic fluid pressures and high temperatures has already progressed slightly further than in the salt-water simulations (Fig. 29.9A and E). After 5000 years, the system also self-organizes into the two distinct fluid domains (Fig. 29.9B). However, the hydrological divide stabilizes at a greater depth. Fluids stay in the single-phase field within the inner part due to the high fluid pressures (Fig. 29.9F).

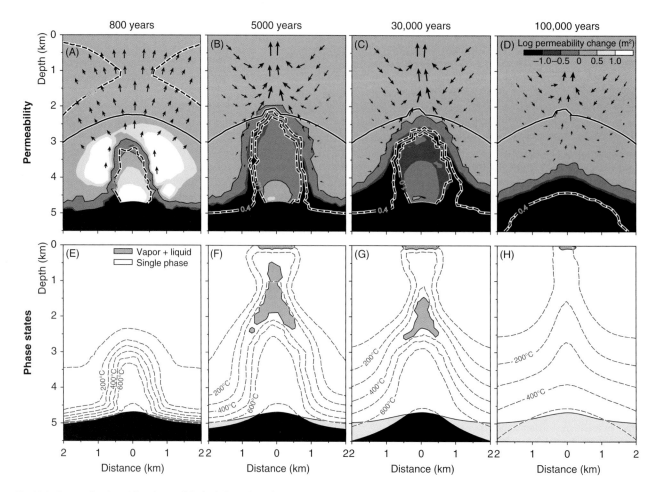

Fig. 29.9. Temporal and spatial evolution of the hydrology of porphyry copper systems without magmatic salt. (A–D) Dynamic permeability changes in reference to the depth-dependent background profile after 800, 5000, 30,000, and 100,000 years of simulation time. The black dashed lines show the pore fluid factor. Black arrows schematically show fluid flow in the dominantly brittle domain. The black solid lines show a vein volume potential of 10%. The black area indicates the magma chamber. (E–H) Temperature distribution and two-phase fluid states with light gray areas indicating liquid–vapor coexistence. The black area indicates the magma chamber and the very light gray area the solidified pluton.

With the transition to the near-hydrostatically pressured domain and with cooling due to mixing with the convecting meteoric waters, the fluids phase-separate into liquid and vapor and develop a boiling zone spanning almost the entire distance from the hydrological divide to the surface.

Again, with further inward crystallization, the hydrological divide retreats to greater depths (Fig. 29.9C and G). After 30,000 years of simulation time, the two-phase area is limited to the center part directly above the divide and directly beneath the surface (Fig. 29.9G). Hydraulic fracturing is less drastic than in the simulation with salt water throughout the simulation and after 30,000 years the background permeability profile is (mostly) sufficient to accommodate the additional amount of magmatic fluids (Fig. 29.9C).

The influence of volcano topography

Fluid expulsion in the simulation with volcano topography follows the same principles as the simulations before. The ascending magmatic fluids develop a similar upflow zone under high temperatures and near-lithostatic pressures as well as the sealing–fracturing cycles described as overpressure–permeability waves (Fig. 29.10). In the brittle domain, however, the higher fluid pressures imposed by the fluid-saturated volcano and the resulting outward-oriented pressure gradient lead to an opposite flow direction for the convecting meteoric fluids.

The round top of the hydrological divide is now shaped by downflowing meteoric fluids that mix with the ascending magmatic fluids and further descend along the hydrological divide. The reversed convection pattern enhances the steepness of the temperature front, as the descending meteoric fluids in the center directly mix with the upflowing hot magmatic fluids without being preheated at the sides of the magmatic fluid plume as in the simulations without volcano topography. Once the fluid mixture is heated to temperatures of 300–400 °C, however, the temperature is maintained as the fluids flow along the transition from brittle to ductile rock behavior. As the convection detaches

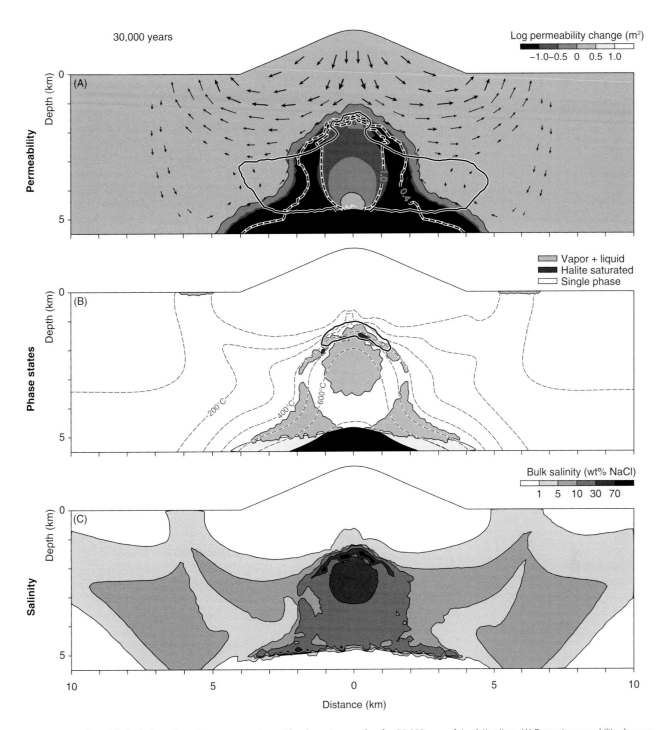

Fig. 29.10. Snapshot of the hydrology of porphyry copper systems with volcano topography after 30,000 years of simulation time. (A) Dynamic permeability changes in reference to the depth-dependent background profile. The black dashed lines show the pore fluid factor. Black arrows schematically show fluid flow in the dominantly brittle domain. The black solid lines show a vein volume potential of 10%. The black area indicates the magma chamber. (B) Temperature distribution and multiphase fluid states with light gray areas indicating liquid–vapor coexistence and dark gray areas halite-saturated fluids. The black solid line shows a copper enrichment potential of 500. The black area indicates the magma chamber and the very light gray area the solidified pluton. (C) Variations in bulk salinity of the fluid mixture including the solid halite phase.

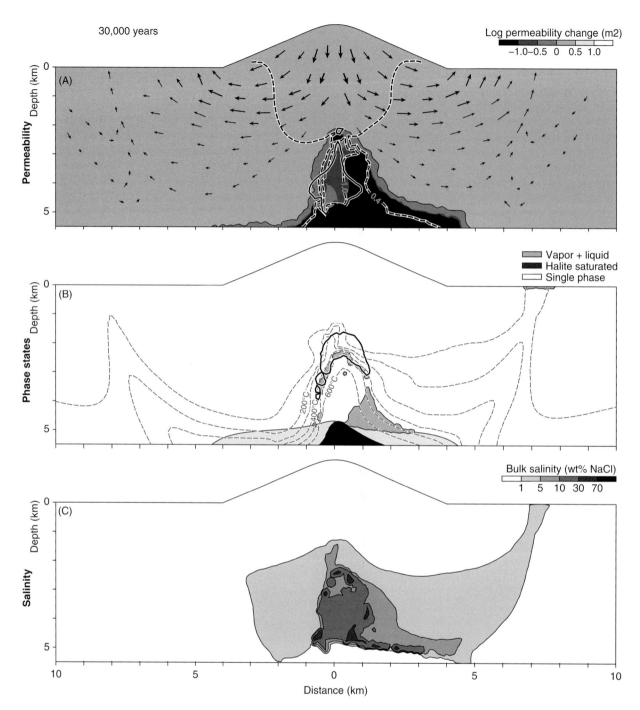

Fig. 29.11. Snapshot of the hydrology of porphyry copper systems with volcano topography and higher permeability profile after 30,000 years of simulation time. (A) Dynamic permeability changes in reference to the depth-dependent background profile of one order of magnitude higher than in Figure 29.1A. The black dashed lines show the pore fluid factor. Black arrows schematically show fluid flow in the dominantly brittle domain. The black solid lines show a vein volume potential of 10%. The black area indicates the magma chamber. (B) Temperature distribution and multiphase fluid states with light gray areas indicating liquid–vapor coexistence. The black solid line shows a copper enrichment potential of 500. The black area indicates the magma chamber and the very light gray area the solidified pluton. (C) Variations in bulk salinity of the fluid mixture including the solid halite phase.

from this transition region and flows upward toward the surface, the fluids cool to temperatures of 100–200 °C. At the base of this topography-driven convection cell, further meteoric convection cells mining heat from the outer parts of the crystallizing magma chamber merge into the main upflow zones.

Salinity variations in the center are similar to the simulation with the same permeability profile in reference to 10^{-14} m^2 at 1 km (Fig. 29.10C). However, in the peripheral parts of the domain, significant amounts of magmatic salt are still retained within the crust. This salt mass has been expelled during the initial phases of fluid release and is now located in areas outside of the main convection cell.

Increasing the permeability profile by an order of magnitude has a narrowing effect on the inner upflow zone (Fig. 29.11) and determines the shape of the predicted ore body. While in the previous simulation copper is enriched within a wider, horizontally oriented shell (Fig. 29.10B), the high-permeability profile produces a narrower, vertically extended shell (Fig. 29.11B). With higher permeabilities, the magma chamber has already crystallized further inward moving down from the volcanic edifice, cooling the top after 30,000 years, and the hydrological divide has retreated to greater depths with decreasing source rate.

Meteoric convection is entirely dominated by fluids moving down from the volcanic edifice, cooling the top and the flanks of the inner, high-pressure part, and flowing outward to rise back to the surface at about 7 km distance from the center. The peripheral convection cells have moved further outward as compared with the lower-permeability simulations and the location where these convection cells merge mark the upflow with the highest temperatures and salinities (Fig. 29.11B and C).

The asymmetric evolution of the hydrothermal system reflects the increasingly dynamic behavior with higher permeabilities. Magmatic fluids traversing the hydrological divide develop a preference toward one side of the two-dimensional modeling domain, which is enhanced by the nonlinear behavior of the hydrothermal system. In the early phases of the system, discharge mainly occurred to the left side, which explains the stronger cooling and the increased copper enrichment along the left limb. At later stages, however, the hydrothermal system switches to preferred discharge at the right side, as indicated by the higher temperatures and salinities of the upflow zone (Fig. 29.11B and C).

DISCUSSION

The distribution of copper enrichment potential and vein volume potential predicted by the model matches with observed copper ore grades and vein densities at porphyry deposits, indicating that porphyry copper ore shells form at stable temperature–pressure fronts at a hydrological divide between a hot overpressured upflow zone of magmatic fluids to a colder

brittle domain with convection of meteoric fluids (see also Weis et al. 2012). Varying rock properties by using overall higher permeability values has a significant effect on the dimensions of the deposit (Figs 29.7 and 29.8). The ore shell and the area of intense quartz veining are much narrower and concentrated. The limbs of the shell at the sides become steeper and partially merge, forming a vertically extended, cylindrical shaped ore body with only a minor barren core at the deeper ends.

Simulations including a volcano at the surface show that topography-driven flow has a significant effect on the hydrology as it reverses the direction of the meteoric convection, leading to discharge locations at epithermal conditions at distances as far as 7 km away from the volcanic center. The predicted deposits in the porphyry environment, however, are comparable with the simulations with flat topography (Figs 29.7–29.11), with only a tendency toward wider ore shells.

The entire ore-forming event is predicted to last between 50,000 and 100,000 years (Fig. 29.12A), which is in good agreement with recent high-precision geochronology and earlier modeling studies (Cathles 1977; Norton & Knight 1977; Hayba & Ingebritsen 1997; Driesner & Geiger 2007; von Quadt et al. 2011).

Saline fluid flow and rock permeability

The interplay between saline fluid flow and rock permeability controls the self-organization of the system. In the simulations, the brittle–ductile transition is assumed to start at a temperature of 360 °C, which is in agreement with borehole measurements at active geothermal systems, where fluid pressures begin to rise at that temperature (Fournier 1991). A similar behavior of rock properties has been inferred to control hydrothermal convection at mid-ocean ridges (Cathles 1993). At the same temperature, fluid properties of (salt) water also develop the strongest buoyancy-driven upward force or fluxibility, leading to an optimized mass and energy transport as can also been shown for mid-ocean-ridge hydrothermal systems (Jupp & Schultz 2000; Coumou et al. 2008, 2009a).

Varying fluid properties by using pure water as a proxy for magmatic fluids expelled from the magma chamber shows that NaCl has an important effect on the dynamics of the system. The increase in fluid volume due to phase separation at high temperatures and fluid pressures during the ascent of salt-water fluids leads to more drastic permeability changes and more intense fracturing, which moves the hydrological divide to much shallower crustal levels than in the pure water case.

The development of a hydrological divide also provides an important mechanism to transport magmatic salt through the crust and to discharge it at the surface (Fig. 29.12). The magma chamber has a volume of about 95 km^3 in three dimensions (Fig. 29.6) and exsolves a total amount of about 13 Gt of fluids. Scaled to the two-dimensional section, a total of about 27 Mt of fluid mass is injected during the simulation (Fig. 29.12A),

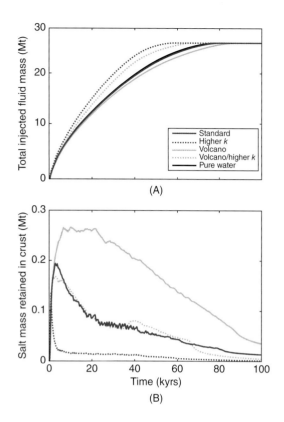

Fig. 29.12. Fluid expulsion and magmatic salt. (A) Temporal evolution of injection of magmatic fluid mass. (B) Temporal evolution of salt mass retained in the modeling domain.

the background permeability profile, these convecting, cooling fluids are initially saline because a significant fraction represents reheated magmatic fluids (see Fig. 29.7J).

After the maximum in salt accumulation, the salt mass retained in the crust decreases steadily even though more dissolved salt is added at the cupola. Due to the higher permeabilities in the uppermost part of the domain, the fluids of the upflow zone stay moderately saline, which means that they are not too dense, so that they can still move upward and transport more salt out of the system than is injected at the cupola. In particular, the high-permeability simulation shows that the meteoric convection is not only an important cooling agent for the porphyry copper system but also an efficient diluting agent that prevents the system from becoming salt oversaturated.

Further dynamics

The dynamic permeability model still uses a number of simplifications and general assumptions. Resolving further dynamics will lead to a more detailed understanding of the system. Both stress field and strain rate may vary in space and time, which will have an impact on failure criteria and the location of the brittle–ductile transition. Preexisting fractures or larger fault zones may play an important role in altering hydrological pathways and stress patterns, especially in the uppermost part of the domain.

The model assumes that the host rock is critically stressed and permeability follows a characteristic profile for tectonically active crust under normal-faulting conditions. However, the crust may also be dominated by less permeable rocks, which may not be critically stressed, or the tectonic setting may lead to strike-slip conditions. These properties of the host rock may further vary in space and time, due to both external changes and the evolution of the porphyry system itself. All of these factors are likely to influence the hydrothermal system, in particular the convection of the meteoric fluids within the brittle domain. Their individual effects are hard to predict but could be investigated with further simulations.

Except for halite precipitation and dissolution, the model currently does not explicitly account for geochemical feedbacks on porosity and permeability. In particular, quartz precipitation and dissolution can have an important effect on fluid flow and hydrofracturing, which is parameterized as an incremental permeability increase here but ultimately requires a description that is able to capture the actual physical process. Similarly, the process of fracture closure and vein sealing requires further constraints. In the nominally ductile domain, the parameterization as immediate closure ensured that overpressure can be released temporarily. This mechanism can be explained by rapid vein sealing with quartz precipitation and matches the observations at porphyry deposits. In the brittle domain, however, permeability decay after a permeability-creating event may be considerably slower (Ingebritsen & Manning 2010). Fluid–rock reaction can also alter the host rock and lead to

including 2.7 Mt of magmatic salt. Considering the porosity of 5% of the modeling domain, this amount of salt would be enough to fill the entire pore space above the cupola in an area of 5 km width, that is, more than the excerpts of Figures 29.7 and 29.8.

The hot, salty fluids will inevitably phase-separate and saturate in solid halite, either by continuous accumulation of a heavy brine phase residing at depth or by phase transition to the vapor–halite coexistence field with decreasing fluid pressures upon ascent of the hot fluids. Simulations with magmatic degassing of H_2O–$NaCl$ fluids with static, uniform permeabilities show that magmatic salt is not easily vented through the crust and that accumulation of solid halite can lead to a pulsating fluid flow in saline magmatic–hydrothermal systems (Weis et al. 2014). However, the presence of a solid halite phase has only been inferred in very few fluid inclusion studies or indirectly shown by the so-called halite trend (Cloke & Kesler 1979; Muntean & Einaudi 2000; Lecumberri-Sanchez et al. 2015).

The salt mass accumulated in the crust over the simulation time for the porphyry model described here increases to a maximum during the first 5000 years (Fig. 29.12B). The maximum is associated with the establishment of the hydrological divide and the change from fluid expulsion to convection in the outer, hydrostatically pressured domain. Depending on changes in rock properties, which may affect the hydrothermal system. Furthermore, an impermeable carapace at the cupola may lead to temporary accumulation of magmatic fluids and episodic injections into the host rock together with magma, which may also result in repeated dike injections as observed at many porphyry deposits (e.g., Dilles 1987; Sillitoe 2010).

The wave-like behavior of fluid flow through the crust has been described in a number of hydrogeological settings. Metamorphic and sedimentary environments may generate compaction-driven porosity waves (Connolly & Podladchikov 1998; Appold & Nunn 2002; Connolly & Podladchikov 2015). Earthquakes have been inferred to trigger fault-valve behavior of fast fluid flow along faults (Sibson 1987; Miller et al. 2004), which may lead to the formation of mesothermal gold deposits (e.g., Sibson et al. 1988; Micklethwaite et al. 2015). Solitary waves may facilitate fluid transport from deep, overpressured regions to shallower levels in hydrocarbon reservoirs and accretionary wedges (e.g., Revil & Cathles 2002; Bourlange & Henry 2007; Joshi et al. 2012). Permeability has also been described as a toggle switch between extremely high and low values depending on whether fluid pressures meet the failure condition or not, leading to the development of an interconnected flow network in fractured rocks (Miller & Nur 2000). All of these processes have in common that they consider permeability and/or porosity as a dynamic parameter.

Relations to active hydrothermal systems

The convection system above the hydrological divide resembles high-enthalpy geothermal systems, where upflow follows a boiling-curve-with-depth profile. However, the relatively high permeability in the uppermost part and the permeability decrease with depth lead to a stronger cooling of the heated fluid due to convection of colder fluids at shallower crustal levels. Numerical simulations have shown that the highest enthalpies are achieved for permeabilities near 10^{-15} m^2 that optimize the balance between heat conduction from the source and advection by the convecting fluids (Hayba & Ingebritsen 1997; Driesner & Geiger 2007). In the presented simulations, episodic magmatic fluid surges can provide additional heat to temporarily maintain the boiling-with-depth profile.

The hydrological divide as described here separates the actual heat source, the magma chamber, from the base of the high-enthalpy system by several kilometers in the vertical direction. This vertical profile is a result of fluid release from the magma chamber, which points to significantly different styles of the driving heat source for hydrous magmas in subduction settings, such as in the Taupo Volcanic Zone, as compared with anhydrous magmas at constructive plate margins, such as in Iceland. Without the release of magmatic fluids, the base of the hydrothermal system will remain closer to the magma chamber, as documented by simulations of subaerial and submarine hydrothermal systems (e.g., Cathles 1977; Norton & Knight

1977; Hayba & Ingebritsen 1997; Coumou et al. 2009b; Gruen et al. 2014; Weis 2014; Weis et al. 2014).

At the Taupo Volcanic Zone, the brittle–ductile transition is generally inferred to be located at 6–7 km depth, while the hydrothermal systems are predominantly located within the uppermost 3 km of the crust (Bertrand et al. 2012). In between, at 3–7 km depths, recent magnetotelluric data show vertically extended low-resistivity zones with a width of a few kilometers, interpreted to represent convection plumes connecting the near-surface hydrothermal activity with an underlying magmatic system at depths of about 10 km (Bertrand et al. 2012). The dimensions and depth ranges of these structures resemble the hydrology of the porphyry systems described in this study, where the high-temperature, near-lithostatically pressured part forms the base of the high-enthalpy system at shallow depths.

The different styles of heat source may also shed light on the so-called supercritical geothermal resources (referring to the critical point of pure water), which aim at tapping fluids with enthalpies of 3 MJ kg^{-1} instead of 1.5 MJ kg^{-1} typical for high-enthalpy systems. Supercritical geothermal resources are currently explored in volcanic areas such as the Krafla site in Iceland in the roots of high-enthalpy systems (Fridleifsson & Elders 2005; Elders et al. 2011), which are likely to be located close to the magma chamber in tectonic settings with predominantly anhydrous magmatism. Also, the inferred brittle–ductile transition may be associated with higher temperatures for basaltic rocks than the 360–500 °C assumed here (Fournier 1999; Scott et al. 2015). In contrast, according to the simulations presented here, supercritical conditions in association with more felsic, hydrous magmatism only develop below the hydrological divide and the targeted semi-brittle, semi-ductile area is limited to a narrow region dividing the two hydrological systems (Weis & Driesner 2013). The simulations suggest that the rock in the upflow zone in the roots of high-enthalpy systems may behave in a ductile (plastic, viscous) manner, which relates to further geothermal projects, where the feasibility of injecting fluids into ductile rock is being investigated (Japan Beyond Brittle Project 2014).

Episodic ground deformations (bradyseism) at active volcanic regions are also closely related to the processes described here (Bodnar et al. 2007). Deformation is not modeled in our simulations, but the dynamics of the high-temperature, high-pressure domain with their overpressure–permeability waves and episodic surges of magmatic fluids are likely to also cause ground deformations, which have been quantified in other approaches that assumed comparable driving forces (e.g. Hutnak et al. 2009).

Excess degassing at active volcanoes refers to the observation that the amount of volatiles released through venting is larger than the volatile contents of the erupted magma and the volcanic conduit (Shinohara 2008). The simulations provide a mechanism for keeping the location of an initial cupola by continuous heating and thereby locally preventing cooling from the outside, which may help accumulate fluids from a larger source pluton. The modeling also describes how the fluids can ascend through the overlying rock within a focused, vertically extended upflow zone.

CONCLUSIONS

The dynamic interplay between saline fluid flow and rock permeability controls the evolution of magmatic–hydrothermal ore deposits. Numerical simulations indicate that porphyry copper ore shells form at a stable temperature–pressure front, dividing a hot inner upflow zone of magmatic fluids under near-lithostatic fluid pressures in a nominally ductile domain from colder surroundings dominated by meteoric convection under near-hydrostatic fluid pressures in a brittle domain. This hydrological divide provides both a cooling and diluting mechanism for saline magmatic fluids expelled from a crystallizing upper crustal magma chamber.

The permeability range influences the timescale of ore formation as well as the shape of the ore body and the distribution of vein density. With higher permeability profiles, the model predicts narrower, vertically extended ore shells. The salinity of the fluids influences the intensity of hydraulic fracturing due to phase separation within the upflow zone and affects the depth of ore formation. The self-organization into the hydrological divide also ensures that magmatic salt can be transported out of the domain with convection of meteoric fluids instead of being accumulated in heavy brines and solid halite within the crust.

With a volcano topography above the hydrothermal system, the direction of flow in the meteoric convection cell reverses, leading to an even sharper temperature gradient at the top of the porphyry system and diverting the upflow zone at epithermal conditions to the flanks of the volcano, at distances of up to 7 km from the volcanic center.

The simulations further suggest that a hydrological divide maintained by magmatic fluid release in tectonic settings with felsic, hydrous magmatism can separate the base of high-enthalpy geothermal systems from the magma chamber as a driving heat source by several kilometers in the vertical direction. This upflow zone is characterized by fluid flow through nominally ductile rock with temporary embrittlement due to hydraulic fracturing and could serve as a source region for the so-called supercritical geothermal systems. In general, the simulations support the concept of crustal permeability as a dynamic property evolving from an intimate interplay of fluid flow and rock permeability.

ACKNOWLEDGMENTS

Many thanks to Stephen Cox and a second reviewer, the editors of *Geofluids* and of the special thematic issue on crustal permeability as well as to Thomas Driesner and Christoph Heinrich for helpful comments and discussions on the manuscript. The work has been supported by the Swiss National Science Foundation within various funding periods.

PART IV

Conclusion

CHAPTER 30

Toward systematic characterization

TOM GLEESON[1] AND STEVEN E. INGEBRITSEN[2]

[1] *Department of Civil Engineering, University of Victoria, Victoria, BC, Canada;* [2] *U.S. Geological Survey, Menlo Park, CA, USA*

Key words: static permeability, dynamic permeability, permeability-depth relations

As noted in the introduction, the measured permeability of the shallow continental crust is so highly variable that it is often considered to defy systematic characterization. Although some order has been revealed in globally compiled data sets, including postulated relations between permeability and depth on a whole-crust scale (i.e., to approximately 30 km depth; e.g., Manning & Ingebritsen 1999; Ingebritsen & Manning 2010), the recognized limitations of such empirical relations helped to inspire this book. For instance, whereas a wide variety of *k–z* relations have been suggested (Fig. 30.1), certain chapters in this book highlight the fallacy of extrapolating crustal-scale *k–z* relations to the uppermost crust, or perhaps even defining such relations (e.g., Burns *et al.* this book; Ranjram *et al.* this book), which can significantly impact the parameterization of permeability in hydrologic models. The permeability structure of the shallow (approximately <1 km) crust is highly heterogeneous, and dominant controls on local permeability include the primary lithology, porosity, rheology, geochemistry, and tectonic and time–temperature histories of the rocks.

The permeability of clastic sediments in the cool, shallow crust is often well predicted as a function of mechanical compaction and consequent porosity–permeability relations (e.g., Daigle & Screaton this book; Luijendijk & Gleeson this book). However, this predictability diminishes at depths where diagenetic process becomes important (e.g., Fig. 30.2A); in the North Sea basin, for instance, pressure solution begins to affect porosity–permeability relations at approximately 2 km depth (Fig. 30.2B). Similarly, hydrothermal alteration of volcanic rocks tends to cause significant reduction of permeability at temperatures in excess of approximately 40–50 °C (e.g., Burns *et al.* this book). Systematic permeability differences among original lithologies persist to contact-metamorphic depths of 3–10 km, but are not evident at regional metamorphic depths of 10–30+ km (Fig. 30.3) – presumably because, at greater depths, the metamorphic textures are largely independent of the original lithology.

Temporal changes in permeability can be gradual or abrupt. Streamflow responses to moderate-to-large earthquakes

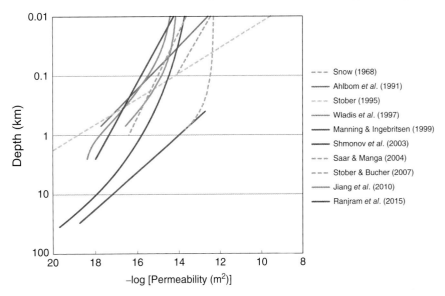

Fig. 30.1. Proposed permeability–depth (*k–z*) relations. Of these, only Manning & Ingebritsen (1999), Shmonov *et al.* (2003), and Saar & Manga (2004) are conditioned on data from depths exceeding a few kilometers. (*See color plate section for the color representation of this figure.*)

Legend (Fig. 30.1):
- Snow (1968)
- Ahlbom *et al.* (1991)
- Stober (1995)
- Wladis *et al.* (1997)
- Manning & Ingebritsen (1999)
- Shmonov *et al.* (2003)
- Saar & Manga (2004)
- Stober & Bucher (2007)
- Jiang *et al.* (2010)
- Ranjram *et al.* (2015)

Crustal Permeability, First Edition. Edited by Tom Gleeson and Steven E. Ingebritsen.
© 2017 John Wiley & Sons, Ltd. Published 2017 by John Wiley & Sons, Ltd.
Companion Websites: www.wiley.com/go/gleeson/crustalpermeability/
http://crustalpermeability.weebly.com/

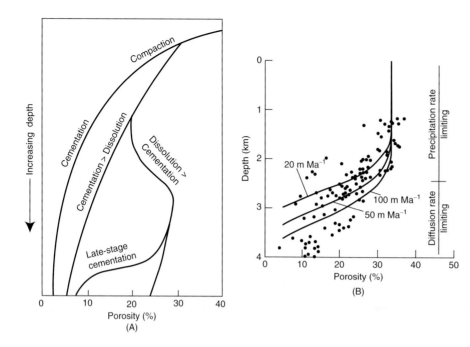

Fig. 30.2. (A) Hypothetical curves of porosity versus depth in siliciclastic basin sediments, showing the effects of simultaneous compaction, cementation, and dissolution at different rates (after Loucks *et al.* 1984) and (B) porosity–depth variation due to pressure solution in the North Sea; here, porosity data are from Ramm (1992), and numerical modeling predictions of porosity loss with depth due to pressure solution alone for three different burial rates are from Renard *et al.* (2000a).

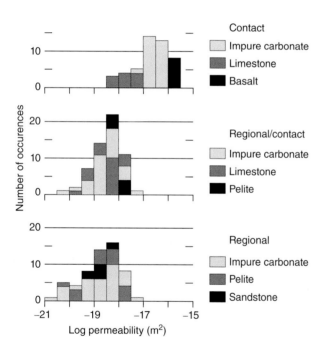

Fig. 30.3. Histograms of metamorphic permeabilities for contact, regional/contact, and regional metamorphic settings, showing differences encountered in different lithologies. "Regional/contact" refers to continental volcanic arcs with elevated geothermal gradients due to heat transported by magma. In contact-metamorphic environments, basalts and impure carbonates record the highest permeability, whereas carbonates record distinctly lower permeability. Lithological controls are less distinct in the deeper environments of regional/contact and regional metamorphism. In the deepest (regional) settings, all lithologies appear to have similar mean permeability. (Adapted from Manning & Ingebritsen 1999.)

demonstrate that dynamic stresses can instantaneously change permeability on a regional scale (e.g., Rojstaczer & Wolf 1992); large (1 mm) fractures can be sealed by silica precipitation within 10 years (Lowell *et al.* 1993); and simulations of calcite dissolution in coastal carbonate aquifers suggest significant changes in porosity and permeability over timescales of 10^4 to 10^5 years (Sanford & Konikow 1989). At the other end of the spectrum, the reduction of pore volume during sediment burial modifies permeability very slowly. For example, shale permeabilities from the US Gulf Coast vary from about 10^{-18} m^2 near the surface to about 10^{-20} m^2 at 5 km depth (Neglia 1979), and the natural subsidence rate is 0.1–10 mm year^{-1} (Sharp & Domenico 1976), so we can infer that it takes perhaps 10^7 years for the permeability of a subsiding package of shale to decrease by a factor of 10. These various observations are consistent with suggestions that crustal-scale permeability is a dynamically self-adjusting property, reflecting a competition between permeability destruction by processes such as compaction and permeability creation by processes such as fluid sourcing (e.g., Connolly & Podladchikov this book) and tectonically driven fracturing and faulting.

Nonetheless, given the highly variable rates and scales of permeability creation and decay, considering permeability as a static parameter can be a reasonable assumption for a wide range of research problems and applications. For example, for typical low-temperature hydrogeologic investigations with timescales of days to decades, permeability may be considered static in the absence of seismicity. Similarly, if it takes perhaps 10^7 years for the permeability of a subsiding package of shale to decrease by a factor of 10, permeability in sedimentary basins may be

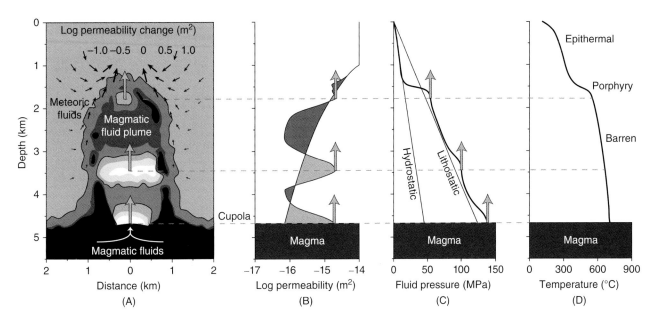

Fig. 30.4. (A) Simulated porosity/permeability waves driven by injection of magmatic volatiles from a cupola at 5 km depth; wave velocity in this example is approximately 3 km year^{-1}; (B) simulated permeability at any given depth varies by a factor of 10^1 to 10^2 as the waves pass; (C) rock failure is assumed to occur at near-lithostatic fluid pressures below the brittle–ductile transition and near-hydrostatic fluid pressures above the brittle–ductile transition. The stable interface between the magmatic fluid plume and meteoric convection seen in (A) acts to localize mineralization (D). (Adapted from Weis *et al.* 2012; see also Connolly & Podladchikov (this book) and Weis (this book).)

considered static for investigations on much shorter timescales. Whether the dynamic variation of permeability is important to include in analyses and quantitative models depends upon its rapidity and magnitude relative to the requirements of the problem at hand.

Recent research on enhanced geothermal reservoirs, ore-forming systems (Fig. 30.4), and the hydrologic effects of earthquakes yields broadly consistent results regarding permeability enhancement by dynamic stresses. Shear dislocation caused by tectonic forcing or fluid injection can increase near- to intermediate-field permeability by factors of 100–1000. Dynamic stresses (shaking) in the intermediate to far field corresponding to seismic energy densities >0.01 J m^{-3} also increase permeability, albeit often by ≪10 and at most by a factor of approximately 20 (e.g., Wang & Manga 2010; Manga *et al.* 2012). These permeability increases are transient, tending to return to preseismic values over timescales on the order of months to decades (e.g., Elkhoury *et al.* 2006; Kitagawa *et al.* 2007; Xue *et al.* 2013). There is reasonable agreement between the magnitude of near- to intermediate-field permeability increases (10^2 to 10^3 fold) directly measured at enhanced geothermal sites (e.g., Evans *et al.* 2005; Haring *et al.* 2008), inferred from field evidence (e.g., Howald *et al.* this book; Saffer this book), invoked in simulations of transient hydrothermal circulation (e.g., Weis *et al.* 2012; Taron *et al.* 2014; Howald *et al.* this book; Weis this book), and

inferred from seismic and metamorphic data (Ingebritsen & Manning 2010). We note that enhanced geothermal systems, geologic carbon sequestration (Lucier & Zoback 2008), and deep injection of waste fluid (Hsieh & Bredehoeft 1981; Frohlich *et al.* 2014; Weingarten *et al.* 2015) all entail similar stimuli, namely fluid-injection rates on the order of 10 s of kg s^{-1}, as do the simulations of ore-forming systems by Weis (this book). In North America, fluid-injection practices have caused a recent and dramatic increase in M_w >3 seismic events on a nearly continental scale (e.g., Hitzman *et al.* 2012; Ellsworth 2013). This ongoing injection experiment, although poorly constrained, represents an opportunity to explore and assess dynamic crustal permeability to depths of perhaps 10 km.

Even the most intensive campaigns to measure permeability in the very shallow crust cannot yield unambiguous determination of the large-scale trends that govern, for instance, transport behavior (e.g., Eggleston & Rojstaczer 1998). Thus, it is likely that models of large-scale fluid transport will continue to depend on inferences based on geophysical imaging and on improved understanding of the thermal, mechanical, and geochemical factors that control the overall permeability structure of the crust. Our hope and expectation is that this book and the associated data at http://crustalpermeability.weebly.com/ will enhance our ability to quantify or predict permeability and its variability with space, direction, and time.

References

Abelin H, Neretnieks I, Tunbrant S, Moreno L (1985) Migration in a single fracture: Experimental results and evaluation. Final report, Stripa Project, SKB, Stockholm.

Acocella V, Gudmundsson A, Funiciello R (2000) Interaction and linkage of extension fractures and normal faults: examples from the rift zone of Iceland. *Journal of Structural Geology*, **22**, 1233–1246.

Adamides NG (1990) Hydrothermal circulation and ore deposition in the Troodos ophiolite, Cyprus. In: *Ophiolites: Oceanic Crustal Analogues* (eds Malpas J, Moores EM, Panayiotou A, Xenophontos C), pp. 685–704, Geological Survey Department, Nicosia, Cyprus.

Adams AL, Germaine JT, Flemings PB, Day-Stirrat RJ (2013) Stress induced permeability anisotropy of resedimented Boston Blue Clay. *Water Resources Research*, **49**, doi:10.1002/wrcr .20470.

Adler PM (1997) Fracture deformation and influence on permeability. *Physical Review*, **56**, 3167–3184.

Ague JJ (2014) Fluid flow in the deep crust. In: *Treatise on Geochemistry*, 2nd edn (eds Holland HD, Turekian KK), pp. 203–247. Elsevier, Oxford.

Ahlbom K, Andersson J-E, Nordqvist R, Ljunggren C, Tiren S, Voss C (1991) Fjallveden study site – Scope of activities and main results. *Swedish Nuclear Fuel Waste Management Company (SKB) Technical Report*, **SKB-TR-91-52**.

Aki K, Richards PG (2002) *Quantitative Seismology*, 2nd edn University Science Books, Sausalito.

Akinfiev NN, Diamond LW (2009) A simple predictive model of quartz solubility in water-salt-C02 systems at temperatures up to 1000°C and pressures up to 1000 MPa. *Geochimica Cosmochimica Acta*, **76**, 1597–1608.

Allen DM, Grasby SE, Voormeij DA (2006) Determining the circulation depth of thermal springs in the southern Rocky Mountain Trench, south-eastern British Columbia, Canada using geothermometry and borehole temperature logs. *Hydrogeology Journal*, **14**, 159–172.

Allen SK, Cox SC, Owens IF (2011) Rock avalanches and other landslides in the central Southern Alps of New Zealand: a regional study considering possible climate change impacts. *Landslides*, **8**, 33–48.

Alley WM, Healy RW, LaBaugh JW, Reilly TE (2002) Flow and storage in groundwater systems. *Science*, **296**, 1985–1990.

Allis RG, Henley RW, Carman AF (1979) The thermal regime beneath the Southern Alps. In: *The Origin of the Southern Alps* (eds Walcott RI, Cresswell MM), *Bulletin of the Royal Society of New Zealand*, **18**, 79–85.

Allis RG, Shi Y (1995) New insights to temperature and pressure beneath the central Southern Alps, New Zealand. *New Zealand Journal of Geology and Geophysics*, **38**, 585–592.

Al-Tabbaa A, Wood DM (1987) Some measurements of the permeability of kaolin. *Geotechnique*, **37**, 499–514.

Alt-Epping P, Diamond LW, Häring MO (2013) Prediction of water-rock interaction and porosity evolution in a granitoid-hosted enhanced geothermal system, using constraints from the 5 km Basel-1 well. *Applied Geochemistry*, **38**, 121–133.

Alves MA, Oliveira PJ, Pinho FT (2003) A convergent and universally bounded interpolation scheme for the treatment of advection. *International Journal for Numerical Methods in Fluids*, **41**, 47–75.

Amann-Hildenbrand A, Bertier P, Busch A, Krooss BM (2013) Experimental investigation of the sealing capacity of generic clay-rich caprocks. *International Journal of Greenhouse Gas Control*, **19**, 620–641.

Ames LL, McGarrah JE, Walker BA (1983) Sorption of trace constituents from aqueous solutions onto secondary minerals, I, Uranium. *Clays and Clay Minerals*, **31**, 321–334.

Anderholm SK (2001) Mountain-front recharge along the east side of the Albuquerque Basin, Central New Mexico. *U.S. Geological Survey Water-Resources Investestigations Report*, **00–4010**.

Anderson A, Blackwell D, Chickering C, Boyd T, Home R, Mackenzie M, Moore J, Nickull D, Richard S, Shevenell L (2013) National Geothermal Data System (NGDS) geothermal data domain: assessment of geothermal community data needs. *Proceedings of the Stanford Geothermal Workshop, 2013, Stanford University, California, paper SGP-TR-198*, http://www.geothermal-energy.org/pdf/IGAstandard/SGW/ 2013/Anderson.pdf (accessed 04 May 2016).

Anderson EM (1951) *The Dynamics of Faulting*. Oliver & Boyd, Edinburgh.

Anderson MP (2005) Heat as a ground water tracer. *Ground Water*, **43**, 951–968.

Anderson RN, Zoback MD, Hickman SH, Newmark RL (1985) Permeability versus depth in the upper oceanic crust: in situ measurements in DSDP hole 504B, eastern equatorial Pacific. *Journal of Geophysical Research*, **90**, 3659–3669.

Anderson-Sprecher R (1994) Model comparisons and R^2. *The American Statistician*, **48**, 113–117.

Ankit K, Nestler B, Selzer M, Reichardt M (2013) Phase-field study of grain boundary tracking behavior in crack-seal microstructures. *Contributions to Mineralogy and Petrology*, **166**, 1709–1723.

Anonymous (1988) Granulometrische analyse Asten 02. Technical report, Rijks Geologische Dienst, Haarlem.

Antonioli A, Piccininni D, Chiaraluce L, Cocco M (2005) Fluid flow and seismicity pattern: evidence from the 1997 Umbria-March (central Italy) seismic sequence. *Geophysical Research Letters*, **32**, doi:10.1029/2004GL022256.

Aplin AC, Fleet AJ, MacQuaker JHS (1999) Muds and mudstones: physical and fluid flow properties. *Geological Society of London Special Publication*, **158**, 1–8.

Appold MS, Garven G, Boles JR, Eichhubl P (2007) Numerical modeling of the origin of calcite mineralization in the Refugio-Carneros fault, Santa Barbara Basin, California. *Geofluids*, **7**, 79–95.

Appold MS, Nunn JA (2002) Numerical models of petroleum migration via buoyancy-driven porosity waves in viscously deformable sediments. *Geofluids*, **2**, 233–247.

Aquilina L, Genter A, Elsass P, Pribnow D (2000) Evolution of fluid circulation in the Rhine graben: constraints from the chemistry of present fluids. In: *Hydrogeology of Crystalline Rocks* (eds Stober I, Bucher K), pp. 177–203. Kluwer Academic Publishers, Dordrecht.

Araneda OA, Morales RF, Rojas EG, Henríquez JO, Molina RE (2007) Rock preconditioning application in virgin caving condition in a panel caving mine, CODELCO Chile El Teniente Division. In: *Proceedings International Symposium, Deep and High Stress Mining*, pp. 111–120. Australian Centre for Geomechanics, Perth, Australia.

Arch J, Maltman A (1990) Anisotropic permeability and tortuosity in deformed wet sediments. *Journal of Geophysical Research*, **95**, 9035–9045.

Arehart GB, Coolbaugh MF, Poulson SR (2003) Evidence for a magmatic source of heat for the Steamboat Springs geothermal system using trace elements and gas geochemistry. *Geothermal Resources Council Transactions*, **27**, 269–274.

Armand G (2000) Contribution à la caractérisation en laboratoire et à la modélisation constitutive du comportement mécanique des joints rocheux. PhD thesis, Université Joseph Fourier, Grenoble, France.

Armbruster T, Kohler T, Meisel T, Nagler TF (1996) The zeolite, fluorite, quartz assemblage of the fissure at Gibelsbach, Fiesch (Valais, Switzerland): crystal chemistry, REE patterns, and genetic speculations. *Schweizerische Mineralogische und Petrographische Mitteilungen*, **76**, 131–146.

Arzi AA (1978) Critical phenomena in rheology of partially melted rocks. *Tectonophysics*, **44**, 173–184.

Ashby MF (1988) The modeling of hot isostatic pressing. In: *Proceedings HIP: Hot Isostatic Pressing − Theories and Applications* (ed Garvare T), pp. 29–40, Centek, Lulea, Sweden.

Asmerom Y, Polyak V, Burns S (2010) Variable winter moisture in the southwestern United States linked to rapid glacial climate shifts. *Nature Geoscience*, **3**, 114–117.

ASTM International (2004) *Standard Test Methods for Measurement of Hydraulic Conductivity of Saturated Porous Materials Using a Flexible Wall Permeameter, D5084–03*. ASTM International, West Conshohocken, PA.

ASTM International (2006) *Standard Test Method for One-dimensional Consolidation Properties of Saturated Cohesive Soils Using Controlled-Strain Loading, D4186–06*. ASTM International, West Conshohocken, PA.

Audet DM, Fowler AC (1992) A mathematical model for compaction in sedimentary basins. *Geophysical Journal International*, **110**, 577–590.

Auer F, Berckhemer H, Oehlschlegel G (1981) Steady-state creep of fine-grain granite at partial melting. *Journal of Geophysics-Zeitschrift Tur Geophysik*, **49**, 89–92.

Aydin A, Schultz R (1990) Effect of mechanical interaction on the development of strike-slip faults with echelon patterns. *Journal of Structural Geology*, **12**, 123–129.

Bächler D (2003) Coupled thermal-hydraulic-chemical modelling at the Soultz-sous-Forêts HDR reservoir (France). PhD thesis, ETH Zurich, No. 15044.

Bäckblom G, Martin CD (1999) Recent experiments in hard rock to study the excavation response: implication of the performance of a nuclear waste geological repository. *Tunneling and Underground Space Technology*, **14**, 377–394.

Baertschi P, Silverman SR (1951) The determination of relative abundances of the oxygen isotopes in silicate rocks. *Geochimica et Cosmochimica Acta*, **1**, 317–328.

Baghbanan A, Jing L (2008) Stress effects on permeability in fractured rock masses with correlated fracture length and aperture. *International Journal of Rock Mechanics and Mining Sciences*, **45**, 1320–1334.

Baker ET, German CR (2004) On the global distribution of hydrothermal vent fields. In: *Mid-Ocean Ridges: Hydrothermal Interactions between the Lithosphere and Oceans* (eds German CR, Lin J, Parson LM), pp. 245–266. American Geophysical Union, Washington, DC.

Baker ET, Massoth GJ, Feely RA (1987) Cataclysmic hydrothermal venting on the Juan de Fuca Ridge. *Nature*, **329**, 149–151.

Balashov VN, Yardley BWD (1998) Modeling metamorphic fluid flow with reaction-compaction-permeability feedbacks. *American Journal of Science*, **298**, 441–470.

Baldridge WS, Keller GR, Haak V, Wendlandt E, Jiracek GR, Olsen KH (1995) The Rio Grande Rift. In: *Continental Rifts: Evolution, Structure, Tectonics* (ed Olsen KH), *Developments in Geotectonics*, **25**, 233–265, Elsevier, Amsterdam.

Bandis S, Lumsden AC, Barton NR (1983) Fundamentals of rock joint deformation. *International Journal of Rock Mechanics and Mining Sciences*, **20**, 249–268.

Bandstra JL, Buss HL, Campen RK, Liermann LJ, Noore J, Hausrath EM, Navarre-Sitchler AK, Jang J-H, Brantley SL, 2008, Compilation of mineral dissolution rates. In: *Kinetics of Water-Rock Interaction* (eds Brantley SL, Kubicki JD, White AF), pp. 737–824, Springer Verlag, New York.

Banerjee NR, Gillis KM (2001) Hydrothermal alteration in a modern suprasubduction zone: the Tonga forearc crust. *Journal of Geophysical Research*, **106**, 21737–21750.

Banerjee NR, Gillis KM, Muehlenbachs K (2000) Discovery of epidosites in a modern oceanic setting, the Tonga forearc. *Geology*, **28**, 151–154.

Bangs NL, Shipley TH, Moore JC, Moore GF (1999) Fluid accumulation and channeling along the northern Barbados Ridge decollement thrust. *Journal of Geophysical Research*, **104**, 20399–20414

Banwart SA, Chorover J, Gaillardet J, Sparks D, White T, Anderson S, Aufdenkampe A, Bernasconi S, Brantley SL, Chadwick O, Dietrich WE, Duffy C, Goldhaber M, Lehnert K, Nikolaidis NP, Ragnarsdottir KV (2013) *Sustaining Earth's Critical Zone Basic Science and Interdisciplinary Solutions for Global Challenges*. University of Sheffield, Sheffield.

Baragar WRA, Lambert MB, Baglow N, Gibson IL (1990) The sheeted dike zone in the Troodos ophiolite. In: *Ophiolites: Oceanic Crustal Analogues* (eds Malpas J, Moores EM, Panayiotou A, Xenophontos C), pp. 37–52. Geological Survey Department, Nicosia, Cyprus.

Barcilon V, Lovera OM (1989) Solitary waves in magma dynamics. *Journal of Fluid Mechanics*, **204**, 121–133.

Barcilon V, Richter FM (1986) Nonlinear-waves in compacting media. *Journal of Fluid Mechanics*, **164**, 429–448.

Bargar KE (1988) Secondary mineralogy of core from geothermal drill hole CTGH-1, Cascade Range, Oregon. In: Geology and geothermal resources of the Breitenbush-Austin Hot Springs area, Clackamas and Marion Counties (ed Sherrod DR), pp. 39–45. *State of Oregon Department of Geology and Mineral Industries Open-File Report*, **O-88-5**.

Bargar KE, Keith TEC (1999) Hydrothermal mineralogy of core from geothermal drill holes at Newberry volcano, Oregon. *U.S. Geological Survey Professional Paper*, **1578**.

Barnes I, Downes CJ, Hulston JR (1978) Warm springs, South Island, New Zealand, and their potential to yield laumontite. *American Journal of Science*, **278**, 1412–1427.

Barroll MW, Reiter M (1990) Analysis of the Socorro hydrogeothermal system: central New Mexico. *Journal of Geophysical Research*, **95**, 21949–21963.

Bart M, Shao JF, Lydzba D, Haji-Sotoudeh M (2004) Coupled hydromechanical modeling of rock fractures under normal stress. *Canadian Geotechnical Journal*, **41**, 686–697.

Barton N (1982) *Modelling Rock Joint behaviour from in situ Block Tests: Implication for Nuclear Waste Repository Design*. Office of Nuclear Waste Isolation, Columbus, Ohio, ONWI-308.

Barton N (2007) Thermal over-closure of joints and rock masses and implications for HLW repositories. *Proceedings of the 11th Congress of the International Society for Rock Mechanics*, Lisbon, Portugal, 9–13 June 2007.

Barton N, Bandis S, Bakhtar K (1985) Strength, deformation and conductivity coupling of rock joints. *International Journal of Rock Mechanics and Mining Sciences*, **22**, 121–140.

Barton N, Choubey V (1977) The shear strength of rock joints in theory and practice. *Rock Mechanics*, **10**, 1–54.

Bateman R, Ayer JA, Dubé B (2008) The Timmins-Porcupine gold camp, Ontario: anatomy of an Archean greenstone belt and ontogeny of gold mineralization. *Economic Geology*, **103**, 1285–1308.

Batjes NH (1996) Total carbon and nitrogen in the soils of the world. *European Journal of Soil Science*, **47**, 151–163.

Batzle M, Wang Z (1992) Seismic properties of pore fluids. *Geophysics*, **57**, 1396–1408.

Bauer F (1987) Die Kristallinen Gesteine aus der Bohrlochvertiefung Urach 3 und ihre fluiden Einschlüsse: eine Interpretation der hydrothermalen Überprägung anhand der Fluid-Daten aus EinschluBmessungen. Dissertation at Universitat (T.H.) Fridericiana Karlsruhe.

Bauer HH, Vaccaro JJ (1990) Estimates of ground-water recharge to the Columbia Plateau regional aquifer system, Washington, Oregon, and Idaho, for pre development and current land-use conditions. *U.S. Geological Survey Water-Resources Investigations Report*, **88–4108**, http://pubs.er.usgs.gov/usgspubs/wri/wri884108 (accessed 04 May 2016).

Bear J (1972) *Dynamics of Fluids in Porous Media*. American Elsevier, New York.

Bear J (1979) *Hydraulics of Groundwater*. McGraw-Hill, New York.

Bear LM (1963) *The Mineral Resources and Mining Industry of Cyprus*. Geological Survey Department, Nicosia, Cyprus.

Beavan RJ, Denys P, Denham M, Hager B, Herring T, Molnar P (2010a) Distribution of present-day vertical deformation across the Southern Alps, New Zealand, from 10 years of GPS data. *Geophysical Research Letters*, **37**, L16305.

Beavan RJ, Ellis SM, Wallace LM, Denys P (2007) Kinematic constraints from GPS on oblique convergence of the Pacific and Australian Plates, central South Island, New Zealand. In: *A Continental Plate Boundary: Tectonics at South Island, New Zealand* (eds Okaya DA, Stern TA, Davey FJ), *American Geophysical Union Geophysical Monograph*, **175**, pp. 75–94, AGU.

Beavan RJ, Fielding E, Motagh M, Samsonov S, Donnelly N (2011) Fault location and slip distribution of the 22 February 2011 M_w6.2 Christchurch, New Zealand, earthquake from geodetic data. *Seismological Research Letters*, **82**, 789–799.

Beavan RJ, Samsonov S, Denys P, Sutherland R, Palmer NG, Denham M (2010b) Oblique slip on the Puysegur subduction interface in the 2009 July M_w7.8 Dusky Sound earthquake from GPS and InSAR observations: implications for the tectonics of southwestern New Zealand. *Geophysical Journal International*, **183**, 1265–1286.

Beavan RJ, Samsonov S, Motagh M, Wallace L, Ellis S, Palmer NG (2010c) The Darfield (Canterbury) earthquake: geodetic observations and preliminary source model. *Bulletin of the New Zealand Society for Earthquake Engineering*, **43**, 228–235.

Becker JA, Bickle MJ, Galy A, Holland TJB (2008) Himalayan metamorphic CO_2 fluxes: quantitative constraints from hydrothermal springs. *Earth and Planetary Science Letters*, **265**, 616–629.

Becker K (1990) A guide to formation testing using ODP drillstring packers. *Ocean Drilling Program Technical Note*, **14**, College Station, Texas.

Becker K, Davis EE (2004) In situ determinations of the permeability of the igneous oceanic crust. In: *Hydrogeology of the Oceanic Lithosphere* (eds Davis E, Elderfield H), pp. 189–224. Cambridge University Press, Cambridge, UK.

Beeler NM, Hickman SH (2004), Stress-induced, time-dependent fracture closure at hydrothermal conditions. *Journal of Geophysical Research*, **109**, B02211, doi:10.1029/2002JB001782.

Bekins BA, Dreiss SJ (1992) A simplified analysis of parameters controlling dewatering in accretionary prisms. *Earth and Planetary Science Letters*, **109**, 275–287.

Bekins B, Matmon D, Screaton EJ, Brown KM (2011) Reanalysis of in situ permeability measurements in the Barbados decollement. *Geofluids*, **11**, 57–70.

Bekins B, McCaffrey AM, Dreiss SJ (1994) Influence of kinetics on the smectite to illite transition in the Barbados accretionary prism. *Journal of Geophysical Research*, **99**, 18147–18158.

Bekins B, McCaffrey AM, Dreiss SJ (1995) Episodic and constant flow models for the origin of low-chloride waters in a modern accretionary complex. *Water Resources Research*, **31**, 3205–3215.

Bekins BA, Screaton EJ (2007) Pore pressure and fluid flow in the northern Barbados accretionary complex. In: *The Seismogenic Zone of Subduction Thrust Faults* (ed Dixon T), pp. 148–170, Columbia University Press, New York.

Belanger DW, Freeze GA, Lolcama JL, Pickens JF (1989) Interpretation of hydraulic testing in crystalline rock at the Leuggern borehole. In: Technical Report 87-19 (ed Nagra), SKB, Baden, Switzerland, http://www.nagra.ch/data/documents/database/dokumente/%24default/Default%20Folder/Publikationen/NTBs%201987-1988/d_ntb87-19.pdf (accessed 04 May 2016).

Belcher WR (2004) Death Valley regional ground-water flow system, Nevada and California—Hydrogeologic framework and transient ground-water flow model. *US Geological Survey Scientific Investigations Report*, **1171**, 408.

Bendall B, Hogarth R, Holl H, McMahon A, Larking A, Reid P (2014) Australian experience in EGS permeability enhancement – A review of 3 case studies. *Proceedings, Thirty-Ninth Workshop on Geothermal Reservoir Engineering, Stanford University, Stanford, California*.

Benjelloun ZH (1993) *Étude Experimentale et Modelisation du Comportement Hydromécanique des Joints Rocheux*. PhD thesis, Université Joseph Fourier, Grenoble, France.

Bennett RH, Fischer KM, Lavoie DL, Bryant WR, Rezak R (1989) Porometry and fabric of marine clay and carbonate sediments: determinants of permeability. *Marine Geology*, **89**, 127–152.

Bense V, Gleeson T, Loveless S, Bour O, Scibek J (2013) Fault zone hydrogeology. *Earth-Science Reviews*, **127**, 171–192.

Benson SM, Cole DR (2008) CO_2 sequestration in deep sedimentary formations. *Elements*, **4**, 325–331.

Berckhemer H, Rauen A, Winter H, Kern H (1997) Petrophysical properties of the 9-km-deep crustal section at KTB. *Journal of Geophysical Research*, **102**, 18337–18361.

Bercovici D, Ricard Y, Schubert G (2001) A two-phase model for compaction and damage, 1, General theory. *Journal of Geophysical Research*, **106**, 8887–8906.

Bergamaschi L, and Putti M (1999) Mixed finite elements and Newton-type linearizations for the solution of Richards' equation. *International Journal of Numerical Methods in Engineering*, **45**, 1026–1045.

Berkowitz B (2002) Characterizing flow and transport in fractured geological media: a review. *Advances in Water Resources*, **25**, 861–884.

Bernabe Y (1987) A wide range permeameter for use in rock physics. *International Journal of Rock Mechanics and Mining Sciences*, **24**, 309–315.

Berner EK, Berner RA (1996) *Global Environment: Water, Air, and Geochemical Cycles*. Prentice-Hall, Upper-Saddle River, NJ.

Berner RA (1978) Rate control of mineral dissolution under Earth surface conditions. *American Journal of Science*, **278**, 1235–1252.

Berner RA (2004) *The Phanerozoic Carbon Cycle: CO$_2$ and O$_2$*. Oxford University Press, New York.

Berryman JG (2003) Dynamic permeability in poroelasticity. *Stanford Exploration Project*, **113**, 443–453.

Berryman KR, Cochran UA, Clark KJ, Biasi GP, Langridge RM, Villamor P (2012) Major earthquakes occur regularly on an isolated plate boundary fault. *Science*, **336**, 1690–1693.

Bertrand EA, Caldwell TG, Hill GJ, Wallin EL, Bennie SL, Cozens N, Onacha SA, Ryan GA, Walter C, Zaino A, Wameyo P (2012) Magnetotelluric imaging of upper-crustal convection plumes beneath the Taupo Volcanic Zone, New Zealand. *Geophysical Research Letters*, **39**, L02304.

Bettison-Varga L, Schiffman P, Janecky DR (1995) Fluid-rock interaction in the hydrothermal upflow zone of the Solea graben, Troodos ophiolite, Cyprus. *Geological Society of America Special Papers*, **296**, 81–111.

Bettison-Varga L, Varga RJ, Schiffman P. (1992) Relation between ore-forming hydrothermal systems and extensional deformation in the Solea graben spreading center, Troodos ophiolite, Cyprus. *Geology*, **20**, 987–990.

Bhatnagar PL, Gross EP, Krook M (1954) A model for collision processes in gases, I, Small amplitude processes in charged and neutral one-component systems. *Physical Review Letters*, **94**, 511–525.

Bickle MJ, Teagle DAH (1992) Strontium alteration in the Troodos ophiolite – Implications for fluid fluxes and geochemical transport in mid-ocean ridge hydrothermal systems. *Earth and Planetary Science Letters*, **113**, 219–237.

Bickle MJ, Teagle DAH, Beynon J, Chapman HJ (1998) The structure and controls on fluid-rock interactions in ocean ridge hydrothermal systems: constraints from the Troodos ophiolite. In: *Modern Ocean Floor Processes and the Geological Record*, **148** (eds Mills RA, Harrison K), pp. 127–152. Geological Society of London, London.

Biot MA (1941) General theory of three-dimensional consolidation. *Journal of Applied Physics*, **12**, 155–164.

Biot MA (1956a) Theory of propagation of elastic waves in a fluid-saturated porous solid, I, Low-frequency range. *Journal of the Acoustical Society of America*, **28**, 168–178.

Biot MA (1956b) Theory of propagation of elastic waves in a fluid-saturated porous solid, II, Higher frequency range. *Journal of the Acoustical Society of America*, **28**, 179–191.

Biot MA (1962) Generalized theory of acoustic propagation in porous dissipative media. *Journal of the Acoustical Society of America*, **34**, 1254–1264.

Birchwood RA, Turcotte DL (1994) A unified approach to geopressuring, low-permeability zone formation, and secondary porosity generation in sedimentary basins. *Journal of Geophysical Research*, **99**, 20051–20058.

Bittleston SH, Ferguson J, Frigaard IA (2002) Mud removal and cement placement during primary cementing of an oil well - Laminar non-Newtonian displacements in an eccentric annular Hele-Shaw cell. *Journal of Engineering Mathematics*, **43**, 229–253.

Bjørlykke K (1999) Principal aspects of compaction and fluid flow in mudstones. *Geological Society of London Special Publication*, **158**, 73–78.

Bjornsson G, Bodvarsson G (1990) A survey of geothermal reservoir properties. *Geothermics*, **19**, 17–27.

Blackwell DD (1983) Heat flow in the northern Basin and Range province. In *The Role of Heat in the Development of Energy and Mineral Resources in the Northern Basin and Range Province, Geothermal Resources Council Special Report*, **13**, 81–93.

Blackwell DD (1994) A summary of deep thermal data from the Cascade Range and analysis of the "rain curtain" effect. *Oregon Department of Geology and Mineral Industries Open-File Report*, **O-94-07**, 75–131.

Blanc G, Doussan C, Thomas C, Boulegue J (1991) Non-steady state diffusion and advection model of transient concentration-depth profiles from the Barbados accretionary complex. *Oceanologica Acta*, **16**, 363–372.

Blankennagel RK, Weir JE Jr (1973) Geohydrology of the eastern part of Pahute Mesa, Nevada Test Site, Nye County, Nevada. *US Geological Survey Professional Paper*, **712-B**.

Blümling P, Bernier F, Lebon P, Martin D (2007) The excavation damaged zone in clay formations time-dependent behaviour and influence on performance assessment. *Physics and Chemistry of the Earth*, **32**, 588–599.

Boano F, Harvey JW, Marion A, Packman AI, Revelli R, Ridolfi L, Wörman A (2014) Hyporheic flow and transport processes: mechanisms, models, and biogeochemical implications. *Reviews of Geophysics*, **52**, 603–679, doi:10.1002/2012RG000417.

Bodnar RJ, Azbej T, Becker SP, Cannatelli C, Fall A, Severs MJ (2013) Whole earth geohydrologic cycle, from the clouds to the core: the distribution of water in the dynamic Earth system. *The Geological Society of America Special Paper*, **500**, 431–461.

Bodnar RJ, Cannatelli C, De Vivo B, Lima A, Belkin HE, Milia A (2007) Quantitative model for magma degassing and ground deformation (bradyseism) at Campi Flegrei, Italy: implications for future eruptions. *Geology*, **35**, 791–794.

Bodnar RJ, Lecumberri-Sanchez P, Moncada D, Steele-MacInnis M (2014) Fluid inclusions in hydrothermal ore deposits. In: *Treatise on Geochemistry*, 2nd edn (eds Holland HD, Turekian KK), pp. 119–142, Elsevier, Oxford.

Boese C, Jacobs K, Smith EGC, Stern TA, Townend J (2014) Background and delayed-triggered swarms in the central Southern Alps, South Island, New Zealand. *Geochemistry, Geophysics, Geosystems*, **15**, 945–964.

Boese CM, Stern TA, Townend J, Bourguignon S, Sheehan A, Smith EGC (2013) Sub-crustal earthquakes within the Australia-Pacific plate boundary zone beneath the Southern Alps, New Zealand. *Earth and Planetary Science Letters*, **376**, 212–219.

Boese C, Townend J, Smith EGC, Stern TA (2012) Microseismicity and stress in the vicinity of the Alpine Fault, central Southern Alps, New Zealand. *Journal of Geophysical Research*, **117**, B02302, doi:10.1029/2011JB008460.

Boles JR, Eichhubl P, Garven G, Chen J (2004) Evolution of a hydrocarbon migration pathway along basin-bounding faults: evidence from fault cement. *American Association of Petroleum Geologists Bulletin*, **88**, 947–970.

Boles JR, Grivetti M (2000) Calcite cementation along the Refugio/Carneros Fault, coastal California: a link between deformation, fluid movement and fluid-rock interaction at a basin margin. *Journal of Geochemical Exploration*, **69–70**, 313–316.

Bolton AJ, Maltman AJ, Clennell MB (1999) Nonlinear stress-dependence of permeability: a mechanism for episodic fluid flow in accretionary wedges. *Geology*, **27**, 239–242.

Borchardt R, Emmermann R (1993) Vein minerals in KTB rocks. *KTB Report*, **2**, 481–488.

Borchardt R, Zulauf G, Emmermann R, Hoefs J, Simon K (1990) Abfolge und Bildungsbedingungen von Sekundarmineralien in der KTB-Vorbohrung. *KTB Report*, **90-4**, 76–88.

Borthwick J, Harmon RS (1982) A note regarding CIF$_3$ as an alternative to BrF$_5$ for oxygen isotope analysis. *Geochimica et Cosmochimica Acta*, **46**, 1665–1668.

Boulon MJ (1995) A 3-D direct shear device for testing the mechanical behaviour and the hydraulic conductivity of rock joints. *Proceedings of the MJFR-2 Conference*, Vienna, Austria, Balkema, Rotterdam, 407–413.

Boulon MJ, Selvadurai APS, Benjelloun H, Feuga B (1993) Influence of rock joint degradation on hydraulic conductivity. *International Journal of Rock Mechanics and Mining Sciences*, **30**, 1311–1317.

Boulton C, Moore DE, Lockner DA, Toy VG, Townend J, Sutherland R (2014) Frictional properties of exhumed fault gouges in DFDP-1 cores, Alpine Fault, New Zealand. *Geophysical Research Letters*, **41**, 356–362.

Bourbie T, Zinszner B (1985) Hydraulic and acoustic properties as a function of porosity in Fontainebleau Sandstone. *Journal of Geophysical Research*, **90**, 11524–11532.

Bourlange S, Henry P (2007) Numerical model of fluid pressure solitary wave propagation along the decollement of an accretionary wedge: application to the Nankai wedge. *Geofluids*, **7**, 323–334.

Bourlange S, Henry P, Moore JC, Mikada H, Klaus A (2003) Fracture porosity in the décollement zone of Nankai accretionary wedge using logging while drilling resistivity data. *Earth and Planetary Science Letters*, **209**, 103–112.

Boutt DF, Saffer D, Doan M, Lin W, Ito T, Kano Y, Flemings P, McNeill LC, Byrne T, Hayman NW, Moe KT (2012) Scale dependence of in situ permeability measurements in the Nankai accretionary prism: the role of fractures. *Geophysical Research Letters*, **39**, L07302.

Bowman JR, Willlett SD, Cook SJ (1994) Oxygen isotopic transport and exchange during fluid flow: one-dimensional models and applications. *American Journal of Science*, **294**, 1–55.

Boyer F, Guazzelli E, Pouliquen O (2011) Unifying suspension and granular rheology. *Physical Review Letters*, **107**, 188301.

Brace WF (1980) Permeability of crystalline and argillaceous rocks. *International Journal of Rock Mechanics and Mining Sciences*, **17**, 241–251.

Brace WF (1984) Permeability of crystalline rocks: new in situ measurements. *Journal of Geophysical Research*, **89**, 4327–4330.

Brace WF, Walsh JB, Frangos WT (1968) Permeability of granite under high pressure. *Journal of Geophysical Research*, **73**, 2225–2236.

Bradley BA, Cubrinovski M (2011) Near-source strong ground motions observed in the 22 February 2011 Christchurch earthquake. *Seismological Research Letters*, **82**, 853–865.

Brandon MT, Vance JA (1992) Tectonic evolution of the Cenozoic Olympic subduction complex, Washington state, as deduced from fission track ages for detrital zircons. *American Journal of Science*, **292**, 565–636.

Brantley SL (2004) Reaction kinetics of primary rock-forming minerals under ambient conditions. In: *Surface and Ground Water, Weathering, and Soils* (eds Drever JI, Holland HD, Turekian KK), pp. 73–117, Treatise on Geochemistry, Elsevier, Amsterdam.

Brantley SL, Chen Y (1995) Chemical weathering rates of pyroxenes and amphiboles. *Mineralogical Society of America Reviews in Mineralogy and Geochemistry*, **31**, 119–172.

Brantley SL, Kubicki JD, White AF (2008) *Kinetics of Water-Rock Interaction*. Springer Verlag, New York.

Brantley SL, Megonigal JP, Scatena FN, Balogh-Brunstad Z, Barnes RT, Bruns MA, Van Cappellen P, Dontsova K, Hartnett HE, Hartshorn AS, Heimsath A, Herndon E, Jin L, Keller CK, Leake JR, Mcdowell WH, Meinzer FC, Mozdzer TJ, Petsch S, Pett-Ridge J, Pregitzer KS, Raymond PA, Riebe CS, Shumaker K, Sutton-Grier A, Walter R, Yoo K (2011) Twelve testable hypotheses on the geobiology of weathering. *Geobiology*, **9**, 140–165.

Brantley SL, Mellott N (2000) Surface area and porosity of primary silicate minerals. *American Mineralogist*, **85**, 1767–1783.

Bray CJ, Karig DE (1985) Porosity of sediments in accretionary prisms and some implications for dewatering processes. *Journal of Geophysical Research*, **90**, 768–778.

Bredehoeft JD (1967) Response of well-aquifer systems to Earth tides. *Journal of Geophysical Research*, **72**, 3075–3087.

Bredehoeft JD (1997) Fault permeability near Yucca Mountain. *Water Resources Research*, **33**, 2459–2463.

Bredehoeft JD, Papadopulos IS (1965) Rates of vertical ground water flow estimated from the Earth's thermal profile. *Water Resources Research*, **1**, 325–328.

Bredehoeft JD, Papadopulos SS (1980) A method for determining the hydraulic properties of tight formations. *Water Resources Research*, **16**, 223–238.

Brodsky EE, Karakostas V, Kanamori H (2000) A new observation of dynamically triggered regional seismicity: earthquakes in Greece following the August, 1999 Izmit, Turkey earthquake. *Geophysical Research Letters*, **27**, 27410–27414.

Brodsky EE, Roeloffs E, Woodcock D, Gall I, Manga M (2003) A mechanism for sustained groundwater pressure changes induced by distant earthquakes. *Journal of Geophysical Research*, **108**, doi:10.1029/2002JB002321.

Brown ET, Hoek E (1978) Trends in relationships between measured in-situ stresses and depth. *International Journal of Rock Mechanics and Mining Sciences*, **15**, 211–215.

Brown KL (1986) Gold deposition from geothermal discharges in New Zealand. *Economic Geology*, **81**, 979–983.

Brown KM (1995) The variation of the hydraulic conductivity structure of an overpressured thrust zone with effective stress. In: *Proceedings of the Ocean Drilling Program, Scientific Results*, **146** (eds Carson B, Westbrook GK, Musgrave RJ, Suess E), pp. 281–289, Texas A & M University, Ocean Drilling Program, College Station, TX.

Brown KM, Bekins BA, Clennell B, Dewhurst D, Westbrook GK (1994) Heterogeneous hydrofracture development and accretionary fault dynamics. *Geology*, **22**, 259–262.

Brown KM, Saffer DM, Bekins BA (2001) Smectite diagenesis, pore-water freshening, and fluid flow at the toe of the Nankai wedge. *Earth and Planetary Science Letters*, **194**, 97–109.

Brown KM, Tryon MD, DeShon HR, Dorman LM, Schwartz SY (2005) Correlated transient fluid pulsing and seismic tremor in the Costa Rica subduction zone. *Earth and Planetary Science Letters*, **238**, 189–203.

Brown M (2010) The spatial and temporal patterning of the deep crust and implications for the process of melt extraction. *Philosophical Transactions of the Royal Society of London, Series A, Mathematical, Physical and Engineering Sciences*, **368**, 11–51.

Brown SR (1987a) A note on the description of surface roughness using fractal dimension. *Geophysical Research Letters*, **14**, 1095–1098.

Brown SR (1987b) Fluid flow thorough rock joints: the effect of surface roughness. *Journal of Geophysical Research*, **92**, 1337–1347.

Brown SR (1989) Transport of fluid and electric current through a single fracture. *Journal of Geophysical Research*, **94**, 9429–9438.

Brown SR, Scholz CH (1985) Closure of random elastic surfaces in contact. *Journal of Geophysical Research*, **90**, 5531–5545.

Brown SR, Scholz CH (1986) Closure of rock joints. *Journal of Geophysical Research*, **91**, 4939–4948.

Brown SR, Stockman HW, Reeves SJ (1995) Applicability of Reynolds equation for modeling fluid flow between rough surfaces. *Geophysical Research Letters*, **22**, 2537–2540.

Browne PRL, Courtney SS, Wood CP (1989) Formation rates of calc-silicate minerals deposited inside drillhole casing, Ngatamariki geothermal field, New Zealand. *American Mineralogist*, **74**, 759–763.

Brückmann W, Moran K, MacKillop AK (1997) Permeability and consolidation characteristics from Hole 949B, northern Barbados Ridge. In: *Proceedings of the Ocean Drilling Program, Scientific Results*, **156** (eds Shipley TH, Ogawa Y, Blum P, Bahr JM), pp. 109–114, Ocean Drilling Program, College Station, TX.

Bruel D (2007) Using the migration of the induced seismicity as a constraint for fractured Hot Dry Rock reservoir modelling. *International Journal of Rock Mechanics and Mining Sciences*, **44**, 1106–1117.

Brunner P, Kinzelbach W (2005) Sustainable groundwater management. In: *Encyclopedia of Hydrological Sciences* (ed Anderson MG), Part 13. Wiley, New York.

Brush DJ, Thomson NR (2003) Fluid flow in synthetic rough-walled fractures: Navier-Stokes, Stokes, and local cubic law assumptions. *Water Resources Research*, **39**, doi:10.1029/2002WR001346.

Bryant WR (2002) Permeability of clays, silty-clays and clayey-silts. *Gulf Coast Association of Geological Societies Transactions*, **52**, 1069–1077.

Bryant WR, Hottman WE, Trabant PK (1974) Permeability of unconsolidated and consolidated marine sediments, Gulf of Mexico. *AAPG Bulletin*, **58**, 2207.

Bucher K, Stober I (2010) Fluids in the upper continental crust. *Geofluids*, **10**, 241–253.

Bucher K, Stober I, Seelig U (2012) Water deep inside the mountains: unique water samples from the Gotthard rail base tunnel, Switzerland. *Chemical Geology*, **334**, 240–253.

Bucher K, Zhang L, Stober I (2009) A hot spring in granite of the western Tianshan, China. *Applied Geochemistry*, **24**, 402–410.

Bucher-Nurminen K (1981) The formation of metasomatic reaction veins in dolomitic marble roof pendants in the Bergell intrusion (Province Sondrio, Northern Italy). *American Journal of Science*, **281**, 1197–1222.

Bucher-Nurminen K (1989) Reaction veins in marbles formed by a fracture-reaction-seal mechanism. *European Journal of Mineralogy*, **1**, 701–714.

Baumgartner LP, Valley JV (2001) Stable isotope transport and contact metamorphic fluid flow. In: *Stable Isotope Geochemistry* (eds Valley JW, Cole DR), *Reviews in Mineralology*, **43**, 415–467, University of Wisconsin, Madison, WI.

Baumgartner LP, Gerdes ML, Person MA, Roselle, GT (1997) Porosity and permeability of carbonate rocks during contact metamorphism. In: *Fluid Flow and Transport in Rocks: Mechanisms and Effects* (eds Jamveit B, Yardley BWD), Chapman & Hall, London, 83–98.

Bunger AP, Jeffrey RG, Kear JP, Zhang XP, Morgan M (2011) Experimental investigation of the interaction among closely spaced hydraulic fractures. In: 45th US Rock Mechanics/ Geomechanics Symposium, June 26–29, 2011, San Francisco, California.

Burchfiel BC, Royden LH, van der Hilst RD, Hager BH (2008) A geological and geophysical context for the Wenchuan earthquake of 12 May 2008, Sichuan, People's Republic of China. *GSA Today*, **18**, 4–11.

Burgmann R, Pollard DD (1994) Strain accommodation about strike-slip fault discontinuities in granitic rock under brittle-to-ductile conditions. *Journal of Structural Geology*, **16**, 1655–1674.

Burns ER, Morgan DS, Lee KK, Haynes JV, Conlon TD (2012a) Evaluation of long-term water-level declines in basalt aquifers near Mosier, Oregon. *U.S. Geological Survey Scientific Investigations Report*, **2012–5002**, http://pubs.er.usgs.gov/ publication/sir20125002 (accessed 04 May 2016).

Burns ER, Morgan DS, Peavler RS, Kahle SC (2011) Three-dimensional model of the geologic framework for the Columbia Plateau Regional Aquifer System, Idaho, Oregon, and Washington. *U.S. Geological Survey Scientific Investigations Report*, **2010–5246**, http://pubs.er.usgs.gov/publication/ sir20105246 (accessed 04 May 2016).

Burns ER, Snyder DT, Haynes JV, Waibel MS (2012b) Groundwater status and trends for the Columbia Plateau Regional Aquifer System, Washington, Oregon, and Idaho. *U.S. Geological Survey Scientific Investigations Report*, **2012–5261**, http://pubs.er .usgs.gov/publication/sir20125261 (accessed 04 May 2016).

Burns E, Williams CF, Ingebritsen SE, Voss CI, Spane FA, DeAngelo J (2015) Understanding heat and groundwater flow through continental flood basalt provinces: insights gained from alternative models of permeability/depth relationships for the Columbia Plateau, USA. *Geofluids*, **15**, 120–138.

Butler GA, Cauffman TL, Lolcama JL, Longsine DE, McNeish JA (1989) Interpretation of hydraulic testing at the Weiach borehole. In: *SKB Technical Report*, **87-20** (ed SKB), Nagra, Baden, Switzerland, http://www.nagra.ch/data/documents/database/ dokumente/$default/Default%20Folder/Publikationen/ NTBs%201987-1988/e_ntb87-20.pdf (accessed 04 May 2016).

Butler JJ Jr (1998) *The Design, Performance, and Analysis of Slug Tests*. Lewis Publishers, New York.

Byerlee J (1978) Friction of rocks. *Pure and Applied Geophysics*, **116**, 615–626.

Caine JS, Evans JP, Forster CB (1996) Fault zone architecture and permeability structure. *Geology*, **24**, 1025–1028.

Candela T, Renard F, Klinger Y, Mair K, Schmittbuhl J, Brodsky EE (2012) Roughness of fault surfaces over nine decades of length scales. *Journal of Geophysical Research*, **117**, B08409, doi:10.1029/2011JB009041.

Canfield DE, Kump LR (2013) Carbon cycle makeover. *Science*, **339**, 533–534.

Cann JR, Gillis KM (2004) Hydrothermal insights from the Troodos ophiolite, Cyprus. In: *Hydrogeology of the Oceanic Lithosphere* (eds Davis E, Elderfield H), pp. 274–310. Cambridge University Press, Cambridge.

Cann JR, McCaig AM, Yardley BWD (2015) Rapid generation of reaction permeability in the roots of black smoker systems, Troodos ophiolite, Cyprus. *Geofluids*, **15**, 179–192.

Cann JR, Strens MR (1989) Modeling periodic megaplume emission by black smoker systems. *Journal of Geophysical Research*, **94**, 12227–12237.

Cappa F (2006) Role of fluids in the hydromechanical behavior of heterogeneous fractured rocks: *in situ* characterization and numerical modelling. *Bulletin of Engineering Geology and the Environment*, **65**, 321–337.

Cappa F, Guglielmi Y, Gaffet S, Lancon H, Lamarque I (2006a) Use of in situ fiber optic sensors to characterize highly heterogeneous elastic displacement fields in fractured rocks. *International Journal of Rock Mechanics and Mining Sciences*, **43**, 647–654.

Cappa F, Guglielmi Y, Rutqvist J, Tsang C-F, Thoraval A (2006b) Hydromechanical modeling of pulse tests that measure both fluid pressure and fracture-normal displacement at the Coaraze Laboratory site, France. *International Journal of Rock Mechanics and Mining Sciences*, **43**, 1062–1082.

Cappa F, Rutqvist J (2011) Modeling of coupled deformation and permeability evolution during fault reactivation induced by deep underground injection of CO_2. *International Journal of Greenhouse Gas Control*, **5**, 336–346.

Cappa F, Rutqvist J, Yamamoto K (2009) Modeling crustal deformation and rupture processes related to upwelling of deep CO_2-rich fluids during the 1965–1967 Matsushiro earthquake swarm in Japan. *Journal of Geophysical Research*, **114**, 1–20.

Cardwell WJ, Parsons R (1945) Average permeabilities of heterogeneous oil sands. *American Institute of Mining and Metallurgical Engineers Technical Publication*, **1852**, 1–9.

Carman PC (1937) Fluid flow through granular beds. *Transactions Institution of Chemical Engineers*, **15**, 150–166.

Carman PC (1939) Permeability of saturated sands, soils and clays. *Journal of Agricultural Science*, **29**, 262–273.

Carman PC (1956) *Flow of Gases Through Porous Media*. Academic Press, New York.

Carrigan CR, King GCP, Barr GE, Bixler NE (1991) Potential for water-table excursions induced by seismic events at Yucca Mountain, Nevada. *Geology*, **19**, 1157–1160.

Carslaw HS, Jaeger JC (1959) *Conduction of Heat in Solids*, 2nd edn Oxford University Press, London.

Carson B, Screaton EJ (1998) Fluid flow in accretionary prisms: evidence for focused, time-variable discharge. *Reviews of Geophysics*, **36**, 329–351.

Carson B, Seke E, Paskevich V, Holmes ML (1994) Fluid expulsion sites on the Cascadia accretionary prism: mapping diagenetic deposits with processed GLORIA imagery. *Journal of Geophysical Research*, **99**, 11959–11969.

Cassar C, Nicolas M, Pouliquen O (2005) Submarine granular flows down inclined planes. *Physics of Fluids*, **17**, 103301.

Catalli F, Meier M-A, Wiemer S (2013) The role of Coulomb stress changes for injection-induced seismicity: the Basel enhanced geothermal system. *Geophysical Research Letters*, **40**, doi:10.1029/2012GL054147.

Cathles LM (1977) An analysis of the cooling of intrusives by ground-water convection which includes boiling. *Economic Geology*, **72**, 804–826.

Cathles LM (1993) A capless 350°C flow zone model to explain megaplumes, salinity variations, and high-temperature veins in ridge axis hydrothermal systems. *Economic Geology*, **88**, 1977–1988.

Cathles LM, Adams JJ (2005) Fluid flow and petroleum and mineral resources in the upper (<20 km) continental crust. *Economic Geology 100th Anniversary Volume*, 77–110.

Cathles LM, Erendi HJ, Barrie T (1997) How long can a hydrothermal system be sustained by a single intrusive event? *Economic Geology*, **92**, 766–771.

Cattaneo CR (1958) Sur une forme de l'equation de la chaleur eliminant le paradoxe d'une propagation instantanee. *Comptes Rendus*, **247**, 431–433.

Cattin R, Chamot-Rooke N, Pubellier M, Rabaute A, Delescluse M, Vigny C, Fleitout L, Dubernet P (2009) Stress change and effective friction coefficient along the Sumatra-Andaman-Sagaing fault system after the 26 December 2004 (M_w = 9.2) and the 28 March 2005 (M_w = 8.7) earthquakes. *Geochemistry, Geophysics, Geosystems*, **10**, Q03011.

Cederbom CE, van der Beek P, Schlunegger F, Sinclair HD, Oncken O (2011) Rapid extensive erosion of the North Alpine foreland basin at 5-4 Ma. *Basin Research*, **23**, 528–550.

Chabora E, Zemach E, Spielman P, Drakos P, Hickman S, Lutz S, Boyle K, Falconer A, Robertson-Tait A, Davatzes NC, Rose P, Majer E, Jarpe S (2012) Hydraulic stimulation of well 27-15, Desert Peak Geothermal Field, Nevada USA. *Proceedings Thirty-Seventh Workshop on Geothermal Reservoir Engineering*, Stanford University.

Chan L-H, Kastner M (2000) Lithium isotopic compositions of pore fluids and sediments in the Costa Rica subduction zone: implications for fluid processes and sediment contribution to the arc volcanoes. *Earth and Planetary Science Letters*, **183**, 275–290.

Chan T, Christiansson R, Boulton GS, Ericsson LO, Hartikainen J, Jensen MR, Mas Ivars D, Stanchell FW, Vistrand P, Wallroth T (2005) DECOVALEX III BMT 3/BENCHPAR WP4: the thermo-hydro-mechanical responses to a glacial cycle and their potential implications for deep geological disposal of nuclear fuel waste in a fractured crystalline rock mass. *International Journal of Rock Mechanics and Mining Sciences*, **42**, 805–827.

Chapuis RP (2012) Predicting the saturated hydraulic conductivity of soils: a review. *Bulletin of Engineering Geology and the Environment*, **71**, 401–434.

Chapuis RP, Aubertin M (2003) On the use of the Kozeny Carman equation to predict the hydraulic conductivity of soils. *Canadian Geotechnical Journal*, **628**, 616–628.

Chauveau B, Kaminski E (2008) Porous compaction in transient creep regime and implications for melt, petroleum, and CO_2 circulation. *Journal of Geophysical Research*, **113**, B09406.

Chebotarev II (1955) Metamorphism of natural waters in the crust of weathering, 1. *Geochimica et Cosmochimica Acta*, **8**, 22–48.

Chen X, Shearer PM, Abercrombie RE (2012) Spatial migration of earthquakes within seismic clusters in Southern California: evidence for fluid diffusion. *Journal of Geophysical Research*, **117**, B04301, doi:10.1029/2011JB008973.

Chen Y, Brantley SL (1998) Diopside and anthophyllite dissolution at 25 degrees and 90 degrees C and acid pH. *Chemical Geology*, **147**, 233–248.

Chen Z, Narayan SP, Yang Z, Rahman SS (2000) An experimental investigation of hydraulic behavior of fractures and joints in granitic rock, *International Journal of Rock Mechanics and Mining Sciences*, **37**, 1061–1071.

Chester FM, Evans JP, Biegel RL (1993) Internal structure and weakening mechanisms of the San Andreas fault. *Journal of Geophysical Research*, **98**, 771–786.

Chia Y, Chiu JJ, Chiang Y-H, Lee T-P, Liu C-W (2008) Spatial and temporal changes of groundwater level induced by thrust faulting. *Pure and Applied Geophysics*, **165**, 5–16.

Chia Y, Wang Y-S, Chiu JJ, Liu C-W (2001) Changes of groundwater level due to the 1999 Chi-Chi earthquake in the Choshui River alluvial fan in Taiwan. *Bulletin of the Seismological Society of America*, **91**, 1062–1068.

Chiodini G, Frondini F, Marini L (1995) Theoretical geothermometers and P_{CO2} indicators for aqueous solutions coming from hydrothermal systems of medium-low temperature hosted in carbonate-evaporite rocks. Application to the thermal springs of the Etruscan Swell, Italy. *Applied Geochemistry*, **10**, 337–346.

Christian GD, Dasgupta P, Schug K. (2014) *Analytical Chemistry*. John Wiley and Sons, Inc., Hoboken, NJ.

Claesson L, Skelton A, Graham C, Morth CM (2007) The timescale and mechanisms of fault sealing and water-rock interaction after an earthquake. *Geofluids*, **7**, 427–440.

Clauser C (1992) Permeability of crystalline rocks. *Eos Transactions American Geophysical Union*, **73**, 233, 237–238.

Clavier C, Coates G, Dumanoir J (1984) Theoretical and experimental bases for the dual-water model for interpretation of shaly sands. *SPE Journal*, **24**, 153–168.

Clift P, Vannucchi P (2004) Controls on tectonic accretion versus erosion in subduction zones: implications for the origin and recycling of the continental crust. *Reviews of Geophysics*, **42**, RG2001.

Cline JS, Bodnar RJ (1991) Can economic porphyry copper mineralization be generated by a typical calc-alkaline melt? *Journal of Geophysical Research*, **96**, 8113–8126.

Cloke PL, Kesler SE (1979) Halite trend in hydrothermal solutions. *Economic Geology*, **74**, 1823–1831.

Cloos M (1984) Landward-dipping reflectors in accretionary wedges: active dewatering conduits? *Geology*, **12**, 519–522.

Clowes RM, Brandon MT, Green AG, Yorath CJ, Brown AS, Kanasewich ER, Spencer C (1987) LITHOPROBE - southern Vancouver Island: Cenozoic subduction complex imaged by deep seismic reflections. *Canadian Journal of Earth Sciences*, **24**, 31–51.

Coggon RM, Teagle DAH, Smith-Duque CE, Alt JC, Cooper MJ (2010) Reconstructing past seawater Mg/Ca and Sr/Ca from mid-ocean ridge flank carbonates. *Science*, **327**, 1114–1117.

Committee on Fracture Characterization and Fluid Flow (1996) *Rock Fractures and Fluid Flow*. National Academy Press, Washington, DC.

Conin M, Bourlange S, Henry P, Boiselet A, Gaillot P (2013) Distribution of resistive and conductive structures in Nankai accretionary wedge reveals contrasting stress paths. *Tectonophysics*, **611**, 181–191.

Connell J (2000) Generation and destruction of porosity during the hydrothermal alteration of the sheeted dike complex in the Troodos ophiolite, Cyprus. MSc (Geochemistry) thesis, University of Leeds.

Connolly JAD (1997) Devolatilization-generated fluid pressure and deformation-propagated fluid flow during prograde regional metamorphism. *Journal of Geophysical Research*, **102**, 18149–18173.

Connolly JAD (2010) The mechanics of metamorphic fluid expulsion. *Elements*, **6**, 165–172.

Connolly JAD, Podladchikov YY (1998) Compaction-driven fluid flow in viscoelastic rock. *Geodinamica Acta*, **11**, 55–84.

Connolly JAD, Podladchikov YY (2000) Temperature-dependent viscoelastic compaction and compartmentalization in sedimentary basins. *Tectonophysics*, **324**, 137–168.

Connolly JAD, Podladchikov YY (2004) Fluid flow in compressive tectonic settings: implications for mid-crustal seismic reflectors and downward fluid migration. *Journal of Geophysical Research*, **109**, B04201.

Connolly JAD, Podladchikov YY (2007) Decompaction weakening and channeling instability in ductile porous media: implications for asthenospheric melt segregation. *Journal of Geophysical Research*, **112**, B10205, doi:10.1029/2005JB004213.

Connolly JAD, Podladchikov YY (2013) A hydromechanical model for lower crustal fluid flow. In: *Metasomatism and the Chemical Transformation of Rock* (eds Harlov DE, Austrheim H), pp. 599–658. Springer, Berlin.

Connolly JAD, Podladchikov YY (2015) An analytical solution for solitary porosity waves: implications for dynamic permeability and fluidization of nonlinear viscous and viscoplastic rock. *Geofluids*, **15**, 269–292.

Connolly P, Cosgrove J (1999) Prediction of static and dynamic fluid pathways within and around dilational jogs. *Geological Society of London Special Publications*, **155**, 105–121.

Constantinou G, Govett GJS (1972) Genesis of sulphide deposits, ochre and umber of Cyprus. *Transactions of the Institution of Mining and Metallurgy*, **81**, B34–B46.

Constantinou G, Govett GJS (1973) Metallogensis associated with the Troodos ophiolite. *Economic Geology*, **68**, 843–858.

Coogan LA (2008) Reconciling temperatures of metamorphism, fluid fluxes and heat transport in the upper crust at intermediate to fast-spreading mid-ocean ridges. *Geochemistry, Geophysics, Geosystems*, **9**, doi:10.1029/2007/GC001787.

Coogan LA (2009) Altered oceanic crust as an inorganic record of palaeoseawater Sr concentration. *Geochemistry, Geophysics, Geosystems*, **10**, doi:10.1029/2008/GC002341.

Cook HE, Corboy J (2004) Great Basin Paleozoic carbonate platform – Facies, facies transitions, depositional models, platform architecture, sequence stratigraphy, and predictive mineral hosts models. *US Geological Survey Open-File Report*, **2004-1078**.

Cook NGW (1992) Natural joints in rock: mechanical, hydraulic and seismic behaviour and properties under normal stress. *International Journal of Rock Mechanics and Mining Sciences*, **29**, 198–223.

Coolbaugh MF, Arehart GB, Faulds JE, Garside LJ (2005) Geothermal systems in the Great Basin, western United States: modern analogues to the roles of magmatism, structure, and regional tectonics in the formation of gold deposits. In: *Geological Society of Nevada Symposium 2005: Window to the World* (eds Rhoden HN, Steininger RC, Vikre PG), pp. 1063–1082. Nevada Geological Society, Reno, NV.

Cooper HH, Bredehoeft JD, Papadopulos IS, Bennett RR (1965) The response of well-aquifer systems to seismic waves. *Journal of Geophysical Research*, **70**, 3915–3926.

Cooper HH, Jacob CE (1946) A generalized graphical method for evaluating formation constants and summarizing well-field history. *Transactions American Geophysical Union*, **27**, 526–534.

Coumou D, Driesner T, Geiger S, Paluszny A, Heinrich CA (2009a) High-resolution three-dimensional simulations of mid-ocean ridge hydrothermal systems. *Journal of Geophysical Research*, **114**, B07104.

Coumou D, Driesner T, Heinrich CA (2008) The structure and dynamics of mid-ocean ridge hydrothermal systems. *Science*, **321**, 1825–1828.

Coumou D, Driesner T, Weis P, Heinrich CA (2009b) Phase separation, brine formation, and salinity variation at Black Smoker hydrothermal systems. *Journal of Geophysical Research*, **114**, B03212.

Cousins WJ, McVerry GH (2010) Overview of strong-motion data from the Darfield earthquake. *Bulletin of the New Zealand Society for Earthquake Engineering*, **43**, 222–227.

Cowie PA, Scholz CH (1992) Displacement-length scaling relationship for faults: data synthesis and discussion. *Journal of Structural Geology*, **14**, 1149–1156.

Cox SC, Barrell DJA (2007) Geology of the Aoraki area: scale 1:250,000. *Institute of Geological & Nuclear Sciences 1:250,000 Geological Map*, **15**.

Cox SC, Craw D, Chamberlain CP (1997) Structure and fluid migration in a late Cenozoic duplex system forming the main divide in the central Southern Alps, New Zealand. *New Zealand Journal of Geology and Geophysics*, **40**, 359–373.

Cox SC, Menzies C, Sutherland R, Denys P, Chamberlain C, Teagle D (2015) Changes in hot spring temperature and hydrogeology of the Alpine Fault hanging wall, New Zealand, induced by South Island earthquakes. *Geofluids*, **15**, 216–239.

Cox SC, Rutter HK, Sims A, Manga M, Wier JJ, Ezzy T, White PA, Horton TW, Scott D (2012a) Hydrological effects of the M_w7.1 Darfield (Canterbury) earthquake, 4 September 2010,

New Zealand. *New Zealand Journal of Geology and Geophysics*, **55**, 231–247.

Cox SC, Song SH, White PA, Davidson P, Strong DT (2010) The Canterbury and other earthquakes: far field effects on groundwater. In: Water: The Blue Gold. *Proceedings of the New Zealand Hydrological Society Conference 2010*, 7–10 December, University of Otago, Dunedin, 175–176.

Cox SC, Stirling MW, Herman F, Gerstenberger M, Ristau J (2012b) Potentially active faults in the rapidly eroding landscape adjacent to the Alpine Fault, central Southern Alps, New Zealand. *Tectonics*, **31**, TC2011.

Cox SC, Sutherland R (2007) Regional geological framework of South Island, New Zealand, and its significance for understanding the active plate boundary. In: *A Continental Plate Boundary: Tectonics at South Island, New Zealand* (eds Okaya DA, Stern TA, Davey FJ). *American Geophysical Union Geophysical Monograph*, **175**, 19–46, AGU.

Cox SF (2005) Coupling between deformation, fluid pressures, and fluid flow in ore-producing hydrothermal systems at depth in the crust. *Economic Geology,* **100**th Anniversary Volume, 39–75.

Cox SF (2010) The application of failure mode diagrams for exploring the roles of fluid pressure and stress states in controlling styles of fracture-controlled permeability enhancement in faults and shear zones. *Geofluids*, **10**, 217–233.

Cox SF, Knackstedt MA, Braun J (2001) Principles of structural control on permeability and fluid flow in hydrothermal systems. *Reviews in Economic Geology*, **14**, 1–24.

Craig H, Lupton JE (1981) Helium and mantle volatiles in the ocean and the oceanic crust. In: *The Sea*, Vol. 7 (ed Emiliani E), pp. 391–428, Wiley Interscience, New York.

Craig RF (2004) *Craig's Soil Mechanics*, 7th edn Spon, London.

Crane K (1987) Structural evolution of the East Pacific Rise axis from 13°10′N to 10°35′N: interpretations from SeaMARC I data. *Tectonophysics*, **136**, 65–92.

Crawford BR, Faulkner DR, Rutter EH (2008) Strength, porosity, and permeability development during hydrostatic and shear loading of synthetic quartz-clay fault gouge. *Journal of Geophysical Research*, **113**, B03207, doi:10.1029/2006JB004634.

Curewitz D, Karson J (1994) Structural settings of hydrothermal outflow: fracture permeability maintained by fault propagation and interaction. *Journal of Volcanology and Geothermal Research*, **79**, 149–168.

Curtis CD, Lipshie SR, Oertel G, Pearson MJ (1980) Clay orientation in some Upper Carboniferous mudrocks, its relationship to quartz content and some inferences about fissility, porosity and compactional history. *Sedimentology*, **27**, 333–339.

Cutillo PA, Screaton EJ, Ge S (2003) Three-dimensional numerical simulation of fluid flow and heat transport within the Barbados Ridge accretionary complex. *Journal of Geophysical Research*, **108**, doi:10.1029/2002JB002240.

Daigle H, Dugan B (2011) Permeability anisotropy and fabric development: a mechanistic explanation. *Water Resources Research*, **47**, W12517, doi:10.1029/2011WR011110.

Daigle H, Dugan B (2014) Data report: permeability, consolidation, stress state, and pore system characteristics of sediments from Sites C0011, C0012, and C0018 of the Nankai Trough. In: *Proceedings of the Integrated Ocean Drilling Program*, **333**,

(eds Henry P, Kanamatsu T, Moe K, and the Expedition 333 Scientists), 1–23.

Daigle H, Screaton E (2015a) Evolution of sediment permeability during burial and subduction. *Geofluids*, **15**, 84–105.

Daigle H, Screaton EJ (2015b) Predicting the permeability of sediments entering subduction zones. *Geophysical Research Letters*, **42**, 5219–5226.

Daigle H, Thomas B, Rowe H, Nieto M (2014) Nuclear magnetic resonance characterization of shallow marine sediments from the Nankai Trough, Integrated Ocean Drilling Program Expedition 333. *Journal of Geophysical Research*, **119**, 2631–2650.

Danielopol DL, Griebler C, Gunatilaka A, Notenboom J (2003) Present state and future prospects for groundwater ecosystems. *Environmental Conservation*, **30**, 104–130.

Davis D, Suppe J, Dahlen FA (1983) Mechanics of fold-and-thrust belts and accretionary wedges. *Journal of Geophysical Research*, **88**, 1153–1172.

Davis EE, Becker K, Wang K, Carson B (1995) Long-term observations of pressure and temperature in Hole 892B, Cascadia accretionary prism. In: *Proceedings of the Ocean Drilling Program, Scientific Results*, **146**, (eds Carson B, Westbrook GK, Musgrave RJ, Suess E), 299–311, ODP.

Davis EE, Heesemann M, Wang K (2011) Evidence for episodic aseismic slip across the subduction seismogenic zone off Costa Rica: CORK borehole pressure observations at the subduction prism toe. *Earth and Planetary Science Letters*, **306**, 299–305.

Day GA (1987) Source of recharge to the Beowawe geothermal system, Nevada. MS thesis, University of Nevada-Reno.

Day-Stirrat RJ, Aplin AC, Srodoh J, van der Pluijm BA (2008) Diagenetic reorientation of phyllosilicate minerals in Paleogene mudstones of the Podhale Basin, southern Poland. *Clays and Clay Minerals*, **56**, 100–111.

Day-Stirrat RJ, Dutton SP, Milliken KL, Loucks RG, Aplin AC, Hillier S, van der Pluijm BA (2010) Fabric anisotropy induced by primary depositional variations in the silt:clay ratio in two fined-grained slope fan complexes: Texas Gulf Coast and northern North Sea. *Sedimentary Geology*, **226**, 42–53.

De Borst, R (1987) Integration of plasticity equations for singular yield functions. *Computers & Structures*, **26**, 823–829.

de Dreuzy JR, de Boiry P, Pichot G, Davy P (2010) Use of power averaging for quantifying the influence of structure organization on permeability upscaling in on-lattice networks under mean parallel flow. *Water Resources Research*, **46**, doi:10.1029/2009WR008769.

Deichmann N, Ernst J (2009) Earthquake focal mechanisms of the induced seismicity in 2006 and 2007 below Basel (Switzerland). *Swiss Journal of Geosciences*, **102**, 457–466.

Deichmann N, Giardini D (2009) Earthquakes induced by the stimulation of an Enhanced Geothermal System below Basel (Switzerland). *Seismological Research Letters*, **80**, 784–798.

Delepine N, Cuenot N, Rothert E, Parotidis M, Rentsch S, Shapiro SA (2004) Characterization of fluid transport properties of the Hot Dry Rock reservoir Soultz-2000 using induced microseismicity. *Journal of Geophysics and Engineering*, **1**, 77–83.

DeMets C, Gordon RG, Argus DF (2010) Geologically current plate motions. *Geophysical Journal International*, **181**, 1–80.

Deming D (1993) Regional permeability estimates from investigations of coupled heat and groundwater flow, North Slope Alaska. *Journal of Geophysical Research*, **98**, 16271–16286.

Deming D, Craganu C, Lee Y (2002) Self-sealing in sedimentary basins. *Journal of Geophysical Research*, **107**, doi:10.1029/2001JB000504.

Dempsey D, Kelkar S, Lewis K, Hickman S, Davatzes N, Moos D, Zemach E (2013) Modeling shear stimulation of the EGS well Desert Peak 27-15 using a coupled thermal-hydrological-mechanical simulator. *47th US Rock Mechanics/Geomechanics Symposium*, American Rock Mechanics Association.

Derode B, Cappa F, Guglielmi Y, Rutqvist J (2013) Coupled seismo-hydromechanical monitoring of inelastic effects on injection-induced fracture permeability. *International Journal of Rock Mechanics and Mining Sciences*, **61**, 266–274.

Detournay E, Cheng AH-D (1993) Fundamentals of poroelasticity. In: *Comprehensive Rock Engineering: Principles, Practice and Projects, Vol. II, Analysis and Design Method* (ed Fairhurst C), pp. 113–171. Pergamon Press, Oxford.

Detrick RS, Buhl P, Mutter J, Orcutt J, Madsen J, Brocher T (1987) Multi-channel seismic imaging of a crustal magma chamber along the East Pacific Rise. *Nature*, **326**, 35–41.

Detwiler RL (2008) Experimental observations of deformation caused by mineral dissolution in variable-aperture fractures. *Journal of Geophysical Research*, **113**, B08202.

Detwiler RL, Pringle SE, Glass RJ (1999) Measurement of fracture aperture fields using transmitted light: an evaluation of measurement errors and their influence on simulations of flow and transport through a single fracture. *Water Resources Research*, **35**, 2605–2617.

Dewhurst DN, Aplin AC, Sarda JP (1999a) Influence of clay fraction on pore-scale properties and hydraulic conductivity of experimentally compacted mudstones. *Journal of Geophysical Research*, **104**, 29261–29274.

Dewhurst DN, Aplin AC, Sarda JP, Yang Y (1998) Compaction-driven evolution of porosity and permeability in natural mudstones: an experimental study. *Journal of Geophysical Research*, **103**, 651–661.

Dewhurst DN, Brown KM, Clennell MB, Westbrook GK (1996) A comparison of the fabric and permeability anisotropy of consolidated and sheared silty clay. *Engineering Geology*, **42**, 253–267.

Dewhurst DN, Yang Y, Aplin AC (1999b) Permeability and fluid flow in natural mudstones. *Geological Society of London Special Publication*, **158**, 23–43.

Dieterich JH (1992) Earthquake nucleation on faults with rate-and state-dependent friction. *Tectonophysics*, **211**, 115–134.

Dietrich HG (1982) Geological results of the Urach 3 Borehole and the correlation with other boreholes. In: *The Urach Geothermal Project* (ed Haenel R), pp. 49–58. Schweizerbart'sche Verlagsbuchhandlung, Stuttgart.

Di Federico V (1997) Estimates of equivalent aperture for non-Newtonian flow in a rough-walled fracture. *International Journal of Rock Mechanics and Mining Sciences*, **34**, 1133–1137.

Di Iorio D, Lavelle JW, Rona PA, Bemis K, Xu G, Germanovich LN, Lowell RP, Gene G (2012) Measurements and models of heat flux and plumes from hydrothermal discharges near the deep seafloor. *Oceanography*, **25**, 168–179.

Dilles JH (1987) The petrology of the Yerington Batholith, Nevada: evidence for the evolution of porphyry copper ore fluids. *Economic Geology*, **82**, 1750–1789.

Dingman SL (2002) *Physical Hydrology*, 2nd edn Waveland Press, Long Grove, IL.

Dixon TH, Moore JC (2007) The seismogenic zone of subduction thrust faults: introduction. In: *The Seismogenic Zone of Subduction Thrust Faults* (eds Dixon TH, Moore JC), pp. 42–85. Columbia University Press, New York.

Dobrovolsky IP, Zubkov SI, Miachkin VI (1979) Estimation of the size of earthquake preparation zones. *Pure and Applied Geophysics*, **117**, 1025–1044.

Doe TW, Korbin GE (1987) A comparison of hydraulic fracturing and hydraulic jacking stress measurements. *Proceedings of 28th U.S. Rock Mechanics Symposium*, Tucson, 283–90.

Domenico PA, Swartz FW (1990) *Physical and Chemical Hydrogeology*. John Wiley & Sons, New York.

Doyen PM (1988) Permeability, conductivity, and pore geometry of sandstone. *Journal of Geophysical Research*, **93**, 7729–7740.

Dowrick DJ (1996) The modified Mercalli earthquake intensity scale; revisions arising from recent studies of New Zealand earthquakes. *Bulletin of the New Zealand National Society for Earthquake Engineering*, **29**, 92–106.

Driesner T (2007) The system H_2O-NaCl, Part II, Correlations for molar volume, enthalpy, and isobaric heat capacity from 0 to 1000 degrees C, 1 to 5000 bar, and 0 to 1 X-NaCl. *Geochimica et Cosmochimica Acta*, **71**, 4902–4919.

Driesner T (2010) The interplay of permeability and fluid properties as a first-order control of heat transport, venting temperatures and venting salinities at mid-ocean ridge hydrothermal systems. *Geofluids*, **10**, 132–141.

Driesner T, Geiger S (2007) Numerical simulation of multiphase fluid flow in hydrothermal systems. In: *Fluid-Fluid Interactions*, Vol. 65 (eds Liebscher A, Heinrich CA), pp. 187–212, Mineralogical Society of America, Chantilly, VA.

Driesner T, Heinrich CA (2007) The system H_2O-NaCl, Part I, Correlation formulae for phase relations in temperature-pressure-composition space from 0 to 1000 degrees C, 0 to 5000 bar, and 0 to 1 X-NaCl. *Geochimica et Cosmochimica Acta*, **71**, 4880–4901.

Drost BW, Whiteman KJ, Gonthier JB (1990) Geologic framework of the Columbia Plateau Aquifer System, Washington, Oregon, and Idaho. *U.S. Geological Survey Water-Resources Investigations Report*, **87-4238**, http://pubs.er.usgs.gov/usgspubs/wri/wri874238 (accessed 06 May 2016).

du Pont SC, Gondret P, Perrin B, Rabaud M, (2003) Granular avalanches in fluids. *Physical Review Letters*, **90**, 044301.

Duba AG, Durham WB, Handin JW, Wang HF (eds) (1990) The brittle-ductile transition in rocks – The Heard volume. *American Geophysical Union Geophysical Monograph*, **56**.

Dugan B, Daigle H (2011) Data report: Permeability, compressibility, stress state, and grain size of shallow sediments from Sites C0004, C0006, C0007, and C0008 of the Nankai accretionary complex. In Kinoshita M, Tobin H, Ashi J, Kimura G, Lallemant S, Screaton EJ, Curewitz D, Masago H, Moe KT, and the Expedition 314/315/316 Scientists, *Proceedings of the Integrated Ocean Drilling Program*, **314/315/316**,1–11.

Dugan B, Sheahan TC (2012) Offshore sediment overpressures of passive margins: mechanisms, measurement, and models. *Reviews of Geophysics*, **50**, RG3001.

Dugan B, Zhao X (2013) Data report: Permeability of sediments from Sites C0011 and C0012, NanTroSEIZE, Stage 2, Subduction inputs. In Saito S, Underwood MB, Kubo Y, and the Expedition 322 Scientists, *Proceedings of the Integrated Ocean Drilling Program*, **322**, 1–14.

Dunbar N (2005) Quaternary volcanism in New Mexico. *New Mexico Museum of Natural History and Science Bulletin*, **28**, 95–106.

Durham WB (1997) Laboratory observations of the hydraulic behavior of a permeable fracture from 3800 m depth in the KTB pilot hole. *Journal of Geophysical Research*, **102**, 18405–18416.

Durham WB, Bonner BP (1994) Self-propping and fluid flow in slightly offset joints at high effective pressures. *Journal of Geophysical Research*, **99**, 9391–9399.

Durham WB, Bourcier WL, Burton EA (2001) Direct observation of reactive flow in a single fracture. *Water Resources Research*, **37**, doi:10.1029/2000WR900228.

Durney DW, Ramsay JG (1973) Incremental strains measured by syntectonic crystal growths. In: *Gravity and Tectonics* (eds DeJong KA, Scholten R), pp. 67–96. Wiley, New York.

Dussan EB, Sharma Y (1992) Analysis of the pressure response of a single-probe formation tester. *SPE Formation Evaluation*, **7**, 151–156.

Dusseault M, McLennan J (2011) Massive multistage hydraulic fracturing: where are we? In: 45th US Rock Mechanics/Geomechanics Symposium, June 26–29, 2011, San Francisco, California.

Dziak RP, Chadwick WW, Christopher GG, Embley RW (2003) Hydrothermal temperature changes at the southern Juan de Fuca Ridge associated with MW 6.2 Blanco Transform earthquake. *Geology*, **31**, 119–122.

Earnest E, Boutt D (2014) Investigating the role of hydromechanical coupling on flow and transport in shallow, fractured rock aquifers. *Hydrogeology Journal*, **22**, 1473–1491.

Eaton D, van der Baan M, Birkelo B, Tary J-B (2014) Scaling relations and spectral characteristics of tensile microseisms: evidence for opening/closing cracks during hydraulic fracturing. *Geophysical Journal International*, **196**, 1844–1857.

Edmunds WM, Savage D (1991) Geochemical characteristics of groundwater in granites and related crystalline rocks. In: *Applied Groundwater Hydrology, a British Perspective* (eds Downing RA, Wilkinson WB), pp. 199–216. Clarendon Press, Oxford.

Edwards S, Hudson-Edwards K, Cann J, Malpas J, Xenophontos C (2010) *Classic Geology in Europe, 7, Cyprus*. Terra Publishing, Harpenden, UK.

Eggleston C, Rojstaczer S (1998) Identification of large-scale hydraulic conductivity trends and the influence of trends on contaminant transport. *Water Resources Research*, **34**, 2155–2168.

Ehrenberg SN, Nadeau PH (2005) Sandstone vs. carbonate petroleum reservoirs: a global perspective on porosity-depth and porosity-permeability relationships. *AAPG Bulletin*, **89**, 435–445.

Ekinci MK (2012) Permeability, clay mineralogy, and microfabric of fine grained sediments from the Nankai Trough and Shikoku Basin, offshore southwest Japan. MS thesis, University of Missouri.

Ekinci MK, Likos WJ, Underwood MB, Guo J (2011) Data report: Permeability of mud(stone) samples from IODP Sites C0006 and C0007, Nankai Trough Seismogenic Zone Experiment. In Kinoshita M, Tobin H, Ashi J, Kimura G, Lallemant S, Screaton EJ, Curewitz D, Masago H, Moe KT, and the Expedition 314/315/316 Scientists, *Proceedings of the Integrated Ocean Drilling Program*, **314/315/316**, 1–40.

Elders WA, Frioleifsson GO, Zierenberg RA, Pope EC, Mortensen AK, Guomundsson A, Lowenstern JB, Marks NE, Owens L, Bird DK, Reed M, Olsen NJ, Schiffman P (2011) Origin of a rhyolite that intruded a geothermal well while drilling at the Krafla volcano, Iceland. *Geology*, **39**, 231–234.

Elkhoury JE, Brodsky EE, Agnew DC (2006) Seismic waves increase permeability. *Nature*, **441**, 1135–1138.

Elkhoury JE, Niemeijer A, Brodsky EE, Marone C (2011) Laboratory observations of permeability enhancement by fluid pressure oscillation of in situ fractured rock. *Journal of Geophysical Research*, **116**, B02311, doi:10.1029/2010JB007759.

Elliot GM, Brown ET, Boodt PI, Hudson JA (1985) Hydrochemical behaviour of joints in the Carmenelis granite SW England. *Proceedings of the International Symposium on Fundamentals of Rock Joints*, Bjorkliden, Sweden, A. A. Balkema, Rotterdam, 249–258.

Ellis AJ, Mahon WAJ (1964) Natural hydrothermal systems and experimental hot water/rock interactions. *Geochimica et Cosmochimica Acta*, **28**, 1323–1357.

Ellis AJ, Mahon WAJ (1967) Natural hydrothermal systems and experimental hot water/rock interactions, Part II. *Geochimica et Cosmochimica Acta*, **31**, 519–538.

Ellis S, Beavan J, Eberhart-Phillips D, Stöckhert B (2006) Simplified models of the Alpine Fault seismic cycle: stress transfer in the mid-crust. *Geophysical Journal International*, **166**, 386–402.

Ellsworth WL (2013) Injection-induced earthquakes. *Science*, **341**, 142.

Elsworth D, Goodman RE (1986) Characterization of rock fissure hydraulic conductivity using idealized wall roughness profiles. *International Journal of Rock Mechanics and Mining Sciences*, **23**, 233–243.

Ely DM, Burns ER, Morgan DS, Vaccaro JJ (2014) Numerical simulation of groundwater flow in the Columbia Plateau Regional Aquifer System, Idaho, Oregon, and Washington. *U.S. Geological Survey Scientific Investigations Report*, **2014–5127**.

Emter D, Wenzel HG, Zürn W (1999) Das Observatorium Schiltach. *DGG Mitteilungen*, **3**, 1–15.

Engvik A., Ihlen PM, Austrheim A. (2014) Characterisation of Na-metasomatism in the Sveconorwegian Bamble sector of south Norway. *Geoscience Frontiers*, **5**, 659–672.

Erzinger J, Stober I (2005) Introduction to special issue: long-term fluid production in the KTB pilot hole, Germany. *Geofluids*, **5**, 1–7.

Esaki T, Du S, Mitani Y, Ikusada K, Jing L (1999) Development of a shear-flow test apparatus and determination of coupled properties for a single rock joint. *International Journal of Rock Mechanics and Mining Sciences*, **36**, 641–650.

Etheridge MA, Wall VJ, Cox SF, Vernon RH (1984) High fluid pressures during regional metamorphism and deformation – Implications for mass-transport and deformation mechanisms. *Journal of Geophysical Research*, **89**, 4344–4358.

Evans JP, Forster CB, Goddard JV (1997) Permeability of fault-related rocks, and implications for hydraulic structure of fault zones. *Journal of Structural Geology*, **19**, 1393–1404.

Evans K, Wyatt F (1984) Water table effects on the measurement of earth strain. *Tectonophysics*, **108**, 323–337.

Evans KF (2005) Permeability creation and damage due to massive fluid injections into granite at 3.5 km at Soultz, 2, Critical stress and fracture strength. *Journal of Geophysical Research*, **110**, B04204, doi:10.1029/2004JB003169.

Evans KF, Genter A, Sausse J (2005) Permeability creation and damage due to massive fluid injections into granite at 3.5 km at Soultz, 1, Borehole observations. *Journal of Geophysical Research*, **110**, B04203, doi:10.1029/2004JB003168.

Fairhurst C (2013) Fractures and fracturing: hydraulic fracturing in jointed rock. In: *Effective and Sustainable Hydraulic Fracturing* (eds Bunger A, McLennan J, Jeffrey R), pp. 47–79, InTech, Rijeka, Croatia.

Fairley JP, Ingebritsen SE, Podgorney RK (2010) Challenges for numerical modeling of enhanced geothermal systems. *Ground Water*, **48**, 482–483.

Fan Y, Richard S, Bristol RS, Peters SE, Ingebritsen SE, Moosdorf N, Packman A, Gleeson T, Zaslavsky I, Peckham S, Murdoch L, Fienen M, Cardiff M, Tarboton D, Jones N, Hooper R, Arrigo J, Gochis D, Olson J, Wolock D (2015) DigitalCrust – A 4D data system of material properties for transforming research on crustal fluid flow. *Geofluids*, **15**, 372–379.

Fardin PC (2003) The effect of scale on the morphology, mechanics and transmissivity of single rock fractures, structural non-stationarity on the surface roughness and the shear strength of rock fractures. PhD thesis, Royal Institute of Technology (KTH), Stockholm, Sweden.

Faulder DD, Johnson SD, Benoit WR (1997) Flow and permeability structure of the Beowawe, Nevada hydrothermal system. Proceedings, Twenty-Second Workshop on Geothermal Research and Engineering, Stanford, California, *Stanford Geothermal Program Technical Report*, **SGP-TR-155**, 63–73.

Faulds JE, Hinz NH, Coolbaugh MF, Cashman PH, Kratt C, Dering G, Edwards J, Mayhew B, McLachlan H (2011) Assessment of favorable structural settings of geothermal systems in the Great Basin, western USA. *Geothermal Research Council Transactions*, **35**, 777–783.

Faulkner DR, Jackson CAL, Lunn RJ, Schlische RW, Shipton ZK, Wibberley CAJ, Withjack MO (2010) A review of recent developments concerning structure, mechanics and fluid flow properties of fault zones. *Journal of Structural Geology*, **32**, 1557–1575.

Faulkner DR, Rutter EH (2001) Can the maintenance of overpressured fluids in large strike-slip fault zones explain their apparent weakness? *Geology*, **29**, 503–506.

Faust CR, Mercer JW (1979) Geothermal reservoir simulation, 1, Mathematical models for liquid- and vapor-dominated hydrothermal systems. *Water Resources Research*, **15**, 23–29.

Favara R, Grassa F, Inguaggiato S, Valenza M (2001) Hydrogeochemistry and stable isotopes of thermal springs: earthquake-related chemical changes along Belice Fault (western Sicily). *Applied Geochemistry*, **16**, 1–17.

Ferguson CA, Osburn GR, McIntosh WC (2012) Oligocene calderas in the San Mateo Mountains, Mogollon-Datil volcanic field, New Mexico. *New Mexico Geological Society Guidebook*, **63**, 74–77.

Ferris JG (1951) Cyclic fluctuations of water level as a basis for determining aquifer transmissivity. *International Association of Hydrological Sciences (IAHS)*, **33**, 148–155.

Ferry JM (1987) Metamorphic hydrology at 13-km depth and 400–500°C. *American Mineralogist*, **72**, 39–58.

Ferry JM (1994) Overview of the petrologic record of fluid flow during regional metamorphism in northern New England. *American Journal of Science*, **294**, 905–988.

Fetter CW (1994) *Applied Hydrogeology*, 3rd edn Prentice Hall, Englewood Cliffs, NJ.

Filgris MN (2001) Römische Baderuine Badenweiler. Historische Wurzeln des Kurortes neu prasentiert. *Denkmalsplege in Baden-Wurttemberg*, **4**, 166–175.

Finnegan S, Peters SE, Fischer WW (2011) Late Ordovician-Early Silurian selective extinction patterns in Laurentia and their relationship to climate change. In: *Ordovician of the World* (eds Gutiérrez-Marco JC, Rábano I, Garcia-Bellido D), pp. 155–159. *Cuadernos del Museo Geominera*, **14**, Madrid, Spain.

Fisher AT, Hounslow MW (1990) Transient fluid flow through the toe of the Barbados accretionary complex: constraints from Ocean Drilling Program Leg 110, heat flow studies and simple models. *Journal of Geophysical Research*, **95**, 8845–8858.

Fisher AT, Zwart G (1997) Packer experiments along the décollement of the Barbados accretionary complex: measurements of in situ permeability. In: *Proceedings of the Ocean Drilling Program, Scientific Results*, **156**, (eds Shipley TH, Ogawa Y, Blum P, Bahr JM), pp. 199–218, Ocean Drilling Program, College Station, TX.

Fisher AT, Zwart G, the Ocean Drilling Program Leg 156 Scientific Party (1996) Relation between permeability and effective stress along a plate boundary fault. Barbados accretionary complex. *Geology*, **24**, 307–310.

Fitts TG, Brown KM (1999) Stress-induced smectite dehydration: ramifications for patterns of freshening and fluid expulsion in the N. Barbados accretionary wedge. *Earth and Planetary Science Letters*, **172**, 179–197.

Fogg G, Senger RL (1985) Automatic generation of flow nets with conventional ground-water modeling algorithms. *Groundwater*, **23**, 336–344.

Fogg GE, LaBolle EM (2006) Motivation of synthesis, with an example on groundwater quality sustainability. *Water Resources Research*, **42**, W03S05.

Folk RL (1966) A review of grain-size parameters. *Sedimentology*, **6**, 73–93.

Fonneland-Jorgensen H, Furnes H, Muehlenbachs K, Dilek Y (2005) Hydrothermal alteration and tectonic evolution of an intermediate- to fast-spreading back-arc oceanic crust: late Ordovician Solund-Stavfjord ophiolite, western Norway. *Island Arc*, **14**, 517–541.

Fontaine FJ, Wilcock WSD, Foustoukos DE, Butterfield DA (2009) A Si-Cl geothermobarometer for the reaction zone of high-temperature, basaltic-hosted midocean ridge hydrothermal systems. *Geochemistry, Geophysics, Geosystems*, **10**, Q05009.

Ford A, Blenkinsop TG, McLellan JG (2009) Factors affecting fluid flow in strike-slip fault systems: coupled deformation and fluid flow modeling with application to the western Mount Isa inlier, Australia. *Geofluids*, **9**, 2–23.

Fornari DJ, Shank T, Von Damm KL, Gregg TKP, Lilley M, Levai G, Bray A, Haymon RM, Perfit MR, Lutz R (1998) Time-series temperature measurements at high-temperature hydrothermal vents, East Pacific Rise 9°49'-51'N: evidence for monitoring a crustal cracking event. *Earth and Planetary Science Letters*, **160**, 419–431.

Forster C, Smith L (1988a) Groundwater flow systems in mountainous terrain, 1, Numerical modeling technique. *Water Resources Research*, **24**, 999–1010.

Forster C, Smith L (1988b) Groundwater flow systems in mountainous terrain, 2, Controlling factors. *Water Resources Research*, **24**, 1011–1023.

Forster C, Smith L (1989) The influence of groundwater flow on thermal regimes in mountainous terrain: a model study. *Journal of Geophysical Research*, **94**, 9439–9451.

Forsyth DW, Scheirer DS, Webb SC, Dorman LM, Orcutt JA, Harding AJ, Blackman DK, Morgan JP, Detrick RS, Shen Y, Wolfe CJ, Canales JP, Toomey DR, Sheehan AF, Solomon SC, Wilcock WSD, Team MS (1998) Imaging the deep seismic structure beneath a mid-ocean ridge: the MELT experiment. *Science*, **280**, 1215–1218.

Fossen H, Schultz RA, Runhovde E, Rotevatn A, Buckley SJ (2010) Fault linkage and graben stopovers in the Canyonlands (Utah) and the North Sea Viking Graben, with implications for hydrocarbon migration and accumulation. *American Association of Petroleum Geologists Bulletin*, **94**, 597–613.

Foster SSD, Chilton PJ (2003) Groundwater: the processes and global significance of aquifer degradation. *Philosophical Transactions of the Royal Society of London. Series B, Biological Sciences*, **358**, 1957–1972.

Fournier RO (1979) A revised equation for the Na/K geothermometer. *Geothermal Resources Council Transactions*, **3**, 221–224.

Fournier RO (1981) Application of water geochemistry to geothermal exploration and reservoir engineering. In: *Geothermal Systems: Principles and Case Histories* (eds Ryback L, Muffler LJP), pp. 109–143. John Wiley and Sons, New York.

Fournier RO (1989) Lectures on geochemical interpretation of hydrothermal waters. *UNU Geothermal Training Programme, Reykjavik Iceland*, **10**.

Fournier RO (1991) The transition from hydrostatic to greater than hydrostatic fluid pressure in presently active continental hydrothermal systems in crystalline rock. *Geophysical Research Letters*, **18**, 955–958.

Fournier RO (1999) Hydrothermal processes related to movement of fluid from plastic into brittle rock in the magmatic-epithermal environment. *Economic Geology*, **94**, 1193–1211.

Fournier RO, Potter RW II (1982) A revised and expanded (quartz) geothermometer. *Geothermal Resources Council Bulletin*, **11**, 3–12.

Fowler AC (1984) A mathematical model of magma transport in the asthenosphere. *Geophysical and Astrophysical Fluid Dynamics*, **33**, 155–190.

Fowler AC, Yang X (1999) Pressure solution and viscous compaction in sedimentary basins. *Journal of Geophysical Research*, **104**, 12989–12997.

Francheteau J, Armijo R, Cheminee JL, Hekinian R, Lonsdale P, Blum N (1992) Dike complex of the East Pacific Rise exposed in the walls of Hess Deep and the structure of the upper oceanic crust. *Earth and Planetary Science Letters*, **111**, 109–121.

Frape SK, Fritz P (1987) Geochemical trends for groundwaters from the Canadian shield. In: *Saline Water and Gases in Crystalline Rocks* (eds Fritz P, Frape SK), pp. 19–38. *Geological Association of Canada Special Paper*, **33**, The Runge Press Limited, Ottawa, ON.

Freeze RA, Cherry JA (1979) *Groundwater*. Prentice-Hall, Englewood Cliffs, NJ.

Fridleifsson GO, Elders WA (2005) The Iceland Deep Drilling Project: a search for deep unconventional geothermal resources. *Geothermics*, **34**, 269–285.

Friedrich AM, Lee J, Wernicke BP, Sieh K (2004) Geologic context of geodetic data across a Basin and Range normal fault, Crescent Valley, Nevada. *Tectonics*, **23**, 1–24.

Frisch U, d'Humieres D, Hasslacher B, Lallemand P, Pomeau Y, Rivet J-P (1987) Lattice gas hydrodynamics in two and three dimensions. *Complex Systems*, **1**, 649–707.

Frisch U, Hasslacher B, Pomeau Y (1986) Lattice-gas automata for the Navier-Stokes equation. *Physical Review Letters*, **56**, 1505–1508.

Fritz DE, Farmer GL, Verplanck EP (2006) Application of Sr isotopes in secondary silicate minerals to paleogroundwater hydrology: an example from K-metasomatized rocks in the western US. *Chemical Geology*, **235**, 276–285.

Frohlich C, Ellsworth W, Brown WA, Brunt M, Luetgert J, MacDonald T, Walter S (2014) The 17 May 2012 M4.8 earthquake near Timpson, East Texas: an event possibly triggered by fluid injection. *Journal of Geophysical Research*, **119**, 581–593.

Frost BR, Bucher K (1994) Is water responsible for geophysical anomalies in the deep continental crust? A petrological perspective. *Tectonophysics*, **231**, 293–309.

Fry B, Bannister SC, Beavan RJ, Bland L, Bradley BA, Cox SC, Cousins WJ, Gale NH, Hancox GT, Holden C, Jongens R, Power WL, Prasetya G, Reyners ME, Ristau J, Robinson R, Samsonov S, Wilson KJ, team G (2010) The M_w 7.6 Dusky Sound earthquake of 2009: preliminary report. *Bulletin of the New Zealand Society for Earthquake Engineering*, **43**, 24–40.

Fuis GS, Moore TE, Plafker G, Brocher TM, Fisher MA, Mooney WD, Nokleberg WJ, Page RA, Beaudoin BC, Christensen NI, Levander AR, Lutter WJ, Saltus RW, Ruppert NA (2008) Trans-Alaska crustal transect and continental evolution involving subduction underplating and synchronous foreland thrusting. *Geology*, **36**, 267–270.

Furlong KP, Hanson RB, Bowers JR (1991) Modeling thermal regimes. *Reviews in Mineralogy*, **26**, 437–505.

Fusseis F, Handy MR (2008) Micromechanisms of shear zone propagation at the brittle-viscous transition. *Journal of Structural Geology*, **30**, 1242–1253.

Fusseis F, Handy, MR, Schrank C (2006) Networking of shear zones at the brittle-to-viscous transition (Cap de Creus, NE Spain). *Journal of Structural Geology*, **28**, 1228–1243.

Fyfe WS, Price NJ, Thompson AB (1978) *Fluids in the Earth's Crust*. Elsevier Scientific, New York.

Gale J (1990) Hydraulic behaviour of rock joints. *Conference on Rock Joints, Proceedings of the ISRM*, 351–362.

Gale JE, Wilson CR, Witherspoon PA, Wilson CR (1982) Swedish-American cooperative program on radioactive waste storage in mined caverns in crystalline rock. In: *SKB Technical Report*, 49 (ed Swedish Nuclear Fuel Supply Co & Lawrence Berkeley Laboratory), National Technical Information Service, U.S. Department of Commerce, Springfield, VA, http://esd.lbl.gov/files/publications/stripa_reports/Technical_Project_Report_No_49.pdf (accessed 06 May 2016).

Galet RM (2011) Thermo-hydro-mechanical study of deformable porous media with double porosity in local thermal non-equilibrium. PhD thesis, University of New South Wales.

Gallahan WE, Duncan RA (1994) Spatial and temporal variability in crystallisation of celadonites within the Troodos ophiolite, Cyprus: implications for low-temperature alteration of the oceanic crust. *Journal of Geophysical Research*, **99**, 3147–3161.

Galvan B, Miller SA (2013) A full GPU simulation of evolving fracture networks in a heterogeneous poro-elasto-plastic medium with effective-stress-dependent permeability. In: *GPU Solutions to Multi-scale Problems in Science and Engineering* (eds Yuen DA, Wong L, Chi X, Johnsson L, Ge W, Shi Y), pp. 305–319, Springer, Berlin, Heidelberg.

Gamage K, Bekins B, Screaton E (2005) Data report: Permeabilities of eastern equatorial Pacific and Peru margin sediments. *Proceedings of the Ocean Drilling Program, Scientific Results*, **201**, 1–18.

Gamage K, Screaton E (2003) Data report: Permeabilities of Nankai accretionary prism sediments. *Proceeding of the Ocean Drilling Program, Scientific Results*, **190/196**, 1–21.

Gamage K, Screaton E (2006) Characterization of excess pore pressures at the toe of the Nankai accretionary complex, Ocean Drilling Program Sites 1173, 1174, and 808: results of one-dimensional modeling. *Journal of Geophysical Research*, **111**, B04103.

Gamage K, Screaton E, Bekins B, Aiello I (2011) Permeability-porosity relationships of subduction zone sediments. *Marine Geology*, **279**, 19–36.

Gangi AF (1978) Variation of whole and fractured porous rock permeability with confining pressure. *International Journal of Rock Mechanics and Mining Sciences*, **15**, 249–257.

Ganguly J (2002) Diffusion kinetics in minerals: principles and applications to tectono-metamorphic processes. *EMU Notes in Mineralogy*, **4**, 271–309.

Garg SK, Pritchett JW (1977) On pressure-work, viscous dissipation and the energy balance relation for geothermal reservoirs. *Advances in Water Resources*, **1**, 41–47.

Garg SK, Pritchett JW, Wannamaker PE, Combs J (2007) Characterization of geothermal reservoirs with electrical surveys: Beowawe geothermal field. *Geothermics*, **36**, 487–517.

Garrels RM, Howard PF (1959) Reactions of feldspar and mica with water at low temperature and pressure. *Clays and Clay Minerals*, **6**, 68–88.

Garven G, Bull SW, Large RR (2001) Hydrothermal fluid flow models of stratiform ore genesis in the McArthur Basin, Northern Territory, Australia. *Geofluids*, **1**, 289–311.

Garza RAP, Titley SR, Pimentel FB (2001) Geology of the Escondida porphyry copper deposit, Antofagasta region, Chile. *Economic Geology*, **96**, 307–324.

Gassiat C, Gleeson T, Luijendijk E (2013) The location of old groundwater in hydrogeologic basins and layered aquifer systems. *Geophysical Research Letters*, **40**, 3042–3047.

Gavrilenko P, Gueguen Y (1993) Fluid overpressures and pressure solution in the crust. *Tectonophysics*, **21**, 91–110.

Ge S (1997) A governing equation for fluid flow in rock fractures. *Water Resources Research*, **33**, 53–61.

Ge S, Stover SC (2000) Hydrodynamic response to strike- and dip-slip faulting in a half-space. *Journal of Geophysical Research*, **105**, 25513–25524.

Genser J, Cloetingh SAPL, Neubauer F (2007) Late orogenic rebound and oblique Alpine convergence: new constraints from subsidence analysis of the Austrian Molasse basin. *Global and Planetary Change*, **58**, 214–223.

Genter A, Evans K, Cuenot N, Fritsch D (2010) Contribution of the exploration of deep crystalline fractured reservoir of Soultz to the knowledge of enhanced geothermal systems (EGS). *Comptes Rendus Geoscience*, **342**, 502–516.

Geological Survey of Japan (2013) Volcanoes of Japan. Retrieved January 10, 2014, from https://gbank.gsj.jp/volcano/.

GeoNet (2014) Earthquake strong motion data and processing. http://info.geonet.org.nz/display/appdata/Strong-Motion+Data, accessed 15 February 2014.

Gerdes M, Baumgartner L, Person M (1995) Permeability heterogeneity in metamorphic rocks: implications from stochastic modeling. *Geology*, **23**, 945–948.

German CR, Lin J (2004) The thermal structure of the oceanic crust, ridge-spreading and hydrothermal circulation: how well do we understand their inter-connections? In: *Mid-Ocean Ridges: Hydrothermal Interactions between the Lithosphere and Oceans* (eds German CR, Lin J, Parson LM), *AGU Geophysical Monograph Series*, **148**, 1–18.

Giacomini A, Buzzi O, Ferrero AM, Migliazza M, Giani GP (2007) Numerical study of flow anisotropy within a single natural rock joint. *International Journal of Rock Mechanics and Mining Sciences*, **45**, 47–58.

Gieskes JM, Vrolijk P, Blanc G (1990) Hydrogeochemistry, ODP Leg 110: an overview. In: *Proceedings of the Ocean Drilling Program, Scientific Results*, **110**, (eds Moore JC, Mascle A, Leg 110 Scientific Party), pp. 395–408, Ocean Drilling Program, College Station, TX.

Giger SB, Tenthorey E, Cox SF, Fitz Gerald JD (2007) Permeability evolution in quartz fault gouges under hydrothermal conditions. *Journal of Geophysical Research*, **112**, doi:10.1029/2006JB004828.

Giggenbach WF (1986) Graphical techniques for the evaluation of water/rock equilibrium conditions by use of Na, K, Mg and Ca contents of discharge waters. *Proceedings of the Eighth New Zealand Geothermal Workshop*, Auckland, New Zealand, 37–43.

Giggenbach WF (1988) Geothermal solute equilibria: derivation of Na-K-Mg-Ca geoindicators. *Geochimica Cosmochimica Acta*, **52**, 2749–2765.

Giggenbach WF (1991) Chemical techniques in geothermal exploration. *UNITAR/UNDP Guidebook: Application of Geochemistry in Resources Development*, 119–144.

Giggenbach WF, Sano Y, Wakita H (1993) Isotopic composition of helium, and CO_2 and CH_4 contents in gases produced along the New Zealand part of a convergent plate boundary. *Geochimica et Cosmochimica Acta*, **57**, 3427–3455.

Gilmer A, Mauldin R, Keller G (1986) A gravity study of the Jornado del Muerto and Palomas Basins. *New Mexico Geological Society Guidebook*, **37**, 131–134.

Gingerich SB, Voss CI (2005) Three-dimensional variable-density flow simulation of a coastal aquifer in southern Oahu, Hawaii, USA. *Hydrogeology Journal*, **13**, 436–450.

Giordano M (2009) Global groundwater? Issues and solutions. *Annual Review of Environment and Resources*, **34**, 153–178.

Gischig VS, Wiemer S (2013) A stochastic model for induced seismicity based on non-linear pressure diffusion and irreversible permeability enhancement. *Geophysical Journal International*, **194**, 1229–1249.

Gledhill KR, Ristau J, Reyners ME, Fry B, Holden C (2011) The Darfield (Canterbury, New Zealand) M_w7.1 earthquake of September 2010: a preliminary seismological report. *Seismological Research Letters*, **82**, 378–386.

Gleeson SA, Yardley BWD, Munz IA (2003) Infiltration of basinal fluids into high-grade basement, South Norway: sources and behaviour of waters and brines. *Geofluids*, **3**, 33–48.

Gleeson T, Alley WM, Allen DM, Sophocleous MA, Zhou Y, Taniguchi M, Van der Steen J (2012a) Towards sustainable groundwater use: setting long-term goals, backcasting, and managing adaptively. *Ground Water*, **50**, 19–26.

Gleeson T, Moosdorf N, Hartmann J, van Beek LPH (2014) A glimpse beneath Earth's surface: GLobal HYdrogeology MaPS (GLHYMPS) of permeability and porosity. *Geophysical Research Letters*, **41**, 3891–3898.

Gleeson T, Smith L, Moosdorf N, Hartmann J, Dürr H, Manning A, Beek L, Jellinek A (2011) Mapping permeability over the surface of the Earth. *Geophysical Research Letters*, **38**, doi:10.1029/2010GL045565.

Gleeson T, Van der Steen J, Sophocleous MA, Taniguchi M, Alley WM, Allen DM, Zhou Y (2010) Groundwater sustainability strategies. *Nature Geoscience*, **3**, 378–379.

Gleeson T, Wada Y, Bierkens MFP, van Beek LPH (2012b) Water balance of global aquifers revealed by groundwater footprint. *Nature*, **488**, 197–200.

Gliko AO, Singh RN, Swathi PS (1999) Physical approach to the problem of origin of charnockitic rocks of southern India: mechanisms of crustal heating and transfer of carbon dioxide. *Russian Journal of Earth Sciences*, **1**, 409–421.

Goertz-Allmann BP, Goertz A, Wiemer S (2011) Stress drop variations of induced earthquakes at the Basel geothermal site. *Geophysical Research Letters*, **38**, doi:10.1029/2011GL047498.

Gómez-Hernández JJ, Gorelick SM (1989) Effective groundwater model parameter values: influence of spatial variability of hydraulic conductivity, leakance, and recharge. *Water Resources Research*, **25**, 405–419.

Goodman, R (1976) *Methods of Geological Engineering in Discontinuous Rocks*. West Publishing, Eagan, MN.

Gorokhovich Y, Fleeger G (2007) Pymatuning earthquake in Pennsylvania and Late Minoan crisis on Crete. *Water Science & Technology: Water Supply*, **7**, 245–251.

Graham DW, Christie DM, Harpp KS, Lupton JE (1993) Mantle-plume helium in submarine basalts from the Galapagos Platform. *Science*, **262**, 2023–2026.

Gratier JP, Favreau P, Renard F (2003) Modeling fluid transfer along California faults when integrating pressure solution crack sealing and compaction processes. *Journal of Geophysical Research*, **108**, B02104, doi:10.1029/2001JB000380.

Griffiths FJ, Joshi RC (1989) Change in pore size distribution due to consolidation of clays. *Géotechnique*, **39**, 159–167.

Gruen G, Weis P, Driesner T, Heinrich CA, de Ronde CEJ (2014) Hydrodynamic modeling of magmatic–hydrothermal activity at submarine arc volcanoes, with implications for ore formation. *Earth and Planetary Science Letters*, **404**, 307–318.

Gueguen Y, Dormieux L, Bouteca M (2004) Fundamentals of poromechanics. In: *Mechanics of Fluid-Saturated Rocks* (eds Gueguen Y, Bouteca M), pp. 55–79. Elsevier Academic Press, Burlington, MA.

Guglielmi Y, Cappa F, Lane H, Janowczyk JB, Rutqvist J, Tsang C-F, Wang JSY (2014) ISRM suggested method for step-rate injection method for fracture in-situ properties (SIMFIP): using a 3-component borehole deformation sensor. *Rock Mechanics and Rock Engineering*, **47**, 303–311.

Gulley AK, Ward NFD, Cox SC, Kaipio JP (2013) Groundwater responses to the recent Canterbury earthquakes: a comparison. *Journal of Hydrology*, **504**, 171–181.

Guo J, Underwood MB (2014) Data report: Consolidation characteristics of sediments from Sites C0011, C0012, and C0018, IODP Expeditions 322 and 333, NanTroSEIZE Stage 2. In: Saito S, Underwood MB, Kubo Y, and the Expedition 322 Scientists, *Proceedings of the Integrated Ocean Drilling Program*, **322**.

Gürbüz A (2010) Geometric characteristics of pull-apart basins. *Lithosphere*, **2**, 199–206.

Gustafson LB, Hunt JP (1975) Porphyry copper deposit at El Salvador, Chile. *Economic Geology*, **70**, 857–912.

Haar L, Gallagher JS, Kell GS (1984) *NBS/NRC Steam Tables: Thermodynamic and Transport Properties and Computer Programs for Vapor and Liquid States of Water in SI Units*. Hemisphere Publishing, New York.

Hacker BR (2008) H_2O subduction beyond arcs. *Geochemistry, Geophysics, Geosystems*, **9**, Q03001.

Häfner F, Lauterbach M, Bamberg HF (1985) Physikalische Eigenschaften von Grund- und mineralisiertem Wasser, von Erdöl und Erdgas. *Wissenschaftlich-Technischer Informationsdienst*, **26**, 18–27.

Hagemann SG, Groves DI, Ridley JR, Vearncombe JR (1992) The Archean lode gold deposits at Wiluna, Western Australia: high-level, brittle-style mineralization in a strike-slip regime. *Economic Geology*, **87**, 1022–1053.

Haimson BC (1975) The state of stress in the Earth's crust. *Reviews of Geophysics*, **13**, 350–352.

Haines SH, van der Pluijm BA, Ikari MJ, Saffer DM, Marone C (2009) Clay fabric intensity in natural and artificial fault gouges: implications for brittle fault zone processes and sedimentary basin clay fabric evolution. *Journal of Geophysical Research*, **114**, B05406.

Hakami E, Larsson E (1996) Aperture measurement and flow experiments on a single natural fracture. *International Journal of Rock Mechanics and Mining Sciences*, **33**, 395–404.

Halevy I, Peters SE, Fischer WW (2012) Sulfate burial constraints on the Phanerozoic sulfur cycle. *Science*, **337**, 331–334.

Hammerschmidt SB, Davis EE, Hupers A, Kopf A (2013) Limitation of fluid flow at the Nankai Trough megasplay fault zone. *Geo-Marine Letters*, **33**, 405–408.

Hancox GT, Cox SC, Turnbull IM, Crozier MJ (2003) Reconnaissance studies of landslides and other ground damage caused by the M_w7.2 Fiordland earthquake of 22 August 2003. *Institute of Geological & Nuclear Sciences Science Report*, **2003/30**.

Hanks TC, Kanamori H (1979) A moment magnitude scale. *Journal of Geophysical Research*, **84**, 2348–2350.

Hannisdal B, Peters SE (2011) Phanerozoic earth system evolution and marine biodiversity. *Science*, **334**, 1121–1124.

Hans J, Boulon MJ (2003) A new device for investigating the hydro-mechanical properties of rock joints. *International Journal for Numerical and Analytical Methods in Geomechanics*, **27**, 513–548.

Hansen AJ, Vaccaro JJ, Bauer HH (1994) Ground-water flow simulation of the Columbia Plateau Regional Aquifer System, Washington, Oregon, and Idaho. *U.S. Geological Survey Water-Resources Investigations Report*, **91–4187**, http://pubs.er.usgs.gov/usgspubs/wri/wri914187 accessed 06 May 2016.

Hanson RB (1992) Effects of fluid production on fluid-flow during regional and contact metamorphism. *Journal of Metamorphic Geology*, **10**, 87–97.

Hanson RB (1995) The hydrodynamics of contact metamorphism. *Geological Society of America Bulletin*, **107**, 595–611.

Hanson RB (1997) Hydrodynamics of regional metamorphism due to continental collision. *Economic Geology*, **92**, 880–891.

Hao KX, Si H, Fujiwara H, Owaze T (2009) Coseismic surface-ruptures and crustal deformations of the 2008 Wenchuan earthquake M_w7.9, China. *Geophysical Research Letters*, **36**, L11303.

Hardebeck JL (2012) Fluid-driven seismicity response of the Rinconada fault near Paso Robles, California, to the 2003 M6.5 San Simeon earthquake. *Bulletin of the Seismological Society of America*, **102**, 377–390.

Harder H (1959) Contribution to the geochemistry of boron, part II, Boron in sediments. *Nachrichten der Akademie der Wissenschaften in Göttingen, II, Mathematisch-Physikalische Klasse*, **6**, 123–183.

Hardin E, Barton N, Voegele M, Board M, Lingle R, Pratt H, Ubbes W (1982) Measuring the thermomechanical and transport properties of a rockmass using the heated block test. *Proceedings of the 23th U.S. Symposium on Rock Mechanics*, University of California, Berkeley, California, August 25–27, 1982, 802–13.

Hardy JR Jr (1991) *Laboratory Tests Conducted on Barre Granite: Compressive Strength, Flexural Strength, Modulus of Rupture, Adsorption and Bulk Specific Gravity and Petrographic Analysis*. Department of Mining Engineering, Pennsylvania State University, PA.

Häring MO, Schanz U, Ladner F, Dyer BC (2008) Characterisation of the Basel 1 enhanced geothermal system. *Geothermics*, **37**, 469–495.

Harlow FH, Ellison MA, Reid JH (1964) The particle-in-cell computing method for fluid dynamics. *Methods in Computational Physics*, **3**, 319–343.

Harper GD, Bowman JR, Kuhns R (1988) A field, chemical and stable isotope study of subseafloor metamorphism of the Josphine ophiolite, California-Oregon. *Journal of Geophysical Research*, **93**, 4625–4656.

Harris C (1989) Oxygen-isotope zonation of agates from Karoo volcanics of the Skeleton Coast, Namibia. *American Mineralogist*, **74**, 476–481.

Harrison R, Cather S (2004) The hot springs fault systems of south-central New Mexico—evidence for northward translation of the Colorado Plateau during the Laramide orogeny. *New Mexico Bureau of Geology and Mineral Resources Bulletin*, **160**, 161–180.

Harrison R, Lozinsky R, Eggleston T, McIntosh W (1993) Geologic map of the Truth or Consequences 30×60 minute quadrangle. *New Mexico Bureau of Mines and Mineral Resources Open-File Report*, **390**, scale 1:100 000.

Harvey C, Gorelick SM (2000) Rate-limited mass transfer or macrodispersion: which dominates plume evolution at the macrodispersion experiment (MADE) site? *Water Resources Research*, **36**, 637–650.

Hasegawa A, Nakajima J, Uchida N, Okada T, Zhao D, Matsuzawa T, Umino N (2009) Plate subduction, and generation of earthquakes and magmas in Japan as inferred from seismic observations: an overview. *Gondwana Research*, **16**, 370–400.

Hasegawa A, Umino N, Takagi A (1978) Double-planed structure of the deep seismic zone in the northeastern Japan arc. *Tectonophysics*, **47**, 43–58.

Hasegawa A, Yoshida K, Asano Y, Okada T, Iinuma T, Ito Y (2012) Change in stress field after the 2011 great Tohoku-Oki earthquake. *Earth and Planetary Science Letters*, **355–356**, 231–243.

Hasegawa A, Yoshida K, Okada T (2011) Nearly complete stress drop in the 2011 M_w9.0 off the Pacific coast of Tohoku earthquake. *Earth, Planets and Space*, **63**, 703–707.

Hatherton T, Leopard AE (1964) The densities of New Zealand rocks. *New Zealand Journal of Geology and Geophysics*, **7**, 605–625.

Haw MD (2004) Jamming, two-fluid behavior, and "self-filtration" in concentrated particulate suspensions. *Physical Review Letters*, **92**, 342–360.

Hayba DO, Ingebritsen SE (1994) The computer model HYDROTHERM, a three-dimensional finite-difference model to simulate ground-water flow and heat transport in the temperature range of 0 to 1,200°C. *U.S. Geological Survey Water-Resources Investigations Report*, **94–4045**.

Hayba DO, Ingebritsen SE (1997) Multiphase groundwater flow near cooling plutons. *Journal of Geophysical Research*, **102**, 12235–12252.

Haymon RM (1996) The response of ridge-crest hydrothermal systems to segmented, episodic magma supply. *Geological Society of London Special Publication*, **118**, 157–168.

Headquarters for Earthquake Research Promotion (2008) Evaluation of the Aizu-Bonchi Toen and Seien faults (in Japanese). Retrieved from http://www.jishin.go.jp/main/chousa, accessed 06 May 2016.

Hedenquist JW, Arribas A, Reynolds TJ (1998) Evolution of an intrusion-centered hydrothermal system: Far Southeast-Lepanto porphyry and epithermal Cu-Au deposits, Philippines. *Economic Geology*, **93**, 373–404.

Hedenquist JW, Lowenstern JB (1994) The role of magmas in the formation of hydrothermal ore-deposits. *Nature*, **370**, 519–527.

Heederik JP (1988) Geothermische Reserves Centrale Slenk, Nederland: Exploratie en evaluatie. Technical report, TNO, Utrecht.

Heimgartner M, Louie JN, Scott JB, Thelen W, Lopez CT, Coolbaugh M (2006) The crustal thickness of the great basin: using seismic refraction to assess regional geothermal potential. *Geothermal Resources Council Transactions*, **30**, 83–86.

Heinrich CA (2006) How fast does gold trickle out of volcanoes? *Science*, **314**, 263–264.

Heinrich CA, Candela PA (2014) Fluids and ore formation in the earth's crust. In: *Treatise on Geochemistry*, 2nd edn (eds Holland HD, Turekian KK), pp. 1–28, Elsevier, Oxford.

Heinrich W, Hoffbauer R, Hubberten HW (1995) Contrasting fluid-flow patterns at the Bufa del Diente contact aureole, northeast Mexico – Evidence from stable isotopes. *Contributions to Mineralogy and Petrology*, **119**, 362–376.

Helfferich F (1966) Ion exchange kinetics. *Ion Exchange*, **1**, 65–100.

Helfrich KR, Whitehead JA (1990) Solitary waves on conduits of buoyant fluid in a more viscous-fluid. *Geophysical and Astrophysical Fluid Dynamics*, **51**, 35–52.

Hendriks B, Andriessen P, Huigen Y, Leighton C, Redfield T, Murrell G, Gallagher K, Nielsen SB (2007) A fission track data compilation for Fennoscandia. *Norsk Geologisk Tidsskrift*, **87**, 143–155.

Henley RW, Ellis AJ (1983) Geothermal systems ancient and modern – A geochemical review. *Earth-Science Reviews*, **19**, 1- 50.

Henry P (2000) Fluid flow at the toe of the Barbados accretionary wedge constrained by thermal, chemical, and hydrogeologic observations and models. *Journal of Geophysical Research*, **105**, 25855–25872.

Henry P, Foucher J-P, Le Pichon X, Sibuet M, Kobayashi K, Tarits P, Chamot-Rooke N, Furuta T, Schultheiss P (1992) Interpretation of temperature measurements from the Kaiko-Nankai cruise: modeling of fluid flow in clam colonies. *Earth and Planetary Science Letters*, **109**, 355–371.

Henry P, Lallemant S, Nakamura K, Tsunogai U, Mazzotti S, Kobayashi K (2002) Surface venting at the toe of the Nankai wedge and implications for flow paths. *Marine Geology*, **187**, 119–143.

Henry P, Le Pichon X (1991) Fluid flow along a décollement layer: a model applied to the 16°N section of the Barbados accretionary wedge. *Journal of Geophysical Research*, **96**, 6507–6528.

Hensen C, Wallmann K, Schmidt M, Ranero C, Suess E (2004) Fluid expulsion related to mud extrusion off Costa Rica – A window to the subducting slab. *Geology*, **32**, 201–204

Herdianita NR, Rodgers KA, Browne PRL (2000) Routine instrumental procedures to characterise the mineralogy of modern and ancient silica sinters. *Geothermics*, **29**, 65–81.

Herman F, Cox SC, Kamp PJJ (2009) Low-temperature thermochronology and thermokinematic modelling of deformation, exhumation, and development of topography in the central Southern Alps, New Zealand. *Tectonics*, **28**, TC5011.

Herman F, Seward D, Valla PG, Carter A, Kohn B, Willett SD, Ehlers TA (2013) Worldwide acceleration of mountain erosion under a cooling climate. *Nature*, **504**, 423–426.

Hersum TG, Marsh BD (2006) Igneous microstructures from kinetic models of crystallization. *Journal of Volcanology and Geothermal Research*, **154**, 34–47.

Hezarkhani A, Williams-Jones AE (1998) Controls of alteration and mineralization in the Sungun porphyry copper deposit, Iran: evidence from fluid inclusions and stable isotopes. *Economic Geology*, **93**, 651–670.

Hezarkhani A, Williams-Jones AE, Gammons CH (1999) Factors controlling copper solubility and chalcopyrite deposition in the Sungun porphyry copper deposit, Iran. *Mineralium Deposita*, **34**, 770–783.

Hickey KA, Barker SLL, Dipple GM, Arehart GB, Donelick RA (2014) The brevity of hydrothermal fluid flow revealed by thermal halos around giant gold deposits: implications for Carlin-type gold systems. *Economic Geology*, **109**, 1461–1487.

Hickman S, Davatzes N (2010) In-situ stress and fracture characterization for planning of an EGS stimulation in the Desert Peak Geothermal Field, Nevada. *Proceedings Thirty-Fifth Workshop on Geothermal Reservoir Engineering*, Stanford University.

Hicks DD, Hill J, Shankar U (1996) Variation of suspended sediment yields around New Zealand: the relative importance of rainfall and geology. *International Association of Hydrological Sciences Publication*, **236**, 149–156.

Hiederer R, Kochy M (2011) Global soil organic carbon estimates and the harmonized world soil database. *European Commission Joint Research Center Scientific and Technical Report*, **EUR 25225 EN**, http://eusoils.jrc.ec.europa.eu/esdb_archive/eusoils_docs/other/EUR25225.pdf accessed 06 May 2016.

Higuera FJ, Jimenez J (1989) Boltzmann approach to lattice gas simulations. *Europhysics Letters*, **9**, 663.

Hildenbrand A, Schlomer S, Kross BM (2002) Gas breakthrough experiments on fine-grained sedimentary rocks. *Geofluids*, **2**, 3–23.

Hildenbrand A, Schlömer S, Kross BM, Littke R (2004) Gas breakthrough experiments on pelitic rocks: comparative study with N_2, CO_2 and CH_4. *Geofluids*, **4**, 61–80.

Hill DP (1977) A model for earthquake swarms. *Journal of Geophysical Research*, **82**, 1347–1352.

Hill DP, Prejean S (2005) Magmatic unrest beneath Mammoth Mountain, California. *Journal of Volcanology and Geothermal Research*, **146**, 257–283.

Hill DP, Reasenberg PA, Michael A, Arabaz WJ, Beroza G, Brumbaugh D, Brune JN, Castro R, Davis S, dePolo D, Ellsworth WL, Gomberg J, Harmsen S, House L, Jackson SM, Johnston MJS, Jones L, Keller R, Malone S, Mungaia L, Nava S, Pechmann JC, Sanford A, Simpson RW, Smith RB, Stark M, Stickney M, Vidal S, Walter S, Wong V, Zollweg J (1993) Seismicity remotely triggered by the magnitude 7.3 Landers, California, earthquake. *Science*, **260**, 1617–1623.

Hinz N, Siler D, Faulds J (2012) 3D Geologic mapping-structural studies of geothermal systems in the Basin and Range, *Digital Mapping Techniques 2012 Proceedings*, http://ngmdb.usgs.gov/Info/dmt/DMT12presentations.html, accessed 06 May 2016.

Hinzen K-G (2003) Stress field in the northern Rhine area, central Europe, from earthquake fault plane solutions. *Tectonophysics*, **377**, 325–356.

Hiramatsu Y, Honma H, Saiga A, Furumoto M, Ooida T (2005) Seismological evidence on characteristic time of crack healing in the shallow crust. *Geophysical Research Letters*, **32**, doi:10.1029/2005GL022657.

Hirose F, Miyaoka K, Hayashimoto N, Yamazaki T, Nakamura M (2011) Outline of the 2011 off the Pacific coast of Tohoku earthquake (M_w9.0)—Seismicity: foreshocks, mainshock, aftershocks, and induced activity. *Earth, Planets and Space*, **63**, 513–518.

Hitzman MW, many others (Committee on Induced Seismicity Potential in Energy Technologies) (2012) *Induced Seismicity Potential in Energy Technologies*. National Academy Press, Washington, DC.

Hobday C, Worthington MH (2012) Field measurements of normal and shear fracture compliance. *Geophysical Prospecting*, **60**, 488–499.

Hofstra AH, Cline JS (2000) Characteristics and models for Carlin-type gold deposits. *Reviews in Economic Geology*, **13**, 163–220.

Holdich RG (2002) *Fundamentals of Particle Technology*. Midland Information Technology and Publishing, Loughborough.

Holland HD (2005) Sea level, sediments and the composition of seawater. *American Journal of Science*, **305**, 220–239.

Holness MB (2003) Growth and albitization of K-feldspar in crystalline rocks in the shallow crust: a tracer for fluid circulation during exhumation? *Geofluids*, **3**, 89–102.

Holtzman BK, Groebner NJ, Zimmerman ME, Ginsberg SB, Kohlstedt DL (2003) Stress-driven melt segregation in partially molten rocks. *Geochemistry Geophysics Geosystems*, **4**, doi:10.1029/2001GC000258.

Hooft EEE, Detrick RS (1993) The role of density in the accumulation of basaltic melts at mid-ocean ridges. *Geophysical Research Letters*, **20**, 423–426.

Hori S, Hasegawa A (1991) Location of a mid-crustal magma body beneath Mt. Moriyoshi, northern Akita Prefecture, as estimated from reflected S×S phases. *Zisin (Journal of the Seismological Society of Japan 2nd series)*, **44**, 39–48.

Horne RN (1996) *Modern Well Test Analysis, A Computer Aided Approach*, 2nd edn Petroway Inc., Palo Alto, CA.

Horsrud P, Sønstebø EF, Bøe R (1998) Mechanical and petrophysical properties of North Sea shales. *International Journal of Rock Mechanics and Mining Sciences*, **35**, 1009–1020.

Horton DG (1991) Secondary minerals in Columbia River Basalt, Pasco basin, Washington, Rep. WHC-SA-1352-FP, Pacific Northwest National Laboratory, Richland, WA.

Howald T, Person M, Campbell A, Lueth V, Hofstra A, Sweetkind D, Gable C, Banerjee A, Luijendijk E, Crossey L, Karlstrom K, Kelley S, Phillips F (2015) Evidence for long-time scale ($>10^3$ years) changes in hydrothermal activity induced by seismic events. *Geofluids*, **15**, 252–268.

Howarth JD, Fitzsimons SJ, Norris RJ, Jacobsen GE (2012) Lake sediments record cycles of sediment flux driven by large earthquakes on the Alpine fault, New Zealand. *Geology*, **40**, 1091–1094.

Hsieh PA, Bredehoeft JD (1981) A reservoir analysis of the Denver earthquakes: a case of induced seismicity. *Journal of Geophysical Research*, **86**, 903–920.

Hsieh PA, Bredehoeft JD, Rojstaczer SA (1988) Response of well aquifer systems to Earth tides: problem revisited. *Water Resources Research*, **24**, 468–472.

Hsieh PA, Tracy JV, Neuzil CE, Bredehoeft JD, Silliman SE (1981) A transient laboratory method for determining the hydraulic properties of 'tight' rocks, I, Theory. *International Journal of Rock Mechanics and Mining Sciences*, **18**, 245–252.

Hsiung SM, Chowdhury AH, Nataraja MS (2005) Numerical simulation of thermal-mechanical processes observed at the Drift-Scale Heater Test at Yucca Mountain, Nevada, USA. *International Journal of Rock Mechanics and Mining Sciences*, **42**, 652–666.

Hu DW, Zhu QZ, Zhou H, Shao JF (2010) A discrete approach for anisotropic plasticity and damage in semi-brittle rocks. *Computational Geotechnics*, **37**, 658–666.

Hu K, Issler D (2009) A comparison of core petrophysical data with well log parameters, Beaufort-Mackenzie Basin. *Geological Survey of Canada Open File*, **6042**.

Huang F, Jian C, Tang Y, Xu G, Deng Z, Chi G-C, Farrar CD (2004) Response changes of some wells in the mainland subsurface fluid monitoring network of China, due to the September 21 1999, M_s7.6 Chi-Chi earthquake. *Tectonophysics*, **390**, 217–234.

Huang R, Audétat A (2012) The titanium-in-quartz (TitaniQ) thermobarometer: a critical examination and re-calibration. *Geochimica et Cosmochimica Acta*, **84**, 75–89.

Hubbert MK (1940) The theory of ground-water motion. *The Journal of Geology*, **48**, 785–944.

Huber C, Chopard B, Manga M (2010) A lattice Boltzmann model for coupled diffusion. *Journal of Computational Physics*, **229**, 7956–7976.

Huber C, Parmigiani A, Chopard B, Manga M, Bachmann O (2008) Lattice Boltzmann model for melting with natural convection. *International Journal of Heat and Fluid Flow*, **29**, 1469–1480.

Huber C, Su Y (2015) A pore-scale investigation of the dynamic response of saturated porous media to transient stresses. *Geofluids*, **15**, 11–23.

Hudson JA, Bäckstrom A, Rutqvist J, Jing L (2008) DECOVALEX-THMC Project, Task B, Understanding and characterizing the excavation disturbed zone, Final Report (EDZ Guidance Document) characterizing and modelling the Excavation damaged Zone (EDZ) in crystalline rocks in the context of radioactive waste disposal. *SKI Report*, **43**.

Huenges E, Erzinger J, Kück J, Engeser B, Kessels W (1997) The permeable crust: geohydraulic properties down to 9101 m depth. *Journal of Geophysical Research*, **102**, 18255–18265.

Hulen JB, Lutz SJ (1999) Altered volcanic rocks as hydrologic seals on the geothermal system of Medicine Lake volcano, California. *Geothermal Resources Council Bulletin*, **7**, 217–222.

Humphris SE, Cann JR (2000) Constraints on the energy and chemical balances of the modern TAG and ancient Cyprus seafloor sulfide deposits. *Journal of Geophysical Research*, **105**, 28477–28488.

Hunt JM (1990) Generation and migration of petroleum from abnormally pressured fluid compartments. *American Association of Petroleum Geologists Bulletin*, **74**, 1–12.

Hüpers A, Kopf AJ (2012) Data report: Consolidation properties of silty claystones and sandstones sampled seaward of the Nankai Trough subduction zone, IODP Sites C0011 and C0012. In Saito S, Underwood MB, Kubo Y, and the Expedition 322 Scientists, *Proceedings of the Integrated Ocean Drilling Program*, **322**, 1–23.

Hurwitz S, Lowenstern JB, Heasler H (2007) Spatial and temporal geochemical trends in the hydrothermal system of Yellowstone National Park: inferences from river solute fluxes. *Journal of Volcanology and Geothermal Research*, **162**, 149–171.

Husen S, Taylor R, Smith RB, Healser H (2004a) Changes in geyser eruption behavior and remotely triggered seismicity in Yellowstone National Park produced by the 2002 M_w7.9 Denali fault earthquake, Alaska. *Geology*, **32**, 537–540.

Husen S, Wiemer S, Smith RB (2004b) Remotely triggered seismicity in the Yellowstone National Park region by the 2002 M_w7.9 Denali fault earthquake, Alaska. *Bulletin of the Seismological Society of America*, **94**, S317–S331.

Hutnak M, Hurwitz S, Ingebritsen SE, Hsieh PA (2009) Numerical models of caldera deformation: effects of multiphase and multicomponent hydrothermal fluid flow. *Journal of Geophysical Research*, **114**, B04411, doi:10.1029/2008JB006151.

Huuse M, Jackson CAL, Van Rensbergen P, Davies RJ, Flemings PB, Dixon RJ (2010) Subsurface sediment remobilization and fluid flow in sedimentary basins: an overview. *Basin Research*, **22**, 342–360.

IAPWS (1997) Release on the IAPWS industrial formulation 1997 for the thermodynamic properties of water and steam. The International Association for the Properties of Water and Steam, Lucerne, Switzerland.

Ichikawa Y, Selvadurai APS (2012) *Transport Phenomena in Porous Media: Aspects of Micro/Macro Behaviour*. Springer Verlag, Berlin.

Ienaga M, McNeill LC, Mikada H, Saito S, Goldberg D, Moore JC (2006) Borehole image analysis of the Nankai accretionary wedge, ODP Leg 196: structural and stress studies. *Tectonophysics*, **426**, 207–220.

Ikari MJ, Saffer DM (2012) Permeability contrasts between sheared and normally consolidated sediments in the Nankai accretionary prism. *Marine Geology*, **295–298**, 1–13.

Illies JH (1972) The Rhine graben rift system-plate tectonics and transform faulting. *Geophysical Surveys*, **1**, 27–60.

Ingebritsen SE, Geiger S, Hurwitz S, Driesner T (2010) Numerical simulation of magmatic hydrothermal systems. *Reviews of Geophysics*, **48**, RG1002.

Ingebritsen SE, Gleeson T (2015) Crustal permeability: introduction to the special issue. *Geofluids*, **15**, 1–10.

Ingebritsen SE, Manning CE (1999) Geological implications of a permeability-depth curve for the continental crust. *Geology*, **27**, 1107–1110.

Ingebritsen SE, Manning CE (2002) Diffuse fluid flux through orogenic belts: implications for the world ocean. *Proceedings of the National Academy of Sciences USA*, **99**, 9113–9116.

Ingebritsen SE, Manning CE (2010) Permeability of the continental crust: dynamic variations inferred from seismicity and metamorphism. *Geofluids*, **10**, 193–205.

Ingebritsen SE, Sanford W, Neuzil CE (2006) *Groundwater in Geologic Processes*, 2nd edn Cambridge University Press, Cambridge, UK.

Institute of Geophysics-CAS (China Earthquake Administration) (1976) *China Earthquake Catalog*. Center for Chinese Research Material, Washington, DC (in Chinese).

Ishibashi T, McGuire TP, Watanabe N, Tsuchiya N, Elsworth D (2013) Permeability evolution in carbonate fractures: competing roles of confining stress and fluid pH. *Water Resources Research*, **49**, 2828–2842, doi:10.1002/wrcr.20253.

Ishibashi T, Watanabe N, Hirano N, Okamoto A, Tsuchiya N (2012) GeoFlow: a novel model simulator for prediction of the 3-D channeling flow in a rock fracture network. *Water Resources Research*, **48**, W07601, doi:10.1029/2011WR011226.

Ishibashi T, Watanabe N, Hirano N, Okamoto A, Tsuchiya N (2015) Beyond-laboratory-scale prediction for channeling flows through subsurface rock fractures with heterogeneous aperture distributions revealed by laboratory evaluation. *Journal of Geophysical Research*, **120**, doi:10.1002/2014JB011555.

Itaba S, Koizumi N, Matsumoto N, Takahashi M, Sato T, Ohtani R, Kitagawa Y, Kuwahara Y, Ozawa K (2008) Groundwater level changes related to the ground shaking of the Noto Hanto Earthquake in 2007. *Earth, Planets and Space*, **60**, 1153–1159.

Itasca (2013) *UDEC 5.0 Universal Distinct Element Code*. Itasca Consulting Group Incorporated, Minneapolis, Minnesota.

Iwai K (1976) Fundamental studies of fluid flow through a single fracture. PhD thesis, University of California, Berkeley.

Iwatsuki T, Yoshida H (1999) Groundwater chemistry and fracture mineralogy in the basement granite rock in the Tono uranium mine area, Gifu Prefecture, Japan – Groundwater composition, Eh evolution analysis by fracture filling minerals. *Geochemical Journal*, **33**, 19–32.

Izawa E, Urashima Y, Ibaraki K, Suzuki R, Yokoyama T, Kawasaki K, Koga A, Taguchi S (1990) The Hishikari gold deposit – High-grade epithermal veins in Quaternary volcanics of southern Kyushu, Japan. *Journal of Geochemical Exploration*, **36**, 1–56.

Jaeger JC (1971) Friction of rocks and stability of rock slopes. *Geotechnique*, **21**, 97–134.

Jaeger JC, Cook NGW, Zimmerman RW (2007) *Fundamentals of Rock Mechanics*, 4th edn Blackwell, Maiden, MA.

Jafari MK, Pellet F, Boulon MJ, Amini Hosseini K (2004) Experimental study of mechanical behaviour of rock joints under cyclic loading. *Rock Mechanics and Rock Engineering*, **37**, 3–23.

Jakob A, Mazurek M, Heer W (2003) Solute transport in crystalline rocks at Aspo, II, Blind predictions, inverse modelling and lessons learnt from test STT1. *Journal of Contaminant Hydrology*, **6**, 175–190.

Jamtveit B, Putnis CV, Malthe-Sorenssen A (2009) Reaction induced fracturing during replacement processes. *Contributions to Mineralogy and Petrology*, **157**, 127–133.

Japan Beyond Brittle Project (2014) Japan Beyond Brittle Project. URL: http://jbbp.kankyo.tohoku.ac.jp/JBBP/outline.html accessed 06 May 2016.

Japsen P, Dysthe DK, Hartz EH, Stipp SLS, Yarushina VM, Jamtveit B (2011) A compaction front in North Sea chalk. *Journal of Geophysical Research*, **116**, B11208.

Jeffrey RG, Bunger A, Lecampion B, Zhang X, Chen ZR, van As A, Allison DP, de Beer W, Dudley JW, Siebrits E, Thiercelin M, Mainguy M (2009) Measuring hydraulic fracture growth in naturally fractured rock. SPE Annual Technical Conference and Exhibition, 4-7 October, New Orleans, Louisiana, http://dx.doi.org/10.2118/124919-MS, accessed 06 May 2016.

Jenner GA, Cawood PA, Rautenschlein M, White WM (1987) Composition of back-arc basin volcanics, Valu Fa Ridge, Lau Basin – Evidence for a slab-derived component in their mantle source. *Journal of Volcanology and Geothermal Research*, **32**, 209–222.

Jerram DA, Widdowson M (2005) The anatomy of continental flood basalt provinces: geological constraints on the processes and products of flood volcanism. *Lithos*, **79**, 385–405.

Jiang T, Peng Z, Wang W, Chen Q-F (2010a) Remotely triggered seismicity in continental China following the 2008 M_w7.9 Wenchuan earthquake. *Bulletin of the Seismological Society of America*, **100**, 2574–2589.

Jiang X-W, Wan L, Cardenas MB, Ge S, Wang X-S (2010b) Simultaneous rejuvenation and aging of groundwater in basins due to depth-decaying hydraulic conductivity and porosity. *Geophysical Research Letters*, **37**, L05403.

Jiang X-W, Wan L, Wang XS, Liang SH, Hu BX (2009) Estimation of fracture normal stiffness using a transmissivity-depth correction. *International Journal of Rock Mechanics and Mining Sciences*, **46**, 51–58.

Jiang X-W, Wang X-S, Wan L (2010c) Semi-empirical equations for the systematic decrease in permeability with depth in porous and fractured media. *Hydrogeology Journal*, **18**, 839–850.

Jing Z, Richards JW, Watanabe K, Hashida T (2000) A three-dimensional stochastic rock mechanics model of engineered geothermal systems in fractured crystalline rock. *Journal of Geophysical Research*, **105**, 23663–23679, doi:10.1029/2000JB900202.

John DA, Hofstra AH, Fleck RJ, Brummer JE, Saderholm EC (2003) Geologic setting and genesis of the mule canyon low-sulfidation epithermal gold-silver deposit, north-central Nevada. *Economic Geology*, **98**, 425–463.

John DA, Wrucke CT (2003) Geologic map of the Mule Canyon quadrangle, Lander County, Nevada. *Nevada Bureau of Mines and Geology Map*, **144**.

Johnson DL, Koplik J, Dashen R (1987) Theory of dynamic permeability and tortuosity in fluid-saturated porous media. *Journal of Fluid Mechanics*, **176**, 379–402.

Jones BL (1987) Conventional core analysis studies for TNO-DGV, Institute of Applied Geoscience on well Asten 2. Technical report, Redwood Corex, Sassenheim.

Jones JB, Mulholland PJ (eds) (2000) *Streams and Ground Waters*. Academic Press, San Diego.

Jones ME, Addis MA (1985) On changes in porosity and volume during burial of argillaceous sediments. *Marine and Petroleum Geology*, **2**, 247–253.

Jónsson S, Segall P, Pedersen R, Björnsson G (2003) Post-earthquake ground movements correlated to pore-pressure transients. *Nature*, **424**, 179–183.

Joshi A, Appold MS, Nunn JA (2012) Evaluation of solitary waves as a mechanism for oil transport in poroelastic media: a case study of the South Eugene Island field, Gulf of Mexico basin. *Marine and Petroleum Geology*, **37**, 53–69.

Jowitt SM, Jenkin GRT, Coogan LA, Naden J (2012) Quantifying the release of base metals from source rocks for volcanogenic massive sulfide deposits: effects of protolith composition and alteration mineralogy. *Journal of Geochemical Exploration*, **118**, 47–59.

Juhlin C, Sandstedt H (1989) Storage of nuclear waste in very deep boreholes: feasibility study and assessment of economic potential. In: *SKB Technical Report*, **89-39** (ed SKB), SKB, Stockholm, Sweden, http://www.skb.se/upload/publications/pdf/TR89-39webb.pdf accessed 06 May 2016.

Juhlin C, Wallroth T, Smellie J, Eliasson T, Ljunggren C, Leijon B, Beswick J (1998) The very deep hole concept – Geoscientific appraisal of conditions at great depth. *SKB Technical Report*, **98-05**.

Jung R (2013) EGS – Goodbye or Back to the Future. In: *Effective and Sustainable Hydraulic Fracturing* (eds Bunger A, McLennan J, Jeffrey R), pp. 95–121, InTech, Rijeka, Croatia.

Jupp T, Schultz A (2000) A thermodynamic explanation for black smoker temperatures. *Nature*, **403**, 880–883.

Juteau T, Manac'h O, Lécuyer C, Ramboz C (2000) The high-temperature reaction zone of the Oman ophiolite: new field data, micro thermometry of fluid inclusions, PIXE analyses and oxygen isotopic ratios. *Marine Geophysical Researches*, **21**, 351–385.

Kahle SC, Morgan DS, Welch WB, Ely DM, Hinlde SR, Vaccaro JJ, Orzol LL (2011) Hydrogeologic framework and hydrologic budget components of the Columbia Plateau Regional Aquifer System, Washington, Oregon, and Idaho. *U.S. Geological Survey Scientific Investigations Report*, **2011–5124**, http://pubs.er.usgs.gov/publication/sir20115124 accessed 06 May 2016.

Kahle SC, Olsen TD, Morgan DS (2009) Geologic setting and hydrogeologic units of the Columbia Plateau Regional Aquifer System, Washington, Oregon, and Idaho. *U.S. Geological Survey Scientific Investigations Map*, **3088**, http://pubs.er.usgs.gov/usgspubs/sim/sim3088 accessed 06 May 2016.

Kaieda H, Jones R, Moriya H, Sasaki S, Ushijima K (2005) Ogachi HDR reservoir evaluation by AE and geophysical methods, In: *Proceedings of World Geothermal Congress 2005*, 24–29 April, Antalya, Turkey.

Kaiser AE, Holden C, Beavan RJ, Beetham RD, Benites RA, Celentano A, Collet D, Cousins WJ, Cubrinovski M, Dellow GD, Denys P, Fielding E, Fry B, Gerstenberger MC, Langridge RM, Massey CI, Motagh M, Pondard N, McVerry GH, Ristau J, Stirling MW, Thomas J, Uma SR, Zhao JX (2012) The M_w6.2 Christchurch Earthquake of February 2011: preliminary report. *New Zealand Journal of Geology and Geophysics*, **55**, 67–90.

Kaiser P, Valley B, Dusseault M, Duff D (2013) Hydraulic fracturing mine back trials – Design rationale and project status, http://dx.doi.org/10.5772/56260 accessed 06 May 2016.

Kameyama M, Yuen DA, Karato SI (1999) Thermal-mechanical effects of low-temperature plasticity (the Peierls mechanism) on the deformation of a viscoelastic shear zone. *Earth and Planetary Science Letters*, **168**, 159–172.

Kanamori H, Anderson DL (1975) Theoretical basis of some empirical relations in seismology. *Bulletin of the Seismological Society of America*, **65**, 1073–1095.

Kano Y, Yanagidani T (2006) Broadband hydroseismograms observed by closed borehole wells in the Kamioka mine, central Japan: response of pore pressure to seismic waves from 0.05 to 2 Hz. *Journal of Geophysical Research*, **111**, B03410.

Karig DE (1990) Experimental and observational constraints on the mechanical behaviour in the toes of accretionary prisms. *Geological Society London Special Publication*, **54**, 383–398.

Karig DE, Lundberg N (1990) Deformation bands from the toe of the Nankai accretionary prism. *Journal of Geophysical Research*, **95**, 9099–9109.

Karingithi C (2009) Chemical geothermometers for geothermal exploration. *Short Course IV on Exploration for Geothermal Resources, organized by UNU-GTP, KenGen and GDC*.

Karnis A, Goldsmith HL, Mason SG (1966) Kinetics of flowing dispersions, I, Concentrated suspensions of rigid particles. *Journal of Colloid and Interface Science*, **22**, 531–553.

Kastner M, Elderfield H, Martin JB (1991) Fluids in convergent margins: what do we know about their composition, origin, role in diagenesis and importance for oceanic chemical fluxes? *Philosophical Transactions of the Royal Society Series A*, **335**, 243–259.

Kaszuba J, Yardley BWD, Andreani M (2013) Experimental perspectives of mineral dissolution and precipitation due to carbon dioxide-water-rock interactions. *Reviews in Mineralogy & Geochemistry*, **77**, 153–188.

Kato A, Igarashi T, Obara K, Sakai S, Takeda T, Saiga A, Iidaka T, Iwasaki T, Hirata N, Goto K, Miyamachi H, Matsushima T, Kubo A, Katao H, Yamanaka Y, Terakawa T, Nakamichi H, Okuda T, Horikawa S, Tsumura N, Umino N, Okada T, Kosuga M, Takahashi H, Yamada T (2013) Imaging the source regions of normal faulting sequences induced by the 2011 M9.0 Tohoku-Oki earthquake. *Geophysical Research Letters*, **40**, 273–278.

Katz RF (2008) Magma dynamics with the enthalpy method: benchmark solutions and magmatic focusing at mid-ocean ridges. *Journal of Petrology*, **49**, 2099–2121.

Kay MA, Main IG, Elphick SC, Ngwenya BT (2006) Fault gouge diagenesis at shallow burial depth: solution-precipitation reactions in well-sorted and poorly sorted powders of crushed sandstone. *Earth and Planetary Science Letters*, **243**, 607–614.

Keller GV, Grose LT, Murray JC, Skokan CK (1979) Results of an experimental drill hole at the summit ol Kilauea Volcano, Hawaii. *Journal of Volcanology and Geothermal Research*, **5**, 345–385.

Kelley DS, Carbotte SM, Caress DW, Clague DA, Delaney JR, Gill JB, Hadaway H, Holden JF, Hooft EEE, Kellogg JP, Lilley MD, Stoermer M, Toomey D, Weeldy R, Wilcock WSD (2012) Endeavour Segment of the Juan de Fuca Ridge: one of the most remarkable places on Earth. *Oceanography*, **25**, 44–61.

Kelley SA, Chapin CE (1997) Cooling histories of mountain ranges in the southern Rio Grande Rift based on apatite fission-track analysis – a reconnaissance survey. *New Mexico Geology*, **19**, 1–14.

Kempe S (1979) Carbon in the rock cycle. In: *The Global Carbon Cycle* (eds Bolin B, Degens ET, Kempe S, Ketner P), pp. 343–375. Scientific Committee on Problems of the Environment (SCOPE), Old Woking, UK.

Kemp SJ, Wagner D (2006) The mineralogy, geochemistry and surface area of mudrocks from the London Clay Formation of southern England. *British Geological Survey Physical Hazards Programme Internal Report*, **IR/06/060**, Nottingham, UK.

Kennedy BM, van Soest MC (2007) Flow of mantle fluids through the ductile lower crust: helium isotope trends. *Science*, **318**, 1433–1436.

Kennett JP (1982) *Marine Geology*. Prentice Hall, Englewood Cliffs, NJ.

Kern LR, Perkins TK, Wyant RE (1959) The mechanics of sand movement in fracturing. *Transactions of the American Institute of Mining and Metallurgical Engineers*, **216**, 403–405.

Kesler SE (1994) *Mineral Resources, Economics and the Environment*. Macmillan Publishing, New York.

Kestin J, Khalifa HE, Correia RJ (1981) Tables of the dynamic and kinematic viscosity of aqueous KCl solutions in the temperature range 25–l50°C and the pressure range 0.1–35 MPa. *Journal of Physical and Chemical Reference Data*, **10**, 57–70.

Khilar KC, Fogler HS (1998), *Migrations of Fines in Porous Media*. Kluwer Academic Publishers, Boston, Dordrecht.

Kidd RGW (1977) A model for the process of formation of the upper oceanic crust. *Geophysical Journal of the Royal Astronomical Society*, **50**, 149–183.

Kidd RGW, Cann JR (1974) Chilling statistics indicate an ocean-floor spreading origin for the Troodos complex, Cyprus. *Earth and Planetary Science Letters*, **24**, 151–155.

Kim ST, O'Neil JR (1997) Equilibrium and nonequilibrium oxygen isotope effects in synthetic carbonates. *Geochimica et Cosmochimica Acta*, **61**, 3461–3475.

Kim Y-S, Peacock DCP, Sanderson DJ (2004) Fault damage zones. *Journal of Structural Geology*, **26**, 503–517.

King C-Y, Azuma S, Igarashi G, Ohno M, Saito H, Wakita H (1999) Earthquake-related water level changes at 16 closely clustered wells in Tono, central Japan. *Journal of Geophysical Research*, **104**, 13073–13082.

King C-Y, Basler D, Presser TS, Evans CW, White LD, Minissale AD (1994a) In search of earthquake-related hydrologic and chemical changes along the Hayward fault. *Applied Geochemistry*, **9**, 83–91.

King GC, Stein RS, Lin J (1994b) Static stress changes and the triggering of earthquakes. *Bulletin of the Seismological Society of America*, **84**, 567–585.

Kinoshita M, Tobin H, Ashi J, Kimura G, Lallemant S, Screaton EJ, Curewitz D, Masago H, Moe KT, the Expedition 314/315/316 Scientists (2009) Proceedings of the Integrated Ocean Drilling Program, **314/315/316**.

Királý L (1969) Anisotropy and heterogeneity within jointed limestone. *Eclogae Geologicae Helvetiae*, **62**, 613–619.

Kita I, Taguchi S, Matsubaya O (1985) Oxygen isotope fractionation between amorphous silica and water at 34–93°C. *Nature*, **314**, 83–84.

Kitagawa Y, Fujimoro K, Koizumi N (2007) Temporal change in permeability of the Nojima fault zone by repeated water injection experiments. *Tectonophysics*, **443**, 183–192.

Kitajima H, Chester FM, Biscontin G (2012) Mechanical and hydraulic properties of Nankai accretionary prism sediments: effect of stress path. *Geochemistry, Geophysics, Geosystems*, **13**, Q0AD27.

Kitajima H, Saffer DM (2012) Elevated pore pressure and anomalously low stress in regions of low frequency earthquakes along the Nankai Trough subduction megathrust. *Geophysical Research Letters*, **39**, L23301.

Kiyama T, Kita H, Ishijima Y, Yanagidani T, Akoi K, Sato T (1996) Permeability in anisotropic granite under hydrostatic compression and tri-axial compression including post-failure region. *Proceedings 2nd North American Rock Mechanics Symposium*, 1643-50.

Knoll MD (1996) A petrophysical basis for ground penetrating radar and very early time electromagnetics: Electrical properties of sand-clay mixtures. PhD thesis, The University of British Columbia.

Knoll MD, Knight R (1994) Relationships between dielectric and hydrogeologic properties of sand-clay mixtures. *Proceedings of the Fifth International Conference on Ground Penetrating Radar*, **1**, 12–16.

Kodaira S, Iidaka T, Kato A, Park JO, Iwasaki T, Kaneda Y (2004) High pore fluid pressure may cause silent slip in the Nankai trough. *Science*, **304**, 1295–1298.

Koerner A, Kissling E, Miller SA (2004) A model of deep crustal fluid flow following the $M_w = 8.0$ Antofagasta, Chile, earthquake. *Journal of Geophysical Research*, 109, doi:10.1029/2003JB002816.

Koh J, Rosnhan H, Rahman SS (2011) A numerical study on the long term thermo-poroelastic effects of cold water injection into naturally fractured geothermal reservoirs. *Computers and Geotechnics*, **38**, 669–682.

Kohl T, Megel T (2005), Coupled hydro-mechanical modelling of the GPK3 reservoir stimulation at the European EGS site, Soultz-sous-Forêts. *Proceedings of the Thirtieth Workshop on Geothermal Reservoir Engineering*, Stanford University, Stanford, California.

Kohlstedt DL, Evans B, Mackwell SJ (1995) Strength of the lithosphere – constraints imposed by laboratory experiments. *Journal of Geophysical Research*, **100**, 17587–17602.

Koiter WT (1953) Stress-strain relations, uniqueness and variational theorems for elastic-plastic materials with a singular yield surface. *Quarterly of Applied Mathematics*, **11**, 350–354.

Koizumi N, Lai WC, Kitagawa Y, Matsumoto N (2004) Comment on "Coseismic hydrological changes associated with dislocation of the September 21, 1999 Chichi earthquake, Taiwan" by Min Lee et al. *Geophysical Research Letters*, **31**, L13603, doi:10.1029/2004GL019897.

Kolditz O (1995) Modeling flow and heat transfer in fractured rocks – Conceptual model of a 3-D Deterministic fracture network. *Geothermics*, **24**, 451–470.

Koltermann CE, Gorelick SM (1995) Fractional packing model for hydraulic conductivity derived from sediment mixtures. *Water Resources Research*, **31**, 3283–3297.

Kondo H, Kimura G, Masago H, Ohmori-Ikehara K, Kitamura Y et al. (2005) Deformation and fluid flow of a major out-of-sequence thrust located at seismogenic depth in an accretionary complex: Nobeoka thrust in the Shimanto Belt, Kyushu, Japan. *Tectonics*, **24**, TC6008.

Konikow LF, Kendy E (2005) Groundwater depletion: a global problem. *Hydrogeology Journal*, **13**, 317–320.

Konzuk JS, Kueper BH (2004) Evaluation of cubic law based models describing single-phase flow through a rough-walled fracture. *Water Resources Research*, **40**, W02402.

Koons PO (1987) Some thermal and mechanical consequences of rapid uplift: an example from the Southern Alps, New Zealand. *Earth and Planetary Science Letters*, **86**, 307–319.

Koons PO (1989) The topographic evolution of collisional mountain belts: a numerical look at the Southern Alps, New Zealand. *American Journal of Science*, **289**, 1041–1069.

Koons PO (1994) Three-dimensional critical wedges: tectonics and topography in oblique collisional zones. *Journal of Geophysical Research*, **99**, 12301–12305.

Koons PO, Craw D (1991) Evolution of fluid driving forces and composition within collisional orogens. *Geophysical Research Letters*, **18**, 935–938.

Koons PO, Craw D, Cox SC, Upton P, Templeton AS, Chamberlain CP (1998) Fluid flow during active oblique convergence: a Southern Alps model from mechanical and geochemical observations. *Geology*, **26**, 159–162.

Koons PO, Norris RJ, Craw D, Cooper AF (2003) Influence of exhumation on the structural evolution of transpressional plate boundaries: an example from the Southern Alps, New Zealand. *Geology*, **31**, 3–6.

Kopf A, Strasser M, Monsees N, Underwood MB, Guo J (2011) Data report: Particle size analysis of sediments recovered during IODP Expeditions 315 and 316, Sites C0001–C0008, Nankai Trough forearc, off Japan. In Kinoshita M, Tobin H, Ashi J, Kimura G, Lallemant S, Screaton EJ, Curewitz D, Masago H, Moe KT, and the Expedition 314/315/316 Scientists, *Proceeding Integrated Ocean Drilling Program*, **314/315/316**, 1–19.

Kopp C, Fruehn J, Flueh ER, Reichert C, Kukowski N, Bialas J, Klaeschen D (2000) Structure of the Makran subduction zone from wide-angle and reflection seismic data. *Tectonophysics*, **329**, 171–191.

Korzhinskii DS (1959) *Physicochemical Basis of the Analysis of the Paragenesis of Minerals*. Consultants Bureau, New York.

Korzhinskii DS (1965) The theory of systems with perfectly mobile components and processes of mineral formation. *American Journal of Science*, **263**, 193–205.

Korzhinskii DS (1970) *Theory of Metasomatic Zoning*. Oxford University Press, Oxford.

Kosuga M, Watanabe K, Hashimoto K, Kasai H (2012) Seismicity in the northern part of Tohoku District induced by the 2011 off the Pacific Coast of Tohoku earthquake. *Zisin (Journal of the Seismological Society of Japan 2nd series)*, **65**, 69–83.

Kothe DB, Rider WJ (1995) A comparison of interface tracking methods. *Proceedings 26th American Institute of Aeronautics and Astronautics (AIAA) Computational Fluid Dynamics Conference*, 19–22 June, San Diego, California.

Kozeny J (1927) Ueber kapillare leitung des wassers im boden. *Sitzungsberichte Akademie der Wissenschaften Wien*, **136**, 271–306.

Krieger IM, Dougherty TJ (1959) A mechanism for non-Newtonian flow in suspensions of rigid spheres. *Transactions of the Society of Rheology*, **3**, 137–152.

Krusemann GP, de Ridder NA (1991) Analysis and evaluation of pumping test data. *International Institute for Land Reclamation and Improvement Publication*, **47**.

Kühn M (2004) *Reactive Flow Modeling of Hydrothermal Systems.* Springer, Berlin, Heidelberg, New York.

Kukkonen IK (1995) Thermal aspects of groundwater circulation in bedrock and its effect on crustal geothermal modelling in Finland, the central Fennoscandian Shield. *Tectonophysics,* **244,** 119–136.

Kumar S, Bodvarsson GS (1990) Fractal study and simulation of fracture roughness. *Geophysical Research Letters,* **17,** 701–704.

Kwon O, Kronenberg AK, Gangi AF, Johnson B, Herbert BE (2004) Permeability of illite-bearing shale, 1, Anisotropy and effects of clay content and loading. *Journal of Geophysical Research,* **109,** B10205.

Lachenbruch AH, Sass JH (1977) Heat flow in the United States and the thermal regime of the crust. *American Geophysical Union Geophysical Monograph Series,* **20,** 626–675.

Labaume P, Kastner M, Trave A, Henry P (1997a) Carbonate veins from the décollement zone at the toe of the northern Barbados accretionary prism: microstructure, mineralogy, geochemistry, and relations with prism structures and fluid regime. In: *Proceedings of the Ocean Drilling Program, Scientific Results,* **156,** (eds Shipley TH, Ogawa Y, Blum P, Bahr JM), pp. 79–96, Ocean Drilling Program, College Station, TX.

Labaume P, Maltman AJ, Bolton A, Tessier D, Ogawa Y, Takizawa S (1997b) Scaly fabrics in sheared clays from the décollement zone of the Barbados accretionary prism. In: *Proceedings of the Ocean Drilling Program, Scientific Results,* **156,** (eds Shipley TH, Ogawa Y, Blum P, Bahr JM), pp. 59–77, Ocean Drilling Program, College Station, TX.

Ladanyi B, Archambault G (1970) Simulation of shear behaviour of a jointed rock mass. *Proceedings 11th Symposium on Rock Mechanics,* American Institute of Mechanical Engineers, New York.

Lai G, Ge H, Xue L, Brodsky EE, Huang F, Wang W (2014) Tidal response variation and recovery following the Wenchuan earthquake from water level data of multiple wells in the near field. *Tectonophysics,* **619–620,** 115–122.

Lai WC, Koizumi N, Matsumoto N, Kitagawa Y, Lin CW, Shieh CL, Lee YP (2004) Effects of seismic ground motion and geological setting on the coseismic groundwater level changes caused by the 1999 Chi-Chi earthquake, Taiwan. *Earth, Planets, and Space,* **56,** 873–880.

Landtwing MR, Pettke T, Halter WE, Heinrich CA, Redmond PB, Einaudi MT, Kunze K (2005) Copper deposition during quartz dissolution by cooling magmatic-hydrothermal fluids: the Bingham porphyry. *Earth and Planetary Science Letters,* **235,** 229–243.

Lauer RM, Saffer DM (2012) Fluid budgets of subduction zone forearcs: the contribution of splay faults. *Geophysical Research Letters,* **39,** LI3604.

Lecampion B, Garagash D (2014) Confined flow of suspensions modeled by a frictional rheology. *Journal of Fluid Mechanics,* **759,** 199–235.

Lecumberri-Sanchez P, Steele-MacInnis M, Weis P, Driesner T, Bodnar RJ (2015) Salt precipitation in magmatic-hydrothermal systems associated with upper crustal plutons. *Geology,* **43,** G37163-1, doi:10.1130/G37163.1.

Lee HS, Cho TF (2002) Hydraulic characteristics of rough fractures in linear flow under normal and shear load. *Rock Mechanics and Rock Engineering,* **35,** 299–318.

Lee S-G, Kim T-K, Lee TJ (2011) Strontium isotope geochemistry and its geochemical implication from hot spring waters in South Korea. *Journal of Volcanology and Geothermal Research,* **208,** 12–22.

Leonard M (2010) Earthquake fault scaling: self-consistent relating of rupture length, width, average displacement, and moment release. *Bulletin of the Seismological Society of America,* **100,** 1971–1988.

Le Pichon X, Henry P, Lallemant S (1990) Water flow in the Barbados accretionary complex. *Journal of Geophysical Research,* **95,** 8945–8967.

Leroueil S, Bouclin G, Tavenas F, Bergeron L, La Rochelle P (1990) Permeability anisotropy of clays as a function of strain. *Canadian Geotechnical Journal,* **27,** 568–579.

Leroy P, Revil A (2004) A triple-layer model of the surface electrochemical properties of clay minerals. *Journal of Colloid and Interface Science,* **270,** 371–380.

Li L, Unsworth MJ, Booker JR, Wei W, Tan H, Jones AG (2003) Partial melt or aqueous fluid in the mid-crust of Southern Tibet? Constraints from INDEPTH magnetotelluric data. *Geophysical Journal International,* **153,** 289–304.

Li Y-G, Ellsworth WL, Thurber CH, Malin PE, Aid K (1997) Fault-zone guided waves from explosions in the San Andreas fault at Parkfield and Cienega Valley, California. *Bulletin of the Seismological Society of America,* **87,** 210–221.

Lilliefors HW (1967) On the Kolmogorov-Smirnov test for normality with mean and variance unknown. *Journal of the American Statistical Association,* **62,** 399–402.

Lin J, Stein RS (2004) Stress triggering in thrust and subduction earthquakes and stress interaction between the southern San Andreas and nearby thrust and strike-slip faults. *Journal of Geophysical Research,* **109,** B02303.

Lin W, Takahashi M, Nakamura T, Fujii Y (2008) Tensile strength and deformability of Inada Granite and their anisotropy: comparison between uniaxial tension test and Brazilian test. *Japanese Geotechnical Journal,* **3,** 165–173.

Lin WN, Daily W (1990) Hydrological properties of Topopah spring tuff under a thermal-gradient-laboratory results. *International Journal of Rock Mechanics and Mining Sciences,* **27,** 373–386.

Lindberg A, Siitari-Kauppi M (1998) Shear zone-related hydrothermal alteration in Proterozoic rocks in Finland. *9th International Symposium on Water-Rock Interaction,* **WRI-9,** 413-6.

Lister CRB (1974) On the penetration of water into hot rock. *Geophysical Journal International,* **39,** 465–509.

Litchfield NJ, Van DR, Sutherland R, Barnes PM, Cox SC, Norris RJ, Beavan RJ, Langridge RM, Villamor P, Berryman KR, Stirling MW, Nicol A, Nodder SD, Lamarche G, Barrell DJA, Pettinga JR, Little TA, Pondard N, Clark KJ (2013) A model of active faulting in New Zealand. *New Zealand Journal of Geology and Geophysics,* **57,** 32–56.

Little CTS, Cann JR, Herrington RJ, Morisseau M (1999) Late Cretaceous hydrothermal vent communities from the Troodos ophiolite, Cyprus. *Geology,* **27,** 1027–1030.

Little TA, Cox SC, Vry JK, Batt G (2005) Variations in exhumation level and uplift rate along the oblique-slip Alpine Fault, central Southern Alps, New Zealand. *Geological Society of America Bulletin,* **117,** 707–723.

Liu G, Zheng C, Gorelick SM (2007) Evaluation of the applicability of the dual-domain mass transfer model in porous media containing connected high-conductivity channels. *Water Resources Research*, **43**, W12407.

Liu H-H, Rutqvist J, Berryman JC (2009) On the relationship between stress and elastic strain for porous and fractured rock. *International Journal of Rock Mechanics and Mining Sciences*, **46**, 289–296.

Liu H-H, Wei M-Y, Rutqvist J (2013) Normal-stress dependence of fracture hydraulic properties including two-phase flow properties. *Hydrogeology Journal*, **21**, 371–382.

Liu LB, Roeloffs E, Zheng XY (1989) Seismically induced water level fluctuations in the Wali well, Beijing, China. *Journal of Geophysical Research*, **94**, 9453–9462.

Liu WQ, Manga M (2009) Changes in permeability caused by dynamic stresses in fractured sandstone. *Geophysical Research Letters*, **36**, L20307.

Liu WQ, Wang CH, Hwang LS (2010) Temporal variation of seepage water chemistry before and after the Hengchun M_s7.2 earthquake in south Taiwan. *Geoderma*, **155**, 107–114.

Liu Y, Rice JR (2007) Spontaneous and triggered aseismic deformation transients in a subduction fault model. *Journal of Geophysical Research*, **112**, B09404.

Long H, Flemings PB, Germaine JT, Saffer DM (2011) Consolidation and overpressure near the seafloor in the Ursa Basin, deepwater Gulf of Mexico. *Earth and Planetary Science Letters*, **305**, 11–20.

Long JCS, Remer JS, Wilson CR, Witherspoon PA (1982a) Porous media equivalents for networks of discontinuous fractures. *Water Resources Research*, **18**, 645–658.

Long JJ, Imber J (2011) Geological controls on fault relay zone scaling. *Journal of Structural Geology*, **33**, 1790–1800.

Long PE, Apted MJ, Spane FA, Kim K (1982b) Geologic, geochemical, rock mechanics, and hydrologic characteristics of candidate repository horizons. Session II-B, BWIP Technology Review, Proceedings of the 1982 National Waste Terminal Storage Program Information Meeting. U.S. Department of Energy, DOE/NWTS-30, pp. 29–36.

Lopez DL, Smith L (1995) Fluid flow in fault zones: analysis of the interplay of convective circulation and topographically driven groundwater flow. *Water Resources Research*, **31**, 1489–1503.

Loucks RG, Dodge MM, Galloway WE (1984) Regional controls of diagenesis and reservoir quality in Lower Tertiary sandstones along the Texas Gulf Coast. In: *Clastic Diagenesis* (eds McDonald DA, Surdam RC), *American Association of Petroleum Geologists Memoir*, **37**, pp. 15–46, AAPG.

Louis C (1969) A study of groundwater flow in jointed rock and its influence on the stability of rock masses. Technical Report 9, Rock Mechanics, Imperial College, London, United Kingdom.

Louis C (1974) Rock hydraulics. In: *Rock Mechanics* (ed Muller L), pp. 299–387. Springer, Vienna.

Louis C, Dessenne JL, Feuga B (1977) Interaction between water flow phenomena and the mechanical behavior of soil or rock masses. In: *Finite Elements in Geomechanics* (ed Gudehus G), pp. 479–511. John Wiley & Sons, New York.

Lowell RP (1991) Modeling continental and submarine hydrothermal systems. *Reviews of Geophysics*, **29**, 457–476.

Lowell RP, Germanovich LN (2004) Hydrothermal processes at mid-ocean ridges: results from scale analysis and single-pass models. In: *Mid-Ocean Ridges: Hydrothermal Interactions between the Lithosphere and Oceans*, **148** (eds German CR, Lin J, Parson LM), pp. 219–244. American Geophysical Union, Washington, DC.

Lowell RP, Van Cappellen P, Germanovich LN (1993) Silica precipitation in fractures and the evolution of permeability in hydrothermal upflow zones. *Science*, **260**, 192–194.

Lowenstein TK, Timofeeff MO, Brennan ST, Hardie LA (2001) Oscillations in Phanerozoic seawater chemistry: evidence from fluid inclusions. *Science*, **294**, 1086–1088.

Lozinsky RP (1987) Cross-section across the Jornado del Muerto, Engle, and Northern Palomas Basins, south-central New Mexico. *New Mexico Geology*, **9**, 55–57.

Lucier A, Zoback M (2008) Assessing the economic feasibility of regional deep saline aquifer CO_2 injection and storage: a geomechanics-based workflow applied to the Rose Run sandstone in eastern Ohio, USA. *International Journal of Greenhouse Gas Control*, **2**, 230–247.

Luijendijk E (2012) The role of fluid flow in the thermal history of sedimentary basins: inferences from thermochronology and numerical modeling in the Roer Valley Graben, southern Netherlands. PhD thesis, Vrije Universiteit, Amsterdam.

Luijendijk E, Gleeson T (2015) How well can we predict permeability in sedimentary basins? Deriving porosity-permeability algorithms for non-cemented sand and clay mixtures. *Geofluids*, **15**, 67–83.

Luijendijk E, Ter Voorde M, Van Balen RT, Verweij H, Simmelink E (2011) Thermal state of the Roer Valley Graben, part of the European Cenozoic Rift System. *Basin Research*, **23**, 65–82.

Luthi S, Souhaité P (1990) Fracture apertures from electrical borehole scans. *Geophysics*, **55**, 821–833.

Lynn BA, Campbell KA, Moore J, Brown PRL (2008) Origin and evolution of the Steamboat Springs siliceous sinter deposit, Nevada, U.S.A. *Sedimentary Geology*, **210**, 111–131.

Lyon MK, Leal LG (1998a) An experimental study of the motion of concentrated suspensions in two-dimensional channel flow, Part 1, Monodisperse systems. *Journal of Fluid Mechanics*, **363**, 25–56.

Lyon MK, Leal LG (1998b) An experimental study of the motion of concentrated suspensions in two-dimensional channel flow, Part 2, Bidisperse systems. *Journal of Fluid Mechanics*, **363**, 57–77.

Lyubetskaya T, Ague JJ (2009) Modeling the magnitudes and directions of regional metamorphic fluid flow in collisional orogens. *Journal of Petrology*, **50**, 1505–1531.

Mader HM, Llewellin EW, Mueller SP (2013) The rheology of two-phase magmas: a review and analysis. *Journal of Volcanology and Geothermal Research*, **257**, 135–158.

Maher K, Chamberlain CP (2014) Hydrologic regulation of chemical weathering and the geologic carbon cycle. *Science*, **343**, 1502–1504.

Mahesh P, Kundu B, Catherine JK, Gahalaut VK (2011) Anatomy of the 2009 Fiordland earthquake (M_w7.8), South Island, New Zealand. *Geoscience Frontiers*, **2**, 17–22.

Mahyari AT, Selvadurai APS (1998) Enhanced consolidation in brittle geomaterials susceptible to damage. *Mechanics of Cohesive Frictional Materials*, **3**, 291–303.

Mailloux B, Person M, Kelley S, Dunbar N, Cather S, Strayer L, Hudleston P (1999) Tectonic controls on the hydrogeology of the Rio Grande Rift, New Mexico. *Water Resources Research*, **35**, 2641–2659.

Makurat A, Barton N, Rad NS, Bandis S (1990a) Joint conductivity variation due to normal and shear deformation. In: *Proceedings of the International Symposium on Rock Joints*, Loen, Norway, 4–7 June 1990 (eds Barton N, Stephansson O), pp. 535–540. Balkema, Rotterdam.

Makurat A, Barton N, Tunbridge L, Vik G (1990b) The measurements of the mechanical and hydraulic properties of rock joints at different scale in the Stripa project. In: *Proceedings of the International Symposium on Rock Joints*, Loen, Norway, 4–7 June 1990 (eds Barton N, Stephansson O), pp. 541–548. Balkema, Rotterdam.

Mallon AJ, Swarbrick RE (2002) A compaction trend for non-reservoir North Sea chalk. *Marine and Petroleum Geology*, **19**, 527–539.

Mallon AJ, Swarbrick RE (2008) How should permeability be measured in fine-grained lithologies? Evidence from the chalk. *Geofluids*, **8**, 35–45.

Mallon AJ, Swarbrick RE, Katsube TJ (2005) Permeability of fine-grained rocks: new evidence from chalks. *Geology*, **33**, 21–24.

Maloney SM, Kaiser PK, Vorauer A (2006) A Re-assessment of in situ stresses in the Canadian Shield. In: *The 41st U.S. Symposium on Rock Mechanics* (ed Yale DP) pp. 1494–1503. American Rock Mechanics Association, Golden, Colorado.

Maltman A, Labaume P, Housen B (1997) Structural geology of the décollement at the toe of the Barbados accretionary prism. In: *Proceedings of the Ocean Drilling Program, Scientific Results*, **156** (eds Shipley TH, Ogawa Y, Blum P, Bahr JM), 279–292, Ocean Drilling Program, College Station, TX.

Manga M (1996) Hydrology of spring-dominated streams in the Oregon Cascades. *Water Resources Research*, **32**, 2435–2439.

Manga M (1997) A model for discharge in spring-dominated streams and implications for the transmissivity and recharge of quaternary volcanics in the Oregon Cascades. *Water Resources Research*, **33**, 1813–1822.

Manga M, Beresnev I, Brodsky EE, Elkhoury JE, Elsworth D, Ingebritsen SE, Mays DC, Wang C-Y (2012) Changes in permeability caused by transient stresses: field observations, experiments, and mechanisms. *Reviews of Geophysics*, **50**, RG2004.

Manga M, Brodsky E (2006) Seismic triggering of eruptions in the far field: volcanoes and geysers. *Annual Review of Earth and Planetary Sciences*, **34**, 263–291.

Manga M, Brodsky EE, Boone M (2003) Response of streamflow to multiple earthquakes. *Geophysical Research Letters*, **30**, 1214.

Manga M, Kirchner JW (2004) Interpreting the temperature of water at cold springs and the importance of gravitational potential energy. *Water Resources Research*, **40**, W05110, doi:10.1029/2003WR002905.

Manga M, Rowland JC (2009) Response of Alum Rock springs to the October 30, 2007 earthquake and implications for the origin of increased discharge after earthquakes. *Geofluids*, **9**, 237–250.

Manga M, Wang C-Y (2007) Earthquake hydrology. In: *Treatise on Geophysics* (ed Schubert GS), pp. 293–320. Elsevier, Amsterdam.

Manighetti I, Campillo M, Sammis C, Mai PM, King G (2005) Evidence for self-similar, triangular slip distributions on earthquakes: implications for earthquake and fault mechanics. *Journal of Geophysical Research*, **110**, doi:10.1029/2004JB003174.

Manning C, Ingebritsen SE (1999) Permeability of the continental crust: the implications of geothermal data and metamorphic systems. *Reviews of Geophysics*, **37**, 127–150.

March A (1932) Mathematische theorie der regelung nach der korngestelt bei affiner deformation. *Zeitschrift fur Kristallographie, Mineralogie und Petrographie*, **81**, 285–297.

Marchildon N, Dipple GM (1998) Irregular isograds, reaction instabilities, and the evolution of permeability during metamorphism. *Geology*, **26**, 15–18.

Mariner RH, Presser TS, Evans WC (1983) Geochemistry of active geothermal systems in the Northern Basin and Range Province. *Geothermal Resources Council Special Report*, **13**, 95–119.

Marion DP (1990) Acoustical, mechanical, and transport properties of sediments and granular materials. PhD thesis, Stanford University.

Markl G, Bucher K (1998) Composition of fluids in the lower crust inferred from metamorphic salt in lower crustal rocks. *Nature*, **391**, 781–783.

Marschall P, Horseman S, Gimmi T (2005) Characterisation of gas transport properties of the Opalinus Clay, a potential host rock formation for radioactive waste disposal. *Oil & Gas Science Technology - Revue d'IFP Energies Nouvelles*, **60**, 121–139.

Marshall TJ (1958) A relation between permeability and size distribution of pores. *Journal of Soil Science*, **9**, 1–8.

Martin CD, Chandler NA (1994) The progressive fracture of Lac du Bonnet Granite. *International Journal of Rock Mechanics and Mining Sciences*, **31**, 643–659.

Martino JB, Chandler NA (2004) Excavation-induced studies at the Underground Research Laboratory. *International Journal of Rock Mechanics and Mining Sciences*, **41**, 1413–1426.

Mascle A, Moore JC, ODP Leg 110 Scientific Party (1988) *Proceedings of the Ocean Drilling Program, Initial Reports (Pt. A)*, **110**.

Massart TJ, Selvadurai APS (2012) Stress-induced permeability evolution in quasi-brittle geomaterials. *Journal of Geophysical Research*, **117**, doi:10.1029/2012JB009251.

Massart TJ, Selvadurai APS (2014) Computational modelling of crack-induced permeability evolution in granite with dilatant cracks. *International Journal of Rock Mechanics and Mining Sciences*, **70**, 593–604.

Matmon D, Bekins BA (2006) Hydromechanics of a high taper angle, low-permeability prism: a case study from Peru. *Journal of Geophysical Research*, **111**, B07101.

Matsuki K, Chida Y, Sakaguchi K, Glover PWJ (2006) Size effect on aperture and permeability of a fracture as estimated in large synthetic fractures. *International Journal of Rock Mechanics and Mining Sciences*, **43**, 726–755.

Matthai SK, Geiger S, Roberts SG, Paluszny A, Belayneh M, Burri A, Mezentsev A, Lu H, Coumou D, Driesner T, Heinrich CA (2007) Numerical simulation of multi-phase fluid flow in structurally complex reservoirs. In: *Structurally Complex Reservoirs*, vol. **292** (eds Jolley SJ, Barr D, Walsh JJ, Knipe RJ), pp. 405–429. The Geological Society, London.

Matthews CS, Russell DG (1967) *Pressure Buildup and Flow Tests in Wells*. Society of Petroleum Engineers, Dallas, TX.

Maurer DK, Plume RW, Thomas JM, Johnson AK (1996) Water resources and effects of changes in ground-water use along the Carlin Trend, north-central Nevada. *U.S. Geological Survey Water-Resources Investigations Report*, **96–4134**.

Mavko G, Nur A (1997) The effect of a percolation threshold in the Kozeny-Carman relation. *Geophysics*, **62**, 1480–1482.

Mazurek M (2000) Geological and hydraulic properties of water-conducting features in crystalline rocks. In: *Hydrogeology of Crystalline Rocks* (eds Stober I, Bucher K), pp. 3–26. Kluwer Academic Publishers, Dordrecht.

Mazurek M, Hurford AJ, Leu W (2006) Unravelling the multi-stage burial history of the Swiss molasse basin: integration of apatite fission track, vitrinite reflectance and biomarker isomerisation analysis. *Basin Research*, **18**, 27–50.

Mazurek M, Jakob A, Bossart P (2003) Solute transport in crystalline rocks at Aspo, I, Geological basis and model calibration. *Journal of Contaminant Hydrology*, **61**, 157–174.

McCaig AM (1997) The geochemistry of volatile flow in shear zones. In: *Deformation-Enhanced Fluid Transport in the Earth's Crust and Mantle* (ed Holness MB) *Mineralogical Society Series*, **8**, pp. 227–266, Chapman & Hall, London.

McCarthy JF (1991) Analytical models of the effective permeability of sand-shale reservoirs. *Geophysical Journal International*, **105**, 513–527.

McCord JP, Moe H (1990) Interpretation of hydraulic testing at the Kaisten borehole. In: *Nagra Technical Report*, **89-18** (ed Nagra), Nagra, Baden, Switzerland, http://www.nagra.ch/data/documents/database/dokumente/$default/Default%20Folder/Publikationen/NTBs%201989-1990/e_ntb89-18.pdf accessed 06 May 2016.

McDonald MG, Harbaugh AW (1984) A modular three-dimensional finite-difference groundwater flow model. *U.S. Geological Survey Open-File Report*, **83–875**.

McGrail BP, Sullivan EC, Spane FA, Bacon DH, Hund G, Thome PD, Thompson CJ, Reidel SP, Colwell FS (2009) *Preliminary Hydrogeologic Characterization Results from the Wallula Basalt Pilot Study*. PNWD-4129, Battelle-Pacific Northwest Division, Richland, Washington.

McGuire KJ, McDonnell JJ, Weiler M, Kendall C, McGlynn BL, Welker JM, Seibert J (2005) The role of topography on catchment-scale water residence time. *Water Resources Research*, **41**, W05002, doi:10.1029/2004WR003657.

Mclnerney P, Guillen A, Courrioux G, Calcagno P, Lees T (2005) Building 3D geological models directly from the data? A new approach applied to Broken Hill, Australia, *U.S. Geological Survey Open-File Report*, **2005-1428**, http://pubs.usgs.gov/of/2005/1428/mcinerney/ accessed 06 May 2016.

McKenna JR, Blackwell DD, Richards MC (2005) Natural state modeling, structure, preliminary temperature and chemical synthesis of the Dixie Valley, Nevada Geothermal System. Proceedings Thirtieth Workshop on Geothermal Reservoir Engineering, Stanford University, Stanford, California, January 31-February 2, 2005, *Stanford Geothermal Program Technical Report*, **SGP-TR-176**.

McKenzie D (1984) The generation and compaction of partially molten rock. *Journal of Petrology*, **2**, 713–765.

McKenzie D (1987) The compaction of igneous and sedimentary rocks. *Journal of the Geological Society*, **144**, 299–307.

McKenzie JM, Voss CI (2013) Permafrost thaw in a nested groundwater-flow system. *Hydrogeology Journal*, **21**, 299–316.

McKiernan AW, Saffer DM (2006) Data report: Permeability and consolidation properties of subducting sediments off Costa Rica, ODP Leg 205. *Proceedings of the Ocean Drilling Program, Scientific Results*, **205**, 1–24.

McNeill LC, Ienaga M, Tobin H, Saito S, Goldberg D, Moore JC, Mikada H (2004) Deformation and in situ stress in the Nankai Accretionary prism from resistivity-at-bit images, ODP Leg 196. *Geophysical Research Letters*, **31**, L02602.

Medina R, Elkhoury J, Morris J, Prioul R, Desroches J, Detwiler R (2015) Flow of dense suspensions through fractures: significant in-plane velocity variations caused by small variations in solid concentration. *Geofluids*, **15**, 24–36.

Méheust Y, Schmittbuhl J (2003) Scale effects related to flow in rough fractures. *Pure and Applied Geophysics*, **160**, 1023–1050.

Menegon L, Fusseis F, Stunitz H, Xiao XH (2015) Creep cavitation bands control porosity and fluid flow in lower crustal shear zones. *Geology*, **43**, 227–230.

Menzies CD (2012) Fluid flow associated with the Alpine Fault, South Island, New Zealand. PhD thesis, University of Southampton.

Menzies CD, Teagle DAH, Craw D, Cox SC, Boyce AJ, Barrie D (2014) Incursion of meteoric waters into the ductile regime in an active orogen. *Earth and Planetary Science Letters*, **399**, 1–14.

Mesri G, Olson RE (1971) Mechanisms controlling the permeability of clays. *Clays and Clay Minerals*, **19**, 151–158.

Meyer H, Hetzel R, Fügenschuh B, Strauss H (2010) Determining the growth rate of topographic relief using in situ-produced ^{10}Be: a case study in the Black Forest, Germany. *Earth and Planetary Science Letters*, **290**, 391–402.

Michaels AS, Lin C (1954) Permeability of kaolinite. *Industrial and Engineering Chemistry*, **46**, 1239–1246.

Micklethwaite S (2008) Optimally oriented "fault-valve" thrusts: evidence for aftershock-related fluid pressure pulses? *Geochemistry, Geophysics, Geosystems*, **9**, doi:10.1029/2007GC001916.

Micklethwaite S, Cox SF (2006) Progressive fault triggering and fluid flow in aftershock domains: examples from mineralized Archaean fault systems. *Earth and Planetary Science Letters*, **250**, 318–330.

Micklethwaite S, Ford A, Witt W, Sheldon H (2015) The where and how of faults, fluids, and permeability – insights from fault stepovers, scaling properties, and gold mineralization. *Geofluids*, **15**, 240–251.

Micklethwaite S, Sheldon HA, Baker T (2010) Active fault and shear processes and their implications for mineral deposit formation and discovery. *Journal of Structural Geology*, **32**, 151–165.

Micucki EJ (1998) Hydrothermal transport and depositional processes in Archean lode-gold systems: a review. *Ore Geology Reviews*, **13**, 307–321.

Mikada H, Becker K, Moore JC, Klaus A, ODP Leg 196 Scientific Party (2002) *Proceedings of the Ocean Drilling Program, Initial Reports*, **196**, doi:10.2973/odp.proc.ir.196.2002.

Miller SA (2013) The role of fluids in tectonic and earthquake processes. *Advances in Geophysics*, **54**, 1–46.

Miller SA (2015) Modeling enhanced geothermal systems and the essential nature of large-scale changes in permeability at the onset of slip. *Geofluids*, **15**, 338–349.

Miller SA, Collettini C, Chiaraluce L, Cocco M, Barchi M, Kaus BJP (2004) Aftershocks driven by a high-pressure CO_2 source at depth. *Nature*, **427**, 724–727.

Miller SA, Nur A (2000) Permeability as a toggle switch in fluid-controlled crustal processes. *Earth and Planetary Science Letters*, **183**, 133–146.

Miller SA, Nur A, Olgaard DL (1996) Earthquakes as a coupled shear stress high pore pressure dynamical system. *Geophysical Research Letters*, **23**, 197–200.

Milliken KL, Esch WL, Reed RM, Zhang T (2012) Grain assemblages and strong diagenetic overprinting in siliceous mudrocks, Barnett Shale (Mississippian), Fort Worth Basin, Texas. *AAPG Bulletin*, **96**, 1553–1578.

Milliken KL, Reed RM (2010) Multiple causes of diagenetic fabric anisotropy in weakly consolidated mud, Nankai accretionary prism, IODP Expedition 316. *Journal of Structural Geology*, **32**, 1887–1898.

Milodowski AE, Gillespie MR, Naden J, Fortey NJ, Shepherd TJ, Pearce JM, Metcalfe R (1998) The petrology and paragenesis of fracture mineralization in the Sellafield area, west Cumbria. *Proceedings of the Yorkshire Geological Society*, **52**, 215–241.

Milord I, Sawyer EW, Brown M (2001) Formation of diatexite migmatite and granite magma during anatexis of semi-pelitic metasedimentary rocks: an example from St. Malo, France. *Journal of Petrology*, **42**, 487–505.

Milsch H, Heinrich W, Dreisen G (2003) Reaction-induced flow in synthetic quartz-bearing marbles. *Contributions to Mineralogy and Petrology*, **146**, 286–296.

Min K-B, Rutqvist J, Elsworth D (2009) Chemically and mechanically mediated influences on the transport and mechanical characteristics of rock fractures. *International Journal of Rock Mechanics and Mining Sciences*, **46**, 80–89.

Min K-B, Rutqvist J, Tsang C-F, Jing L (2004) Stress-dependent permeability of fractured rock masses: a numerical study. *International Journal of Rock Mechanics and Mining Sciences*, **41**, 1191–1210.

MINEDEX. Geological Survey of Western Australia, http://www.dmp.wa.gov.au/3970.aspx, accessed 2014.

Miyazawa M (2011) Propagation of an earthquake triggering front from the 2011 Tohoku-Oki earthquake. *Geophysical Research Letters*, **38**, L23307.

Moe H, McNeish JA, McCord JP, Andrews RW (1990) Interpretation of hydraulic testings at the Schafisheim borehole. In: *Nagra Technical Report*, **89-09** (ed Nagra), Nagra, Baden, Switzerland, http://www.nagra.ch/data/documents/database/dokumente/$default/Default%20Folder/Publikationen/NTBs%201989-1990/e_ntb89-09.pdf accessed 06 May 2016.

Mogi K, Mochizuki H, Kurokawa Y (1989) Temperature changes in an artesian spring at Usami in the Izu Peninsula (Japan) and their relation to earthquakes. *Tectonophysics*, **159**, 95–108.

Monaghan JJ (2012) Smoothed particle hydrodynamics and its diverse applications. *Annual Review of Fluid Mechanics*, **44**, 323–346.

Mondol NH, Bjorlykke K, Jahren J (2008) Experimental compaction of clays: relationship between permeability and petrophysical properties in mudstones. *Petroleum Geoscience*, **14**, 319–337.

Montgomery C (2013). Fracturing fluids. *Proceedings International Conference for Effective and Sustainable Hydraulic Fracturing (HF2013)*, Brisbane, Australia, May 20–22, 2013, InTech, pp. 3–24.

Montgomery DR, Manga M (2003) Streamflow and water well responses to earthquakes. *Science*, **300**, 2047–2049.

Moore DE, Lockner DA, Byerlee JD (1994) Reduction of permeability in granite at elevated temperatures. *Science*, **265**, 1558–1561.

Moore DE, Morrow CA, Byerlee JD (1982) Use of swelling clays to reduce permeability and its potential application to nuclear waste repository sealing. *Geophysical Research Letters*, **9**, 1009–1012.

Moore DE, Morrow CA, Byerlee JD (1983) Chemical reactions accompanying fluid flow through granite held in a temperature gradient. *Geochimica et Cosmochimica Acta*, **47**, 445–453.

Moore GF, Taira A, Klaus A, ODP Leg 190 Scientific Party (2001) *Proceedings of the Ocean Drilling Program, Initial Reports*, **190**, doi:10.2973/odp.proc.ir.l90.2001.

Moore JC (1989) Tectonics and hydrogeology of accretionary prisms: role of the décollement zone. *Journal of Structural Geology*, **11**, 95–106.

Moore JC, Moore GF, Cochrane GR, Tobin HJ (1995) Negative-polarity seismic reflections along faults of the Oregon accretionary prism: indicators of overpressuring. *Journal of Geophysical Research*, **100**, 12895–12906.

Moore JC, ODP Leg 110 Scientific Party (1987) Expulsion of fluids from depth along a subduction-zone décollement horizon. *Nature*, **326**, 785–788.

Moore JC, Orange D, Kulm LD (1990) Interrelationship of fluid venting and structural evolution: *Alvin* observations from the frontal accretionary prism, Oregon. *Journal of Geophysical Research*, **95**, 8795–8808.

Moore JC, Saffer D (2001) Updip limit of the seismogenic zone beneath the accretionary prism of southwest Japan: an effect of diagenetic to low-grade metamorphic processes and increasing effective stress. *Geology*, **29**, 183–186.

Morad S, Ketzer JM, de Ros LF (2000) Spatial and temporal distribution of diagenetic alterations in siliciclastic rocks: implications for mass transfer in sedimentary basins. *Sedimentology*, **47**, 95–120.

Moran K, Gray WGD, Brown KM (1995) Permeability and stress history of sediment from the Cascadia margin. *Proceedings of the Ocean Drilling Program, Scientific Results*, **146-1**, 275–280.

Morgan JK, Ramsey EB, Ask MVS (2008) Deformation and mechanical strength of sediments at the Nankai subduction zone. In: *The Seismogenic Zone of Subduction Thrust Faults* (eds Dixon TH, Moore JC), pp. 210–256. Columbia University Press, New York.

Morgan P, Seager WR, Golombek MP (1986) Cenozoic thermal mechanical and tectonic evolution of the Rio Grande Rift. *Journal of Geophysical Research*, **91**, 6263–6276.

Morris BL, Lawrence ARL, Chilton PJC, Adams B, Calow RC, Klinck BA (2003a) Groundwater and its susceptibility to degradation: a global assessment of the problem and options for management. *United Nations Environment Programme (UNEP) Early Warning and Assessment Report Series*, **RS 03-3**.

Morris JD, Villinger HW, Klaus A, ODP Leg 205 Scientific Party (2003b) *Proceedings of the Ocean Drilling Program, Initial Reports*, **205**, doi:10.2973/odp.proc.ir.205.2003.

Morris JP, Monaghan JJ (1997) A switch to reduce SPH viscosity. *Journal of Computational Physics*, **136**, 41–50.

Morrow CA, Lockner DA (1997) Permeability and porosity of the Illinois UPH 3 drillhole granite and a comparison with other deep drillhole rocks. *Journal of Geophysical Research*, **102**, 3067–3075.

Morrow CA, Lockner DA, Moore DE, Byerlee JD (1981) Permeability of granite in a temperature gradient. *Journal of Geophysical Research*, **86**, 3002–3008.

Mortensen AK, Axelsson G (2013) Developing a Conceptual Model of a Geothermal System: presented at "Short Course on Conceptual Modelling of Geothermal Systems", Santa Tecla, El Salvador, February 24 - March 2, 2013, accessed at http://www.os.is/gogn/unu-gtp-sc/UNU-GTP-SC-16-29.pdf.

Mortimer N (1993) Jurassic tectonic history of the Otago schist, New Zealand. *Tectonics*, **12**, 237–244.

Mottl MJ, Wheat CG, Fryer P, Gharib J, Martin JB (2004) Chemistry of springs across the Mariana forearc shows progressive devolatilization of the subducting plate. *Geochimica et Cosmochimica Acta*, **68**, 4915–4933.

Muir-Wood R, King G (1993) Hydrological signatures of earthquake strain. *Journal of Geophysical Research*, **98**, 22035–22068.

Müller TM, Sahay PN (2011) Stochastic theory of dynamic permeability in poroelastic media. *Physical Review E*, **84**, 026329.

Munier R, Talbot CJ (1993) Segmentation, fragmentation and jostling of cratonic basement in and near Äspö, southeast Sweden. *Tectonics*, **12**, 713–727.

Muntean JL, Einaudi MT (2000) Porphyry gold deposits of the Refugio district, Maricunga belt, northern Chile. *Economic Geology*, **95**, 1445–1472.

Munz IA, Yardley BWD, Banks DA, Wayne D (1995) Deep penetration of sedimentary fluids into basement rocks from southern Norway: evidence from hydrocarbon and brine inclusions in quartz veins. *Geochimica et Cosmochimica Acta*, **59**, 239–254.

Munz IA, Yardley BWD, Gleeson SA (2002) Petroleum infiltration of high-grade basement, South Norway: pressure-temperature-time (P-T-t-X) constraints. *Geofluids*, **2**, 41–53.

Murdoch LC, Germanovich LN (2012) Storage change in a flat-lying fracture during well tests. *Water Resources Research*, **48**, W12528.

Murdoch LC, Richardson JR, Tan QF, Malin SC, Fairbanks C (2006) Forms and sand transport in shallow hydraulic fractures in residual soil. *Canadian Geotechnical Journal*, **43**, 1061–1073.

Nabelek PI (2002) Calc-silicate reactions and bedding-controlled isotopic exchange in the Notch Peak aureole, Utah: implications for differential fluid fluxes with metamorphic grade. *Journal of Metamorphic Geology*, **20**, 429–440.

Nabelek PI (2009) Numerical simulation of kinetically-controlled calc-silicate reactions and fluid flow with transient permeability around crystallizing plutons. *American Journal of Science*, **309**, 517–548.

Nagra (1985) Sondierbohrung Böttstein, Untersuchungsbericht. *Nagra Technischer Bericht (Technical Report of the Swiss National Cooperative for the Disposal of Radioactive Waste)*, **85–01**.

Nagra (1992) Sondierbohrung Schafisheim, Untersuchungsbericht. *Nagra Technischer Bericht (Technical Report of the Swiss National Cooperative for the Disposal of Radioactive Waste)*, **88-11**.

Najari M, Selvadurai APS (2014) Thermo-hydro-mechanical response of granite to temperature changes. *Environmental Earth Sciences*, **72**, 189–198.

Nakagawa M (1983) Geology and petrology of Moriyoshi volcano. *Journal of the Japanese Association of Mineralogy, Petrology and Economic Geology*, **78**, 197–210.

Nakajima J, Yoshida K, Hasegawa A (2013) An intraslab seismic sequence activated by the 2011 Tohoku-oki earthquake: evidence for fluid-related embrittlement. *Journal of Geophysical Research*, **118**, 3492–3505.

Nakazato H, Oba T, Itaya T (1996) The geology and K-Ar ages of the Gassan volcano, northeast Japan. *Journal the Japanese Association of Mineralogy, Petrology and Economic Geology*, **91**, 1–10.

National Research Council (1996) *Rock Fractures and Fluid Flow*. National Academies Press, Washington, DC.

National Research Council (2001) *Basic Research Opportunities in the Earth Sciences*. National Academies Press, Washington, DC.

Nazareth JJ, Hauksson E (2004) The seismogenic thickness of the southern California crust. *Bulletin of the Seismological Society of America*, **94**, 940–960.

Neglia S (1979) Migration of fluids in sedimentary basins. *American Association of Petroleum Geologists Bulletin*, **63**, 573–597.

Nehlig P, Juteau T (1988) Flow porosities, permeabilities and preliminary data on fluid inclusions and fossil thermal gradients in the crustal sequence of the Semail ophiolite (Oman). *Tectonophysics*, **151**, 199–221.

Nehlig P, Juteau T, Bendel V, Cotten J (1994) The root zones of oceanic hydrothermal systems – Constraints from the Samail Ophiolite (Oman). *Journal of Geophysical Research*, **99**, 4703–4713.

Nelson PH, Kibler JE (2001) A catalog of porosity and permeability from core plugs in siliciclastic rocks. *U.S. Geological Survey Open-File Report*, **03–420**.

Nemcok M, Henk A, Gayer RA, Vandycke S, Hathaway TM (2002) Strike-slip fault bridge fluid pumping mechanism: insights from field-based palaeostress analysis and numerical modelling. *Journal of Structural Geology*, **24**, 1885–1902.

Nemoto K, Watanabe N, Hirano N, Tsuchiya N (2009) Direct measurement of contact area and stress dependence of anisotropic flow through rock fracture with heterogeneous aperture distribution. *Earth and Planetary Science Letters*, **281**, 81–87.

Neuman SP (1994) Generalized scaling of permeabilities: validation and effect of support scale. *Geophysical Research Letters*, **21**, 349–352.

Neuman SP (2005) Trends, prospects and challenges in quantifying flow and transport through fractured rocks. *Hydrogeology Journal*, **13**, 124–147.

Neuzil C (1994) How permeable are clays and shales? *Water Resources Research*, **30**, 145–150.

Neuzil CE (1995) Abnormal pressures as hydrodynamic phenomena. *American Journal of Science*, **295**, 742–786.

Neuzil CE (2003) Hydromechanical coupling in geologic processes. *Hydrogeology Journal*, **11**, 41–83.

Newcomb RC (1959) Some preliminary notes on ground water in the Columbia River Basalt. *Northwest Science*, **33**, 1–18.

Nguyen TS, Börgesson L, Chijimatsu M, Hernelind J, Jing L, Kobayashi A, Fujita T, Jussila P, Rutqvist J, Jing L (2009) A case study on the influence of THM coupling on the near field safety of a spent fuel repository in sparsely fractured granite. *Environmental Geology*, **57**, 1239–1254.

Nguyen TS, Jing L (ed) (2008) *DECOVALEX-THMC Project, Task A, Influence of near field coupled THM phenomena on the performance of a spent fuel repository*. Report of Task *A2*, SKI Report, **44**.

Nguyen TS, Selvadurai APS (1995) Coupled thermal-mechanical-hydrological behaviour of sparsely fractured rock: implications for nuclear fuel waste disposal. *International Journal of Rock Mechanics and Mining Sciences*, **32**, 465–479.

Nguyen TS, Selvadurai APS (1998) A model for coupled mechanical and hydraulic behaviour of a rock joint. *International Journal for Numerical and Analytical Methods in Geomechanics*, **22**, 29–48.

Nielsen KA (2007) *Fractured Aquifers: Formation Evaluation by Well Testing*. Trafford Publishing, Victoria, BC.

Nishimoto S, Yoshida H (2010) Hydrothermal alteration of deep fractured granite: effects of dissolution and precipitation. *Lithos*, **115**, 153–162.

Noir J, Jacques E, Bekri S, Adler PM, Tapponnier P, King GCP (1997) Fluid flow triggered migration of events in the 1989 Dobi earthquake sequence of Central Afar. *Geophysical Research Letters*, **24**, 2335–2338.

Nolan TB, Anderson GH (1934) The geyser area near Beowawe, Eureka County, Nevada. *American Journal of Science*, **27**, 215–229.

Noorishad J, Tsang CF, Witherspoon PA (1984) Coupled thermo-hydraulic-mechanical phenomena in saturated fractured porous rocks: numerical approach. *Journal of Geophysical Research*, **89**, 10365–10373.

Nordgård Bolås HM, Hermanrud C, Schutter TA, Grimsmo Teige GM (2008) Is stress-insensitive chemical compaction responsible for high overpressures in deeply buried North Sea chalks? *Marine and Petroleum Geology*, **25**, 565–587.

Norris RJ, Cooper AF (2001) Late Quaternary slip rates and their significance for slip partitioning on the Alpine Fault, New Zealand. *Journal of Structural Geology*, **23**, 507–520.

Norton D, Knapp R (1977) Transport phenomena in hydrothermal systems – Nature of porosity. *American Journal of Science*, **277**, 913–936.

Norton D, Knight J (1977) Transport phenomena in hydrothermal systems: cooling plutons. *American Journal of Science*, **277**, 937–981.

Nunn JA, Deming D (1991) Thermal constraints on basin-scale flow systems. *Geophysical Research Letters*, **18**, 967–970.

Nye JF (1953) The flow law of ice from measurements in glacier tunnels, laboratory experiments and the Jungfraufirn borehole experiment. *Proceedings of the Royal Society of London, Series A, Mathematical and Physical Sciences*, **219A**, 477–489.

Oelkers EH, Schott J (2001) An experimental study of enstatite dissolution rates as a function of pH, temperature, and aqueous Mg and Si concentration, and the mechanism of pyroxene/pyroxenoid dissolution. *Geochimica et Cosmochimica Acta*, **65**, 1219–1231.

Ohzono M, Yabe Y, Iinuma T, Ohta Y, Miura S, Tachibana K, Sato T, Demachi T (2013) Strain anomalies induced by the 2011 Tohoku Earthquake (M_w 9.0) as observed by a dense GPS network in northeastern Japan. *Earth, Planets and Space*, **64**, 1231–1238.

Okada T, Matsuzawa T, Umino N, Yoshida K, Hasegawa A, Takahashi H, Yamada T, Kosuga M, Takeda T, Kato A, Igarashi T, Obara K, Sakai S, Saiga A, Iidaka T, Iwasaki T, Hirata N, Tsumura N, Yamanaka Y, Terakawa T, Nakamichi H, Okuda T, Horikawa S, Katao H, Miura T, Kubo A, Matsushima T, Goto K, Miyamachi H (2015) Hypocenter migration and crustal seismic velocity distribution observed for the inland earthquake swarms induced by the 2011 Tohoku earthquake in NE Japan: implications for crustal fluid distribution and crustal permeability. *Geofluids*, **15**, 293–309.

Okada T, Umino N, Hasegawa A (2003) Rupture process of the July 2003 northern Miyagi earthquake sequence, NE Japan, estimated from double-difference hypocenter locations. *Earth, Planets, and Space*, **55**, 741–750.

Okada T, Umino N, Hasegawa A, Group for the aftershock observations of the Iwate-Miyagi Nairiku Earthquake in 2008 (2012) Hypocenter distribution and heterogeneous seismic velocity structure in and around the focal area of the 2008 Iwate-Miyagi Nairiku Earthquake, NE Japan—Possible seismological evidence for a fluid driven compressional inversion earthquake. *Earth, Planets, and Space*, **64**, 717–728.

Okada T, Yoshida K, Ueki S, Nakajima J, Uchida N, Matsuzawa T, Umino N, Hasegawa A, Group for the aftershock observations of the 2011 off the Pacific coast of Tohoku earthquake (2011) Shallow inland earthquakes in NE Japan possibly triggered by the 2011 off the Pacific coast of Tohoku Earthquake. *Earth, Planets and Space*, **63**, 749–754.

Okada Y (1992) Internal deformation due to shear and tensile faults in a half-space. *Bulletin of the Seismological Society of America*, **82**, 1018–1040.

Oliver NHS (1995) Hydrothermal history of the Mary Kathleen fold belt, Mt. Isa Block, Queensland. *Australian Journal of Earth Sciences*, **42**, 267–280.

Oliver NHS (1996) Review and classification of structural controls on fluid flow during regional metamorphism. *Journal of Metamorphic Geology*, **14**, 477–492.

Oliver NH, Butera KM, Rubenach MJ, Marshall LJ, Cleverley, JS, Mark G, Esser D (2008). The protracted hydrothermal evolution of the Mount Isa Eastern Succession: A review and tectonic implications. *Precambrian Research*, **163**, 108–130.

Olmsted FH, Rush FE (1987) Hydrogeologic reconnaissance of the Beowawe geysers geothermal area, Nevada. *Geothermics*, **16**, 27–46.

Olsen HW (1966) Darcy's law in saturated kaolinite. *Water Resources Research*, **2**, 287–295.

Olson P, Christensen U (1986) Solitary wave-propagation in a fluid conduit within a viscous matrix. *Journal of Geophysical Research*, **91**, 6367–6374.

Olsson R, Barton N (2001) An improved model for hydromechanical coupling during shearing of rock joints. *International Journal of Rock Mechanics and Mining Sciences*, **38**, 317–329.

Ordonez-Miranda J, Alvarado-Gil J (2011) On the stability of the exact solutions of the dual-phase lagging model of heat conduction. *Nanoscale Research Letters*, **6**, 327.

Ordonez-Miranda J, Alvarado-Gil JJ, Yang R (2012) Effective thermal conductivity of metal-dielectric composites at the non-dilute limit. *International Journal of Thermophysics*, **33**, 2118–2124.

Orense RP, Kiyota T, Yamada S, Cubrinovski M, Hosono Y, Okamura M, Yasuda S (2011) Comparison of liquefaction features observed during the 2010 and 2011 Canterbury earthquakes. *Seismological Research Letters*, **82**, 905–918.

Ortiz AER, Renner J, Jung R (2011) Hydromechanical analyses of the hydraulic stimulation of borehole Basel 1. *Geophysical Journal International*, **185**, 1266–1287.

Ostrowski LP, Kloska MB (1989) Final interpretation of hydraulic testing at the Siblingen borehole. In: *Nagra Technical Report*, **89-10** (ed Nagra), Nagra, Baden, Switzerland. http://www.timeride.ch/data/documents/database/dokumente/$default/Default%20Folder/Publikationen/NTBs%201989-1990/e_ntb89-10.pdf, accessed 06 May 2016.

Owens L (2013) Geochemical investigation of hydrothermal and volcanic systems in Iceland, New Mexico and Antarctica. PhD thesis, New Mexico Institute of Mining and Technology, http://www.ees.nmt.edu/outside/alumni/papers/2013d_owens_lb.pdf accessed 06 May 2016.

Ozisik MN, Tzou DY (1994) On the wave theory in heat conduction. *Journal of Heat Transfer*, **116**, 526–535.

Papadopulos SS, Bredehoeft JD, Cooper HH (1973) On the analysis of 'slug test' data. *Water Resources Research*, **9**, 1087–1089.

Parsons B, Sclater JG (1977) An analysis of the variation of ocean floor bathymetry and heat flow with age. *Journal of Geophysical Research*, **82**, 803–827.

Parsons T, Sliter R, Geist EL, Jachens RC, Jaffe BE, Foxgrover A, Hart PE, McCarthy J (2003) Structure and mechanics of the Hayward-Rodgers Creek Fault step-over, San Francisco Bay, California. *Bulletin of the Seismological Society of America*, **93**, 2187–2200.

Paschke SS, Banta ER, Dupree JA, Capesius JP, Litke DW (2011) Groundwater availability of the Denver Basin aquifer system, Colorado, *U.S. Geological Survey Professional Paper*, **1770**.

Patrinos GP, Cooper DN, van Mulligen E, Gkantouna V, Tzimas G, Tatum Z, Schultes E, Roos M, Barend M (2012) Microattribution and nanopublication as means to incentivize the placement of human genome variation data into the public domain. *Human Mutation*, **33**, 1503–1512.

Patton FD (1966) Multiple modes of shear failure in rock. *Proceedings 1st Congress of International Society of Rock Mechanics*, Lisbon, 509-13.

Pazdniakou A, Adler PM (2013) Dynamic permeability of porous media by the lattice Boltzmann method. *Advances in Water Resources*, **62B**, 292–302.

Pearce JA, Lippard SJ, Roberts S (1984) Characteristics and tectonic significance of supra-subduction zone ophiolites. *Geological Society of London Special Publications*, **16**, 77–94.

Peltzer G, Rosen P, Rogez F, Hudnut K (1996) Postseismic rebound in fault step-overs caused by pore fluid flow. *Science*, **273**, 1202–1204.

Pepin J, Person M, Phillips F, Kelley S, Timmons S, Owens L, Witcher J, Gable C (2015) Deep fluid circulation within crystalline basement rocks and the role of hydrologic windows in the formation of the Truth or Consequences, New Mexico low-temperature geothermal system. *Geofluids*, **15**, 139–160.

Person M, Banerjee A, Hofstra D, Sweetkind D, Gao Y (2008) Hydrologic models of modern and fossil geothermal systems within the Great Basin: implications for Carlin-type gold mineralization. *Geosphere*, **4**, 888–917.

Person M, Hofstra A, Sweetkind D, Stone W, Cohen D, Gable C, Banerjee A (2012) Analytical and numerical models of hydrothermal fluid flow at fault intersections. *Geofluids*, **12**, 312–326.

Person M, Taylor J, Dingman S (1998) Sharp interface models of salt water intrusion and wellhead delineation on Nantucket Island, Massachusetts. *Groundwater*, **36**, 731–742.

Pester NJ, Reeves EP, Rough ME, Ding K, Seewald JS, Seyfried WE (2012) Subseafloor phase equilibria in high-temperature hydrothermal fluids of the Lucky Strike Seamount (mid-Atlantic Ridge, 37°17'N). *Geochimica et Cosmochimica Acta*, **90**, 303–322.

Peters C (2008a) Accessibilities of reactive minerals in consolidated sedimentary rock: an imaging study of three sandstones. *Chemical Geology*, **265**, 198–208.

Peters EJ (2012) *Advanced Petrophysics, Volume 1, Geology, Porosity, Absolute Permeability, Heterogeneity, and Geostatistics*. Live Oak Book Company, Palo Alto, CA.

Peters SE (2005) Geologic constraints on the macroevolutionary history of marine animals. *Proceedings of the National Academy of Sciences USA*, **102**, 12326–12331.

Peters SE (2006) Macrostratigraphy of North America. *The Journal of Geology*, **114**, 391–412.

Peters SE (2008b) Environmental determinants of extinction selectivity in the fossil record. *Nature*, **454**, 626–629.

Peters SE, Kelly DC, Fraass A (2013) Oceanographic controls on the diversity and extinction of planktonic foraminifera. *Nature*, **4932**, 398–401.

Peters SE, Zhang C, Livny M, Ré C (2014) A machine-compiled macroevolutionary history of Phanerozoic life. ArXiv Preprint: 1406.2963.

Petrovitch CL, Nolte DD, Pyrak-Nolte LJ (2013) Scaling of fluid flow versus fracture stiffness. *Geophysical Research Letters*, **40**, 2076–2080.

Phillips OM (1991) *Flow and Reactions in Permeable Rocks*. Cambridge University Press, Cambridge.

Piggott AR, Elsworth D (1993), Laboratory assessment of the equivalent apertures of a rock fracture. *Geophysical Research Letters*, **20**, 1387–1390.

Pine RJ, Batchelor AS (1984) Downward migration of shearing in jointed rock during hydraulic injections. *International Journal of Rock Mechanics and Mining Sciences*, **21**, 249–263.

Platt JP, Leggett JK, Young J, Raza H, Alam S (1985) Large-scale sediment underplating in the Makran accretionary prism, southwest Pakistan. *Geology*, **13**, 507–511.

Plesha ME (1987) Constitutive models for rock discontinuities with dilatancy and surface degradation. *International Journal for Numerical and Analytical Methods in Geomechanics*, **11**, 345–362.

Plouraboué F, Kurowski P, Boffa JM, Hulin JP, Roux S (2000) Experimental study of the transport properties of rough self-affine fractures. *Journal of Contaminant Hydrology*, **46**, 295–318.

Plume RW, Carlton SM (1988) Hydrogeology of the Great Basin region of Nevada, Utah, and adjacent states. *U.S. Geological Survey Hydrologic Investigations Atlas*, **HA-694-A**, scale 1:100,000.

Plummer LN, Bexfield LM, Anderholm SK, Sanford WE, Eurybiades B (2004) Geochemical characterization of groundwater flow in the Santa Fe Group aquifer system, Middle Rio Grande Basin, New Mexico. *U.S. Geological Survey Water-Resources Investigations Report*, **03–4131**.

Poiseuille J-M (1844) *Recherches Expérimentales sur le Mouvement des Liquides Dans les Tubes de Très-petits Diamètres*. Imprimerie Royale, Paris.

Pokrovski GS, Borisova AY, Harrichoury JC (2008) The effect of sulfur on vapor-liquid fractionation of metals in hydrothermal systems. *Earth and Planetary Science Letters*, **266**, 345–362.

Polak A, Elsworth D, Yasuhara A, Grader AS, Halleck PM (2003) Permeability reduction of a natural fracture under net dissolution by hydrothermal fluids. *Geophysical Research Letters*, **30**, doi:10.1029/2003GL017575.

Pollack HN, Hurter SJ, Johnson JR (1993) Heat flow from the Earth's interior: analysis of the global data set. *Reviews of Geophysics*, **31**, 267–280.

Polski Y, Capuano L, Finger J, Huh M, Knudsen S, Chip Mansure A, Raymond D, Swanson R (2008) Enhanced Geothermal Systems (EGS) well construction technology evaluation report. *Sandia Report*, **SAND2008-7866**, Sandia National Laboratories, Albuquerque, New Mexico.

Porcello JJ, Tolan TL, Lindsey KA (2009) Groundwater level declines in the Columbia River Basalt Group and their relationship to mechanisms for groundwater recharge—A conceptual groundwater system model, Columbia Basin Ground Water Management Area of Adams, Franklin, Grant, and Lincoln Counties. Othello, Washington, prepared by the Columbia Basin Ground Water Management Area of Adams, Franklin, Grant, and Lincoln Counties, June 2009.

Porta G, Chaynikov S, Riva M, Guadagnini A (2013) Upscaling solute transport in porous media from the pore scale to dual-and multicontinuum formulations. *Water Resources Research*, **49**, 2025–2039.

Powell T, Cumming W (2010) Spreadsheets for geothermal water and gas geochemistry. *Proceedings of the 35th Workshop on Geothermal Reservoir Engineering, Stanford University*, **SGP-TR-188**.

Powell WC (1929) Report of an investigation of the Hot Springs artesian basin, Hot Springs, New Mexico. *Ninth Biennial Report of the State Engineer of New Mexico*, 1929–1930, **120-9**.

Power IM, Wilson SA, Dipple GM (2013) Serpentinite carbonation for CO_2 sequestration. *Elements*, **9**, 115–121.

Power WL, Durham WB (1997) Topography of natural and artificial fractures in granitic rocks: implications for studies of rock friction and fluid migration. *International Journal of Rock Mechanics and Mining Sciences*, **34**, 979–989.

Power WL, Tullis TE, Brown SR, Boitnott GN, Scholz CH (1987) Roughness of natural fault surfaces. *Geophysical Research Letters*, **14**, 29–32.

Pratt H, Swolfs H, Brace W, Black A (1977) Elastic and transport properties of in situ jointed granite. *International Journal of Rock Mechanics and Mining Sciences*, **14**, 35–45.

Preisig G, Cornaton FJ, Perrochet P (2012) Regional flow simulation in fractured aquifers using stress-dependent parameters. *Groundwater*, **50**, 376–385.

Preisig G, Eberhardt E, Gischig V, Roche V, van der Baan M, Valley B, Kaiser P, Duff D, Lowther R (2015) Development of connected rock mass permeability in massive crystalline rocks through hydraulic fracture propagation and shearing accompanying fluid injection. *Geofluids*, **15**, 321–337.

Pride SR, Tromeur E, Berryman JG (2002) Biot slow-wave effects in stratified rock. *Geophysics*, **67**, 271–281.

PRISM Climate Group (2004) *PRISM climate group: Oregon State University*. Accessed October 1, 2009, http://www.prismclimate.org.

Priyatkina N, Kullerud K, Bergh S, Armitage P, Ravna E (2011) CO_2 sequestration during interactions between fluid and mafic to intermediate intrusive rocks on Vannoya Island, West Troms Basement Complex, North Norway. *Proceedings 10th International Congress of Applied Mineralogy, Trondheim (ICAM)*, pp. 549–553.

Putnis A, John T (2010) Replacement processes in the Earth's crust. *Elements*, **6**, 159–164.

Pyrak-Nolte LJ, Cook NGW, Nolte DD (1988) Fluid percolation through single fractures, *Geophysical Research Letters*, **15**, 1247–1250.

Pyrak-Nolte LJ, Morris JP (2000) Single fractures under normal stress: the relation between fracture specific stiffness and fluid flow. *International Journal of Rock Mechanics and Mining Sciences*, **37**, 245–262.

Pyrak-Nolte LJ, Myer LR, Cook NGW, Witherspoon PA (1987) Hydraulic and mechanical properties of natural fractures in low permeability rock. *Proceedings International Congress on Rock Mechanics of ISRM*, Balkema, Rotterdam, 225–231.

Quigley M, Van DR, Litchfield NJ, Villamor P, Duffy B, Barrell DJA, Furlong K, Stahl T, Bilderback E, Noble D (2012) Surface rupture during the 2010 M_w7.1 Darfield (Canterbury) earthquake: implications for fault rupture dynamics and seismic-hazard analysis. *Geology*, **40**, 55–58.

Quigley M, Van DR, Villamor P, Litchfield NJ, Barrell DJA, Furlong K, Stahl T, Duffy B, Bilderback E, Noble D, Townsend DB, Begg JG, Jongens R, Ries W, Claridge J, Klahn A, Mackenzie H, Smith A, Hornblow S, Nicol R, Cox SC, Langridge RM, Pedley K (2010) Surface rupture of the Greendale Fault during the Darfield (Canterbury) Earthquake, New Zealand: initial findings. *Bulletin of the New Zealand Society for Earthquake Engineering*, **43**, 236–242.

Quon SH, Ehlers EG (1963) Rocks of the northern part of the mid-Atlantic Ridge. *Geological Society of America Bulletin*, **74**, 1–7.

Rabinowicz M, Ricard Y, Gregoire M (2002) Compaction in a mantle with a very small melt concentration: implications for the generation of carbonatitic and carbonate-bearing high alkaline mafic melt impregnations. *Earth and Planetary Science Letters*, **203**, 205–220.

Rabinowicz M, Vigneresse JL (2004) Melt segregation under compaction and shear channeling: application to granitic magma segregation in a continental crust. *Journal of Geophysical Research*, **109**, B04407.

Raffensperger JP, Garven G (1995) The formation of unconformity-type uranium ore deposits, 1, Coupled groundwater flow and heat transport modeling. *American Journal of Science*, **295**, 581–636.

Ramm M (1992) Porosity-depth trends in reservoir sandstones: theoretical models related to Jurassic sandstones, offshore Norway. *Marine and Petroleum Geology*, **9**, 553–567.

Ramsay JG (1980) The crack-seal mechanism of rock deformation. *Nature*, **284**, 135–139.

Ranalli G (1995) *Rheology of the Earth*. Springer-Verlag, New York.

Ranero CR, Grevemeyer I, Sahling H, Barckhausen U, Hensen C, Wallmann K, Weinrebe W, Vannucchi P, von Huene R, McIntosh K (2008) Hydrological system of erosional convergent margins and its influence on tectonics and interplate seismogenesis. *Geochemistry, Geophysics, Geosystems*, **9**, Q03S04.

Ranero CR, Morgan JP, McIntosh K, Reichert C (2003) Bending-related faulting and mantle serpentinization at the Middle America trench. *Nature*, **425**, 367–373.

Ranjram M, Gleeson T, Luijendijk E (2015) Is the permeability of crystalline rock in the shallow crust related to depth, lithology, or tectonic setting? *Geofluids*, **15**, 106–119.

Raven KG, Gale JE (1985) Water flow in a natural rock fracture as a function of stress and sample size. *International Journal of Rock Mechanics and Mining Sciences*, **22**, 251–261.

Redmond PB, Einaudi MT, Inan EE, Landtwing MR, Heinrich CA (2004) Copper deposition by fluid cooling in intrusion-centered systems: new insights from the Bingham porphyry ore deposit, Utah. *Geology*, **32**, 217–220.

Reece JS, Flemings PB, Germaine JT (2013) Data report: Permeability, compressibility, and microstructure of resedimented mudstone from IODP Expedition 322, Site C0011. In Saito S, Underwood MB, Kubo Y, and the Expedition 322 Scientists, *Proceedings of the Integrated Ocean Drilling Program*, **322**, 1–26.

Reidel SP (1983) Stratigraphy and petrogenesis of the Grande Ronde Basalt from the deep canyon country of Washington, Oregon, and Idaho. *Geological Society of America Bulletin*, **94**, 519–542.

Reidel SP, Camp VE, Tolan TL, Martin BS (2013) The Columbia River flood basalt province: stratigraphy, areal extent, volume, and physical volcanology. *Geological Society of America Special Paper*, **497**, 1–43.

Reidel SP, Johnson VG, Spane FA (2002) Natural gas storage in basalt aquifers of the Columbia Basin, Pacific Northwest USA— A guide to site characterization. Pacific Northwest National Laboratory, Richland, Washington, PNNL-13962, http://www.pnl.gov/main/publications/external/technical_reports/PNNL-13962.pdf, accessed 06 May 2016.

Reinecker J, Tingay M, Muller B, Heidbach O (2010) Present-day stress orientation in the Molasse Basin. *Tectonophysics*, **482**, 129–138.

Reiter M, Eggleston R, Broadwell B, Minier J (1986) Estimates of terrestrial heat flow from deep petroleum tests along the Rio Grande Rift in central and southern New Mexico. *Journal of Geophysical Research*, **91**, 6225–6245.

Rejeb A, Rouabhi A, Millard A, Maßmann J, Uehara S (2008) *DECOVALEX-THMC Project, Task C, Hydromechanical response of the Tournemire Argillite to the underground openings excavation: unsaturated zones and mine-by-test experiment*. Final Report, SKI Report, **44**.

Renard F, Brosse E, Gratier JP (2000a) The different processes involved in the mechanism of pressure-solution in quartz-rich rocks and their interactions. In: *Quartz Cementation of Sandstones* (eds Worden RH, Morad S). *International Association of Sedimentologists Special Publication*, **29**, pp. 67–78, IAS.

Renard F, Gratier JP, Jamtveit B (2000b) Kinetics of crack-sealing, intergranular pressure solution, and compaction around active faults. *Journal of Structural Geology*, **22**, 1395–1407.

Renner JL, White DE, Williams DL (1975) Hydrothermal convection systems. In: Assessment of geothermal resources of the United States 1975. *United States Geological Survey Circular*, **726**, 5–57.

Renshaw CE (1995) On the relationship between mechanical and hydraulic apertures in rough-walled fractures. *Journal of Geophysical Research*, **100**, 24629–24636.

Reuschle T (2011) Data report: Permeability measurements under confining pressure, Expeditions 315 and 316, Nankai Trough. In: Kinoshita M, Tobin H, Ashi J, Kimura G, Lallement S, Screaton EJ, Curewitz D, Masago H, Moe KT, and the Expedition 314/315/316 Scientists, *Proceedings of the Integrated Ocean Drilling Program*, **314/315/316**, 1–17.

Revil A (2002) Mechanical compaction of sand/clay mixtures. *Journal of Geophysical Research*, **107**, doi:10.1029/2001JB000318.

Revil A, Cathles LM (1999) Permeability of shaly sands. *Water Resources Research*, **35**, 651–662.

Revil A, Cathles LM (2002) Fluid transport by solitary waves along growing faults – A field example from the South Eugene Island Basin, Gulf of Mexico. *Earth and Planetary Science Letters*, **202**, 321–335.

Revil A, Florsch N (2010) Determination of permeability from spectral induced polarization in granular media. *Geophysical Journal International*, **181**, 1480–1498.

Revil A, Leroy P, Titov K (2005) Characterization of transport properties of argillaceous sediments: application to the Callovo-Oxfordian argillite. *Journal of Geophysical Research*, **110**, B06202, doi:10.1029/2004JB003442.

Reyes AG, Christenson BW, Faure K (2010) Sources of solutes and heat in low-enthalpy mineral waters and their relation to tectonic setting, New Zealand. *Journal of Volcanology and Geothermal Research*, **192**, 117–141.

Reyners ME (2009) Large subduction thrust earthquake shakes southern New Zealand. *Eos Transactions American Geophysical Union*, **90**, 282.

Reyners ME (2011) Lessons from the destructive M_w6.3 Christchurch, New Zealand, earthquake. *Seismological Research Letters*, **82**, 371–372.

Rice JR (1992) Fault stress states, pore pressure distributions, and the weakness of the San Andreas fault. In: *Fault Mechanics and Transport Properties of Rocks* (eds Evans B, Wong T-F), pp. 475–503. Academic Press, San Diego.

Richard GC, Kanjilal S, Schmeling H (2012) Solitary-waves in geophysical two-phase viscous media: a semi-analytical solution. *Physics of the Earth and Planetary Interiors*, **198**, 61–66.

Richards JP (2003) Tectono-magmatic precursors for porphyry Cu-(Mo-Au) deposit formation. *Economic Geology*, **98**, 1515–1533.

Richards JP (2013) Giant ore deposits formed by optimal alignments and combinations of geological processes. *Nature Geoscience*, **6**, 911–916.

Richardson CJ, Cann JR, Richards HG, Cowan JG (1987) Metal-depleted root zones of the Troodos ore-forming hydrothermal systems, Cyprus. *Earth and Planetary Science Letters*, **84**, 243–253.

Richter FM, McKenzie D (1984) Dynamical models for melt segregation from a deformable rock matrix. *Journal of Geology*, **92**, 729–740.

Rider MH (2002) *The Geological Interpretation of Well Logs*. Rider-French Consulting Ltd., Sutherland.

Rimstidt JD, Barnes HL (1980) The kinetics of silica-water reactions. *Geochimica et Cosmochimica Acta*, **44**, 1683–1699.

Rimstidt JD, Cole DR (1983) Geothermal mineralization, I, The mechanism of formation of the Beowawe, Nevada, siliceous sinter deposit. *American Journal of Science*, **283**, 861–875.

Rinaldi AP, Rutqvist J, Cappa F (2013) Geomechanical effects on CO_2 leakage through fault zones during large-scale underground injection. *International Journal of Greenhouse Gas Control*, **20**, 117–131.

Robert F, Boullier A-M, Firdaous K (1995) Gold-quartz veins in metamorphic terranes and their bearing on the role of fluids in faulting. *Journal of Geophysical Research*, **100**, 861–881.

Roberts SJ, Nunn JA, Cathles LM, Cipriani F-D (1996) Expulsion of abnormally pressured fluids along faults. *Journal of Geophysical Research*, **101**, 231–252.

Roeloffs EA (1988) Hydrologic precursors to earthquakes: a review. *Pure and Applied Geophysics*, **126**, 177–209.

Roeloffs EA (1996) Poroelastic techniques in the study of earthquake related hydrologic phenomena. *Advances in Geophysics*, **37**, 135–195.

Roeloffs EA (1998) Persistent water level changes in a well near Parkfield, California, due to local and distant earthquakes. *Journal of Geophysical Research*, **103**, 868–889.

Rogers PSZ, Pitzer KS (1982) Volumetric properties of aqueous sodium chloride solutions. *Journal of Physical and Chemical Reference Data*, **11**, 15–81.

Rojstaczer SA, Ingebritsen SE, Hayba DO (2008) Permeability of continental crust influenced by internal and external forcing. *Geofluids*, **8**, 128–139.

Rojstaczer SA, Wolf S (1992) Permeability changes associated with large earthquakes: an example from Loma Prieta, California. *Geology*, **20**, 211–214

Rojstaczer SA, Wolf S, Michel R (1995) Permeability enhancement in the shallow crust as a cause of earthquake-induced hydrological changes. *Nature*, **373**, 237–239.

Rolandone F, Bürgmann R, Nadeau RM (2004) The evolution of the seismic-aseismic transition during the earthquake cycle: constraints from the time-dependent depth distribution of aftershocks. *Geophysical Research Letters*, **31**, doi:10.1029/2004GL021379.

Rona PA, Hannington MD, Raman CV, Thompson G, Tivey MK, Humphris SE, Lalou C, Petersen S (1993) Active and relict sea-floor hydrothermal mineralization at the TAG hydrothermal field, mid-Atlantic Ridge. *Economic Geology*, **88**, 1989–2017.

Ronov AB (1978) The Earth's sedimentary shell. *International Geology Review*, **24**, 1313–1363.

Roselle GT, Baumgartner LP, Valley JW (1999) Stable isotope evidence of heterogeneous fluid infiltration at the Ubehebe Peak contact aureole, Death Valley National Park, California. *American Journal of Science*, **299**, 93–138.

Rowe CD, Meneghini F, Moore JC (2009) Fluid-rich damage zone of an ancient out-of-sequence thrust, Kodiak Islands, Alaska. *Tectonics*, **28**, TC1006.

Rowe K, Screaton EJ, Ge S (2012) Coupled fluid-flow and deformation modeling of the frontal thrust region of the Kumano Basin transect, Japan: implications for fluid pressures and decollement downstepping. *Geochemistry, Geophysics, Geosystems*, **13**, Q0AD23.

Rowe K, Screaton E, Guo J, Underwood MB (2011) Data report: Permeabilities of sediments from the Kumano Basin transect off Kii Peninsula, Japan. In: Kinoshita M, Tobin H, Ashi J, Kimura G, Lallemant S, Screaton EJ, Curewitz D, Masago H, Moe KT, and the Expedition 314/315/316 Scientists, *Proceedings of the Integrated Ocean Drilling Program*, **314/315/316**, 1–24.

Rowland JC, Manga M, Rose TP (2008) The influence of poorly interconnected fault zone flow paths on spring geochemistry. *Geofluids*, **8**, 93–101.

Rowland JV, Simmons SF (2012) Hydrologic, magmatic and tectonic controls on hydrothermal flow, Taupo Volcanic Zone, New Zealand: implications for the formation of epithermal vein deposits. *Economic Geology*, **107**, 427–457.

Rubenach M (2013) Structural controls of metasomatism on a regional scale. In: *Metasomatism and the Chemical Transformation of Rocks* (eds Harlov DE, Austrheim H), Springer-Verlag, Berlin, Heidelburg, 93–140.

Rutqvist J (1995) Determination of hydraulic normal stiffness of fractures in hard rock from hydraulic well testing. *International Journal of Rock Mechanics and Mining Sciences*, **32**, 513–523.

Rutqvist J (2004) *Drift Scale THM Model*. MDL-NBS-HS-000017 REV 0, Bechtel SAIC Company, Las Vegas, Nevada.

Rutqvist J (2011) Status of the TOUGH-FLAC simulator and recent applications related to coupled fluid flow and crustal deformations. *Computers and Geoscience*, **37**, 739–750.

Rutqvist J (2012) The geomechanics of CO_2 storage in deep sedimentary formations. *International Journal of Geotechnical and Geological Engineering*, **30**, 525–551.

Rutqvist J (2015) Fractured rock stress-permeability relationships from in situ data and effects of temperature and chemical-mechanical couplings. *Geofluids*, **15**, 48–66.

Rutqvist J, Barr D, Birkholzer JT, Fujisaki K, Kolditz O, Liu Q-S, Fujita T, Wang W, Zhang C-Y (2009a) A comparative simulation study of coupled THM processes and their effect on fractured rock permeability around nuclear waste repositories. *Environmental Geology*, **57**, 1347–1360.

Rutqvist J, Börgesson L, Chijimatsu M, Hernelind J, Jing L, Kobayashi A, Nguyen S (2009b) Modeling of damage, permeability changes and pressure responses during excavation of the TSX tunnel in granitic rock at URL, Canada. *Environmental Geology*, **57**, 1263–1274.

Rutqvist J, Chijimatsu M, Jing L, De Jonge J, Kohlmeier M, Millard A, Nguyen TS, Rejeb A, Souley M, Sugita Y, Tsang C-F (2005) Numerical study of the THM effects on the near-field safety of a hypothetical nuclear waste repository – BMT1 of the DECOVALEX III project, Part 3, Effects of THM coupling in fractured rock. *International Journal of Rock Mechanics and Mining Sciences*, **42**, 745–755.

Rutqvist J, Freifeld B, Min K-B, Elsworth D, Tsang Y (2008) Analysis of thermally induced changes in fractured rock permeability during eight years of heating and cooling at the Yucca Mountain Drift Scale Test. *International Journal of Rock Mechanics and Mining Sciences*, **45**, 1373–1389.

Rutqvist J, Leung C, Hoch A, Wang Y, Wang Z (2013a) Linked multicontinuum and crack tensor approach for modeling of coupled geomechanics, fluid flow and transport in fractured rock. *International Journal of Rock Mechanics and Geotechnical Engineering*, **5**, 18–31.

Rutqvist J, Rinaldi AP, Cappa F, Moridis GJ (2013b) Modeling of fault reactivation and induced seismicity during hydraulic fracturing of shale-gas reservoirs. *Journal of Petroleum Science and Technology*, **107**, 31–44.

Rutqvist J, Stephansson O (1996) A cyclic hydraulic jacking test to determine the in situ stress normal to a fracture. *International Journal of Rock Mechanics and Mining Sciences*, **33**, 695–711.

Rutqvist J, Stephansson O (2003) The role of hydromechanical coupling in fractured rock engineering. *Hydrogeology Journal*, **11**, 7–40.

Rutqvist J, Tsang C-F (2002) A study of caprock hydromechanical changes associated with CO_2 injection into a brine formation. *Environmental Geology*, **42**, 296–305.

Rutqvist J, Tsang C-F (2003) Analysis of thermal-hydrologic-mechanical behavior near an emplacement drift at Yucca Mountain. *Journal of Contaminant Hydrology*, **62–63**, 637–652.

Rutqvist J, Tsang C-F (2012) Multiphysics processes in partially saturated fractured rock: experiments and models from Yucca Mountain. *Reviews of Geophysics*, **50**, RG3006.

Rutqvist J, Tsang C-F, Ekman D, Stephansson O (1997) Evaluation of in situ hydromechanical properties of rock fractures at Laxemar in Sweden. *Proceedings of 1st Asian Rock Mechanics Symposium ARMS 97*, Seoul, South Korea, 619–624.

Rutqvist J, Wu Y-S, Tsang C-F, Bodvarsson G (2002) A modeling approach for analysis of coupled multiphase fluid flow, heat transfer, and deformation in fractured porous rock. *International Journal of Rock Mechanics and Mining Sciences*, **39**, 429–442.

Saar MO (2011) Geothermal heat as a tracer of large-scale groundwater flow and as a means to determine permeability fields. *Hydrogeology Journal*, **19**, 31–52.

Saar MO, Manga M (2004) Depth dependence of permeability in the Oregon Cascades inferred from hydrogeologic, thermal, seismic, and magmatic modeling constraints. *Journal of Geophysical Research*, **109**, B04204.

Sadalage PJ, Fowler M (2012) *NoSQL Distilled: A Brief Guide to the Emerging World of Polyglot Persistence*. Addison-Wesley, Upper Saddle River, NJ.

Saffer DM (2003) Pore pressure development and progressive dewatering in underthrust sediments at the Costa Rican subduction margin: comparison with Northern Barbados and Nankai. *Journal of Geophysical Research*, **108**, doi:10.1029/2002JB001787.

Saffer DM (2007) Pore pressure within underthrust sediments in subduction zones. In: *The Seismogenic Zone of Subduction Thrust Faults* (eds Dixon T, Moore JC), pp. 171–209. Columbia University Press, New York.

Saffer DM (2010) Hydrostratigraphy as a control on subduction zone mechanics through its effects on drainage: an example from the Nankai Margin, SW Japan. *Geofluids*, **10**, 114–131.

Saffer DM (2015) The permeability of active subduction plate boundary faults. *Geofluids*, **15**, 193–215.

Saffer DM, Bekins BA (1998a) Episodic fluid flow in the Nankai accretionary complex: timescale, geochemistry, flow rates, and fluid budget. *Journal of Geophysical Research*, **103**, 30351–30370.

Saffer D, Bekins BA (1998b) Fluid budgets and pore pressures in the shallow Subduction Zone: a comparison of the Nankai and Cascadia accretionary systems (abstract). *Eos Transactions American Geophysical Union*, **79**, Fall Meeting Supplement, F899.

Saffer DM, Bekins BA (1999) Fluid budgets at convergent plate margins: implications for the extent and duration of fault zone dilation. *Geology*, **27**, 1095–1098.

Saffer DM, Bekins BA (2002) Hydrologic controls on the morphology and mechanics of accretionary wedges. *Geology*, **30**, 271–274.

Saffer DM, Bekins BA (2006) An evaluation of factors influencing pore pressure in accretionary complexes: implications for taper angle and wedge mechanics. *Journal of Geophysical Research*, **111**, B04101.

Saffer DM, Guo J, Underwood MB, Likos W, Skarbek RM, Song I, Gildow M (2011) Data report: Consolidation, permeability, and fabric of sediments from the Nankai continental slope, IODP Sites C0001, C0008, and C0004. In: Kinoshita M, Tobin H, Ashi J, Kimura G, Lallemant S, Screaton EJ, Curewitz D, Masago H, Moe KT, and the Expedition 314/315/316 Scientists, *Proceedings of the Integrated Ocean Drilling Program*, **314/315/316**, 1–61.

Saffer DM, McKiernan AW (2009) Evaluation of in situ smectite dehydration as a pore-water freshening mechanism in the Nankai Trough, offshore southwest Japan. *Geochemistry, Geophysics, Geosystems*, **10**, Q02010.

Saffer DM, Screaton EJ (2003) Fluid flow pathways at the toe of convergent margins: interpretation of sharp geochemical gradients. *Earth and Planetary Science Letters*, **213**, 261–270.

Saffer DM, Silver EA, Fisher AT, Tobin H, Moran K (2000) Inferred pore pressures at the Costa Rica subduction zone: implications for dewatering processes. *Earth and Planetary Science Letters*, **177**, 193–207.

Saffer DM, Tobin H (2011) Hydrogeology and mechanics of subduction zone forearcs: fluid flow and pore pressure. *Annual Review of Earth and Planetary Sciences*, **39**, 157–186.

Sahling H, Masson DG, Ranero C, Huhnerbach V, Weinrebe W et al. (2008) Fluid seepage at the continental margin offshore Costa Rica and southern Nicaragua. *Geochemistry, Geophysics, Geosystems*, 9, Q05S05.

Sample J (1996) Isotopic evidence from authigenic carbonates for rapid upward fluid flow in accretionary wedges. *Geology*, 24, 897–900.

Sanematsu K, Watanabe K, Duncan RA, Izawa E (2006) The history of vein formation determined by $^{40}Ar/^{39}Ar$ dating of adularia in the Hosen-1 vein at the Hishikari epithermal deposit, Japan. *Economic Geology*, 101, 685–698.

Sanford RM, Bowers RL, Combs J (1979) Rio Grande Rift geothermal exploration case history, Elephant Butte prospect, south-central New Mexico. *Transactions Geothermal Resources Council*, 3, 609–612.

Sanford WE, Konikow LF (1989) Simulation of calcite dissolution and porosity changes in salt water mixing zones in coastal aquifers. *Water Resources Research*, 25, 655–667.

Sangawa A (1987) Damage of the 1611 Aizu earthquake in relation to surface faulting. *Zisin (Journal of the Seismological Society of Japan 2nd series)*, 40, 235–245.

Sanner B (2000) Baden-Baden, a famous thermal spa with a long history. *Geo-Heat Center Quarterly Bulletin*, 21, 16–22.

Santagata M, Germaine JT (2002) Sampling disturbance effects in normally consolidated clays. *Journal of Geotechnical and Geoenvironmental Engineering*, 128, 997–1006.

Santagata M, Germaine JT (2005) Effect of OCR on sampling disturbance of cohesive soils and evaluation of laboratory reconsolidation procedures. *Canadian Geotechnical Journal*, 42, 459–474.

Santagata M, Kang YI (2007) Effects of geologic time on the initial stiffness of clays. *Engineering Geology*, 89, 98–111.

Santamarina JC, Klein KA, Wang YH, Prencke E (2002) Specific surface: determination and relevance. *Canadian Geotechnical Journal*, 39, 233–241.

Sardini, P. Ledésert, B., Touchard, G. (1997) Quantification of microscopic pore networks by image analysis and measurements of permeability in the Soultz-sous-Forêts granite (Alsace, France). In *Fluid Flow and Transport in Rocks* (eds Jamtveit B, Yardley BWD), pp. 171–189, Chapman & Hall, London.

Sass JH, Lachenbruch AH, Munroe RJ, Green GW, Moses TH Jr (1971) Heat flow in the western United States. *Journal of Geophysical Research*, 76, 6356–6431.

Sassa K, He B, Miyagi T, Strasser M, Konagai K, Ostric M, Setiawan H, Takara K, Nagai O, Yamashiki Y, Tutumi S (2012) A hypothesis of the Senoumi submarine megaslide in Suruga Bay in Japan – based on the undrained dynamic-loading ring shear tests and computer simulation. *Landslides*, 9, 439–455.

Sato H, Hirata N, Iwasaki T, Matsubara M, Ikawa T (2002) Deep seismic reflection profiling across the Ou Backbone range, northern Honshu Island, Japan. *Tectonophysics*, 355, 41–52.

Sato H, Yoshida T, Iwasaki T, Sato T, Ikeda Y, Umino N (2004) Late Cenozoic tectonic development of the back arc region of central northern Honshu, Japan, revealed by recent deep seismic profiling. *Journal of the Japanese Association for Petroleum Technology*, 69, 145–154.

Sato T, Sakai R, Furuya K, Kodama T (2000) Coseismic spring flow changes associated with the 1995 Kobe earthquake. *Geophysical Research Letters*, 27, 1219–1222.

Sausse J, Genter A (2005) Types of permeable fractures in granite. *Geological Society of London Special Publication*, 240, 1–14.

Sawyer AH, Flemings PB, Elsworth D, Kinoshita M (2008) Response of submarine hydrologic monitoring instruments to formation pressure changes: theory and application to Nankai advanced CORKs. *Journal of Geophysical Research*, 113, B01102.

Sawyer EW (1998) Formation and evolution of granite magmas during crustal reworking: the significance of diatexites. *Journal of Petrology*, 39, 1147–1167.

Schiffman P, Smith BM (1988) Petrology and oxygen isotope geochemistry of a fossil seawater hydrothermal system within the Solea graben, northern Troodos ophiolite, Cyprus. *Journal of Geophysical Research*, 93, 4612–4624.

Schiffman P, Smith BM, Varga RJ, Moores EM (1987) Geometry, conditions and timing of off-axis hydrothermal metamorphism and ore-deposition in the Solea graben. *Nature*, 325, 423–425.

Schlische RW, Young SS, Ackermann RV, Gupta A (1996) Geometry and scaling relations of a population of very small rift-related normal faults. *Geology*, 24, 683–686.

Schloemer S, Krooss BM (1997) Experimental characterisation of the hydrocarbon sealing efficiency of cap rocks. *Marine and Petroleum Geology*, 14, 565–580.

Schlumberger Limited (1981) *RFT – Essentials of Pressure Test Interpretation*. Schlumberger, Paris.

Schneider J, Flemings PB, Day-Stirrat RJ, Germaine JT (2011) Insights into pore-scale controls on mudstone permeability through resedimentation experiments. *Geology*, 39, 1011–1014.

Schneider Reece J, Flemings PB, Dugan B, Long H, Germaine JT (2012) Permeability-porosity relationships of shallow mudstones in the Ursa Basin, northern deepwater Gulf of Mexico. *Journal of Geophysical Research*, 117, B12102.

Scholz CH (1998) Earthquakes and friction laws. *Nature*, 391, 37–42.

Schroder RA, Strait SR (1987) Fluid temperature data from selected boreholes on the Hanford site. Rockwell Hanford Operations BWIP Supporting Document, SD-BWI-DP-065, Rev. 0.

Schultz RA, Soliva R, Fossen H, Okubo CH, Reeves DM (2008) Dependence of displacement-length scaling relations for fractures and deformation bands on the volumetric changes across them. *Journal of Structural Geology*, 30, 1405–1411.

Schweisinger T, Murdoch LC, Huey CO Jr (2007) Design of a removable borehole extensometer. *Geotechnical Testing Journal*, 30, 202–211.

Schweisinger T, Svenson EJ, Murdoch LC (2009) Introduction to hydromechanical well tests in fractured rock aquifers. *Ground Water*, 47, 69–79.

Schweisinger T, Svenson EJ, Murdoch LC (2011) Hydromechanical behavior during constant-rate pumping tests in fractured gneiss. *Hydrogeology Journal*, 19, 963–980.

Scott DR, Stevenson DJ (1984) Magma solitons. *Geophysical Research Letters*, 11, 1161–1164.

Scott DR, Stevenson DJ (1986) Magma ascent by porous flow. *Journal of Geophysical Research*, 91, 9283–9296.

Scott DR, Stevenson DJ, Whitehead JA (1986) Observations of solitary waves in a viscously deformable pipe. *Nature*, **319**, 759–761.

Scott RB, Rona PA, McGregor BA, Scott MR (1974) The TAG hydrothermal field. *Nature*, **251**, 301–302.

Scott S, Driesner T, Weis P (2015) Geologic controls on supercritical geothermal resources above magmatic intrusions. *Nature Communications*, **6**, 7837, doi:10.1038/ncomms8837.

Screaton EJ (2010) Recent advances in subseafloor hydrogeology: focus on basement-sediment interactions, subduction zones, and continental slopes. *Hydrogeology Journal*, **18**, 1547–1570.

Screaton EJ, Carson B, Davis EE, Becker K (2000) Permeability of a décollement zone: results from a two-well experiment in the Barbados accretionary complex. *Journal of Geophysical Research*, **105**, 21403–21410.

Screaton EJ, Carson B, Lennon GP (1995) Hydrogeologic properties of a thrust fault within the Oregon accretionary prism. *Journal of Geophysical Research*, **100**, 20025–20035.

Screaton EJ, Fisher AT, Carson B, Becker K (1997) Barbados Ridge hydrogeologic tests: implications for fluid migration along an active décollement. *Geology*, **25**, 239–242.

Screaton EJ, Gamage K, James S (2014) Data report: Permeabilities of Expedition 320/321 sediments from the Pacific Equatorial Transect. *Proceedings of the Integrated Ocean Drilling Program*, **320/321**.

Screaton EJ, Ge S (2012) The impact of megasplay faulting and permeability contrasts on Nankai Trough subduction zone pore pressures. *Geophysical Research Letters*, **39**, L22301.

Screaton EJ, Hays T, Gamage K, Martin JM (2006) Data report: Permeabilities of Costa Rica subduction zone sediments. *Proceeding of the Ocean Drilling Program, Scientific Results*, **205**.

Screaton EJ, Rowe K, Sutton J, Atalan G (2013) Data report: Permeabilities of Expedition 322 and 333 sediments from offshore the Kii Peninsula, Japan. *Proceedings of the Integrated Ocean Drilling Program*, **322**.

Screaton EJ, Saffer DM (2005) Fluid expulsion and overpressure development during initial subduction at the Costa Rica convergent margin. *Earth and Planetary Science Letters*, **233**, 361–374.

Screaton EJ, Wuthrich DR, Dreiss SJ (1990) Permeabilities, fluid pressures, and flow rates in the Barbados Ridge complex. *Journal of Geophysical Research*, **95**, 8997–9007.

Seager WR, Mack GH (2003) Geology of the Caballo Mountains, New Mexico. *New Mexico Bureau of Geology and Mineral Resources Memoir*, **49**.

Seager WR, Morgan P (1979) Rio Grande Rift in southern New Mexico, west Texas, and northern Chihuahua. In: *Rio Grande Rift-Tectonics and Magmatism* (ed Riecker RE), pp. 87–106, American Geophysical Union, Washington, DC

Segall P, Rice JR (1995) Dilatancy, compaction, and slip instability of a fluid-infiltrated fault. *Journal of Geophysical Research*, **100**, 22155–22171.

Selvadurai APS (ed) (1996) *Mechanics of Poroelastic Media*. Kluwer Academic Publishers, Dordrecht, The Netherlands.

Selvadurai APS (2000) *Partial Differential Equations in Mechanics Vol. 1: Fundamentals, Laplace's Equation, the Diffusion Equation, the Wave Equation*. Springer-Verlag, Berlin.

Selvadurai APS (2004) Stationary damage modelling of poroelastic contact. *International Journal of Solids and Structures*, **41**, 2043–2064.

Selvadurai APS (2007) The analytical method in geomechanics. *Applied Mechanics Reviews*, **60**, 87–106.

Selvadurai APS (2009) Fragmentation of ice sheets during impact. *Computer Modeling in Engineering and Science*, **52**, 259–277.

Selvadurai APS (2015) Normal stress-induced permeability hysteresis of a fracture in a granite cylinder. *Geofluids*, **15**, 37–47.

Selvadurai APS, Boulon MJ (eds) (1995) *Mechanics of Geomaterial Interfaces*. Studies in Applied Mechanics, **42**, Elsevier, Amsterdam.

Selvadurai APS, Boulon MJ, Nguyen TS (2005) The permeability of an intact granite. *Pure and Applied Geophysics*, **162**, 373–407.

Selvadurai APS, Carnaffan P (1997) A transient pressure pulse technique for the measurement of permeability of a cement grout. *Canadian Journal of Civil Engineering*, **24**, 489–502.

Selvadurai APS, Glowacki A (2008) Evolution of permeability hysteresis of Indiana Limestone during isotropic compression. *Ground Water*, **46**, 113–119.

Selvadurai APS, Ichikawa Y (2013) Some aspects of air-entrainment on decay rates in hydraulic pulse tests. *Engineering Geology*, **165**, 38–45.

Selvadurai APS, Jenner L (2012) Radial flow permeability testing of an argillaceous limestone. *Ground Water*, **51**, 100–107.

Selvadurai APS, Letendre A, Hekimi B (2011) Axial flow hydraulic pulse testing of an argillaceous limestone. *Environmental Earth Sciences*, **64**, 2047–2058.

Selvadurai APS, Najari M (2013) On the interpretation of hydraulic pulse tests on rock specimens. *Advances in Water Resources*, **53**, 139–149.

Selvadurai APS, Najari M (2015) Laboratory-scale hydraulic pulse testing: influence of air fraction in the fluid-filled cavity in the estimation of permeability. *Geotechnique*, **65**, 124–134.

Selvadurai APS, Nguyen TS (1995) Computational modeling of isothermal consolidation of fractured porous media. *Computers and Geotechnics*, **17**, 39–73.

Selvadurai APS, Nguyen TS (1997) Scoping analyses of the coupled thermal-hydrological-mechanical behaviour of the rock mass around a nuclear fuel waste repository. *Engineering Geology*, **47**, 379–400.

Selvadurai APS, Selvadurai PA (2010) Surface permeability tests: experiments and modelling for estimating effective permeability. *Proceedings of the Royal Society, Mathematics and Physical Sciences Series A*, **466**, 2819–2846.

Selvadurai APS, Sepehr K (1999a) Discrete element modelling of fragmentable geomaterials with size dependent strength. *Engineering Geology*, **53**, 235–241.

Selvadurai APS, Sepehr K (1999b) Two dimensional discrete element simulation of ice-structure interaction. *International Journal of Solids and Structures*, **36**, 4919–4940.

Selvadurai APS, Shirazi A (2004) Mandel-Cryer effects in fluid inclusions in damage susceptible poroelastic media. *Computers and Geotechnics*, **37**, 285–300.

Selvadurai APS, Shirazi A (2005) An elliptical disc anchor in a damage-susceptible poroelastic medium. *International Journal of Numerical Methods in Engineering*, **16**, 2017–2039.

Selvadurai APS, Suvorov AP, Selvadurai PA (2015) Thermo-hydro-mechanical processes in fractured rock formations during glacial advance. *Geoscientific Model Development*, **8**, 2167–2185.

Selvadurai APS, Yu Q (2005) Mechanics of a discontinuity in a geomaterial. *Computers and Geotechnics*, **32**, 92–106.

Selvadurai APS, Yue ZQ (1994) On the indentation of a poroelastic layer. *International Journal of Numerical and Analytical Methods in Geomechanics*, **18**, 161–175.

Senger RK, Fogg GE (1990) Stream functions and equivalent freshwater heads for modeling regional flow of variable-density ground water, 2, Application and implications for modeling strategy. *Water Resources Research*, **26**, 2097–2106.

Serra O (1982) *Fundamentals of Well-log Interpretation, 1, The Acquisition of Logging Data*. Elsevier, Amsterdam.

Shand SJ (1949) Rocks of the mid-Atlantic ridge. *Journal of Geology*, **57**, 89–92.

Shao JF, Hoxha D, Bart M, Homand F, Duveau G, Souley M, Hoteit N (1999) Modelling of induced anisotropic damage in granites. *International Journal of Rock Mechanics and Mining Sciences*, **36**, 1001–1012.

Shapiro SA, Dinske C (2009a) Scaling of seismicity induced by nonlinear fluid-rock interaction. *Journal of Geophysical Research*, **114**, doi:10.1029/2008JB006145.

Shapiro SA, Dinske C (2009b) Fluid-induced seismicity: pressure diffusion and hydraulic fracturing. *Geophysical Prospecting*, **57**, 301–310.

Shapiro SA, Huenges E, Borm G (1997) Estimating the crust permeability from fluid-injection-induced seismic emission at the KTB site. *Geophysical Journal International*, **131**, F15–F18.

Shapiro SA, Patzig R, Rothert E, Rindschwentner J (2003) Triggering of seismicity by pore pressure perturbations: permeability signatures of the phenomenon. *Pure and Applied Geophysics*, **160**, 1051–1066.

Shapiro SA, Rothert E, Rath V, Rindschwentner J (2002) Characterization of fluid transport properties of reservoirs using induced microseismicity. *Geophysics*, **67**, 212–220.

Sharp JM Jr, Domenico PA (1976) Energy transport in thick sequences of compacting sediment. *Geological Society of America Bulletin*, **87**, 390–400.

Sheldon HA, Micklethwaite S (2007) Damage and permeability around faults: implications for mineralization. *Geology*, **34**, 903–906.

Shen Z, Sun J, Zhang P, Wan Y, Wang M, Biirgmann R (2009) Slip maxima at fault junctions and rupturing of barriers during the 2008 Wenchuan earthquake. *Nature Geoscience*, **2**, 718–724.

Sheng P, Zhou M-Y (1988) Dynamic permeability in porous media. *Physical Review Letters*, **61**, 1591–1594.

Sheppard SMF (1986) Characterization and isotopic variations in natural waters. In: *Stable Isotopes in High Temperature Geologic Processes* (eds Valley JW, Taylor HP, Jr, O'Neil JR). *Reviews of Mineralogy*, **16**, pp. 165–184, Mineralogical Society of America.

Sheppard SMF, Gilg HA (1996) Stable isotope geochemistry of clay minerals. *Clay Mineralogy*, **31**, 1–24.

Shi Y, Wang CY (1986) Pore pressure generation in sedimentary basins: overloading versus aquathermal. *Journal of Geophysical Research*, **91**, 2153–2162.

Shi YL, Wang CY (1988) Generation of high pore pressures in accretionary prisms – Inferences from the Barbados subduction complex. *Journal of Geophysical Research*, **93**, 8893–8910.

Shi Z, Wang G, Liu C (2013a) Advances in research on earthquake fluids hydrogeology in China: a review. *Earthquake Science*, **26**, 415–425.

Shi Z, Wang G, Liu C (2013b) Co-seismic groundwater level changes induced by the May 12, 2008 Wenchuan earthquake in the near field. *Pure and Applied Geophysics*, **170**, 1773–1783.

Shi Z, Wang G, Liu C, Mei J, Wang J, Fang H (2013c) Coseismic response of groundwater level in the Three Gorges well network and its relationship to aquifer parameters. *Chinese Science Bulletin*, **58**, 3080–3087.

Shi Z, Wang G, Manga M, Wang C-Y (2015) Continental-scale water-level response to a large earthquake. *Geofluids*, **15**, 310–320.

Shi Z, Wang G, Wang C-Y, Manga M, Liu C (2014) Comparison of hydrological responses to the Wenchuan and Lushan earthquakes. *Earth and Planetary Science Letters*, **391**, 193–200.

Shinohara H (2008) Excess degassing from volcanoes and its role on eruptive and intrusive activity. *Reviews of Geophysics*, **46**, RG4005.

Shipboard Scientific Party (1994) Site 892. In: *Proceedings of the Ocean Drilling Program, Initial Reports*, **146** (eds Carson B, Westbrook GK, Musgrave RJ, Suess E), 301–378.

Shipboard Scientific Party (1995a) Site 948. *Proceedings of the Ocean Drilling Program, Initial Reports*, **156**, 87–192.

Shipboard Scientific Party (1995b) Site 949. *Proceedings of the Ocean Drilling Program, Initial Reports*, **156**, 193–257.

Shiping L, Yushou L, Yi L, Zhenye W, Gang Z (1994) Permeability-strain equations corresponding to the complete stress-strain path of Yinzhuang Sandstone. *International Journal of Rock Mechanics and Mining Sciences*, **31**, 383–391.

Shipley TH, ODP Leg 156 Scientific Party (1994) Seismically inferred dilatancy distribution, northern Barbados Ridge décollement: implications for fluid migration and fault strength. *Geology*, **22**, 411–414.

Shipley TH, Ogawa Y, Blum P, ODP Leg 156 Scientific Party (1995) Proceedings of the Ocean Drilling Program, Initial Reports, **156**.

Shmonov VM, Vitiovtova VM, Zharikov AV, Grafchikov AA (2003) Permeability of the continental crust: implications of experimental data. *Journal of Geochemical Exploration*, **78–79**, 697–699.

Shrag D (2007) Preparing to capture carbon. *Science*, **315**, 812–813.

Sibson R (1981a) A brief description of natural neighbor interpolation. In: *Interpolating Multivariate Data* (ed Barnett V), pp. 21–36, John Wiley & Sons, New York.

Sibson RH (1981b) Fluid flow accompanying faulting: field evidence and models. In: *Earthquake Prediction: An International Review*, Maurice Ewing Ser., Vol. 4 (eds Simpon DW, Richards PG), pp. 593–603. American Geophysical Union, Washington, DC.

Sibson RH (1987) Earthquake rupturing as a mineralizing agent in hydrothermal systems. *Geology*, **15**, 701–704.

Sibson RH (1992) Fault-valve behavior and the hydrostatic lithostatic fluid pressure interface. *Earth-Science Reviews*, **32**, 141–144.

Sibson RH (1996) Structural permeability of fluid-driven fault-fracture meshes. *Journal of Structural Geology*, **18**, 1031–1042.

Sibson RH (2001) Seismogenic framework for hydrothermal transport and ore deposition. *Society of Economic Geologists Reviews*, **14**, 25–50.

Sibson RH (2007) An episode of fault-valve behavior during compressional inversion? The 2004 M6.8 Mid-Niigata Prefecture, Japan, earthquake sequence. *Earth and Planetary Science Letters*, **257**, 188–199.

Sibson RH (2013) Stress switching in subduction forearcs: implications for overpressure containment and strength cycling on megathrusts. *Tectonophysics*, **600**, 142–152.

Sibson RH, Moore JMM, Rankin AH (1975) Seismic pumping – A hydrothermal fluid transport mechanism. *Journal of the Geological Society of London*, **131**, 653–659.

Sibson RH, Robert F, Poulsen KH (1988) High-angle reverse faults, fluid-pressure cycling, and mesothermal gold-quartz deposits. *Geology*, **16**, 551–555.

Sibson RH, Rowland JV (2003) Stress, fluid pressure and structural permeability in seismogenic crust, North Island, New Zealand. *Geophysical Journal International*, **154**, 584–594.

Sil S, Freymueller JT (2006) Well water level changes in Fairbanks, Alaska, due to the great Sumatra-Andaman earthquake. *Earth, Planets, and Space*, **58**, 181–184.

Sillitoe RH (2010) Porphyry copper systems. *Economic Geology*, **105**, 3–41.

Silver EA, Kastner M, Fisher AT, Morris JD, McIntosh KD, Saffer DM (2000) Fluid flow paths in the crust of the Middle America Trench, Costa Rica margin. *Geology*, **28**, 679–682.

Simmons SF, Brown KL (2007) The flux of gold and related metals through a volcanic arc, Taupo Volcanic Zone, New Zealand. *Geology*, **35**, 1099–1102.

Simpson GDH (1998) Dehydration-related deformation during regional metamorphism, NW Sardinia, Italy. *Journal of Metamorphic Geology*, **16**, 457–472.

Skarbek RM, Saffer DM (2009) Pore pressure development beneath the decollement at the Nankai subduction zone: implications for plate boundary fault strength and sediment dewatering. *Journal of Geophysical Research*, **114**, B07401.

SKB (2008) Site description of Forsmark at completion of the site investigation phase. In: *SKB Technical Report*, **08-05** (ed SKB), SKB, Stockholm, Sweden, http://www.skb.se/upload/publications/pdf/TR-08-05.pdf accessed 06 May 2016.

Skelton ADL, Graham CM, Bickle, MJ (1995) Lithological and structural controls on regional 3-D fluid flow patterns during greenschist facies metamorphism of the Dalradian of the SW Scottish Highlands. *Journal of Petrology*, **36**, 563–586.

Skelton ADL, Valley JV, Graham CM, Bickle, MJ, Fallick AE (2000) The correlation of reaction and isotope fronts and the mechanism of metamorphic fluid flow. *Contributions to Mineralogy and Petrology*, **138**, 364–375.

Slack TZ, Murdoch LC, Germanovich LN, Hisz DB (2013) Reverse water-level change during interference slug tests in fractured rock. *Water Resources Research*, **49**, 1552–1567.

Sloan SW, Booker R (1986) Removal of singularities in Tresca and Mohr-Coulomb yield functions. *Communications in Applied Numerical Methods*, **2**, 173–179.

Smeulders DMJ, Eggels RLGM, Dongen MEHV (1992) Dynamic permeability: reformulation of theory and new experimental and numerical data. *Journal of Fluid Mechanics*, **245**, 211–227.

Smith C (1983) Thermal hydrology and heat flow of Beowawe geothermal area, Nevada. *Geophysics*, **48**, 618–626.

Smith GI, Friedman I, Veronda G, Johnson CA (2002) Stable isotope compositions of water in the Great Basin, United States, 3, Comparison of ground waters with modern precipitation. *Journal of Geophysical Research*, **107**, doi:10.1029/2001JD000567.

Smith L, Chapman DS (1983) On the thermal effects of groundwater flow, 1, Regional scale systems. *Journal of Geophysical Research*, **88**, 593–608.

Smith-Konter B, Sandwell D (2009) Stress evolution of the San Andreas fault system: recurrence interval versus locking depth. *Geophysical Research Letters*, **36**, doi:10.1029/2009GL037235.

Snow DT (1968a) Hydraulic character of fractured metamorphic rocks of the Front Range and implications to the Rocky Mountain Arsenal Well. *Quarterly of the Colorado School of Mines*, **63**, 167–200.

Snow DT (1968b) Rock fracture spacings, openings, and porosities. *Journal of the Soil Mechanics and Foundations Division, ASCE*, **94**, 73–91.

Snow DT (1970) The frequency and apertures of fractures in rock. *International Journal of Rock Mechanics and Mining Sciences*, 7, 23–40.

Snyder DT, Haynes JV (2010) Groundwater conditions during 2009 and changes in groundwater levels from 1984 to 2009, Columbia Plateau Regional Aquifer System, Washington, Oregon, and Idaho. *U.S. Geological Survey Scientific Investigations Report*, **2010-5040**.

Sohn RA (2007) Stochastic analysis of exit fluid temperature records from the active TAG hydro thermal mound (Mid-Atlantic Ridge, 26°N), 1, Modes of variability and implications for subsurface flow. *Journal of Geophysical Research*, **112**, B07101, doi:10.1029/2006JB004435.

Soliva R, Benedicto A (2004) A linkage criterion for segmented normal faults. *Journal of Structural Geology*, **26**, 2251–2267.

Solomon EA, Kastner M, Wheat G, Jannasch HW, Robertson G Davis EE, Morris JD (2009) Long-term hydrogeochemical records in the oceanic basement and forearc prism at the Costa Rica subduction zone. *Earth and Planetary Science Letters*, **282**, 240–251.

Solum JG, van der Pluijm BA (2009) Quantification of fabrics in clay gouge from the Carboneras fault, Spain and implications for fault behavior. *Tectonophysics*, **475**, 554–562.

Song C, Ekinci MK, Underwood MB, Henry P (2015) Data report: Permeability and microfabric of mud (stone) samples from IODP Sites C0011 and C0012, NanTroSEIZE subduction inputs. In: Saito S, Underwood MB, Kubo Y, and the Expedition 322 Scientists, Proceedings of the Integrated Ocean Drilling Program, 322, doi:10.2204/iodp.proc.322.211.2015.

Sophocleous M (2010) Groundwater management practices, challenges, and innovations in the High Plains aquifer, USA—lessons and recommended actions. *Hydrogeology Journal*, **18**, 559–575.

Souley M, Homand F, Pepa S, Hoxha D (2001) Damage-induced permeability changes in granite: a case example at the URL in Canada. *International Journal of Rock Mechanics and Mining Sciences*, **38**, 297–310.

Spane FA (1982) Hydrologic studies within the Pasco Basin. Proceedings of the 1982 National Waste Terminal Storage Program Information Meeting, U.S. Department of Energy, DOE/NWTS-30, p. 23–8.

Spane FA (2013) *Preliminary Analysis of Grande Ronde Basalt Formation Flow Top Transmissivity as It Relates to Assessment and Site Selection Applications for Fluid/Energy Storage and Sequestration Projects*. Pacific Northwest National Laboratory, U.S. Department of Energy, PNNL-22436.

Spane FA, Bonneville A, McGrail BP, Thorne PD (2012) *Hydrologic Characterization Results and Recommendations for the Wallula Basalt Pilot Well*. PNWD-4368, Battelle-Pacific Northwest Division, Richland, Washington.

Spencer DW (1963) The interpretation of grain size distribution curves of clastic sediments. *Journal of Sedimentary Petrology*, **33**, 180–190.

Spiegelman M (1993) Physics of melt extraction – Theory, implications and applications. *Philosophical Transactions of the Royal Society of London, Series A, Mathematical, Physical and Engineering Sciences*, **342**, 23–41.

Spinelli GA, Mozley PS, Tobin HJ, Underwood MB, Hoffman NW, Bellew GM (2007) Diagenesis, sediment strength, and pore collapse in sediment approaching the Nankai Trough subduction zone. *GSA Bulletin*, **119**, 377–390.

Spinelli GA, Saffer DM (2004) Along-strike variations in underthrust sediment dewatering on the Nicoya margin, Costa Rica related to the updip limit of seismicity. *Geophysical Research Letters*, **31**, L04613.

Spinelli G, Saffer DM, Underwood MB (2006) Effects of along-strike variability in temperature on the hydrogeology of the Nicoya margin subduction zone, Costa Rica. *Journal of Geophysical Research*, **111**, B04403.

Standards New Zealand (2004) NZS1170.5:2004, Structural Design Actions, Part 5, Earthquake Actions New Zealand.

Staude S, Bons PD, Markl G (2009) Hydrothermal vein formation by extension-driven dewatering of the middle crust: an example from SW Germany. *Earth and Planetary Science Letters*, **286**, 387–395.

Staudigel H, Gillis K, Duncan R (1986) K/Ar and Rb/Sr ages of celadonites from the Troodos ophiolite, Cyprus. *Geology*, **14**, 72–75.

Stauffer P, Bekins BA (2001) Modeling consolidation and dewatering near the toe of the northern Barbados accretionary complex. *Journal of Geophysical Research*, **106**, 6369–6383.

Steefel CI, Maher K (2009) Fluid-rock interaction: a reactive transport approach. In: *Thermodynamics and Kinetics of Water-Rock Interaction*, vol. **70** (eds Oelkers EH, Schott J), pp. 87–124, Mineralogical Society of America, Chantilly, VA.

Steele-MacInnes M, Han L, Lowell RP, Rimstidt JD, Bodnar RJ (2012a) The role of fluid phase immiscibility in quartz dissolution and precipitation in sub-seafloor hydrothermal systems. *Earth Planetary Science Letters*, **321–322**, 139–151.

Steele-MacInnes M, Han L, Lowell RP, Rimstidt JD, Bodnar RJ (2012b) Quartz precipitation and fluid inclusion characteristics in sub-seafloor hydrothermal systems associated with volcanogenic massive sulphide deposits. *Central European Journal of Geosciences*, **4**, 275–286.

Stenger R (1982) Petrology and geochemistry of the basement rocks of the research drilling project Urach 3. In: *The Urach Geothermal Project* (ed Haenel R), pp. 41–48, Schweizerbart'sche Verlagsbuchhandlung, Stuttgart.

Stephansson O (ed) (1985) *Proceedings of the International Symposium on Fundamentals of Rock Joints*. Björkliden, Norway.

Steurer JF, Underwood MB (2003) Data report: The relation between physical properties and grain-size variations in hemipelagic sediments from Nankai Trough. *Proceedings of the Ocean Drilling Program, Scientific Results*, **190/196**, 1–25.

Stevenson D (1989) Spontaneous small-scale melt segregation in partial melts undergoing deformation. *Geophysical Research Letters*, **16**, 1067–1070.

Stirling MW, McVerry GH, Berryman KR (2002) A new seismic hazard model for New Zealand. *Bulletin of the Seismological Society of America*, **92**, 1878–1903.

Stirling MW, McVerry GH, Gerstenberger MC, Litchfield NJ, Van DR, Berryman KR, Barnes P, Wallace LM, Villamor P, Langridge RM, Lamarche G, Nodder S, Reyners ME, Bradley B, Rhoades DA, Smith WD, Nicol A, Pettinga J, Clark KJ, Jacobs K (2012) National seismic hazard model for New Zealand: 2010 update. *Bulletin of the Seismological Society of America*, **102**, 1514–1542.

Stober I (1986) Strömungsverhalten in Festgesteinsaquiferen mit Hilfe von Pump- und Injektionsversuchen. *Geologisches Jahrbuch, Reihe C*, **42**, 204.

Stober I (1995) *Die Wasserführung des kristallinen Grundgebirges [Water in the Crystalline Basement]*. Ferdinand Enke Verlag, Stuttgart.

Stober I (1996) Researchers study conductivity of crystalline rock in proposed radioactive waste site. *Eos, Transactions American Geophysical Union*, **77**, 93–94.

Stober I (2011) Depth- and pressure-dependent permeability in the upper continental crust: data from the Urach 3 geothermal borehole, southwest Germany. *Hydrogeology Journal*, **19**, 685–699.

Stober I, Bucher K (2004) Fluid sinks within the Earth's crust. *Geofluids*, **4**, 143–151.

Stober I, Bucher K (2005a) The upper continental crust, an aquifer and its fluid: hydraulic and chemical data from 4 km depth in fractured crystalline basement rocks at the KTB test site. *Geofluids*, **5**, 8–19.

Stober I, Bucher K (2005b) Deep-fluids: Neptune meets Pluto. In: *The Future of Hydrogeology* (ed Voss C), *Hydrogeology Journal*, **13**, 112–115.

Stober I, Bucher K (2007a) Hydraulic properties of the crystalline basement. *Hydrogeology Journal*, **15**, 213–224.

Stober I, Bucher K (2007b) Erratum to: hydraulic properties of the crystalline basement. *Hydrogeology Journal*, **15**, 1643.

Stober I, Bucher K (2015) Hydraulic conductivity of fractured upper crust: insights from hydraulic tests in boreholes and fluid-rock interaction in crystalline basement rocks. *Geofluids*, **15**, 161–178.

Stober I, Richter A, Brost E, Bucher K (1999) The Ohlsbach Plume: natural release of deep saline water from the crystalline basement of the black forest. *Hydrogeology Journal*, 7, 273–283.

Strasser M, Henry P, Kanamatsu T, Moe KT, Moore GF, IODP Expedition 333 Scientists (2012) Scientific drilling of masstransport deposits in the Nankai accretionary wedge: first results from IODP Expedition 333. In: *Submarine Mass Movements and Their Consequences* (eds Yamada Y, Kawamura K, Ikehara K, Ogawa Y, Urgeles R, Mosher D, Chaytor J, Strasser M), pp. 671–681. Springer, Dordrecht.

Struhsacker EM (1980) The geology of the Beowawe geothermal system, Eureka and Lander Counties, Nevada. *University of Utah Research Institute Report*, **ESL-37**.

Stuyfzand PJ (1989) An accurate, relatively simple calculation of the saturation index of calcite for fresh to salt water. *Journal of Hydrology*, **105**, 95–107.

Suetnova EI, Carbonell R, Smithson SB (1994) Bright seismic reflections and fluid movement by porous flow in the lower crust. *Earth and Planetary Science Letters*, **126**, 161–169.

Sumita I, Yoshida S, Kumazawa M, Hamano Y (1996) A model for sedimentary compaction of a viscous medium and its application to inner-core growth. *Geophysical Journal International*, **124**, 502–524.

Summers WK (1976) Catalog of thermal waters in New Mexico. *New Mexico Bureau of Mines and Mineral Resources Hydrologic Report*, **4**.

Sun H, Feistel R, Koch M, Markoe A (2008) New equations for density, entropy, heat capacity and potential temperature of a saline thermal fluid. *Deep-Sea Research, Part I, Oceanographic Research Papers*, **55**, 1304–1310

Sundaram PN, Watkins DJ, Ralph WE (1987) Laboratory investigation of coupled stress-deformation-hydraulic flow in a natural rock fracture. *Proceedings of the 28th U.S. Symposium on Rock Mechanics*, University of Arizona, Tuscon, 29 June-1 July, 1987, 593–600.

Sussman M, Smereka P, Osher S (1994) A level set approach for computing solutions to incompressible 2-phase flow. *Journal of Computational Physics*, **114**, 146–159.

Sutherland R, Berryman KR, Norris R (2006) Quaternary slip rate and geomorphology of the Alpine Fault: implications for kinematics and seismic hazard in southwest New Zealand. *Geological Society of America Bulletin*, **118**, 464–474.

Sutherland R, Eberhart-Phillips D, Harris RA, Stern TA, Beavan RJ, Ellis SM, Henrys SA, Cox SC, Norris RJ, Berryman KR, Townend J, Bannister SC, Pettinga J, Leitner B, Wallace LM, Little TA, Cooper AF, Yetton M, Stirling MW (2007) Do great earthquakes occur on the Alpine Fault in central South Island, New Zealand? In: *A Continental Plate Boundary: Tectonics at South Island, New Zealand* (eds Okaya DA, Stern TA, Davey FJ). *American Geophysical Union Geophysical Monograph*, **175**, pp. 235–251, AGU.

Sutherland R, Toy VG, Townend J, Cox SC, Eccles JD, Faulkner DR, Prior DJ, Norris RJ, Mariani E, Boulton C, Carpenter BM, Menzies CD, Little TA, Hastings M, De PG, Langridge RM, Scott HR, Lindroos ZR, Fleming B, Kopf AJ (2012) Drilling reveals fluid control on architecture and rupture of the Alpine fault, New Zealand. *Geology*, **40**, 1143–1146.

Svenson E, Schweisinger T, Murdoch LC (2008) Field evaluation of the hydromechanical behavior of flat-lying fractures during slug tests. *Journal of Hydrology*, **359**, 30–45.

Swanberg CA, Walkey WC, Combs J (1988) Core hole drilling and the "rain curtain" phenomenon at Newberry Volcano, Oregon. *Journal of Geophysical Research*, **93**, 10163–10173.

Tada R, Siever R (1989) Pressure solution during diagenesis. *Annual Reviews of Earth and Planetary Science*, **17**, 89–118.

Tadokoro K, Ando M (2002) Evidence for rapid fault healing derived from temporal changes in S wave splitting. *Geophysical Research Letters*, **29**, doi:10.1029/2001GL013644.

Taira A, Hill I, Firth J, Berner U, Brückmann W, Byrne T, Chabernaud T, Fisher A, Foucher JP, Gamo T, Gieskes J, Hyndman R, Karig D, Kastner M, Kato Y, Lallemant S, Lu R, Maltman A, Moore G, Moran K, Olaffson G, Owens W, Pickering K, Siena F, Taylor E, Underwood M, Wilkinson C, Yamano M, Zhang J (1992) Sediment deformation and hydrogeology of the Nankai Trough accretionary prism: synthesis of shipboard results of ODP Leg 131. *Earth and Planetary Science Letters*, **109**, 431–450.

Taira T, Silver PG, Niu F, Nadeau RM (2009) Remote triggering of fault-strength changes on the San Andreas fault at Parkfield. *Nature*, **461**, 636–677.

Tait A, Henderson R, Turner R, Zheng X (2006) Thin plate smoothing spline interpolation of daily rainfall for New Zealand using a climatological rainfall surface. *International Journal of Climatology*, **26**, 2097–2115.

Takahashi M, Hirata A, Koide H (1990) Effect on confining pressure and pore pressure on permeability of Inada granite. *Journal of the Japan Society of Engineering Geology*, **31**, 105–114 (in Japanese with English abstract).

Tanaka A, Ishikawa Y (2005) Crustal thermal regime inferred from magnetic anomaly data and its relationship to seismogenic layer thickness. *Physics of the Earth and Planetary Interiors*, **152**, 257–266.

Tanikawa W, Hirose T, Mukoyoshi H, Tadai O, Lin W (2013) Fluid transport properties in sediments and their role in large slip near the surface of the plate boundary fault in the Japan Trench. *Earth and Planetary Science Letters*, **382**, 150–160.

Tanikawa W, Tadai O, Mukoyoshi H (2014) Permeability changes in simulated granite faults during and after frictional sliding. *Geofluids*, **14**, 481–494.

Tanner WF (1964) Modification of sediment size distributions. *Journal of Sedimentary Research*, **34**, 156–164.

Taron J, Elsworth D (2009) Thermal-hydrologic-mechanical-chemical processes in the evolution of engineered geothermal reservoirs. *International Journal of Rock Mechanics and Mining Sciences*, **46**, 855–864.

Taron J, Elsworth D (2010a) Coupled mechanical and chemical processes in engineered geothermal reservoirs with dynamic permeability. *International Journal of Rock Mechanics and Mining Sciences*, **47**, 1339–1348.

Taron J, Elsworth D (2010b) Constraints on the compaction rate and equilibrium in the pressure solution creep of quartz aggregates and fractures: controls of aqueous concentration. *Journal of Geophysical Research*, **115**, B07211, doi:10.1029/2009JB007118.

Taron J, Elsworth D, Min K-B (2009) Numerical simulation of thermal-hydrologic-mechanical-chemical processes in deformable, fractured porous media. *International Journal of Rock Mechanics and Mining Sciences*, **46**, 842–854.

Taron J, Hickman S, Ingebritsen SE, Williams C (2014) Using a fully coupled, open-source THM simulator to examine the role of thermal stresses in shear stimulation of enhanced geothermal systems. 48th US Rock Mechanics/Geomechanics Symposium, Minneapolis, Minnesota, 1–4 June 2014.

Tavenas F, Jean P, Leblond P, Leroueil S (1983) The permeability of natural soft clays, Part II, Permeability characteristics. *Canadian Geotechnical Journal*, **20**, 645–660.

Taylor B, Zellmer K, Martinez F, Goodliffe Andrew (1996) Sea-floor spreading in the Lau back-arc basin. *Earth and Planetary Science Letters*, **144**, 35–40.

Taylor DW (1948) *Fundamentals of Soil Mechanics*. John Wiley and Sons, New York.

Taylor E, Leonard J (1990) Sediment consolidation and permeability at the Barbados forearc. *Proceedings of the Ocean Drilling Program, Scientific Results*, **110**, 289–308.

Teagle DAH, Alt JC, Halliday AN (1998) Tracing the evolution of hydrothermal fluids in the upper oceanic crust: Sr-isotopic constraints from DSDP/ODP Holes 504B and 896A. In: *Modern Ocean-Floor Processes and the Geological Record*, 148 (eds Mills RA, Harrison K), pp. 81–97. Geological Society of London, London.

Teichert BMA, Torres ME, Bohrmann G, Eisenhauer A (2005) Fluid sources, fluid pathways and diagenetic reactions across an accretionary prism revealed by Sr and B geochemistry. *Earth and Planetary Science Letters*, **239**, 106–121.

Tenthorey E, Cox SF (2006) Cohesive strengthening of fault zones during the interseismic period: an experimental study. *Journal of Geophysical Research*, **11**, B09202.

Tenthorey E, Fitz Gerald J (2006) Feedbacks between deformation, hydrothermal reaction and permeability evolution in the crust: experimental insights. *Earth and Planetary Science Letters*, **247**, 117–129.

Terakawa T, Hashimoto C, Matsu'ura M (2013) Changes in seismic activity following the 2011 Tohoku-oki earthquake: effects of pore fluid pressure. *Earth and Planetary Science Letters*, **365**, 17–24.

Terakawa T, Miller SA, Deichmann N (2012) High fluid pressure and triggered earthquakes in the enhanced geothermal system in Basel, Switzerland. *Journal of Geophysical Research*, **117**, doi:10.1029/2011JB008980.

Terzaghi C (1925) Principles of soil mechanics. *Engineering News-Record*, **95**, 19–27.

Tester JW (2006) *The Future of Geothermal Energy, Part 1, Summary and Part 2, Full Report*. Massachusetts Institute of Technology, Cambridge, MA.

Theis CV (1963) Estimating the transmissivity of a water-table aquifer from the specific capacity of a well. *U.S. Geological Survey Water Supply Paper*, **1536-1**, 332–336.

Theis CV, Taylor GC, Murray CR (1941) Thermal waters of the Hot Springs artesian basin, Sierra County, New Mexico. *Fourteenth and Fifteenth Biennial Reports of the State Engineer of New Mexico*, 419–492.

Thielicke W, Stamhuis E (2012) PIVLab - Time-resolved digital particle image velocimetry tool for MATLAB. Ver. 1.32.

Thompson TB, Teal L, Meeuwig RO (2002) Gold deposits of the North Carlin Trend. *Nevada Bureau of Mines and Geology Bulletin*, **111**.

Thorpe RK, Watkins DJ, Ralph WE, Hsu R, Flexser R (1982) Strength and permeability test on ultra-large Stripa Granite core. *Lawrence Berkeley Laboratory, Berkeley, California, Report*, **LBL-11203**.

Thorwart M, Dzierma Y, Rabbel W, Hensen C (2013) Seismic swarms, fluid flow and hydraulic conductivity in the forearc offshore North Costa Rica and Nicaragua. *International Journal of Earth Sciences*, **103**, 1789–1799.

Tian M, Ague JJ (2014) The impact of porosity waves on crustal reaction progress and CO_2 mass transfer. *Earth and Planetary Science Letters*, **390**, 80–92.

Timar-Geng Z, Fügenschuh B, Wetzel A, Dresmann H (2006) Low-temperature thermochronology of the flanks of the southern Upper Rhine Graben. *International Journal of Earth Sciences*, **95**, 685–702.

Tivey MK, Humphris SE, Thompson G, Hannington MD, Rona P (1995) Deducing patterns of fluid flow and mixing within the TAG active hydrothermal mound using mineralogical and geochemical data. *Journal of Geophysical Research*, **100**, 12527–12555.

Tobin HJ, Kinoshita M (2006) NanTroSEIZE: the IODP Nankai Trough seismogenic zone experiment. *Scientific Drilling*, **2**, 23–27.

Tobin HJ, Moore JC, Moore GF (1994) Fluid pressure in the frontal thrust of the Oregon accretionary prism: experimental constraints. *Geology*, **22**, 979–982.

Tobin HJ, Saffer DM (2009) Elevated fluid pressure and extreme mechanical weakness of a plate boundary thrust, Nankai Trough subduction zone. *Geology*, **37**, 679–682.

Tobin HJ, Vannucchi P, Meschede M (2001) Structure, inferred mechanical properties, and implications for fluid transport in the décollement zone, Costa Rica convergent margin. *Geology*, **29**, 907–910.

Toda S, Stein RS, Lin J (2011) Widespread seismicity excitation throughout central Japan following the 2011 M = 9.0 Tohoku earthquake and its interpretation by Coulomb stress transfer. *Geophysical Research Letters*, **38**, doi:10.1029/2011GL047834.

Toda S, Stein RS, Richards-Dinger K, Bozkurt SB (2005) Forecasting the evolution of seismicity in southern California: animations built on earthquake stress transfer. *Journal of Geophysical Research*, **110**, B05S16.

Tolan TL, Reidel SP, Beeson MH, Anderson JL, Fecht KR, Swanson DA (1989) Revisions to the estimates of the areal extent and volume of the Columbia River Basalt Group. In: *Volcanism and Tectonism in the Columbia River Flood-Basalt Province* (eds Reidel SP, Hooper PR) *Geological Society of America Special Paper*, **239**, pp. 1–20, Geological Society of America.

Tokunaga T, Hosoya S, Tosaka H, Kojima K (1998) An estimation of the intrinsic permeability of argillaceous rocks and the effects on long-term fluid migration. *Geological Society of London Special Publication*, **141**, 83–94.

Tong P, Zhao D, Yang D (2012) Tomography of the 2011 Iwaki earthquake (M 7.0) and Fukushima nuclear power plant area. *Solid Earth*, **3**, 43–51.

Töth J (1962) A theory of groundwater motion in small drainage basins in central Alberta, Canada. *Journal of Geophysical Research*, **67**, 4375–4388.

Töth J (1978) Gravity-induced cross-formational flow of formation fluids, Red Earth region, Alberta, Canada: analysis, patterns, and evolution. *Water Resources Research*, **14**, 805–843.

Townend J, Sherburn S, Arnold R, Boese C, Woods L (2012) Three-dimensional variations in present-day tectonic stress along the Australia-Pacific plate boundary in New Zealand. *Earth and Planetary Science Letters*, **353–354**, 47–59.

Townend J, Zoback MD (2000) How faulting keeps the crust strong. *Geology*, **28**, 399–402.

Truesdell AH (1976) Summary of section III: geochemical techniques in exploration. *Proceedings of the 2nd U.N. Symposium on the Development and Use of Geothermal Resources, San Francisco 1975.* Vol. I, pp. liii–lxxx. U.S. Government Printing Office, Washington, DC.

Tsang C-F (1991) Coupled thermomechanical hydrochemical processes in rock fractures. *Reviews of Geophysics*, **29**, 537–551.

Tsang C-F (1999) Linking thermal, hydrological and mechanical processes in fractured rocks. *Annual Review of Earth and Planetary Sciences*, **27**, 359–384.

Tsang C-F, Bernier F, Davies C (2005) Geohydromechanical processes in the excavation damaged zone in crystalline rock, rock salt, and indurated and plastic clays – in the context of radioactive waste disposal. *International Journal of Rock Mechanics and Mining Sciences*, **42**, 109–125.

Tsang C-F, Neretnieks I (1998) Flow channeling in heterogeneous fractured rocks. *Reviews of Geophysics*, **36**, 275–298.

Tsang YW, Birkholzer JT, Muldiopadhyay S (2009) Modeling of thermally driven hydrological processes in partially saturated fractured rock. *Reviews of Geophysics*, **47**, RG3004.

Tsang YW, Witherspoon P (1981) Hydromechanical behaviour of a deformable rock fracture subject to normal stress. *Journal of Geophysical Research*, **86**, 9287–9298.

Tsuji T, Kawamura K, Kanamatsu T, Kasaya T, Fujikura K, Ito Y, Tsuru T, Kinoshita M (2013) Extension of continental crust by anelastic deformation during the 2011 Tohoku-oki earthquake: the role of extensional faulting in the generation of a great tsunami. *Earth and Planetary Science Letters*, **364**, 44–58.

Tumarkina E, Misra S, Burlini L, Connolly JAD (2011) An experimental study of the role of shear deformation on partial melting of a synthetic metapelite. *Tectonophysics*, **503**, 92–99.

Uehara SI, Shimamoto T (2004) Gas permeability evolution of cataclasite and fault gouge in triaxial compression and implications for changes in fault-zone permeability structure through the earthquake cycle. *Tectonophysics*, **378**, 183–195.

Ujiie K, Hisamitsu T, Taira A (2003) Deformation and fluid pressure variation during initiation and evolution of the plate boundary décollement zone in the Nankai accretionary prism. *Journal of Geophysical Research*, **108**, doi:10.1029/2002JB002314.

Ulven OI, Jamtveit B, Malthe-Sorenssen A (2014) Reaction-driven fracturing of porous rock. *Journal of Geophysical Research*, **119**, 7473–7486, doi:10.1002/2014JB011102.

Umino N, Okada T, Hasegawa A (2002) Foreshock and aftershock sequence of the 1998 M 5.0 Sendai, northeastern Japan, earthquake and its implications for earthquake nucleation. *Bulletin of the Seismological Society of America*, **92**, 2465–2477.

Underwood MB (2007) Sediment inputs to subduction zones: why lithostratigraphy and clay mineralogy matter. In: *The Seismogenic Zone of Subduction Thrusts* (eds Dixon TH, Moore JC), pp. 42–85. Columbia University Press, New York.

Upton P, Sutherland R (2014) High permeability and low temperature correlates with proximity to brittle failure within mountains at an active tectonic boundary, Manapouri tunnel, Fiordland, New Zealand. *Earth and Planetary Science Letters*, **389**, 176–187.

Valley B, Evans KF (2009) Stress orientation to 5 km depth in the basement below Basel (Switzerland) from borehole failure analysis. *Swiss Journal of Geosciences*, **102**, 467–480.

Valsami E, Cann JR (1992) Mobility of rare earth elements in zones of intense hydro- thermal alteration in the *Pindos ophiolite*, Greece. In: *Ophiolites and Their Modern Oceanic Analogues*, vol. **60** (ed Parson LM), pp. 219–232, Geological Society of London, London.

Van Ark E, Detrick RS, Canales JP, Carbotte SM, Harding AJ, Kent GM, Nedimovic MR, Wilcock WSD, Diebold JB, Babcock J (2007) Seismic structure of the Endeavour segment, Juan de Fuca Ridge: correlations with seismicity and hydrothermal activity. *Journal of Geophysical Research*, **112**, B02401.

Vannucchi P, Leoni L (2007) Structural characterization of the Costa Rica décollement: evidence for seismically induced fluid pulsing. *Earth and Planetary Science Letters*, **262**, 413–428.

Varga RJ, Gee JS, Bettison-Varga L, Anderson RS, Johnson CE (1999) Early establishment of hydrothermal systems during structural extension; palaeomagnetic evidence from the Troodos ophiolite, Cyprus. *Earth and Planetary Science Letters*, **171**, 221–235.

Vasseur G, Djeran-Maigre I, Grunberger D, Rousset G, Tessier D, Velde B (1995) Evolution of structural and physical parameters of clays during experimental compaction. *Marine and Petroleum Geology*, **12**, 941–954.

Vermilye JM, Scholz CH (1995) Relation between vein length and aperture. *Journal of Structural Geology*, **17**, 423–434.

Vernotte P (1958) Les paradoxes de la theorie continue de l'equation de la chaleur. *Comptes Rendus*, **246**, 3154–3155.

Vidale JE, Li Y-G (2003) Damage to the shallow Landers Fault from the nearby Hector Mine earthquake. *Nature*, **421**, 524–526.

Vigneresse JL, Barbey P, Cuney M (1996) Rheological transitions during partial melting and crystallization with application to felsic magma segregation and transfer. *Journal of Petrology*, **37**, 1579–1600.

Vila M, Fernandez M, Jiminez-Munt I (2010) Radiogenic heat production variability of some common lithological groups and its significance to lithospheric thermal modeling. *Tectonophysics*, **490**, 152–164.

Vincent M (2013) Five things you didn't want to know about hydraulic fractures. In: *Effective and Sustainable Hydraulic Fracturing* (eds Bunger A, McLennan J, Jeffrey R), pp. 81–93, InTech, Rijeka, Croatia.

Viola G, Venvik Ganerød G, Wahlgren CH (2009) Unraveling 1.5 Ga of brittle deformation history in the Laxemar-Simpevarp area, southeast Sweden: a contribution to the Swedish site investigation study for the disposal of highly radioactive nuclear waste. *Tectonics*, **28**, TC5007.

Von Damm K (1995) Controls on the chemistry and temporal variability of seafloor hydrothermal fluids. In: *Seafloor Hydrothermal Systems*, vol. **91** (eds Humphris SE, Zierenberg RA, Mullineaux LS, Thomson RE), pp. 222–247, American Geophysical Union, Washington, DC.

Von Damm KL, Bischoff JL, Rosenbauer RJ (1991) Quartz solubility in hydrothermal seawater; an experimental study and equation describing quartz solubility for up to 0.5 M NaCl solutions. *American Journal of Science*, **291**, 977–1007.

von Huene R, Lee H (1982) The possible significance of pore fluid pressures in subduction zones. In: *Studies in Continental Margin Geology* (eds Watkins JS, Drake CL), *American Association of Petroleum Geologists Memoir,* **34**, pp. 781–791, American Association of Petroleum Geologists.

von Quadt A, Erni M, Martinek K, Moll M, Peytcheva I, Heinrich CA (2011) Zircon crystallization and the lifetimes of ore-forming magmatic-hydrothermal systems. *Geology*, **39**, 731–734.

Voss CI, Provost AM (2002) SUTRA, A model for saturated-unsaturated variable-density ground-water flow with solute or energy transport. *U.S. Geological Survey Water-Resources Investigations Report*, **02-4231**.

Vrolijk P, Fisher AT, Gieskes J (1991) Geochemical and thermal evidence for fluid migration in the Barbados accretionary prism (ODP Leg 110). *Geophysical Research Letters*, **18**, 947–950.

Vrolijk P, Sheppard SMF (1987) Syntectonic veins from the Barbados accretionary prism (ODP Leg 110): record of paleohydrology. *Sedimentology*, **38**, 671–690.

Wagner GA, Reimer GM (1972) Fission track tectonics: the tectonic interpretation of fission track apatite ages. *Earth and Planetary Science Letters*, **14**, 263–268.

Wagner W, Kretschmar H-J (2008) *International Steam Tables, Properties of Water and Steam,* 2nd edn Springer-Verlag, Berlin, Heidelberg.

Wakita H (1975) Water wells as possible indicators of tectonic strain. *Science*, **189**, 553–555.

Walder J, Nur A (1984) Porosity reduction and crustal pore pressure development. *Journal of Geophysical Research*, **89**, 1539–1548.

Walderhaug O, Eliassen A, Aase NE (2012) Prediction of permeability in quartz-rich sandstones: examples from the Norwegian continental shelf and the Fontainebleau Sandstone. *Journal of Sedimentary Research*, **82**, 899–912.

Waldhauser F, Ellsworth B (2000) A double-difference earthquake location algorithm: method and application to the northern Hayward Fault, California. *Bulletin of the Seismological Society of America*, **90**, 1353–1368.

Waldhauser F, Schaff DP, Diehl T, Engdahl ER (2012) Splay faults imaged by fluid-driven aftershocks of the 2004 M_w9.2 Sumatra-Andaman earthquake. *Geology*, **40**, 243–246.

Walker CT (1975) Geochemistry of boron. *Benchmark Papers in Geology*, **23**.

Walker D, Rhen I, Gurban I (1997) Summary of hydrogeologic conditions at Aberg, Bebergy and Ceberg. In: *SKB Technical Report*, **97-23** (ed SKB), SKB, Stockholm, Sweden, http://www.skb.se/upload/publications/pdf/TR97-23webb.pdf, accessed 07 May 2016.

Walker JD, Kirby E, Andrew JE (2005) Strain transfer and partitioning between the Panamint Valley, Searles Valley, and Ash Hill fault zones, California. *Geosphere*, **1**, 111–118.

Wallace LM, Beavan RJ, McCaffrey R, Berryman KR, Denys P (2007) Balancing the plate motion budget in the South Island, New Zealand using GPS, geological and seismological data. *Geophysical Journal International*, **168**, 332–352.

Walsh JJ, Bailey WR, Childs C, Nicol A, Bonson CG (2003) Formation of segmented normal faults: a 3-D perspective. *Journal of Structural Geology*, **25**, 1251–1262.

Wang C (1985) Ground-water studies for earthquake prediction in China. *Pure and Applied Geophysics*, **122**, 215–217.

Wang C-Y (2001) Coseismic hydrologic response of an alluvial fan to the 1999 Chi-Chi earthquake, Taiwan. *Geology*, **29**, 831–834.

Wang C-Y, Chia Y (2008) Mechanism of water level changes during earthquakes: near field versus intermediate field. *Geophysical Research Letters*, **35**, L12402.

Wang C-Y, Chia Y, Wang P-L, Dreger D (2009) Role of S waves and Love waves in coseismic permeability enhancement. *Geophysical Research Letters*, **36**, L09404.

Wang C-Y, Dreger DS, Wang CH, Mayeri D, Berryman JG (2003) Field relations among coseismic ground motion, water level change and liquefaction for the 1999 Chi-Chi ($M_w = 7.5$) earthquake, Taiwan. *Geophysical Research Letters*, **30**, doi:10.1029/2003GL017601.

Wang C-Y, Manga M (2010a) *Earthquakes and Water* (Lecture Notes in Earth Sciences, **114**). Springer-Verlag, Berlin, Heidelberg.

Wang C-Y, Manga M (2010b) Hydrologic responses to earthquakes and a general metric. *Geofluids*, **10**, 210–216.

Wang C-Y, Wang CH, Kuo CH (2004a) Temporal change in groundwater level following the 1999 ($M_w = 7.5$) Chi-Chi earthquake, Taiwan. *Geofluids*, **4**, 210–220.

Wang C-Y, Wang CH, Manga M (2004b) Coseismic release of water from mountains: evidence from the 1999 ($M_w = 7.5$) Chi-Chi, Taiwan, earthquake. *Geology*, **32**, 769–772.

Wang C-Y, Wang LP, Manga M, Wang CH, Chen CH (2013) Basin-scale transport of heat and fluid induced by earthquakes. *Geophysical Research Letters*, **40**, 3893–3897, doi:10.1002/grl.50738

Wang H (2000) *Theory of Linear Poroelasticity with Applications to Geomechanics and Hydrogeology*. Princeton University Press, Princeton, NJ.

Wang J, Cook P, Trautz R, Flexser S, Hu Q, Salve R, Hudson D, Conrad M, Tsang Y, Williams K, Soll W, Turin J (2001) *In-Situ Field Testing of Processes*. ANL-NBS-HS-000005 REV01, Bechtel SAIC Company, Las Vegas, Nevada.

Wang K (1994) Kinematic models of dewatering accretionary prisms. *Journal of Geophysical Research*, **99**, 4429–4438.

Wang K, Davis EE (1996) Theory for the propagation of tidally induced pore pressure variations in layered subseafloor formations. *Journal of Geophysical Research*, **101**, 11483–11495.

Wark DA, Watson EB (2006) TitaniQ: a titanium-in-quartz geothermometer. *Contributions to Mineralogy and Petrology*, **152**, 743–754.

Warren JE, Price HS (1961) Flow in heterogeneous porous media. *SPE Journal*, **1**, 153–169.

Watanabe N, Hirano N, Tsuchiya N (2008) Determination of aperture structure and fluid flow in a rock fracture by high-resolution numerical modeling on the basis of a flow-through experiment under confining pressure. *Water Resources Research*, **44**, W06412, doi:10.1029/2006WR005411.

Watanabe N, Hirano N, Tsuchiya N (2009) Diversity of channeling flow in heterogeneous aperture distribution inferred from integrated experimental-numerical analysis on flow through shear fracture in granite. *Journal of Geophysical Research*, **114**, B04208, doi:10.1029/2008JB005959.

Watanabe N, Ishibashi T, Hirano N, Tsuchiya N, Ohsaki Y, Tamagawa T, Tsuchiya Y, Okabe H (2011a) Precise 3D numerical modeling of fracture flow coupled with X-ray Computed tomography for reservoir core samples. *SPE Journal*, **16**, 683–691.

Watanabe N, Ishibashi T, Ohsaki Y, Tsuchiya NY, Tamagawa T, Hirano N, Okabe H, Tsuchiya N (2011b) X-ray CT based numerical analysis of fracture flow for core samples under various confining pressure. *Engineering Geology*, **123**, 338–346.

Watanabe N, Ishibashi T, Tsuchiya N, Ohsaki Y, Tamagawa T, Tsuchiya Y, Okabe H, Ito H (2012) Geologic core holder with CFR PEEK body for the X-ray CT based numerical analysis of fracture flow under confining pressure. *Rock Mechanics and Rock Engineering*, **46**, 413–418.

Watt JT, Glen JMG, John D, Ponce DA (2007) Three-dimensional geologic model of the northern Nevada rift and the Beowawe geothermal system, north-central Nevada. *Geosphere*, **3**, 667–682.

Wei ZQ, Hudson JA (1988) Permeability of jointed rock masses. In: *Rock Mechanics and Power Plants* (ed Romana M), pp. 613–626. Balkema, Rotterdam.

Weinberg RF, Hodkiewicz PF, Groves DI (2004) What controls gold distribution in Archean terranes? *Geology*, **32**, 545–548.

Weingarten M, Ge S, Godt JW, Bekins BA, Rubinstein JL (2015) High-rate injection is associated with the increase in U.S. mid-continent seismicity. *Science*, **348**, 1336–1340.

Weis P (2014) The physical hydrology of ore-forming magmatic-hydrothermal systems. *Society of Economic Geologists Special Publication*, **18**, 59–75.

Weis P (2015) The dynamic interplay between saline fluid flow and rock permeability in magmatic -hydrothermal systems. *Geofluids*, **15**, 350–371.

Weis P, Driesner T (2013) The interplay of non-static permeability and fluid flow as a pre-requisite for supercritical geothermal resources. *Energy Procedia*, **40**, 102–106.

Weis P, Driesner T, Coumou D, Geiger S (2014) Hydrothermal, multi-phase convection of H_2O-NaCl fluids from ambient to magmatic temperatures: a new numerical scheme and benchmarks for code comparison. *Geofluids*, **14**, 347–371.

Weis P, Driesner T, Heinrich CA (2012) Porphyry-copper ore shells form at stable pressure-temperature fronts within dynamic fluid plumes. *Science*, **338**, 1613–1616.

Weisenberger T, Bucher K (2010) Zeolites in fissures of granites and gneisses of the central Alps. *Journal of Metamorphic Geology*, **28**, 825–847.

Weisenberger T, Bucher K (2011) Mass transfer and porosity evolution during low temperature water-rock interaction in gneisses of the Simano nappe: Arvigo, Val Calanca, Swiss Alps. *Contributions to Mineralogy and Petrology*, **162**, 61–81.

Weiss E (1982) A computer program for calculating relative-transmissivity input arrays to aid model calibration. *U.S. Geological Survey Open-File Report*, **82–447**.

Weissmann GS, Zhang Y, LaBolle EM, Fogg GE (2002) Dispersion of groundwater age in an alluvial aquifer system. *Water Resources Research*, **38**, doi:10.1029/2001WR000907.

Welch AH, Bright DJ, Knochenmus LA (2007) Water resources of the Basin and Range carbonate-rock aquifer system in White Pine County, Nevada, and adjacent areas in Nevada and Utah. *U.S. Geological Survey Scientific Investigations Report*, **2007–5261**.

Welhan JA, Poredai RJ, Rison W, Craig H (1988) Helium isotopes in geothermal and volcanic gases of the western United States, I, Regional variability and magmatic origin. *Journal of Volcanology and Geothermal Research*, **34**, 185–199.

Wells A, Yetton MD, Duncan RP, Stewart GH (1999) Prehistoric dates of the most recent Alpine fault earthquakes, New Zealand. *Geology*, **27**, 995–998.

Wells DL, Coppersmith KJ (1994) New empirical relationships among magnitude, rupture length, rupture width, rupture area, and surface displacement. *Bulletin of the Seismological Society of America*, **84**, 974–1002.

Wells JT, Ghiorso MS (1991) Coupled fluid flow and reaction in mid-ocean ridge hydrothermal systems; the behavior of silica. *Geochimica et Cosmochimica Acta*, **55**, 2467–2481.

Wells NS, Clough TJ, Condron LM, Baisden WT, Harding JS, Dong Y, Lewis GD, Lear G (2013) Biogeochemistry and community ecology in a spring-fed urban river following a major earthquake. *Environmental Pollution*, **182**, 190–200.

Wells SG, Granzow H (1981) Hydrogeology of the thermal aquifer near Truth or Consequences, New Mexico. In: State-Coupled Low-Temperature Geothermal Resource Assessment Program, Fiscal Year 1980, 3–5 to 3–51.

Weltje GJ, Prins MA (2003) Muddled or mixed? Inferring palaeoclimate from size distributions of deep-sea clastics. *Sedimentary Geology*, **162**, 39–62.

Wesnousky SG, Barron AD, Briggs RW, Caskey SJ, Kumar S, Owen L (2005) Paleoseismic transect across the northern Great Basin. *Journal of Geophysical Research*, **110**, doi:10.1029/2004JB003283.

West AJ (2012) Thickness of the chemical weathering zone and implications for erosional and climatic drivers of weathering and for carbon-cycle feedbacks. *Geology*, **40**, 811–814.

Westbrook GK, Carson B, Musgrave RJ, ODP Leg 146 Scientific Party (1994) *Proceedings of the Ocean Drilling Program, Initial Reports*, **146** (Pt. 1).

Westergaard HM (1952) *Theory of Elasticity and Plasticity*. Harvard University Press, Cambridge.

White DE (1998) The Beowawe geysers, Nevada, before geothermal development. *United States Geological Survey Bulletin*, **1998**.

White R, Spinelli GA, Mozley PS, Dunbar NW (2010) Importance of volcanic glass alteration to sediment stabilization: offshore Japan. *Sedimentology*, **58**, 1138–1154.

Wiggins C, Spiegelman M (1995) Magma migration and magmatic solitary waves in 3-D. *Geophysical Research Letters*, **22**, 1289–1292.

Wilkinson BH, McElroy BJ, Kesler SE, Peters SE, Rothman ED (2009) Global geologic maps are tectonic speedometers – rates of rock cycling from area-age frequencies. *Geological Society of America Bulletin*, **121**, 760–779.

Wilkinson DS, Ashby MF (1975) Pressure sintering by power law creep. *Acta Metallurgica*, **23**, 1277–1285.

Wilkinson JJ (2013) Triggers for the formation of porphyry ore deposits in magmatic arcs. *Nature Geoscience*, **6**, 917–925.

Wilkinson WB, Shipley EL (1972) Vertical and horizontal laboratory measurements in clay soils. *Developments in Soil Science*, **2**, 285–298.

Willemse EJM (1997) Segmented normal faults: correspondence between three-dimensional mechanical models and field data. *Journal of Geophysical Research*, **102**, 675–692.

Williams CF, DeAngelo J (2008) Mapping geothermal potential in the western United States. *Transactions of the Geothermal Resources Council*, **32**, 181–188.

Williams CF, DeAngelo J (2011) Evaluation of approaches and associated uncertainties in the estimation of temperatures in the upper crust of the Western United States. *Transactions of the Geothermal Resources Council*, **35**, 1599–1605.

Williams GP, Troutman BM (1987) Algebraic manipulation of equations for best-fit straight lines. In: *Use and Abuse of Statistical Methods in the Earth Sciences* (ed Size WB), pp. 129–141. Oxford University Press, Oxford.

Wilson AM, Huettel M, Klein S (2008) Grain size and depositional environment as predictors of permeability in coastal marine sands. *Estuarine, Coastal and Shelf Science*, **80**, 193–199.

Wilson RAM (1959) The geology of the Xeros-Troodos area. *Geological Survey Department Memoir*, **1**, Geological Survey Department, Nicosia, Cyprus.

Winograd IJ (1971) Hydrogeology of ash-flow tuff: a preliminary statement. *Water Resources Research*, **7**, 994–1006.

Winograd IJ, Thordarson W (1975) Hydrogeologic and hydrochemical framework, south-central Great Basin, Nevada-California, with special reference to the Nevada Test Site. *U.S. Geological Survey Professional Paper*, **712**-C.

Winter TC, Harvey JW, Franke OL, Alley WA (1998) Ground water and surface water: a single resource. *U.S. Geological Survey Circular*, **1139**.

Wissa AEZ, Christian JT, Davis EH, Heiberg S (1971) Consolidation at a constant rate of strain. *Journal of the Soil Mechanics and Foundations Division ASCE*, **97**, 1393–1413.

Witcher JC (1988) Geothermal resources of southwestern New Mexico and southeastern Arizona. *New Mexico Geological Society Guidebook*, **39**, 191–198.

Witherspoon PA, Amick C, Gale J, Iwai K (1979) Observation of a potential size effect in experimental determination of the hydraulic properties of fractures. *Water Resources Research*, **15**, 1142–1146.

Witherspoon PA, Wang JSY, Iwai K, Gale JE (1980) Validity of cubic law for fluid-flow in a deformable rock fracture. *Water Resources Research*, **16**, 1016–1024.

Witt WK, Ford A, Hanrahan B, Mamuse A (2013) Regional-scale targeting for gold in the Yilgarn Craton: part 1 of the Yilgarn Gold Exploration Targeting Atlas. *Western Australia Geological Survey Report*, **125**.

Wittwer C (1986) Probenahmen und chemische analysen von grundwassern aus den Sondierbohrungen. In: *Nagra Technical Report*, **85–49** (ed Nagra), Nagra, Baden, Switzerland, http://www.nagra.cn/data/documents/database/dokumente/%24default/Default%20Folder/Publikationen/NTBs%201985-1986/d_ntb85-49.pdf, accessed 07 May 2016.

Wladis D, Jönsson P, Wallroth T (1997) Regional characterization of hydraulic properties or rock using well test data. *Swedish Nuclear Fuel Waste Management Company (SKB) Technical Report*, **SKB-TR- 97-2** (ed SKB), SKB, Stockholm, Sweden.

Wolf-Gladrow D (2000) *Lattice-Gas Cellular Automata and Lattice Boltzmann Models – An Introduction*. Springer, Berlin.

Wong A, Wang C-Y (2007) Field relations between the spectral composition of ground motion and hydrological effects during the 1999 Chi-Chi (Taiwan) earthquake. *Journal of Geophysical Research*, **112**, B10305.

Worthington M, Lubbe R (2007) The scaling of fracture compliance. *Geological Society London Special Publication*, **270**, 73–82.

Xu X, Wen X, Yu G, Chen G, Klinger Y, Hubbard J, Shaw J (2009) Coseismic reverse-and oblique-slip surface faulting generated by the 2008 M_w7.9 Wenchuan earthquake, China. *Geology*, **37**, 515–518.

Xue L, Li H-B, Brodsky EE, Xu Z-Q, Kano Y, Wang H, Mori JJ, Si J-L, Pei J-L, Zhang W, Yang G, Sun Z-M, Huang Y (2013) Continuous permeability measurements record healing inside the Wenchuan earthquake fault zone. *Science*, **340**, 1555–1559.

Yadav SK, Chakrapani GJ (2006) Dissolution kinetics of rock-water interactions and its implications. *Current Science*, **90**, 932–937.

Yamamoto K, Yoshida H, Akagawa F, Nishimoto S, Metcalfe R (2013) Redox front penetration in the fractured Toki Granite, central Japan: an analogue for redox buffering in fractured crystalline host rocks for repositories of long-lived radioactive waste. *Applied Geochemistry*, **35**, 75–87.

Yang Y, Aplin AC (1998) Influence of lithology and compaction on the pore size distribution and modelled permeability of some mudstones from the Norwegian margin. *Marine and Petroleum Geology*, **15**, 163–175.

Yang Y, Aplin AC (2007) Permeability and petrophysical properties of 30 natural mudstones. *Journal of Geophysical Research*, **112**, B03206.

Yang Y, Aplin AC (2010) A permeability-porosity relationship for mudstones. *Marine and Petroleum Geology*, **27**, 1692–1697.

Yang Z, Peng X-F, Lee D-J, Chen M-Y (2009) An image-based method for obtaining pore-size distribution of porous media. *Environmental Science & Technology*, **43**, 3248–3253.

Yardley BWD (1981) The effect of cooling on the water content and mechanical properties of metamorphosed rocks. *Geology*, **9**, 405–408.

Yardley BWD (1986) Fluid migration and veining in the Connemara Schists. In: *Fluid-Rock Reactions During Metamorphism, Advances in Physical Geochemistry*, Vol. 5 (eds Walther JV, Wood BJ), pp. 109–131. Springer-Verlag, New York.

Yardley BWD (2009) The role of water in the evolution of the continental crust. *Journal of the Geological Society*, **166**, 585–600.

Yardley BWD (2013) The chemical composition of metamorphic fluids in the crust. In: *Metasomatism and the Chemical Transformation of Rock* (eds Harlov DE, Austrheim H), Springer-Verlag, Berlin, Heidelberg, 17–51.

Yardley BWD, Baumgartner LP (2007) Fluid processes in deep crustal fault zones. In: *Tectonic Faults–Agents of Change on a*

Dynamic Earth (eds. Handy MR, Hirth G, Hovius N), pp. 295–318. The MIT Press, Boston.

Yardley BWD, Bodnar RJ (2014) Fluids in the continental crust. *Geochemical Perspectives*, **3**, 127.

Yardley BWD, Bottrell SH, Cliff RA (1991a) Evidence for a regional-scale fluid loss event during mid-crustal metamorphism. *Nature*, **490**, 151–154.

Yardley BWD, Cleverley JS (2013) The role of metamorphic fluids in the formation of ore deposits. *Geological Society of London Special Publication*, **393**, doi:10.1144/SP393.5.

Yardley BWD, Harlov DE, Heinrich W (2010) Rates of retrograde metamorphism and their implications for crustal rheology. *Geofluids*, **10**, 234–240.

Yardley BWD, Lloyd GE (1989) An application of cathodoluminescence microscopy to the study of textures and reactions in high grade marbles from Connemara, Ireland. *Geological Magazine*, **126**, 333–337.

Yardley BWD, Rhede D, Heinrich W (2014) Rates of retrograde metamorphism and their implications for the rheology of the crust: an experimental study. *Journal of Petrology*, **65**, 623–641.

Yardley BWD, Rochelle CA, Barnicoat AC, Lloyd GE (1991b) Oscillatory zoning in metamorphic minerals: an indicator of infiltration metasomatism. *Mineralogical Magazine*, **55**, 357–365.

Yardley BWD, Valley JW (1997) The petrologic case for a dry lower crust. *Journal Geophysical Research*, **102**, 12173–12185.

Yarushina VM (2009) (De)compaction waves in porous viscoelastoplastic media. PhD thesis, University of Oslo, Oslo.

Yasuhara H, Kinoshita N, Ohfuji H, Lee DS, Nakashima S, Kishida K (2011) Temporal alteration of fracture permeability in granite under hydrothermal conditions and its interpretation by coupled chemo-mechanical model. *Applied Geochemistry*, **26**, 2074–2088.

Yasuhara HD, Elsworth D, Polak A (2004) Evolution of permeability in a natural fracture: significant role of pressure solution. *Journal of Geophysical Research*, **109**, B3204.

Yasuhara HD, Polak A, Mitani Y, Grader AS, Halleck PM, Elsworth D (2006) Evolution of fracture permeability through fluid-rock reaction under hydrothermal conditions. *Earth and Planetary Science Letters*, **244**, 186–200.

Yeo IW, de Freitas MH, Zimmermann RW (1998) Effect of shear displacement on the aperture and permeability of a rock fracture. *International Journal of Rock Mechanics and Mining Sciences*, **35**, 1051–1070.

Yoshida H, Metcalfe R, Ishibashi M, Minami M (2013) Long-term stability of fracture systems and their behaviour as flow paths in uplifting granitic rocks from the Japanese orogenic field. *Geofluids*, **13**, 45–55.

Yoshida K, Hasegawa A, Okada T, Iinuma T, Ito Y, Asano Y (2012) Stress before and after the 2011 great Tohoku-oki earthquake and induced earthquakes in inland areas of eastern Japan. *Geophysical Research Letters*, **39**, L03302.

Yoshinaka R, Yoshida J, Arai H, Arisaka S (1993) Scale effects on shear and deformability of rock joints. In: *Scale Effects in Rock Masses 93* (ed Pinto da Chunha A), pp. 143–149. Balkema, Rotterdam.

You CF, Castillo PR, Gieskes JM, Chan LH, Spivack AJ (1996) Trace element behavior in hydrothermal experiments: implications for fluid processes at shallow depths in subduction zones. *Earth and Planetary Science Letters*, **140**, 41–52.

Youngs DL (1984) An interface tracking method for a 3D Eulerian hydrodynamics code. *Atomic Weapons Research Establishment (AWRE) Technical Report*, **44/92** (AWRE/44/92/35).

Yue L, Likos WJ, Guo J, Underwood MB (2012) Data report: Permeability of mud(stone) samples from Site C0001, IODP Expedition 315, Nankai Trough – NanTroSEIZE, Stage 1. In Kinoshita M, Tobin H, Ashi J, Kimura G, Lallemant S, Screaton EJ, Curewitz D, Masago H, Moe KT, and the Expedition 314/315/316 Scientists, *Proceedings of the Integrated Ocean Drilling Program*, **314/315/316**, 1–41.

Yukutake Y, Ito H, Honda R, Harada M, Tanada T, Yoshida A (2011) Fluid-induced swarm earthquake sequence revealed by precisely determined hypocenters and focal mechanisms in the 2009 activity at Hakone volcano, Japan. *Journal of Geophysical Research*, **116**, B04308.

Zakharova NV, Goldberg DS, Sullivan EC, Herron MM, Grau JA (2012) Petrophysical and geochemical properties of Columbia River flood basalt: implications for carbon sequestration. *Geochemistry, Geophysics, Geosystems*, **13**, Q11001.

Zangeneh N, Eberhardt E, Bustin R (2012) Application of the distinct-element method to investigate the influence of natural fractures and in situ stresses on hydrofrac propagation. In: 46th US Rock Mechanics/Geomechanics Symposium, Chicago, Illinois.

Zangerl C, Evans KF, Eberhardt E, Loew S (2008) Normal stiffness of fractures in granitic rock: a compilation of laboratory and in-situ experiments. *International Journal of Rock Mechanics and Mining Sciences*, **45**, 1500–1507.

Zhang H, Thurber CH (2003) Double-difference tomography: the method and its application to the Hayward Fault, California. *Bulletin of the Seismological Society of America*, **93**, 1875–1889.

Zhang LG, Jingxiu L, Zhensheng C, Huanbo Z (1994) Experimental investigations of oxygen isotope fractionation in cassiterite and wolframite. *Economic Geology*, **89**, 150–157.

Zhang P, Deng Q, Zhang G, Ma J, Gan W, Min W, Mao F, Wang Q (2003) Active tectonic blocks and strong earthquakes in the continent of China. *Science in China, Series D, Earth Sciences*, **46**, 13–24.

Zhao Z, Rutqvist J, Leung C, Hokr M, Neretnieks I, Hoch A, Havlicek J, Wang Y, Wang Z, Zimmerman R (2013) Stress effects on solute transport in fractured rocks: a comparison study. *Journal of Rock Mechanics and Geotechnical Engineering*, **5**, 110–123.

Zheng YF (1999) Oxygen isotope fractionations in carbonate and sulfate minerals. *Journal of Geochemistry*, **33**, 109–126.

Zhou JJ, Shao JF, Xu WY (2006) Coupled modeling of damage growth and permeability variation in brittle rocks. *Mechanics Research Communications*, **33**, 450–459.

Zhu W, David C, Wong T-F (1995) Network modeling of permeability evolution during cementation and hot isostatic pressing. *Journal of Geophysical Research*, **100**, 15451–15464.

Zhu W, Evans B, Bernabe Y (1999) Densification and permeability reduction in hot-pressed calcite: a kinetic model. *Journal of Geophysical Research*, **104**, 25501–25511.

Zhu W, Wong TF (1997) The transition from brittle faulting to cataclastic flow: permeability evolution. *Journal of Geophysical Research*, **102**, 3027–3041.

Ziagos JP, Blackwell DD (1986) A model for the transient temperature effects of horizontal fluid flow in geothermal systems. *Journal of Volcanology and Geothermal Research*, **27**, 371–397.

Zienkiewicz OC, Taylor RL (2005) *The Finite Element Method for Solid and Structural Mechanics*, 6th edn Elsevier, Oxford.

Zimmerman RM, Wilson ML, Board MP, Hall ME, Schuch RL (1985) Thermal-cycle testing of the G-tunnel heated block. *Proceedings of the 26th US Symposium on Rock Mechanics*, Rapid City, South Dakota, 26–28 June, pp. 749–58. AA Balkema, Rotterdam.

Zimmerman RW, Bodvarsson G (1996), Hydraulic conductivity of rock fractures. *Transport in Porous Media*, **23**, 1–30.

Zimmerman RW, Chen DW, Cook NGW (1992) The effect of contact area on the permeability of fractures. *Journal of Hydrology*, **139**, 79–96.

Zoback MD (2010) *Reservoir Geomechanics*. Cambridge University Press, Cambridge.

Zoback MD, Byerlee JD (1975) The effect of micro-crack dilatancy on the permeability of Westerly Granite. *Journal of Geophysical Research*, **80**, 752–755.

Zoback MD, Harjes H-P (1997) Injection-induced earthquakes and crustal stress at 9 km depth at the KTB deep drilling site, Germany. *Journal of Geophysical Research*, **102**, 18477–18491.

Zoback MD, Townend J (2001) Implications of hydrostatic pore pressures and high crustal strength for the deformation of intraplate lithosphere. *Tectonophysics*, **336**, 19–30.

Zoback MD, Townend J, Grollimund B (2002) Steady-state failure equilibrium and deformation of intraplate lithosphere. *International Geology Review*, **44**, 383–401.

Zoback ML (1979) Geologic and geophysical investigation of the Beowawe geothermal area, north-central Nevada. *Stanford University School of Earth Sciences*, **16**.

Zwart G, Bruckmann W, Moran K, MacKillop AK, Maltman AJ, Bolton A, Vrolijk P, Miller T, Gooch MJ, Fisher A (1997) Evaluation of hydrogeologic properties of the Barbados accretionary prism: a synthesis of Leg 156 results. In: *Proceedings of the Ocean Drilling Program, Scientific Results*, **156**, (eds Shipley TH, Ogawa Y, Blum P, Bahr JM), pp. 303–310, Ocean Drilling Program, College Station, TX.

Zwart G, Moore JC, Cochrane GR (1996) Variations in temperature gradients identify active faults in the Oregon accretionary prism. *Earth and Planetary Science Letters*, **139**, 485–495.

Index

Crustal Permeability, First Edition. Edited by Tom Gleeson and Steven E. Ingebritsen.
© 2017 John Wiley & Sons, Ltd. Published 2017 by John Wiley & Sons, Ltd.
Companion Websites: www.wiley.com/go/gleeson/crustalpermeability/
http://crustalpermeability.weebly.com/